Principles of Lasers

FIFTH EDITION

Principles of Lasers

FIFTH EDITION

Orazio Svelto
Polytechnic Institute of Milan
and National Research Council
Milan, Italy

Translated from Italian and edited by

David C. Hanna
Southampton University
Southampton, England

Springer

Orazio Svelto
Politecnico di Milano
Dipto. Fisica
Piazza Leonardo da Vinci, 32
20133 Milano
Italy

ISBN 978-1-4899-7713-7 ISBN 978-1-4419-1302-9 (eBook)
DOI 10.1007/978-1-4419-1302-9
Springer New York Dordrecht Heidelberg London

Printed on acid-free paper

Springer is part of Springer Science+Business Media (www.springer.com)

To my wife Rosanna
and to my sons Cesare and Giuseppe

Preface

This book is motivated by the very favorable reception given to the previous editions as well as by the considerable range of new developments in the laser field since the publication of the third edition in 1989. These new developments include, among others, Quantum-Well and Multiple-Quantum Well lasers, diode-pumped solid-state lasers, new concepts for both stable and unstable resonators, femtosecond lasers, ultra-high-brightness lasers etc. The basic aim of the book has remained the same, namely to provide a *broad* and *unified description* of laser behavior at the simplest level which is compatible with a correct physical understanding. The book is therefore intended as a text-book for a senior-level or first-year graduate course and/or as a reference book.

This edition corrects several errors introduced in the previous edition. The most relevant *additions* or *changes* to since the third edition can be summarized as follows:

1. A much-more detailed description of Amplified Spontaneous Emission has been given [Chapt. 2] and a novel simplified treatment of this phenomenon both for homogeneous or inhomogeneous lines has been introduced [Appendix C].
2. A major fraction of a chapter [Chapt. 3] is dedicated to the interaction of radiation with semiconductor media, either in a bulk form or in a quantum-confined structure (quantum-well, quantum-wire and quantum dot).
3. A modern theory of stable and unstable resonators is introduced, where a more extensive use is made of the ABCD matrix formalism and where the most recent topics of dynamically stable resonators as well as unstable resonators, with mirrors having Gaussian or super-Gaussian transverse reflectivity profiles, are considered [Chapt. 5].
4. Diode-pumping of solid-state lasers, both in longitudinal and transverse pumping configurations, are introduced in a unified way and a comparison is made with corresponding lamp-pumping configurations [Chapt. 6].
5. Spatially-dependent rate equations are introduced for both four-level and quasi-three-level lasers and their implications, for longitudinal and transverse pumping, are also discussed [Chapt. 7].

6. Laser mode-locking is considered at much greater length to account for e.g. new mode-locking methods, such as Kerr-lens mode-locking. The effects produced by second-order and third-order dispersion of the laser cavity and the problem of dispersion compensation to achieve the shortest pulse-durations are also discussed at some length [Chapt. 8].
7. New tunable solid-state lasers, such as Ti: sapphire and Cr: LISAF, as well as new rare-earth lasers such as Yb^{3+}, Er^{3+}, and Ho^{3+} are also considered in detail [Chapt. 9].
8. Semiconductor lasers and their performance are discussed at much greater length [Chapt. 9].
9. The divergence properties of a multimode laser beam as well as its propagation through an optical system are considered in terms of the M^2-factor and in terms of the embedded Gaussian beam [Chapt. 11 and 12].
10. The production of ultra-high peak intensity laser beams by the technique of chirped-pulse-amplification and the related techniques of pulse expansion and pulse compression are also considered in detail [Chapt. 12].

The book also contains numerous, thoroughly developed, examples, as well as many tables and appendixes. The examples either refer to real situations, as found in the literature or encountered through my own laboratory experience, or describe a significative advance in a particular topic. The tables provide data on optical, spectroscopic and nonlinear-optical properties of laser materials, the data being useful for developing a more quantitative context as well as for solving the problems. The appendixes are introduced to consider some specific topics in more mathematical detail. A great deal of effort has also been devoted to the *logical organization* of the book so as to make its content more accessible.

The *basic philosophy* of the book is to resort, wherever appropriate, to an intuitive picture rather than to a detailed mathematical description of the phenomena under consideration. Simple mathematical descriptions, when useful for a better understanding of the physical picture, are included in the text while the discussion of more elaborate analytical models is deferred to the appendixes. The *basic organization* starts from the observation that a laser can be considered to consists of three elements, namely the active medium, the resonator, and the pumping system. Accordingly, after an introductory chapter, Chapters 2–3, 4–5 and 6 describe the most relevant features of these elements, separately. With the combined knowledge about these constituent elements, chapters 7 and 8 then allow a discussion of continuos-wave and transient laser behavior, respectively. Chapters 9 and 10 then describe the most relevant types of laser exploiting high-density and low-density media, respectively. Lastly, chapters 11 and 12 consider a laser beam from the user's view-point examining the properties of the output beam as well as some relevant laser beam transformations, such as amplification, frequency conversion, pulse expansion or compression.

With so many topics, examples, tables and appendixes, it is clear that the entire content of the book could not be covered in only a one semester-course. However the organization of the book allows several different learning paths. For instance, one may be more interested in learning the *Principles of Laser Physics*. The emphasis of the study should then be mostly concentrated on the first section of the book [Chapt. 1–5 and Chapt. 7–8]. If, on the other hand, the reader is more interested in the *Principles of Laser Engineering*, effort should mostly be concentrated on the second part of the book Chap. 6 and 9–12. The *level of understanding* of a given topic may also be suitably *modulated* by e.g. considering, in more or less detail, the numerous examples, which often represent an extension of a given topic, as well as the numerous appendixes.

Writing a book, albeit a satisfying cultural experience, represents a heavy intellectual and physical effort. This effort has, however, been gladly sustained in the hope that this edition can serve the pressing need for a general introductory course to the laser field.

ACKNOWLEDGMENTS. I wish to acknowledge the following friends and colleagues, whose suggestions and encouragement have certainly contributed to improving the book in a number of ways: Christofer Barty, Vittorio De Giorgio, Emilio Gatti, Dennis Hall, Günther Huber, Gerard Mourou, Colin Webb, Herbert Welling. I wish also to warmly acknowledge the critical editing of David C. Hanna, who has acted as much more than simply a translator. Lastly I wish to thank, for their useful comments and for their critical reading of the manuscript, my former students: G. Cerullo, S. Longhi, M. Marangoni, M. Nisoli, R. Osellame, S. Stagira, C. Svelto, S. Taccheo, and M. Zavelani.

Milano Orazio Svelto

Contents

List of Examples

Chapter 2

Chapter 3

Chapter 4

Chapter 5

Chapter 6

Chapter 7

Chapter 8

Chapter 9

Chapter 11

Chapter 12

1

Introductory Concepts

In this introductory chapter, the fundamental processes and the main ideas behind laser operation are introduced in a very simple way. The properties of laser beams are also briefly discussed. The main purpose of this chapter is thus to introduce the reader to many of the concepts that will be discussed later on, in the book, and therefore help the reader to appreciate the logical organization of the book.

1.1. SPONTANEOUS AND STIMULATED EMISSION, ABSORPTION

To describe the phenomenon of spontaneous emission, let us consider two energy levels, 1 and 2, of some atom or molecule of a given material, their energies being E_1 and E_2 ($E_1 < E_2$) (Fig. 1.1a). As far as the following discussion is concerned, the two levels could be any two out of the infinite set of levels possessed by the atom. It is convenient, however, to take level 1 to be the ground level. Let us now assume that the atom is initially in level 2. Since $E_2 > E_1$, the atom will tend to decay to level 1. The corresponding energy difference, $E_2 - E_1$, must therefore be released by the atom. When this energy is delivered in the form of an electromagnetic (e.m. from now on) wave, the process will be called *spontaneous* (or *radiative*) *emission*. The frequency ν_0 of the radiated wave is then given by the well known expression

$$\nu_0 = (E_2 - E_1)/h \tag{1.1.1}$$

where h is Planck's constant. Spontaneous emission is therefore characterized by the emission of a photon of energy $h\nu_0 = E_2 - E_1$, when the atom decays from level 2 to level 1 (Fig. 1.1a). Note that radiative emission is just one of the two possible ways for the atom to decay. The decay can also occur in a nonradiative way. In this case the energy difference $E_2 - E_1$ is delivered in some form of energy other than e.m. radiation (e.g. it may go into kinetic or internal energy of the surrounding atoms or molecules). This phenomenon is called *non-radiative decay*.

O. Svelto, *Principles of Lasers*,
DOI: 10.1007/978-1-4419-1302-9_1,

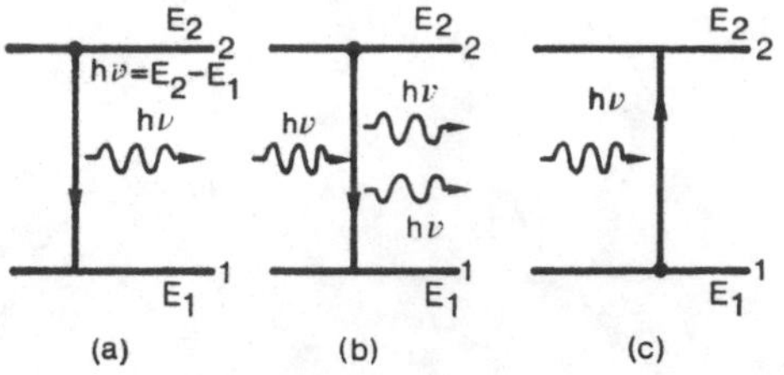

FIG. 1.1. Schematic illustration of the three processes: (a) spontaneous emission; (b) stimulated emission; (c) absorption.

Let us now suppose that the atom is found initially in level 2 and that an e.m. wave of frequency $\nu = \nu_o$ (i.e., equal to that of the spontaneously emitted wave) is incident on the material (Fig. 1.1b). Since this wave has the same frequency as the atomic frequency, there is a finite probability that this wave will force the atom to undergo the transition $2 \rightarrow 1$. In this case the energy difference $E_2 - E_1$ is delivered in the form of an e.m. wave that adds to the incident one. This is the phenomenon of *stimulated emission*. There is a fundamental difference between the spontaneous and stimulated emission processes. In the case of spontaneous emission, the atoms emits an e.m. wave that has no definite phase relation with that emitted by another atom. Furthermore, the wave can be emitted in any direction. In the case of stimulated emission, since the process is forced by the incident e.m. wave, the emission of any atom adds in phase to that of the incoming wave and along the same direction.

Let us now assume that the atom is initially lying in level 1 (Fig. 1.1c). If this is the ground level, the atom will remain in this level unless some external stimulus is applied to it. We shall assume, then, that an e.m. wave of frequency $\nu = \nu_o$ is incident on the material. In this case there is a finite probability that the atom will be raised to level 2. The energy difference $E_2 - E_1$ required by the atom to undergo the transition is obtained from the energy of the incident e.m. wave. This is the *absorption* process.

To introduce the probabilities for these emission and absorption phenomena, let N be the number of atoms (or molecules) per unit volume which, at time t, are lying in a given energy level. From now on the quantity N will be called the *population* of the level.

For the case of spontaneous emission, the probability for the process to occur can be defined by stating that the rate of decay of the upper state population, $(dN_2/dt)_{sp}$, must be proportional to the population N_2. We can therefore write

$$\left(\frac{dN_2}{dt}\right)_{sp} = -AN_2 \tag{1.1.2}$$

where the minus sign accounts for the fact that the time derivative is negative. The coefficient A, introduced in this way, is a positive constant and is called the rate of spontaneous emission or the Einstein A coefficient (an expression for A was in fact first obtained by Einstein from thermodynamic considerations). The quantity $\tau_{sp} = 1/A$ is called the spontaneous emission (or radiative) lifetime. Similarly, for non-radiative decay, we can often write

$$\left(\frac{dN_2}{dt}\right)_{nr} = -\frac{N_2}{\tau_{nr}} \tag{1.1.3}$$

where τ_{nr} is referred to as the non-radiative decay lifetime. Note that, for spontaneous emission, the numerical value of A (and τ_{sp}) depends only on the particular transition considered. For non-radiative decay, τ_{nr} depends not only on the transition but also on the characteristics of the surrounding medium.

We can now proceed, in a similar way, for the stimulated processes (emission or absorption). For stimulated emission we can write

$$\left(\frac{dN_2}{dt}\right)_{st} = -W_{21}N_2 \tag{1.1.4}$$

where $(dN_2/dt)_{st}$ is the rate at which transitions $2 \rightarrow 1$ occur as a result of stimulated emission and W_{21} is called the rate of stimulated emission. Just as in the case of the A coefficient defined by Eq. (1.1.2) the coefficient W_{21} also has the dimension of $(\text{time})^{-1}$. Unlike A, however, W_{21} depends not only on the particular transition but also on the intensity of the incident e.m. wave. More precisely, for a plane wave, it will be shown that we can write

$$W_{21} = \sigma_{21}F \tag{1.1.5}$$

where F is the photon flux of the wave and σ_{21} is a quantity having the dimension of an area (the stimulated emission *cross section*) and depending on the characteristics of the given transition.

In a similar fashion to Eq. (1.1.4), we can define an absorption rate W_{21} by means of the equation

$$\left(\frac{dN_1}{dt}\right)_{a} = -W_{12}N_1 \tag{1.1.6}$$

where $(dN_1/dt)_a$ is the rate of the $1 \rightarrow 2$ transitions due to absorption and N_1 is the population of level 1. Furthermore, just as in Eq. (1.1.5), we can write

$$W_{12} = \sigma_{12}F \tag{1.1.7}$$

where σ_{12} is some characteristic area (the *absorption cross section*), which depends only on the particular transition.

In what has just been said, the stimulated processes have been characterized by the stimulated emission and absorption cross-sections, σ_{21} and σ_{12}, respectively. Now, it was shown by Einstein at the beginning of the twentieth century that, if the two levels are non-degenerate, one always has $W_{21} = W_{12}$ and $\sigma_{21} = \sigma_{12}$. If levels 1 and 2 are g_1-fold and g_2-fold degenerate, respectively one has instead

$$g_2W_{21} = g_1W_{12} \tag{1.1.8}$$

i.e.

$$g_2\sigma_{21} = g_1\sigma_{12} \tag{1.1.9}$$

Note also that the fundamental processes of spontaneous emission, stimulated emission and absorption can readily be described in terms of absorbed or emitted photons as follows (see Fig. 1.1). (1) In the spontaneous emission process, the atom decays from level 2 to level 1 through the emission of a photon. (2) In the stimulated emission process, the incident photon stimulates the $2 \rightarrow 1$ transition and we then have two photons (the stimulating plus the stimulated one). (3) In the absorption process, the incident photon is simply absorbed to produce the $1 \rightarrow 2$ transition. Thus we can say that each stimulated emission process creates while each absorption process annihilates a photon.

1.2. THE LASER IDEA

Consider two arbitrary energy levels 1 and 2 of a given material and let N_1 and N_2 be their respective populations. If a plane wave with a photon flux F is traveling along the z direction in the material (Fig. 1.2), the elemental change, dF, of this flux along the elemental length, dz, of the material will be due to both the stimulated and emission processes occurring in the shaded region of Fig. 1.2. Let S be the cross sectional area of the beam. The change in number between outgoing and incoming photons, in the shaded volume per unit time, will thus be SdF. Since each stimulated process creates while each absorption removes a photon, SdF must equal the difference between stimulated emission and absorption events occurring in the shaded volume per unit time. From (1.1.4) and (1.1.6) we can thus write $SdF = (W_{21}N_2 - W_{12}N_1)(Sdz)$ where Sdz is, obviously, the volume of the shaded region. With the help of Eqs. (1.1.5), (1.1.7) and (1.1.9) we obtain

$$dF = \sigma_{21} F\,[N_2 - (g_2 N_1/g_1)]\,dz \tag{1.2.1}$$

Note that, in deriving Eq. (1.2.1), we have not taken into account the radiative and non-radiative decays. In fact, non-radiative decay does not add any new photons while the photons created by the radiative decay are emitted in any direction and do not contribute to the incoming photon flux F.

Equation (1.2.1) shows that the material behaves as an amplifier (i.e., $dF/dz > 0$) if $N_2 > g_2N_1/g_1$, while it behaves as an absorber if $N_2 < g_2N_1/g_1$. Now, at thermal equilibrium, the populations are described by Boltzmann statistics. So, if N_1^e and N_2^e are the thermal equilibrium

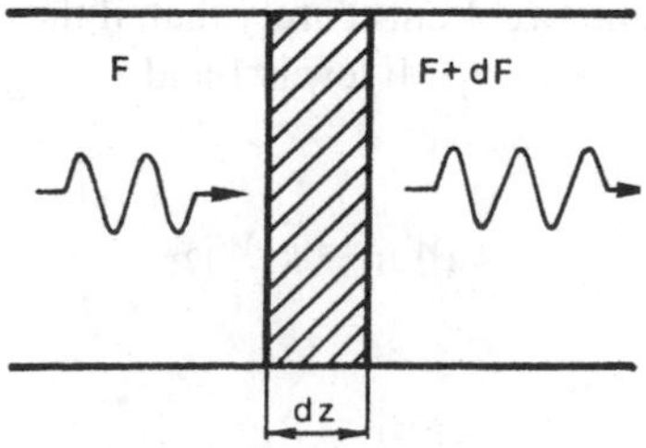

FIG. 1.2. Elemental change dF in the photon flux F fro a plane e.m. wave in traveling a distance dz through the material.

populations of the two levels, we have

$$\frac{N_2^e}{N_1^e} = \frac{g_2}{g_1} \exp - \left[\frac{E_2 - E_1}{kT}\right] \tag{1.2.2}$$

where k is Boltzmann's constant and T the absolute temperature of the material. In thermal equilibrium we thus have $N_2^e < g_2 N_1^e / g_1$. According to Eq. (1.2.1), the material then acts as an absorber at frequency ν. This is what happens under ordinary conditions. If, however, a non-equilibrium condition is achieved for which $N_2 > g_2 N_1 / g_1$ then the material will act as an amplifier. In this case we will say that there exists a *population inversion* in the material, by which we mean that the population difference $N_2 - (g_2 N_1 / g_1)$ is opposite in sign to that which exists under thermodynamic equilibrium $[N_2 - (g_2 N_1 / g_1) < 0]$. A material in which this population inversion is produced will be called an *active material*.

If the transition frequency $\nu_0 = (E_2 - E_1)/ kT$ falls in the microwave region, this type of amplifier is called a *maser* amplifier. The word *maser* is an acronym for "*m*icrowave *a*mplification by *s*timulated *e*mission of *r*adiation." If the transition frequency falls in the optical region, the amplifier is called a *laser* amplifier. The word *laser* is again an acronym, with the letter *l* (*l*ight) substituted for the letter *m* (*m*icrowave).

To make an oscillator from an amplifier, it is necessary to introduce a suitable positive feedback. In the microwave region this is done by placing the active material in a resonant cavity having a resonance at frequency ν_0. In the case of a laser, the feedback is often obtained by placing the active material between two highly reflecting mirrors (e.g. plane parallel mirrors, see Fig. 1.3). In this case, a plane e.m. wave traveling in the direction perpendicular to the mirrors will bounce back and forth between the two mirrors and be amplified on each passage through the active material. If one of the two mirrors is made partially transparent, a useful output beam is obtained from this mirror. It is important to realize that, for both masers and lasers, a certain threshold condition must be reached. In the laser case, for instance, the oscillation will start when the gain of the active material compensates the losses in the laser (e.g. the losses due to the output coupling). According to Eq. (1.2.1), the gain per pass in the active material (i.e. the ratio between the output and input photon flux) is $\exp\{\sigma[N_2 - (g_2 N_1 / g_1)]l\}$ where we have denoted, for simplicity, $\sigma = \sigma_{21}$, and where l is the length of the active material. Let R_1 and R_2 be the power reflectivity of the two mirrors (Fig. 1.3) and let L_i be the internal loss per pass in the laser cavity. If, at a given time, F is the photon flux in the cavity, leaving mirror 1 and traveling toward mirror 2, then the photon flux, F', again leaving mirror 1 after one round trip will be $F' = F \exp\{\sigma[N_2 - (g_2 N_1/g_1)]l\} \times (1 - L_i) R_2 \times \exp\{\sigma[N_2 - (g_2 N/g_1)]l\} \times (1 - L_i) R_1$. At threshold we must have $F' = F$, and therefore $R_1 R_2 (1 - L_i)^2 \exp\{2\sigma[N_2 - (g_2 N_1/g_1)]l\} = 1$. This equation shows that threshold is reached when the population inversion, $N = N_2 - (g_2 N_1 / g_1)$, reaches a critical value, known as the *critical inversion*, given by

$$N_c = -[\ln R_1 R_2 + 2 \ln (1 - L_i)] / 2\sigma l \tag{1.2.3}$$

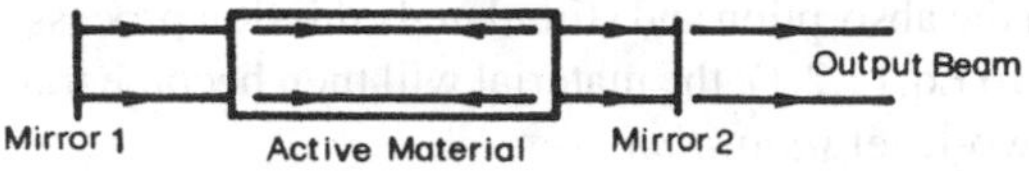

FIG. 1.3. Scheme of a laser.

The previous expression can be put in a somewhat simpler form if we define

$$\gamma_1 = -\ln R_1 = -\ln(1 - T_1) \tag{1.2.4a}$$

$$\gamma_2 = -\ln R_2 = -\ln(1 - T_2) \tag{1.2.4b}$$

$$\gamma_i = -\ln(1 - L_i) \tag{1.2.4c}$$

where T_1 and T_2 are the two mirror transmissions (for simplicity mirror absorption has been neglected). The substitution of Eq. (1.2.4) in Eq. (1.2.3) gives

$$N_c = \gamma/\sigma l \tag{1.2.5}$$

where we have defined

$$\gamma = \gamma_i + (\gamma_1 + \gamma_2)/2 \tag{1.2.6}$$

Note that the quantities γ_i, defined by Eq. (1.2.4c), may be called the logarithmic internal loss of the cavity. In fact, when $L_i \ll 1$ as usually occurs, one has $\gamma_i \cong L_i$. Similarly, since both T_1 and T_2 represent a loss for the cavity, γ_1 and γ_2, defined by Eq. (1.2.4a and b), may be called the logarithmic losses of the two cavity mirrors. Thus, the quantity γ defined by Eq. (1.2.6) will be called the single pass loss of the cavity.

Once the critical inversion is reached, oscillation will build up from spontaneous emission. The photons that are spontaneously emitted along the cavity axis will, in fact, initiate the amplification process. This is the basis of a laser oscillator, or laser, as it is more simply called. Note that, according to the meaning of the acronym laser as discussed above, the word should be reserved for lasers emitting visible radiation. The same word is, however, now commonly applied to any device emitting stimulated radiation, whether in the far or near infrared, ultraviolet, or even in the X-ray region. To be specific about the kind of radiation emitted one then usually talks about infrared, visible, ultraviolet or X-ray lasers, respectively.

1.3. PUMPING SCHEMES

We will now consider the problem of how a population inversion can be produced in a given material. At first sight, it might seem that it would be possible to achieve this through the interaction of the material with a sufficiently strong e.m. wave, perhaps coming from a sufficiently intense lamp, at the frequency $\nu = \nu_0$. Since, at thermal equilibrium, one has $g_1 N_1 > g_2 N_2 g_1$, absorption will in fact predominate over stimulated emission. The incoming wave would produce more transitions $1 \rightarrow 2$ than transitions $2 \rightarrow 1$ and we would hope in this way to end up with a population inversion. We see immediately, however, that such a system would not work (at least in the steady state). When in fact the condition is reached such that $g_2 N_2 = g_1 N_1$, then the absorption and stimulated emission processes will compensate one another and, according to Eq. (1.2.1), the material will then become transparent. This situation is often referred to as two-level *saturation*.

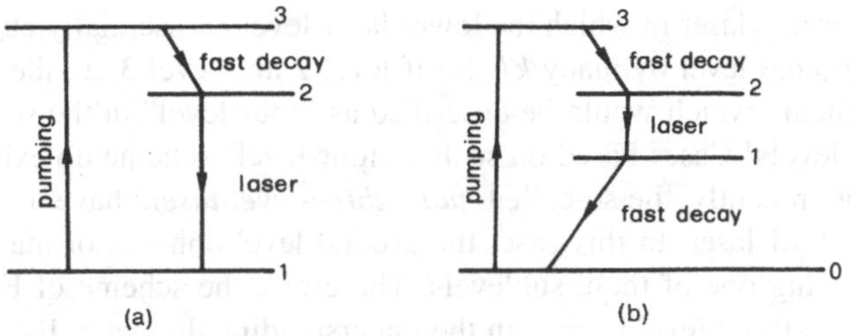

FIG. 1.4. (a) Three-level and (b) four-level laser schemes.

With just two levels, 1 and 2, it is therefore impossible to produce a population inversion. It is then natural to question whether this is possible using more than two levels out of the infinite set of levels of a given atomic system. As we shall see, the answer is in this case positive, and we will accordingly talk of a *three-level laser* or *four-level laser*, depending upon the number of levels used (Fig. 1.4). In a three-level laser (Fig. 1.4a), the atoms are in some way raised from the ground level 1 to level 3. If the material is such that, after an atom has been raised to level 3, it decays rapidly to level 2 (perhaps by a rapid nonradiative decay), then a population inversion can be obtained between levels 2 and 1. In a four-level laser (Fig. 1.4b), atoms are again raised from the ground level (for convenience we now call this level 0) to level 3. If the atom then decays rapidly to level 2 (e.g. again by a fast nonradiative decay), a population inversion can again be obtained between levels 2 and 1. Once oscillation starts in such a four-level laser, however, the atoms will then be transferred to level 1, through stimulated emission. For continuos wave (henceforth abbreviated as cw) operation it is therefore necessary that the transition 1 $\rightarrow$ 0 should also be very fast (this again usually occurs by a fast nonradiative decay).

We have just seen how to make use of a three or four levels of a given material to produce population inversion. Whether a system will work in a three- or four-level scheme (or whether it will work at all!) depends on whether the various conditions given above are fulfilled. We could of course ask why one should bother with a four level scheme when a three-level scheme already seems to offer a suitable way of producing a population inversion. The answer is that one can, in general, produce a population inversion much more easily in a four-level than in a three-level laser. To see this, we begin by noting that the energy difference among the various levels of Fig. 1.4 are usually much greater than kT. According to Boltzmann statistics [see, e.g., Eq. (1.2.2)] we can then say that essentially all atoms are initially (i.e., at equilibrium) in the ground level. If we now let N_t be the atom density in the material, these will initially all be in level 1 from the three-level case. Let us now begin raising atoms from level 1 to level 3. They will then decay to level 2 and, if this decay is sufficiently fast, level 3 will remain more or less empty. Let us now assume, for simplicity, that the two levels are either non-degenerate (i.e. $g_1 = g_2 = 1$) or have the same degeneracy. Then, according to Eq. (1.2.1), the absorption losses will be compensated by the gain when $N_2 = N_1$. From this point on, any further atom that is raised will then contribute to population inversion. In a four-level laser, however, since level 1 is also empty, any atom that has been raised to level 2 immediately produces population inversion. The above discussion shows that, whenever possible, we should look for a material that can be operated as a four-level rather than a three-level system. The use of more than four levels is, of course, also possible. It should be noted that the term "four-level laser" has come to be used for any laser in which the lower laser level is essentially empty, by virtue of being above the ground level by many kT. So if level 2 and level 3 are the same level, then one has a level scheme which would be described as "four-level" in the sense above, while only having three levels! Cases based on such a "four-level" scheme do exist. It should also be noted that, more recently, the so-called *quasi-three-level lasers* have also become a very important cathegory of laser. In this case, the ground level consists of many sublevels, the lower laser level being one of these sublevels. Therefore, the scheme of Fig. 1.4b can still be applied to a quasi-three-level laser with the understanding that level 1 is a sublevel of the ground level and level 0 is the lowest sublevel of the ground level. If all ground state sublevels are strongly coupled, perhaps by some fast non-radiative decay process, then the populations of these sublevels will always be in thermal equilibrium. Let us further assume that the energy separation between level 1 and level 0 (see Fig. 1.4b) is comparable to kT. Then, according to Eq. (1.2.2), there will always be some population present in the lower laser level and the laser system will behave in a way which is intermediate between a three- and a four-level laser.

The process by which atoms are raised from level 1 to level 3 (in a three-level scheme), from 0 to 3 (in a four-level scheme), or from the ground level to level 3 (in a quasi-three-level scheme) is known as *pumping*. There are several ways in which this process can be realized in practice, e.g., by some sort of lamp of sufficient intensity or by an electrical discharge in the active medium. We refer to Chap. 6 for a more detailed discussion of the various pumping processes. We note here, however, that, if the upper pump level is empty, the rate at which the upper laser level becomes populated by the pumping, $(dN_2/dt)_p$, can in general be written as $(dN_2/dt)_p = W_p N_g$ where W_p is a suitable rate describing the pumping process and N_g is the population of the ground level for either a three- or four-level laser while, for a quasi-three-level laser, it can be taken to be the total population of all ground state sublevels. In what follows, however, we will concentrate our discussion mostly on four level or quasi-three-level lasers. The most important case of three-level laser, in fact, is the Ruby laser, a historically important laser (it was the first laser ever made to operate) although no longer so widely used. For most four-level and quasi-three-level lasers in commun use, the depletion of the ground level, due to the pumping process, can be neglected.* One can then write N_g = const and the previous equation can be written, more simply, as

$$(dN_2/dt)_p = R_p \tag{1.3.1}$$

where R_p may be called the pump rate per unit volume or, more briefly, the *pump rate*. To achieve the threshold condition, the pump rate must reach a threshold or critical value, R_{cp}. Specific expressions for R_{cp} will be obtained in Chap. 6 and Chap. 7.

1.4. PROPERTIES OF LASER BEAMS

Laser radiation is characterized by an extremely high degree of (1) monochromaticity, (2) coherence, (3) directionality, and (4) brightness. To these properties a fifth can be added,

* One should note that, as a quasi-3-level laser becomes progressively closer to a pure 3-level laser, the assumption that the ground state population is changed negligibly by the pumping process will eventually not be justified. One should also note that in fiber lasers, where very intense pumping is readily achieved, the ground state can be almost completely emptied.

viz., (5) short time duration. This refers to the capability for producing very short light pulses, a property that, although perhaps less fundamental, is nevertheless very important. We shall now consider these properties in some detail.

1.4.1. Monochromaticity

Briefly, we can say that this property is due to the following two circumstances: (1) Only an e.m. wave of frequency ν_0 given by (1.1.1) can be amplified. (2) Since the two-mirror arrangement forms a resonant cavity, oscillation can occur only at the resonance frequencies of this cavity. The latter circumstance leads to the laser linewidth being often much narrower (by as much as to ten orders of magnitude!) than the usual linewidth of the transition $2 \rightarrow 1$ as observed in spontaneous emission.

1.4.2. Coherence

To first order, for any e.m. wave, one can introduce two concepts of coherence, namely, spatial and temporal coherence.

To define spatial coherence, let us consider two points P_1 and P_2 that, at time $t = 0$, lie on the same wave-front of some given e.m. wave and let $E_1(t)$ and $E_2(t)$ be the corresponding electric fields at these two points. By definition, the difference between the phases of the two field at time $t = 0$ is zero. Now, if this difference remains zero at any time $t > 0$, we will say that there is a perfect coherence between the two points. If this occurs for any two points of the e.m. wave-front, we will say that the wave has *perfect spatial coherence*. In practice, for any point P_1, the point P_2 must lie within some finite area around P_1 if we want to have a good phase correlation. In this case we will say that the wave has a *partial spatial coherence* and, for any point P, we can introduce a suitably defined coherence area $S_c(P)$.

To define temporal coherence, we now consider the electric field of the e.m. wave at a given point P, at times t and $t + \tau$. If, for a given time delay τ, the phase difference between the two field remains the same for any time t, we will say that there is a temporal coherence over a time τ. If this occurs for any value of τ, the e.m. wave will be said to have perfect time coherence. If this occurs for a time delay τ such that $0 < \tau < \tau_0$, the wave will be said to have partial temporal coherence, with a coherence time equal to τ_0. An example of an e.m wave with a coherence time equal to τ_0 is shown in Fig. 1.5. The figure shows a sinusoidal electric field undergoing random phase jumps at time intervals equal to τ_0. We see that the concept of temporal coherence is, at least in this case, directly connected with that of monochromaticity. We will show, in fact, in Chap. 11, that any stationary e.m. wave with coherence time τ_0 has a bandwidth $\Delta\nu \cong 1/\tau_0$. In the same chapter it will also be shown that, for a non-stationary but repetitively reproducing beam (e.g., a repetitively Q-switched or a mode-locked laser beam) the coherence time is not related to the inverse of the oscillation bandwidth $\Delta\nu$ and may actually be much longer than $1/\Delta\nu$.

It is important to point out that the two concepts of temporal and spatial coherence are indeed independent of each other. In fact, examples can be given of a wave having perfect spatial coherence but only limited temporal coherence (or vice versa). If, for instance, the wave shown in Fig. 1.5 were to represent the electric fields at points P_1 and P_2 considered earlier,

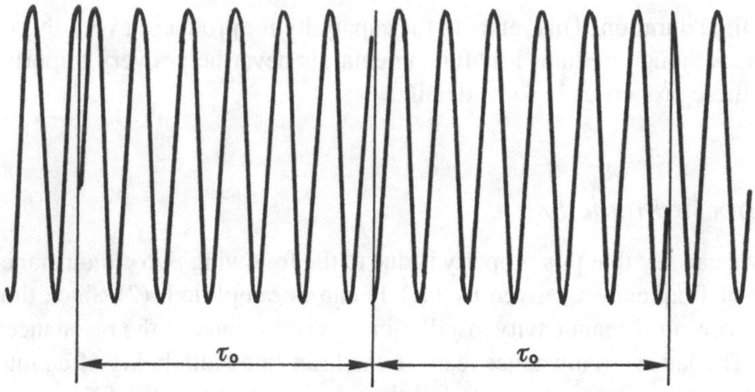

FIG. 1.5. Example of an e.m. wave with a coherence time of approximately τ_0.

the spatial coherence between these two points would be complete still the wave having a limited temporal coherence.

We conclude this section by emphasizing that the concepts of spatial and temporal coherence provide only a first-order description of the laser's coherence. Higher order coherence properties will in fact discussed in Chap. 11. Such a discussion is essential for a full appreciation of the difference between an ordinary light source and a laser. It will be shown in fact that, by virtue of the differences between the corresponding higher-order coherence properties, a laser beam is fundamentally different from an ordinary light source.

1.4.3. Directionality

This property is a direct consequence of the fact that the active medium is placed in a resonant cavity. In the case of the plane parallel one of Fig. 1.3, for example, only a wave propagating in a direction orthogonal to the mirrors (or in a direction very near to it) can be sustained in the cavity. To gain a deeper understanding of the directional properties of a laser beam (or, in general, of any e.m. wave), it is convenient to consider, separately, the case of a beam with perfect spatial coherence and the case of partial spatial coherence.

Let us first consider the case of perfect spatial coherence. Even for this case, a beam of finite aperture has unavoidable divergence due to diffraction. This can be understood with the help of Fig. 1.6, where a monochromatic beam of uniform intensity and plane wave-front is assumed to be incident on a screen S containing an aperture D. According to Huyghens' principle the wave-front at some plane P behind the screen can be obtained from the superposition of the elementary waves emitted by each point of the aperture. We thus see that, on account of the finite size D of the aperture, the beam has a finite divergence θ_d. Its value can be obtained from diffraction theory. For an arbitrary amplitude distribution we get

$$\theta_d = \beta\, \lambda / D \tag{1.4.1}$$

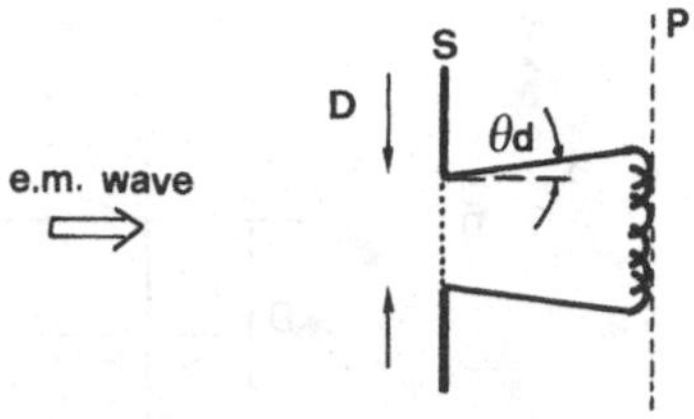

FIG. 1.6. Divergence of a plane e.m. wave due to diffraction.

where λ and D are the wavelength and the diameter of the beam. The factor β is a numerical coefficient of the order of unity whose value depends on the shape of the amplitude distribution and on the way in which both the divergence and the beam diameter are defined. A beam whose divergence can be expressed as in Eq. (1.4.1) is described as being *diffraction limited.*

If the wave has only a partial spatial coherence, its divergence will be larger than the minimum value set by diffraction. Indeed, for any point P' of the wave-front, the Huygens' argument of Fig. 1.6 can only be applied for points lying within the coherence area S_c around point P'. The coherence area thus acts as a limiting aperture for the coherent superposition of the elementary wavelets. The beam divergence will now be given by

$$\theta = \beta\lambda / [S_c]^{1/2} \tag{1.4.2}$$

where. again, β is a numerical coefficient of the order of unity whose exact value depends on the way in which both the divergence θ and the coherence area S_c are defined.

We conclude this general discussion of the directional properties of e.m. waves by pointing out that, given suitable operating conditions, the output beam of a laser can be made diffraction limited.

1.4.4. Brightness

We define the brightness of a given source of e.m. waves as the power emitted per unit surface area per unit solid angle. To be more precise, let dS be the elemental surface area at point O of the source (Fig. 1.7a). The power dP emitted by dS into a solid angle $d\Omega$ around direction OO' can be written as

$$dP = B \cos\theta \, dS \, d\Omega \tag{1.4.3}$$

where θ is the angle between OO' and the normal $\mathbf{n}$ to the surface. Note that the factor $\cos\theta$ arises simply from the fact that the physically important quantity for the emission along the OO' direction is the projection of dS on a plane orthogonal to the OO' direction, i.e. $\cos\theta \, dS$. The quantity B defined through Eq. (1.4.3) is called the source brightness at the point O in the direction OO'. This quantity will generally depend on the polar coordinates θ and ϕ of the direction OO' and on the point O. When B is a constant, the source is said to be isotropic (or a Lambertian source).

Let us now consider a laser beam of power P, with a circular cross section of diameter D and with a divergence θ (Fig. 1.7b). Since θ is usually very small, we have $\cos\theta \cong 1$. Since

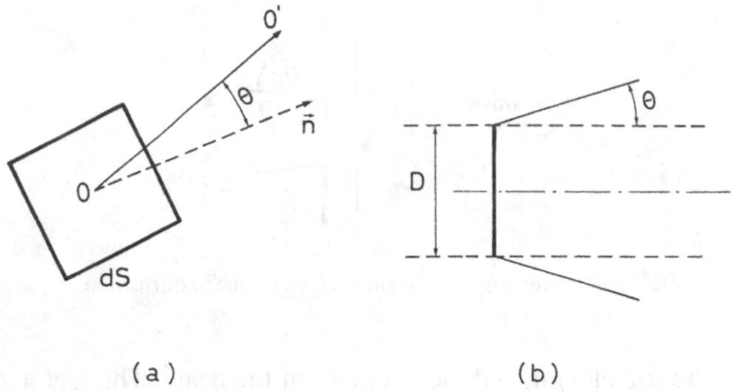

FIG. 1.7. (a) Surface brightness at the point O for a general source of e.m. waves. (b) Brightness of a laser beam of diameter D and divergence θ.

the area of the beam is equal to $\pi D^2/4$ and the emission solid angle is $\pi\theta^2$, then, according to Eq. (1.4.3), we obtain the beam brightness as

$$B = 4P/(\pi D\theta)^2 \tag{1.4.4}$$

Note that, if the beam is diffraction limited, we have $\theta = \theta_d$ and, with the help of Eq. (1.4.1), we obtain from Eq. (1.4.4)

$$B = \left(\frac{2}{\beta\pi\lambda}\right)^2 P \tag{1.4.5}$$

which is the maximum brightness that a beam of power P can have.

Brightness is the most important parameter of a laser beam and, in general, of any light source. To illustrate this point we first recall that, if we form an image of any light source through a given optical system and if we assume that object and image are in the same medium (e.g. air), then the following property holds: The brightness of the image is always less than or equal to that of the source, the equality holding when the optical system provides lossless imaging of the light emitted by the source. To further illustrate the importance of brightness, let us consider the beam of Fig. 1.7b, having a divergence equal to θ, to be focused by a lens of focal length f. We are interested in calculating the peak intensity of the beam in the focal plane of the lens (Fig. 1.8a). To make this calculation we recall that the beam can be decomposed into a continuous set of plane waves with an angular spread of approximately θ around the propagation direction. Two such waves, making an angle θ' are indicated by solid and dashed lines, respectively, in Fig. 1.8b. The two beams will each be focused to a distinct spot in the focal plane and, for small angle θ', the two spots are transversely separated by a distance $r = f\theta'$. Since the angular spread of the plane waves which make up the beam of Fig. 1.8a is equal to the beam divergence θ, we arrive at the conclusion that the diameter, d, of the focal spot in Fig. 1.8a is approximately equal to $d = 2f\theta$. For an ideal, lossless, lens the overall power in the focal plane equals the power, P, of the incoming wave. The peak intensity in the focal plane is thus found to be $I_p = 4P/\pi d^2 = P/\pi(f\theta)^2$. In terms of beam brightness, according to (1.4.4) we then have $I_p = (\pi/4)B(D/f)^2$. Thus I_p increases with increasing beam

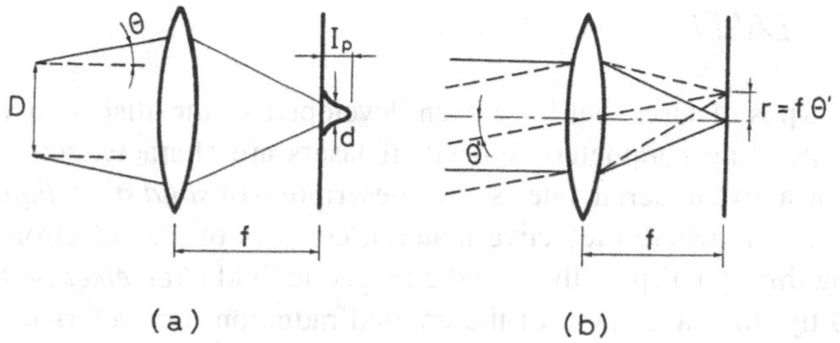

FIG. 1.8. (a) Intensity distribution in the focal plane of a lens for a beam of divergence θ. (b) Plane-wave decomposition of the beam of a.

diameter D. The maximum value of I_p is then attained when D is made equal to the lens diameter D_L. In this case we obtain

$$I_p = (\pi/4)\,(N.A.)^2\,B \tag{1.4.6}$$

where $N.A. = \sin[\tan^{-1}(D_L/f)] \cong (D_L/f)$ is the lens numerical aperture. Equation (1.4.6) then shows that, for a given numerical aperture, the peak intensity in the focal plane of a lens depends only on the beam brightness.

A laser beam of even moderate power (e.g. a few milliwatts) has a brightness that is several orders of magnitude greater than that of the brightest conventional sources (see, e.g., problem 1.7). This is mainly due to the highly directional properties of the laser beam. According to Eq. (1.4.6), this means that the peak intensity produced in the focal plane of a lens can be several order of magnitude larger for a laser beam compared to that of a conventional source. Thus the focused intensity of a laser beam can reach very large values, a feature which is exploited in many applications of lasers.

1.4.5. Short Time Duration

Without going into any detail at this stage, we simply mention that by means of a special technique called mode locking, it is possible to produce light pulses whose duration is roughly equal to the inverse of the linewidth of the $2 \rightarrow 1$ transition. Thus, with gas lasers, whose linewidth is relatively narrow, the pulse-width may be of ~ 0.1–1 ns. Such pulse durations are not regarded as particularly short and indeed even some flashlamps can emit light pulses with a duration of somewhat less than 1 ns. On the other hand, the linewidth of some solid state and liquid lasers can be 10^3–10^5 times larger than that of a gas laser, and, in this case, much shorter pulses may be generated (down to $\sim$ 10 fs). This opens up exciting new possibilities for laser research and applications.

Notice that the property of short time duration, which implies energy concentration in time, can, in a sense, be considered to be the counterpart of monochromaticity, which implies energy concentration in wavelength. Short time duration would, however, perhaps be regarded as a less fundamental property than monochromaticity. While in fact all lasers can, in principle, be made extremely monochromatic, only lasers with a broad linewidth, i.e. solid state and liquid lasers, may produce pulses of very short time duration.

1.5. TYPES OF LASERS

The various types of laser that have been developed so far, display a very wide range of physical and operating parameters. Indeed, if lasers are characterized according to the physical state of the active material, one uses the description of *solid state*, *liquid* or *gas lasers*. A rather special case is where the active material consists of free electrons, at relativistic velocities, passing through a spatially periodic magnetic field (*free-electron lasers*). If lasers are characterized by the wavelength of the emitted radiation, one refers to *infrared lasers*, *visible lasers*, *UV* and *X-ray lasers*. The corresponding wavelength may range from ≈ 1 mm (i.e. millimeter waves) down to ≈ 1 nm (i.e. to the upper limit of hard X-rays). The span in wavelength can thus be a factor of $\approx 10^6$ (we recall that the visible range spans less than a factor 2, roughly from 700 to 400 nm). Output powers cover an even larger range of values. For cw lasers, typical powers go from a few mW, in lasers used for signal sources (e.g. for optical communications or for bar-code scanners), to tens of kW in lasers used for material working, to a few MW (≈ 5 MW so far) in lasers required for some military applications (e.g. for directed energy weapons). For pulsed lasers the peak power can be much higher than for cw lasers and can reach values as high as 1 PW (10^{15} W)! Again for pulsed lasers, the pulse duration can vary widely from the ms level typical of lasers operating in the so-called *free-running* regime (i.e. without any *Q-switching* or *mode-locking* element in the cavity) down to about 10 fs (1 fs $= 10^{-15}$ s) for some mode locked lasers. The physical dimensions can also vary widely. In terms of cavity length, for instance, the length can be as small as ~ 1 μm for the shortest lasers up to some km for the longest (e.g. a laser 6.5 km long, which was set up in a cave for geodetic studies).

This wide range of physical or operating parameters represent both a strength and a weakness. As far as applications are concerned, this wide range of parameters offers enormous potential in several fields of fundamental and applied sciences. On the other hand, in terms of markets, a very wide spread of different devices and systems can be an obstacle to mass production and its associated price reduction.

1.6. ORGANIZATION OF THE BOOK

The organization of the book is based on the fact that, as indicated in our discussion so far, a laser can be considered to consist of three elements: (1) an active material, (2) a pumping scheme, (3) a resonator. Accordingly, the next two chapters deal with the interaction of radiation with matter, starting from the simplest cases, i.e. atoms or ions in an essentially isolated situation, (Chap. 2), and going on to the more complicated cases, i.e. molecules and semiconductors, (Chap. 3). As an introduction to optical resonators, the next Chapter (Chap. 4) considers some topics relating to ray and wave propagation in particular optical elements such as free-space, optical lens-like media, Fabry-Perot interferometers and multi-layer dielectric coatings. Chapter 5 then deals with the theory of optical resonators while the next Chapter (Chap. 6) deals with the pumping processes. The concepts introduced in these chapters are then used in next two chapters (Chap. 7 and 8) where the theory is developed for continuous wave and transient laser behavior, respectively. The theory is based on the lowest order approximation, i.e. using the rate equation approach. This treatment is, in fact, capable of describing most laser characteristics. Obviously, lasers based upon different types of active media have significant differences in their characteristics. So, the next two chapters (Chap. 9 and 10) discuss the characteristic properties of a number of types of laser. Thus Chap. 9 covers ionic crystal, dye and semiconductor lasers, these having a number of common features, while Chap. 10 considers gas, chemical and free-electron lasers. By this point, the reader should have acquired sufficient understanding of laser behavior to go on to a study of the properties of the output beam (coherence, monochromaticity, brightness, noise). These properties are considered in Chap. 11. Finally, the theme of Chap. 12 is based on the fact that, before being put to use, a laser beam is generally transformed in some way. This includes: (1) spatial transformation of the beam due to its propagation through e.g. a lens system; (2) amplitude transformation as a result of passing through an amplifier; (3) wavelength transformation, or frequency conversion, via a number of nonlinear phenomena (second harmonic generation, parametric processes); (4) time transformation by e.g. pulse compression.

PROBLEMS

1.1. The part of the e.m. spectrum that is of interest in the laser field starts from the submillimiter wave region and goes down in wavelength to the X-ray region. This covers the following regions in succession: (1) far infrared; (2) near infrared; (3) visible; (4) ultraviolet (uv); (5) vacuum ultraviolet (vuv); (6) soft X-ray; (7) X-ray: From standard textbooks find the wavelength intervals of the above regions. Memorize or record these intervals since they are frequently used in this book.

1.2. As a particular case of Problem 1.1, memorize or record the wavelengths corresponding to blue, green, and red light.

1.3. If levels 1 and 2 of Fig. 1.1 are separated by an energy E_2-E_1 such that the corresponding transition frequency falls in the middle of the visible range, calculate the ratio of the populations of the two levels in thermal equilibrium at room temperature.

1.4. When in thermal equilibrium at $T = 300\,\mathrm{K}$, the ratio of the level populations N_2/N_1 for some particular pair of levels is given by $1/e$. Calculate the frequency ν for this transition. In what region of the e.m. spectrum does this frequency fall?

1.5. A laser cavity consists of two mirrors with reflectivities $R_1 = 1$ and $R_2 = 0.5$ while the internal loss per pass is $L_i = 1\%$. Calculate the total logarithmic losses per pass. If the length of the active material is $l = 7.5\,\mathrm{cm}$ and the transition cross section is $\sigma = 2.8 \times 10^{-19}\,\mathrm{cm}^2$, calculate then the threshold inversion.

1.6. The beam from a ruby laser ($\lambda \cong 694\,\mathrm{nm}$) is sent to the moon after passing through a telescope of 1 m diameter. Calculate the approximate value of beam diameter on the moon assuming that the beam has perfect spatial coherence (the distance between earth and moon is approximately 384,000 km).

1.7. The brightness of probably the brightest lamp so far available (PEK Labs type 107/109, excited by 100 W of electrical power) is about 95 $\mathrm{W/cm^2}$ sr in its most intense green line ($\lambda = 546\,\mathrm{nm}$). Compare this brightness with that of a 1 W Argon laser ($\lambda = 514.5\,\mathrm{nm}$), which can be assumed to be diffraction limited.

2

Interaction of Radiation with Atoms and Ions

2.1. INTRODUCTION

This chapter deals with the interaction of radiation with atoms and ions which are weakly interacting with any surrounding species, such as atoms or ions in a gas phase or impurity ions in an ionic crystal. The somewhat more complicated case of interaction of radiation with molecules or semiconductors will be considered in the next chapter. Since the subject of radiation interaction with matter is, of course, very wide, we will limit our discussion to those phenomena which are relevant for atoms and ions acting as active media. So, after an introductory section dealing with the theory of blackbody radiation, a milestone for the whole of modern physics, we will consider the elementary processes of absorption, stimulated emission, spontaneous emission, and nonradiative decay. They will first be considered within the simplifying assumptions of a dilute medium and a low intensity. Following this, situations involving a high beam intensity and a medium that is not dilute (leading, in particular, to the phenomena of saturation and amplified spontaneous emission) will be considered. A number of very important, although perhaps less general, topics relating to the photophysics of dye lasers, free-electron lasers, and X-ray lasers will be briefly considered in Chaps. 9 and 10 immediately preceding the discussion of the corresponding laser.

2.2. SUMMARY OF BLACKBODY RADIATION THEORY[1]

Let us consider a cavity filled with a homogeneous and isotropic medium. If the walls of the cavity are kept at a constant temperature, T, they will continuously emit and receive power in the form of electromagnetic (e.m.) radiation. When the rates of absorption and emission

O. Svelto, *Principles of Lasers*,
DOI: 10.1007/978-1-4419-1302-9_2, © Springer Science+Business Media LLC 2010

becomes equal, an equilibrium condition is established at the walls of the cavity as well as at each point of the dielectric. This situation can be described by introducing the energy density ρ, which represents the electromagnetic energy contained in unit volume of the cavity. This energy density can be expressed as a function of the electric field, $E(t)$, and magnetic field, $H(t)$, according to the formula

$$\rho = <\frac{1}{2}\varepsilon E^2> + \frac{1}{2}\mu H^2> \tag{2.2.1}$$

where ε and μ are, respectively, the dielectric constant and the magnetic permeability of the medium inside the cavity and where the symbol $< >$ indicates a time average over a cycle of the radiation field. We can then represent the spectral energy distribution of this radiation by the function ρ_ν, which is a function of frequency ν. This is defined as follows: $\rho_\nu d\nu$ represents the energy density of radiation in the frequency range from ν to $\nu + d\nu$. The relationship between ρ and ρ_ν is obviously

$$\rho = \int_0^\infty \rho_\nu d\nu \tag{2.2.2}$$

Suppose now that a hole is made in the wall of the cavity. If we let I_ν be the spectral intensity of the light escaping from the hole, one can show that I_ν is proportional to ρ_ν obeying the simple relation

$$I_\nu = (c/4n)\rho_\nu \tag{2.2.3}$$

where c is the velocity of light in the vacuum and n is the refractive index of the medium inside the cavity. We can now show that I_ν and hence ρ_ν are universal functions, independent of either the nature of the walls or the cavity shape, and dependent only on the frequency ν and temperature T of the cavity. This property of ρ_ν can be proven through the following simple thermodynamic argument. Let us suppose we have two cavities of arbitrary shape, whose walls are at the same temperature T. To ensure that the temperature remains constant, we may imagine that the walls of the two cavities are in thermal contact with two thermostats at temperature T. Let us suppose that, at a given frequency ν, the energy density ρ'_ν in the first cavity is greater than the corresponding value ρ''_ν in the second cavity. We now optically connect the two cavities by making a hole in each and then imaging, with some optical system, each hole onto the other. We also insert an ideal filter in the optical system, which lets through only a small frequency range around the frequency ν. If $\rho'_\nu > \rho''_\nu$ then, according to Eq. (2.2.3), one will have $I'_\nu > I''_\nu$ and there will be a net flow of electromagnetic energy from cavity 1 to cavity 2. Such a flow of energy, however, would violate the second law of thermodynamics, since the two cavities are at the same temperature. Therefore one must have $\rho'_\nu = \rho''_\nu$ for all frequencies.

The problem of calculating this universal function $\rho_\nu(\nu, T)$ was a very challenging one for the physicists of the time. Its complete solution was provided by Planck, who, in order to find a correct solution of the problem, had to introduce the so-called hypothesis of light quanta. The blackbody theory is therefore one of the fundamental bases of modern physics.[(1)] Before going further into it, we first need to consider the electromagnetic modes of a blackbody cavity. Since the function ρ_ν is independent of the cavity shape or the nature of the dielectric medium, we choose to consider the relatively simple case of a rectangular cavity uniformly filled with dielectric and with perfectly conducting walls.

2.2.1. Modes of a Rectangular Cavity

Let us consider the rectangular cavity of Fig. 2.1. To calculate ρ_ν, we begin by calculating the standing e.m. field distributions that can exist in this cavity. According to Maxwell's equations, the electric field $\mathbf{E}(x, y, z, t)$ must satisfy the wave equation

$$\nabla^2 \mathbf{E} - \frac{1}{c_n^2}\frac{\partial^2 \mathbf{E}}{\partial t^2} = 0 \tag{2.2.4}$$

where ∇^2 is the Laplacian operator and c_n is the velocity of light in the medium considered. In addition, the field must satisfy the following boundary condition at each wall:

$$\mathbf{E} \times \mathbf{n} = 0 \tag{2.2.5}$$

where $\mathbf{n}$ is the normal to the particular wall under consideration. This condition expresses the fact that, for perfectly conducting walls, the tangential component of the electric field must vanish on the walls of the cavity.

It can be easily shown that the problem is soluble by separation of the variable. Thus, if we put

$$\mathbf{E} = \mathbf{u}(x, y, z)E(t) \tag{2.2.6}$$

and substitute Eq. (2.2.6) in Eq. (2.2.4), we have

$$\nabla^2 \mathbf{u} = -k^2 \mathbf{u} \tag{2.2.7a}$$

$$\frac{d^2 E}{dt^2} = -(c_n k)^2 E \tag{2.2.7b}$$

where k is a constant. Equation (2.2.7b) has the general solution

$$E = E_0 \cos(\omega t + \phi) \tag{2.2.8}$$

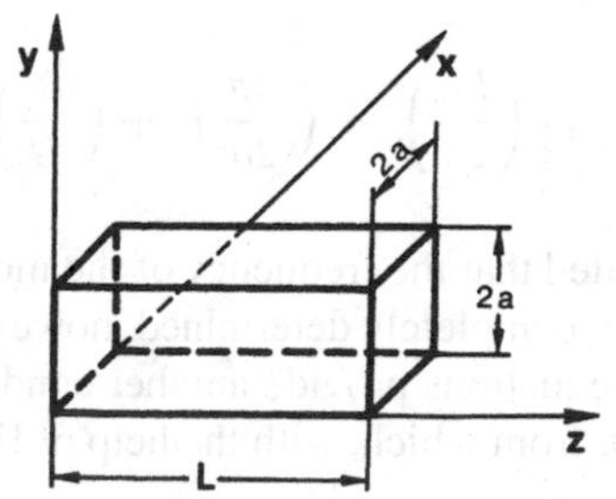

FIG. 2.1. Rectangular cavity with perfectly conducting walls kept at temperature T.

where E_0 and ϕ are arbitrary constant and where

$$\omega = c_n k \tag{2.2.9}$$

With $E(t)$ given by Eq. (2.2.8), we see that the solution Eq. (2.2.6) can be written as

$$\mathbf{E}(x, y, z, t) = E_0 \mathbf{u}(x, y, z) \exp(j\omega t + \phi) \tag{2.2.9a}$$

and thus corresponds to a standing wave configuration of the e.m. field within the cavity. In fact the amplitude of oscillation at a given point of the cavity is constant in time. A solution of this type is referred to as a an e.m. *mode* of the cavity.

We are now left with the task of solving Eq. (2.2.7a), known as the Helmholtz equation, subject to the boundary condition given by Eq. (2.2.5). It can readily be verified that the expressions

$$\begin{aligned} u_x &= e_x \cos k_x x \sin k_y y \sin k_z z \\ u_y &= e_y \sin k_x x \cos k_y y \sin k_z z \\ u_z &= e_z \sin k_x x \sin k_y y \cos k_z z \end{aligned} \tag{2.2.10}$$

satisfy Eq. (2.2.7a) for any value of e_x, e_y, e_z, provided that

$$k_x^2 + k_y^2 + k_z^2 = k^2 \tag{2.2.11}$$

Furthermore, the solution Eq. (2.2.10) already satisfies the boundary condition Eq. (2.2.5) on the three planes $x = 0, y = 0, z = 0$. If we now impose the condition that Eq. (2.2.5) should also be satisfied on the other walls of the cavity, we obtain

$$\begin{aligned} k_x &= l\pi/2a \\ k_y &= m\pi/2a \\ k_z &= n\pi/L \end{aligned} \tag{2.2.12}$$

where l, m, and n are positive integers. Their physical significance can be seen immediately: they represent the number of nodes that the standing wave mode has along the directions x, y, and z, respectively. For fixed values of l, m, and n it follows that k_x, k_y, and k_z will also be fixed and, according to Eqs. (2.2.9) and (2.2.11), the angular frequency ω of the mode will also be fixed and given by

$$\omega_{lmn} = c_n \left[\left(\frac{l\pi}{2a}\right)^2 + \left(\frac{m\pi}{2a}\right)^2 + \left(\frac{n\pi}{L}\right)^2 \right]^{1/2} \tag{2.2.13}$$

where we have explicitly indicated that the frequency of the mode will depend on the indices l, m, and n. The mode is still not completely determined, however, since e_x, e_y, and e_z are still arbitrary. However, Maxwell's equations provide another condition that must be satisfied by the electric field, i.e., $\nabla \cdot \mathbf{u} = 0$, from which, with the help of Eq. (2.2.10), we get

$$\mathbf{e} \cdot \mathbf{k} = 0 \tag{2.2.14}$$

In Eq. (2.2.14) we have introduced the two vectors **e** and **k**, whose components along x, y, and z axes are respectively, e_x, e_y, and e_z and k_x, k_y, and k_z. Equation (2.2.14) therefore shows that, out of the three quantities e_x, e_y, and e_z, only two are independent. In fact, once we fix l, m, and n (i.e., once **k** is fixed), the vector **e** is bound to lie in a plane perpendicular to **k**. In this plane, only two degree of freedom are left for the choice of the vectors **e**, and only two independent modes are thus present. Any other vector, **e**, lying in this plane can in fact be obtained as a linear combination of the previous two vectors.

Let us now calculate the number of resonant modes, N_ν, whose frequency lies between 0 and ν. This will be the same as the number of modes whose wave vector **k** has a magnitude, k, between 0 and $2\pi\nu/c_n$. From Eq. (2.2.12) we see that, in a system coordinate k_x, k_y, k_z, the possible values for **k** are given by the vectors connecting the origin with the nodal points of the three-dimensional lattice shown in Fig. 2.2. Since, however, k_x, k_y, and k_z are positive quantities, we must count only those points lying in the positive octant. It can furthermore be easily shown that there is a one to one correspondence between these points and the unit cell of dimensions $(\pi/2a, \pi/2a, \pi/L)$. The number of points having k between 0 and $(2\pi\nu/c_n)$ can thus be calculated as (1/8) times the volume of the sphere, centered at the origin, and of radius $(2\pi\nu/c_n)$ divided by the volume of the unit cell of dimensions $(\pi/2a, \pi/2a, \pi/L)$. Since, as previously noted, there are two modes possible for each value of k, we have

$$N_\nu = 2\frac{\frac{1}{8}\frac{4}{3}\pi\left(\frac{2\pi\nu}{c_n}\right)^3}{\frac{\pi}{2a}\frac{\pi}{2a}\frac{\pi}{L}} = \frac{8\pi\nu^3}{3c_n^3}V \qquad (2.2.15)$$

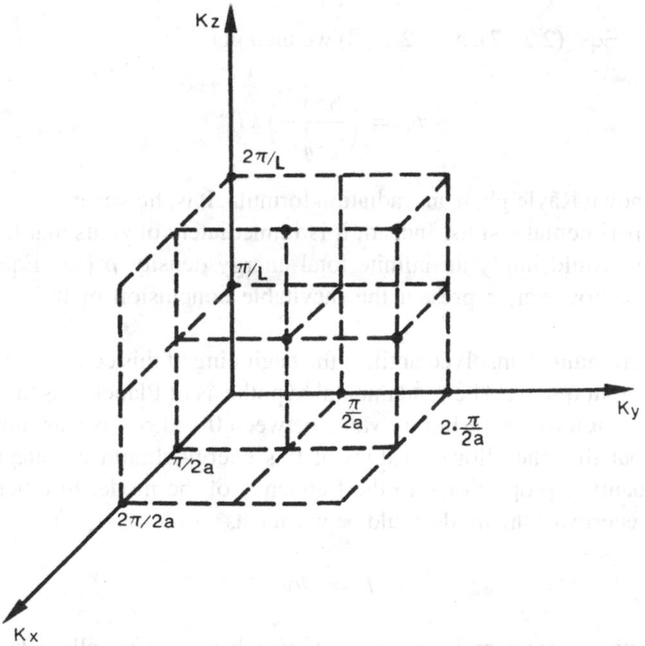

FIG. 2.2. Pictorial illustration of the density of modes in the cavity of Fig. 2.1. Each point of the lattice corresponds to two cavity modes.

where V is the total volume of the cavity. If we now define $p(\nu)$ as the number of modes per unit volume and per unit frequency range, we have

$$p(\nu) = \frac{1}{V}\frac{dN}{d\nu} = \frac{8\pi\nu^2}{c_n^3}. \tag{2.2.16}$$

2.2.2. The Rayleigh-Jeans and Planck Radiation Formula

Having calculated the quantity $p(\nu)$ we can now proceed to calculate the energy density ρ_ν. We can begin by writing ρ_ν as the product of the number of modes per unit volume per unit frequency range, $p(\nu)$, multiplied by the average energy $<E>$ contained in each mode, i.e.

$$\rho_\nu = p_\nu <E> \tag{2.2.17}$$

To calculate $<E>$ we assume that the cavity walls are kept at a constant temperature T. According to Boltzmann's statistics, the probability dp that the energy of a given cavity mode lies between E and $E + dE$ is expressed by $dp = C\exp[-(E/kT)]dE$, where C is a constant to be established by the condition $\int_0^\infty C\exp[-(E/kT)dE = 1$. The average energy of the mode $<E>$ is therefore given by

$$<E> = \frac{\int_0^\infty E\exp[-(E/kT)]dE}{\int_0^\infty \exp[-(E/kT)]dE} = kT \tag{2.2.18}$$

From Eq. (2.2.16), Eqs. (2.2.17), and (2.2.18) we then get

$$\rho_\nu = \left(\frac{8\pi\nu^2}{c_n^3}\right)kT \tag{2.2.19}$$

This is the well known Rayleigh-Jeans radiation formula. It is, however, in complete disagreement with the experimental results. Indeed, it is immediately obvious that Eq. (2.2.19) must be wrong since it would imply an infinite total energy density ρ [see Eq. (2.2.2)]. Equation (2.2.19) does, however, represent the inevitable conclusion of the previous classical arguments.

The problem remained unsolved until, at the beginning of this century, Planck introduced the hypothesis of light quanta. The fundamental hypothesis of Planck was that the energy in a given mode could not have any arbitrary value between 0 and ∞, as was implicitly assumed in Eq. (2.2.18), but that the allowed values of this energy should be integral multiples of a fundamental quantity, proportional to the frequency of the mode. In other words, Planck assumed that the energy of the mode could be written as

$$E = nh\nu \tag{2.2.20}$$

where n is a positive integer and h a constant (which was later called Planck's constant). Without entering into too many details, here, on this fundamental hypothesis, we merely wish to note that this essentially implies that energy exchange between the inside of the cavity and its walls must involve a discrete amount of energy $h\nu$. This minimum quantity that can be exchanged is called a light quantum or photon. According to this hypothesis, the average energy of the mode is now given by

$$E = \frac{\sum_0^\infty {}_n nh\nu \exp[-(nh\nu/kT)]}{\sum_0^\infty {}_n \exp[-(nh\nu/kT)]} = \frac{h\nu}{\exp(h\nu/kT) - 1} \tag{2.2.21}$$

This formula is quite different from the classical expression Eq. (2.2.18). Obviously, for $h\nu \ll kT$, Eq. (2.2.21) reduces to Eq. (2.2.18). From Eq. (2.2.16), Eqs. (2.2.17), and (2.2.21) we now obtain the Planck formula,

$$\rho_\nu = \frac{8\pi\nu^2}{c_n^3} \frac{h\nu}{\exp(h\nu/kT) - 1} \tag{2.2.22}$$

which is in perfect agreement with the experimental results, provided that we choose for h the value $h = 6.62 \times 10^{-34}$ J × s. For example, Fig. 2.3 shows the behavior predicted by Eq. (2.2.22) for ρ_ν vs frequency ν for two values of temperature T.

Lastly, we may notice that the ratio

$$<\phi> = \frac{<E>}{h\nu} = \frac{1}{\exp(h\nu/kT) - 1} \tag{2.2.23}$$

gives the average number of photons $<\phi>$ for each mode. If we now consider a frequency ν in the optical range ($\nu \approx 4 \times 10^{14}$ Hz), we get $h\nu \approx 1\,eV$. For $T \cong 300$ K we have $kT \cong (1/40)$ eV, so that from Eq. (2.2.23) it is $<\phi> \cong \exp(-40)$. We thus see that the average number of photons per mode, for blackbody radiation at room temperature, is very much smaller than unity. This value should be compared with the number of photons ϕ_0 that can be obtained in a laser cavity for a single mode (see Chap. 7).

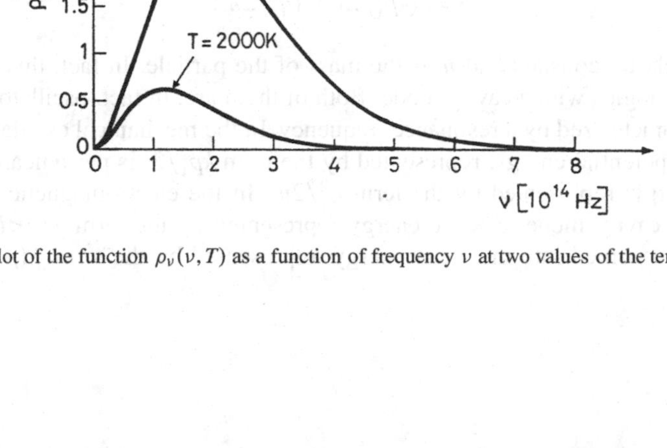

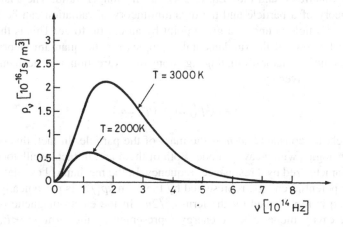

FIG. 2.3. Plot of the function $\rho_\nu(\nu, T)$ as a function of frequency ν at two values of the temperature T.

2.2.3. Planck's Hypothesis and Field Quantization

The fundamental assumption of Planck given by Eq. (2.2.20) was considered with a degree of caution if not suspicion for some time after the idea was proposed. Someone even considered it as a mere mathematical trick to transform an integral [Eq. (2.2.18)] into a summation [Eq. (2.2.21)] to get, by luck, a result in agreement with experiments. However, the theory of the photoelectric effect, due to Einstein [1904], which was essentially based upon Planck's hypothesis, soon provided further evidence that the fundamental assumption of Planck was indeed correct. It was many years later, however, before this assumption received its complete conceptual justification by the quantum field theory of Dirac [1927]. Although a detailed description of field quantization is beyond the scope of this book, it is worth devoting a little space to indicate how field quantization arises.[(2)] This will also help provide a deeper understanding of some topics to be considered later on in the book.

Consider a mode of the e.m. cavity, i.e. characterized by a given standing wave pattern, and let ν be its resonance frequency. If $E_x(\mathbf{r}, t)$ and $H_y(\mathbf{r}, t)$ are the transverse components of its electric and magnetic fields, the corresponding energy density ρ will be given by Eq. (2.2.1) and its energy will be equal to

$$E = \int \rho dV \tag{2.2.24}$$

where V is the volume of the cavity. A possible starting point to understand the basis of quantum field theory is a recognition that, by analogy with the case of a particle, the pair of quantities $E_x(\mathbf{r}, t)$ and $H_y(\mathbf{r}, t)$ cannot be measured simultaneously with arbitrary precision.[(2)] In other words, there is a form of Heisenberg uncertainty relation between $E_x(\mathbf{r}, t)$ and $H_y(\mathbf{r}, t)$ analogous to that which exists between the position p_x and momentum q_x of a particle moving e.g. in the x direction. Note that the Heisenberg uncertainty relation between p_x and q_x can provide the starting point for the quantum theory of a particle. It indicates, in fact, that the equations of classical mechanics, which are essentially based on the canonical variables p_x and q_x, are no longer valid. Likewise, the uncertainty relation between $E_x(\mathbf{r}, t)$ and $H_y(\mathbf{r}, t)$ can provide for the starting point of the quantum theory of radiation in the sense that they show that Maxwell's equations, and thus Eq. (2.2.4), are no longer valid. The analogy between the quantum theory of a particle and the quantum theory of radiation can be taken further by considering a particle bound to a given point by an elastic force. This is the case of the harmonic oscillator, one of the fundamental examples for the quantum theory of a bound particle. A harmonic oscillator oscillating e.g. along the x direction, is a mechanical oscillator whose total energy is given by

$$E = (kp_x^2/2) + (q_x^2/2m) \tag{2.2.25}$$

where k is the elastic constant and m is the mass of the particle. In fact, this oscillator provides several analogies with a cavity mode. Both of them are, in fact, oscillators in the sense that they are characterized by a resonance frequency. In the mechanical oscillator, oscillation occurs because potential energy, represented by the term $kp_x^2/2$, is periodically transformed into kinetic energy, represented by the term $q_x^2/2m$. In the electromagnetic oscillator represented by the cavity mode, electric energy represented by the term $\int(\varepsilon{<}E_x{}^2{>}/2)dV$, is periodically transformed into magnetic energy, represented by the term $\int(\mu{<}H_y{}^2{>}/2)dV$.

Based on this close analogy, one can then look for similar quantization rules. The appropriate quantization procedure leads to the fundamental result that the energy of the given cavity mode is quantized in exactly the same way as the harmonic oscillator. Namely, the eigenvalues for the mode energy are given by

$$E = (1/2)h\nu + nh\nu \qquad (2.2.26)$$

where n is an integer value. The first term, the zero point energy, has a similar origin to that of the harmonic oscillator. In the latter case, in fact, it arises because the energy cannot be zero since, according to Eq. (2.2.25), this would require that both p_x and q_x are zero, which is contrary to the Heisenberg uncertainty principle. Likewise, for the cavity mode, the energy cannot be zero because, according to Eq. (2.2.1), this would require both E_x and H_y to be zero, which, by the same argument is again impossible. Thus field quantization predicts that the energy levels of a given cavity mode of frequency ν are given by Eq. (2.2.26), a conclusion which coincides with the Planck's assumption [Eq. (2.2.20)] apart from the zero point energy term. The results of field quantization thus provide a framework wherein Planck's assumption is given a more fundamental justification. Needless to say, Maxwell's equations, as seen in Sect. 2.2.1, do not impose any condition on the total energy density of a cavity mode. Thus, according to these equations, the mode energy could have any value covering the range between 0 and ∞, *continuously*.

As a closing comment to this section we note that, according to Eq. (2.2.26), the energy levels of a cavity mode, just like those of the harmonic oscillator, can be displayed as in Fig. 2.4. In the lowest, zero-point energy, level both $<E_x^2>$ and $<H_y^2>$ are different from zero and are referred to as the zero-point fluctuations of the electric and magnetic field, respectively. Note also that the zero point energy value of $(h\nu/2)$ has really no physical significance. If, instead of Eq. (2.2.24), one were to define the energy of the mode as

$$E = \left(\int \rho dV\right) - (h\nu/2) \qquad (2.2.27)$$

then one would have a zero value for the lowest energy state. It can be shown, however, that this state would still include, at the same level as before, the zero-point field fluctuations of both $<E_x^2>$ and $<H_y^2>$, these fluctuations being the quantities which actually characterize the zero-point energy state.

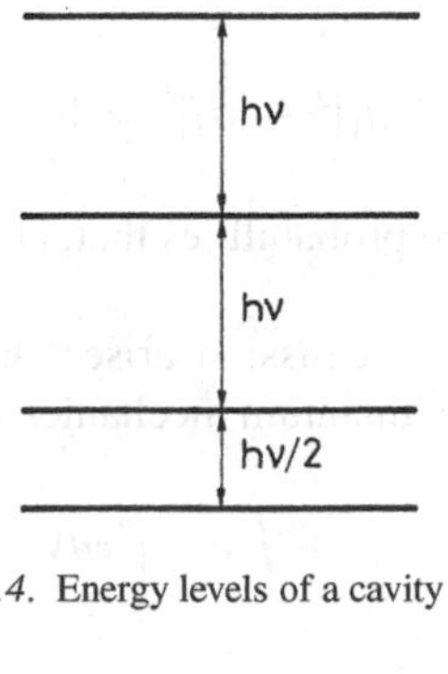

FIG. 2.4. Energy levels of a cavity mode.

2.3. SPONTANEOUS EMISSION

As a first attempt at describing spontaneous emission, we will follow a *semiclassical approach* where the atoms are treated as quantized (i.e. treated according to quantum mechanics) while the fields are treated classically (i.e. treated through Maxwell's equations). As we shall see, this attempt fails to describe the phenomenon of spontaneous emission in a correct way (i.e. in agreement with experiment). The approach will turn out to be very instructive, however. The results obtained will then be compared with the correct ones, i.e. those predicted by a full *quantum theory* where both atoms and fields are quantized, the former by quantum mechanics and the latter by quantum field theory. Thus, to correctly describe the phenomenon of spontaneous emission, a very common phenomenon of every day experience (the light from the sun or from ordinary lamps arises from spontaneous emission), we will need to introduce sophisticated concepts of quantum theory.

2.3.1. Semiclassical Approach

Let us assume that a given atom, initially raised to its excited level 2, of energy E_2, is decaying by spontaneous emission to level 1, of energy E_1 (Fig. 1.1a). We will assume that the two levels are non-degenerate and so let

$$\psi_1(\mathbf{r}, t) = u_1(\mathbf{r}) \exp[-j(E_1/\hbar)t] \tag{2.3.1a}$$

and

$$\psi_2(\mathbf{r}, t) = u_2(\mathbf{r}) \exp[-j(E_2/\hbar)t] \tag{2.3.1b}$$

be the corresponding wave functions, where $u_{1,2}(\mathbf{r})$ are the eigenfunctions of the two stationary states, $\mathbf{r}$ denotes the co-ordinate of the electron undergoing the transition, the origin being taken at the nucleus, and $\hbar = h/2\pi$. When the atom is undergoing the $2 \rightarrow 1$ transition due to spontaneous emission, its wave function can be expressed as a linear combination of the wave functions of the two states, i.e.

$$\psi = a_1(t)\psi_1 + a_2(t)\psi_2 \tag{2.3.2}$$

where a_1 and a_2 are time-dependent complex numbers. Note that, according to quantum mechanics, we have

$$|a_1|^2 + |a_2|^2 = 1 \tag{2.3.3}$$

and thus $|a_1|^2$ and $|a_2|^2$ represent the probabilities that, at time t, the atom is found in state 1 or 2, respectively.

To understand how spontaneous emission arises, let us calculate the electric dipole moment of the atom $\boldsymbol{\mu}$. According to quantum mechanics we have

$$\boldsymbol{\mu} = -\int e\,|\psi|^2 \mathbf{r}\,dV \tag{2.3.4}$$

where e is the magnitude of the electron charge and the integral is taken over the whole volume of the atom. The form of Eq. (2.3.4) can be readily understood by noting that $e|\psi|^2 dV$ is the elemental charge expected in the volume dV at position $\mathbf{r}$ and that this charge produces an elemental dipole moment $d\boldsymbol{\mu} = -(e|\psi|^2 dV)\mathbf{r}$. The substitution of Eq. (2.3.2) into Eq. (2.3.4) with the help of Eq. (2.3.1) gives

$$\boldsymbol{\mu} = \int e\mathbf{r}|a_1|^2|u_1|^2 dV + \int e\mathbf{r}|a_2|^2|u_2|^2 dV + \\ + \int e\mathbf{r}\left[a_1 a_2^* u_1 u_2^* \exp j(\omega_0 t) + a_1^* a_2 u_1^* u_2 \exp -j(\omega_0 t)\right] dV \tag{2.3.5}$$

where * stands for complex conjugate and $\omega_0 = (E_2 - E_1)/\hbar$. Equation (2.3.5) shows that $\boldsymbol{\mu}$ has a term $\boldsymbol{\mu}_{osc}$, oscillating at the frequency ω_0, which can be written as

$$\boldsymbol{\mu}_{osc} = \mathrm{Re}[2a_1/a_2^* \boldsymbol{\mu}_{21} \exp j(\omega_0 t)] \tag{2.3.6}$$

where we have defined a time-independent dipole moment $\boldsymbol{\mu}_{21}$ given by

$$\boldsymbol{\mu}_{21} = \int u_2^* e\mathbf{r}\, u_1 dV. \tag{2.3.7}$$

The vector $\boldsymbol{\mu}_{21}$ is referred to as the matrix element of the electric dipole moment operator or, in short, the electric dipole moment of the atom. Equation (2.3.6) shows that, during the $2 \rightarrow 1$ transition, the atom acquires a dipole moment, $\boldsymbol{\mu}_{osc}$, which is oscillating at frequency ω_o and whose amplitude is proportional to the vector $\boldsymbol{\mu}_{21}$ given by Eq. (2.3.7). Now, from classical electrodynamics we know that any oscillating dipole moment must radiate power into the surrounding space. Accordingly, from a semiclassical standpoint, the process of spontaneous emission can be identified as arising from this radiated power. To be more specific, let us write the oscillating dipole moment as $\boldsymbol{\mu} = \boldsymbol{\mu}_0 \cos(\omega_0 t + \phi) = \mathrm{Re}[\boldsymbol{\mu}_0' \exp(i\omega_0 t)]$, where $\boldsymbol{\mu}_0$ is a real vector describing the amplitude of the dipole moment, Re stands for real part and $\boldsymbol{\mu}_0'$ is a complex vector† given by $\boldsymbol{\mu}_0' = \boldsymbol{\mu}_0 \exp(j\phi)$. According to classical electrodynamics, we know that this oscillating dipole moment will radiate into the surrounding space a power P_r given by[(3)]

$$P_r = \frac{n\mu^2 \omega_0{}^4}{12\pi\varepsilon_0 c_0{}^3} \tag{2.3.8}$$

where $\mu = |\boldsymbol{\mu}_0| = |\boldsymbol{\mu}_0'|$ is the amplitude of the electric dipole moment, n is the refractive index of the medium surrounding the dipole, and c_0 is the light velocity in the vacuum. In the present case we can still use Eq. (2.3.8) provided that μ is now taken to be $\mu = 2|a_1 a_2^* \boldsymbol{\mu}_{21}|$, i.e. it is the magnitude of the complex vector $2a_1 a_2^* \boldsymbol{\mu}_{21}$. We thus see that the radiated power can be written as

$$P_r = P_r' |a_1|^2 |a_2|^2 \tag{2.3.9}$$

† We recall that a complex vector **A**, is a vector whose components e.g. A_x, A_y, A_z are complex numbers. The magnitude A of a complex vector is a real quantity given by $A = [\mathbf{A}\,\mathbf{A}^*]^{1/2}$ where $\mathbf{A}^*$ is the vector conjugate to **A** (i.e. with components $A^*{}_x$, $A^*{}_y$ and $A^*{}_z$ which are the complex conjugated of those for **A**).

where P'_r is a time independent quantity given by

$$P'_r = \frac{16\pi^3 n|\mu|^2 \nu_0^4}{3\varepsilon_0 c_0^3} \tag{2.3.10}$$

and where $|\mu| = |\boldsymbol{\mu}_{21}|$ is the magnitude of the complex vector $\boldsymbol{\mu}_{21}$. To calculate the atom's rate of decay we use an energy balance argument to write

$$\frac{dE}{dt} = -P_r \tag{2.3.11}$$

where the atom energy is now given by

$$E = |a_1|^2 E_1 + |a_2|^2 E_2 \tag{2.3.12}$$

With the help of Eq. (2.3.3), Eq. (2.3.12) can be readily transformed to

$$E = E_1 + h\nu_0|a_2|^2 \tag{2.3.13}$$

where $\nu_0 = (E_2 - E_1)/h$ is the transition frequency. Equation (2.3.11) with the help of Eq. (2.3.9), Eqs. (2.3.10) and (2.3.13) can then be written as

$$\frac{d|a_2|^2}{dt} = -\frac{1}{\tau_{sp}}|a_1|^2|a_2|^2 = -\frac{1}{\tau_{sp}}\left(1 - |a_2|^2\right)|a_2|^2 \tag{2.3.14}$$

where we have defined a characteristic time $\tau_{sp} = h\nu_0/P'_r$ as

$$\tau_{sp} = \frac{3h\varepsilon_0 c_0^3}{16\pi^3 \nu_0^3 n|\mu|^2} \tag{2.3.15}$$

which is called the spontaneous-emission (or radiative) lifetime of level 2. The solution of Eq. (2.3.14) can be conveniently written in the form

$$|a_2|^2 = \frac{1}{2}\left[1 - \tanh\left(\frac{t - t_0}{2\tau_{sp}}\right)\right] \tag{2.3.16}$$

where t_0 is set by the initial condition, i.e. by the value $|a_2(0)|^2$ Indeed, from Eq. (2.3.16) one gets that

$$|a_2(0)|^2 = \frac{1}{2}\left[1 - \tanh\left(-\frac{t_0}{2\tau_{sp}}\right)\right] \tag{2.3.17}$$

which, for a given value of $|a_2(0)|^2$ (provided it is smaller than 1) yields a unique value of t_0. As an example, Fig. 2.5 shows the time behavior of $|a_2(t)|^2$ for the initial condition $|a_2(0)|^2 = 0.96$. Note that, by choosing a different value of $|a_2(0)|^2$ one merely changes the value of t_0 in Eq. (2.3.16), i.e. one only changes the origin of the time axis. Assuming for

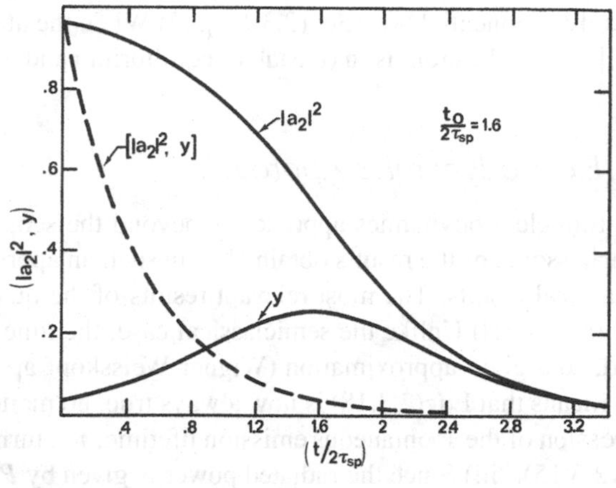

FIG. 2.5. Time behavior of the upper state occupation probability, $|a_2|^2$, and of the normalized radiated power, $y = \tau_{sp}P_r/h\nu_0$. Solid lines: semiclassical theory; dashed line: quantum electrodynamics theory.

instance $|a_2(0)|^2 = 0.8$, the curve of $|a_2(t)|^2$ is obtained simply by horizontally shifting the curve of Fig. 2.5 to the left until it crosses the vertical $t = 0$ axis at the value 0.8. This shows the advantage of expressing the decay of $|a_2(t)|^2$ in the form of Eq. (2.3.16). Once $|a_2(t)|^2$ has been calculated, the radiated power, P_r, according to Eqs. (2.3.11) and (2.3.13), is obtained as $P_r = -h\nu_0 d|a_2|^2/dt$. The time behavior of the normalized radiated power, $y = \tau_{sp}P_r/h\nu_0$, is also shown in Fig. 2.5. For the discussion that follows it is important to notice that the time behavior of $|a_2(t)|^2$ can be approximated by an exponential law, i.e.

$$|a_2(t)|^2 = |a_2(0)|^2 \exp[-(t/\tau_{sp})] \tag{2.3.18}$$

only when $|a_2(0)|^2 \ll 1$. In this case, in fact, we can put $|a_1|^2 \cong 1$ into Eq. (2.3.14) thus readily obtaining Eq. (2.3.18).

A particularly important case occurs when $|a_2(0)|^2 = 1$. In this case we find from Eq. (2.3.17) that $t_0 = \infty$, which means that, according to this semiclassical theory, the atom should not decay. Indeed, when $|a_2(0)|^2 = 1$, then $|a_1(0)|^2 = 0$, and from Eq. (2.3.14) one gets $d|a_2|^2/dt = 0$. Another way of looking at this case is to observe that, when $a_1(0) = 0$, μ_{osc} given by Eq. (2.3.6) vanishes. Since the atom does not have an oscillating dipole moment, it cannot radiate power and it is therefore in an equilibrium state. Let us now investigate the stability of this equilibrium. To do this, we assume the atom to be perturbed so that $|a_2| < 1$ at $t = 0$. Physically, this means that, as a result of this perturbation, there is a finite probability $|a_1|^2$ of finding the atom in level 1. Equation (2.3.6) then shows that a dipole moment oscillating at frequency ω_0 is now produced. This moment will radiate into the surrounding space and the atom will tend to decay to level 1. This implies a decrease of $|a_2|^2$ and the atom moves further away from equilibrium. The atom is therefore in unstable equilibrium.

Before going further, it is worth summarizing the main results obtained with this semiclassical approach: (1) The time behavior of $|a_2|^2$ can generally be described in terms of an hyperbolic tangent equation, Eq. (2.3.16), but, for very weak excitation (i.e. for $|a_2|^2 \ll 1$), it follows an approximately exponential law [Eq. (2.3.18)]. (2) When the atom is initially in the upper state (i.e. $|a_2(0)|^2 = 1$), the atom is in (unstable) equilibrium and no radiation occurs.

2.3.2. Quantum Electrodynamics Approach

Although a quantum electrodynamics approach is beyond the scope of this book, it is worthwhile to summarize some of the results obtained from such an approach and comparing them with the semiclassical results. The most relevant results of the quantum approach can be summarized as follows:[(4,5)] (i) Unlike the semiclassical case, the time behavior of $|a_2|^2$ is now always described, to a good approximation (Wigner-Weisskopf approximation), by an exponential law. This means that Eq. (2.3.18) is now always true, no matter what the value of $|a_2(0)|^2$. (ii) The expression of the spontaneous emission lifetime, τ_{sp}, turns out to be given, in this case too, by Eq. (2.3.15). (iii) Since the radiated power is given by $P_r = -h\nu_0 d|a_2|^2/dt$, this power will also decay exponentially with a time constant τ_{sp}. We see that the semiclassical and the quantum electrodynamics approaches give completely different predictions for the phenomenon of spontaneous emission (see Fig. 2.5). All available experimental results* confirm that the quantum electrodynamics approach gives the correct answer to the problem. From Eq. (2.3.15) we can then write the rate of spontaneous emission, $A = 1/\tau_{sp}$, as

$$A = \frac{16\pi^3 \nu_0^3 n|\mu|^2}{3h\varepsilon_0 c_0^3} \tag{2.3.19}$$

The above remarks imply that, according to quantum electrodynamics, an atom in the upper level is not in a state of unstable equilibrium and the physical reason for the disappearance of this unstable state on passing from the semiclassical to the quantum electrodynamics approach deserves some further discussion. In the semiclassical case, the atom's wave function was generally written as in Eq. (2.3.2) and this implies that the atom is not in a stationary state. According to quantum mechanics, this can only occur when some sort of perturbation is already applied to the atom. Furthermore, to remove the unstable equilibrium position discussed before, we need again to assume that the atom is somewhat perturbed, and we now look for some cause for this perturbation. At first sight we may be tempted to say that there will always be enough stray radiation around the material to perturb the atom. To be more specific, let us suppose that the material is contained in a blackbody cavity whose walls are kept at temperature T. We might then imagine this stray radiation to be provided by the blackbody radiation within the cavity. This conclusion would be wrong, however, since the radiation produced in this way would actually be due to the process of stimulated emission, i.e., stimulated by the blackbody radiation. The phenomenon of spontaneous emission would then depend upon the wall temperature and would cease at $T = 0$. The correct form of perturbation needed to describe the phenomenon of spontaneous emission is actually provided by the quantum electrodynamics approach. In fact, according to the discussion presented in

* Of these, we should like to mention the very accurate measurements of the so-called Lamb shift, another phenomenon that occurs during spontaneous emission. The center frequency of the spontaneously emitted light does not occur at frequency ν_0 (the transition frequency) but at slightly different value. Lamb-shift measurements for hydrogen are among the most careful measurements so far made in physics, and they are always exactly agreed (within the experimental errors) with the predictions of the quantum electrodynamic approach.

Sect. 2.2.3, the mean square values $<E^2>$ and $<H^2>$ of both the electric and magnetic fields of a given cavity mode are different from zero even at $T = 0$ (*zero-point field fluctuations*). We may therefore consider these fluctuations as the perturbation acting on the atom and which, in particular, upsets the unstable equilibrium predicted by the semiclassical treatment. Correspondingly we may think of the spontaneous emission process as originating from these zero-point fluctuations.

2.3.3. Allowed and Forbidden Transitions

Equation (2.3.19) shows that, to have $A \neq 0$, we must have $|\mu| \neq 0$. In this case the spontaneous emission process arises from the power radiated by the electric dipole of the atom and the transition is said to be electric dipole allowed. By contrast, when $|\mu| = 0$, we have $A = 0$ and the transition is said to be electric dipole forbidden. In this case the transition may occur via other multipole radiation processes e.g. through the oscillating magnetic dipole moment of the atom (*magnetic dipole transitions*). This is usually a much weaker process, however.

Let us now consider the situation when the transition is electric dipole forbidden i.e., when $|\mu| = 0$. Since $|\mu| = |\boldsymbol{\mu}_{21}|$, Eq. (2.3.7) shows that this occurs when the eigenfunctions u_1 and u_2 are either both symmetric or both anti symmetric*. In fact, in this case, the two contributions from the integrand of Eq. (2.3.7) at points $\mathbf{r}$ and $-\mathbf{r}$, respectively, are equal and opposite. It is therefore of interest to see when the wave functions $u(\mathbf{r})$ are either symmetric or anti symmetric. This occurs when the Hamiltonian $H_o(\mathbf{r})$ of the system is unchanged by changing $\mathbf{r}$ into $-\mathbf{r}$, i.e. when†

$$H_o(-\mathbf{r}) = H_o(\mathbf{r}) \tag{2.3.20}$$

In this case, in fact, for any eigenfunction $u_n(\mathbf{r})$, one has

$$H_o(\mathbf{r})u_n(\mathbf{r}) = E_n u_n(\mathbf{r}) \tag{2.3.21}$$

From Eq. (2.3.21), changing $\boldsymbol{r}$ into $-\boldsymbol{r}$ and using Eq. (2.3.20), one gets

$$H_o(\mathbf{r})u_n(-\mathbf{r}) = E_n u_n(-\mathbf{r}) \tag{2.3.22}$$

Equations (2.3.21) and (2.3.22) show that $u_n(\mathbf{r})$ and $u_n(-\mathbf{r})$ are both eigenfunctions of the Hamiltonian H_o with the same eigenvalue E_n. For non degenerate energy levels, there is, by definition only one eigenfunction for each eigenvalue, apart from an arbitrary choice of sign, so that

$$u_n(-\mathbf{r}) = \pm u_n(\mathbf{r}) \tag{2.3.23}$$

Therefore, if $H_o(\mathbf{r})$ is symmetric, the eigenfunctions must be either symmetric or antisymmetric. In this case, it is usually said that the eigenfunctions have a well defined parity.

* It may be recalled here that a function $f(\mathbf{r})$ is symmetric (or of even parity) if $f(-\mathbf{r}) = f(\mathbf{r})$, while it is antisymmetric (or of odd parity) if $f(-\mathbf{r}) = -f(\mathbf{r})$.

† If the Hamiltonian H_0 is a function of more than one coordinate $\mathbf{r}_1, \mathbf{r}_2, \ldots,$ the inversion operation must be simultaneously applied to all these coordinates.

Example 2.1. *Estimate of τ_{sp} and A for electric-dipole allowed and forbidden transitions* For an electric-dipole allowed transition at a frequency corresponding to the middle of the visible range, an estimate the order of magnitude of A is obtained from Eq. (2.3.19) by there substituting the values $\lambda = c/\nu = 500\,\text{nm}$ and $|\mu| \cong ea$, where a is the atomic radius ($a \cong 0.1\,\text{nm}$). We therefore get $A \cong 10^8\,\text{s}^{-1}$ (i.e. $\tau_{sp} \cong 10\,\text{ns}$). For magnetic dipole transitions A is approximately 10^5 times smaller, and therefore $\tau_{sp} \cong 1\,\text{ms}$. Note that, according to Eq. (2.3.19), A increases as the cube of the frequency, so that the importance of the process of spontaneous emission increases rapidly with frequency. In fact spontaneous emission is often negligible in the middle to far infrared where nonradiative decay usually dominates. On the other hand, when one considers the x-ray region (say $\lambda \le 5\,\text{nm}$), τ_{sp} becomes very short (10–100 fs) a feature that constitutes a major problem for achieving a population inversion in X-ray lasers.

It remains now to see when the Hamiltonian satisfies Eq. (2.3.20), i.e., when it is invariant under inversion. Obviously this occurs when the system has a center of symmetry. Another important case is that of an isolated atom. In this case, the potential energy of the k-th electron of the atom is given by the sum of -the potential energy due to the nucleus (which is symmetric) and that of all other electrons. For the i-th electron, this energy will depend on $|\mathbf{r}_i - \mathbf{r}_k|$, i.e., on the magnitude of the distance between the two electrons. Therefore, these terms will also be invariant under inversion. An important case where Eq. (2.3.20) is not valid is where an atom is placed in an external electric field (e.g., a crystal's electric field) that does not possess a center of inversion. In this case the wave functions will not have a definite parity.

To sum up, we can say that electric dipole transitions only occur between states of opposite parity, and that the states have a well-defined parity if the Hamiltonian is invariant under inversion.

2.4. ABSORPTION AND STIMULATED EMISSION

In this section, we will study in some detail the processes of absorption and stimulated emission induced in a two-level system for a single atom interacting with a monochromatic electromagnetic (e.m.) wave. In particular, our aim is to calculate the rates of absorption W_{12} and stimulated emission W_{21} [see Eqs. (1.1.4) and (1.1.6)]. We will follow the semiclassical approximation, wherein, as already explained, the atom is quantized while the e.m. radiation is treated classically. It can be shown, in fact, that the quantum electrodynamics approach gives the same result as the semiclassical treatment when the number of photons in a given radiation mode is much greater than unity. Since this applies to any system other than an exceedingly weak e.m. wave, we can dispense with the complication of the full quantum treatment. We will at first assume the two levels to be non-degenerate and treat the case of degenerate levels later in this Chapter.

2.4.1. Rates of Absorption and Stimulated Emission

Let us first consider the case of absorption and assume that, for time $t \ge 0$, a monochromatic e.m. wave is incident on the atom so that the atomic wave-function can be described as in Eq. (2.3.2) where we will assume the initial conditions $|a_1(0)|^2 = 1$ and $|a_2(0)|^2 = 0$.

As a result of the interaction with the e.m. wave, the atom will acquire an interaction energy H'. In the treatment that follows this energy H' is considered to be due to the interaction of the electric dipole moment of the atom with the electric field $\mathbf{E}(\mathbf{r},t)$ of the e.m. wave (*electric dipole interaction*), where the origin of the **r** coordinate is taken at the nucleus. The electric field at the nuclear position can then be written as

$$\mathbf{E}(0,t) = \mathbf{E}_0 \sin(\omega t) \tag{2.4.1}$$

where ω is the angular frequency of the wave. We will also assume that the wavelength of the e.m. radiation is much greater than the atom's dimension so that the phase-shift of the e.m. wave over an atomic dimension is very small. Then Eq. (2.4.1) can be taken to give the value of the electric field for any location in the atom (*electric dipole approximation*). We will also assume the frequency ω to be close to the resonant frequency, ω_0, of the transition.

Classically, for a given position **r** of the electron within the atom, the atom would exhibit an electric dipole moment $\boldsymbol{\mu} = -e\mathbf{r}$, where e is the magnitude of the electronic charge. The interaction energy H' resulting from the external electric field would then be

$$H' = \boldsymbol{\mu} \cdot \mathbf{E} = -e \cdot \mathbf{E}_0 \sin \omega t \tag{2.4.2}$$

In a quantum mechanical treatment, this sinusoidally time-varying interaction energy is treated as a sinusoidally time-varying interaction Hamiltonian $\mathcal{H}'(t)$, which is then inserted into the time-dependent Schrödinger wave equation. Since $\omega \cong \omega_o$, this interaction Hamiltonian results in the transition of the atom from one level to the other. This implies that, for $t > 0$, $|a_1(t)|^2$ will decrease from its initial value $|a_1(0)|^2 = 1$ and $|a_2(t)|^2$ will increase correspondingly. To derive an expression for $a_2(t)$ we will additionally assume that the transition probability is weak, so that a perturbation analysis can be used (*time dependent perturbation theory*), and that the interaction occurs for a long time after $t = 0$.

Given the above assumptions, the time behaviour of $|a_2(t)|^2$ is shown, in Appendix *A*, to be given by

$$|a_2(t)|^2 = \frac{\pi^2}{3h^2} |\mu_{21}|^2 E_0^2 \delta(\nu - \nu_0) t \tag{2.4.3}$$

where $\nu = \omega/2\pi$, $\nu_0 = \omega_0/2\pi$, δ is the Dirac delta function, E_0 is the amplitude of the vector $\mathbf{E}_0$ and $|\boldsymbol{\mu}_{21}|$ is the amplitude of the complex vector $\boldsymbol{\mu}_{21}$ given by Eq. (2.3.7). Equation (2.4.3) shows that, for $t > 0$, $|a_2(t)|^2$ increases linearly with time. We can then define the transition rate W_{12}^{sa} as

$$W_{12}^{sa} = d|a_2|^2/dt \tag{2.4.4}$$

From Eq. (2.4.3) we then get

$$W_{12}^{sa} = \frac{\pi^2}{3h^2} |\mu_{21}|^2 \, E_0^2 \delta(\nu - \nu_0) \tag{2.4.5}$$

Note that the transition rate defined by Eq. (2.4.4) refers to the case of a single atom interacting with monochromatic radiation and this situation is denoted by the superscript *sa* (single atom) added to W_{12}.

To gain a more physical understanding of the absorption phenomenon, we begin by noticing that, for $t > 0$, the wave function can be described as in Eq. (2.3.2). For $t > 0$ the atom thus acquires an oscillating dipole moment, μ_{osc}, given by Eq. (2.3.6). In contrast to the case of spontaneous emission, however, since $a_1(t)$ and $a_2(t)$ are now driven by the electric field of the e.m. wave, the phase of μ_{osc} turns out to be correlated to that of the wave. In particular, for absorption, i.e. when one starts with the initial conditions $a_1(0) = 1$ and $a_2(0) = 0$, the phase of the dipole is such that the dipole absorbs power from the e.m. wave. The interaction phenomenon is thus seen to be very similar to that of a classical oscillating dipole moment driven by an external field.[(6)]

Equation (2.4.5) can also be expressed in terms of the energy density of the e.m. wave. Since

$$\rho = n^2 \varepsilon_0 E_0^2/2 \tag{2.4.6}$$

where n is the refractive index of the medium and ε_o is the vacuum permittivity, we obtain

$$W_{12}^{sa} = \frac{2\pi^2}{3n^2\varepsilon_0 h^2}|\mu_{21}|^2 \rho\delta(\nu - \nu_0) \tag{2.4.7}$$

Note that W_{12}^{sa} is proportional to the Dirac δ function. This implies the unphysical result that $W = 0$ for $\nu \neq \nu_0$ and $W_{12} = \infty$ when $\nu = \nu_0$, i.e. when the frequency of the e.m. wave is exactly coincident with the frequency of the atomic transition. The reason for this unphysical result can be traced back to the assumption that the interaction of the e.m. wave with the atom could continue undisturbed for an indefinite time. Indeed, from a classical viewpoint, if a sinusoidal electric field at frequency ν drives a (lossless) oscillating dipole moment at frequency ν_0, there would only be an interaction, i.e. a net energy transfer, if $\nu = \nu_0$. Actually, there are a number of perturbation phenomena (such as collisions with other atoms or with lattice phonons) that prevent this interaction from continuing undisturbed indefinitely. These phenomena will be discussed at some length in a later section, but the general result they lead to can be expressed in a very simple way: Eq. (2.4.7) remains valid provided the Dirac δ function – an infinitely sharp function centred at $\nu = \nu_0$ and of unit area, i.e., such that $\int \delta(\nu - \nu_0)d\nu = 1$ – be replaced by a new function $g(\nu - \nu_0)$, symmetric about $\nu = \nu_0$ again of unit area, i.e. such that $\int g(\nu - \nu_0)d\nu = 1$, and generally given by

$$g(\nu - \nu_0) = \frac{2}{\pi \Delta\nu_0} \frac{1}{1 + [2(\nu - \nu_0)/\Delta\nu_0]^2} \tag{2.4.8}$$

where $\Delta\nu_o$ depends on the particular broadening mechanism involved. We can therefore write for W_{12}^{sa} the expression

$$W_{12}^{sa} = \frac{2\pi^2}{3n^2\varepsilon_0 h^2}|\mu_{21}|^2 \rho g(\nu - \nu_0) \tag{2.4.9}$$

The normalized function $[g(\nu - \nu_0)\Delta\nu_0]$ is plotted in Fig. 2.6 vs the normalized frequency difference $(\nu - \nu_0)/(\Delta\nu_0/2)$. The full width of the curve between the two points having

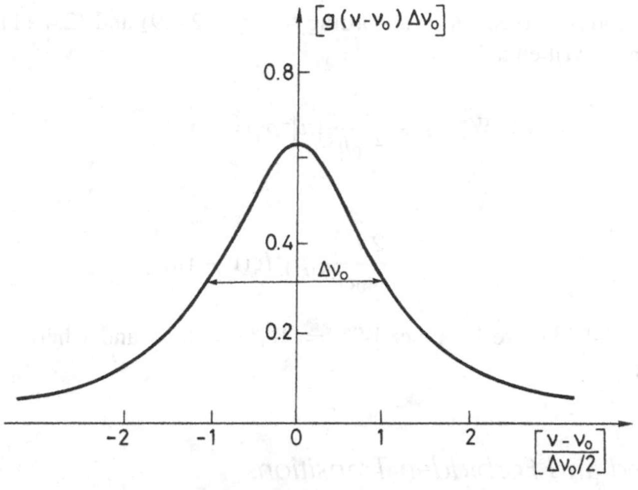

FIG. 2.6. Normalized plot of a Lorentzian line.

half the maximum value (FWHM, from Full Width at Half Maximum) is simply $\Delta\nu_0$. The maximum of $g(\nu - \nu_0)$ occurs for $\nu = \nu_0$ and its value is given by

$$g(0) = 2/\pi\Delta\nu_0 = 0.637/\Delta\nu_0 \tag{2.4.9a}$$

A curve of the general form described by Eq. (2.4.8) is called *Lorentzian* after H.E. Lorentz who first derived it in his theory of the electron oscillator.[(6)] For a plane e.m. wave it is often useful to express W_{12}^{sa} in terms of the intensity, I, of the incident radiation. Since

$$I = c_0\rho/n \tag{2.4.10}$$

where n is the refractive index of the medium, we find from Eq. (2.4.9) that

$$W_{12}^{so} = \frac{2\pi^2}{3n\varepsilon_0 ch^2}|\mu_{21}|^2 I g(\nu - \nu_0) \tag{2.4.11}$$

We consider next the case of stimulated emission. The starting points, namely the wavefunction of the two-level system [Eq. (2.3.2)] and the interaction energy H' [Eq. (2.4.2)] remain unchanged. Thus, the corresponding pair of equations describing the evolution, with time, of $|a_2(t)|^2$ and $|a_1(t)|^2$ (see Appendix *A*) also remain unchanged. The only difference arises from the fact that the initial condition is now given by $|a_2(0)|^2 = 1$ and thus $|a_1(0)|^2 = 0$. It can be readily seen that the new equations for stimulated emission are then obtained from those corresponding to absorption by simply interchanging the indices 1 and 2. Thus the transition rate W_{21}^{sa} is obtained from Eq. (2.4.5) by interchanging the two indices. From to Eq. (2.3.7) one immediately see that $\mu_{12} = \mu_{21}^*$, implying that $|\mu_{12}| = |\mu_{21}|$. Therefore we have

$$W_{12}^{sa} = W_{21}^{sa} \tag{2.4.12}$$

showing that the probabilities of absorption and stimulated emission are equal in this case [compare with Eq. (1.1.8)].

As a conclusion to this section, according to Eqs. (2.4.9) and (2.4.11), the stimulated transition rate can be written as*

$$W^{sa} = \frac{2\pi^2}{3n^2\varepsilon_0 h^2}|\mu|^2 \rho g(\nu - \nu_0). \tag{2.4.13}$$

$$W^{sa} = \frac{2\pi^2}{3n\varepsilon_0 ch^2}|\mu|^2 I g(\nu - \nu_0) \tag{2.4.13a}$$

where, from Eq. (2.4.12), we have set $W^{sa} = W^{sa}_{12} = W^{sa}_{21}$. and where we have also set $|\mu| = |\mu_{12}| = |\mu_{21}|$.

2.4.2. Allowed and Forbidden Transitions

Equations (2.4.13) and (2.3.19) show that the transition rate W^{sa}, and the spontaneous emission rate A are proportional to $|\mu|^2$. This indicates that the two phenomena must obey the same selection rule. Thus the stimulated transition via electric dipole interaction (electric dipole transition) only occurs between states, u_1 and u_2, of opposite parity. The transition is then said to be electric dipole allowed. Conversely, if the parity of the two states is the same, then $W^{sa} = 0$ and the transition is said to be electric-dipole forbidden. This does not mean, however, that the atom cannot pass from level 1 to level 2 through the influence of an incident e.m. wave. In this case, the transition can occur, for instance, as a result of the interaction of the magnetic field of the e.m. wave with the magnetic dipole moment of the atom. For the sake of simplicity, we will not consider this case any further (*magnetic dipole interaction*), but limit ourselves to observing that the analysis can be carried out in a similar manner to that used to obtain Eq. (2.4.11). We may also point out that a magnetic dipole transition is allowed between states of equal parity (even-even or odd-odd transitions). Therefore, a transition that is forbidden by electric dipole interaction is, however, allowed for magnetic dipole interaction and vice versa. It is also instructive to calculate the order of magnitude of the ratio of the electric dipole transition probability, W_e, to the magnetic dipole transition probability, W_m. Obviously the calculation refers to two different transitions, one being allowed for electric dipole and the other for magnetic dipole interaction. We shall assume that the intensity of the e.m. wave is the same for the two cases. For an allowed electric dipole transition, according to Eq. (2.4.5), we can write $W_e \propto (\mu_e E_0)^2 \approx (eaE_0)^2$, where E_0 is the electric field amplitude and where the electric dipole moment of the atom μ_e has been approximated (for an allowed transition) by the product of the electron charge

* It should be noted that the factor 3 appearing in the denominator of Eq. (2.4.3) and, hence, of Eq. (2.4.5), Eq. (2.4.7), Eq. (2.4.9), Eq. (2.4.11), Eqs. (2.4.13), and (2.4.13a), refers to the case of a linearly polarized wave interacting with randomly oriented atoms (such as in a gas). In this case, in fact, we have $W \propto <|\boldsymbol{\mu}_{21} \cdot \mathbf{E}_0|^2> = |\mu_{21}|^2 = |\mu_{21}|^2 E_0^2 <\cos^2\theta> = |\mu_{21}|^2 E_0^2/3$, where θ is the angle between $\boldsymbol{\mu}_{21}$ and $\mathbf{E}_0$ and the average is taken over all atom-field orientations. Indeed, for randomly oriented $\boldsymbol{\mu}_{21}$ vectors, one has $<\cos^2\theta> = 1/3$ where the average is taken in the three dimensional space, i.e. $<\cos^2\theta> = \int \cos^2\theta \, d\Omega/4\pi$. For different cases of atom/field orientation, the factor $|\mu_{21}|^2/3$ should be changed appropriately. Thus, for aligned ions (such as in ionic crystals) and a linearly polarized wave, the factor 3 should be dropped and $|\mu_{21}|$ should represent the magnitude of the component of $\boldsymbol{\mu}_{21}$ along $\mathbf{E}_0$.

e and the radius a of the atom. For a magnetic dipole interaction, it can likewise be shown that $W_m \propto (\mu_m B_0)^2 \approx (\beta B_0)^2$, where B_0 is the magnetic field amplitude of the wave and where the magnetic dipole moment of the atom, μ_m, has been approximated (for an allowed transition) by the Bohr magneton β ($\beta = 9.27 \times 10^{-24}$ A $\times$ m^2). Thus we get

$$(W_e/W_m) = (eaE_0/\beta B_0)^2 = (eac/\beta)^2 \cong 10^5 \tag{2.4.14}$$

To obtain the numerical result of Eq. (2.4.14) we have made use of the fact that, for a plane wave it is $E_0/B_0 = c$ (where c is the light velocity), and we have assumed that $a \cong 0.5$ nm. The probability of an electric dipole transition is thus much greater than that of a magnetic dipole.

2.4.3. Transition Cross Section, Absorption and Gain Coefficient

In Sect. 2.4.1, the transition rate has been calculated for the case of a single atom interacting with an incident e.m. wave and whose linewidth is determined by some line-broadening mechanism. We now consider an ensemble of N_t atoms per unit volume and we want to calculate the corresponding, average, transition rate.

The first case we will consider is where both the resonance frequency ν_0 and the line shape are the same for every atom (the case of *homogeneous broadening*). The transition rate, W_h, for this homogeneous case will then be the same for every atom, so that we can simply write

$$W_h(\nu - \nu_0) = W^{sa}(\nu - \nu_0) \tag{2.4.15}$$

If we now let all atoms be in the ground state, the power absorbed per unit volume, dP_a/dV, will then be given by

$$(dP_a/dV) = W_h N_t h\nu \tag{2.4.16}$$

Since W_h is proportional to the wave intensity, hence to the photon flux $F = I/h\nu$, we can define an absorption cross section, σ_h, as

$$\sigma_h = W_h/F \tag{2.4.17}$$

From Eqs. (2.4.13a) and (2.4.17), σ_h is seen to be given by

$$\sigma_h = \frac{2\pi^2}{3n\varepsilon_0 ch}|\mu|^2 \nu g(\nu - \nu_0) \tag{2.4.18}$$

With the help of the same argument used in connection with Fig. 1.2, we obtain from Eqs. (2.4.16) and (2.4.17) the equation describing the variation of the photon flux along the z direction as [compare with Eq. (1.2.1)]

$$dF = -\sigma N_t F dz \tag{2.4.19}$$

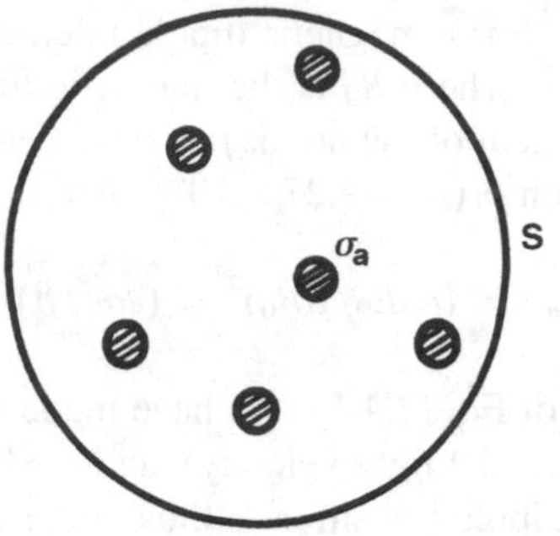

FIG. 2.7. Effective absorption cross section, σ_a, of atoms in a light beam of cross section S.

Examination of Eq. (2.4.19) leads to a simple physical interpretation of this transition cross section. First, let us suppose that we can associate with each atom an effective absorption cross section σ_a, in the sense that, if a photon enters this cross section, it will be absorbed by the atom (Fig. 2.7). If we let S be the cross-sectional area of the e.m. wave, the number of atoms in the element dz of the material (see also Fig. 1.2) is $N_t S dz$, thus giving a total absorption cross section of $\sigma_a N_t S dz$. The fractional change (dF/F) of photon flux in the element dz of the material is therefore

$$(dF/F) = -(\sigma_a N_t S dz/S) \tag{2.4.20}$$

A comparison of Eq. (2.4.20) with Eq. (2.4.19) shows that $\sigma_h = \sigma_a$, so that the meaning we can attribute to σ_h is that of an effective absorption cross section as defined above.

A somewhat different case occurs when the resonance frequencies ν_0' of the atoms are distributed around some central frequency ν_0 (case of *inhomogeneous broadening*). This distribution will be described by the function $g^*(\nu_0' - \nu_0)$ whose definition is such that $dN_t = N_t g^*(\nu_0' - \nu_0)\, d\nu_0'$ gives the elemental number of atoms with resonance frequency between ν_0' and $\nu_0' + d\nu_0'$. According to Eq. (2.4.16), the elemental power absorbed by this elemental number of atoms, dN_t, is given by $d(dP_a/dV) = (N_t h\nu) W_h(\nu - \nu_0')\, g^*(\nu_0' - \nu_0)\, d\nu_0'$, where $W_h(\nu - \nu_0')$ is the transition rate for those atoms with resonance frequency ν_0'. The total power absorbed per unit volume is then given by

$$(dP_a/dV) = N_t h\nu \int W_h(\nu - \nu_0')g^*(\nu_0' - \nu_0)d\nu_0' \tag{2.4.21}$$

A comparison of Eq. (2.4.21) with Eq. (2.4.16) shows that we can define an inhomogeneous transition rate, W_{in}, as

$$W_{in} = \int W_h(\nu - \nu_0')g^*(\nu_0' - \nu_0)d\nu_0' \tag{2.4.22}$$

According to Eq. (2.4.17) we can now define an inhomogeneous cross-section σ_{in} as $\sigma_{in} = W_{in}/F$. Upon dividing both sides of Eq. (2.4.22) by F and using Eq. (2.4.17) we then get

$$\sigma_{in} = \int \sigma_h(\nu - \nu_0')g^*(\nu_0' - \nu_0)d\nu_0' \tag{2.4.23}$$

Following the argument presented in connection with Fig. 2.7 one can see that σ_{in} is an effective absorption cross section that one can associate with a single atom, so that a photon will be absorbed if it enters this cross section. Note, however, that, in this case each atom has, in reality, a different cross section $\sigma_h(\nu - \nu_0')$ at the frequency of the incoming radiation and σ_{in} is just an effective average cross section. Note also that, according to Eq. (2.4.23), the line shape and linewidth of σ_{in} depend on the function $g^*\left(\nu_0' - \nu_0\right)$ i.e. on the distribution of the atomic resonance frequencies. The phenomena leading to this frequency distribution will be discussed at some length in a later section. Here we limit ourselves to pointing out that $g^*\left(\nu_0' - \nu_0\right)$ is generally described by a function of the form

$$g^*(\nu_0' - \nu_0) = \frac{2}{\Delta\nu_0^*}\left(\frac{\ln 2}{\pi}\right)^{1/2} \exp -\left[\frac{4(\nu_0' - \nu_0)^2}{\Delta\nu_0^{*2}} \ln 2\right] \tag{2.4.24}$$

where $\Delta\nu_o^*$ is the transition linewidth (FWHM), its value depending of the particular broadening mechanism under consideration.

With the help of Eq. (2.4.18), Eq. (2.4.23) can be transformed to

$$\sigma_{in} = \frac{2\pi^2}{3n\varepsilon_0 ch}|\mu|^2 \nu g_t(\nu - \nu_0) \tag{2.4.25}$$

In Eq. (2.4.25) we have used the symbol $g_t(\nu - \nu_0)$ for the total line shape function which can be expressed as

$$g_t = \int_{-\infty}^{+\infty} g^*(x) g\left[(\nu - \nu_0) - x\right] dx \tag{2.4.26}$$

where we have put $x = \nu_0' - \nu_0$. The expression of the cross section for inhomogeneous broadening, σ_{in}, is thus obtained from that for homogeneous broadening, given by Eq. (2.4.18), by substituting $g(\nu - \nu_0)$ with $g_t(\nu - \nu_0)$. Note that, according to Eq. (2.4.26), g_t is the convolution of the functions g and g^*. Since both functions are normalised to unity it can be shown that g_t is also normalised to unity, i.e. that $\int g_t(\nu - \nu_o)d\nu = 1$. Note also that Eq. (2.4.25) provides a generalisation of Eq. (2.4.18). Indeed, it is immediately seen from Eqs. (2.4.26) and (2.4.25) that σ_{in} reduces to σ_h when $g^*\left(\nu_0' - \nu_0\right) = \delta\left(\nu_0' - \nu_0\right)$, i.e. when all atoms have the same resonance frequency. Conversely, if the width of the line shape function, $g\left(\nu - \nu_0'\right)$, is much smaller than that due to inhomogeneous broadening, $g^*\left(\nu_0' - \nu_0\right)$, then $g\left(\nu - \nu_0'\right)$ can be approximated by a Dirac δ function in Eq. (2.4.26) and one gets $g_t \cong g^*(\nu - \nu_0)$ (case of *pure inhomogeneous broadening*). In this case from Eq. (2.4.24) we get

$$g_t = g^*(\nu - \nu_0) = \frac{2}{\Delta\nu_0^*}\left(\frac{\ln 2}{\pi}\right)^{1/2} \exp -\left[\frac{4(\nu - \nu_0)^2}{\Delta\nu_0^*} \ln 2\right] \tag{2.4.27}$$

The normalized function $\left[g^*(\nu - \nu_0)\Delta\nu_0^*\right]$ is plotted in Fig. 2.8 vs the normalized frequency difference $(\nu - \nu_0)/\left(\Delta\nu_0^*/2\right)$. According to Eq. (2.4.27), the width of the curve (FWHM) is simply $\Delta\,\nu_0^*$, the maximum of the function occurs for $\nu = \nu_0$ and its value is given by

$$g^*(0) = \frac{2}{\Delta\nu_0^*}\left(\frac{\ln 2}{\pi}\right)^{1/2} = \frac{0.939}{\Delta\nu_0^*} \tag{2.4.28}$$

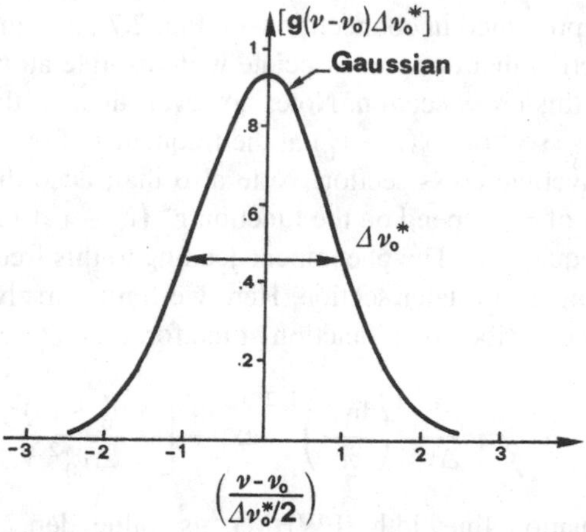

FIG. 2.8. Normalized plot of a Gaussian line.

A curve of the general form described by Eq. (2.4.27) is called *Gaussian*.

Based on the preceding discussion we will, from now on, use the symbol $\sigma = \sigma_{in}$ to indicate the absorption cross section, for which the general expression can be written as

$$\sigma = \frac{2\pi^2}{3n\varepsilon_0 ch}|\mu|^2 \nu g_t(\nu - \nu_0) \tag{2.4.29}$$

The corresponding expression for the absorption rate $W = \sigma F$ can then be written as

$$W = \frac{2\pi^2}{3n^2\varepsilon_0 h^2}|\mu|^2 \rho g_t(\nu - \nu_0) \tag{2.4.30}$$

where $\rho = (nI/c) = (nFh\nu/c)$ is the energy density of the e.m. wave.

One could now repeat the same arguments for the case of stimulated emission. According to Eq. (2.4.12) one readily sees that, for non degenerate levels, the general expressions for the stimulated emission cross section and for the rate of stimulated emission are again given by Eqs. (2.4.29) and (2.4.30) respectively.

It should be emphasized that, according to Eq. (2.4.29), σ depends only on material parameters $[|\mu|^2$, g_t, and $\nu_0]$ and on the frequency of the incident wave. A knowledge of σ as a function of ν is therefore all that is needed to describe the interaction process. The transition cross section σ is therefore a very important and widely used parameter of the transition. It should also be observed that, for the case where the populations of the two levels are N_1 and N_2, Eq. (2.4.19) generalises to

$$dF = -\sigma(N_1 - N_2)Fdz \tag{2.4.31}$$

This has the same form as that originally derived in Chap. 1 [see Eq. (1.2.1) with $g_1 = g_2$]. The discussion presented in this section, however, provides a deeper understanding of the meaning of the (effective) cross section σ.

Another way of describing the interaction of radiation with matter involves defining a quantity α as

$$\alpha = \sigma(N_1 - N_2) \tag{2.4.32}$$

If $N_1 > N_2$ then α is positive and is referred to as the absorption coefficient of the material. Using Eq. (2.4.29) the following expression is obtained for α:

$$\alpha = \frac{2\pi^2}{3n\varepsilon_0 ch}(N_1 - N_2)|\mu|^2 \nu g_t(\nu - \nu_0) \tag{2.4.33}$$

Since α depends on the populations of the two levels, it is not the most suitable parameter for describing a situation where the level populations are changing, such as, for example, in a laser. Its usefulness, however, lies in the fact that the absorption coefficient α can often be directly measured. From Eqs. (2.4.31) and (2.4.32) we get in fact

$$dF = -\alpha F dz \tag{2.4.34}$$

The ratio between the photon flux after traversing a length l of the material and the incident flux is therefore $[F(l)/F(0)] = \exp(-\alpha l)$. By experimentally measuring this ratio with a sufficiently monochromatic radiation, we can obtain the value of α for that particular wavelength. The corresponding value of the transition cross section is then obtained from Eq. (2.4.32) once N_1 and N_2 are known. If the medium is in thermodynamic equilibrium, N_1 and N_2 can be obtained from Eq. (1.2.2) once the total population $N_t = N_1 + N_2$ and the level's degeneracies are known. The instrument used the measure the absorption coefficient α is known as an absorption spectrophotometer. We note, however, that an absorption measurement obviously cannot be performed for a transition in which level 1 is empty. This situation, for instance, occurs when level 1 is not the ground level and its energy above the ground level is much larger than kT. As a final observation we note that if $N_2 > N_1$, the absorption coefficient α, defined by Eq. (2.4.32), becomes negative and, of course, the wave gets amplified rather than absorbed in the material. In this case it is customary to define the new quantity g, as

$$g = \sigma(N_2 - N_1) \tag{2.4.35}$$

which is positive and is called the gain coefficient.

2.4.4. Einstein Thermodynamic Treatment

In this section we will describe a treatment, given by Einstein,[7] of both spontaneous and stimulated transitions (absorption and emission). In this treatment the concept of stimulated emission was first clearly established and the correct relationship between spontaneous and stimulated transition rates was derived well before the formulation of quantum mechanics and quantum electrodynamics. The calculation makes use of an elegant thermodynamic argument. To this end, let us assume that the material is placed in a blackbody cavity whose walls are kept at a constant temperature T. Once thermodynamic equilibrium is reached, an e.m. energy density with a spectral distribution ρ_ν given by Eq. (2.2.22) will be established and the material will be immersed in this radiation. In this material, both stimulated-emission and absorption processes will occur, in addition to the spontaneous-emission process. Since the system is in thermodynamic equilibrium, the number of transitions per second from level 1 to level 2 must be equal to the number of transitions from level 2 to level 1. We now set

$$W_{21} = B_{21}\rho_{\nu 0} \tag{2.4.36}$$

$$W_{12} = B_{12}\rho_{\nu 0} \tag{2.4.37}$$

where B_{21} and B_{12} are constant coefficients (the *Einstein B coefficients*), and let N_1^e and N_2^e be the equilibrium populations of levels 1 and 2, respectively. We can then write

$$AN_2^e + B_{21}\rho_{\nu 0}N_2^e = B_{12}\rho_{\nu 0}N_1^e \tag{2.4.38}$$

From Boltzmann statistics we also know that, for non degenerate levels, one has

$$N_2^e/N_1^e = \exp(-h\nu_0/kT) \tag{2.4.39}$$

From Eqs. (2.4.38) and (2.4.39) it then follows that

$$\rho_{\nu 0} = \frac{A}{B_{12}\exp(h\nu_0/kT) - B_{21}} \tag{2.4.40}$$

A comparison of Eq. (2.4.40) with Eq. (2.2.22), when $\nu = \nu_0$, leads to the following relations:

$$B_{12} = B_{21} = B \tag{2.4.41}$$

$$\frac{A}{B} = \frac{8\pi h\nu_0^3 n^3}{c^3} \tag{2.4.42}$$

Equation (2.4.41) shows that the probabilities of absorption and stimulated emission due to blackbody radiation are equal. This relation is therefore analogous to that established, in a completely different way, in the case of monochromatic radiation [see Eq. (2.4.12)]. Equation (2.4.42), on the other hand, allows the calculation of A, once B, i.e., the coefficient for stimulated emission due to blackbody radiation, is known. This coefficient can easily be obtained from Eq. (2.4.30) once we remember that this equation was established for monochromatic radiation. For blackbody radiation, we can write $\rho_\nu d\nu$ for the energy density of radiation whose frequency lies between ν and $\nu + d\nu$ and simulate this elemental radiation by a monochromatic wave. The corresponding elemental transition probability dW is then obtained from Eq. (2.4.30) by substituting $\rho_\nu d\nu$ for ρ. Upon integration of the resulting equation with the assumption that $g_t(\nu - \nu_0)$ can be approximated by a Dirac δ function in comparison with ρ_ν (see Fig. 2.3), we get

$$W = \frac{2\pi^2}{3n^2\varepsilon_0 h^2}|\mu|^2\rho_{\nu 0} \tag{2.4.43}$$

The comparison of Eq. (2.4.43) with Eqs. (2.4.36) or (2.4.37) then gives

$$B = \frac{2\pi^2|\mu|^2}{3n^2\varepsilon_0 h^2} \tag{2.4.44}$$

so that, from Eq. (2.4.42), we obtain

$$A = \frac{16\pi^3 \nu_0^3 n|\mu|^2}{3 h\varepsilon_0 c^3} \tag{2.4.45}$$

It should be noted that the expression for A that we have just obtained is exactly the same as that obtained by a quantum electrodynamics approach [see Eq. (2.3.19)]. Its calculation is in fact based on thermodynamics and the use of Planck's law (which is quantum electrodynamically correct).

2.5. LINE BROADENING MECHANISMS

In this section we will discuss, in some detail, the various line broadening mechanisms mentioned in previous sections. According to the earlier discussion, there is an important distinction to be made from the outset between homogeneous and inhomogeneous line-broadening. A line-broadening mechanism is referred to as *homogeneous* when it broadens the line of each atom in the same way. In this case the line-shape of the single-atom cross section and that of the overall absorption cross section would be identical. Conversely, a line-broadening mechanism is said to be *inhomogeneous* when it distributes the atomic resonance frequencies over some spectral range. Such a mechanism thus broadens the overall line of the system (i.e. that of α) without broadening the lines of individual atoms.

Before proceeding, it is worth recalling that the shape of the function $g_t(\nu - \nu_0)$ can be determined in two ways: (a) By an absorption experiment, with the help of a spectrophotometer. In this case one measures the absorption coefficient as a function of frequency ν, using the spectrophotometer to select the light frequency. From Eq. (2.4.33) one sees that $\alpha \propto \nu g_t(\nu - \nu_0)$. Since the linewidth of the function $g_t(\nu - \nu_0)$ is, typically, much smaller than ν_0, we can approximately write $\alpha \propto \nu_0 g_t(\nu - \nu_0)$. Thus, to a very good approximation, the shape of the α vs ν curve coincides with that of the function $g_t(\nu - \nu_0)$. (b) By an emission experiment, in which one passes the spontaneously emitted light trough a spectrometer of sufficiently high resolution and one determines $g_t(\nu - \nu_0)$ by measuring the shape of the spectral emission. It can be shown that, for any transition, the lineshapes obtained by these two approaches are always the same. So, in the discussion that follows, we will consider the lineshape function either in absorption or in emission, whichever is the more convenient.

2.5.1. Homogeneous Broadening

The first homogeneous line-broadening mechanism we consider is one due to collisions and is known as *collision broadening*. In a gas, it is due to the collision of an atom with other atoms, ions, free electrons, etc. or with the walls of the container. In a solid it is due to the interaction of the atom with the phonons of the lattice. After a collision the two level wavefunctions ψ_1 and ψ_2 of the atom [see Eq. (2.3.1)] will undergo a random phase jump. This means that the phase of the oscillating dipole moment μ_{osc} [see Eq. (2.3.6)] will undergo a random jump compared to that of the incident e.m. wave. These collisions thus interrupt

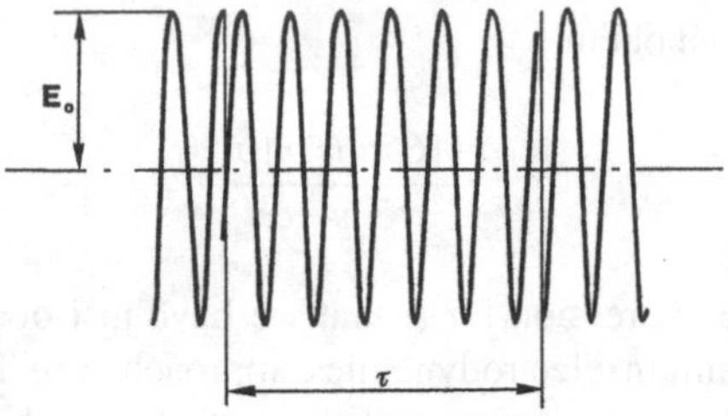

FIG. 2.9. Time behavior of the electric field of an e.m. wave, $E(t)$, as seen from an atom undergoing collisions. Note that, in actual cases, there may 10^7 or more collisions during the collision time τ.

the process of coherent interaction between the atom and the incident e.m. wave. Since it is the relative phase which is important during the interaction process, an equivalent way of treating this problem is to assume that it is the phase of the electric field rather than that of μ_{osc} that undergoes a jump at each collision. The electric field will therefore no longer appear sinusoidal but will instead appear as shown in Fig. 2.9, where each phase jump occurs at the time of a collision. It is therefore clear that, under these conditions, the atom no longer sees a monochromatic wave. In this case, if we write $d\rho = \rho_{\nu'} d\nu'$ for the energy density of the wave in the frequency interval between ν' and $\nu' + d\nu'$, we can use this elemental energy density in the formula valid for monochromatic radiation, i.e., Eq. (2.4.7), which gives

$$dW_{12} = \frac{2\pi^2}{3\,n^2 \varepsilon_0\, h^2} |\mu_{21}|^2 \rho_{\nu'} \delta(\nu' - \nu_0) d\nu' \tag{2.5.1}$$

The overall transition probability is then obtained by integrating Eq. (2.5.1) over the entire frequency spectrum of the radiation, thus giving

$$W_{12} = \frac{2\pi^2}{3\,n^2 \varepsilon_0\, h^2} |\mu_{21}|^2 \int_{-\infty}^{+\infty} \rho_{\nu'}\, \delta(\nu' - \nu_0) d\nu' \tag{2.5.2}$$

We can now write $\rho_{\nu'}$ as

$$\rho_{\nu'} = \rho\, g(\nu' - \nu) \tag{2.5.3}$$

where ρ is the energy density of the wave [see Eq. (2.4.6)], and $g(\nu' - \nu)$ describes the spectral distribution of $\rho_{\nu'}$. Since one obviously has $\rho = \int \rho_{\nu'} d\nu'$, the integration of both sides of Eq. (2.5.3) then shows that $g(\nu' - \nu)$ must satisfy the normalization condition

$$\int_{-\infty}^{+\infty} g(\nu' - \nu) d\nu' = 1 \tag{2.5.4}$$

Upon substituting Eq. (2.5.3) into Eq. (2.5.2) and using a well known mathematical property of the δ function we get

$$W_{12} = \frac{2\pi^2}{3\,n^2 \varepsilon_0\, h^2} |\mu_{21}|^2 \rho\, g(\nu - \nu_0) \tag{2.5.5}$$

As anticipated in Sect. 2.4.1, it is seen that W_{12} is indeed obtained by substituting $g(\nu - \nu_0)$ to $\delta(\nu - \nu_0)$ in Eq. (2.4.7). Note that, according to Eq. (2.5.4), we also have

$$\int_{-\infty}^{+\infty} g(\nu - \nu_0) d\nu = 1 \tag{2.5.6}$$

There now remains the problem of calculating the normalized spectral density of the incident radiation $g(\nu' - \nu)$. This will depend on the time interval, τ, between collisions (Fig. 2.9), which will obviously be different for each collision. We will assume that the distribution of the values of τ can be described by a probability density

$$p_\tau = [\exp(-\tau/\tau_c)]/\tau_c \tag{2.5.7}$$

Here $p_\tau\, d\tau$ is the probability that the time interval between two successive collisions lies between τ and $\tau + d\tau$. Note that τ_c has the physical meaning of the average time $<\tau>$ between collisions. It is easy, in fact, to see that

$$<\tau> = \int_0^{\infty} \tau\, p_\tau d\tau = \tau_c \tag{2.5.8}$$

At this point the mathematical problem to be solved is well defined. We need to obtain the normalized spectral lineshape of a wave as in Fig. 2.9 for which the time τ between two successive collisions has the statistical distribution p_τ given by Eq. (2.5.7). Referring to the Appendix B for the mathematical details we merely quote the final result here. The required normalized spectral lineshape is given by

$$g(\nu' - \nu) = 2\tau_c \frac{1}{[1 + 4\pi^2 \tau_c^2 (\nu' - \nu)^2]} \tag{2.5.9}$$

According to Eq. (2.5.5) the line shape of the transition is obtained from Eq. (2.5.9) by substituting ν' by ν_0. We then get

$$g(\nu - \nu_0) = 2\tau_c \frac{1}{[1 + 4\pi^2 \tau_c^2 (\nu - \nu_0)^2]} \tag{2.5.10}$$

which is our final result. We thus obtain a function with a Lorentzian lineshape, as generally described by Eq. (2.4.8) [see also Fig. 2.6], where the peak value is now $2\tau_c$ and the linewidth $\Delta\nu_0$ is

$$\Delta\nu_0 = 1/\pi\tau_c \tag{2.5.11}$$

Example 2.2. *Collision broadening of a He-Ne laser* As a first example of collision broadening, we consider the case of a transition for an atom, or ion, in a gas at pressure p. An estimate of τ_c is, in this case, given by $\tau_c = l/\upsilon_{th}$ where l is the mean free path of the atom in the gas and υ_{th} is its average thermal velocity. Since $\upsilon_{th} = (3kT/M)^{1/2}$ where M is the atomic mass and taking l to be given by the expression resulting from the hard-sphere model of a gas, we obtain

$$\tau_c = \left(\frac{2}{3}\right)^{1/2} \frac{1}{8\pi} \frac{(MkT)^{1/2}}{pa^2} \qquad (2.5.12)$$

where a is the radius of the atom and p is the gas pressure. For a gas of Neon atoms at room temperature and at a pressure $p \cong 0.5$ Torr (typical pressure in a He-Ne gas laser) using Eq. (2.5.12) with $a \cong 0.1$ nm and $\tau_c \cong 0.5\ \mu$s, we find from Eq. (2.5.11) that $\Delta\nu_o = 0.64$ MHz. Note that τ_c is inversely proportional, and hence $\Delta\nu_0$ directly proportional, to p. As a rough "rule of thumb" we can say that, for any atom, collisions in a gas contribute to the line broadening by an amount $(\Delta\nu_0/p) \cong 1$ MHz/Torr, comparable to that shown in the example of Ne atoms. Note also that, during the collision time τ_c the number of cycles of the e.m. wave is equal to $m = \nu\tau_c$ For a wave whose wavelength falls in the middle of the visible range we have $\nu = 5 \times 10^{14}$ Hz and thus the number of cycles is 5×10^8. This emphasizes the fact that Fig. 2.9 is not to scale since the number of cycles in the time τ is much larger than suggested in the figure.

Example 2.3. *Linewidth of Ruby and Nd:YAG* As a third example of collision broadening, we will consider an impurity ion in an ionic crystal. In this case the collisions of the ion occur with the lattice phonons. Since the number of phonons in a given lattice vibration is a strong function of the lattice temperature, we expect the transition linewidth to show a strong dependence on temperature. As a representative example, Fig. 2.10 shows the linewidth versus temperature for both Nd:YAG and ruby, the linewidth being expressed in wavenumbers [cm^{-1}], a quantity widely used by spectroscopists rather than actual frequency.* At 300 K the laser transition linewidths are seen to be $\Delta\nu_0 \cong 4\ cm^{-1} \cong 120$ GHz for Nd:YAG and $\Delta\nu_0 \cong 11\ cm^{-1} = 330$ GHz for ruby.

A second homogeneous line-broadening mechanism has its origin in the phenomenon of spontaneous emission. Since this emission is an inevitable feature of any transition, the corresponding broadening is called *natural* or *intrinsic*. In the case of natural broadening, it is easiest to consider the behavior in terms of the spectrum of the emitted radiation. It should be noted however that, as pointed out in Sect. 2.3.2, spontaneous emission is a purely quantum phenomenon, i.e. it can only be correctly explained by quantizing both matter and radiation. It follows therefore that a correct description of the lineshape of the emitted radiation also needs a quantum electrodynamics treatment. We will therefore limit ourselves to quoting the final result, which happens to be very simple, and to justifying it by some simple arguments. The quantum electrodynamics theory of spontaneous emission[(8)] shows that the spectrum $g(\nu - \nu_0)$ is again described by a Lorentzian line whose shape can be obtained from Eq. (2.5.10) by replacing τ_c by $2\tau_{sp}$, where τ_{sp} is the decay time of the spontaneous emission. Thus, in particular, the full width of the line (FWHM) is given by

$$\Delta\nu_0 = 1/2\pi\tau_{sp} \qquad (2.5.13)$$

* For a given wave of frequency ν, the corresponding frequency in wave numbers (e.g. in cm^{-1}) is given by $w = 1/C$, where c is the velocity of the wave in a vacuum (in cm/s). The true frequency ν is then obtained from the frequency in wave numbers by the simple relation $\nu = cw$ while the corresponding wavelenght is simply given by $\lambda = c/\nu = 1/w$ (in cm). This illustrates the advantages of the wave number notation. The trem wave number arises from the fact that w gives the number of wave periods, n, comprised in a given unitary length l (e.g. in 1 cm). The number n is in fact given by $n = l/\lambda$ so that $n/l = 1\lambda = w$.

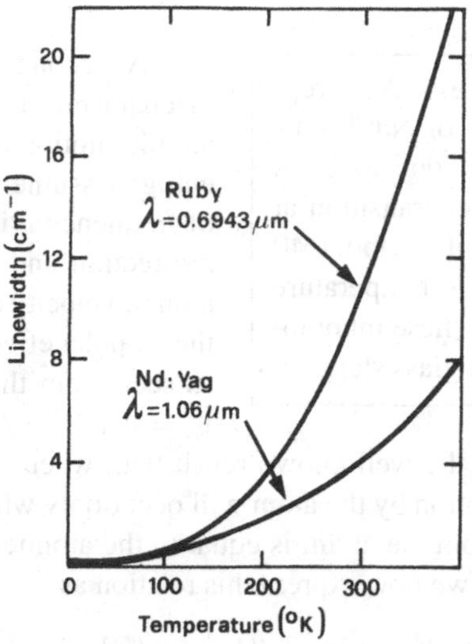

FIG. 2.10. Laser linewidth vs temperature for ruby and Nd:YAG, as determined by phonon broadening.

To justify this result we notice that, since the power emitted by the atom decays in time as $\exp(-t/\tau_{sp})$, the corresponding electric field can be thought as decaying according to the relationship $E(t) = \exp(-t/2\tau_{sp}) \times \cos\omega_0 t$. The decay of emitted intensity [which is proportional to $<E^2(t)>$] would then show the correct temporal behavior, namely, $\exp(-t/\tau_{sp})$. We can now easily calculate the power spectrum corresponding to such a field $E(t)$ and verify that the line shape is Lorentzian and that its width is given by Eq. (2.5.13).

Example 2.4. *Natural linewidth of an allowed transition* As a representative example we can find an order of magnitude estimate for $\Delta\nu_{nat}$ for an electric-dipole allowed transition. Assuming $|\mu| = ea$ with $a \cong 0.1$ nm and $\lambda = 500$ nm (green light) we already obtained in example 2.1 that $\tau_{sp} \cong 10$ ns. From Eq. (2.5.13) we then get $\Delta\nu_{nat} \cong 16$ MHz. Note that $\Delta\nu_{nat}$, just as $A = 1/\tau_{sp}$, is expected to increase with frequency as ν_0^3. Therefore the natural linewidth increases very rapidly for transitions at shorter wavelengths (down to the UV or X-ray region).

2.5.2. Inhomogeneous Broadening

We will now consider some mechanisms where the broadening arises from the distribution of the atomic resonance frequencies (inhomogeneous broadening).

As a first case of inhomogeneous broadening we consider that which occurs for ions in ionic crystals or glasses. Ions will experience a local electric field produced by the surrounding atoms of the material and, due to material inhomogeneities which are particularly significant in glass medium, these fields will be different from ion to ion. These local field variations will then produce, via the Stark effect, local variation of the energy levels and thus of the transition frequencies of the ions (the term inhomogeneous broadening originates from this case). For random local field variations, the corresponding distribution of the transition frequencies $g^*(\nu_0' - \nu_0)$ turns out to be given by a Gaussian function, i.e. by the general expression Eq. (2.4.27) where the linewidth $\Delta\nu_0^*$ (FWHM) will depend upon the extent of variation of transition frequencies in the material and hence upon the amount of field inhomogeneity within the crystal or glass.

Example 2.5. *Linewidth of a Nd:glass laser* As a representative example we consider the case of Nd^{3+} ions doped into a silicate glass. In this case, due to glass inhomogeneities, the linewidth of the laser transition at $\lambda = 1.05\,\mu m$ is $\Delta\nu_0^* \cong 5.4\,THz$ i.e. it is about 40 times broader than that of Nd:YAG at room temperature (see Example 2.3). It should be noted that these inhomogeneities are an unavoidable feature of the glass state.

A second inhomogeneous broadening mechanism, typical of gas, arises from atomic motion and is called Doppler broadening. Assume that an incident e.m. wave of frequency ν is propagating in the positive z direction and let υ_z be the component of atomic velocity along this axis. According to the Doppler effect, the frequency of the wave, as seen from the rest frame of the atom, is $\nu' = \nu[1 - (\upsilon_z/c)]$ where c is the velocity of light in the medium. Notice the well known result that, when $\upsilon_z > 0$, we have $\nu' < \nu$ and vice versa. Of course, absorption by the atom will occur only when the apparent frequency ν' of the e.m. wave, as seen from the atom, is equal to the atomic transition frequency ν_0, i.e., when $\nu[1 - (\upsilon_z/c)] = \nu_0$. If we now express this relation as

$$\nu = \nu_0/[1 - (\upsilon_z/c)] \tag{2.5.14}$$

we can arrive at a different interpretation of the process. As far as the interaction of the e.m. radiation with the atom is concerned, the result would be the same if the atom were not moving but instead had a resonant frequency ν_0' given by

$$\nu_0' = \nu_0/[1 - (\upsilon_z/c)] \tag{2.5.15}$$

where ν_0 is the true transition frequency. Indeed, following this interpretation, absorption is expected to occur when the frequency ν of the e.m. wave is equal to ν_0' i.e. when $\nu = \nu_0'$, in agreement with Eq. (2.5.14) when the expression Eq. (2.5.15) for ν_0' is used. When looked at in this way, one can see that this broadening mechanism does indeed belong to the inhomogeneous category as defined at the beginning of this section.

To calculate the corresponding line shape $g^*(\nu_0' - \nu_0)$ it is now sufficient to remember that, if we let $p_\upsilon d\upsilon_z$ be the probability that an atom of mass M in a gas at temperature T has a velocity component between υ_z and $\upsilon_z + d\upsilon_z$, then p_υ is given by the Maxwell distribution

$$p_\upsilon = \left(\frac{M}{2\pi kT}\right)^{1/2} \exp -(M\upsilon_z^2/2kT) \tag{2.5.16}$$

From Eq. (2.5.15), since $|\upsilon_z| \ll c$, we get $\nu_0' \cong \nu_0[1 + (\upsilon_z/c)]$ and thus $\upsilon_z = c(\nu_0' - \nu_0)/\nu_0$. From Eq. (2.5.16) one then obtains the desired distribution upon recognizing that one must have $g^*(\nu_0' - \nu_0)d\nu_0' = p_\upsilon d\upsilon_z$. One then gets

$$g^*(\nu_0' - \nu_0) = \frac{1}{\nu_0}\left(\frac{Mc^2}{2\pi kT}\right)^{1/2} \exp -\left[\frac{Mc^2}{2kT}\frac{(\nu_0' - \nu_0)^2}{\nu_0^2}\right] \tag{2.5.17}$$

Thus one again obtains a Gaussian function whose FWHM linewidth (Doppler linewidth) is now readily found from a comparison of Eq. (2.5.17) with Eq. (2.4.24), giving

$$\Delta\nu_0^* = 2\nu_0\left[2kT\ln 2/Mc^2\right]^{1/2} \tag{2.5.18}$$

For the purely inhomogeneous case the lineshape will be given by the general expression of Eq. (2.4.27) where $\Delta\nu_0^*$ is given by Eq. (2.5.18).

Example 2.6. *Doppler linewidth of a He-Ne laser* Consider the Ne line at the wavelength $\lambda = 632.8$ nm (the red laser line of a He-Ne laser) and assume T = 300 K. Then from Eq. (2.5.18), using the appropriate mass for Ne, we get $\Delta\nu_0^* \cong 1.7$ GHz. A comparison of this value with those obtained for collision broadening, see example 2.2, and natural broadening, see Example 2.4 (the transition is allowed by electric dipole), shows that Doppler broadening is the predominant line broadening mechanism in this case.

2.5.3. Concluding Remarks

According to the previous discussion, we have seen that the shape of a homogeneous line is always Lorentzian while that of an inhomogeneous line is always Gaussian. When two mechanisms contribute to line broadening, the overall line shape turns out to be always given by the convolution of the corresponding line-shape functions, as indicated in Eq. (2.4.26) for the case of one line being homogeneously and the other inhomogeneously broadened. It can now be shown that the convolution of a Lorentzian line, of width $\Delta\nu_1$, with another Lorentzian line, of width $\Delta\nu_2$, again gives a Lorentzian line whose width is now $\Delta\nu = \Delta\nu_1 + \Delta\nu_2$. The convolution of a Gaussian line, of width $\Delta\nu_1$, with another Gaussian line, of width $\Delta\nu_2$, is again a Gaussian line, this time of width $\Delta\nu = \left(\Delta\nu_1^2 + \Delta\nu_2^2\right)^{1/2}$. For any combination of broadening mechanisms, it is therefore always possible, to reduce the problem to a convolution of a single Lorentzian line with a single Gaussian line and this integral (which is known as the Voigt integral[9]) is tabulated. Sometimes, however, (e.g. as in the previously discussed cases for Ne), one mechanism predominates. In this case, it is then possible to talk of a pure Lorentzian or Gaussian line.

We conclude this section by showing, in Table 2.1, the actual range of linewidths for the various line-broadening mechanisms considered. Note that, in the middle of the visible range, we have $\tau_{sp} \cong 10$ ns and hence $\Delta\nu_{nat} \cong 10$ MHz for an electric dipole allowed transition. For an electric dipole forbidden transition, on the other hand, one has $\tau_{sp} \cong 1$ ms and hence $\Delta\nu_{nat} \cong 1$ kHz. Note also that, in the case of a liquid, collision broadening and local field inhomogeneous broadening are the predominant broadening mechanisms. In this case, the average time between two consecutive collisions is indeed much shorter than in the gas phase [$\tau_c \cong 0.1$ ps] and hence we have $\Delta\nu_c = 1/\pi\tau_c \cong 100\ \text{cm}^{-1}$. Inhomogeneous broadening arises from the local density variations associated with a given temperature and may produce a value for the linewidth $\Delta\nu_0^*$ comparable to that of collision broadening. In a solid, inhomogeneous broadening due to local field variations may be as high as 300 cm^{-1} for a glass and as low as 0.5 cm^{-1}, or even lower, for a good quality crystal such as in presently available Nd:YAG crystals.

TABLE 2.1. Typical magnitude of frequency broadening for the various line-broadening mechanisms

	Type	Gas	Liquid	Solid
Homogeneous	Natural	1 kHz ÷ 10 MHz	Negligible	Negligible
	Collisions	5 ÷ 10 MHz/Torr	~300 cm^{-1}	–
	Phonons	–	–	~10 cm^{-1}
Inhomogeneous	Doppler	50 MHz ÷ 1 GHz	Negligible	–
	Local field	–	~500 cm^{-1}	1 ÷ 500 cm^{-1}

2.6. NONRADIATIVE DECAY AND ENERGY TRANSFER

Besides decaying via radiative emission, an excited species can also undergo nonradiative decay. There exists a variety of ways by which this can occur and the detailed description of the various physical phenomena can often be quite complicated. We shall therefore limit ourselves to a qualitative discussion with the main aim being to elucidate the physical phenomena involved. We will then consider the combined effect of radiative and non-radiative decay processes.

2.6.1. Mechanisms of Nonradiative Decay[10]

First we consider a nonradiative decay mechanism which arises from collisions, sometimes called *collisional deactivation*. In this case, for a gas or a liquid, the transition energy is released as excitation and/or as kinetic energy of the colliding species or given to the walls of the container. In the case of a solid, such as an ionic crystal or glass, the energy of the excited ion is taken up by the lattice phonons or by the glass vibrational modes.

The collisional deactivation process, for the case where the energy of an excited species B^* is released as kinetic energy of a colliding species A, can be expressed in the form

$$B^* + A \rightarrow B + A + \Delta E \tag{2.6.1}$$

where ΔE is equal to the excitation energy. Since ΔE ends up as kinetic energy of the colliding partners, the process is also referred to as a *superelastic collision* or a *collision of second kind*. For a process of the form shown in Eq. (2.6.1), the rate of change of B^* population, N_{B^*}, can be written as

$$\frac{dN_{B^*}}{dt} = -k_{B^*A} N_{B^*} N_A \tag{2.6.2}$$

where N_A is the population of species A and k_{B^*A} is a coefficient which depends on the transition of species B and on species A. The process is particularly effective, i.e. k_{B^*A} is particularly large, when A has a very small mass (e.g. the He in the gas of a CO_2 laser) so it can more readily take-up the surplus energy ΔE, from the collision process, as kinetic energy. For the same reason the process can readily occur in a gas discharge when A is a discharge electron (e.g. deactivation of the 2^3S state of He in a He-Ne laser). According to Eq. (2.6.2), we can now define a nonradiative decay rate

$$W_{nr} = k_{B^*A} N_A \tag{2.6.3}$$

From Eqs. (2.6.2) and (2.6.3) we then get

$$\left(\frac{dN_2}{dt}\right) = -\frac{N_2}{\tau_{nr}} \tag{2.6.4}$$

where, to conform with previous notations, we have let N_2 be the population of the species undergoing collisional deactivation and where we have defined a nonradiative decay time as $\tau_{nr} = (1/W_{nr})$.

It should be observed that, in writing Eq. (2.6.2), we have neglected the reverse process of that given by Eq. (2.6.1) i.e.,

$$B + A \rightarrow B^* + A - \Delta E \tag{2.6.5}$$

where species B is excited at the expense of the kinetic energy ΔE of the two colliding partners (*thermal activation* or *collision of first kind*). If this process were taken into account, one should write, instead of Eq. (2.6.2), the following equation

$$(dN_{B^*}/dt) = -k_{B^*A}N_{B^*}N_A + k_{BA}N_BN_A \tag{2.6.6}$$

where k_{BA} is a coefficient describing the process of thermal activation. To find the relationship between k_{BA} and k_{B^*A} we can consider species B in thermal equilibrium with species A and then apply the so-called *principle of detailed balance*. This principle can generally be formulated by requiring that, in thermodynamic equilibrium, the rate of *any* process must be exactly balanced by the rate of the corresponding reverse process*. Thus in this case, according to Eq. (2.6.6), we require

$$k_{B^*A}N_{B^*}N_A = k_{BA}N_BN_A \tag{2.6.7}$$

In thermal equilibrium and for nondegenerate levels we have $N_{B^*} = N_B \exp(-\Delta E/kT)$ where ΔE is the excitation energy of species B and T is the temperature of the ensemble of species B and A. From Eq. (2.6.7) we then get

$$k_{B^*A} = k_{BA} \exp(\Delta E/kT) \tag{2.6.8}$$

which shows that the rate coefficient k for the exothermic reaction Eq. (2.6.1) is always larger than that of the endothermic reaction Eq. (2.6.5). Actually, for electronic and for most vibrational transitions, ΔE is much larger than kT. Thus, according to Eq. (2.6.8), we have $k_{B^*A} \gg k_{BA}$. It is very important also to realize that, although Eq. (2.6.8) has been derived for thermal equilibrium conditions, the same relation still holds if the population of species B is maintained in a non equilibrium state of excitation, e.g. by some pumping process, provided that the translational degrees of freedom of both species B and A are still in thermal equilibrium. In fact, the quantum mechanical calculation of the rate coefficient k does not depend on the population of B but only upon the eigenfunctions of the two species involved, and on their relative velocities. For a steady excitation of species B away from the Boltzmann equilibrium, i.e. when N_{B^*} is of the same order of N_B, we thus have $k_{B^*A}N_{B^*} \gg k_{BA}N_B$ and Eq. (2.6.6) reduces to Eq. (2.6.2). Thus, to conclude, collisional deactivation takes the simple form given by Eq. (2.6.4) only when $\Delta E >> kT$ so that thermal activation may be neglected, which is the case for electronic transitions and for most vibrational transitions. For deactivation of the lowest energy vibrational levels of some molecules [e.g. the (010) state of CO_2] and for rotational transitions, thermal excitation must however be taken into account.

* Note that the equation expressing the balance of processes between a two level atom and the blackbody radiation, established in Sect. 2.4.4., is another example of the principle of detailed balance.

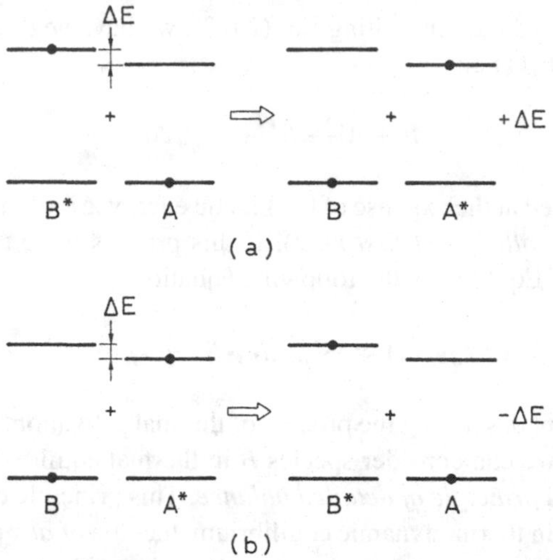

FIG. 2.11. (a) Nonradiative decay of a species B by near resonant energy transfer to a species A. (b) Reverse, back-transfer, process.

When the electronic energy of species B is released in the form of internal energy of some other species A, we can represent this with an equation of the form (collision of the second kind*)

$$B^* + A \rightarrow B + A^* + \Delta E \tag{2.6.9}$$

where $\Delta E = E_B - E_A$ is the difference between the internal energies of the two species (see Fig. 2.11a). The quantum mechanical calculation of the corresponding transition rate is beyond the scope of this book and we refer the reader elsewhere for details.[11] Here we limit ourselves to pointing out that, since ΔE must be added to or removed from the kinetic energy of the two colliding partners, the process turns out to be particularly effective when ΔE is appreciably smaller than kT. Therefore, the process is also called *near-resonant energy transfer* and often plays an important role as a pumping mechanism in gas lasers (e.g. energy transfer between excited He and ground state Ne, in a He-Ne laser, or between excited N_2 and ground state CO_2 in a CO_2 laser). The process also results in an effective deactivation channel for species B. To consider the dynamics of this deactivation process, we must also take into account the reverse process (*back-transfer*, see Fig. 2.11b)

$$B + A^* \rightarrow B^* + A - \Delta E \tag{2.6.10}$$

Actually, again applying the principle of detailed balance one can now show that, e.g., for the case of exact resonance (i.e., $\Delta E = 0$), one has $k_{B^*A} = k_{BA^*}$, where k_{B^*A} and k_{BA^*} are the

* Collisions of the *first kind* involve conversion of the kinetic energy of one species into internal energy of another species [see Eq. (2.6.5)]. In collisions of the *second kind*, internal energy is converted into some other form of energy (other than radiation) such as kinetic energy [see Eq. (2.6.1)], or is transferred into internal energy of another species (same or different species) [see Eq. (2.6.9)]. Collisions of the second kind thus also include, for instance, the conversion of excitation energy into chemical energy.

rate constants of the two processes described by Eqs. (2.6.9) and (2.6.10), respectively. This indicates that the back-transfer reaction often plays a very important role. This process can however be neglected when the decay of species A from its excited state is very fast, as it may occur by the onset of stimulated emission. In this case one has $(N_{A^*}/N_A) \ll (N_{B^*}/N_B)$, the back transfer may be neglected, and the rate of decay of the excited species, B^*, can simply be written as

$$(dN_{B^*}/dt) = -k_{B^*A}N_{B^*}N_A \tag{2.6.11}$$

We again obtain an equation of the general form given by Eq. (2.6.4) where now $(1/\tau_{nr}) = k_{B^*A}N_A$.

Finally we consider the case where collisional deactivation of species B (e.g. an active ion in an ionic crystal) occurs through interaction with lattice phonons or with glass vibrational modes*. In many cases, except for some nonradiative decay processes occurring in tunable solid state lasers (see Chap. 9), we are dealing with electronic transitions and thus with transition energies of species B which are many times (typically at least 3 to 4 times) larger than that of the most energetic phonon. This means that, to conserve energy, the transition energy must be released in the form of many phonons (*multiphonon deactivation*). Thus, in this case, the deactivation process can be represented in the form

$$B^* \rightarrow B + \sum_{1}^{n}{}_i(h\nu_i) \tag{2.6.12}$$

where ν_i are the frequencies of the phonons involved and the sum is extended over all phonons created in this resonant or near-resonant process. Again we can define a transition rate W_{nr} according to the relation

$$\frac{dN_{B^*}}{dt} = -W_{nr}B^* \tag{2.6.13}$$

In this case, since many phonons are involved, the quantum mechanical calculation for the process would involve a higher order perturbation theory. It is therefore not considered here in any detail. We simply limit ourselves to pointing out that, if only a phonon of frequency ν is involved, W_{nr} can be written as $W_{nr} = A\exp(-B\Delta E/h\nu)$, where A and B are host-dependent constants and ΔE is the transition energy of species B. We thus see that the transition rate rapidly decreases with the increasing number, $n = \Delta E/h\nu$, of phonons involved i.e. with the increasing order of the multiphonon process. The dominant contribution to the nonradiative process thus comes from the lattice phonon of the highest energy, since this means that the lowest order process is then involved. The large variation in vibrational spectra shown by different material then makes W_{nr} extremely host dependent. By contrast, the rate is found to be relatively independent of the actual electronic state or even the particular active ion involved.

* The absence of translational invariance in glass means that, strictly speaking, one should not talk in terms of phonons, in this case, as one does for a crystal. For now on, however, for brevity we will refer to phonons even in this case.

As a conclusion to this discussion on collisional deactivation, we note that, while the process can take a variety of forms, the decay behavior of the excited state can, subjects to limits that we have discussed, always be described by the general relation Eq. (2.6.4), where the value of τ_{nr} will depend upon the particular process under consideration. It should be noted explicitly, however, that there is a fundamental difference between the nonradiative decay time, τ_{nr}, discussed here, and the collision time τ_c discussed in Sect. 2.5.1, although they both originate from collisions. In fact, a nonradiative decay process requires an inelastic collision since the decaying species gives up its energy to its surroundings. By contrast, τ_c is the average time between two consecutive dephasing collisions and thus arises from elastic collisions only. Note that, in general, elastic collisions are more likely than inelastic ones and thus τ_c is smaller, and often much smaller, than τ_{nr}.

A kind of nonradiative decay that does not rely on collisions, arises from *dipole-dipole interaction* between an excited species that we shall call the donor, D, and, e.g., a ground state species that we shall call the acceptor, A. The interaction results in energy being transferred between donor and acceptor. This process has been extensively studied by Förster for liquids(12) and by Dexter for solids.(13) It plays a very important role e.g. for active ions in crystals or glasses and for mixtures of organic dyes in solution. Consider the donor, undergoing the downward transition, while at a distance R from the acceptor. During the transition, the donor will develop a dipole moment, $\boldsymbol{\mu}_D$, oscillating at its transition frequency. From the theory of electric dipole radiation,(14) it is known that this moment generates, at a distance R, a nonradiating electric field (the so called *near-zone field*) whose magnitude, $\mathbf{E}_D(t)$, as for an electrostatic dipole, is equal to $\boldsymbol{\mu}_D/4\pi\varepsilon_0 R^3$. Under these conditions, nonradiative decay may occur by energy transfer arising from the interaction of the near-zone field $\mathbf{E}_D(t, R)$ at the position of the acceptor with the oscillating dipole moment of the acceptor, $\boldsymbol{\mu}_A$. The interaction energy, H, can then be written as

$$H \propto |\mathbf{E}_D \cdot \boldsymbol{\mu}_A| \propto |\boldsymbol{\mu}_D \cdot \boldsymbol{\mu}_A|/R^3 \tag{2.6.14}$$

Of course, the interaction will have a significant strength only if the oscillation frequency of $\boldsymbol{\mu}_D$ is nearby resonant with that of $\boldsymbol{\mu}_A$. This means that there should be a good overlap between the emission spectrum of the donor and the absorption spectrum of the acceptor from its initial state (often not necessarily being the ground state). The detailed calculation shows that, for a single donor and acceptor separated by a distance R, the rate of energy transfer can be written as(12)

$$W_{DA} = \left(\frac{3}{64\pi^5}\right)\left(\frac{1}{R^6}\right)\left[\frac{1}{\tau_{sp}}\int_0^\infty \left(\frac{c}{n\nu}\right)^4 g_D(\nu)\sigma_A(\nu)d\nu\right] \tag{2.6.15}$$

where τ_{sp} is the spontaneous lifetime of the donor, n is the refractive index of the surrounding medium g_D is the line-shape function of the donor and σ_A is the absorption cross-section of the acceptor. Note that since, as usual, the dependence of W_{DA} with the interaction energy H is $W_{DA} \propto |H|^2$, from Eq. (2.6.14) we expect $W_{DA} \propto |\mu_D|^2|\mu_A|^2/R^6$. By this relation one can understand the dependence of W_{DA} not only from R^{-6} but also from $(1/\tau_{sp})$ [remember that $1/\tau_{sp} \propto |\boldsymbol{\mu}_D|^2$, see Eq. (2.3.15)] and to the cross section σ_A of the acceptor [remember that $\sigma_A \propto |\boldsymbol{\mu}_A|^2$, see Eq. (2.4.29)].

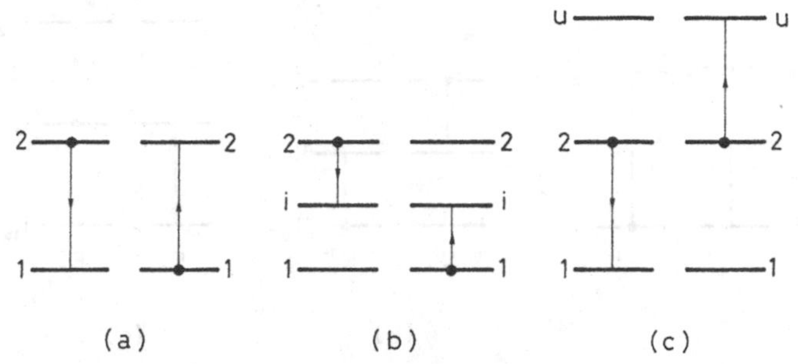

FIG. 2.12. Different forms of energy transfer by dipole-dipole interaction within the same species: (a) Migration of excitation. (b) Cross relaxation. (c) Cooperative up-conversion.

It should be noted finally that, dipole-dipole interactions may take somewhat different forms from the classical donor-acceptor situation just considered. For instance, it may occur between members of a single species (referred to as species D) and thus lead for example to energy transfer between an excited and an unexcited atom [Fig. 2.12a]. Such a resonant energy transfer usually can, for example, lead simply to spatial *excitation migration* within the same species D. It may also lead, however, to a nonradiative decay if the excitation eventually reaches a D site which is close to a, nearly resonant, acceptor site which itself has a fast decay. Energy transfer may also occur into an intermediate level, i, as shown in Fig. 2.12b (*cross relaxation*). This process will be particularly effective for near resonant transfer i.e. when $\Delta E_{2i} \cong \Delta E_{i1}$. Finally, energy transfer may occur throughout an upper level, with both donor and acceptor initially excited and the acceptor then being excited to a higher level u (Fig. 2.12c). This process, known as *cooperative up-conversion*, is particularly effective again for near resonant energy transfer i.e. when $\Delta E_{2u} \cong \Delta E_{21}$.

Example 2.7. *Energy transfer in the Yb^{3+} : Er^{3+}: glass laser system*[15] Donor-acceptor type of energy transfer is very effective in the Yb^{3+} : Er^{3+}:glass laser system (see Chap. 9) in transferring the excitation from the Yb^3 ion, initially excited to its $^2F_{5/2}$ state, to the $^4I_{11/2}$ excited level of Er^{3+} (Fig. 2.13a). This energy transfer, besides being an effective nonradiative decay mechanism for the Yb ion, constitutes a very effective way of pumping the active Er ion. Note that, at high Yb ion concentrations, this energy transfer is assisted by energy migration between Yb ions until the excitation reaches a closely spaced Yb-Er pair.

Example 2.8. *Nonradiative decay from the $^4F_{3/2}$ upper laser level of Nd:YAG* Cross relaxation turns out to be the main nonradiative decay mechanism for the Nd:YAG $^4F_{3/2}$ upper laser level. In this case the intermediate level i of Fig. 2.12b is the $^4I_{15/2}$ level of the Nd ion (Fig. 2.13b). The excitation energy of this level is then rapidly lost via multiphonon relaxation and the ion passes successively through the $^4I_{13/2}$ and the $^4I_{11/2}$ lower levels (not shown in the figure, see however Fig. 2.15) to the $^4I_{9/2}$ ground level. The energy difference between these sublevels (e.g. $^4I_{13/2} \rightarrow {}^4I_{11/2}$) is, in fact, about $2,000\,\text{cm}^{-1}$ (the Stark sublevels are even closer) i.e. only 4 times larger than the highest vibrational frequency of YAG crystal ($\sim 450\,\text{cm}^{-1}$). This mechanism limits the optimal concentration of Nd ions in a YAG crystal to about 1%.

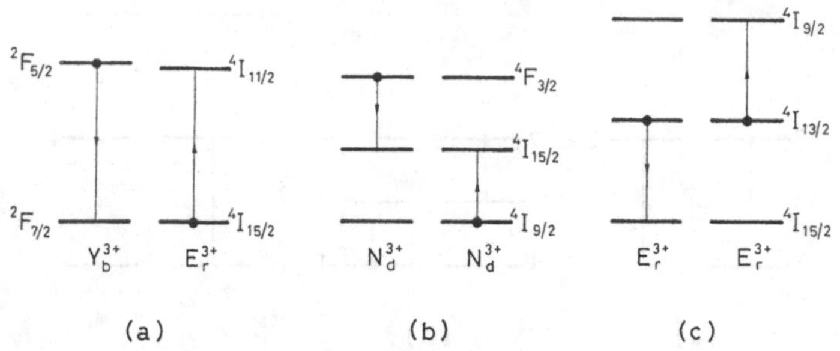

FIG. 2.13. Examples of energy transfer by dipole-dipole interaction: (a) $Yb^{3+} - Er^{3+}$ energy transfer in an Yb:Er laser or amplifier. (b) Nonradiative decay of Nd:YAG by cross relaxation. (c) Cooperative upconversion in an Er^{3+} laser or amplifier.

Example 2.9. *Cooperative upconversion in Er^{3+} lasers and amplifiers*[15] Cooperative up-conversion is believed to be the major cause of inefficiency for Er^{3+} lasers or amplifiers (Fig. 2.13c). In this case, out of two neighboring Er ions initially excited to the $^4I_{13/2}$ laser level, one decays to the ground $^4I_{15/2}$ level while the other is raised to the $^4I_{9/2}$ level. From this level the ion, when in an oxide glass host, then decays rapidly back to the $^4I_{13/2}$ level by multiphonon decay. The net result of this cooperative up-conversion is that one Er ion, initially excited to the laser $^4I_{13/2}$ level, is effectively quenched to the ground level i.e., one looses 50% of the population.

2.6.2. Combined Effects of Radiative and Nonradiative Processes

Let us first consider the case where the nonradiative decay can be described by an equation of the general form Eq. (2.6.4). The time variation of the upper state population N_2 can then be written as

$$\frac{dN_2}{dt} = -\left(\frac{N_2}{\tau_r} + \frac{N_2}{\tau_{nr}}\right) \tag{2.6.16}$$

Equation (2.6.16) can be put in the simpler form

$$dN_2/dt = -(N_2/\tau) \tag{2.6.17}$$

provided that one defines an overall decay time τ given by

$$\frac{1}{\tau} = \frac{1}{\tau_r} + \frac{1}{\tau_{nr}} \tag{2.6.18}$$

The population $N_2(t)$ at time t is then obtained by integrating Eq. (2.6.17). We get

$$N_2(t) = N_2(0) \exp -(t/\tau) \tag{2.6.19}$$

where $N_2(0)$ is the population at $t = 0$. To calculate the time behavior of the spontaneously emitted light, we notice that, according to Eq. (2.6.16), N_2/τ_r gives the number of atoms decaying radiatively per unit volume and unit time. Assuming, for simplicity, that radiative decay occurs to one lower level only, say level 1, and letting ν_0 be the corresponding transition frequency, the spontaneously emitted power at time t will then be

$$P(t) = N_2(t)h\nu_0 V/\tau_r \tag{2.6.20}$$

where V is the volume of the material. With the help of Eq. (2.6.19), Eq. (2.6.20) gives

$$P(t) = [N_2(0)h\nu_0 V/\tau_r] \exp -(t/\tau) \tag{2.6.21}$$

Note that the time decay of the emitted light is exponential with a time constant τ rather than τ_r as one, perhaps, might have expected at a first sight. By monitoring the decay of the spontaneously emitted light from a sample having, at $t = 0$, an initial upper state population $N_2(0)$, one thus measures the overall lifetime τ. To obtain τ_r, let us first define the fluorescence quantum yield ϕ as the ratio of the number of emitted photons to the number of atoms initially raised to level 2. Using Eq. (2.6.21), we have

$$\phi = \frac{\int (P(t)/h\nu_0)dt}{N_2(0)V} = \frac{\tau}{\tau_r} \tag{2.6.22}$$

Note that one can easily show that the above relation remains true also when the decay is into a number of lower levels provided that, in defining the quantum efficiency ϕ, one includes the photons emitted in all of these transitions. The measurement of ϕ thus allows calculation of τ_r once τ is known from the decay measurement of the emitted radiation. This measurement, however, is sometimes not an easy one especially when τ is very short (picoseconds or even less), i.e. when ϕ is very small.

We now briefly consider the case in which nonradiative decay occurs via a dipole-dipole mediated energy transfer. According to Eq. (2.6.15), the transition rate W_{DA} is strongly dependent upon the donor to acceptor distance, R. For a population of N_D donors and N_A acceptors, due to the different distances between donors and acceptors, the decay rate will be different from each donor to acceptor couple and the resulting overall decay will show a non-exponential behavior, the initial faster decay corresponding to sites with the smallest separation R. A particularly important case occurs where there are random values of the donor-acceptor spacing and where this distance is either fixed, as in a solid, or slowly varying over a spontaneous decay time, as often occurs for liquids (*Förster regime*). In this case, when both radiative and nonradiative decay channels are taken into account, the overall decay is given by

$$N_2(t) = N_2(0) \exp -\left[(t/\tau_r) + Ct^{1/2}\right] \tag{2.6.23}$$

The time behavior of the radiated power is then obtained by substituting Eq. (2.6.23) into Eq. (2.6.20) and thus follows the same non-exponential decay as that of the population $N_2(t)$.

2.7. DEGENERATE OR STRONGLY COUPLED LEVELS

So far we have considered only the simplest case in which both levels 1 and 2 are non-degenerate. We will now briefly consider the case when the two levels are degenerate or are made up of a number of strongly coupled levels. The situation is depicted in Fig. 2.14 where levels 1 and 2 are assumed to consists of g_1 and g_2 sublevels which are either degenerate (i.e. have the same energy) or so close in energy to be strongly coupled. We will let N_1 and N_2 be the total populations of levels 1 and 2 and use N_{1i} and N_{2j} to indicate the population of a particular sublevel of the lower and upper manifolds respectively.

2.7.1. Degenerate Levels

We first look at the degenerate case and begin by considering the thermal equilibrium situation. In this case, the population of each sublevel of both upper and lower state will obey the usual Boltzmann equation, thus

$$N_{2_j}^e = N_{1_i}^e \exp\left[-\left(E_2 - E_1\right)/kT\right] \tag{2.7.1}$$

Since, however, the sublevels of e.g. level 1 are also in thermal equilibrium, their population must all be equal, thus

$$N_{1_i}^e = N_1^e/g_1 \tag{2.7.2a}$$

Similarly we have

$$N_{2_j}^e = N_2^e/g_2 \tag{2.7.2b}$$

From Eqs. (2.7.1) and (2.7.2b) we then get

$$N_2^e = N_1^e(g_2/g_1)\exp\left[-\left(E_2 - E_1\right)/kT\right] \tag{2.7.3}$$

Let us now see how the expressions for transition cross section, gain, and absorption coefficient need to be modified in the case of degenerate levels. For this purpose we consider an e.m. wave passing through a material with given overall populations, N_1 and N_2, in the two

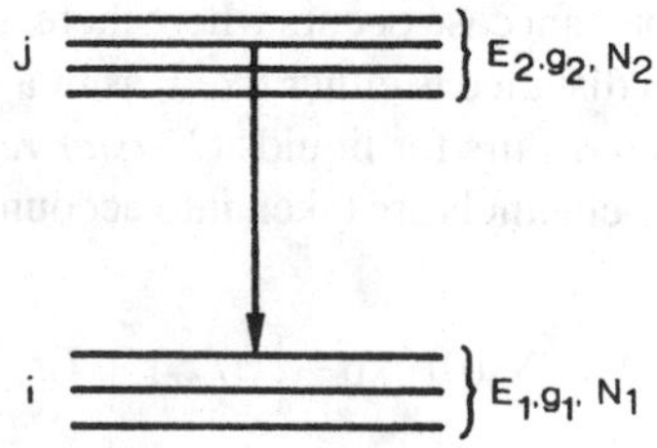

FIG. 2.14. Two level system in which the two levels comprise many sublevels which are either degenerate or strongly coupled.

levels, and we calculate the rate of change of the overall population N_2 due to all radiative and nonradiative transitions between sublevels j and i. We therefore write

$$\left(\frac{dN_2}{dt}\right) = -\sum_1^{g_1}{}_i \sum_1^{g_2}{}_j \left(W_{ji}N_{2j} - W_{ij}N_{1i} + \frac{N_{2j}}{\tau_{ji}}\right) \tag{2.7.4}$$

where W_{ji} is the rate of stimulated transition between j and i sublevels, W_{ij} is the rate of absorption and $(1/\tau_{ji})$ is the rate of spontaneous decay, radiative and nonradiative, between the same two sublevels. Note that W_{ji} and W_{ij} are obtained from Eq. (2.4.30) by substituting the dipole moments between j and i sublevels, $|\mu_{ij}|^2$ and $|\mu_{ji}|^2$, for $|\mu|^2$. These dipole moments can in turn be readily obtained from Eq. (2.3.7). For instance $|\mu_{ij}|$ is obtained from Eq. (2.3.7) by substituting u_i, the eigenfunction of the i-th lower level, for u_1 and u_j, the eigenfunction of the j-th upper level, for u_2. It then follows that:

$$W_{ji} = W_{ij} \tag{2.7.5}$$

If a rapid relaxation towards thermal equilibrium occurs between the sublevels within each level, then all sublevels of the upper level will again be equally populated, and the same will occur to the sublevels of the lower level. Therefore

$$N_{2j} = N_2/g_2 \tag{2.7.6a}$$

$$N_{1i} = N_1/g_1 \tag{2.7.6b}$$

Upon substitution of Eq. (2.7.6) into Eq. (2.7.4) we then get

$$\frac{dN_2}{dt} = -W\left(\frac{N_2}{g_2} - \frac{N_1}{g_1}\right) - \frac{N_2}{\tau} \tag{2.7.7}$$

where, with the help of Eq. (2.7.5), we have defined

$$W = \sum_1^{g_1}{}_i \sum_1^{g_2}{}_j W_{ij} = \sum_1^{g_1}{}_i \sum_1^{g_2}{}_j W_{ji} \tag{2.7.8}$$

and

$$\frac{1}{\tau} = \frac{\sum_1^{g_1}{}_i \sum_1^{g_2}{}_j (1/\tau_{ji})}{g_2} \tag{2.7.9}$$

From Eq. (2.7.7) one can observe now that WN_2/g_2 represents the change in the unit time of the total upper state population due to all stimulated emissions processes and, likewise, WN_1/g_1 represents the population change due to all absorption processes. The change in photon flux dF when the beam travels a distance dz in the material (see Fig. 1.2) can then be written as

$$dF = W\left(\frac{N_2}{g_2} - \frac{N_1}{g_1}\right) dz \tag{2.7.10}$$

We can now define a stimulated emission cross section, σ_{21}, and absorption cross section, σ_{12}, as

$$\sigma_{21} = W/(g_2 F) \tag{2.7.11a}$$

$$\sigma_{12} = W/(g_1 F) \tag{2.7.11b}$$

from which we obviously have

$$g_2\sigma_{21} = g_1\sigma_{12} \tag{2.7.12}$$

When $(N_1/g_1) > (N_2/g_2)$ Eq. (2.7.10) with the help of Eq. (2.7.11b) can be put in the familiar form $dF = -\alpha F dz$ provided one defines the absorption coefficient α as

$$\alpha = \sigma_{12}\left(N_1 - N_2\frac{g_1}{g_2}\right) \tag{2.7.13}$$

Similarly, when $(N_2/g_2) > (N_1/g_1)$, Eq. (2.7.10) with the help of Eq. (2.7.11a) can be put in the familiar form $dF = gFdz$ provided one defines the gain coefficient g as

$$g = \sigma_{21}\left(N_2 - N_1\frac{g_2}{g_1}\right) \tag{2.7.14}$$

The reasons for defining σ_{21} and σ_{12}, respectively, as in Eqs. (2.7.11a) and (2.7.11b) is now apparent. When in fact $N_1 \gg N_2$ (as usually applies to absorption measurements involving optical transitions) Eq. (2.7.13) simply reduces to $\alpha = \sigma_{12}N_1$. Conversely, when $N_2 \gg N_1$ (as applies in a four-level laser), Eq. (2.7.14) simply reduces to $g = \sigma_{21}N_2$.

2.7.2. Strongly Coupled Levels

We now turn to the case where the upper level, 2, and lower level, 1, actually consist of g_2 and g_1 sublevels, respectively, with different energies but with very fast relaxation among the sublevels belonging to each particular level (strongly coupled levels). Each sublevel, of both upper and lower levels, may also consists, itself, of many degenerate levels. In this case, thermalization among the sublevels of either lower and upper level will occur rapidly so that we can assume Boltzmann's statistics to be always obeyed. Instead of Eq. (2.7.6), we write now

$$N_{2j} = f_{2j}N_2 \tag{2.7.15a}$$

$$N_{1i} = f_{1i}N_1 \tag{2.7.15b}$$

where f_{2j} (f_{1i}) is the fraction of total population of level 2 (level 1) that, at thermal equilibrium, is found in sublevel $j(i)$. According to Boltzmann's statistics, we then have

$$f_{2j} = \frac{g_{2j}\exp-(E_{2j}/kT)}{\sum_{1}^{g_2}{}_m g_{2m}\exp-(E_{2m}/kT)} \tag{2.7.16a}$$

$$f_{1i} = \frac{g_{1i}\exp-(E_{1i}/kT)}{\sum_{1}^{g_1}{}_l g_{1l}\exp-(E_{1l}/kT)} \tag{2.7.16b}$$

where E_{2m} and E_{1l} are the energies of the sublevels in the upper and lower level, respectively and g_{2m} and g_{1l} are the corresponding degeneracies.

Let us now assume that the stimulated transition occurs between a given sublevel (say l) of level 1 to a given sublevel (say m) of level 2. Equation (2.7.4) then simplifies to

$$\left(\frac{dN_2}{dt}\right) = -W_{ml}N_{2m} + W_{lm}N_{1l} - \sum_{1}^{g_1}{}_i\sum_{1}^{g_2}{}_j\left(\frac{N_{2j}}{\tau_{ji}}\right) \tag{2.7.17}$$

With the help of Eq. (2.7.15), Eq. (2.7.17) can be written as

$$(dN_2/dt) = -W_{ml}^e N_2 + W_{lm}^e N_1 - (N_2/\tau) \tag{2.7.18}$$

where we have defined the effective rates of stimulated emission, W_{ml}^e, stimulated absorption, W_{lm}^a, and spontaneous decay, $(1/\tau)$, respectively as

$$W_{ml}^e = f_{2m}W_{ml} \tag{2.7.19a}$$

$$W_{lm}^a = f_{1l}W_{lm} \tag{2.7.19b}$$

$$(1/\tau) = \sum_{1}^{g_1}{}_i\sum_{1}^{g_2}{}_j(f_{2j}/\tau_{ji}) \tag{2.7.19c}$$

According to Eq. (2.7.18), the change in photon flux dF when the beam travels a distance dz in the material is given now

$$dF = \left(W_{ml}^e N_2 - W_{lm}^a N_1\right)dz \tag{2.7.20}$$

We can then define an effective stimulated emission cross section, σ_{ml}^e, and an effective absorption cross section, σ_{lm}^a, as

$$\sigma_{ml}^e = W_{ml}^e/F = f_{2m}\sigma_{lm} \tag{2.7.21a}$$

$$\sigma_{ml}^a = W_{ml}^a/F = f_{1l}\sigma_{lm} \tag{2.7.21b}$$

where Eqs. (2.7.19a) and (2.7.19b) have been used and where $\sigma_{lm} = W_{lm}/F$ and $\sigma_{ml} = W_{ml}/F$ are, respectively, the effective cross sections of absorption and stimulated emission for the l

to m transition. Note that, if the two sub-levels l and m are non-degenerate (or have the same degeneracy) one has $\sigma_{lm} = \sigma_{ml}$. Note also that, according to Eqs. (2.7.20) and (2.7.21), the absorption coefficient for the propagating photon flux, can be written as

$$\alpha_{lm} = \sigma^a_{lm} N_1 - \sigma^e_{ml} N_2 \tag{2.7.22}$$

This shows the usefulness of the concepts of effective cross sections: the absorption coefficient, or the gain coefficient when $N_2 > N_1$, is simply obtained upon multiplying the effective cross section with the *total* population of the upper and lower state. In particular, at thermal equilibrium one has $N_2 \cong 0$ and $N_1 \cong N_t$ where N_t is the total population and Eq. (2.7.22) gives

$$\alpha_{lm} = \sigma^a_{lm} N_t \tag{2.7.23}$$

This equation indicates that σ^e_{lm} can readily be obtained from an absorption measurement.

Example 2.10. *Effective stimulated emission cross section for the $\lambda = 1.064\,\mu m$ laser transition of Nd:YAG* The scheme of the relevant energy levels for the Nd:YAG laser is shown in Fig. 2.15. Laser action can occur on the ${}^4F_{3/2} \to {}^4I_{11/2}$ transition ($\lambda = 1.064\,\mu$m), which is the most popular one, as well as on ${}^4F_{3/2} \to {}^4I_{13/2}$($\lambda = 1.32\,\mu$m) and ${}^4F_{3/2} \to {}^4I_{9/2}$ transitions ($\lambda = 0.94\,\mu$m). The 1.064 μm transition occurs between one sublevel, $m = 2$, of the ${}^4F_{3/2}$ level to one sublevel, $l = 3$, of the ${}^4I_{11/2}$ level ($R_2 \to Y_3$ transition). We let $f_{22} = N_{22}/N_2 = N_{22}/(N_{21} + N_{22})$ be the fraction of the total population which is found in the upper laser level, where N_{22} and N_{21} are the populations of the two sublevels of the ${}^4F_{3/2}$ level and N_2 is the total population of this level. Since the two sub-levels are each two-times degenerate, then, according to Eq. (2.7.3) one has $N_{22} = N_{21} \exp -(\Delta E/kT)$ where ΔE is the energy separation between the two sublevels. Form the previous expression of f_{22} we then obtain $f_{22} = 1/[1 + \exp(\Delta E/kT)]$. For $\Delta E = 84\,\text{cm}^{-1}$ and $kT = 208\,\text{cm}^{-1}$[$T = 300\,$K], we get $f_{22} = 0.4$. From measured spectroscopic data on the $R_2 \to Y_3$ transition, the actual peak cross section of the transition has been deduced as $\sigma_{23} = 6.5 \times 10^{-19}\,\text{cm}^2$.[20] The effective cross section of the $R_2 \to Y_3$ transition, σ^e_{23}, is then obtained from Eq. (2.7.21a) as $\sigma^e_{23} = f_{22}\sigma_{23} \cong 2.8 \times 10^{-19}\,\text{cm}^2$.

Example 2.11. *Effective stimulated emission cross section and radiative lifetime in Alexandrite* The relevant energy levels of Alexandrite are shown in Fig. 2.16. The upper laser level is the ${}^4T_2$ state and the laser transition occurs to a vibronic level of the ${}^4A_2$ ground state ($\lambda \cong 730 \div 800\,nm$). Since the ${}^4T_2$ level is strongly coupled to the ${}^2E$ level, the fraction of the total population which is found in the ${}^4T_2$ state, f_{2T}, is given by $f_{2T} = N_{2T}/(N_{2E} + N_{2T})$ where N_{2E} and N_{2T} are the populations of the two levels. At thermal equilibrium, we also have $N_{2T} = N_{2E} \exp -(\Delta E/kT)$, where ΔE is the energy separation between the two levels. From the previous expressions we obtain $f_{2T} = \exp -(\Delta E/kT)/[1 + \exp -(\Delta E/kT)]$. Assuming $\Delta E = 800\,\text{cm}^{-1}$ and $kT = 208\,\text{cm}^{-1}$ [$T = 300\,$K] we get $f_{2T} = 2.1 \times 10^{-2}$. According to Eq. (2.7.21a), the effective stimulated emission cross section is given by $\sigma^e_{TA} = f_{2T}\sigma_{TA}$, where σ_{TA} is the actual cross section. From this last expression, assuming $\sigma_{TA} = 4 \times 10^{-19}\,\text{cm}^2$ at $\lambda = 704\,$nm,[21] we obtain

$f_{2T} \cong 2.1 \times 10^{-2}$ and $\sigma_{TA}^{e} \cong 0.8 \times 10^{-20}$ cm^2. Note the small value of f_{2T} i.e. the small fractional value of the upper laser level population, which results in a strong reduction of the effective stimulated emission cross section. Note also that this cross section increases with increasing temperature as f_{2T} increases with temperature. To calculate the effective lifetime of the upper laser level, τ, we note that the rate of spontaneous decay, $1/\tau_T$, of the ${}^4T_2 \rightarrow {}^4A_2$ laser transition is $(1/\tau_T) = 1.5 \times 10^5$ sec^{-1} ($\tau_T \cong 6.6\,\mu$s) while the rate of the ${}^2E \rightarrow {}^4A_2$ transition is $(1/\tau_E) = 666.6$ sec^{-1} ($\tau_E = 1.5$ ms). From Eq. (2.7.19c) we then get $(1/\tau) = (f_{2E}/\tau_E) + (f_{2T}/\tau_T)$ where $f_{2E} = N_{2E}/(N_{2E} + N_{2T}) = 1 - f_{2T}$ is the fraction of the total population which is found in the ${}^2E$ level. Inserting the appropriate numbers into the previous expression of $(1/\tau)$ we get $\tau = 200\,\mu$s at $T = 300$ K. Thus the effective lifetime is considerably lengthened (from 6.6 to 200 μs) due to the presence of the strongly coupled and long lived ${}^2E$ level which then acts as a storage level or reservoir. Note that the effective lifetime, likewise the effective cross section, is temperature dependent.

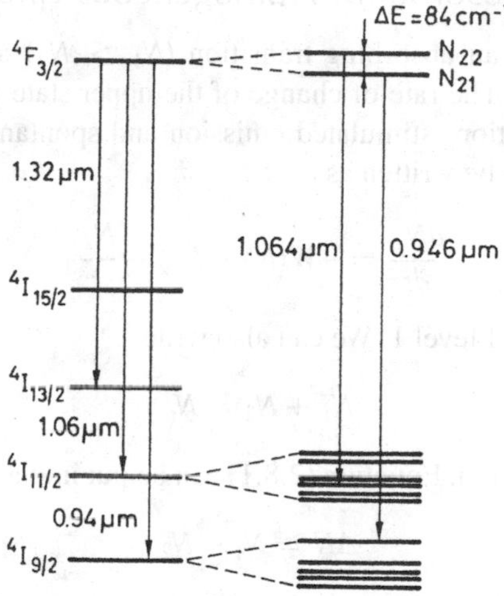

FIG. 2.15. Relevant energy levels for the $\lambda = 1.064\,\mu m$ laser transition of Nd:YAG laser.

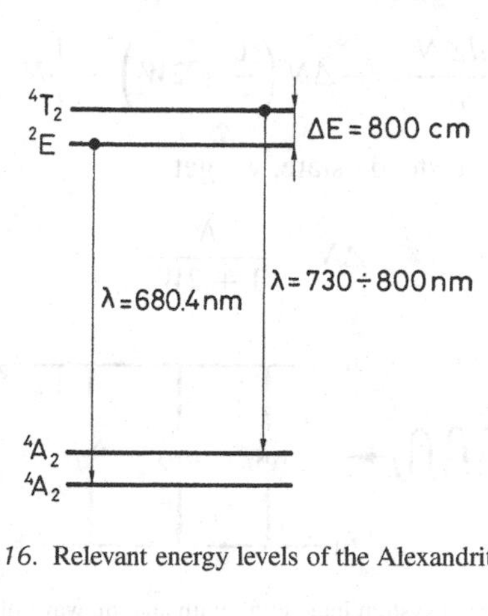

FIG. 2.16. Relevant energy levels of the Alexandrite laser.

2.8. SATURATION

The purpose of this section is to examine the absorption and emission behavior of a transition (of frequency ν_0) in the presence of a strong monochromatic e.m. wave of intensity I and frequency $\nu \cong \nu_0$. For simplicity, we will assume the levels to be non degenerate. Consider first the case where I is sufficiently weak that the populations of the two levels, N_1 and N_2, do not differ significantly from their thermal equilibrium values. One then has $N_1 > N_2$ (often $N_1 \gg N_2$) and the absorption processes, of rate WN_1, will dominate the stimulated emission process, of rate WN_2, i.e. more atoms undergo the $1 \rightarrow 2$ transition than the $2 \rightarrow 1$ transition. Consequently, at sufficiently high values of the intensity I, the two populations will tend to equalize. This phenomenon is referred to as *saturation*.

2.8.1. Saturation of Absorption: Homogeneous Line

We will consider first an absorbing transition ($N_1 > N_2$) and assume the line to be homogeneously broadened. The rate of change of the upper state population, N_2, due to the combined effects of absorption, stimulated emission and spontaneous decay (radiative and nonradiative), Fig. 2.17, can be written as

$$\frac{dN_2}{dt} = -W(N_2 - N_1) - \frac{N_2}{\tau} \tag{2.8.1}$$

where N_1 is the population of level 1. We can also write

$$N_1 + N_2 = N_t \tag{2.8.2}$$

where N_t is the total population. Equation (2.8.1) can be put into a simpler form by defining

$$\Delta N = N_1 - N_2 \tag{2.8.3}$$

Equations (2.8.2) and (2.8.3) then give N_1 and N_2 as a function of ΔN and N_t, and Eq. (2.8.1) becomes

$$\frac{d\Delta N}{dt} = -\Delta N\left(\frac{1}{\tau} + 2W\right) + \frac{1}{\tau}N_t \tag{2.8.4}$$

When $(d\Delta N/dt) = 0$, i.e. in the steady state, we get

$$\Delta N = \frac{N_t}{1 + 2W\tau} \tag{2.8.5}$$

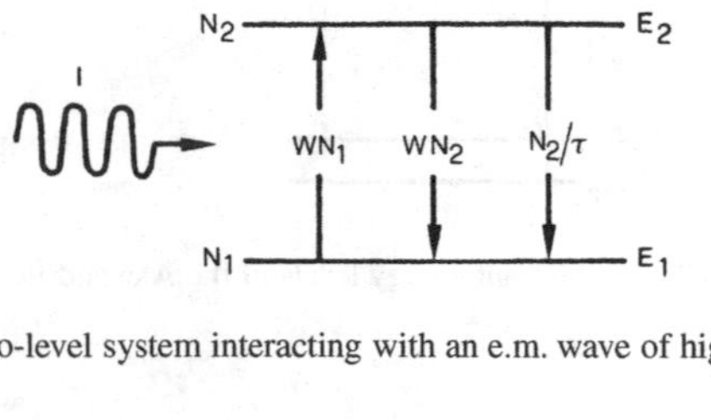

FIG. 2.17. Two-level system interacting with an e.m. wave of high intensity I.

To maintain a given population difference, ΔN, the material needs to absorb from the incident radiation a power per unit volume (dP/dV) given by

$$\frac{dP}{dV} = (h\nu)W\Delta N = (h\nu)\frac{N_t W}{1 + 2W\tau} \tag{2.8.6}$$

which, at saturation, i.e., for $W\tau \gg 1$, becomes

$$(dP/dV)_s = (h\nu)N_t/2\tau \tag{2.8.7}$$

Equation (2.8.7) shows that the power that must be absorbed by the system to keep it in saturation, $(dP/dV)_s$, is, as expected, equal to the power lost by the material due to the decay of the upper state population ($N_t/2$).

It is sometimes useful to have Eqs. (2.8.5) and (2.8.6) rewritten in a more convenient form. To do this we first notice that, according to Eq. (2.4.17), W can be expressed as

$$W = \sigma I/h\nu \tag{2.8.8}$$

where σ is the absorption cross section of the transition. Equations (2.8.5) and (2.8.6) with the help of Eq. (2.8.8) can be recast in the following forms:

$$\frac{\Delta N}{N_t} = \frac{1}{1 + (I/I_s)} \tag{2.8.9}$$

$$\frac{dP/dV}{(dP/dV)_s} = \frac{I/I_s}{1 + (I/I_s)} \tag{2.8.10}$$

where

$$I_s = h\nu/2\sigma\tau \tag{2.8.11}$$

is a parameter that depends on the given material and on the frequency of the incident wave. Its physical meaning is obvious from Eq. (2.8.9). In fact, for $I = I_s$, we get $\Delta N = N_t/2$. When $\nu = \nu_0$, the quantity I_s has a value that depends only on the parameters of the transition. This quantity is called the *saturation intensity*.

Let us now see how the shape of an absorption line changes with increasing value of the intensity, I, of the saturating beam. To do this, consider the idealized experimental situation shown in Fig. 2.18 where the absorption measurements are made using a probe beam of variable frequency ν' and whose intensity I' is small enough so as not to perturb the system appreciably. In practice the beams need to be more or less collinear to ensure that the probe beam interacts only with the saturated region. Under these conditions, the absorption coefficient as seen by the probe beam is obtained from Eq. (2.4.33) by substituting the total lineshape $g_t(\nu - \nu_0)$ with the homogeneous lineshape $g(\nu' - \nu_0)$, where ν has been substituted by ν'. Since $N_1 - N_2 = \Delta N$ is now given by Eq. (2.8.9), we can write

$$\alpha = \frac{\alpha_0}{1 + (I/I_s)} \tag{2.8.12}$$

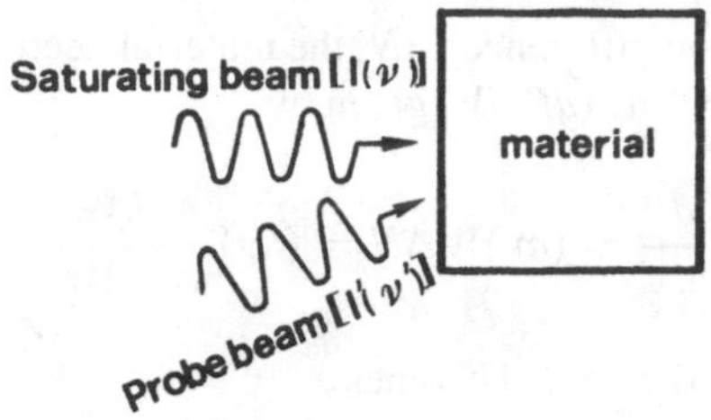

FIG. 2.18. Measurement of the absorption or gain coefficient at frequency ν', by a probe beam of intensity $I'(\nu')$ in the presence of a saturating beam, $I(\nu)$, of intensity I and frequency $\nu[I(\nu) \gg I'(\nu')]$.

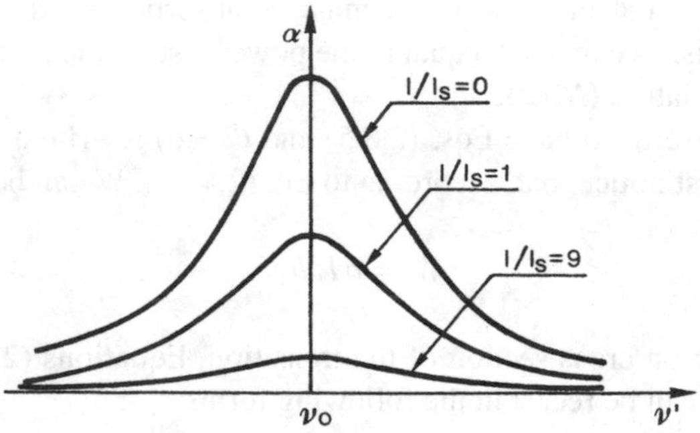

FIG. 2.19. Saturation behavior of the absorption coefficient, α, vs frequency, ν', for increasing values of the intensity, I, of the saturating beam (*homogeneous line*).

where

$$\alpha_0 = \frac{2\pi^2}{3n\varepsilon_0 ch} |\mu|^2 N_t \nu' g(\nu' - \nu_0) \tag{2.8.13}$$

is the absorption coefficient when the saturating wave at frequency ν is absent (*unsaturated absorption coefficient*). Equations (2.8.12) and (2.8.13) show that, when the intensity I of the saturating beam is increased, the absorption coefficient is reduced. The line shape, however, remains the same since it is always described by the function $g(\nu' - \nu_0)$. Figure 2.19 shows three plots of the absorption coefficient α vs ν' at three different values of I/I_s.

We next consider the case where the saturating e.m. wave consists of a light pulse with intensity $I = I(t)$, rather than a cw beam. For simplicity, we will confine ourselves to a comparison between two limiting cases in which the pulse duration is either very long or very short compared to the upper state lifetime τ. If the pulse duration is very long compared to the lifetime, the time evolution of the resulting population difference ΔN, due to saturation, will occur at very slow rate so that we can assume in Eq. (2.8.4) $d\Delta N/dt \ll N_t/\tau$. Accordingly, ΔN turns out to be still given by the steady state Eq. (2.8.9) where now $I = I(t)$. The saturation behavior in this case is essentially the same as for a cw beam. If, on the other hand, the light pulse is very short compared to the lifetime τ, then the absorption term $-2W\Delta N$ in Eq. (2.8.4) dominates the spontaneous decay term $(N_t - \Delta N)/\tau$, i.e. $[(N_t - \Delta N)/\tau] \ll 2W\Delta N$. In this case Eq. (2.8.4) reduces to

$$(d\Delta N/dt) = -2W\Delta N = -(2\sigma/h\nu)I(t)\Delta N \tag{2.8.14}$$

where Eq. (2.8.8) has also been used. Integration of Eq. (2.8.14) with the initial condition $\Delta N(0) = N_t$ gives

$$\Delta N(t) = N_t \exp\left[-(2\sigma/h\nu)\int_0^t I(t)dt\right] \tag{2.8.15}$$

Equation (2.8.15) can be put in a more suggestive form if we define the energy fluence $\Gamma(t)$ as

$$\Gamma(t) = \int_0^t I(t)dt \tag{2.8.16}$$

and the saturation fluence as

$$\Gamma_s = h\nu/2\sigma \tag{2.8.17}$$

From Eq. (2.8.15) we then get

$$\Delta N(t) = N_t \exp - [\Gamma(t)/\Gamma_s] \tag{2.8.18}$$

We see that, in this case, it is the beam energy fluence rather than its intensity that determines the saturation behavior. The population difference ΔN_∞ that is left in the material after the pulse has passed is, according to Eq. (2.8.18), given by

$$\Delta N_\infty = N_t \exp -(\Gamma_t/\Gamma_s) \tag{2.8.19}$$

where Γ_t is the total energy fluence of the light pulse. The material saturation fluence Γ_s can therefore be looked upon as the fluence that the pulse should have to produce a population difference $\Delta N_\infty = N_t/e$. Having calculated the population difference resulting from saturation by a light pulse, the corresponding absorption coefficient of the material can then be obtained, for a homogeneous line, again from Eq. (2.4.33) by substituting $g_t(\nu - \nu_0)$ with $g(\nu' - \nu_0)$. For a light pulse that is, either, slow or fast compared to τ, the value of α is respectively given by Eq. (2.8.12) [with $I = I(t)$] or by

$$\alpha = \alpha_0 \exp[-\Gamma(t)/\Gamma_s] \tag{2.8.20}$$

where α_0 is the unsaturated absorption coefficient. Note that, in the pulsed regime as in the cw regime, the shape of the absorption line remains unchanged when saturation occurs.

2.8.2. Gain Saturation: Homogeneous Line

We now consider the case where the transition $2 \to 1$ exhibits net gain rather than net absorption. We assume that the medium behaves as a four-level system (Fig. 2.20) and the inversion between levels 2 and 1 is produced by some suitable pumping process. We will further assume that the $3 \to 2$ and $1 \to g$ transitions are so fast that we can take $N_3 \cong$

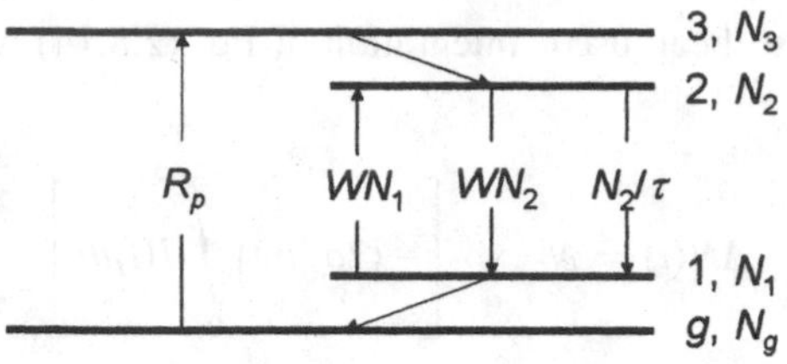

FIG. 2.20. Energy levels and transitions involved in gain saturation of a four-level laser.

$N_1 \cong 0$. With these simplifying assumptions we can write the following rate equation for the population of level 2:

$$(dN_2/dt) = R_p - WN_2 - (N_2/\tau) \tag{2.8.21}$$

where R_p is the pumping rate and N_t is the total population. In the steady state (i.e., for $dN_2/dt = 0$) we find from Eq. (2.8.21)

$$N_2 = \frac{R_p\tau}{1 + W\tau} \tag{2.8.22}$$

With the help of Eq. (2.8.8), Eq. (2.8.22) can be rewritten as

$$N_2 = \frac{N_{20}}{1 + (I/I_s)} \tag{2.8.23}$$

where $N_{20} = R_p\tau$ is the population of level 2 in the absence of the saturating beam (i.e. for $I = 0$) and

$$I_s = h\nu/\sigma\tau \tag{2.8.24}$$

A comparison of Eq. (2.8.24) with Eq. (2.8.11) shows that the expression for the saturation intensity I_s of a four-level system is twice that of the two-level system of Fig. 2.17. The difference arises from the fact that, in a two-level system, a change in population of one level causes an equal and opposite change in population in the other level. Thus ΔN is twice the change in population of each level.

In an experiment such as that shown in Fig. 2.18, the probe beam at frequency ν' will now measure gain rather than absorption. From Eq. (2.4.35), with $N_1 = 0$, using also Eq. (2.8.23), the gain coefficient, g, can be written as

$$g = \frac{g_0}{1 + (I/I_s)} \tag{2.8.25}$$

where $g_0 = \sigma N_{20}$ is the gain coefficient for $I = 0$, i.e. when the saturating beam is absent (*unsaturated gain coefficient*). This gain coefficient, since the line is homogeneously broadened, can be obtained, using Eq. (2.4.18), as

$$g_0 = \frac{2\pi^2}{3n\varepsilon_0 ch}|\mu|^2\nu' N_{20} g(\nu, - \nu_0) \tag{2.8.26}$$

Equations (2.8.25) and (2.8.26) show that, just as in the case of absorption, the saturation again leads to a decrease of g as I increases while the gain profile remains unchanged.

We next consider the case where the saturating e.m. wave consists of a light pulse of intensity $I(t)$. If the pulse duration is very long compared to the lifetime τ, we can neglect the time derivative of N_2 in Eq. (2.8.21) compared to the other terms. Thus we again get Eq. (2.8.23) for the upper state population and Eq. (2.8.25) for the gain coefficient, where I is now function of time. If the light pulse is very short compared to the lifetime τ, then, during the interaction of the light pulse, the pump term R_p and the spontaneous decay term N_2/τ can be neglected compared to the stimulated term WN_2. Thus we get

$$(dN_2/dt) = -(\sigma I/h\nu)N_2 \tag{2.8.27}$$

where Eq. (2.8.8) has been used again. Integration of Eq. (2.8.27) gives

$$N_2(t) = N_{20} \exp\{-[\Gamma(t)/\Gamma_s]\} \tag{2.8.28}$$

where $N_{20} = R_p\tau$ is the population of level 2 before the arrival of the pulse, $\Gamma(t)$ is the energy fluence of the beam [see Eq. (2.8.16)], and

$$\Gamma_s = h\nu/\sigma \tag{2.8.29}$$

is the amplifier saturation fluence. A comparison of Eq. (2.8.29) with Eq. (2.8.17) shows that the saturation fluence of a four-level amplifier is twice that of an absorber. The saturated gain coefficient is given by

$$g = g_0 \exp\{-[\Gamma(t)/\Gamma_s]\} \tag{2.8.30}$$

where $g_0 = \sigma N_{20}$ is the unsaturated gain coefficient and is again given by Eq. (2.8.26). Thus, in the pulsed regime, just as for the cw case, the shape of the gain line remains unchanged when saturation occurs.

2.8.3. Inhomogeneously Broadened Line

When the line is inhomogeneously broadened, the saturation phenomenon is more complicated and we will limit ourselves to just a qualitative discussion (see Problems for further details). For the sake of generality, we will assume that the line is broadened both by homogeneous and inhomogeneous mechanisms so that its shape is expressed as in Eq. (2.4.26): the overall line $g_t(\nu - \nu_0)$ is given by the convolution of the homogeneous contributions $g(\nu - \nu_0')$ of the various atoms. Thus, in the case of absorption, the resulting absorption coefficient can be visualized as shown in Fig. 2.21. In this case, for an experiment such as that envisaged in Fig. 2.18, the saturating beam of intensity $I(\nu)$ will interact with only those atoms whose resonance frequency, ν_0', is in the neighborhood of the frequency, ν. Accordingly, only these atoms will undergo saturation when $I(\nu)$ becomes sufficiently large. The modified shape of the absorption line, for various values of $I(\nu)$, will then be as shown in Fig. 2.22. In this case, as $I(\nu)$ is increased, a hole will be produced in the absorption line at frequency ν. The width of this hole is of the same order as the width of each of the dashed absorption profiles of

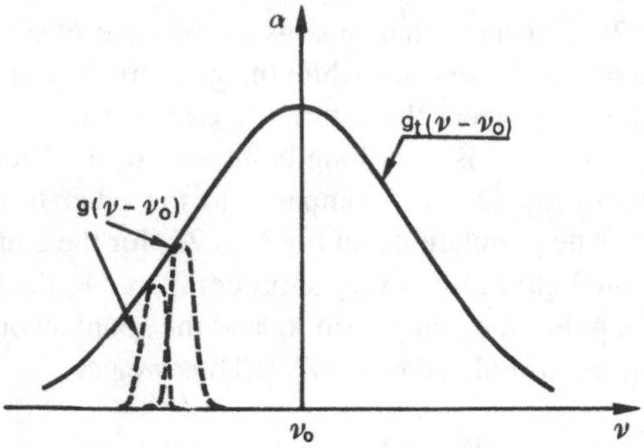

FIG. 2.21. Lineshape of a transition broadened by both homogeneous and inhomogeneous mechanisms. The corresponding line-shape function, $g_t(\nu - \nu_0)$, is obtained from the convolution of the homogeneous lines $g(\nu - \nu_0')$ of the individual atoms.

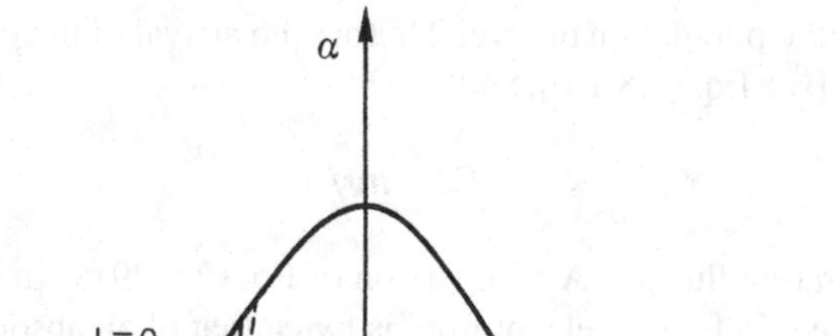

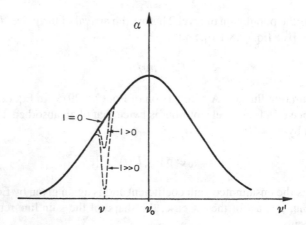

FIG. 2.22. Saturation behavior of the absorption coefficient, α, vs frequency, ν', as measured by the test beam of intensity for increasing values of the intensity, $I'(\nu')$, for increasing values of the intensity, $I(\nu)$, of the saturating beam (*inhomogeneous line*).

Fig. 2.21, i.e. the width of the homogeneous line. A similar argument applies if a transition with net gain rather than absorption is considered. The effect of the saturating beam will, in this case, be to burn holes in the gain profile rather than the absorption profile. Note also that a similar argument can be applied when absorption or gain saturation is produced by a light pulse of sufficiently high energy fluence.

2.9. DECAY OF AN OPTICALLY DENSE MEDIUM

In Sect. 2.3 the decay of an essentially isolated atom or ion has been considered. In a real situation, any atom will be surrounded by many other atoms, some in the ground state and some in the excited state. In this case new phenomena may occur since the decay may be due to the simultaneous occurrence of both spontaneous and stimulated processes. These phenomena will be briefly discussed in this section.

2.9.1. Radiation Trapping

If the fraction of atoms that is raised to the upper level is very small and if the medium is optically dense, the phenomenon known as *radiation trapping* may play a significant role. In this case, a photon that is spontaneously emitted by one atom, instead of escaping from the medium, can be absorbed by another atom which thereby ends up in the excited state. The process therefore has the effect of slowing down the effective rate of spontaneous emission. A detailed discussion about radiation trapping can be found elsewhere.(16) We merely limit to point out here that the lifetime increase depends on the atomic density, on the cross section of the transition involved, and on the geometry of the medium. Radiation trapping may be particularly important for UV transitions with large cross sections [according to Eq. (2.4.29) one has $\sigma \propto |\mu|^2 \nu$ which increases rapidly in going to the UV through the ν term and the increase of $|\mu|^2$ with frequency]. This can result in an increase of the effective life-time of spontaneous emission by as much as a few orders of magnitude.

2.9.2. Amplified Spontaneous Emission

If the fraction of atoms raised to the upper level is very large and if the medium is optically dense the phenomenon known as Amplified Spontaneous Emission (ASE) may play a very important role.

Consider a cylindrically shaped active medium and let Ω be the solid angle subtended by one face of the cylinder as seen from the center O of the other face [Fig. 2.23a]. If the gain of the active material, $G = \exp\sigma(N_2 - N_1)l$, is large enough, the fluorescence power, emitted by atoms around point O into the solid angle Ω, may be strongly amplified by the active medium by a factor that may, in some cases, be as high as 10^4 or even higher. Thus, under suitable conditions, which are considered below, the active medium will preferentially emit its stored energy into the solid angle Ω of Fig. 2.23a and, obviously, along the opposite direction as well. If a totally reflecting mirror ($R = 1$) is placed at one end of the medium (Fig. 2.23b), then an unidirectional output is obtained. This is the basic ASE phenomenon. In contrast to spontaneous emission, ASE possesses some distinctive features which shows some similarity

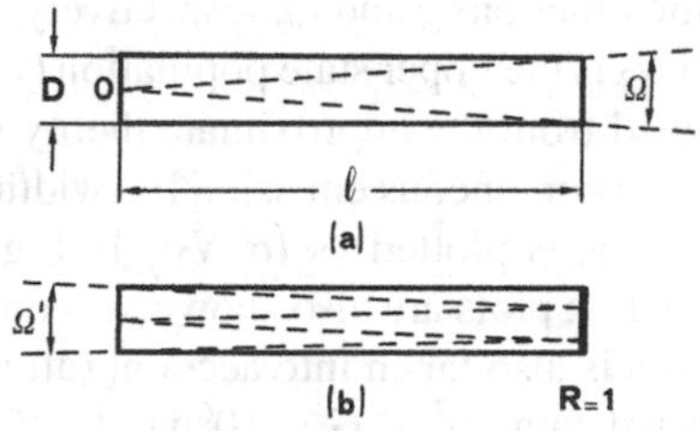

FIG. 2.23. Solid angle of emission in the case of amplified spontaneous emission: (a) Active material without end mirrors. (b) Active material with one end mirror.

to laser action. Indeed ASE has, to some degree, the property of directionality, its bandwidth is appreciably narrower than that of spontaneous emission, it shows a "soft" threshold behavior, and the beam of *ASE* light can be quite intense. We will briefly consider these properties here, while we refer to Appendix C for further details.

The directionality is immediately apparent from Fig. 2.23. For $D \ll l$ the emission solid angle Ω of Fig. 2.23a is, in fact, readily seen to be given by

$$\Omega = \pi D^2/4l^2 \tag{2.9.1}$$

where D is the diameter and l is the length of the active material. Likewise, in the case of Fig. 2.23b, the emission solid angle is

$$\Omega' = \pi D^2/16l^2 \tag{2.9.2}$$

Note that, in both cases, due to refraction taking place at the exit face of the active medium, the external solid angle Ω_n of both Figs. 2.23a and b is obtained from Eqs. (2.9.1) and (2.9.2) by multiplying the right hand side by n^2 where n is the refractive index of the material. In any case, if $D \ll l$, ASE will occur in a narrow cone as can be perhaps best appreciated from the following example.

Example 2.12. *Directional property of ASE* Let us consider the active medium to consists of gaseous Nitrogen, for which there is a laser transition occurring at $\lambda \cong 337$ nm (see Chap. 10). We take $D = 2$ cm and $l = 1$ m and assume that a totally reflecting mirror is placed at one end. From Eq. (2.9.2) we get $\Omega' \cong 0.8 \times 10^{-4}$ sterad which shows that the emission solid angle is very much smaller than the 4π sterad angle into which spontaneous emission occurs. On the other hand, the beam divergence is much larger than would be obtained from the same active medium used in a two-mirror laser resonator. The half-cone divergence angle of the ASE beam, θ', is in fact given by $\theta' = [\Omega'/\pi]^{1/2} \cong 5$ mrad. By comparison, in the case of a laser resonator, the minimum attainable divergence, as set by diffraction, is given by $\theta_d \cong (\lambda/D) \cong 20\,\mu$ rad, i.e. it is 250 times smaller.

The spectral narrowing of ASE can be understood when we notice that the gain experienced by the spontaneously emitted beam will be much higher at the peak, i.e. at $\nu = \nu_0$, than in the wings of the gain line. This situation is illustrated in Fig. 2.24 for a Lorentzian line. The dashed curve shows, in fact, the normalized spectral profile, $g(\nu - \nu_0)/g_p$, of the spontaneously emitted light while the solid lines show the normalized profile, $I_\nu/I_{\nu p}$, of ASE spectral emission at two different values of the peak gain G. In the previous expressions, g_p and $I_{\nu p}$ are the peak values of the functions g and I_ν, respectively, and $G = \exp\left(\sigma_p N_2 l\right)$, where σ_p is the peak cross section and N_2 is the upper state population (we assume $N_1 \cong 0$). The ASE spectral profile has been obtained from the approximate theory of Appendix C. The ratio of the ASE linewidth, $\Delta\nu_{ASE}$, to the spontaneous emission linewidth, $\Delta\nu_o$ (FWHM), as obtained from the same approximate theory, is plotted vs $\left(\sigma_p N_2 l\right)$ in Fig. 2.25 as a dashed curve. In the same figure, three corresponding plots are also shown, as continuous lines, at three values of $(\Omega/4\pi)$, when gain saturation is also taken into account (after reference[17]).Note that, for practical values of the unsaturated gain $10^3 \le G \le 10^6$ i.e. for $7 \le \sigma_p N_2 l \le 14$ and for practical values of the emission solid angle $10^{-3} \le (\Omega/4\pi) \le 10^{-5}$, the reduction in linewidth is roughly in the range between 3 and 4.

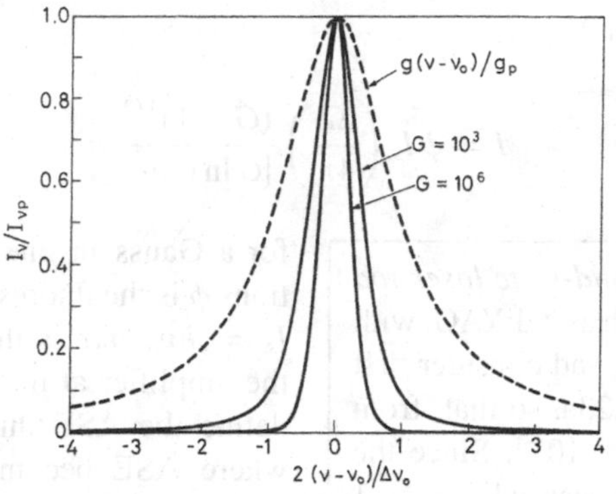

FIG. 2.24. **Normalized ASE spectral emission at two different values of the peak, unsaturated, single pass gain. For comparison, the dashed curve shows the normalized spectrum of spontaneous emission.**

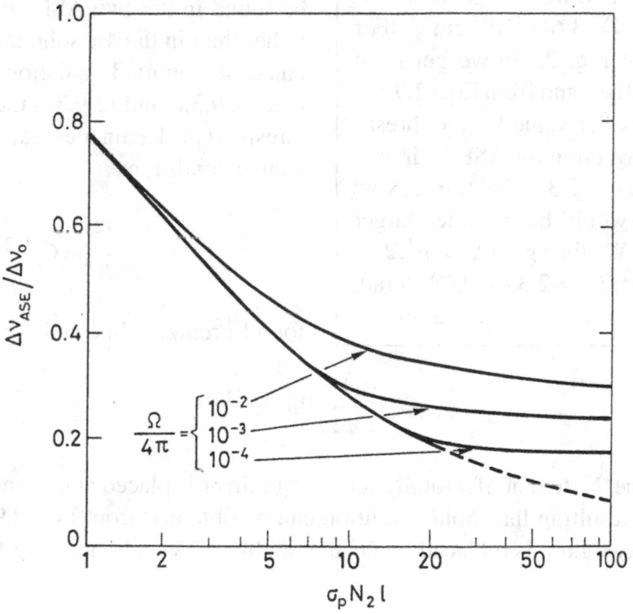

FIG. 2.25. Linewidth of ASE, $\Delta\nu_{ASE}$, normalized to the linewidth of spontaneous emission, $\Delta\nu_0$, as a function of the unsaturated single-pass gain $\sigma_p N_2 l$.

To calculate the "apparent" threshold we begin by pointing out that, according to the theory of Appendix C, the intensity of one of the two ASE beams of Fig. 2.23a is given by

$$I = \phi I_s \left(\frac{\Omega}{4\pi^{3/2}}\right) \frac{(G-1)^{3/2}}{[G \ln G]^{1/2}} \tag{2.9.3a}$$

for a Lorentzian line and by

$$I = \phi I_s \left(\frac{\Omega}{4\pi}\right) \frac{(G-1)^{3/2}}{[G \ln G]^{1/2}} \tag{2.9.3b}$$

Example 2.13. *ASE threshold for a solid-state laser rod* We consider a solid-state laser rod, such as Nd:YAG, with $D = 6\,\text{mm}$, $l = 10\,\text{cm}$ and $n = 1.82$ and consider first the symmetric configuration of Fig. 2.23a, so that, from Eq. (2.9.1) we get $(\Omega/4\pi) = 2.25 \times 10^{-4}$. Since the line of Nd:YAG is Lorentzian and one can take $\phi \cong 1$ for this line, from Eq. (2.9.4a) we obtain $G = 2.5 \times 10^4$ i.e. $\sigma_p N_{th} l = \ln G = 10.12$. Taking $2.8 \times 10^{-19}\,\text{cm}^2$ as the value of the peak stimulated emission cross section σ_p for Nd:YAG [see Example 2.10], we then get a threshold inversion for ASE of $N_{th} = 3.6 \times 10^{18}\,\text{cm}^{-3}$. For the single end configuration of Fig. 2.23b we get from Eq. (2.9.2) $(\Omega'/4\pi) = 5.62 \times 10^{-5}$ and from Eq. (2.9.5a) $G = 6.4 \times 10^2$, i.e. a much smaller value for the threshold peak gain. The threshold inversion for ASE is, in this case, equal to $N_{th} = \ln G/\sigma_p l = 2.3 \times 10^{18}\,\text{cm}^3$. Note that the emission solid angle would be n^2 times larger than the value calculate above. We thus get $\Omega_n = n^2 \Omega = 9.36 \times 10^{-3}$ sterad and $\Omega'_n = n^2 \Omega_n = 2.33 \times 10^{-3}$ sterad, in the two cases respectively.

for a Gaussian line. In both previous equations ϕ is the fluorescence quantum yield and $I_s = h\nu_0/\sigma_p \tau$ is the saturation intensity of the amplifier at the transition peak. We can define the ASE threshold as the condition where ASE becomes the dominant mechanism depopulating the available inversion. We thus require that I becomes comparable to the saturation intensity I_s. In this case, in fact, a sizable fraction of the available energy will be found in the two ASE cones of Fig. 2.23 rather than in the 4π solid angle of the spontaneously emitted radiation. For $G \gg 1$, Eqs. (2.9.3a) and (2.9.3b) then show that the threshold peak gain must satisfy the relatively simple conditions

$$G = \frac{4\pi^{3/2}}{\phi \Omega} [\ln G]^{1/2} \tag{2.9.4a}$$

for a Lorentzian line and

$$G = \frac{4\pi}{\phi \Omega} [\ln G]^{1/2} \tag{2.9.4b}$$

for a Gaussian line. Note that, if a totally reflecting mirror is placed at one end of the medium (Fig. 2.23b), the resulting threshold condition can be obtained from Eq. (2.9.4) provided G, the single-pass peak gain, is replaced by G^2 the double-pass peak gain, and Ω is replaced by Ω'. We thus get

$$G^2 = \frac{4\pi^{3/2}}{\phi \Omega'} [\ln G^2]^{1/2} \tag{2.9.5a}$$

for a Lorentzian line and

$$G^2 = \frac{4\pi}{\phi \Omega'} [\ln G^2]^{1/2} \tag{2.9.5b}$$

for a Gaussian line.

The "soft" threshold which is characteristics of ASE is apparent from Fig. 2.26 where the normalized intensity of one of the two ASE beams of Fig. 2.23a is plotted vs $\sigma_p N\, l$ for

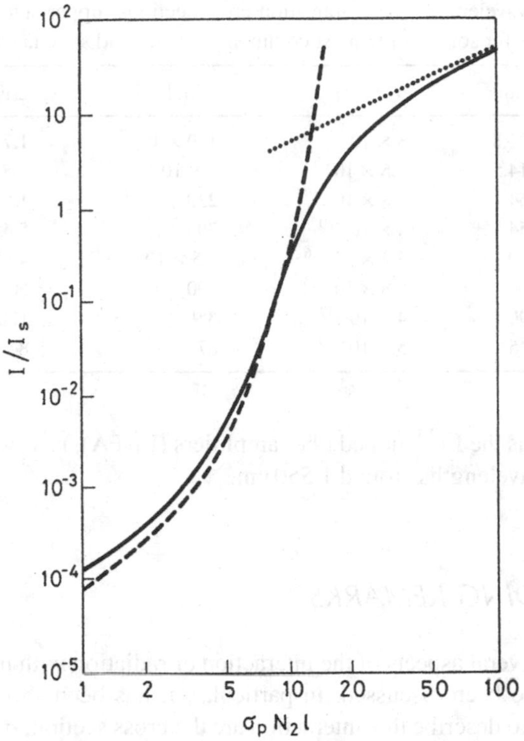

FIG. 2.26. Intensity of ASE emission, I, normalized with respect to the saturation intensity, I_s, as a function of the single-pass gain, $\sigma_p N_2 l$, for an emission solid angle $\Omega = 4\pi \times 10^{-4}$ sterad. The dashed and dotted lines show the results of simplified theories which apply for $(I/I_s) \ll 1$ and $(I/I_s) \gg 1$, respectively. The solid line shows the computed behavior when gain saturation has been taken into account exactly.

$(\Omega/4\pi) = 10^{-4}$ and upon assuming a Lorentzian line and $\phi = 1$. The dashed curve is obtained from Eq. (2.9.3a), which is valid in the limit case $I \ll I_s$. The dotted line, which applies in the other limit case $I \gg I_s$, is obtained from the condition that half of the available fluorescence power is found in the right-propagating ASE beam, i.e. from the equation $(I/I)_s = \sigma_p N_2 l/2$. The solid line is obtained by a more accurate calculation in which the saturation of the upper state population, i.e. gain saturation, has been properly taken into account.[17]

ASE has been used, usually with the configuration of Fig. 2.23b, to obtain directional and narrow bandwidth radiation of high intensity from high gain materials such as nitrogen, excimers or plasmas for X-rays (see Chap. 10). Since one either requires only one mirror or no mirror at all, these systems have been (incongruously) called *mirrorless lasers*. In fact, ASE emission, although having some spatial and temporal coherence, just consists of amplified spontaneous emission noise and should therefore not be confused with laser radiation, whose coherence properties are conceptually different, as will be explained in Chap. 11. In many other situations, ASE is generally a nuisance. For instance it limits the maximum inversion which can be stored in high gain pulsed laser amplifiers. It is also the dominant noise term in

TABLE 2.2. Emission wavelengths, peak transition cross sections, upper-state lifetime and transition linewidths for some of the most common gas and solid state lasers[18–21]

Transition	λ [nm]	σ_p [cm^2]	τ [μs]	$\Delta\nu_0$	Remarks
He-Ne	632.8	3×10^{-13}	150×10^{-3}	1.7 GHz	
Ar^+	514.5	2.5×10^{-13}	6×10^{-3}	3.5 GHz	
Nd:YAG	1,064	2.8×10^{-19}	230	120 GHz	
Nd:Glass	1,054	4×10^{-20}	300	5.4 THz	
Rhod. 6G	570	3.2×10^{-16}	5.5×10^{-3}	46 THz	
Alexandrite	704	0.8×10^{-20}	300	60 THz	T = 300 K
Ti^{3+} : Al_2O_3	790	4×10^{-19}	3.9	100 THz	E||c axis
Cr^{3+} : LiSAF	845	5×10^{-20}	67	84 THz	E||c axis

fiber amplifiers, such as the Er^{3+} doped fiber amplifiers (EDFA), now widely used for optical communications at wavelengths around 1,550 nm.

2.10. CONCLUDING REMARKS

In this Chapter several aspects of the interaction of radiation with matter, mostly relating to atoms or ions, have been discussed. In particular, it has been shown that the two most important parameters to describe this interaction are the cross section, $\sigma = \sigma(\nu - \nu_0)$, and the lifetime, τ, of the upper laser level. In the case of a pure Gaussian or Lorentzian lineshape, one actually needs to know only the peak value, σ_p, of the cross section and the value of the linewidth $(\Delta\nu_0 \text{ or} \Delta\nu_0^*)$. Note also that τ refers to the overall upper level lifetime and, as such, it includes all radiative and nonradiative decay processes which depopulate the upper level. In the case of degenerate or strongly coupled levels, σ_p and τ refer to the effective stimulated emission cross section and upper state lifetime, respectively, as discussed in Sect. 2.7.

A summary of the values of σ_p, τ and $\Delta\nu_0$ $(\text{or } \Delta\nu_0^*)$ for many common laser transitions in gases and ionic crystals are shown in Table 2.2. For comparison, the corresponding values for Rhodamine 6G, a common dye laser material, are also included. Note the very high values of σ_p ($\approx 10^{-13}$ cm^2) for gas lasers as a result of the rather small values of $\Delta\nu_0^*$ (a few GHz) and the rather short lifetimes (a few ns). The lifetime is short because the transitions are electric-dipole allowed. By contrast, for active ions in ionic crystals or glasses such as Nd:YAG or Nd:phosphate glass, σ_p is much smaller ($10^{-20} \div 10^{-19}$ cm^2) and the lifetime is much longer (several hundredths of μs) indicative of a forbidden electric dipole transition. Note also that the linewidths are much larger (from hundredths to thousands of GHz) which also results in a strong reduction of the peak cross section. Dye laser materials, such as Rhodamine 6G, are intermediate between these two cases, showing a fairly high cross section ($\approx 10^{-16}$ cm^2) and also a very short lifetime, a few ns, since, again, the transitions are electric-dipole allowed. The last three laser materials listed in Table 2.2, namely Alexandrite, Ti^{3+} : Al_2O_3 and Cr^{3+}:LISAF, belong to the category of tunable solid state lasers. Indeed, for these materials, the laser linewidths are extremely wide (tens to more than a hundred THz). The cross sections are comparable to those of narrower linewidth materials such as Nd:YAG, while the lifetimes are somewhat shorter.

PROBLEMS

2.1. For a cavity volume $V = 1\,\text{cm}^3$ calculate the number of modes that fall within a bandwidth $\Delta\lambda = 10\,\text{nm}$ centered at $\lambda = 600\,\text{nm}$.

2.2. Instead of ρ_ν, a spectral energy density ρ_λ can be defined, ρ_λ being such that $\rho_\lambda d\lambda$ gives the energy density for the e.m. waves of wavelength lying between λ and $\lambda + d\lambda$. Find the relationship between ρ_λ and ρ_ν.

2.3. For blackbody radiation find the maximum of ρ_λ vs λ. Show in this way that the wavelength λ_M at which the maximum occurs satisfies the relationship $\lambda_M T = hc/ky$ (Wien's law), where the quantity y satisfies the equation $5\,[1 - \exp(-y)] = y$. From this equation find an approximate value of y.

2.4. The wavelength λ_M at which the maximum occurs for the distribution in Fig. 2.3 satisfies the relation $\lambda_M T = 2.9 \times 10^{-3}\,\text{m} \times \text{K}$ (Wien's law). Calculate λ_M for $T = 6{,}000\,\text{K}$. What is the color corresponding to this wavelength?

2.5. The R_1 laser transition of ruby has, to a good approximation, a Lorentzian shape of width (FWHM) 330 GHz at room temperature (see Fig. 2.10). The measured peak transition cross section in $\sigma = 2.5 \times 10^{-20}\,\text{cm}^2$. Calculate the radiative lifetime (the refractive index is $n = 1.76$). Since the observed room temperature lifetime is 3 ms, what is the fluorescence quantum yield?

2.6. Nd:YAG, a typical active laser material, is a crystal of $Y_3Al_5O_{12}$ (yttrium aluminum garnet, YAG) in which some of the Y^{3+} ions are substituted by Nd^{3+} ions. The typical Nd^{3+} atomic concentration used is 1%, i.e. 1% of Y^{3+} ions are replaced by Nd^{3+}. The YAG density is $4.56\,\text{g/cm}^3$. Calculate the Nd^{3+} concentration in the ground ($^4I_{9/2}$) level. This level is actually made up of five (doubly degenerate) levels (see Fig. 2.16), the four higher levels being spaced from the lowest level by 134, 197, 311,and 848 cm^{-1}, respectively. Calculate the Nd^{3+} concentration in the lowest level of the $^4I_{9/2}$ state.

2.7. The neon laser transition at $\lambda = 1.15\,\mu\text{m}$ is predominantly Doppler broadened to $\Delta\nu_0{}^* = 9 \times 10^8\,\text{Hz}$. The upper state lifetime is $\approx 10^{-7}\,\text{s}$. Calculate the peak cross section assuming that the laser transition lifetime is equal to the total upper state lifetime.

2.8. The quantum yield of the $S_1 \rightarrow S_0$ transition (see Chap.9) for Rhodamine 6G is 0.87, and the corresponding lifetime is $\approx 5\,\text{ns}$. Calculate the radiative and nonradiative lifetimes of the S_1 level.

2.9. Calculate the total homogeneous linewidth of the 633 nm laser transition of Ne knowing that $\Delta\nu_{nat} \cong 20\,\text{MHz}$ and $\Delta\nu_c = 0.64\,\text{MHz}$. What is the shape of the overall line?

2.10. Find the relationship between the intensity, I, and the corresponding energy density, ρ, for a plane wave.

2.11. A cylindrical rod of Nd:YAG with diameter of 6.3 mm and length of 7.5 cm is pumped very hard by a suitable flashlamp. The peak cross section for the $1.06\,\mu\text{m}$ laser transition is $\sigma = 2.8 \times 10^{-19}\,\text{cm}^2$ and the refractive index of YAG is $n = 1.82$. Calculate the critical inversion for the onset of the amplified spontaneous emission (ASE) process (the two rod end faces are assumed to be perfectly antireflection coated, i.e., non-reflecting). Also calculate the maximum energy that can be stored in the rod if the ASE process is to be avoided.

2.12. A solution of cryptocyanine (1,1′-diethyl-4,4′-carbocyanine iodiode) in methanol has been used simultaneously to Q-switch and mode-lock (see Chap. 8) a ruby laser. The absorption cross section of cryptocyanine for ruby laser radiation ($\lambda = 694.3\,\text{nm}$) is $8.1 \times 10^{-16}\,\text{cm}^2$. The upper state lifetime is $\tau \cong 22\,\text{ps}$. Calculate the saturation intensity at this wavelength.

2.13. Upon applying the principle of detailed balance to the two near-resonant transfer processes of Eqs. (2.6.9) and (2.6.10), show that, at exact resonance i.e., for $\Delta E = 0$, one has $k_{B^*A} = k_{BA^*}$, where k_{B^*A} and k_{BA^*} are the rate constant of the two processes, respectively

2.14. Instead of observing saturation as in Fig. 2.19, we can also do this by using just the beam $I(\nu)$ and measuring the absorption coefficient for this beam at sufficiently high values of the intensity $I(\nu)$. For a homogeneous line, show that the absorption coefficient is, in this case,

$$\alpha(\nu - \nu_0) = \frac{\alpha_0(0)}{1 + [2(\nu - \nu_0)/\Delta\nu_0]^2 + (I/I_{s0})}$$

where $\alpha_0(0)$ is the unsaturated ($I \ll I_{s0}$) absorption coefficient at $\nu = \nu_0$ and I_{s0} is the saturation intensity, as defined by Eq. (2.8.11), at $\nu = \nu_0$. Hint: begin by showing that

$$\alpha(\nu - \nu_0) = \frac{\alpha_0(0)}{1 + [2(\nu - \nu_0)/\Delta\nu_0]^2} \frac{1}{1 + (I/I_s)}$$

where I_s is the saturation intensity at frequency ν. Continue by expressing I_s in terms of I_{s0}.

2.15. From the expression derived above, find the behavior of the peak absorption coefficient and the linewidth versus I. How would you then measure the saturation intensity I_{s0}?

2.16. Show that, for an inhomogeneous line with line shape function g, the saturated absorption coefficient for an experiment as in Fig. 2.20 can be written as

$$\alpha = \left(\frac{2\pi^2}{3n\varepsilon_0 c_0 h}\right) |\mu|^2 N_t \int \frac{(2/\pi\Delta\nu_0)\, \nu' g*(\nu_0' - \nu_0)}{1 + \left[2\left(\nu' - \nu_0'\right)/\Delta\nu_0\right]^2} \frac{1}{1 + \frac{I}{I_{s0}} \frac{1}{1 + [2(\nu - \nu_0')/\Delta\nu_0]^2}} d\nu_0'$$

where the homogeneous contribution is accounted for by a Lorentzian line. [Hint: begin by calculating the elemental contribution, $d\alpha$, of the saturated absorption coefficient due to the fraction $g^*(\nu_0' - \nu_0)d\nu_0'$ of atoms whose resonant frequencies lie between ν_0' and $\nu_0' + d\nu_0'$]

2.17. Under the assumptions that (1) the homogeneous linewidth is much smaller than the inhomogeneous linewidth and (2) $I \ll I_{s0}$, show that the expression for α given in the previous problem can be approximated as

$$\alpha = \left(\frac{2\pi^2}{3n\varepsilon_0 c_0 h}\right) |\mu|^2 N_t \nu' g*(\nu' - \nu_0) \times$$

$$\times \left[1 - (2/\pi\Delta\nu_0) \frac{I}{I_{s0}} \int \frac{d\nu_0'}{\left\{1 + \left[2\left(\nu' - \nu_0'\right)/\Delta\nu_0\right]^2\right\}\left\{1 + \left[2\left(\nu - \nu_0'\right)/\Delta\nu_0\right]^2\right\}}\right]$$

Since the integral is now the convolution of two Lorentzian lines, what is the width of the hole in Fig. 2.22?

References

1. R. Reiff, *Fundamentals of Statistical and Thermal Physics*, (McGraw-Hill, New York, 1965), Chap. 9
2. W. Heitler, *The Quantum Theory of Radiation*, 3rd edn.(Oxford University Press, London, 1953), Sect. II.9
3. J.A. Stratton, *Electromagnetic Theory*, 1st edn. (McGraw-Hill, New York, 1941), pp. 431–438
4. R.H. Pantell and H.E. Puthoff, *Fundamentals of Quantum Electronics*, (Wiley, New York, 1964), Chap. 6

5. W. Louisell, *Radiation and Noise in Quantum Electronics*, (McGraw-Hill, New York, 1964), Chap. 6
6. H.A. Lorentz, *The Theory of Electrons*, 2nd edn. (Dover, New York 1952), Chap. III
7. A. Einstein, On the Quantum Theory of Radiation, *Z. Phys.* **18**, 121 (1917)
8. *Radiation and Noise in Quantum Electronics*, (McGraw-Hill, New York, 1964), Chap. 5
9. H.G. Kuhn, *Atomic Spectra*, 2nd edn. (Longmans, Green, London 1969), Chap. VII
10. *Radiationless Transitions*, ed. by F.J. Fong (Springer-Verlag, Berlin, 1976) Chap. 4
11. C.K. Rhodes and A. Szoke, Gaseous Lasers: Atomic, Molecular, Ionic in *Laser Handbook* ed. by F.T. Arecchi and E.O. Schultz-DuBois (North Holland, Amsterdam 1972) Vol. 1 pp 265–324
12. J.B. Birks, *Photophysics of Aromatic Molecules* (Wiley-Interscience, New-York, 1970), Sect. 11.9
13. D.L. Dexter, *J. Chem. Phys.* **21**, 836 (1953)
14. J.D. Jackson, *Classical Electrodynamics* (Wiley, New York, 1975) Sect. 9.2
15. W.J. Miniscalco, Optical and Electronic Properties of Rare Earth Ions in Glasses, in *Rare Earth Doped Fiber Lasers and Amplifiers* ed. by M.J.F. Digonnet (Marcel Dekker, New York, 1993), Chap. 2
16. T. Holstein, Imprisonment of Resonant Radiation in Gases, *Phys. Rev.* **72**, 1212 (1947)
17. L.W. Casperson, Threshold Characteristics of Mirrorless Lasers, *J. Appl. Phys.* **48**, 256 (1977)
18. R. Arrathoon, Helium-Neon Lasers and the Positive Column in *Lasers* ed. by A.K. Levine and A.J. DeMaria (Marcel Dekker, New York, 1976), Tab. 2
19. M.H. Dunn and J.N. Ross, The Argon Laser in *Progress in Quantum Electronics*, Vol. 4 ed. by J.H. Sanders and S. Stenholm (Pergamon Press, Oxford 1977), Tab. 2
20. W.F. Krupke, M.D. Shinn, J.E. Marion, J.A. Caird, S.E. Stokowski, Spectroscopic, Optical, and Thermomechanical Properties of Neodymium- and Chromium-Doped Gadolinium Scandium Gallium Garnet, *J. Opt. Soc. Am. B* **3**, 102 (1986)
21. J.C. Walling, O.G. Peterson, J.P. Jennsen, R.C. Morris, and E.W. O'Dell, Tunable Alexandrite Lasers, *IEEE J. Quant. Electr.* **QE-16**, 1302 (1980)

3

Energy Levels, Radiative and Nonradiative Transitions in Molecules and Semiconductors

The purpose of this chapter is to specialize some of the results and considerations of the previous chapter to the somewhat more complicated case of molecules and semiconductors. Particular emphasis will be given to semiconductors, either in bulk or quantum well form, since they play an increasingly important role as laser media.

3.1. MOLECULES

We will first consider the energy levels, and the radiative and nonradiative transitions in molecules. The considerations will be limited to a qualitative description of those features which are particularly relevant for a correct understanding of laser action in active media such as molecular gases or organic dyes. For a more extensive treatment of the wider subject of molecular physics the reader is referred to specialized texts.[(1)]

3.1.1. Energy Levels

The total energy of a molecule consists generally of the sum of four contributions: (1) electronic energy, E_e due to the motion of electrons about the nuclei; (2) vibrational energy E_v, due to the vibrational motion of the nuclei; (3) rotational energy E_r, due to the rotational motion of the molecule; and (4) translational energy. We will not consider the translational energy any further since it is not usually quantized. The other types of energy, however, are quantized and it is instructive to derive, from simple arguments, the order of magnitude of the

O. Svelto, *Principles of Lasers*,
DOI: 10.1007/978-1-4419-1302-9_3,

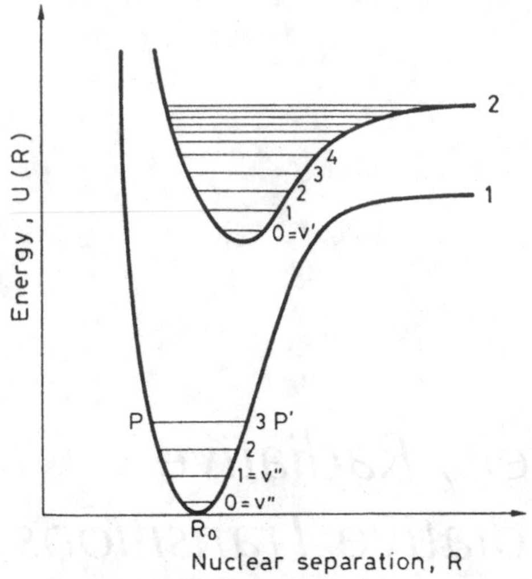

FIG. 3.1. Potential energy curves and vibrational levels of a diatomic molecule.

energy difference between electronic levels (ΔE_e), vibrational levels (ΔE_v), and rotational levels (ΔE_r). The order of magnitude of ΔE_e is given by

$$\Delta E_e \cong \frac{\hbar^2}{ma^2} \tag{3.1.1}$$

where $\hbar = h / 2\pi, m$ is the mass of the electron, and a is the size of molecule. In fact, if we consider an outer electron of the molecule, the uncertainty in its position is of the order of a, then the uncertainty in momentum, via the uncertainty principle, is $\hbar / a$, and the minimum kinetic energy is therefore $\hbar^2 / 2ma^2$. For a diatomic molecule consisting of masses M_1 and M_2 we assume that the corresponding potential energy, U_p, vs internuclear distance R, around the equilibrium distance R_0, can be approximated by the parabolic expression $U_p = k_0(R-R_0)^2/2$ (see Fig. 3.1). Then, the energy difference ΔE_v between two consecutive vibrational levels is given by the well known harmonic oscillator expression

$$\Delta E_v = h\nu_0 = \hbar \left[\frac{k_0}{\mu}\right]^{1/2} \tag{3.1.2}$$

where $\mu = M_1 M_2 / (M_1 + M_2)$ is the reduced mass. For a homonuclear molecule made of two atoms of mass M, the energy difference between two vibrational levels is then

$$\Delta E_v = \hbar \left[\frac{2k_0}{M}\right]^{1/2} \tag{3.1.3}$$

We also expect that a displacement of the two atoms from equilibrium by an amount equal to the size of the molecule would produce an energy change of about ΔE_e since this separation would result in a considerable distortion of the electronic wavefunctions. We can thus write

$$\Delta E_e = k_0 a^2 / 2 \tag{3.1.4}$$

From Eqs. (3.1.1), (3.1.3) and (3.1.4) one can eliminate a^2 and k_0 to obtain

$$\Delta E_v = 2(m / M)^{1/2} \Delta E_e \tag{3.1.5}$$

For a homonuclear diatomic molecule, the rotational energy is then given by $E_r = \hbar^2 J(J+1) / Ma^2$, where J is the rotational quantum number. Therefore, the difference ΔE_r in rotational energy between e.g. the $J = 0$ and $J = 1$ levels is given by $\Delta E_r = 2\hbar^2 / Ma^2 = (m / M)\Delta E_e$, where Eq. (3.1.1) has been used. From Eq. (3.1.5) we then obtain

$$\Delta E_r = (m / M)^{1/2} \Delta E_v \tag{3.1.6}$$

Since $m / M \cong 10^{-4}$, it follows that the separation of rotational levels is about one-hundredth that of the vibrational levels. The spacing of the vibrational levels is, in turn, about one-hundredth of ΔE_e. In fact, as indicated in earlier discussion, the actual frequency ranges, $(\Delta E_e / h)$, for electronic, $(\Delta E_v / h)$, for vibrational and, $(\Delta E_r / h)$, for rotational transitions are found to be roughly $25-50 \times 10^3\,\text{cm}^{-1}$, $500-3000\,\text{cm}^{-1}$ and $1-20\,\text{cm}^{-1}$, respectively.

After these preliminary considerations, we will first consider the simplest case of a molecule consisting of two identical atoms. Since, as already stated, rotations and vibrations occur on a much slower time scale than electronic motion, we can use the Born–Oppenheimer approximation in which the two atoms are first considered to be at a fixed nuclear separation R and non-rotating. By solving Schrödinger's equation for this situation it is then possible to find the dependence of the electronic energy levels on the separation R. Even without actually solving the equation (which is usually very complicated), it is easy to appreciate that, for bound states, the dependence of energy on R must have the form shown in Fig. 3.1, where the ground state, 1, and first excited state, 2, are shown as examples. If the atomic separation is very large ($R \to \infty$), the levels will obviously be the same as those of the single atom. If the separation R is finite, then, as a result of the interaction between the atoms, the energy levels will be displaced. To understand the shape of these curves we note that, with the inclusion of a suitable constant, they can be shown to represent the potential energy of the molecule as a function of the internuclear distance R. In particular, since the minimum energy for curve 1 has been set equal to zero in Fig. 3.1, this curve just represents the potential energy of the ground electronic state. Since the derivative of the potential energy with respect to R gives the force exerted by the atoms on each other, the force is seen to be attractive at large separations and then to become repulsive for small separations. The force is zero for the position corresponding to the minimum of each curve (e.g. R_0), which is, therefore, the separation that the atoms tend to take up (in the absence of oscillation). One notes that the minimum of the curve for the excited state is generally shifted to larger values of R relative to that of the ground-state, owing to the larger orbit occupied by the excited electrons.

So far our discussion has referred to the case in which the two atoms are held fixed at some nuclear separation R. If we now suppose that the molecule is e.g. in its electronic state, 1, and that the two atoms are released at some value R, with $R \neq R_0$, the internuclear force will cause the atoms to oscillate around the equilibrium position R_0. The total energy will then be the sum of the potential energy already discussed, plus the vibrational energy. For small oscillations about the position R_0, curve 1 can be approximated by a parabola and the restoring force between the two atoms is elastic, i.e. is proportional to the displacement from equilibrium. In this case the problem has well-known solutions i.e. those of the harmonic oscillator.

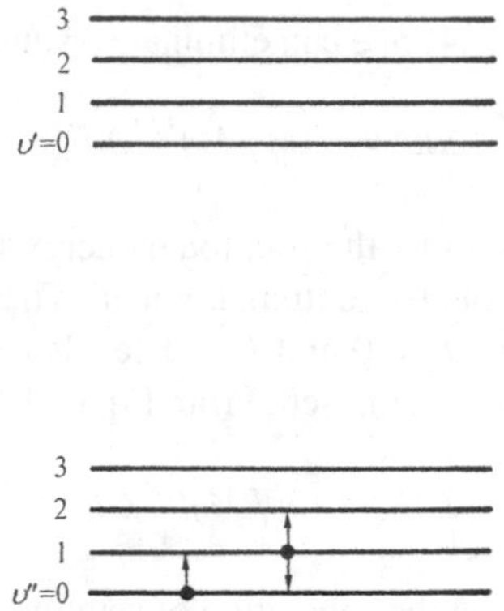

FIG. 3.2. Vibrational energy levels belonging to two consecutive electronic states of a molecule. The arrows indicate allowed transitions starting from the $\upsilon'' = 0$ and $\upsilon'' = 1$ levels.

The energy levels are thus equally spaced by an amount $h\nu_0$ given by Eq. (3.1.2) where the elastic force constant k_0 is equal to the curvature of the parabola. Therefore, when vibrations are taken into account, it is seen that the energy levels (for each of the two electronic states) are given by levels 1,2,3, etc., of Fig. 3.1. We note that the $\upsilon = 0$ level does not coincide with the minimum of the curve because of the well-known zero-point energy, $(h\nu_0 / 2)$, of a harmonic oscillator. Curves 1 and 2 now no longer represent the energy of the system since the atoms are no longer fixed and, instead of Fig. 3.1, the simpler representation of Fig. 3.2 is sometimes used. However, the representation of Fig. 3.1 is, in fact, more meaningful than that of Fig. 3.2. Suppose, for example, that the system is in the $\upsilon'' = 3$ vibrational level of the ground level 1. From Fig. 3.1 one readily sees that the nuclear distance R oscillates between values corresponding to the points P and P' shown in the figure. At these two points, in fact, the vibrational energy coincides with the potential energy, which means that the kinetic energy must be zero. For large oscillations about the equilibrium position R_0, the curve for the potential energy cannot be approximated adequately by a parabola and, in fact, the higher vibrational levels are no longer equally spaced. One can show that the level spacing decreases with increasing energy because the restoring force becomes smaller than that predicted by the parabolic approximation.

We next briefly consider the case of a polyatomic molecule. In this case, the representation given by Fig. 3.1 can still be used provided R is intepreted as some suitable coordinate that can describe the given mode of vibration. Consider for example the SF_6 molecule, which has an octahedral shape (see Fig. 3.3) with the sulfur atom at the center of the octahedron and each of the six fluorine atoms at an apex. For the symmetric mode of vibration shown in the same figure (mode A_{1g}) the coordinate R may be taken as the distance between the sulfur atom and each of the fluorine atoms. Actually SF_6 is seen from Fig. 3.1 to have six independent, nondegenerate modes of vibration. The potential energy U for a general state of the molecule will thus depend upon all six vibrational coordinates of the molecule and should therefore be represented in a seven-dimensional space. So, the representation of Fig. 3.1 can now be regarded as a section of this seven-dimensional function when only one vibrational coordinate is undergoing change.

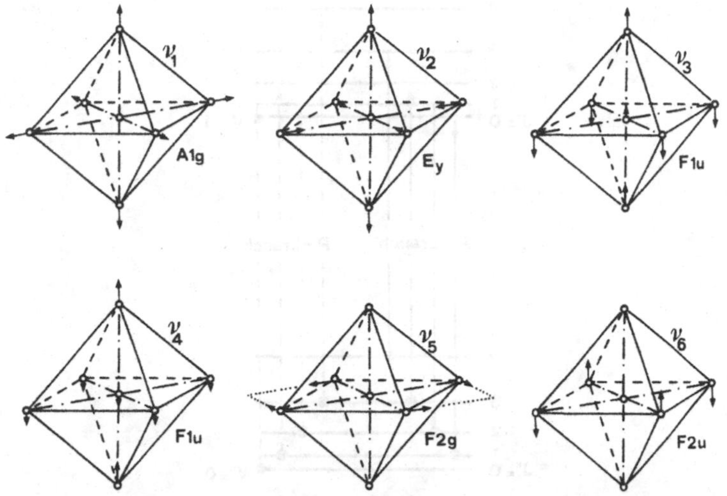

FIG. 3.3. Normal modes of vibration of an octahedral molecule (e.g. SF_6) The sulfur atom occupies the center of the octahedron and the six fluorine atoms are at the corners of the octahedron (after ref.[(2)] by permission).

The description given so far does not give a complete picture of the molecular system since we have ignored the fact that molecule can also rotate. According to quantum mechanics the rotational energy is also quantized, and, for a linear rigid rotator (e.g. a rigid diatomic or linear triatomic molecule), it can be expressed as

$$E_r = BJ(J+1) \tag{3.1.7}$$

where the rotational constant B is given by $\hbar^2 / 2I$ with I being the moment of inertia about an axis perpendicular to the internuclear axis and through the center of mass. Thus, the total energy of the system is given by the sum of the electronic, vibrational, and rotational energies. Accordingly the energy levels of, say, the $\upsilon'' = 0$ and $\upsilon' = 1$ vibrational levels of the ground state will be as indicated in Fig. 3.4. Note that, unlike tha case for vibrational levels, the spacing between consecutive rotational levels is not constant. In fact it increases linearly with the rotational quantum number J, i.e. $[E_r(J) - E_r(J-1)] = 2BJ$.

3.1.2. Level Occupation at Thermal Equilibrium

At thermodynamic equilibrium the population, $N(E_e, E_v, E_r)$, of a rotational-vibrational level belonging to a given electronic state can be written as

$$N(E_e, E_v, E_r) \propto g_e g_v g_r \exp - [(E_e + E_v + E_r) / kT] \tag{3.1.8}$$

where E_e, E_v, and E_r are the electronic, vibrational, and rotational energies of the level and g_e, g_v, and g_r are the corresponding level degeneracies [see Eq. (2.7.3)]. According to the estimates of the previous section, the order of magnitude of E_v / hc is 1,000 cm^{-1} while E_e / hc

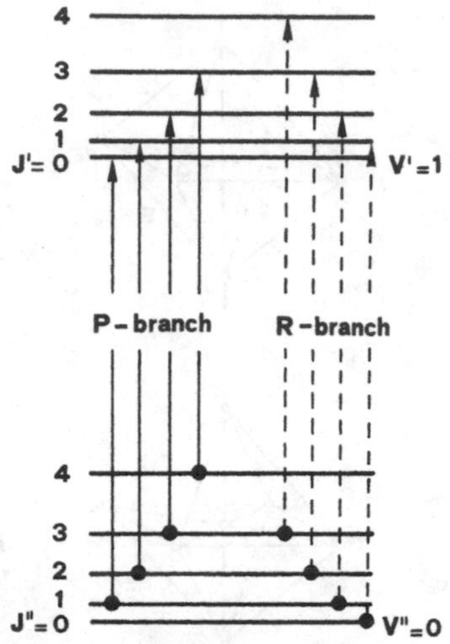

FIG. 3.4. Rotational energy levels belonging to two consecutive vibrational states of a molecule. The arrows indicate allowed transitions belonging to the P – branch and R – branch.

is more than an order of magnitude larger. Since $kT / hc \cong 209\,\text{cm}^{-1}$ (at $T = 300\,\text{K}$), it then follows that both E_e and E_v are appreciably larger than kT. Accordingly we can say as a "rule of thumb" that, at thermal equilibrium at room temperature, a molecule lies in the lowest vibrational level* of the ground electronic state. The probability of occupation of a given rotational level of this lowest vibrational state can then be written, according to Eqs. (3.1.7) and (3.1.8), as

$$p(J) \propto (2J + 1) \exp[-BJ(J + 1) / kT] \qquad (3.1.9)$$

The factor $(2J+1)$ in front of the exponential accounts for level degeneracy: a rotational level of quantum number J is in fact $(2J+1)$-fold degenerate. Taking, as an example, $B = 0.5\,\text{cm}^{-1}$ and assuming $kT = 209\,\text{cm}^{-1}$ (room temperature) we show in Fig. 3.5 the population distribution among the various rotational levels of a given vibrational state (e.g. the ground state). Note that, as a result of the factor $(2J + 1)$ in Eq. (3.1.9), the most heavily populated level is not the ground (i.e. $J = 0$) level but rather the one whose rotational quantum number J satisfies the relation

$$(2J + 1)_m = (2kT / B)^{1/2} \qquad (3.1.10)$$

A conclusion that can be drawn from this section is that for simple molecules at room temperature, the molecular population will be distributed among several rotational levels of the ground vibrational state.

* While this statement is true for diatomic molecules, it is generally not true for polyatomic molecules. In the latter case (e.g., the SF_6 molecule) the spacing between vibrational levels is often appreciably smaller than $1000\,\text{cm}^{-1}$ (down to $\sim 100\,\text{cm}^{-1}$) and many excited vibrational levels of the ground electronic state may have a significant population at room temperature.

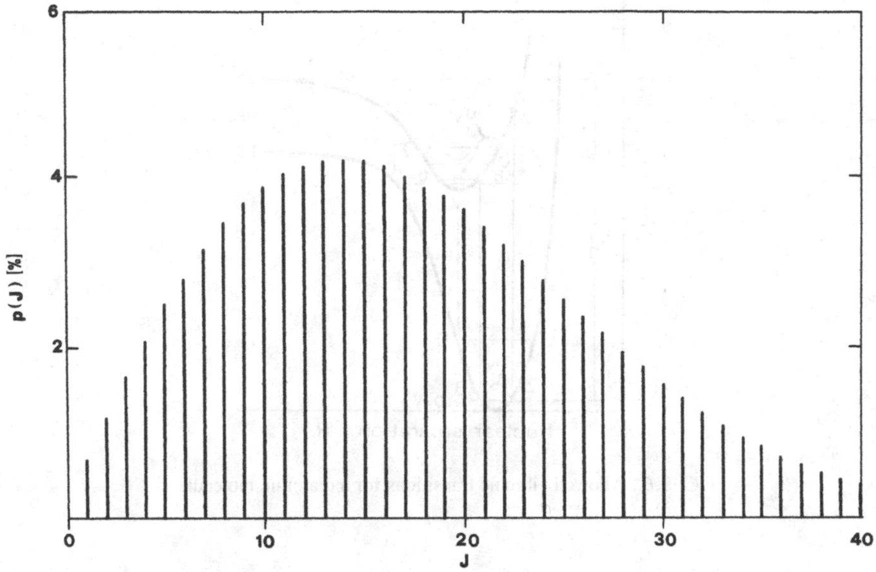

FIG. 3.5. Population distribution among the rotational levels of a given vibrational state.

3.1.3. Stimulated Transitions

According to the earlier discussion, transitions between energy levels of a molecule can be divided into three types:

- Transitions between two rotational-vibrational levels of different electronic states-these are called *vibronic* transitions, a contraction from the words *vibr*ational and electr*onic*. They generally fall in the near-UV spectral region.
- Transitions between two rotational-vibrational levels of the same electronic state (*rotational-vibrational transitions*). They generally fall in the near- to middle-infrared spectral region.
- Transitions between two rotational levels of the same vibrational state, e.g. $\upsilon'' = 0$, of the ground electronic state (*pure rotational transistions*). They generally fall in the far-infrared spectral region.

In the discussion that follows, we briefly consider vibronic and rotational-vibrational transitions, since the most widely used molecular gas lasers are based on these two types of transitions. Lasers based on pure rotational transitions, thus oscillating in the far-IR, also exist, but their use is relatively limited so far (e.g. for spectroscopic applications). In what follows, the quantum mechanical selection rules for these three types of transitions will be briefly considered (see Appendix D for more details).

Consider first a vibronic transition and assume that the symmetry of the electronic wavefunctions in the lower and upper electronic states allows an electric dipole transition. Since the electronic motion occurs at much faster speed than nuclear motion we readily appreciate the so-called *Franck–Condon principle* which states that the nuclear separations do not change during the process of a radiative transition. If we now also assume that all molecules

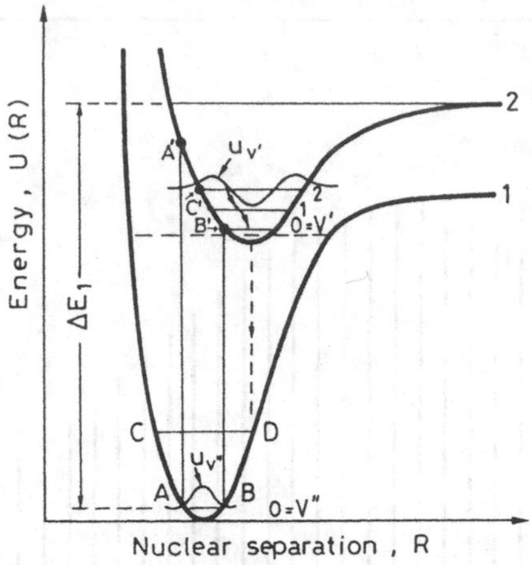

FIG. 3.6. Allowed vibronic transitions for a diatomic molecule.

are in the $\upsilon'' = 0$ level of the ground electronic state,* then, referring to Fig. 3.6, the transition must occur vertically i.e. somewhere between the transitions $A - A'$ and $B - B'$. The Franck–Condon principle can be rephrased in a more precise way by saying that the transition probability between a given vibrational level, υ'', of the ground state and some vibrational level, υ', of the upper electronic state can be written as

$$W_{12} \propto \left| \int u_{\upsilon''} u_{\upsilon'} \, dR \right|^2 \tag{3.1.11}$$

where $u_{\upsilon''}(R)$ and $u_{\upsilon'}(R)$ are the vibrational wave functions of the two levels. Within the harmonic approximation these functions are know to be given by the product of a Gaussian function and a Hermite polynominal. Since the $\upsilon'' = 0$ wavefunction is known to be a Gaussian function., the transition probability, according to Eq. (3.1.11), will be greatest to the vibrational state whose wavefunctions $u_{\upsilon'}$ ensures the best overlap with the function $u_{\upsilon''}$. In the example of Fig. 3.6 the most probable transition will therefore be to the $\upsilon' = 2$ level. Another, simple minded, way of understanding this circumstance follows from noticing that, neglecting zero-point energy, the molecule in the ground state may be considered at rest with a nuclear distance midway between points A and B. Upon absorption, the molecule will pass to an upper vibrational level with the same nuclear separation and still remains at rest (nuclear motion, i.e. position and velocity, cannot change during an electronic transition). This requires the transition to occur toward point C' of level 2. Since the minimum of the potential energy curve for the excited state is shifted toward larger values of the internuclear distance R, the two atoms of the molecule, after absorption, will experience a repulsive force and the molecule will be left in the excited, $\upsilon'' = 2$, vibrational state. As a conclusion, we can say

* When many vibrational levels of the ground electronic state are occupied, transitions may start from any of these levels. Absorption bands originating from $v'' > 0$ are referred to as *hot bands*.

that the transition probability for an electric-dipole allowed vibronic transition is proportional to $\left|\int u_{v''}u_{v'}dR\right|^2$, this quantity being called the *Franck–Condon factor*.

Let us now consider a transition between two vibrational levels of the same electronic state (rotational-vibrational transitions) and assume that the symmetry of the molecule allows for this transition to occur. In this case the transition is said to be *infrared active*.* For such a transition, the quantum mechanical selection rule requires that $\Delta\upsilon = \pm 1$, where $\Delta\upsilon$ is the change in vibrational quantum number. Thus, if we start from the ground state $\upsilon'' = 0$, a transition can only occur to the $\upsilon'' = 1$ state, see Fig. 3.2. If, however, we start from the $\upsilon'' = 1$ level, then the transition may occur to the $\upsilon'' = 2$ (absorption) or to the $\upsilon'' = 0$ level (emission). This result should be contrasted with that for vibronic transitions for which the transition may occur to several vibrational levels, with a probability proportional to the corresponding Franck–Condon factor. It should also be pointed out that the $\Delta\upsilon = \pm 1$ selection rule holds rigorously within the harmonic potential approximation. Since the electronic energy curves of Fig. 3.6 are not exactly parabolic, then it can be shown that transitions obeying the selection rules $\Delta\upsilon = \pm 2, \pm 3$, etc., may also occur as a result of this anharmonicity although with much lower probability (*overtone transitions*).

For both vibronic and vibrational-rotational transitions, we have so far ignored the fact that, corresponding to each vibrational level, there actually exists a whole set of closely spaced rotational levels and these are occupied, at thermal equilibrium, according to Eq. (3.1.9) (see also Fig. 3.5). We thus realize that e.g. the absorption takes place between a given rotational level of the lower vibrational state to some rotational level of the upper vibrational state. For diatomic or linear triatomic molecules the selection rules usually require that $\Delta J = \pm 1$, ($\Delta J = J'' - J'$, where J'' and J' are the rotational numbers of the lower and upper vibrational states). In the case of a rotational-vibrational transition, for instance, a given vibrational transition (e.g. $\upsilon'' = 0 \rightarrow \upsilon'' = 1$ of Fig. 3.2), which, in the absence of rotation, would consist of just a single frequency ν_0, is in fact made up of two sets of lines (Fig. 3.7). The first set, having the lower frequencies, is called the *P* branch and corresponds to the $\Delta J = 1$ transition. The transition frequencies of this branch are lower than ν_0 because the rotational energy of the upper level is smaller than that of the lower level (see Fig. 3.4). The second set, having the higher frequencies, is called the *R* branch and corresponds to $\Delta J = -1$. With the help of Eq. (3.1.7), it can be readily shown that the lines are evenly spaced in frequency by the amount $2B/h$. One also observes from Fig. 3.7 that the amplitudes of the lines are not the same, as a result of the different populations in the rotational levels of the ground state (see Fig. 3.5). Note also that each line is assumed to be broadened by some line-broadening mechanism (e.g. Doppler or collision broadening). For more complex molecules, the selection rule $\Delta J = 0$ also holds and, in this case, the transitions from all the rotational levels of a given vibrational state give a single line centered at frequency ν_0 (*Q* branch). Finally we observe that, when a population inversion is present between the vibrational levels (such as the $\upsilon' = 1$ and $\upsilon'' = 0$ levels of Fig. 3.4) the same spectrum of Fig. 3.7 can be observed in emission rather than in absorption.

* A simple example of infrared inactive transition is that of homonuclear diatomic molecules (e.g. H_2). Ro-vibrational transitions are not allowed in this case, because, on account of symmetry, the molecule cannot develop an electric dipole moment when it vibrates.

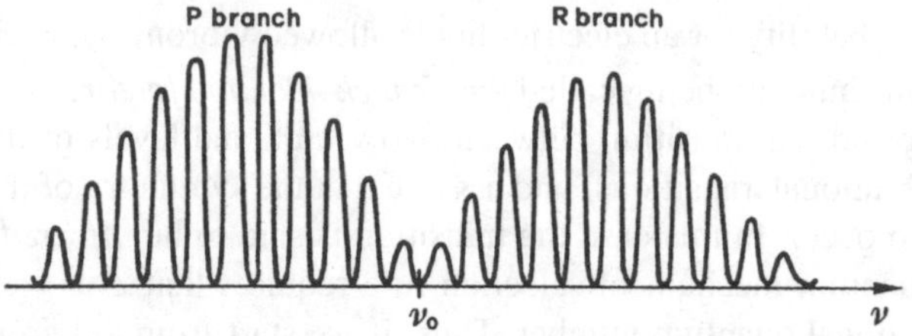

FIG. 3.7. Transitions between two vibrational levels, taking account of the rotational splitting. This transition, which, in the absence of rotational energy, would consist of a single line centered at ν_0, actually consists of two groups of lines: the so-called P – branch, which corresponds to a jump in rotational quantum number of $\Delta J = +1$ and the so-called R – branch, which corresponds to a jump in rotational quantum number of $\Delta J = -1$.

Example 3.1. *Emission spectrum of the CO_2 laser transition at $\lambda = 10.6\,\mu m$.* Here we will consider the $00^\circ 1 \rightarrow 10^\circ 0$ transition (see the section on CO_2 lasers in Chap. 10), whose fundamental frequency, ν_0, in wavenumbers is at $\nu_0 = 960.8\,\text{cm}^{-1}$.[20] The rotational constant of the CO_2 molecule is $B \cong 0.387\,\text{cm}^{-1}$[20] and this value will be taken to be the same for upper ($00^\circ 1$) and lower ($10^\circ 0$) vibrational levels. From previous considerations, the transition energies of the P-branch transitions are given by

$$E = h\,\nu_0 + BJ'(J'+1) - BJ''(J''+1) = h\,\nu_0 - 2BJ'' \tag{3.1.12}$$

where J'' is the rotational quantum number of the lower vibrational state. The rotational number J'_m of the most populated rotational level of the upper vibrational state is given by Eq. (3.1.10). Assuming a rather hot CO_2 molecule, i.e. $T = 450\,\text{K}$, we get $J_{max} \approx 19.6$. For the CO_2 molecule, symmetry dictates that only J'(odd) $\rightarrow J''$(even) transitions can occur. Thus the most populated rotational level in the upper state which is available for the transition is either the $J' = 19$ or the $J' = 21$ level. Assuming that the $J' = 21$ level is the most populated, this level will decay, for a P-branch transition, to the $J'' = 22$ level [P(22) transition]. The corresponding transition frequency, according to Eq. (3.1.12), will then be $\nu = \nu_0 - (2BJ''/h) = 943.8\,\text{cm}^{-1}$, corresponding to a wavelength of $\lambda = (1/943.8)$ cm $\cong 10.6\,\mu\text{m}$. Note that the wavelength corresponding to the fundamental frequency ν_0 is $\lambda = c/\nu_0 \cong 10.4\,\mu\text{m}$. Since only even J'' numbers are involved, the separation between two consecutive P-branch transitions, according to Eq. (3.1.12), will given by $\Delta\nu = 2B\Delta J'' = 4B = 1.55\,\text{cm}^{-1}$.

Example 3.2. *Doppler linewidth of a CO_2 laser.* Consider a CO_2 laser oscillating on the $P(22)$ line at $\lambda = 10.6\,\mu m$ and assume $T = 450\,\text{K}$ [see example 3.1]. Then from Eq. (2.5.18), using the appropriate mass of CO_2 we get $\Delta\,\nu_0^* \cong 50\,\text{MHz}$. Note that, since according to Eq. (2.5.18) one has $\Delta\,\nu_0^* \propto \nu_0$, the calculated linewidth for a CO_2 molecule is much smaller than that of the He-Ne laser in example 2.7 of Chap. 2, essentially because the oscillation frequency ν_0 is now approximately 17 times smaller. Note also that the gas is assumed hotter in this case because, to obtain the high output powers typical of CO_2 lasers, higher pump power are used than in the He–Ne case.

Example 3.3. *Collision broadening of a CO_2 laser.* We will consider a CO_2 laser containing a gas mixture of the H_2, N_2, and CO_2. In this case, the laser linewidth due to collision broadening is found experimentally to be given by $\Delta\nu = 77.58\,(\psi_{CO_2} + 0.73\psi_{N_2} + 0.6\psi_{He}) \times p\,(300/T)^{1/2}$ MHz [compare with Eqs. (2.5.12) and (2.5.11)] where ψ are the fractional partial pressures of the gas mixture, T is the gas temperature and p is the total pressure (in Torr). Taking, as an example, a typical low pressure gas mixture ($p \cong 15$ Torr in a 1:1:8 $CO_2 : N_2$: He mixture) at $T = 450$ K we get $\Delta\nu_c \cong 40$ MHz. A comparison with the result of example 3.2 then shows that, for a low pressure CO_2 laser, collision broadening is comparable to Doppler broadening. However, for higher pressure CO_2 lasers, e.g. atmospheric pressure lasers (see Chap. 10), collision broadening becomes the dominant line broadening mechanism.

Finally, for pure rotational transitions, the selection rule requires that the molecule possess a permanent dipole moment. In fact, considering e.g. the phenomenon of spontaneous emission, the emitted radiation can be looked upon as originating from the rotation of this dipole moment. For a diatomic or linear triatomic molecule the selection rule further requires that $\Delta J = \pm 1$. Thus, in the case of stimulated emission from a given rotational level J, transitions can only occur to the rotational level with quantum number $J-1$.

Before continuing, it is worth summarizing the selection rules which apply for vibronic, rotational-vibrational, and rotational transitions. For an electric dipole allowed vibronic transition one has $\Delta J = \pm 1$ for the change of rotational quantum number while the change in the vibrational quantum number is not strictly established by a precise selection rule. In fact, starting from a given vibrational level υ'' of the lower electronic state the transition may occur to several vibrational levels of the upper electronic state with probabilities proportional to the corresponding Franck–Condon factors. For an infrared-active rotational-vibrational transition, one must have, within the harmonic approximation, $\Delta\upsilon = \pm 1$ for the change of vibrational quantum number and again $\Delta J = \pm 1$ for the change of rotational quantum number. For pure rotational transitions in molecules with a permanent dipole moment one again has $\Delta J = \pm 1$.

3.1.4. Radiative and Nonradiative Decay

Let us first consider spontaneous emission and assume that the molecule is raised to some vibrational level of an excited electronic state (Fig. 3.6). From this state the molecule often decays rapidly by some nonradiative process (e.g. by collision) to the $\upsilon' = 0$ vibrational level.* This is particularly the case for molecules in the liquid phase where collisions occur very frequently. From there the molecule may decay radiatively to a vibrational level of the ground state (*fluorescence* see Fig. 3.6). This transition again occurs vertically and the transition probability from the $\upsilon' = 0$ level to some level of the ground state will again be proportional to the corresponding Franck–Condon factor. Again roughly speaking, the ground state vibrational levels involved will be those nearby the *CD* level of Fig. 3.6. Finally, by nonradiative decay (e.g. by collisions), the molecule rapidly returns to the $\upsilon'' = 0$ level of the ground electronic state (or, more precisely, thermal equilibrium is again established in the ground electronic state). It is now clear from Fig. 3.6 why the fluorescence wavelength is

* Actually, this rapid decay results in a *thermalization* of the molecules in the upper electronic state. The probability of occupation of a given vibrational level of this state is thus given by (3.2.8). For simple molecules, therefore, the lowest vibrational level has the predominant populatation.

longer than that of absorption, a phenomenon referred to as *Stoke's law*. Spontaneous emission may also occur between two ro-vibrational levels of e.g. the ground electronic state and again, for an infrared active transition, the selection rules $\Delta \upsilon = \pm 1$ and $\Delta J = \pm 1$ apply. For pure rotational transitions, spontaneous emission may only occur in molecules having a permanent dipole moment (as explained in the previous section) and the $\Delta J = \pm 1$ selection rule again applies. It should be noted however that, for ro-vibrational transitions and more so for pure rotational transitions, the small value of the transition frequency means that the spontaneous emission lifetime becomes very long i.e. from milliseconds even to seconds (one should remember that $\tau_{sp} \propto 1 / \nu_0^3$). The spontaneous decay of the molecule is then usually dominated by nonradiative processes.

We next briefly consider the phenomena which may cause nonradiative decay. With reference to the more general discussion presented in Sect. 2.6.1, we point out that the main mechanisms are as follows: (a) Collisional deactivation with another like or unlike species. As pointed out before, this occurs particularly for molecules in the liquid phase. In the gas phase, this decay route is particularly effective when the transition energy is small (e.g. for a rotational transition) and when the colliding species have small mass [e.g. deactivation of the $CO_2(0, 1, 0)$ level by He atoms, see Chap. 10]. Collisional deactivation results in a rapid thermalization among the rotational levels of a given vibrational state. (b) By a near-resonant energy transfer to another like or unlike species [see Eq. (2.6.9)]. The phenomenon is particularly effective when the energy imbalance ΔE is appreciably smaller than kT. A notable example of this nonradiative decay process is again found in a CO_2 molecule for the relaxation of the $CO_2(0, 2, 0)$ level to the $CO_2(0, 1, 0)$ level [see Chap. 10]. (c) By *internal conversion* to some other vibrational-rotational level of the same molecule (Fig. 3.8). The process is also called unimolecular decay since it occurs within the same molecule and it is particularly effective when there is a large number of vibrational-rotational modes which are near resonant with the given transition. These modes may also belong to a different electronic state. Thus, for instance, referring again to Fig. 3.6, we note that the molecule, once it is in the lowest vibrational level of the upper electronic state ($\upsilon' = 1$ level of Fig. 3.6), can decay nonradiatively to a nearly isoenergetic vibrational level of the ground electronic state (dotted level in Fig. 3.6). Internal conversion may be particularly effective for large molecules, e.g. dye molecules, which have many modes of vibration. In this case, in fact, the numbers of vibrational modes belonging to the ground electronic level that are in near resonance with the $\upsilon' = 0$ level of Fig. 3.6 can be quite large and the corresponding nonradiative lifetime may even be as short as a few tens of picoseconds.

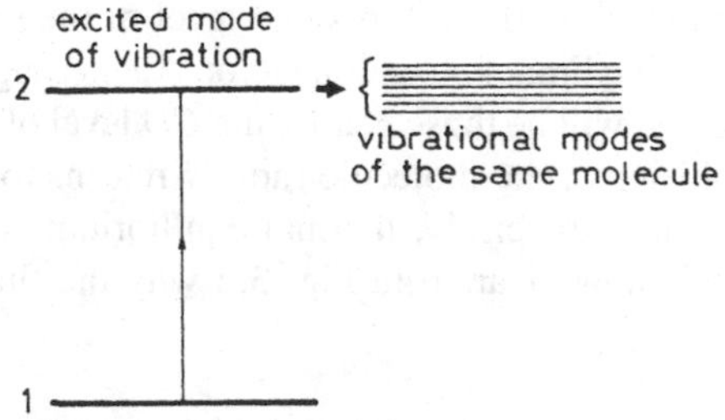

FIG. 3.8. Internal conversion between near-resonant rotational-vibrational modes of the same molecule.

3.2. BULK SEMICONDUCTORS

In this section we will consider the problem of interaction of radiation with matter in the case of a *bulk semiconductor*, i.e. whose physical dimensions are much larger than the de Broglie wavelength of the electrons under consideration. The case of *quantum-confined semiconductors* [quantum wells, quantum wires and quantum dots], wherein one, two, or all three physical dimensions, respectively, are comparable to the De Broglie wavelength and which play an increasingly important role in laser physics, will be considered in the next two sections. Again we will limit our description to the most prominent features of the complex phenomena that occur. For a more extensive treatment of this subject the reader is referred to a specialized text.[3]

3.2.1. Electronic States

The outer electrons of the atom of a semiconductor material are delocalized over the whole crystal and the corresponding wave functions can then be written as *Bloch wave functions*[4]

$$\psi(\mathbf{r}) = u_k(\mathbf{r})[\exp j(\mathbf{k}\cdot\mathbf{r})] \tag{3.2.1}$$

where $u_k(\mathbf{r})$ has the periodicity of the crystalline lattice. The substitution of Eq. (3.2.1) into the Schrödinger wave equation shows that the corresponding eigenvalues of the electron energy, E, are a function of $\mathbf{k}$ and that these values fall within allowed bands. From now on we will limit our considerations to the highest filled band, known as the valence band, and the next higher one, known as the conduction band. Within the *parabolic band approximation*, the E vs k relations can be approximated by a parabola and we then arrive at the picture of Fig. 3.9 for the valence and conduction bands. The energy E_c in the conduction band, measured from the bottom of the band upwards (Fig. 3.9a), can then be written as

$$E_c = \frac{\hbar^2 k^2}{2m_c} \tag{3.2.2a}$$

where $m_c = \hbar^2/[d^2 E_c / d k^2]_{k=0}$ is the effective mass of the electron at the bottom of the conduction band. Likewise, the energy in the valence band, measured from the top of the band downwards (Fig. 3.9a), can be written as

$$E_v = \frac{\hbar^2 k^2}{2m_v} \tag{3.2.2b}$$

where $m_v = \hbar^2/(d^2 E_v / d k^2)_{k=0}$ is the effective mass of the electron at the top of the valence band. In some cases, particularly when dealing with a given transition, it may be more convenient to refer the energy to the same reference level e.g. from the top of the valence band upwards (Fig. 3.9b). If we call E' the energy in this coordinate system, the energies in the conduction and valence bands are now obviously given by

$$E'_c = E_g + E_c \tag{3.2.3a}$$

$$E'_v = -E_v \tag{3.2.3b}$$

where E_g is the energy gap.

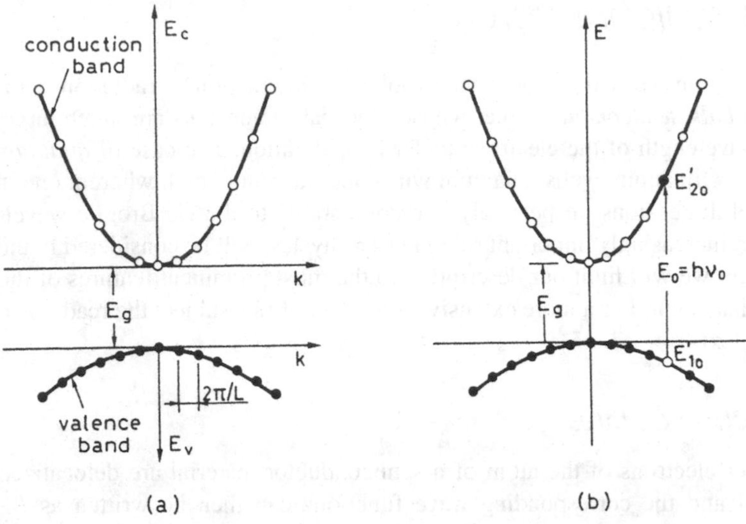

FIG. 3.9. Energy vs k relation for a bulk semiconductor: (a) Energy scale starting from the bottom of the conduction band, upwards, for the conduction band and from the top of the valence band, downwards, for the valence band. (b) Energy scale starting from the top of the valence band, upwards, for both conduction and valence bands.

The above simple one-dimensional model can be readily generalized to the three-dimensional case. If we let k_x, k_y and k_z be the components of the electron's **k** vector, and if we assume that the effective mass, i.e. the band curvature, is the same along x, y, and z directions, we again obtain Eqs. (3.2.2) and (3.2.3) where now $k^2 = k_x^2 + k_y^2 + k_z^2$.

So far we have assumed the semiconductors crystal to be of infinite extent. For a finite sized crystal in the form of a rectangular parallelepiped with dimensions L_x, L_y, and L_z we need to impose the boundary condition that the total phase shift $\mathbf{k}\cdot\mathbf{r}$ across the crystal be some multiple integer of 2π. Thus we get

$$k_i = (2\pi\, l\, /\, L_i) \tag{3.2.4}$$

where $i = x, y, z$ and l is an integer. So, in the one dimensional case, the available states can be indicated as dots, in the valence band, or open circles, in the conduction band, as shown in Fig. 3.9.

The existence of a valence and conduction band can also be explained by a simple physical argument. Consider for simplicity the case of sodium, where each atom contains 11 electrons. Ten of these electrons are tightly bound to the nucleus to form an ion of overall positive charge e. The eleventh electron moves in an orbit around this ion. Let E_1 and E_2 be the energies of this electron in its ground and first excited state, respectively, and ψ_1, ψ_2 the corresponding wave functions. Consider now two sodium atoms at some distance d apart. If d is much larger than the atomic dimensions, the two atoms will not interact with each other and the energy of the two states will remain unchanged. Another way of expressing this is to say that, considering, e. g., the two atoms in their energy state E_1, the one-electron energy

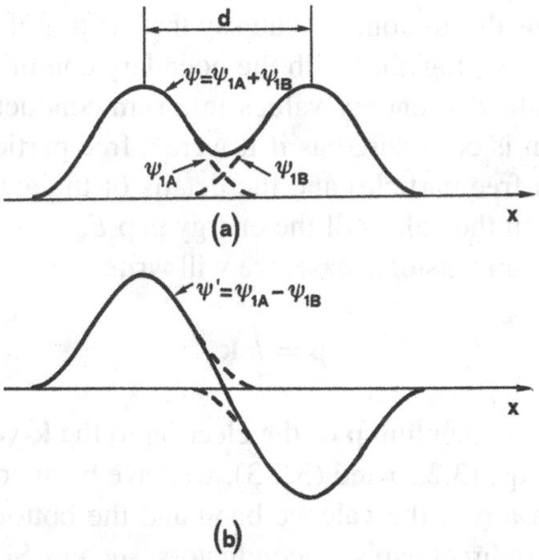

FIG. 3.10. (a) Symmetric and (b) antisymmetric linear combination of the atomic wave functions Ψ_{1A} and Ψ_{1B} of two identical atoms at separation d.

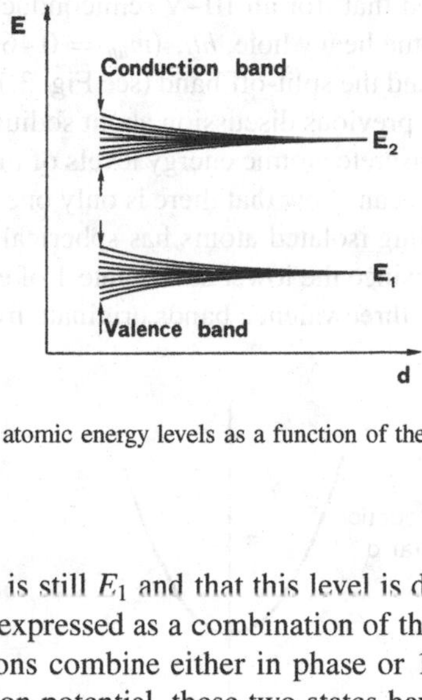

FIG. 3.11. N – fold splitting of the atomic energy levels as a function of the atomic separation d for an N – atom system.

level of the two-atom system is still E_1 and that this level is doubly degenerate. The overall wave function can in fact be expressed as a combination of the two wave functions ψ_{1A} and ψ_{1B} in which the two functions combine either in phase or 180° out of phase (Fig. 3.10). In the absence of an interaction potential, these two states have the same energy E_1. When, however, the atomic separation d becomes sufficiently small, the energies of these two states become slightly different owing to the interaction, and the doubly degenerate level is split into two levels. Likewise, for an N-atom system where the atoms are close enough to interact with each other, the N-fold degenerate level of the state of energy E_1 is split into N closely spaced levels. The state of energy E_1 will thus give rise to the valence band while the state of energy E_2 gives rise, likewise, to the conduction band (Fig. 3.11). From the previous argument it is apparent that each band actually consists of N closely spaced levels, where N is the total number of atoms in the semiconductor crystal. Since N is usually a very large number, the individual energy levels of a semiconductor, in each band, are generally not resolvable.

To sum up the previous discussion, we can say that, within the parabolic band approximation, Eqs. (3.2.2) and (3.2.3) together with the boundary conditions (3.2.4) provide a very simple description of the allowed energy values in a semiconductor. Note that, within this approximation, the electron is considered as if it were a free particle of momentum $p = \hbar k$ (indeed $E = p^2 / 2m$ for a free particle) and the details of the actual quantum system have been reduced to appearing in the values of the energy gap E_g and of the effective masses m_c and m_v. Thus, for the three-dimensional case, we will write

$$\mathbf{p} = \hbar\,\mathbf{k} \tag{3.2.5}$$

as the equation relating the momentum **p** of the electron to the **k**-vector of the wavefunction. Note also that, in writing Eqs. (3.2.2) and (3.2.3), we have been considering only *direct gap* semiconductors, where the top of the valence band and the bottom of the conduction band occur at the same k value. Indirect gap semiconductors, such as Si or Ge, are not considered here since they are not relevant as laser materials. Of the various direct gap semiconductors, we will limit our considerations to the III–V compounds such as GaAs, InGaAs, AlGaAs or InGaAsP. In particular, for GaAs, one has $m_c = 0.067\,m_0$ where m_0 is the rest mass of a free electron. It should also be noted that, for all III–V semiconductors, there are three different types of valence band, namely the heavy hole, *hh*, ($m_{hh} = 0.46\,m_0$ for GaAs), the light hole, *lh*, ($m_{lh} = 0.08\,m_0$ for GaAs), and the split-off band (see Fig. 3.12). This circumstance can be understood when, based on the previous discussion about sodium atoms, we view the energy bands as originating from the discrete atomic energy levels of the isolated atoms which made up the crystal. Accordingly, one can show that there is only one conduction band because the excited state of the corresponding isolated atoms has spherical symmetry like that of the s-state atomic orbitals. Likewise, since the lower state (state 1 of energy E_1 in Fig. 3.11) can be shown to have p-symmetry, the three valence bands originate from a suitable combination of

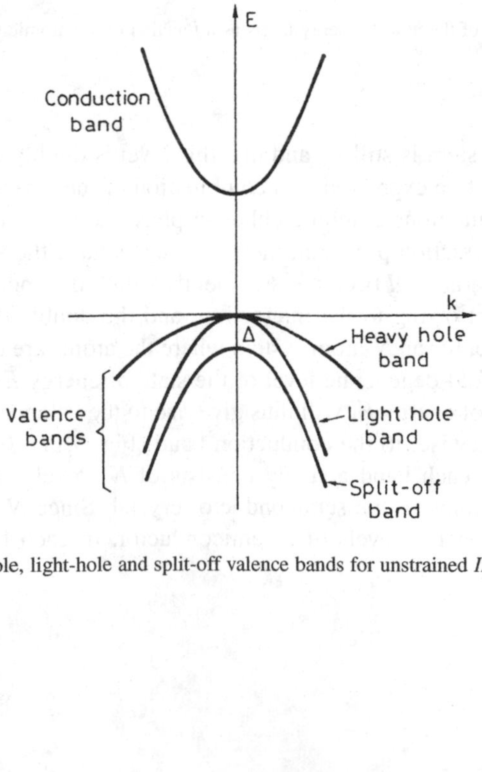

FIG. 3.12. Heavy-hole, light-hole and split-off valence bands for unstrained *III–V* semiconductors.

the p_x, p_y and p_z orbitals of this state, account being taken of the crystal's symmetry. Actually in a crystal of cubic symmetry such as in all unstrained III–V compounds, the three bands are expected to have the same energy at $k = 0$. However, spin-orbit interactions lower one of these bands, the split-off band, by an amount corresponding to $\Delta E = 0.34$ eV for GaAs. Since this value is much larger than $kT[\cong 0.028 \text{ eV}]$, the split-off band will always be filled with electrons and will not participate in radiative and non-radiative transitions. For reasons which will be explained in the next section, the light hole band also makes little contribution to these transitions. Thus, to first order, the valence band of a III–V semiconductor can be thought to be made of only the heavy hole band.

3.2.2. Density of States

Following what was done for the cavity modes in Sect. 2.2.1, we can now proceed to calculating the number of energy states, $p(k)$, whose k value ranges from 0 to k. With reference to Fig. 2.2, since now both positive and negative values of k_i are allowed, $p(k)$ is given by the volume of the sphere of radius k, $4\pi\, k^3 / 3$, divided by the volume of the unit cell, $(2\pi)^3 / L_x L_y L_z$, times a factor 2 to account for the two states arising from the electron spin. Thus

$$p\,(k) = (k^3 V / 3\pi^2) \tag{3.2.6}$$

where $V = L_x\, L_y\, L_z$ is the crystal volume. Since the number of states is very large, we can calculate the density of states per unit volume, $\rho\,(k)$, as

$$\rho_{c,v} = \frac{dp}{V\,dk} = \frac{k^2}{\pi^2} \tag{3.2.7}$$

where Eq. (3.2.6) has been used. Note that this expression is valid for both the valence and conduction bands and, to indicate this, the density of states has been denoted with both indices c and v. We are also interested in calculating the density of states, $\rho\,(E)$, in terms of the electron energy. Since $\rho_{c,v}\,(E)\,dE = \rho_{c,v}\,(k)\,dk$, from Eq. (3.2.2) we obtain

$$\rho_c\,(E_c) = \frac{1}{2\pi^2}\left(\frac{2\,m_c}{\hbar^2}\right)^{3/2} E_c^{1/2} \tag{3.2.8a}$$

$$\rho_v\,(E_v) = \frac{1}{2\pi^2}\left(\frac{2\,m_v}{\hbar^2}\right)^{3/2} E_v^{1/2} \tag{3.2.8b}$$

We recall that E_c and E_v are measured from the bottom of the conduction and the top of the valence bands, upwards and downwards, respectively (Fig. 3.9a). One notes that, since for III–V compounds, one has $m_c << m_v = m_{hh}$, then it follows that $\rho_c << \rho_v$. One also notes that, since $m_{lh} << m_{hh}$, the density of states of light holes is only a small fraction of that of heavy holes. Accordingly, the light holes are very much in a minority for a III–V semiconductor, and their presence can normally be neglected in comparison with heavy holes.

3.2.3. Level Occupation at Thermal Equilibrium

We will first assume that the semiconductor is in overall thermal equilibrium. Since electrons are fermions, i.e. must comply with the Pauli exclusion principle, they must obey Fermi–Dirac statistics rather than Boltzmann statistics. The probability for the electron to occupy a given level of energy E', either in the valence or conduction band, is then given by

$$f(E') = \frac{1}{1 + \exp\left[\left(E' - E'_F\right) / kT\right]} \tag{3.2.9}$$

where E'_F is the Fermi level. In this case, the energy of both valence and conduction bands has been referred to the same reference energy level as in Fig. 3.9b. An interpretation of E'_F is obtained from Eq. (3.2.9) by setting $E' = E'_F$. We get $f\left(E'_F\right) = 1/2$. Another interpretation of the significance of E'_F is also obtained from Eq. (3.2.9) by letting $T \to 0$. We get $f(E') = 1$ for $E' < E'_F$ and $f(E') = 0$ for $E' > E'_F$. Thus, at $T = 0$, the Fermi level separates the filled region from the empty region in a semiconductor. One should now remember that, for undoped semiconductors, E'_F is situated approximately in the middle of the energy gap. Thus, for $T > 0$, the relation $f(E')$ vs E' will be as shown in Fig. 3.13b. This means that, since $E_g >> kT$, the level occupancy in the conduction band is very small i.e. very few electrons are thermally activated to the conduction band. As a consequence of this circumstance, in both Fig. 3.13a and Fig. 3.9a, the available states in the valence band are denoted by a full circle to indicate the presence of an electron. Conversely, the available states in the conduction band are denoted by an open circle to indicate the absence of an electron i.e. the presence of a hole. For n-type-doping, on the other hand, E_F must be displaced toward the conduction band in order

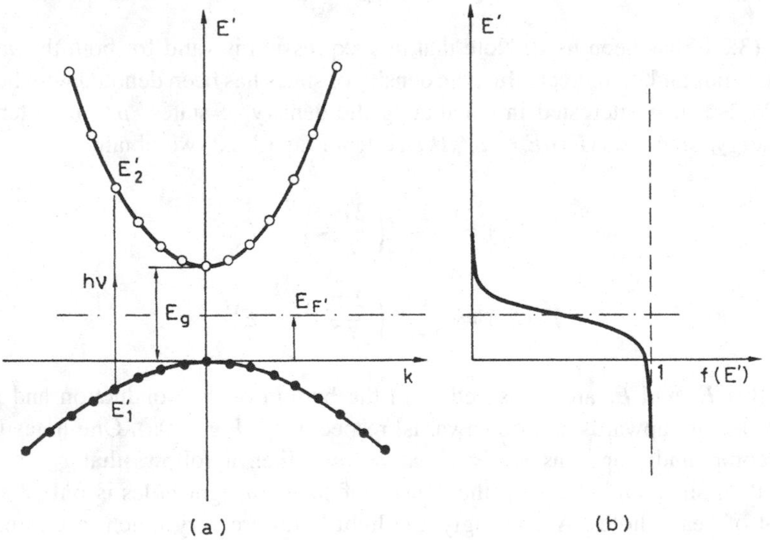

FIG. 3.13. Energy E' vs k relation, (a) and level occupation probability $f(E')$, (b) for both conduction and valence bands under thermal equilibrium.

to accomodate the electrons in this band arising from the dopant ions. Similarly, for p-type-doping, the E_F is displaced toward the valence band. Finally, for very heavy doping (doping level of $\sim 10^{18}\,\text{cm}^{-3}$), E_F is displaced so much that it actually enters the conduction or the valence band, respectively. The semiconductor is then called degenerate since its conductivity becomes similar to that of a metal.

Suppose now that electrons are raised from the valence to the conduction band by some suitable pumping mechanism. The intraband relaxation (whose typical relaxation time τ, as established by electron–phonon collisions, is ≈ 1 ps) is usually much faster than interband relaxation (whose typical relaxation time τ is ≈ 1 ns, due to electron-hole recombination). Thus, a thermal equilibrium will rapidly be established within each band even though there is no overall equilibrium in the semiconductor. One can therefore talk of occupation probabilities f_v and f_c for the valence and conduction band separately. This means that f_c and f_v will be given by expressions of the general form of Eq. (3.2.9) in each band, respectively. More precisely, referring now to the energy coordinate system of Fig. 3.9a, one can write

$$f_c(E_c) = \frac{1}{1 + \exp\left[\left(E_c - E_{F_c}\right)/kT\right]} \tag{3.2.10a}$$

and

$$f_v(E_v) = \frac{1}{1 + \exp\left[\left(E_{F_v} - E_v\right)/kT\right]} \tag{3.2.10b}$$

where E_{F_c} and E_{F_v} are now the energies of the so-called quasi-Fermi levels of the valence and conduction bands, respectively. Thus, for given values of E_{F_c} and E_{F_v}, the plots of $f_c(E_c)$ vs E_c and of $f_v(E_v)$ vs E_v will be as shown in Fig. 3.14b. Note that, following the previous discussion about the Fermi level, the quasi-Fermi levels indicate, in each band, the boundaries between the zones of fully occupied and completely empty states at $T = 0$ K. Accordingly, for $T = 0$ K, the states occupied by an electron (*full circle*) and the states occupied by a hole (*open circle*) will be as shown in Fig. 3.14a. In the same figure, the hatched area thus correspond to states filled with an electron. Sometimes, it is more convenient to express Eq. (3.2.10) using the energy coordinate of Fig. 3.9b. According to Eqs. (3.2.3a) and (3.2.3b) we then obtain

$$f_c\left(E'_c\right) = \frac{1}{1 + \exp\left[\left(E'_c - E'_{F_c}\right)/kT\right]} \tag{3.2.11a}$$

and

$$f_v\left(E'_v\right) = \frac{1}{1 + \exp\left[\left(E'_v - E'_{F_v}\right)/kT\right]} \tag{3.2.11b}$$

As observed above, the quasi-Fermi levels indicate, in each band, the boundaries between occupied and empty states. Consequently, the values of E'_{F_c} and E'_{F_v} in Eq. (3.2.11) must depend of the number of electrons raised to the conduction band. To obtain this dependence, we calculate the electron density in the conduction band, N_e, as

$$N_e = \int_0^\infty \rho_c\left(E_c\right) f_c\left(E_c\right) dE_c \tag{3.2.12}$$

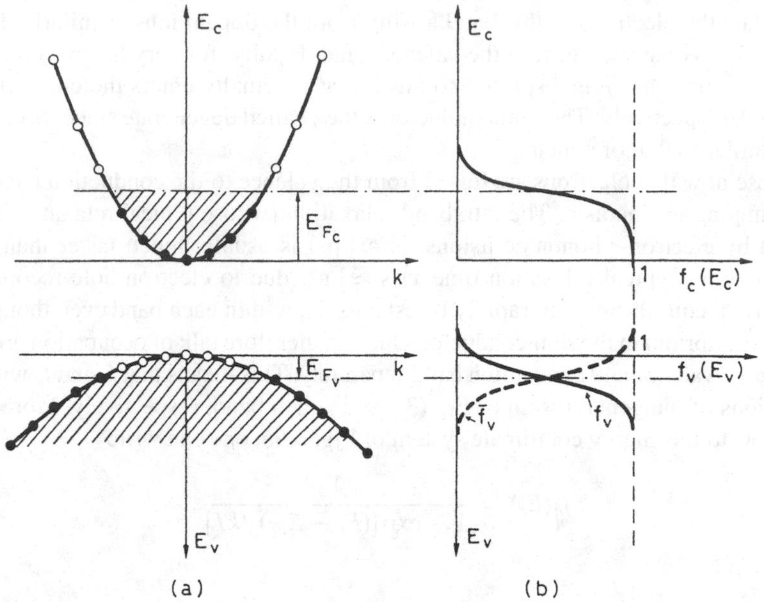

FIG. 3.14. Energy E vs k relation, (a) and level occupation probability $f_{c,v}(E)$, (b) for conduction and valence bands under thermal equilibrium within each band.

To calculate the corresponding hole density, N_h, in the valence band, we notice that $\bar{f}_v(E_v) = 1 - f_v(E_v)$ is the probability that a given state in the valence band is not occupied by an electron and thus filled by a hole. From Eq. (3.2.10b) we then get

$$\bar{f}_v(E_v) = \frac{1}{1 + \exp\left[(E_v - E_{F_v}) \,/\, kT\right]} \tag{3.2.13}$$

Equation (3.2.13) shows that, in the energy coordinate system of Fig. 3.9a, the probability of hole occupation in the valence band takes on the same functional forme as that of electron occupation in the conduction band [compare Eq. (3.2.13) with Eq. (3.2.10a)]. This makes the calculation for the valence band completely symmetrical to that of the conduction band. Thus, for a given value of the quasi-Fermi level in the valence band, the hole density, N_h, is obtained as

$$N_h = \int_0^\infty \rho_v(E_v)\bar{f}_v(E_v)dE_v \tag{3.2.14}$$

Suppose now that a given density of electrons, N, is raised by a suitable pumping process from the valence to the conduction band. The hole density left in the valence band will also be equal to N and the quasi-Fermi levels of both the valence and conduction bands can be obtained from Eqs. (3.2.12) and (3.2.14) by setting the condition $N_e = N_h = N$. In fact, from

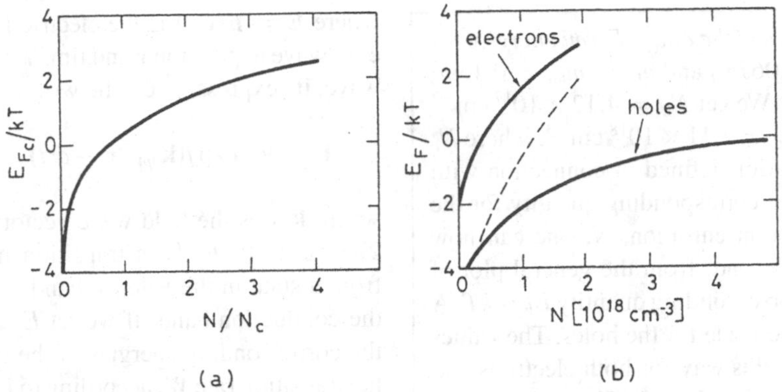

FIG. 3.15. (a) Normalized plot of the quasi-Fermi energy of the conduction band, E_{F_c}, vs the normalized concentration of injected electrons density, N_e. The same normalized relation also holds for holes in the valence band. (b) Normalized plots of the quasi-Fermi levels of both valence and conduction bands, E_F / kT, vs the concentration of injected carriers, N, for GaAs.

Eq. (3.2.12) with the help of Eqs. (3.2.8a) and (3.2.10a), we get

$$N = N_c \frac{2}{\pi^{1/2}} \int_0^\infty \frac{\varepsilon^{1/2}\, d\varepsilon}{1 + \exp\,[\varepsilon - \varepsilon_F]} \tag{3.2.15}$$

where $N_c = 2\left(2\pi\, m_c\, kT/h^2\right)^{3/2}$, $\varepsilon = E_c / kT$ and $\varepsilon_F = E_{F_c} / kT$. From Eq. (3.2.14), with the help of Eqs. (3.2.8b) and (3.2.13), we obtain an expression which is the same as Eq. (3.2.15), provided we interchange suffixes c and v. Equation (3.2.15) then shows that E_{F_c} / kT is a function of only N / N_c and this function is plotted vs N / N_c in Fig. 3.15a. The same figure also holds for the valence band provided we interchange suffix c with υ.

3.2.4. Stimulated Transitions

Let us consider the interaction of a monochromatic e.m. wave of frequency ν with a bulk semiconductor. As for the case of an atomic system, the interaction Hamiltonian, within the electric dipole approximation, can be written as,* [see Eq. (2.4.2)],

$$H' = -e\, \mathbf{E} \cdot \mathbf{r} \tag{3.2.16}$$

* To conform with the treatment of Chapt. 2, the interaction Hamiltonian is written in terms of an electric dipole interaction rather than in terms of the interaction of the vector potential with the electron momentum **p**, as commonly done in many textbooks on semiconductors. The two Hamiltonians can be shown however to lead to the same final results.

Example 3.4. *Calculation of the quasi-Fermi energies for GaAs.* We take $m_c = 0.067m_0$ and $m_v = m_{hh} = 0.46m_0$ and assume $T = 300$ K. We get $N_c = 4.12 \times 10^{17}$ cm^{-3} and $N_v = (m_v / m_c)^{3/2} N_c = 7.41 \times 10^{18}$ cm^{-3}, where N_c is the electron concentration defined in connection with Eq. (3.2.15) and N_v is the corresponding quantity for the holes. At each electron concentration, N, one can now obtain the quantity N / N_c and, from the general plot of Fig. 3.15a, deduce the corresponding quantity E_{F_c} / kT. A similar calculation can be made for the holes. The values of E_F / kT, calculated in this way for both electrons and holes in GaAs, are plotted in Fig. 3.15b vs the carrier concentration, N.

where $\mathbf{E} = \mathbf{E}(\mathbf{r}, t)$ is the electric field of the e.m. wave at position $\mathbf{r}$ and time t. For a plane wave, its expression can be written as

$$\mathbf{E} = \mathbf{E}_0 \exp j(\mathbf{k}_{opt} \cdot \mathbf{r} - \omega t) \qquad (3.2.17)$$

where $\mathbf{k}_{opt}$ is the field wave vector and $\omega = 2\pi\nu$. If $\nu \geq E_g / h$ a transition may occur from a state in the valence band to a state in the conduction band. If we let E_2' and E_1' be the corresponding energies of the two states, the transition rate W, according to Eq. (A. 23) of Appendix A, is

$$W = \frac{\pi^2}{\hbar^2} \left|H_{12}'^o\right|^2 \delta(\nu - \nu_o) \qquad (3.2.18)$$

where $\nu_0 = (E'_2 - E'_1) / h$ and

$$\left|H_{12}'^o\right|^2 = \left| \int \psi_c^* \left(-e\mathbf{r} \cdot \mathbf{E}_0\, e^{j\mathbf{k}_{opt}\cdot\mathbf{r}}\right) \psi_v dV \right|^2 \qquad (3.2.19)$$

Note that ψ_v and ψ_c in Eq. (3.2.19) are the Bloch wave functions of levels 1 and 2 as given by Eq. (3.2.1).

From Eqs. (3.2.18) and (3.2.19) we can now obtain the selection rules for the interaction. From Eq. (3.2.18), noting the δ-Dirac function on the right hand side, one sees that $\nu = \nu_0$. This means that

$$\left(E_2' - E'\right) = h\nu \qquad (3.2.20)$$

which is often referred to as the energy conservation rule for the interaction. Similarly, from Eq. (3.2.19), since $\psi_v \propto \exp(j\mathbf{k}_v \cdot \mathbf{r})$ and $\psi_c \propto \exp(i\mathbf{k}_c \cdot \mathbf{r})$ one can show that the integral is non-vanishing only when

$$\mathbf{k}_c = \mathbf{k}_{opt} + \mathbf{k}_v \qquad (3.2.21)$$

The proof of Eq. (3.2.21) is somewhat involved and requires that the periodic properties of $u_c(\mathbf{r})$ and $u_v(\mathbf{r})$ appearing in Eq. (3.2.1) be properly taken into account.(15) The selection rule Eq. (3.3.21) can however be physically understood when we notice that an exponential factor of the form $\exp j\left[\left(\mathbf{k}_v + \mathbf{k}_{opt} - \mathbf{k}_c\right) \cdot \mathbf{r}\right]$ is present in the integrand of Eq. (3.3.19) and this term, since it oscillates rapidly with $\mathbf{r}$, makes the value of the integral zero unless $\mathbf{k}_v + \mathbf{k}_{opt} - \mathbf{k}_c = 0$. Since $\hbar\mathbf{k}_{c,v}$ is the electron momentum in the conduction or valence band and $\hbar\mathbf{k}_{opt}$ is the photon momentum, Eq. (3.2.21) shows that the total momentum must be conserved in the transition. One should note that one has $k_{opt} = 2\pi n / \lambda$ where n is the semiconductor refractive index and λ is the transition wavelength. Thus, with e.g. $n = 3.5$ and $\lambda \cong 1\ \mu$m, one has $k_{opt} \cong 10^5$ cm^{-1}. On the other hand one typically has $k_{c,v} = 10^6 \div 10^7$ cm^{-1} for an electron or hole of average thermal energy (see Example 3.5). Thus $k_{opt} << k_{c,v}$ and Eq. (3.2.21) simplifies to

$$k_c = k_v \tag{3.2.22}$$

Equation (3.2.22) is often referred to as the **k**-selection or **k**-conservation rule and indicates that stimulated transitions must occur vertically in the E vs k diagram (see Fig. 3.13a). Note finally that the e.m. wave does not interact with the electron's spin or, in other words, that spin is not involved in the interaction Hamiltonian Eq. (3.2.16). The spin, therefore, cannot change in the transition i.e. the selection rule for the change of the electron spin, ΔS, is simply

$$\Delta S = 0 \tag{3.2.23}$$

As for the case of atomic transitions considered in the previous Chapter, Eq. (3.2.18) needs to be modified when line broadening mechanisms are taken into account. For semiconductors, the main broadening mechanism arises from electron-phonon dephasing collisions. Thus the δ-Dirac function in Eq. (3.2.18) must be replaced by a Lorentzian function $g(\nu - \nu_0)$, whose width, according to Eq. (2.5.11) is given by $\Delta\nu_0 = 1/\pi\tau_c$ where τ_c is the average electron-phonon dephasing collision time ($\tau_c \cong 0.1$ ps for GaAs). Proceeding as in Sect. 2.4.4, we can arrive at a definition of a transition cross section which has the same form as for atomic transitions,* namely [see also Eq. (2.4.18)]

> **Example 3.5.** *Calculation of typical values of k for a thermal electron.* For an electron in the conduction band having average thermal velocity υ_{th}, one has $m_c\upsilon_{th}^2 = 3kT$ where T is the electron temperature. We also have $p = \hbar k_c = m_c\upsilon_{th}$. Combining the two previous expressions, we get $k_c = [3m_ckT]^{1/2}/\hbar$. If we take $m_c = 0.067\,m_0$, as for GaAs, and $kT = 0.028$ eV ($T \cong 300$ K) we thus get $k_c = 2.7 \times 10^6\,\mathrm{cm}^{-1}$. Similarly one has $k_v = [3m_vkT]^{1/2}/\hbar$ and thus $k_v = (m_v / m_c)^{1/2} k_c \cong 7 \times 10^6\,\mathrm{cm}^{-1}$ if we take $m_c = m_{hh} = 0.46\,m_0$ for GaAs.

$$\sigma = \frac{2\pi^2\nu}{n\varepsilon_0 ch}\frac{\mu^2}{3}\,g(\nu - \nu_0) \tag{3.2.24}$$

where $\mu = |\boldsymbol{\mu}|$ and

$$\boldsymbol{\mu} = \int u_c e\mathbf{r} u_v dV \tag{3.2.25}$$

where $u_c = u_{ck}$ amd $u_v = u_{vk}$ are the Bloch wavefunctions, appearing in Eq. (3.2.1). Note the factor 3 in the term $\mu^2 / 3$ of Eq. (3.3.24) which arises from averaging the matrix element $\boldsymbol{\mu}$ over all electron **k** vector directions, for a fixed electric field polarization [in this regard, see footnote appearing in connection with Eqs. (2.4.13).

* Note that the concept of cross section as discussed in connection with Fig. 2.7 loses its meaning for a delocalized wavefunction such as the Bloch wavefunction. We nevertheless retain the same symbol σ for a semiconductor to make an easier comparison with the case of isolated atoms or ions. Here σ has the only meaning that the transition rate for a plane wave is $W = \sigma F$, where F is the photon flux of the wave or, alternatively, $W = \sigma\rho\,c/h\nu$, where ρ is the energy density and ν is the frequency of the wave.

3.2.5. Absorption and Gain Coefficients

Consider first two energy levels, E'_2 and E'_1, in the conduction and valence band, respectively, whose energy difference is equal to $E_0 = h\nu_0$, where ν_0 is the frequency of the transition. Under the k-selection rule given by Eq. (3.2.22), the energies E'_2 and E'_1 are uniquely established for a given value of ν_0. In fact, from Eqs. (3.2.3) and (3.2.2), one writes

$$E'_2 = E_g + (\hbar^2 k^2 / 2m_c) \tag{3.2.26a}$$

$$E'_1 = -\hbar^2 k^2 / 2m_v \tag{3.2.26b}$$

where we have set $k = k_c = k_v$. From Eq. (3.2.26), since $E'_2 - E'_1 = E_0 = h\nu_0$, we obtain

$$h\nu_0 = E_g + (\hbar^2 k^2 / 2m_r) \tag{3.2.26c}$$

where m_r is the reduced mass of the semiconductor, given by the relation $m_r^{-1} = m_c^{-1} + m_v^{-1}$. Equation (3.2.26) constitute a set of three equations in the three unknowns E'_2, E'_1 and k.

As a next step, we define the *joint density of states* with respect to the energy variable $E_0 = E'_2 - E'_1$ so that $\rho_j dE_0$ gives the density of transitions with transition energy lying between E_0 and $E_0 + dE_0$. Under the k-selection and spin selection rules given by Eqs. (3.2.22) and (3.2.23), any state in e.g. the valence band, with a given spin, is coupled to only one state in the conduction band with the same spin. The number of transitions is thus equal to the number of corresponding states in either valence or conduction bands. We thus write $\rho_j dE_0 = \rho(k)\, dk$, where $\rho(k) = \rho_{c,v}(k)$ is given by Eq. (3.2.7), so that we obtain

$$\rho_j(E_0) = (k^2 / \pi^2)(dk / dE_0) \tag{3.2.27}$$

With the help of Eq. (3.2.26c), Eq. (3.2.27) gives

$$\rho_j(E_0) = \frac{1}{2\pi^2}\left(\frac{2m_r}{\hbar^2}\right)^{3/2}(E_0 - E_g)^{1/2} \tag{3.2.28}$$

For our purposes it is better to introduce the joint density of states, $\rho_j(\nu_0)$, with respect to the transition frequency $\nu_0 = E_0 / h$. Since $\rho_j(\nu_0)\, d\nu_0 = \rho_j(E_0)\, dE_0$, we get from Eq. (3.2.28)

$$\rho_j(\nu_0) = \frac{4\pi}{h^2}(2m_r)^{3/2}\left[h\nu_0 - E_g\right]^{1/2} \tag{3.2.29}$$

Consider now the elemental number of transitions $dN = \rho_j(\nu_0)\, d\nu_0$ whose transition frequency lies between ν_0 and $\nu_0 + d\nu_0$. For absorption to occur, the lower level, of energy E'_1, must be occupied by an electron while the upper level, of energy E'_2, must be empty. The number of transitions available for absorption will thus be

$$dN_a = (dN)\, f_v(E'_1)\left[1 - f_c(E'_2)\right] \tag{3.2.30}$$

where $f_v(E'_1)$ is the probability that the lower level is full while $\left[1 - f_c(E'_2)\right]$ is the probability that the upper level is empty. Note that a general case of equilibrium within each band is assumed so that $f_v(E'_1)$ and $f_c(E'_2)$ are obtained by Eqs. (3.2.11b) and (3.2.11a) with E'_v and

E'_c being substituted by E'_1 and E'_2 respectively. To calculate the net absorption, we must also take into account the process of stimulated emission between the same two levels. This will occur when the upper state is full while the lower state is empty. The number of transitions available for stimulated emission will then be

$$dN_{se} = (dN) f_c\left(E'_2\right)\left[1 - f_v\left(E'_1\right)\right] \tag{3.2.31}$$

Once the elemental numbers of available transitions for absorption and stimulated emission are calculated, the absorption coefficient at frequency ν is obtained, via (2.4.32), as $d\alpha = \sigma(\nu - \nu_0)\,(dN_a - dN_{se})$, where $\sigma = \sigma_h$ is the homogeneous cross section for the $E'_1 \to E'_2$ transition. From Eq. (3.2.24) we then get

$$d\alpha = \left(\frac{2\pi^2\nu}{n\varepsilon_0 ch}\right)\frac{\mu^2}{3}g(\nu - \nu_0)\,\rho_j(\nu_0)\left[f_v\left(E'_1\right) - f_c\left(E'_2\right)\right] d\nu_0 \tag{3.2.32}$$

The overall absorption coefficient at frequency ν is obtained from Eq. (3.2.32) by integrating over all transition frequencies ν_0. If we assume that $g(\nu - \nu_0)$ vs ν_0 is a much narrower function than both $\rho_j(\nu_0)$ and $(f_c - f_v)$, than $g(\nu - \nu_0)$ can be approximated by a δ function, $\delta = \delta(\nu - \nu_0)$. We then get

$$\alpha = \left(\frac{2\pi^2\nu}{n\varepsilon_0 ch}\right)\frac{\mu^2}{3}\rho_j(\nu)\left[f_v\left(E'_1\right) - f_c\left(E'_2\right)\right] \tag{3.2.33}$$

where E'_2 and E'_1 are now the energies of the two levels whose energy difference is $h\nu$. They can be readily calculated from Eq. (3.2.36) by substituting $h\nu_0$ with $h\nu$.

According to Eq. (3.2.33), the absorption coefficient $\alpha = \alpha(\nu)$ can be written as

$$\alpha = \alpha_0\left[f_v(E'_1) - f_c(E'_2)\right] \tag{3.2.34}$$

where

$$\alpha_0 = \left(\frac{2\pi^2\nu}{n\varepsilon_0 ch}\right)\frac{\mu^2}{3}\rho_j(\nu) \tag{3.2.35}$$

The meaning of $\alpha_0 = \alpha_0(\nu)$ is understood when we consider a semiconductor in overall thermal equilibrium at $T = 0\,\mathrm{K}$. The quasi-Fermi levels coincide in this case with the Fermi-level and, if this level is within the energy gap, one has $f_v\left(E'_1\right) = 1$ and $f_c\left(E'_2\right) = 1$. One then has $\alpha(\nu) = \alpha_0(\nu)$ which is the maximum absorption coefficient that the semiconductor can have at frequency ν. Note that, for an intrinsic semiconductor and assuming $E_g >> kT$ as is the case for all *III–V* semiconductors, we still have $f_v\left(E'_1\right) \cong 1$ and $f_c\left(E'_2\right) \cong 0$ i.e. $\alpha \cong \alpha_0$ even at room temperature. From Eq. (3.2.35), with the help of Eq. (3.2.29) with ν_0 substituted by ν, we then get

$$\alpha \cong \alpha_0 = \frac{\pi^3\nu}{n\varepsilon_0 ch^3}\frac{\mu^2}{3}(2m_r)^{3/2}\left[h\nu - E_g\right]^{1/2} \tag{3.2.36}$$

The frequency behaviour of $\alpha(\nu)$ is then determined, to a good approximation, simply by the frequency behaviour of $\left(h\nu - E_g\right)^{1/2}$.

Example 3.6. *Calculation of the absorption coefficient for GaAs.* As an approximation, we will assume the frequency ν, appearing in the first term of the right hand side of Eq. (3.2.36), to have the value $\nu \cong E_g / h = 3.43 \times 10^{14}$ Hz, where the energy gap E_g is taken to be 1.424 eV. We will also take $m_v = 0.46\, m_0$ and $m_c = 0.067\, m_0$, so that $m_r = 0.059\, m_0 = 5.37 \times 10^{-32}$ Kg. To calculate the average dipole moment $\mu_{av} = \left[\mu^2 / 3\right]^{1/2}$ we note that the accurate value of the average electron moment M_{av} was recently shown to be such that $M_{av}^2 = 3.38\, m_o E_g$.[5] Since the relation between average dipole moment and electron momentum is $M_{av} = m_0\omega\, |\mu_{av}| / e$,[5] we get $\mu_{av} = e\left[3.38 E_g / m_0\right]^{1/2} / 2\pi\nu \cong 0.68 \times 10^{-25}$ C × m. We see that, if we write $r_{av} = \mu_{av} / e$, then $r_{av} \cong 0.426$ nm. Substitution into Eq. (3.2.36) of the values for ν and μ_{av} given above together with the value $n = 3.64$ for the refractive index, gives $\alpha_0 = 19{,}760 \left[h\nu - E_g\right]^{1/2}$, where α_0 is expressed in cm^{-1} and the energy in eV. The absorption coefficient, as calculated from the latter expression, is plotted vs $h\nu - E_g$ in Fig. 3.16. One notes that, when $h\nu$ exceeds the energy gap by only 10 meV, the absorption coefficient already reaches very large value ($\approx$2, 000 cm^{-1}).

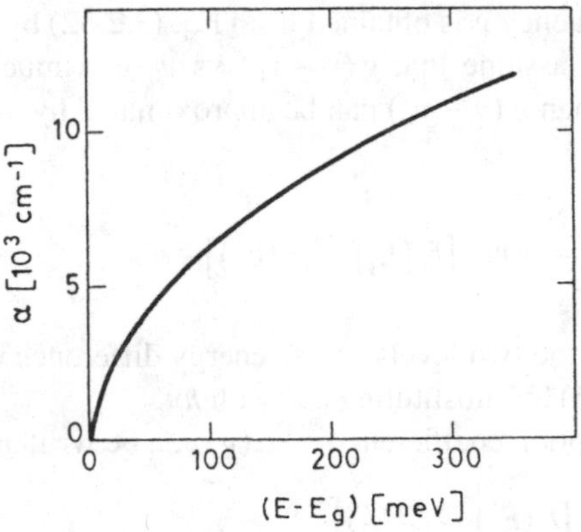

FIG. 3.16. Idealized plot of the absorption coefficient, α, vs the difference between the photon, E, and gap energy, E_g, for an intrinsic GaAs bulk semiconductor.

Consider next the case of the gain coefficient of an "inverted" semiconductor. One can readily see that the previous considerations remain valid provided we interchange the suffices v and c. Thus, from Eq. (3.2.34), the gain coefficient is seen to be given by

$$g = \alpha_0\left[f_c(E_2') - f_v(E_1')\right] \tag{3.2.37}$$

It then follows that, at any transition frequency, the maximun gain coefficient is attained at $T = 0$ K and equals α_0. Note from Eq. (3.2.37) that, for any temperature, the condition for net gain is $f_c\left(E_2'\right) > f_v\left(E_1'\right)$. With the help of Eqs. (3.2.11a) and (3.2.11b) one can readily show that this implies

$$E_2' - E_1' < E_{F_c}' - E_{F_v}' \tag{3.2.38}$$

This is a necessary condition for net gain and was originally derived by Bernard and Duraffourg.[6] One can see that the factor $f_c\left(E_2'\right) - f_v\left(E_1'\right)$ in Eq. (3.2.37) originates from the term

$f_c(E_2')[1-f_v(E_1')]-f_v(E_1')[1-f_c(E_2')]$ which gives the difference in probability between stimulated emission and absorption. Thus the Bernard–Duraffourg condition states that the stimulated events must exceed the absorption events and, in this respect, is seen to be equivalent to the $N_2 > N_1$ condition for a simple two-level atomic system. The relation Eq. (3.2.38) can also be understood graphically if we consider the simple case of $T = 0\,\mathrm{K}$. For a given level of electron-hole injection the position of the quasi-Fermi levels will be as shown in Fig. 3.17, where the dashed zones are filled with electrons and the clear zones are unoccupied by electrons (i.e. full of holes). The condition Eq. (3.2.38) then simply implies that level 2 must belong to the full zone while level 1 must belong to the empty zone in Fig. 3.17. The actual derivation of the Bernard–Duraffourg condition shows however that Eq. (3.2.38) is valid at any temperature.

It is worthwhile remembering, at this point, that $E_2' - E_1' = h\nu$ and that one must also have $h\nu > E_g$. Then from Eq. (3.2.38) we get

$$E_g \le h\nu \le E'_{F_c} - E'_{F_v} \tag{3.2.39}$$

which establishes the gain bandwidth of the semiconductor. According to Eq. (3.2.39), to have gain at any frequency, one must have $E'_{F_c} - E'_{F_v} \ge E_g$ and the limiting case

$$E'_{F_c} - E'_{F_v} = E_g \tag{3.2.40}$$

is called the *transparency condition*. In this case one has $g = 0$ at $\nu = E_g/h$. To achieve this condition we must inject a density of electrons in the conduction band (and holes in the valence band) which is called the *transparency density* and which will be indicated as N_{tr}.

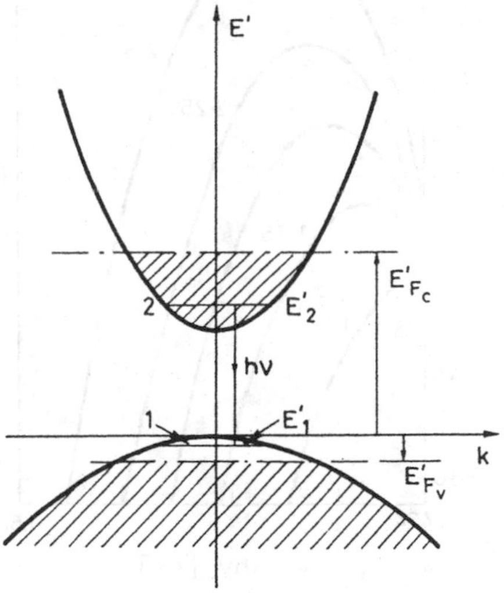

FIG. 3.17. Graphical illustration of the Bernard–Duraffourg condition for achieving net gain in a bulk semiconductor at $T = 0\,\mathrm{K}$.

Example 3.7. *Calculation of the transparency density for GaAs.* It is first convenient to transform Eq. (3.2.40) into the unprimed energy axes of Fig. 3.9a. According to Eq. (3.2.3) one can write $E'_{F_c} = E_g + E_{F_c}$ and $E'_{F_v} = -E_{F_v}$, and Eq. (3.2.40) transforms to $E_{F_c} + E_{F_v} = 0$. From Fig. 3.15a one sees that E_{F_c} / kT is a function of (N/N_c) i.e. one can write $E_{F_c} / kT = f(N / N_c)$. Similarly, one can write $E_{F_v} / kT = f(N / N_v)$ and the transparency condition becomes

$$f(N_{tr} / N_c) + f(N_{tr} / N_v) = 0 \tag{3.2.41}$$

To obtain N_{tr} from Eq. (3.2.41) for GaAs, we have plotted in Fig. 3.15b, as a dashed line, the function $(E_{F_c} / kT) + (E_{F_v} / kT)$ vs N. The curve is obtained as the sum, at each carrier concentration N, of the values given by the two continuous curves of the figure. According to Eq. (3.2.41) we can now say that the transparency density, N_{tr}, is the carrier concentration at which the dashed line of Fig. 3.15b crosses the $(E_{F_c} / kT) + (E_{F_v} / kT) = 0$ horizontal line. From Fig. 3.15b we get $N_{tr} = 1.2 \times 10^{18}\ \mathrm{cm}^{-3}$.

When the density of injected electrons, N, exceeds the transparency density, we have $E'_{F_c} - E'_{F_v} > E_g$, and, according to Eq. (3.2.39), net gain will occur for a photon energy between E_g and $E'_{F_c} - E'_{F_v}$. The gain coefficient vs photon energy, as calculated using Eq. (3.2.37), is shown in Fig. 3.18 for GaAs, using the injected carrier density, N, as a parameter. One notes that, upon increasing the carrier density, the difference in quasi-Fermi energies, $E'_{F_c} - E'_{F_v}$, increases and this results in a corresponding increase of the gain bandwidth. This bandwidth,

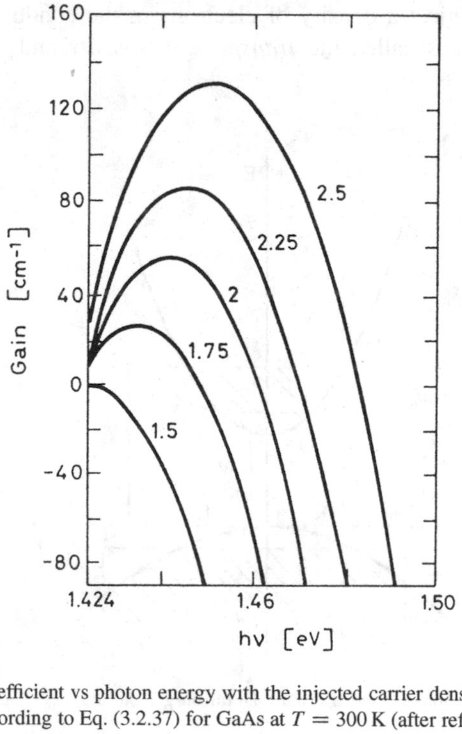

FIG. 3.18. Plot of the gain coefficient vs photon energy with the injected carrier density, N, as a parameter (in units of $10^{18}\ \mathrm{cm}^{-3}$) as expected according to Eq. (3.2.37) for GaAs at $T = 300$ K (after ref.,[15] by permission).

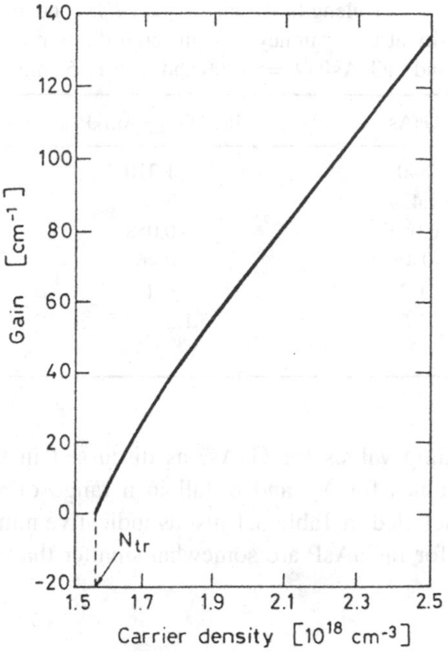

FIG. 3.19. Plot of the peak gain coefficient vs the injected carrier density for GaAs (after ref.,[15] by permission).

even at the highest carrier injections considered in the figure is, however, smaller than 0.07 eV i.e. it is a small fraction of the energy gap. One can also observe from Fig. 3.18 that the peak gain of each curve increases with increasing N. Again for GaAs, Fig. 3.19 shows the plot of this peak gain coefficient vs the density of injected electrons. For typical gain coefficients of interest for semiconductors lasers ($20 \le g \le 80\,\text{cm}^{-1}$), the plot of Fig. 3.19 can be approximated by a linear relation i.e. one can write

$$g = \sigma (N - N_{tr}) \tag{3.2.42}$$

where $\sigma \cong 1.5 \times 10^{-6}\,\text{cm}^2$ for GaAs. It should be noted that σ has some analogy to the gain cross section defined for atomic systems [compare Eqs. (3.2.42) with (2.4.35)]. As already mentioned, however, the concept of cross section is not appropriate for a delocalized wave-function such as that of an electron in a semiconductor. For this reason, since from Eq. (3.2.42) one has $\sigma = dg / dN$, σ is often referred to as the differential gain of the semiconductor. We will still retain the notation of σ for this differential gain, however, as a reminder of the fact that σ has the dimension of an area.

Most of the examples which have been discussed in this section refer to the particular case of a GaAs semiconductor. However, many other materials are also of interest as laser materials, a notable example being the quaternary alloy $In_{1-x}Ga_xAs_yP_{1-y}$ which, depending upon the composition indices x and y, covers the so-called second and third communication windows of optical fibers (1,300 nm $\le \lambda \le$ 1,600 nm). For the purpose of comparison, Table 3.1 shows the values of E_g, m_c / m_0, m_{hh} / m_0, N_{tr} and σ, for $In_{0.75}Ga_{0.25}As_{0.55}P_{0.45}$ ($\lambda \cong$ 1,300 nm) and for $In_{0.6}Ga_{0.4}As_{0.88}P_{0.12}$ ($\lambda \cong$ 1,550 nm),[7]

TABLE 3.1. Values of emission wavelength, λ, energy gap, E_g, conduction band electron mass, m_c, heavy-hole, m_{hh}, carrier density at transparency, N_{tr}, material differential gain, σ, and lifetime τ, for GaAs ($\lambda \cong 850$ nm), and InGaAsP ($\lambda = 1{,}300$ and $\lambda = 1{,}550$ nm) bulk semiconductors

	GaAs	$In_{0.73}Ga_{0.27}As_{0.6}P_{0.4}$	$In_{0.58}Ga_{0.42}As_{0.9}P_{0.1}$
λ [nm]	840	1,310	1,550
E_g [eV]	1.424	0.96	0.81
m_c/m_0	0.067	0.058	0.046
m_{hh}/m_0	0.46	0.467	0.44
N_{tr} [10^{18} cm^{-3}]	1.2	1	1
σ[10^{-16} cm^2]	1.5	$1.2 \div 2.5$	$1.2 \div 2.5$
τ[ns]	3	4.5	4.5

as well as the corresponding values for GaAs, as discussed in this section. It should be noted that the reported values for N_{tr} and σ fall in a range of values reported for these semiconductors and are included in Table 3.1 just as indicative numbers. It does seem, however, that both N_{tr} and σ for InGaAsP are somewhat smaller than the corresponding values for GaAs.

3.2.6. Spontaneous Emission and Nonradiative Decay

Let us first consider the spontaneous emission process and define the spectral rate R_ν so that $R_\nu d\nu$ represents the number of spontaneous emission events per unit time and volume which result in light emitted with frequency between ν and $\nu + d\nu$. To calculate R_ν consider first the transitions, $\rho_j(\nu_0)\, d\nu_0$, whose transition frequencies lie between ν_0 and $\nu_0 + d\nu_0$. They will give an elemental contribution, $d R_\nu$, to R_ν given by $dR_\nu = A_{21} g(\nu - \nu_0) \times \{f_c(E'_{20})\left[1 - f_v(E'_{10})\right]\}\ \rho_j(\nu_0)\, d\nu_0$ where $A_{21} = A_{21}(\nu_0)$ is the rate of spontaneous emission between the two levels and $g(\nu - \nu_0)$ is the lineshape function of the transition. It should be noted that $\rho_j(\nu_0)$ has been multiplied by $f_c(E'_{20})\left[1 - f_v(E'_{10})\right]$ since, as for stimulated emission, spontaneous emission can only occur between an occupied upper state and an empty lower state. The total spectral rate R_ν is then obtained by integrating the above expression over all transition frequencies ν_0. Thus

$$R_\nu = \int A_{21} g(\nu - \nu_0)\{f_c(E'_2)\left[1 - f_v(E'_1)\right]\}\ \rho_j(\nu_0) d\nu_0 \tag{3.2.43}$$

In the limit where $g(\nu - \nu_0)$ can be considered a much narrower function of ν_0 than all other functions in the integrand, $g(\nu - \nu_0)$ can be approximated by a δ function $\delta(\nu - \nu_0)$ and Eq. (3.2.43) reduces to

$$R_\nu = A_{21}\{f_c(E'_2)\left[1 - f_v(E'_1)\right]\}\rho_j(\nu) \tag{3.2.44}$$

where A_{21} is the spontaneous emission rate for $\nu_0 = \nu$ and E'_2 and E'_1 are now the energy levels corresponding to a transition frequency ν. As an example, Fig. 3.20 shows the qualitative behaviour of R_ν vs the photon energy $h\nu$ for an electron injection rate exceeding the rate for

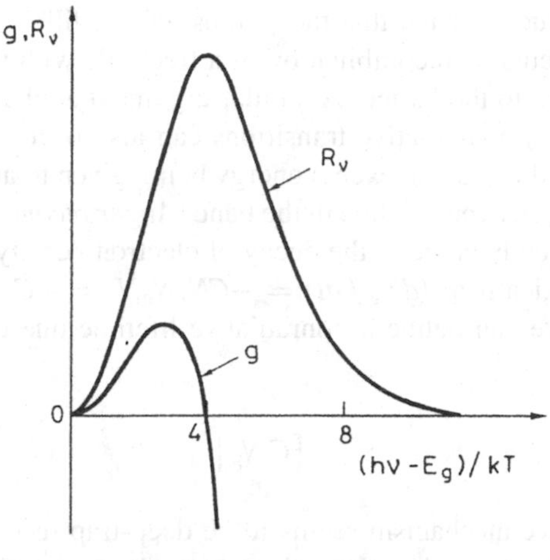

FIG. 3.20. Qualitative behaviour of the spontaneous emission spectra, R_ν, and of optical gain, g, at a given value of the injected carrier density (after ref.[11] by permission).

transparency and by assuming A_{21} independent of ν. In the same figure the gain coefficient, as calculated by Eq. (3.2.37), is also indicated for comparison. One can observe that the emission spectrum, unlike the case for atomic systems, is now different from and generally wider than the gain spectrum. This is because R_ν is proportional to $f_c\left(E_2'\right)\left[1-f_v\left(E_1'\right)\right]$ while α_g is proportional to $f_c\left(E_2'\right)-f_v\left(E_1'\right)$.

Once the spectral rate, R_ν, of spontaneous emission is calculated, the total rate R is obtained by integrating R_ν over all emission frequencies. Thus

$$R=\int A_{21}f_c(E_2')\left[1-f_v(E_1')\right]\rho_j(\nu)\,d\nu \tag{3.2.45}$$

In practice, however, one often makes use of the phenomenological relation

$$R=BN_eN_h\cong BN_e^2 \tag{3.2.46}$$

where B is a suitable constant. A justification of Eq. (3.2.46) can be given by assuming that any electron can recombine with any hole, which implies that the k-selection rule does not strictly hold.[9] We will not give any further discussion of this question which has to do with the so called band tails in a semiconductor,[10] and we will take Eq. (3.2.46) as a phenomenological relation which holds well at the electron and hole densities of interest. We note that, according to the definition of R, one has $(dN_e\,/\,dt)=-R$. We can therefore define a radiative lifetime τ_r such that $R=N_e\,/\,\tau_r$ and thus write

$$\tau_r=(BN_e)^{-1} \tag{3.2.47}$$

Let us next consider nonradiative transitions. They generally occur at deep impurity centers in which a carrier, electron or hole, is trapped (*deep trap recombination*). Consider for instance an n-type semiconductor. At sufficiently high doping values, the Fermi level will be close enough to the conduction band that these traps will be filled with electrons. A nonradiative transition then occurs by recombination of a free hole with this trapped electron, the excess energy being given to the lattice. A similar argument applies for *p*-type-doping. For small gap semiconductors, nonradiative transitions can also occur by direct recombination of untrapped electrons and holes, the excess energy being given to another electron (or hole) which gets excited to a higher energy state in the band (*Auger recombination*[12]). Since Auger recombination is a three-body process, the decay of electron density due to this procees can be written phenomenologically as $(dN_e/dt) = -CN_eN_hN_e = -CN_e^3$, where C is a suitable constant. Accordingly, we can define a nonradiative lifetime due to Auger recombination, τ_A, as

$$\tau_A = \left[CN_e^2\right]^{-1} \tag{3.2.48}$$

The dominant nonradiative mechanism seems to be deep-trap recombination for GaAs and Auger recombination in long wavelength semiconductor laser materials such as InGaAsP.

Example 3.8. *Radiative and nonradiative lifetimes in GaAs and InGaAsP.* For GaAs, we will take $B \cong 1.8 \times 10^{-10}\,\text{cm}^3\,\text{s}^{-1}$ and $N_e \cong N_{tr} = 1.2 \times 10^{18}\,\text{cm}^{-3}$. We then get $\tau_r = 1/BN_{tr} \cong 4.6\,\text{ns}$, to be compared with the measured overall lifetime, at transparency, of $\tau \cong 3\,\text{ns}$ ($T = 300\,\text{K}$). Since $\tau^{-1} = \tau_r^{-1} + \tau_{nr}^{-1}$, where τ_{nr} is the lifetime due to the nonradiative process, we can infer a nonradiative lifetime τ_{nr}, in this case due to deep-trap recombination, of about 9 ns. For InGaAsP at $\lambda = 1,300\,\text{nm}$ we take $B = 2 \times 10^{-10}\,\text{cm}^3\,\text{s}^{-1}$, $N_e \cong N_{tr} \cong 1 \times 10^{18}\,\text{cm}^3$ and $C \cong 3 \times 10^{-29}\,\text{cm}^6\,\text{s}^{-1}$. We get from Eq. (3.2.47) $\tau_r \cong 5\,\text{ns}$ and from Eq. (3.2.48) $\tau_A \cong 33.3\,\text{ns}$ which gives an overall lifetime, τ, in agreement with the measured value [$\tau \cong 5\,\text{ns}$ at $T = 300\,\text{K}$].

3.2.7. Concluding Remarks

We have seen in this section that the phenomena leading to radiative and nonradiative transitions in a bulk semiconductor are notably more complicated than those occurring for isolated atoms or ions, which were considered in the previous Chapter. From a practical viewpoint, however, the most important physical parameters which are needed to predict laser behaviour are the differential gain σ, the transparency density N_{tr}, and the overall lifetime τ for spontaneous decay (resulting from both radiative and non radiative processes). For GaAs and for the InGaAs alloys considered here, these quantities can be obtained from Table 3.1. It should be reminded that the lifetime depends on the carrier concentration and the values reported in the table refer to a concentration equal to the transparency density.

In concluding this section we also recall that the gain value given by Eq. (3.2.37) refers to a semiconductor of large dimensions (bulk semiconductor). For this reason, the gain values quoted in Table 3.1 are often referred to as the *material gain*. The actual gain in a double-heterostructure laser is smaller than this and it is determined by the ratio of the transverse dimension of the active layer to that of the cavity mode. This gain, often called the *modal gain*, thus depends on details of the laser configuration and will be considered in the relevant section of Chap. 9 dealing with semiconductor lasers.

3.3. SEMICONDUCTOR QUANTUM WELLS

In a quantum well (QW) semiconductor, a very thin layer ($L_z \cong 5 \div 20\,\text{nm}$) of a smaller bandgap material, E_{g_1}, is sandwiched between two layers of larger bandgap material, E_{g_2}, (Fig. 3.21a). Technically, this is done by the sophisticated techniques of Molecular Beam Epitaxy (*MBE*) or Metallo-Organic Chemical Vapor Deposition (*MOCVD*). Since $E_{g_1} < E_{g_2}$, potential wells will be established for the electrons at the top of the valence band, *v.b.*, and, for the holes, at the bottom of the conduction band, *c.b.*, (Fig. 3.21b). Due to the electron and hole confinement in these potential wells and since the semiconductor dimension is now comparable to the electron and hole DeBroglie wavelength, the energy levels of electrons and holes show very marked quantum size effects. Furthermore, due to the small thickness of the layer, one can allow the lattice constants for the two materials to differ significantly, resulting in strain being developed within the thin quantum layer. The strain changes the quantum properties of the QW semiconductor considerably and, in particular, it changes the effective masses. The quantum size effects and, for a strained QW, the change in effective masses, result in the optical properties of semiconductor's QW being markedly different from those of the corresponding bulk material. In particular, the material differential gain increases considerably. The transparency electron density remains comparable, for an unstrained QW, while it shows a sizeable decrease for a strained QW. The advantages that these improved properties imply in terms of lowering the laser threshold and increasing the modal gain will be discussed in the relevant section on semiconductor lasers in Chap. 9. Here we merely limit ourselves to pointing out that semiconductor quantum wells, of either strained or unstrained type, have become the most widely used semiconductor laser materials.

3.3.1. Electronic States

To calculate the energy levels of the electrons and holes in the potential wells of Fig. 3.21.b, one needs to know how the difference in bandgap energy $\Delta E_g = E_{g_2} - E_{g_1}$ is partitioned between the well in the conduction band (ΔE_c) and that in the valence band (ΔE_v). This problem (the so-called band offset) involves complicated details of physics of

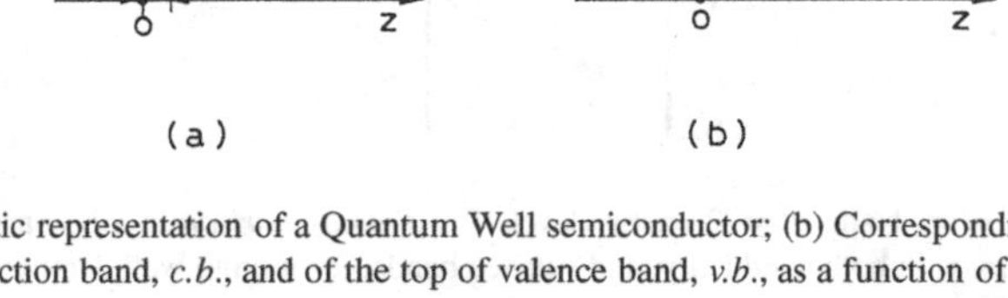

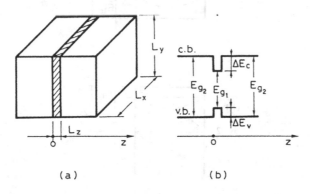

FIG. 3.21. (a) Schematic representation of a Quantum Well semiconductor; (b) Corresponding plot of the energy of the bottom of the conduction band, *c.b.*, and of the top of valence band, *v.b.*, as a function of the *z*-coordinate of (a).

semicondutors. Experimentally, for two of the most important types of QW system, one finds: (1) $\Delta E_c = 0.67\ \Delta E_g$ and $\Delta E_v = 0.33\ \Delta E_g$. for an AlGaAs/GaAs/AlGaAs QW. (2) $\Delta E_c = 0.39\ \Delta E_g$ and $\Delta E_v = 0.61\ \Delta E_g$ for a InP/InGaAsP/InP QW.

To calculate the energy levels of both electrons and holes in the corresponding QW, we will make the much simplified assumption of infinite well depth [i.e. $(\Delta E_c, \Delta E_v) \rightarrow \infty$]. The potential wells will then appear as in Fig. 3.22. We take the z-axis orthogonal to the well with the origin at one well interface. According to Eq. (3.2.1), the Bloch wavefunctions, both in the conduction and valence bands, can be written as

$$\psi_{c,v}(\mathbf{r}) = u(\mathbf{r}_\perp)\, e^{j\mathbf{k}_\perp \cdot \mathbf{r}_\perp} \sin(n\pi z / L_z) \tag{3.3.1}$$

where $\mathbf{r}_\perp$ and $\mathbf{k}_\perp$ are the components of $\mathbf{r}$ and $\mathbf{k}$ in the well plane (the x,y plane) and where n is a positive integer. Note that, written in this way, $\psi_{c,v}$ already satisfies the boundary conditions $\psi_{c,v} = 0$ for $z = 0$ and $z = L_z$ i.e. at the two well boundaries. If we set similar periodic conditions along the x and y axes, we get

$$k_x = (l\pi / L_x) \tag{3.3.2a}$$

$$k_y = (m\pi / L_x) \tag{3.3.2b}$$

where l and m are also positive integers. Note the difference between Eqs. (3.3.2) and (3.2.4) which essentially reflects the fact that we are limiting ourselves, in this case, to positive numbers. Of course, one can also write the boundary condition as in Eq. (3.2.4) i.e. allowing for

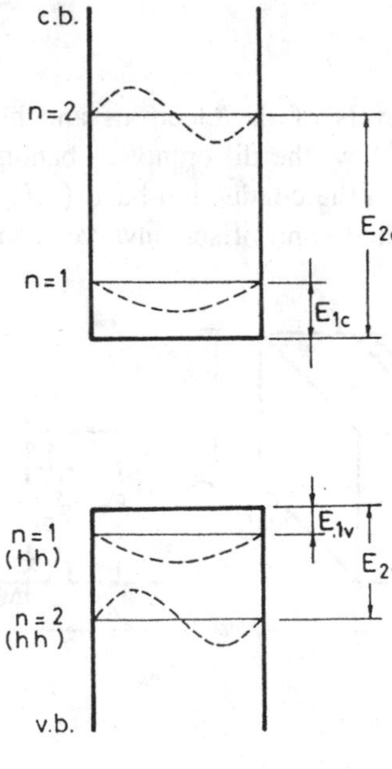

FIG. 3.22. Plots of the $n = 1$ and $n = 2$ energy levels (*continuous horizontal lines*) and of the corresponding eigenfunctions (*dashed lines*) in both conduction and valence bands, for infinite well depths.

both positive and negative integers, and still obtain the same final results. Within the parabolic band approximation, the energy eigenvalues for either valence or conduction bands are

$$E_{c,v} = \frac{\hbar^2 k_\perp^2}{2m_{c,v}} + \frac{n^2\hbar^2\pi^2}{2m_{c,v}L_z^2} = \frac{\hbar^2 k_\perp^2}{2m_{c,v}} + n^2 E_{1c,v} \tag{3.3.3}$$

where $k_\perp^2 = k_x^2 + k_g^2$, $m_{c,v}$, is the electron mass in the conduction band or the hole mass in the valence band (only the heavy hole mass is considered, for simplicity) and where we have indicated by $E_{1c,v}$ the energy of the first quantum-well state ($n = 1$) for either conduction or valence bands, as given by

$$E_{1c,v} = \frac{\hbar^2\pi^2}{2m_{c,v}L_z^2} \tag{3.3.4}$$

One should note that, for both Eqs. (3.3.3) and (3.3.4), the energy is measured from the bottom of the conduction band upwards, for the electrons, and from the top of the valence band, downwards, for the holes. One should also note that, for finite depth of the potential wells, the electrons are not totally reflected at the well interfaces, i.e. the wavefunction is not zero at the interfaces, as assumed in Eq. (3.3.1). The wavefunction will then penetrate into the barrier layer and the expressions for the wavefunctions and energy eigenvalues become more complicated.[14] We will not consider this case any further, since it only produces quantitative rather than qualitative changes to the results which follow.

To discuss Eqs. (3.3.3) and (3.3.4) let us first consider the case of electrons with zero transverse momentum [$k_\perp = 0$]. The first two energy levels ($n = 1$ and $n = 2$) for both the conduction and valence bands are shown as solid horizontal lines in Fig. 3.22 while the corresponding eigenfunctions are shown as dashed lines. One can see that, according to Eq. (3.3.3), one has $E_{2c} = 4E_{1c}$, the same relation holding also for the valence band. If we now consider electrons with $k_\perp > 0$, the energy E vs $k_\perp$ relations, for each of the $n = 1$, $n = 2$, etc., states considered before, will be as shown in Fig. 3.23a. One sees that individual sub-bands are now introduced in the conduction and valence bands. In the same figure, the available states, as obtained via Eq. (3.3.2), are shown as dots in the valence band and open circles in the conduction band. Note finally that, when dealing with transitions between valence and conduction sub-bands, an alternative energy scale, that we shall call E', starting e.g. from the top of the valence band and increasing upwards may, sometimes, be more convenient (see Fig. 3.23b). The transformation between the primed, E', and unprimed, E, energy scales will again be given by Eq. (3.2.3), where E_c and E_v are now expressed by Eq. (3.3.3).

Example 3.9. *Calculation of the first energy levels in a GaAs / AlGaAs quantum well.* Let us take $L_z = 10$ nm and assume that the electron and hole (heavy hole) masses in the GaAs well are the same as those of the bulk material, i.e. $m_c = 0.067\, m_0$ and $m_v = m_{hh} = 0.46\, m_0$. From Eq. (3.3.4) we get $E_{1c} = 56.2$ meV and $E_{1v} = 8$ meV. If the confinment layer, at both sides, is $Al_{0.2}Ga_{0.8}As$ then $E_{g_2} = 1.674$ eV. Since the bandgap of GaAs is $E_{g_1} = 1.424$ eV, we obtain $\Delta E_g = 250$ meV and thus $\Delta E_c = 0.65\, E_g = 162.5$ meV and $\Delta E_v = 0.35\, \Delta E_g = 87.5$ meV. Since E_{1c} is comparable to ΔE_c, the assumption of an infinite well is not a good approximation in this case. Taking barrier penetration into account, the actual values can be obtained from e.g. Fig. 9.1 in ref[3] as $E_{1c} \cong 28$ meV and $E_{1v} \cong 5$ meV.

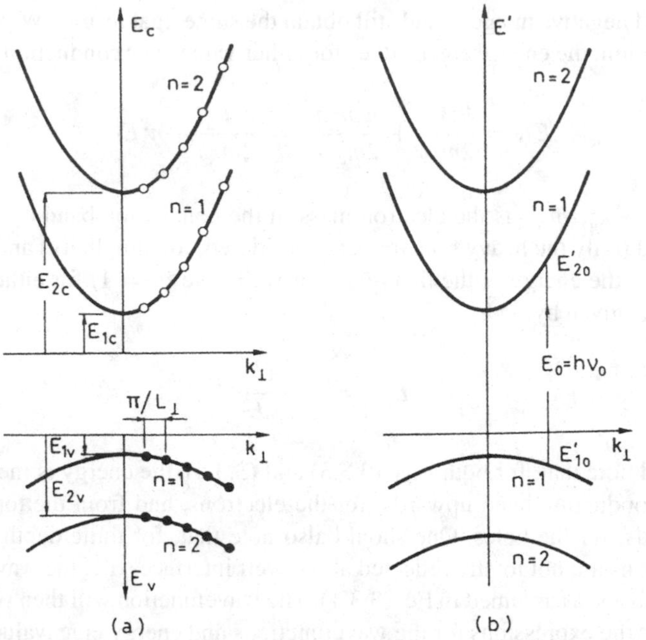

FIG. 3.23. Energy vs $k_\perp$ relations of the $n = 1$ and $n = 2$ subbands of both valence and conduction bands for a quantum well semiconductor. In (a) the origin of the energy axis for the conduction subbands is taken at the bottom of the conduction band of the bulk material and increases upwards. The origin of the energy axis for the valence subbands is taken at the top of the valence band and it increases downwards. In (b) the energy axis is the same for all subbands, the origin is taken at the top of the valence band and energy increases upwards.

3.3.2. Density of States

Let us refer to Fig. 3.24 where the allowed states, as obtained via Eq. (3.3.2), are indicated as dots in the (k_x, k_y) plane (compare with Fig. 2.2). One can see that only the allowed states of the $n = 1$ level are indicated. Indeed, note that one typically has $L_z = 10$ nm while L_x and L_y may range between 10 and 100 μm, i.e. they are 10^3 to 10^4 times larger than L_z. Thus the separation in Δk_z between two successive states along the k_z-axis ($\Delta k_z = \pi / L_z$) is about 10^3 to 10^4 times larger than the separation between successive states along the k_x or k_y direction. The allowed states are now lying in well-separated planes orthogonal to the k_z-axis so it is appropriate now to calculate the density of states in each of these planes. Thus, let $N(k_\perp)$ be the number of states, in each plane e.g. in the $n = 1$ plane of Fig. 3.24, whose transverse vector is between 0 and $k_\perp$. According to the discussion relating to Fig. 2.2, it is readily seen that $N(k_\perp)$ is given by (1/4) the area of the circle of radius $k_\perp$ divided by the area, $\Delta k_x \Delta k_y$, of the unit cell, and then multiplied by a factor 2 to account for the two possible spin orientations in each state. We thus get

$$N(k_\perp) = \frac{2\,(1/4)\,\pi\,k_\perp^2}{\Delta k_x \Delta k_y} = \frac{k_\perp^2}{2\pi} A_\perp \tag{3.3.5}$$

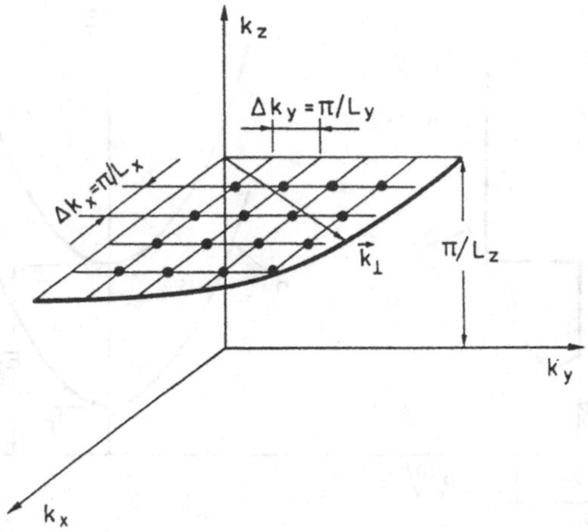

FIG. 3.24. Representation in k_x, k_y, k_z space of the allowed states for the $n = 1$ subband.

where $A_\perp = L_x L_y$ is the transverse area of the quantum well. The number of states per unit $k_\perp$ and per unit area is then obtained as

$$\rho_k^{2D} = \frac{dN(k_\perp)}{A_\perp dk_\perp} = \frac{k_\perp}{\pi} \tag{3.3.6}$$

One can see that, compared to the case of a bulk semiconductor, ρ_k now gives the number of states per unit area rather than per unit volume. As a reminder of this feature, we have used the superscript $2D$ to draw attention to the fact that we are now in a two-dimensional rather than in a three-dimensional situation. One should also note that Eq. (3.3.6) holds both for the valence and conduction bands.

To obtain the density of states in energy coordinates, we write, e.g. for the conduction band, $\rho_c^{2D} dE_c = \rho_k^{2D} dk_\perp$. From Eq. (3.3.6) we then get

$$\rho_c^{2D} = k_\perp dk_\perp / \pi dE_c \tag{3.3.7}$$

From (3.3.3), for the example of the $n = 1$ subband, we have

$$k_\perp^2 = \left(2m_c / \hbar^2\right)(E_c - E_{1c}) \tag{3.3.8}$$

The quantity $k_\perp dk_\perp$ in Eq. (3.3.7) is readily obtained by differentiating both sides of Eq. (3.3.8). Equation (3.3.7) then gives

$$\rho_c^{2D} = m_c / \pi \hbar^2 \tag{3.3.9}$$

One observes that ρ_c^{2D} is independent of the value of $k_\perp$ i.e. of the transverse part of the energy $\hbar^2 k_\perp^2 / 2m_c$ [see Eq. (3.3.3)]. This is shown graphically in Fig. 3.25a where the quantity ρ_c^{2D} / L_z is plotted vs the electron energy E_c (continuous line). Energy is measured from the bottom of the conduction band and the function is plotted for $E_{1c} \le E_c \le E_{2c}$, where E_{2c} is the

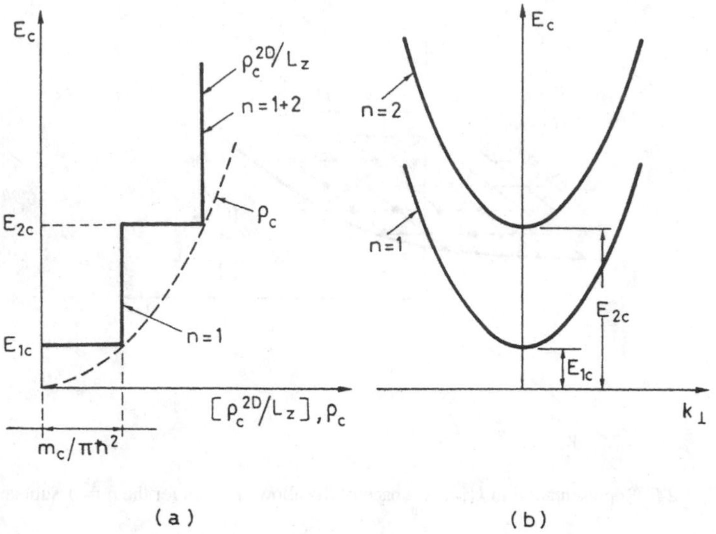

FIG. 3.25. (a) Plot of the quantum well density of states in the conduction band, ρ_c^{2D}, normalized to the well thickness, L_z, as a function of the state energy E_c [*staircase, solid line*]. In the same figure, the plot of the density of states for the corresponding bulk semiconductor, ρ_c, is also shown as a dashed line. (b) Plots of the E_c vs $k_\perp$ relations for the $n = 1$ and $n = 2$ conduction subbands.

energy of the $n = 2$ subband. In fact, for $E_c \geq E_{2c}$, one must also take into account the states lying in the plane $k_z = 2\pi / L_z$ (not shown in Fig. 3.24). The density of these states, however, is the same as that for the $n = 1$ plane i.e. it is again given by Eq. (3.3.9). For $E_c \geq E_{2c}$ the overall density will then be the sum of the densities of both $n = 1$ and $n = 2$ subbands. The corresponding curve is given by the solid-line step function labelled $n = 1 + 2$ in Fig. 3.25a. For the sake of comparison we also show in Fig. 3.25a, as a dashed curve, the density of states for the same semiconductor material in bulk form, ρ_c, as given by Eq. (3.2.8a). One can readily show that the ρ_c curve touches the ρ_c^{2D} / L_z staircase plot at $E_c = E_{1c}$, $E_c = E_{2c}$ and so on. For completeness, we also show in Fig. 3.25b a plot of E_c vs $k_\perp$ [see Fig. 3.23]. Thus, for any value of energy $E_{1c} \leq E_c \leq E_{2c}$, Fig. 3.25b gives directly the corresponding value of the $k_\perp$ component of the electron **k** vector. Similar considerations can also be made for the density of states in the valence band. The corresponding density, ρ_v^{2D}, is simply obtained from Eq. (3.3.9) by substituting m_c with m_v and similar plots to those of Fig. 3.25 can then be made for the valence band. Since for GaAs one has $m_v = m_{hh} \cong 5m_c$, the steps of the staircase in Fig. 3.25a for the valence band are 5 times larger in state density, ρ_v^{2D}, and 5 times smaller in energy, E_v.

3.3.3. Level Occupation at Thermal Equilibrium

Let us first consider the case of overall thermal equilibrium. The probability of occupation of a given state of energy E′ (see Fig. 3.23b), either in the conduction or valence subbands, is again given by Fermi–Dirac statistics as in Eq. (3.2.9) where E'_F is the Fermi energy. Suppose now that some electrons are raised to the conduction subbands $n = 1, n = 2$, etc., and assume a fast relaxation among these subbands (typically $\tau \cong 0.1\,ps$) both in the conduction and in the valence band. The equilibrium situation can then be described by again introducing two quasi-Fermi levels. The probability of occupation of a given level in the conduction or valence subbands will be given by Eqs. (3.2.10a) and (3.2.10b) for the unprimed energy axes of Fig. 3.23a or by Eqs. (3.2.11a) and (3.2.11b) for the primed energy axes of Fig. 3.23b.

Just as for a bulk semiconductor, the values of E_{F_c} and E_{F_v} are established by the number of electrons, N_e, and holes, N_h, which are injected in the corresponding bands. One can indeed calculate N_e and N_h by the relations

$$N_e = \int (\rho_c^{2D}/L_z) f_c \, dE_c \tag{3.3.10a}$$

$$N_h = \int (\rho_v^{2D}/L_z) \bar{f}_v \, dE_v \tag{3.3.10b}$$

In Eq. (3.3.10a) ρ_c^{2D} is the surface density of states, and it is given, for each subband, by Eq. (3.3.9) (see also Fig. 3.25a). In Eq. (3.3.10b) ρ_v^{2D} is the surface density of states for the valence subbands and $\bar{f}_v$ is the occupation probability of the holes as given by Eq. (3.2.13). Since ρ^{2D} is constant in each subband, the integrals in Eq. (3.3.10) can be calculated analytically and the final result can be written as

$$N_e = kT \sum_i \left(\frac{m_{ci}}{\pi \hbar^2 L_z}\right) \ln\left[1 + \exp\left(\frac{E_{F_c} - E_{ic}}{kT}\right)\right] \tag{3.3.11a}$$

$$N_h = kT \sum_i \left(\frac{m_{vi}}{\pi \hbar^2 L_z}\right) \ln\left[1 + \exp\left(\frac{E_{F_v} - E_{iv}}{kT}\right)\right] \tag{3.3.11b}$$

where the sum is taken over all subbands, m_{ci} and m_{vi} are the electron and hole masses in each subband, and E_{ic} and E_{iv} are the minimum energies of each subband. One should note that, by choosing the unprimed energy axes of Fig. 3.23a, the expressions for N_e and N_h take exactly the same functional form.

3.3.4. Stimulated Transitions

Consider a stimulated transition (absorption or stimulated emission) between two given levels, 1 and 2, belonging to a valence and a conduction subband, respectively. Within the electric dipole approximation, the corresponding transition probability, W, will again be proportional to $\left|H_{12}'^{0}\right|^2$ given by

$$\left|H_{12}'^{0}\right|^2 = \left|\int \psi_c^* \left[-e\mathbf{r}\cdot\mathbf{E}(\mathbf{r})\right] \psi_v dV\right|^2 \tag{3.3.12}$$

where ψ_c and ψ_v are now given by Eq. (3.3.1) and $\mathbf{E(r)}$ is the electric field of the e.m. wave at the position $\mathbf{r}$ in the quantum well [compare with Eq. (3.2.19)]. To simplify our considerations, we will take the case of E-field polarization in the plane of the well. Then $e\mathbf{r}\cdot\mathbf{E}(\mathbf{r}) = e\mathbf{r}_\perp\cdot\mathbf{E}(\mathbf{r})$

> **Example 3.10.** *Calculation of the Quasi-Fermi energies for a GaAs/AlGaAs quantum well.* We will take $m_{ci} = m_c = 0.067\,m_0$ and $m_{vi} = 0.46\,m_0$, i.e. we will assume that the masses are the same as those of the bulk material and we will neglect the contribution of the light holes. We will also assume $L_z = 10\,\text{nm}$ and $T = 300\,\text{K}$. From Eq. (3.3.11) one readily obtains the two plots of concentration N vs $(E_F - E_1)/kT$, both for electrons and holes, shown in Fig. 3.26. From this figure, the position of the quasi-Fermi levels for a given injection, N, of electrons and holes, can be readily obtained.

where $\mathbf{r}_\perp$ is the component of the $\mathbf{r}$ vector in the well plane. We further notice that the well thickness is much smaller than the wavelength of light. Hence $\mathbf{E}(\mathbf{r})$ can be taken to be constant along the direction z orthogonal to the well and we can thus write $\mathbf{E} = \mathbf{E}(\mathbf{r}_\perp)$ i.e. a function of only the transverse coordinate $\mathbf{r}_\perp$. It then follows that $e\mathbf{r}_\perp \cdot \mathbf{E}\,(\mathbf{r})$ reduces to $e\mathbf{r}_\perp \cdot \mathbf{E}\,(\mathbf{r}_\perp)$ and Eq. (3.3.12) can be split into two integrals, one over the orthogonal coordinates, x and y, and the other over the longitudinal one z, viz

$$\left|H_{12}^{0}\right|^2 = \left|\int u_c^*\,(\mathbf{r}_\perp)\,e^{-j\mathbf{k}_{c\perp}\cdot\mathbf{r}_\perp}\left[-e\mathbf{r}_\perp\cdot\mathbf{E}_o e^{j\mathbf{k}_{opt}\cdot\mathbf{r}_\perp}\right]u_v(\mathbf{r}_\perp)e^{j\mathbf{k}_{v\perp}\cdot\mathbf{r}_\perp}dxdy\right|^2 \times \qquad (3.3.13)$$

$$\times\left|\int \sin\,(n_c\pi z/L_z)\,\sin\,(n_v\pi z/L_z)dz\right|^2$$

As in the case of a bulk semiconductor, the integral over the transverse coordinates can be shown to vanish unless $\mathbf{k}_{v\perp} + \mathbf{k}_{opt} = \mathbf{k}_{c\perp}$. Since again $\left|\mathbf{k}_{\text{opt}}\right| << (\left|\mathbf{k}_{v\perp}\right|, \left|\mathbf{k}_{c\perp}\right|)$ we obtain the selection rule [compare with Eq. (3.2.22)]

$$k_{c\perp} = k_{v\perp} \qquad (3.3.14)$$

Thus the $\mathbf{k}$-conservation rule still holds for the transverse component, $\mathbf{k}_\perp$, and this implies that transitions must occur vertically in Fig. 3.23. From the second integral of the right hand

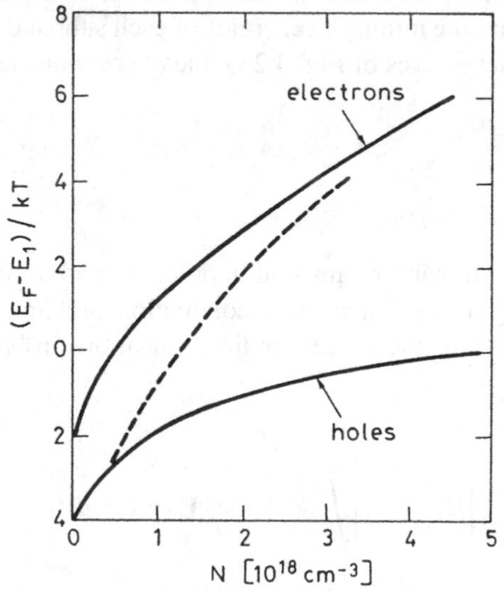

FIG. 3.26. Plots of the normalized difference between the quasi-Fermi energy, E_F, and the energy of the $n = 1$ subband, E_1, vs density of injected carriers, for both electrons and holes, in a 10 nm GaAs/AlGaAs quantum well.

side of Eq. (3.3.13), since n_c and n_v are positive integers we get the selection rule for the quantum number n as

$$\Delta n = n_c - n_v = 0 \tag{3.3.15}$$

which shows that transitions can only occur between two subbands, one in the conduction and the other in the valence band, with the same quantum number n. One should finally note that spin is not involved in the interaction Hamiltonian $e\,\mathbf{r}\cdot\mathbf{E}$, i.e. the e.m. wave does not interact with spin. This implies that spin cannot change in the transition i.e. that

$$\Delta S = 0 \tag{3.3.16}$$

where S is the spin quantum number of the electron involved.

It should be noted that the selection rules Eqs. (3.3.14), (3.3.15), and (3.3.16) have been derived subject to the simplifying assumptions that the **E**-field be polarized in the plane of the well. It can be shown however[13] that the same rules hold in general and, from now on, these rules will be used extensively.

3.3.5. Absorption and Gain Coefficients

To calculate absorption obeying $\mathbf{k}_\perp$-conservation rule it is appropriate to first introduce the joint density of transitions or joint density of states, ρ_{Jk}^{2D}, such that $\rho_{Jk}^{2D}dk_\perp$ gives the number of available transitions or the number of coupled states per unit area, with $k_\perp$ ranging between $k_\perp$ and $k_\perp + dk_\perp$. Since transitions can only occur vertically in Fig. 3.23 and since $\Delta S = 0$, this number is also equal to the number of states in either the valence or conduction bands within the same elemental interval $dk_\perp$. Thus we get

$$\rho_{Jk}^{2D} = \rho_k^{2D} = k_\perp / \pi \tag{3.3.17}$$

where Eq. (3.3.6) has been used. Consider now two given energy levels of energy E'_2 and E'_1 belonging to e.g. the $n = 1$ subbands of the conduction and valence band respectively. From Fig. 3.23b and Eq. (3.3.3), the energy difference, $E_0 = h\nu_0 = E'_2 - E'$, is seen to be given by

$$E_0 = E_g + \frac{\hbar^2 k_\perp^2}{2\,m_r} + \Delta E_1 \tag{3.3.18}$$

where m_r is the reduced mass and $\Delta E_1 = E_{1c} + E_{1v}$. If we now introduce the density of states in the E_0 coordinate, $\rho_{JE_0}^{2D}$, we can write

$$\rho_{JE_0}^{2D}dE_0 = \rho_{Jk}^{2D}dk_\perp = k_\perp dk_\perp / \pi \tag{3.3.19}$$

where Eq. (3.3.17) has been used. The quantity $k_\perp dk_\perp$ is then obtained by differentiating both sides of Eq. (3.3.18). From Eq. (3.3.19) we then get

$$\rho_{JE_0}^{2D} = m_r / \pi\,\hbar^2 \tag{3.3.20}$$

If we now define the density of states with respect to the transition frequency coordinate ν_0, $\rho_{J\nu_0}$, then, since we must have again $\rho_{J\nu_0}^{2D}\, d\nu_0 = \rho_{JE_0}^{2D} dE_0$, we get from Eq. (3.3.20)

$$\rho_{J\nu_0}^{2D} = 4\pi m_r / h \tag{3.3.21}$$

The calculation of the overall absorption at the frequency ν of the incoming e.m. wave can be made in the same way as the calculation performed for bulk material, i.e. through Eqs. (3.2.30)–(3.2.32), provided we substitute $\rho_J(\nu_0)$ with $\left(\rho_{J\nu_0}^{2D}/L_z\right)$ which is the joint density of states for our case. Again assuming infinitely narrow transitions between any two states, the absorption coefficient for the $n = 1 \rightarrow n = 1$ quantum well transition is readily obtained from Eqs. (3.2.34) and (3.2.35) as

$$\alpha_{QW} = \left(\frac{2\pi^2\nu}{n\varepsilon_0 ch}\right)\frac{\mu^2}{3}\left(\frac{\rho_{J\nu}^{2D}}{L_z}\right)\left[f_v\left(E_1'\right) - f_c\left(E_2'\right)\right] \tag{3.3.22}$$

where E_1' and E_2' are now the energies of the two levels whose transition frequency is equal to ν. Under the $\mathbf{k}_\perp$-conservation rule, E_2' and E_1' can readily be obtained from Fig. 3.23a,b with the help of Eq. (3.3.3), by letting $\nu_0 = \nu$.

Example 3.11. *Calculation of the absorption coefficient in a GaAs/AlGaAs quantum well.* We will first consider the case of $T = 0$ K. In this case all valence subbands are full, all conduction subbands are empty and we have $f_v\left(E_1'\right) = 1$ and $f_c\left(E_2'\right) = 0$. The absorption coefficient then has its maximum value, given by

$$\alpha_{QW}^{\max} = \left(\frac{2\pi^2\nu}{n\varepsilon_0 ch}\right)\frac{\mu^2}{3}\frac{\rho_{J\nu}^{2D}}{L_z} \tag{3.3.23}$$

whose dependence upon the photon energy is essentially determined by $\rho_{J\nu}^{2D}$. The absorption coefficient vs the difference, $\left(E - E_g\right)$, between the photon energy and the gap energy as calculated from Eq. (3.3.23) for a $L_z = 10$ nm quantum well is shown in Fig. 3.27. According to Fig. 3.25a, $\rho_{J\nu}^{2D}$ is seen to be zero for a photon energy $E < E_g + E_{1c} + E_{1v} = E_g + \Delta E_1$. No absorption is thus expected for $\left(E - E_g\right) < \Delta E_1$. Assuming again $m_c = 0.067\, m_0$ and $m_v = 0.46\, m_0$, from example 3.8 we get $\Delta E_1 = E_{1c} + E_{1v} \cong 65$ meV. For $\Delta E_1 \le (E - E_g) \le \Delta E_2$, where $\Delta E_2 = E_{2c} + E_{2v}$, $\rho_{J\nu}^{2D}$ is given by Eq. (3.3.21), with $\nu_0 = \nu$, and the absorption coefficient has a constant value given by

$$\alpha_{QW} = \frac{8\pi^3}{n\varepsilon_0 \lambda h^2}\left(\frac{\mu^2}{3}\right)\frac{m_r}{L_z} \tag{3.3.24}$$

where $\lambda = c/\nu$ According to example 3.5, we will take $\left[\mu^2/3\right]^{1/2} = 0.68 \times 10^{-25}$ C × m, $m_r = 5.37 \times 10^{-32}$ Kg, $n = 3.64$ and $\lambda = 833$ nm. From Eq. (3.3.24) we then get $\alpha_{QW} = 5{,}250\,\mathrm{cm}^{-1}$. For $\left(E - E_g\right) \ge \Delta E_2$, transitions between the $n = 2$ subbands also occur, the joint density of states doubles (see also Fig. 3.25a) and the absorption coefficient will also double. Note that, since $E_{2c} = 4E_{1c}$ and $E_{2v} = 4E_{1v}$, one has $\Delta E_2 = 4\Delta E_1 = 260$ meV $= 260$ meV (one can now compare Fig. 3.27 with Fig. 3.16).

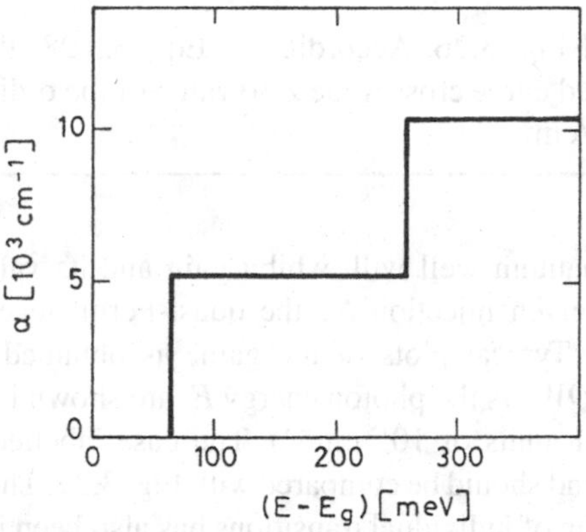

FIG. 3.27. Idealized plot of the absorption coefficient, α, vs the difference between photon energy and gap energy in a 10 nm GaAs/AlGaAs quantum well.

We can proceed in a similar way for the case of stimulated emission. It can readily be seen that the corresponding formula for gain coefficient can be obtained from Eq. (3.3.22) by interchanging the indices c and v and the indices 1 with 2. We then get

$$g_{QW} = \left(\frac{2\pi^2\nu}{n\varepsilon_0 ch}\right)\frac{\mu^2}{3}\left(\frac{\rho_{J\nu}^{2D}}{L_z}\right)\left[f_c\left(E_2'\right) - f_v\left(E_1'\right)\right] \tag{3.3.25}$$

The necessary condition for positive net gain is again that $f_c\left(E_2'\right) \geq f_v\left(E_1'\right)$, which again implies the Bernard–Duraffourg condition $h\nu = E_2' - E_1' \leq E_{F_c}' - E_{F_v}'$. On the other hand, $h\nu$ must be larger than $E_g + \Delta E_1$ so that

$$E_g + \Delta E_1 \leq h\nu \leq E_{F_c}' - E_{F_v}' \tag{3.3.26}$$

which establishes the gain bandwidth. From Eq. (3.3.26), the transparency condition is obtained as

$$E_{Fc}' - E_{Fv}' = E_g + \Delta E_1 \tag{3.3.27}$$

Example 3.12. *Calculation of the transparency density in a GaAs quantum well.* From Fig. 3.23 [see also Eq. (3.2.3)] we have $E_{F_c}' = E_{F_c} + E_g$ and $E_{F_v}' = -E_{F_v}$. Equation (3.3.27), in the new variables E_{F_c} and E_{F_v}, transforms to the simpler expression

$$(E_{F_c} - E_{1c}) + (E_{F_v} - E_{1v}) = 0 \tag{3.3.28}$$

where we have used the relation $\Delta E_1 = E_{1c} + E_{1v}$. To obtain from Eq. (3.3.28) the corresponding value of the transparency density, N_{tr}, we have plotted in Fig. 3.26 the quantity $\left[(E_{F_c} - E_{1c}) + (E_{F_v} - E_{1v})\right]/kT$ as a dashed curve. The curve is obtained by taking, for each carrier concentration N, the sum of the values given by the two continuous curves of Fig. 3.26. According to Eq. (3.3.28) the transparency density corresponds to the point where the dashed curve crosses the zero value of the ordinate. From Fig. 3.26 we thus obtain $N_{tr} \cong 1.25 \times 10^{18}$ electrons/cm^3.

For $N > N_{tr}$, the quantum well will exhibit gain and its value can be obtained from Eq. (3.3.25) once, for a given injection N, the quasi-Fermi levels are calculated (in our example from Fig. 3.26). Typical plots of the gain, as obtained by this procedure for a 8 nm GaAs/$Al_{0.2}Ga_{0.8}As$ QW, vs the photon energy E, are shown in Fig. 3.28 as solid curves for several values of N (in units of 10^{18} cm^{-3}). The case labelled $N = 0$ corresponds to quantum-well absorption and should be compared with Fig. 3.27. The steps are not sharps here because spectral broadening of individual transitions has also been included. One should also note that all possible transitions to the heavy-hole (HH) and light-hole (LH) subbands have been taken into account, in this case, and that transparency occurs for $N_{tr} \cong 2 \times 10^{18}$ cm^{-3}. For $N \geq N_{tr}$, the peak gain coefficient can again be approximated by an expression similar to Eq. (3.2.42), namely

$$g_p = \sigma_{QW}(N - N_{tr}) \tag{3.3.29}$$

with $\sigma_{QW} \cong 7 \times 10^{-16}$ cm^2. Comparing these results with those of the bulk material shows that, while N_{tr} is almost the same for the two cases, the differential gain σ_{QW} for a quantum well is considerably larger (about twice) than that of the bulk semiconductor. The same situation also occurs for $In_{1-x}Ga_xAs_yP_{1-y}$/InP quantum well lasers,(14) and is basically related to the different form for the density of states for the two cases (see Fig. 3.25a).(16) We note however that, both for GaAs/AlGaAs and $In_{1-x}Ga_xAs_yP_{1-y}$/InP quantum wells, a linear relation between α_{gp} and N holds less accurately than for the corresponding bulk materials. In fact, a plot of α_{gp} vs N, at a given temperature, shows that a saturation of α_{gp} occurs at sufficiently

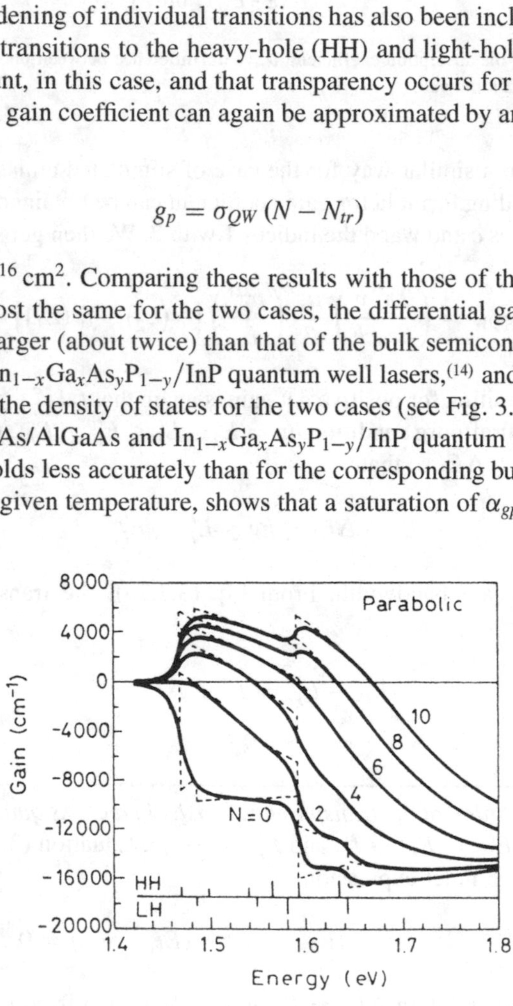

FIG. 3.28. Plots of the gain (or absorption) coefficient vs photon energy with the injected carrier density, N, as a parameter (in units of 10^{18} cm^{-3}) for a 8 nm GaAs/$Al_{0.2}Ga_{0.8}As$ quantum well, in the parabolic band approximation (after ref.(8) by permission).

high values of N,[14] this effect also being basically related to the different form for the density of states for the two cases. Lastly, we recall that the expression in Eq. (3.3.29) gives the material gain coefficient of the quantum-well. How this translates into the modal gain coefficient in a quantum well semiconductor laser and the real advantages of quantum-well lasers will be discussed in Sect. 9.4.4.

3.3.6. Strained Quantum Wells

In a GaAs/$Al_{0.2}Ga_{0.8}As$ quantum well, the lattice constant of GaAs is, to within better than 0.1%, equal to that of $Al_{0.2}Ga_{0.8}As$ (all III–V materials have cubic symmetry). The same situation occurs for the $In_{1-x}Ga_xAs_yP_{1-y}$/InP quantum well if we choose $x \cong 0.45\,y$. Consider now, for example, the case of a $In_xGa_{1-x}As/Al_{0.2}Ga_{0.8}As$ quantum well where, since the substitution of Al with In lowers the bandgap, $In_xGa_{1-x}As$ is the well material. For $0 \leq x \leq 0.5$ one then has $1.424\,\text{eV} \geq E_g \geq 0.9\,\text{eV}$ and the emitted light covers the important wavelength range $840\,\text{nm} \leq \lambda \leq 1{,}330\,\text{nm}$. The lattice constant of InGaAs is now larger than that of AlGaAs [by as much as 3.6% at $x = 0.5$], and, before the well is formed, the situation of the two materials will be as shown in Fig. 3.29a. In forming the quantum well, the two lattice constants must become equal in the QW well plane and this produces a biaxial compression of InGaAs in this plane and a uniaxial tension along the direction orthogonal to the plane (Fig. 3.29b). The InGaAs QW then looses its cubic symmetry and this changes the values of the valence band effective masses and of the band gap.*

What needs to concern us mostly is the heavy hole mass in the plane of the QW as it enters into the expression for the density of states in the valence band (see Sect. 3.3.2). Under compressive strain this mass is greatly reduced (by as much as a factor 2 for $x = 0.2$) becoming closer in value to that of the electron mass in the conduction band. This makes the density of states in the valence band, ρ_v^{2D}, comparable to that, ρ_c^{2D}, in the conduction band. The reduction of heavy hole mass and the corresponding reduction in state density,

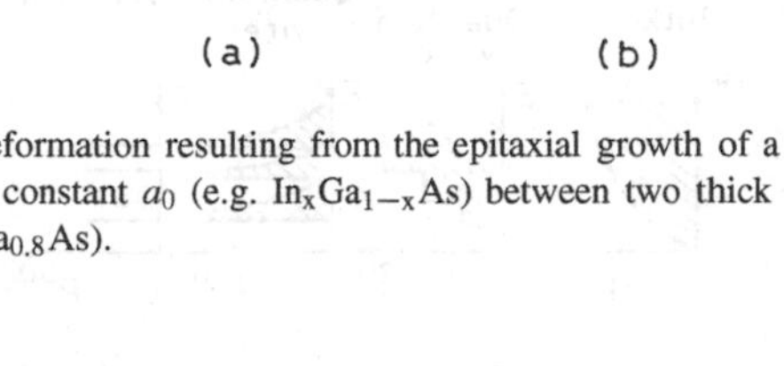

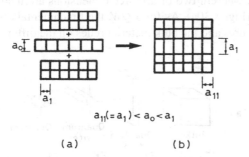

FIG. 3.29. Crystal lattice deformation resulting from the epitaxial growth of a thin quantum-well layer of III–V material with original lattice constant a_0 (e.g. $In_xGa_{1-x}As$) between two thick layers of a material with a lattice constant $a_1 < a_0$ (e.g. $Al_{0.2}Ga_{0.8}As$).

* We recall that, in the parabolic band approximation, all the quantum details are essentially hidden in the values of the effective masses and energy gap.

ρ_v^{2D}, results in two very important advantages compared to typical unstrained QW: (1) The transparency density, N_{tr}, is greatly reduced by an amount as large as a factor 2 [down to $N_{tr} \cong (0.5\text{–}1) \times 10^{18}\,\text{cm}^{-3}$]. (2) The differential gain $d\alpha_g/dN$ is greatly increased by an amount as large as a factor 2 [up to $(15\text{–}30) \times 10^{-16}\,\text{cm}^2$]. The reasons for both circumstances are fundamentally related to the reduced value of ρ_v^{2D} and to the shift in position of the quasi-Fermi levels, at transparency, upon reducing the hole mass.(17) Indeed, the lowest value of N_{tr} and the highest value of dg/dN are attained in the completely symmetrical case $m_v = m_c$.

In concluding this section we can say that there are three main beneficial effects of strained QW lasers: (1) A considerable reduction of N_{tr}. This effect results in a consistent reduction of the threshold current density, J_{th}, since it will be shown in Chap. 9 (see Sect. 9.4.4) that J_{th} is fundamentally related to N_{tr}. (2) An increase of electron-hole recombination time, τ, because both the radiative decay rate, $(1/\tau_r) = BN$, and the Auger rate, $(1/\tau_A) = CN^2$, are reduced as a consequence of the reduction of N_{tr}. This effect also results in a further decrease of J_{th} since $J_{tr} \propto 1/\tau$. (see Sect. 9.4.4 again). (3) A considerable increase of the differential material gain and, hence, of the differential modal gain. It will be shown in Chap. 9, that this effect not only decreases the threshold current density but also increases the laser efficiency. For these reasons, strained QW lasers are becoming increasingly important as laser media.

3.4. QUANTUM WIRES AND QUANTUM DOTS

We have seen in the previous section that the improvement in optical properties obtained on going from bulk material to the corresponding QW material is essentially due to a quantum confinement effect arising from the fact that one dimension of the semiconductor has become comparable to the DeBroglie wavelength. It is therefore natural to expand this idea to consider the other two possible cases of quantum confinment, namely quantum wires, QWR, and quantum dots, QD, wherein two or all three dimensions are made comparable with the DeBroglie wavelength (Fig. 3.30a). As for a QW, the fundamental difference between these quantum confined structures and the bulk material relies on the different forms of the density

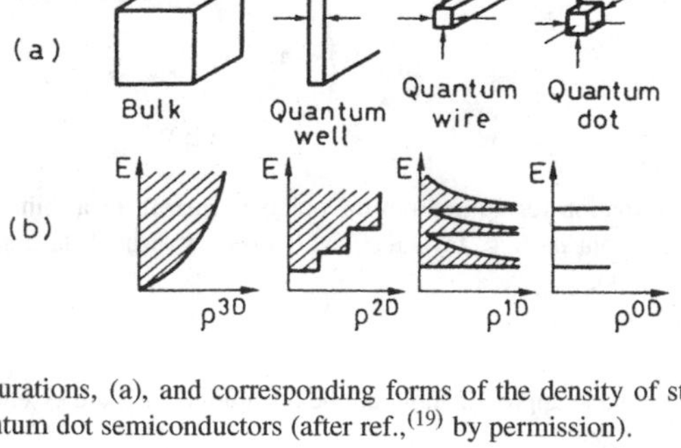

FIG. 3.30. Different configurations, (a), and corresponding forms of the density of states, (b), for bulk, quantum well, quantum wire and quantum dot semiconductors (after ref.,(19) by permission).

of states. Without entering into detail here (for more detail see ref.[(18)]) we show in Fig. 3.30b the qualitative behavior of the density of states for *QWR* and *QD* as compared to bulk and *QW* materials. Using these different forms of state density, one can proceed in a similar way as for a *QW* to calculate the expected gain. We do not pursue this here and, as a representative example, we limit ourselves to showing in Fig. 3.31 the expected material gain vs photon energy for a $Ga_{0.47}In_{0.51}As$/InP system ($Ga_{0.47}In_{0.51}As$ now constitutes the quantum confinment material). In the figure, the curves of predicted material gain vs photon energy for the bulk case, for a 10 nm *QW*, 10 nm × 10 nm *QWR*, and 10 nm × 10 nm × 10 nm *QD* are plotted at the same electron injection $N = 3 \times 10^{18}\,\text{cm}^{-3}$.[(19)] The calculated transparency density is about the same for bulk, *QW* and *QWR* $[N_{tr} \cong 1.3 \times 10^{18}\,\text{cm}^{-3}]$ while it is somewhat higher for *QD* $[N_{tr} \cong 1.8 \times 10^{18}\,\text{cm}^{-3}]$. In agreement with our earlier discussion of *QW* structures, the gain is seen to increase on going from bulk to *QW*, from *QW* to *QWR*, and from *QWR* to *QD*. The gain bandwidth, on the other hand, decreases from *QW* to *QWR* and from *QWR* to *QD*.

As a laser material, quantum wires and dots will perhaps be used in the form of an array, such as the planar ones of Fig. 3.32a, b. Considerable technological difficulties (such as high packing density, low size fluctuations, and low defect density) are still preventing the fabrication of quantum wires and quantum dots having good optical properties. If these difficulties can be solved, semiconductor laser materials of still lower threshold, much higher differential gain and narrower bandwidth will become available.

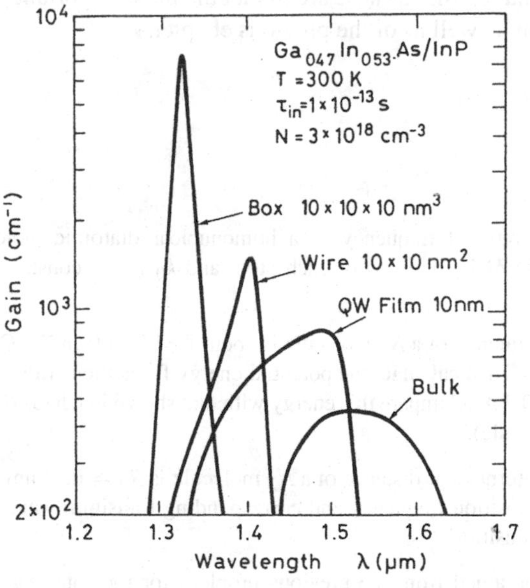

FIG. 3.31. Plot of calculated gain coefficient vs emission wavelenght, at $N = 3 \times 10^{18}\,\text{cm}^{-3}$ electron injection, for a $Ga_{0.47}In_{0.53}As$ bulk semiconductor and for $Ga_{0.47}In_{0.53}As$/InP 10 nm quantum well, 10 nm × 10 nm quantum wire and 10 nm × 10 nm × 10 nm quantum dot. (after ref.[(19)] by permission).

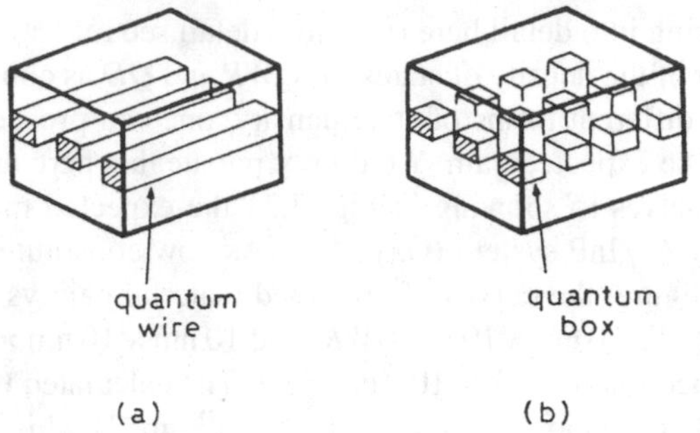

FIG. 3.32. Planar array of multiple quantum wires, (a), and multiple quantum dots, (b).

3.5. CONCLUDING REMARKS

In this chapter, as compared to the previous one, we have progressed from the simple case of atoms to the more complicated cases of molecules and semiconductors and have analyzed these in some detail. As our discussion showed, a physical understanding of the optical properties of these materials requires a rather detailed description of their physical behavior. In doing this we have limited ourselves to the most elementary aspects. From a phenomenological viewpoint, however, as we shall see in Chaps. 7 and 8, only a few physical parameters are needed to predict laser behaviour, namely: (1) The wavelengths and bandwidths of the gain transitions. (2) The transition cross section or, for a semiconductor, the differential gain and the transparency density. (3) The lifetime of the upper state or, for a semiconductor, the electron-hole recombination time. These are in fact the most important physical parameters to come out of the present as well as of the previous chapter.

PROBLEMS

3.1. Show that the vibrational frequency of a homonuclear diatomic molecules is $\nu = (1/2\pi)(2k_0 / M)^{1/2}$ where M is the mass of each atom and k_0 is the constant of the elastic restoring force.

3.2. The vibrational frequency of a N_2 molecule is about $\tilde{\nu} = 2,360\,\text{cm}^{-1}$. Calculate the value of the elastic constant k_0. Then calculate the potential energy for a nuclear distance away from equilibrium of $R - R_0 = 0.3$ Å.(Compare this energy with that shown in Chap. 10 for the potential energy curve of a N_2 molecule).

3.3. The equilibrium internuclear distance of a N_2 molecule is $R_0 \cong 0.11$ nm. Calculate the rotational constant, B, the transition frequency and corresponding transition wavelength for the $J = 0 \rightarrow J = 1$ rotational transition.

3.4. Using the result obtained from the previous problem for the rotational constant, B, of the N_2 molecule, calculate the frequency separation between two consecutive lines of the P-branch of the $\upsilon'' = 0 \rightarrow \upsilon' = 1$ transition. Also calculate the quantum number of the most populated rotational level of the $\upsilon'' = 0$ state.

3.5. The rotational constant of a CO_2 molecule in its 00°1 vibrational level is $B = 0.37\,\text{cm}^{-1}$. Assuming the same value for the rotational constant of the 10°0 level, calculate the P-branch and R-branch spectrum at $T = 450\,\text{K}$ of the 00°1 → 10°0 transition [remember that only rotational levels of the 00°1 upper state of even J number can partecipate in the transition].

3.6. Assuming that the rotational constant of the ground vibrational state of a CO_2 molecule is $B = 0.37\,\text{cm}^{-1}$, i.e. it is equal to that of the (00°1) state, calculate the equilibrium distance R_0 between the carbon and oxygen atoms.

3.7. From the condition $f_c\left(E'_2\right) \geq f_v\left(E'_1\right)$ prove the Bernard–Duraffourg relation $E'_{F_c} - E'_{F_v} \geq h\nu$.

3.8. With the help of Fig. 3.15b calculate for GaAs: (a) The values of E_{F_c} and E_{F_v} at $N = 1.6 \times 10^{18}\,\text{cm}^{-3}$ carrier injection. (a) The overall gain bandwidth at the same injection.

3.9. In the energy reference system of Fig. 3.9a calculate, for GaAs, the energies, E_2 and E_1, of the upper and lower laser levels for a transition energy exceeding the bandgap energy by $0.45\,kT$.

3.10. For a bulk GaAs semiconductor, with the help of Fig. 3.16, calculate the expected gain at a photon energy exceeding the badgap energy by $0.45\,kT$ and for a carrier injection of $N = 1.6 \times 10^{18}\,\text{cm}^{-3}$.

3.11. Assuming that the peak gain in a bulk GaAs semiconductor at a carrier injection $N = 1.6 \times 10^{18}\,\text{cm}^{-3}$ occurs at a photon energy exceeding the gap energy by $0.45\,kT$ and using some of the results obtained in problems 3.9, 3.10 and 3.11, calculate the differential gain $\sigma = d\alpha_g/dN$.

3.12. With the help of Fig. 3.15a, plot on the same figure the quantities E_{F_c}/kT, E_{F_v}/kT, and $\left(E_{F_c} + E_{F_v}\right)/kT$ vs the concentration, N, of electrons and holes for bulk InGaAsP at $\lambda = 1{,}300\,\text{nm}$. From these plots then calculate the transparency density N_{tr} and, from a comparison of this figure with Fig. 3.15b, explain why N_{tr} is, in this case, somewhat smaller than in the GaAs case. From the same plots calculate also the overall gain bandwidth and the values of E_{F_c} and E_{F_v} at $N = 1.6 \times 10^{18}\,\text{cm}^{-3}$ carrier injection. On assuming that the maximum gain occurs at an energy of $\Delta E = 0.65\,kT$ above band gap energy, calculate the corresponding wavelength.

3.13. For a 10 nm GaAs quantum well calculate from Fig. 3.26 the overall bandwidth of the gain curve and the values of the quasi-Fermi levels for an injected carrier density of $N = 2 \times 10^{18}\,\text{cm}^{-3}$. In the ($E$-$E_g$) reference axis of Fig. 3.27 find the energy interval in which positive gain occurs.

3.14. Calculate how the first step of Fig. 3.27 at (E-E_g) $\cong$ 65 meV needs to be modified if a Lorentzian lineshape with a dephasing time $\tau_c = 0.1\,\text{ps}$ is assumed for each transition from an upper level in the first conduction subband to a lower level in the first valence subband.

References

1. G. Herzberg, *Spectra of Diatomic Molecules* (D. Van Nostrand Company, Princeton NJ, 1950).
2. G. Herzberg, *Molecular Spectra and Molecular Structure: Infrared and Raman Spectra of Polyatomic Molecules*, (D. Van Nostrand Company, Princeton NJ, 1968), p. 122, Fig. 51.
3. G.B. Agrawal and N.K. Dutta, *Long Wavelength Semiconductor Lasers* (Chapman and Hall, New York, 1986).
4. C. Kittel, *Introduction to Solid State Physics*, 6th Ed. (Wiley, New York, 1986)
5. R.H. Yan, S.W. Corzine, L.A. Coldren, and I Suemune, Correction for the Expression for Gain in GaAs, *IEEE J. Quantum Electron* **QE–26**, 213–216 (1990).
6. M.G. Bernard and G. Duraffourg, Laser Conditions in Semiconductors, *Phys. Status Solidi* **1**, 699 (1961).
7. Ref. [3] Chap. 3

8. S.W. Corzine, R.H. Yan, and L.A. Coldren, Optical Gain in III-V Bulk and Quantum well Semiconductors, In: *Quantum Well Lasers* ed. by P.S. Zory (Academic, San Diego, CA, 1993) Chap. 1.
9. G.H.B. Thompson, *Physics of Semiconductor Lasers Devices* (Wiley, New York, 1980) sect. 2.5.2.
10. Ref. [9] sect. 2.4.2.
11. Ref. [8] Fig. 2.14.
12. Ref. [3] sect. 3.3.
13. Ref. [8] sect. 3.1.
14. Ref. [3] Chap. 9.
15. A. Yariv, *Quantum Electronics* third ed. (Wiley, New York, 1989) sect. 11.2
16. Ref. [8], sect. 4.1.
17. Ref. [8] Fig. 6.
18. E. Kapon, Quantum Wire Semiconductor Lasers, In: *Quantum Well Lasers*, ed. by P.S. Zory (Academic, San Diego, CA, 1983) Chap. 10.
19. M. Asada, Y. Miyamoto, and Y. Suematsu, Gain and Threshold of Three-Dimensional Quantum-Box Lasers, *IEEE J. Quantum Electron* **QE–22**, 1915–1921 (1986).
20. P.K. Cheo, CO_2 Lasers, In: *Lasers* Vol. 3, Eds. A.K. Levine and A. DeMaria (Marcel Dekker, New York, 1971).

4

Ray and Wave Propagation Through Optical Media

4.1. INTRODUCTION

Before beginning a detailed discussion on optical resonators, which is to be the subject of the next Chapter, we introduce in this Chapter a few topics from geometrical and wave optics. The aim is to introduce some subjects that are not usually covered in elementary optics texts and that constitute a very useful background for the topics to be considered in next chapter. Thus, in particular, the matrix formulation of geometrical optics within the paraxial-ray approximation and wave propagation within the paraxial-wave approximation, leading to the ideas of Gaussian beam propagation, will be discussed here. Situations involving multiple interference such as in a multilayer dielectric coating or in a Fabry-Perot interferometer will also be considered.

4.2. MATRIX FORMULATION OF GEOMETRICAL OPTICS[1]

Consider a ray of light that is either transmitted by or reflected from an optical element which has reciprocal and polarization-independent behavior (e.g. a lens or a mirror). Let z be the optical axis of this element (e.g. the line passing through the centers of curvature of the two spherical surfaces of the lens). Assume that the ray is traveling approximately along the z direction in a plane containing the optical axis. The ray vector $\mathbf{r}_1$ at a given input plane $z = z_1$ of the optical element (Fig. 4.1) can be characterized by two parameters, namely, its radial displacement $r(z_1)$ from the z axis and its angular displacement θ_1. Likewise, the ray-vector $\mathbf{r}_2$ at a given output plane $z = z_2$ can be characterized by its radial, $r_2(z_2)$, and angular θ_2, displacements. Note that the $\mathbf{r}$-axis is taken to be the same for both input and output rays and oriented as in Fig. 4.1. The sign convention for the angles is that the angle is positive if

O. Svelto, *Principles of Lasers*,
DOI: 10.1007/978-1-4419-1302-9_4,

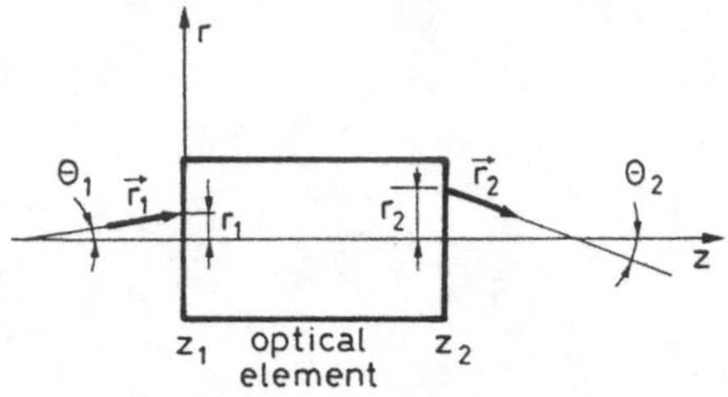

FIG. 4.1. Matrix formulation for the propagation of a ray through a general optical element.

the **r**-vector must be rotated clockwise to make it coincide with the positive direction of the z-axis. Thus, for example, θ_1 is positive while θ_2 is negative in Fig. 4.1.

Within the *paraxial-ray approximation* the angular displacements θ are assumed to be small enough to allow the approximation to be made, $\sin\theta \cong \tan\theta \cong \theta$. In this case the output, $(r_2,\ \theta_2)$, and input, $(r_1,\ \theta_1)$ variables are related by a linear transformation. If we therefore put $\theta_1 \cong (dr_1/dz_1)_{z_1} = r_1'$ and $\theta_2 \cong (dr_2/dz_2)_{z_2} = r_2'$ we can write

$$r_2 = Ar_1 + Br_1' \tag{4.2.1a}$$

$$r_2' = Cr_1 + Dr_1' \tag{4.2.1b}$$

where A, B, C, and D are constants characteristic of the given optical element. In a matrix formulation it is therefore natural to write (4.2.1) as

$$\begin{vmatrix} r_2 \\ r_2' \end{vmatrix} = \begin{vmatrix} A & B \\ C & D \end{vmatrix} \begin{vmatrix} r_1 \\ r_1' \end{vmatrix} \tag{4.2.2}$$

where the $ABCD$ matrix completely characterizes the given optical element within the paraxial ray approximation.

As a first and simplest example we will consider the free-space propagation of a ray along a length $\Delta z = L$ of a material with refractive index n (Fig. 4.2a). If the input and output planes lie just outside the medium, in a medium of refractive index equal to unity, we have, using Snell's law in the paraxial approximation

$$r_2 = r_1 + Lr_1'/n \tag{4.2.3a}$$

$$r_2' = r_1' \tag{4.2.3b}$$

and the corresponding $ABCD$ matrix is therefore

$$\begin{vmatrix} 1 & L/n \\ 0 & 1 \end{vmatrix} \tag{4.2.4}$$

As a next example we consider ray propagation through a lens of focal length f (f is taken to be positive for a converging lens). For a thin lens we obviously have (Fig. 4.2b).

$$r_2 = r_1 \tag{4.2.5a}$$

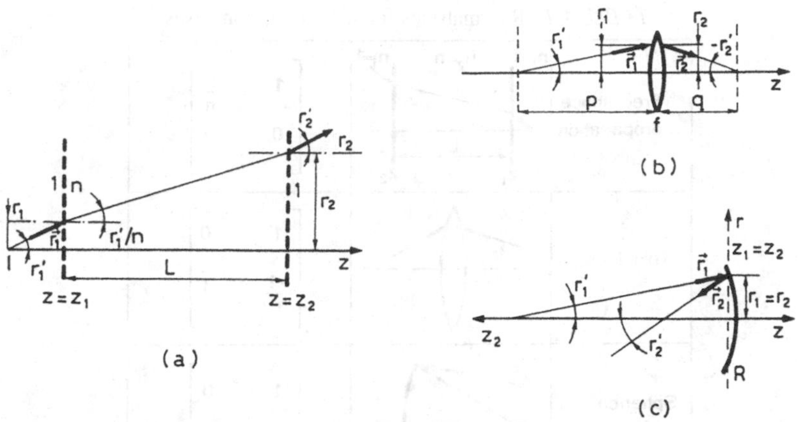

FIG. 4.2. Calculation of the *ABCD* matrix for: (a) free-space propagation, (b) propagation through a thin lens, (c) reflection from a spherical mirror.

The second relation is obtained from the well-known law of geometrical optics, viz., $(1/p) + (1/q) = (1/f)$, and using the fact that $p = r_1/r_1'$ and $q = -r_2/r_2'$. By also using Eq. (4.2.5a) we get

$$r_2' = -(1/f)r_1 + r_1' \tag{4.2.5b}$$

According to Eqs. (4.2.5) the *ABCD* matrix is, in this case,

$$\begin{vmatrix} 1 & 0 \\ -1/f & 1 \end{vmatrix} \tag{4.2.6}$$

As a third example we consider reflection of a ray by a spherical mirror of radius of curvature R (R is taken to be positive for a concave mirror). In this case the z_1 and z_2 planes are taken to be coincident and to be placed just in front of the mirror and the positive direction of the $\boldsymbol{r}$-axis is taken to be the same for incident and reflected rays (Fig. 4.2c). The positive direction of the z axis is taken to be that from left to right for the incident vector and from right to left for the reflected vector. The angle for the incident ray is positive if the $\boldsymbol{r}_1$ vector must be rotated clockwise to make it coincide with the positive z_1 direction while the angle for the reflected ray is positive if the $\boldsymbol{r}_2$-vector must be rotated anticlockwise to make it coincide with the positive z_2 rection of the z-axis. Thus, for example, r_1' is positive while r_2' is negative in Fig. 4.2c. Given these conventions, the ray matrix of a concave mirror of curvature R and, hence, focal length $f = R/2$ can be shown to become identical to that of a positive lens of focal length $f = R/2$. The ray matrix is therefore equal to

$$\begin{vmatrix} 1 & 0 \\ -2/R & 1 \end{vmatrix} \tag{4.2.7}$$

In Table 4.1 we have collected together the ray matrices for the optical elements considered so far as well as for a spherical dielectric interface. We draw attention to the fact that the

TABLE 4.1. Ray matrices for some common cases

Free space propagation	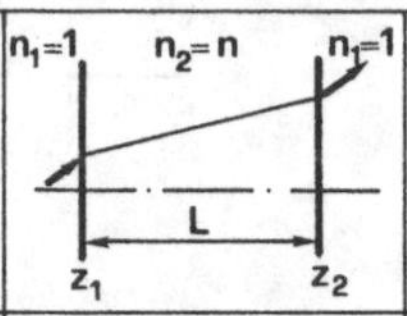	$\begin{bmatrix} 1 & \frac{L}{n} \\ 0 & 1 \end{bmatrix}$
Thin lens	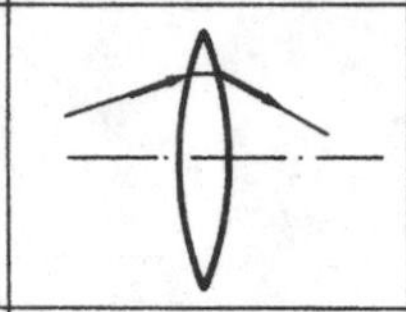	$\begin{bmatrix} 1 & 0 \\ \frac{-1}{f} & 1 \end{bmatrix}$
Spherical mirror	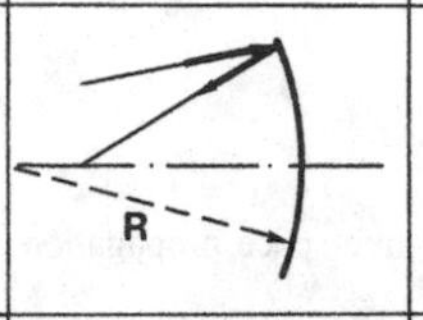	$\begin{bmatrix} 1 & 0 \\ \frac{-2}{R} & 1 \end{bmatrix}$
Spherical dielectric interface	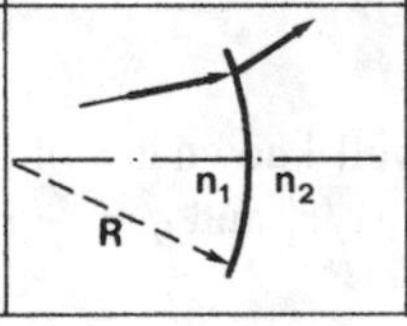	$\begin{bmatrix} 1 & 0 \\ \frac{n_2-n_1}{n_2}\frac{1}{R} & \frac{n_1}{n_2} \end{bmatrix}$

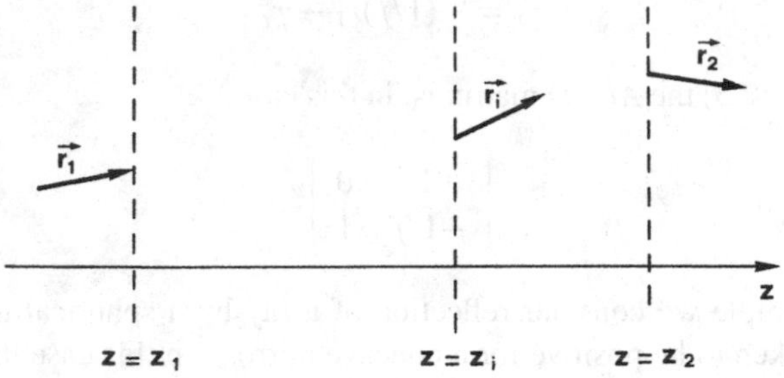

FIG. 4.3. Ray propagation through three distinct planes when the two matrices between planes $z = z_1$ and $z = z_i$ and between $z = z_i$ and $z = z_2$ are known.

determinant of the *ABCD* matrix is unitary i.e.

$$AD - BC = 1 \tag{4.2.8}$$

provided that the input and output planes lye in media of the same refractive index. In fact, this situation holds for the first three cases considered in the table.

Once the matrices of the elementary optical elements are known, one can readily obtain the overall matrix of a more complex optical element by subdividing it into these elementary components. Suppose in fact that, within a given optical element, we can consider an intermediate plane of coordinate z_i (Fig. 4.3) such that the two *ABCD* matrices between planes $z = z_1$ and $z = z_i$ and planes $z = z_i$ and $z = z_2$ are known. If we now call r_i and r_i' the coordinates of the ray vector at plane $z = z_i$, we can obviously write

$$\begin{vmatrix} r_i \\ r_i' \end{vmatrix} = \begin{vmatrix} A_1 & B_1 \\ C_1 & D_1 \end{vmatrix} \begin{vmatrix} r_1 \\ r_1' \end{vmatrix} \tag{4.2.9}$$

$$\begin{vmatrix} r_2 \\ r_2' \end{vmatrix} = \begin{vmatrix} A_2 & B_2 \\ C_2 & D_2 \end{vmatrix} \begin{vmatrix} r_i \\ r_i' \end{vmatrix} \tag{4.2.10}$$

If Eq. (4.2.9) is substituted for the vector $\mathbf{r}_i$ on the right-hand side of Eq. (4.2.10) we obtain

$$\begin{vmatrix} r_2 \\ r_2' \end{vmatrix} = \begin{vmatrix} A_2 & B_2 \\ C_2 & D_2 \end{vmatrix} \begin{vmatrix} A_1 & B_1 \\ C_1 & D_1 \end{vmatrix} \begin{vmatrix} r_1 \\ r_1' \end{vmatrix} \tag{4.2.11}$$

The overall $ABCD$ matrix can thus be obtained by the multiplication of the $ABCD$ matrices of the elementary components. Note that the order in which the matrices appear in the product is the opposite to the order in which the corresponding optical elements are traversed by the light ray.

As a first and perhaps somewhat trivial example of using the above result, we will consider free-space propagation through a length L_1 followed again by free-space propagation through a second length L_2, in a medium with refractive index n. According to Eq. (4.2.4) the overall matrix equation can be written as

$$\begin{vmatrix} r_2 \\ r_2' \end{vmatrix} = \begin{vmatrix} 1 & L_2/n \\ 0 & 1 \end{vmatrix} \begin{vmatrix} 1 & L_1/n \\ 0 & 1 \end{vmatrix} \begin{vmatrix} r_1 \\ r_1' \end{vmatrix} \tag{4.2.12}$$

Using well-known rules of matrix multiplication it can readily be shown that the product of the two square matrices gives an overall matrix

$$\begin{vmatrix} 1 & (L_1 + L_2)/n \\ 0 & 1 \end{vmatrix} \tag{4.2.13}$$

This calculation confirms the obvious result that the overall propagation is equivalent to a free-space propagation over a total length $L = L_1 + L_2$.

As a less trivial and more useful example, we will consider free propagation over a length L (in a medium with refractive index $n = 1$) followed by reflection from a mirror of radius of curvature R. According to Eqs. (4.2.4), (4.2.7), and (4.2.11) the overall $ABCD$ matrix is given by

$$\begin{vmatrix} A & B \\ C & D \end{vmatrix} = \begin{vmatrix} 1 & 0 \\ -(2/R) & 1 \end{vmatrix} \begin{vmatrix} 1 & L \\ 0 & 1 \end{vmatrix} = \begin{vmatrix} 1 & L \\ -(2/R) & 1-(2L/R) \end{vmatrix} \tag{4.2.14}$$

Note that the determinant of the matrix of Eq. (4.2.13) as well as that of Eq. (4.2.14) are again unitary, and this result holds for any arbitrary cascade of optical elements since the determinant of a matrix product is the product of the determinants.

We now address the question of finding the ray-matrix elements A', B', C', D' for reverse propagation through an optical system, in terms of the given matrix elements A, B, C, D for forward propagation. So, referring to Fig. 4.1, one can see that if we take $-\boldsymbol{r}_2$ as the input vector, i.e. if we reverse the propagation direction of the $\boldsymbol{r}_2$ vector, then the output vector must be $-\boldsymbol{r}_1$. For backward propagation we take the same sign conventions as those used for the ray reflected from a spherical mirror (Fig. 4.2c) namely: (a) The z-axis is reversed while the r-axis remains unchanged. (b) The angle between the $\boldsymbol{r}$-vector and the z-axis is positive if the $\boldsymbol{r}$-vector be must rotated anticlockwise to coincide with the z-axis. Given these conventions, it is seen that the rays $-\boldsymbol{r}_1$ and $-\boldsymbol{r}_2$ are described by coordinates $(r_1, -r_1')$ and $(r_2, -r_2')$, respectively. Thus one must have

$$\begin{vmatrix} r_1 \\ -r_1' \end{vmatrix} = \begin{vmatrix} A' & B' \\ C' & D' \end{vmatrix} \begin{vmatrix} r_2 \\ -r_2' \end{vmatrix} \tag{4.2.15}$$

From Eq. (4.2.15) we can obtain r_2 and r_2' as a function of r_1 and r_1'. Since the determinant of the $A'B'C'D'$ matrix is also unitary, we get

$$r_2 = D'r_l + B'r_1' \tag{4.2.16a}$$

$$r_2' = C'r_l + A'r_1' \tag{4.2.16b}$$

A comparison of Eqs. (4.2.16) with Eqs. (4.2.1) then shows that $A' = D,\ B' = B,\ C' = C$, and $D' = A$ so that the overall $A'B'C'D'$ matrix is

$$\begin{vmatrix} A' & B' \\ C' & D' \end{vmatrix} = \begin{vmatrix} D & B \\ C & A \end{vmatrix} \tag{4.2.17}$$

Equation (4.2.17) then shows that the matrix for backward propagation is obtained from that of forward propagation by simply interchanging the matrix elements A and D.

The matrix formulation is not only useful to describe the behavior of a ray as it passes through an optical system, but it can also be used to describe the propagation of a spherical wave. Consider in fact a spherical wave originating from point P_1 of Fig. 4.4 and propagating along the positive z direction. After traversing an optical element described by a given $ABCD$ matrix, this wave will be transformed into a new spherical wave whose center is the point P_2. Consider now two conjugate rays $\mathbf{r}_1$ and $\mathbf{r}_2$ of the two waves, which means that the optical element transforms the incident (or input) ray $\mathbf{r}_1$ into the output ray $\mathbf{r}_2$. The radii of curvature R_1 and R_2 of the two waves at the input plane, z_1, and output plane, z_2, of the optical element are readily obtained as

$$R_1 = r_1/r_1' \tag{4.2.18a}$$

$$R_2 = r_2/r_2' \tag{4.2.18b}$$

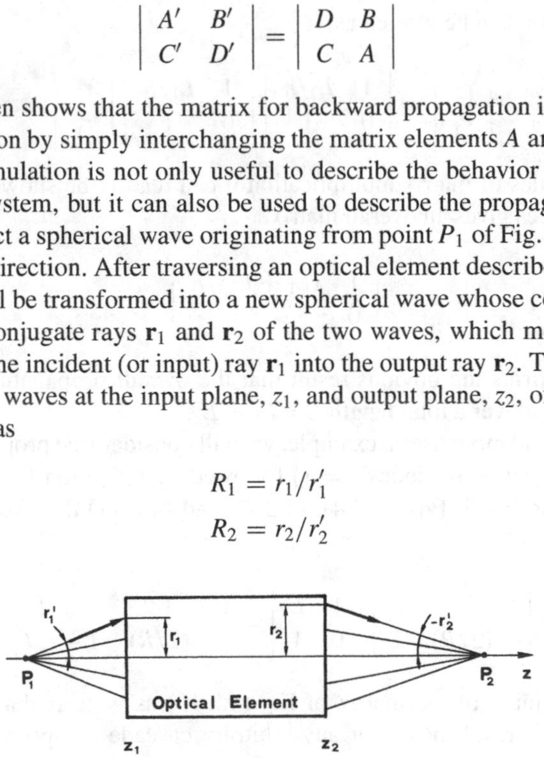

FIG. 4.4. Propagation of a spherical wave emitted from point P_1 through a general optical element described by a given $ABCD$ matrix.

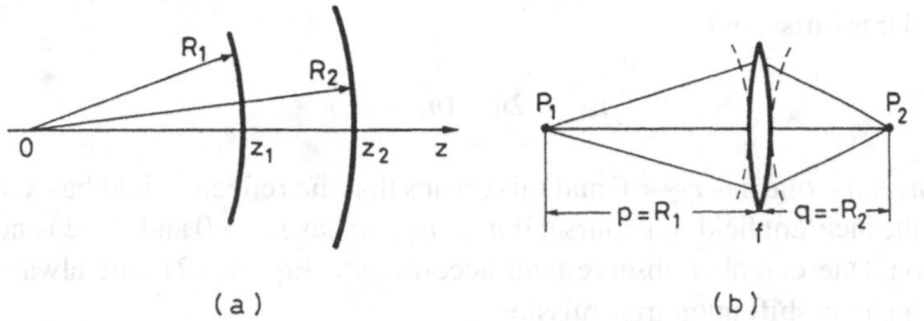

FIG. 4.5. Propagation of a spherical wave: (a) through free space; (b) through a thin lens.

Note that in (4.2.18) we have used the sign convention that R is positive if the center of curvature is to the left of the wave front. From Eqs. (4.2.1) and (4.2.18) we get

$$R_2 = \frac{AR_1 + B}{CR_1 + D} \tag{4.2.19}$$

Equation (4.2.19) is a very important result since it relates, in simple terms, the radius of curvature, R_2, of the output wave to the radius of curvature, R_1, of the input wave via the *ABCD* matrix elements of the given optical component.

As a first and elementary example using this result, consider the free-space propagation of a spherical wave between points having coordinates z_1 and z_2 in Fig. 4.5a. From Eq. (4.2.4), with $n=1$ and $L=z_2-z_1$, and Eq. (4.2.19) we get $R_2=R_1+(z_2-z_1)$ which of course is an obvious result. Consider next the propagation of a spherical wave through a thin lens (Fig. 4.5b). From Eqs. (4.2.6) and (4.2.19) we get

$$\frac{1}{R_2} = \frac{1}{R_1} - \frac{1}{f} \tag{4.2.20}$$

which is simply the familiar law of geometrical optics $p^{-1}+q^{-1}=f^{-1}$.

Although the two examples of Fig. 4.5 are both rather elementary applications of Eq. (4.2.19), the usefulness of this equation can really be appreciated when dealing with a more complicated optical system made up, e.g., of a sequence of lenses and spaces between them. In this case, the overall *ABCD* matrix will be given by the product of the matrices of each optical component and the radius of curvature of the output wave will again be given by Eq. (4.2.19).

4.3. WAVE REFLECTION AND TRANSMISSION AT A DIELECTRIC INTERFACE[2]

Consider a wave which is incident on the plane interface between two media of refractive indices n_1 and n_2. If the wave is initially in the medium of refractive index n_1 and it is normally incident on the surface, the electric field reflectivity is

$$r_{12} = (n_1 - n_2)\,/\,(n_1 + n_2) \tag{4.3.1}$$

while the field transmission is

$$t_{12} = 2n_1/(n_1 + n_2) \tag{4.3.2}$$

Note that, if $n_1 < n_2$, one has $r_{12} < 0$ and this means that the reflected field has a π phase-shift compared to the incident field. Of course, if $n_1 > n_2$ one has $r_{12} > 0$ and there is no phase-shift upon reflection. One can also observe that, according to Eq. (4.3.2), one always has $t_{12} > 0$ i.e. there is no phase-shift upon transmission.

For non-normal incidence, the expressions for electric field reflectivity and transmission are more complicated and depend also on the field polarization. As a representative example Fig. 4.6 shows the plots of the intensity reflectivity, or reflectance, $R = (r_{12})^2$ vs the incidence angle θ for a p-polarized wave (E-field in the plane of incidence) and a s-polarized wave (E-field orthogonal to the plane of incidence), and for $n_1 = 1$ and $n_2 = 1.52$. One can see that, for $\theta = 0$, the two reflectances are obviously equal and, according to Eq. (4.3.1), have the value $R = 4.26\%$. One also notices that, for a p-polarized wave, there is a particular angle ($\theta_B = 56.7°$ in the figure) at which $R = 0$. The situation occurring in this case can be described with the help of Fig. 4.7. Suppose that the incidence angle θ_B is such that the refracted beam is orthogonal to the direction of the reflected beam. The E field in the optical material and hence its polarization vector will therefore be parallel to the direction of reflection. Since the reflected beam may be considered to be produced by radiation emitted by the polarization vector of the medium where refraction occurs, this reflected beam will in this case be zero since an electric dipole does not radiate along its own direction. A straightforward calculation based on geometrical optics can now give the value of the incidence angle θ_B, which is called the *Brewster angle* or polarizing angle. According to the previous discussion we have

$$\theta'_B + \theta_B = \pi/2 \tag{4.3.3}$$

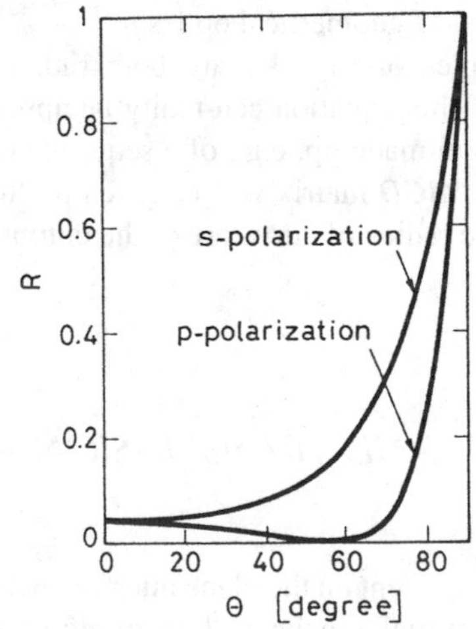

FIG. 4.6. Power reflectivity, R, vs angle of incidence, θ, at an interface between air and a medium of refractive index $n = 1.5$. The two curves refer to the cases of E-field polarization in the plane of incidence (p-polarization) and orthogonal to this plane (s-polarization).

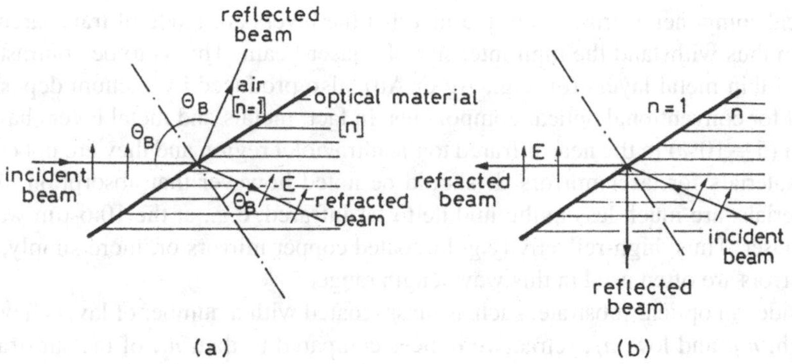

FIG. 4.7. Reflected and refracted beams for incidence at Brewster's angle: (a) incidence from the less dense medium; (b) incidence from the more dense medium.

where θ'_B is the angle of the refracted beam. From Snell's law we also have

$$n \sin \theta'_B = \sin \theta_B \tag{4.3.4}$$

Since according to Eq. (4.3.3) one has $\sin \theta'_B = \cos \theta_B$, from Eq. (4.3.4) we get the following expression for the Brewster angle:

$$\tan \theta_B = n \tag{4.3.5}$$

Note that, if the direction of the rays is reversed (Fig. 4.7), the reflected beam will again be zero since the refracted and reflected beams are again orthogonal. So, if a plane parallel plate of a given optical material is inserted at Brewster's angle into a beam polarized in the plane of the figure, no reflection will occur at the two surfaces of the plate. Let us now assume that a plane parallel plate of e.g. refractive index $n = 1.52$ is inserted, at the Brewster angle, within an optical cavity. According to Fig. 4.6, the reflectance of an *s*-polarized beam at each of the two interfaces will be $R \cong 15\%$. Thus, an *s*-polarized beam would suffer around 30% loss, due to the reflection at the two interfaces. If the laser gain per pass is smaller than 30%, the *s*-polarization will not oscillate and the laser beam will be found to be linearly polarized in the plane of incidence to the plate.

4.4. MULTILAYER DIELECTRIC COATINGS[3,4]

The mirror surfaces, used as high-reflectivity laser mirrors or beam splitters, are commonly fabricated by the technique of deposition of a multilayer dielectric stack on the optical surface, plane or curved, of a substrate material, such as glass. The same technique can also be used to greatly reduce the surface reflectivity of optical components (antireflection coating) or to produce optical elements such as interference filters or polarizers. The coating is usually produced in a vacuum chamber by evaporation of the required dielectric materials, which then condense in a layer on the substrate. The widespread use of multilayer dielectric coatings for laser optical components arises from the fact that the layers are made of transparent materials and can thus withstand the high intensity of a laser beam. This is to be contrasted to the behavior of thin metal layers (of, e.g., Ag or Au), also produced by vacuum deposition and often used for conventional optical components. In fact, metals and metal layers have a large absorption ($5 \div 10\%$) in the near-infrared to the ultraviolet region and they are not communly used as materials for laser mirrors. It should be noted however that absorption losses, for these materials, are much less in the middle to far infrared, e.g., at the 10.6-μm wavelength of a CO_2 laser. Thus, high-reflectivity gold-coated copper mirrors or, more simply, polished copper mirrors are often used in this wavelength range.

Consider an optical substrate, such as glass, coated with a number of layers having alternately high, n_H, and low, n_L, refractive indices compared to that, n_s, of the substrate. If the thickness of the layers l_H and l_L are such that $n_H\, l_H = n_L\, l_L = \lambda_0/4$ where λ_0 is a specified wavelength, the electric field reflections at all layer interfaces, for an incident beam of wavelength $\lambda = \lambda_0$, will add in phase. Consider, for instance, the two interfaces of a high-index layer (Fig. 4.8a). According to Eq. (4.3.1), the electric field reflectivity at the low-to-high index interface has a negative sign and the electric field undergoes a phase shift of $\phi_1 = \pi$ upon reflection. Conversely, the reflectivity at the high-to-low index interface is positive and no phase shift of the reflected wave will occur there. If now the optical thickness, $n_H\, l_H$, of the layer is equal to $\lambda_0/4$, the phase shift after the round trip in the high refractive index layer will be $\phi_2 = 2kl_H = (4\pi n_H/\lambda)l_H = \pi$. This means that the two reflected waves have the same phase and the corresponding fields will add. One can easily show that the same conclusion applies for the two interfaces of a low-index layer. It then follows that all reflected beams in a multilayer dielectric coating, as well as their multiple reflections, add in phase. If therefore a sufficient number of $(\lambda/4)$ layers of alternating low and high indices are deposited, the overall reflectivity, due to all of the multiple reflections, can reach a very high value. If the multilayer stack starts and ends with a high-index layer, so that there is an odd number, J, of layers, the resulting power reflectivity (at $\lambda = \lambda_0$) turns out to be

$$R(\lambda_0) = \left(\frac{n_H^{J+1} - n_L^{J-1} n_s}{n_H^{J+1} + n_L^{J-1} n_s} \right)^2 \tag{4.4.1}$$

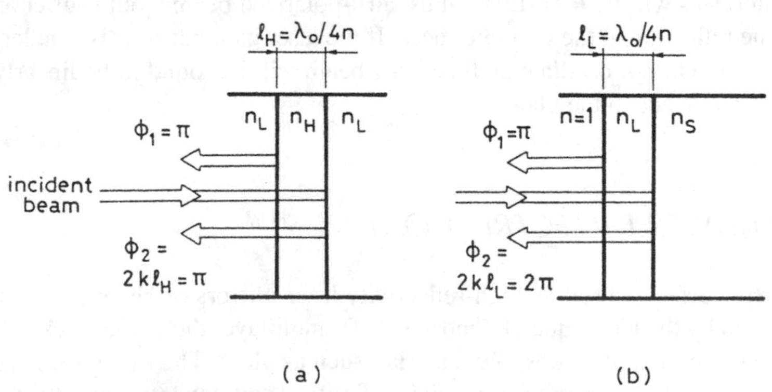

FIG. 4.8. (a) First two reflections at the two interfaces of a high index layer in a multilayer dielectric coating. (b) First two reflections at the two interfaces of a low index layer in a single layer antireflection coating. Multiple reflections also occur [see e.g. the case of a Fabry-Perot interferometer], but are not shown in the figures.

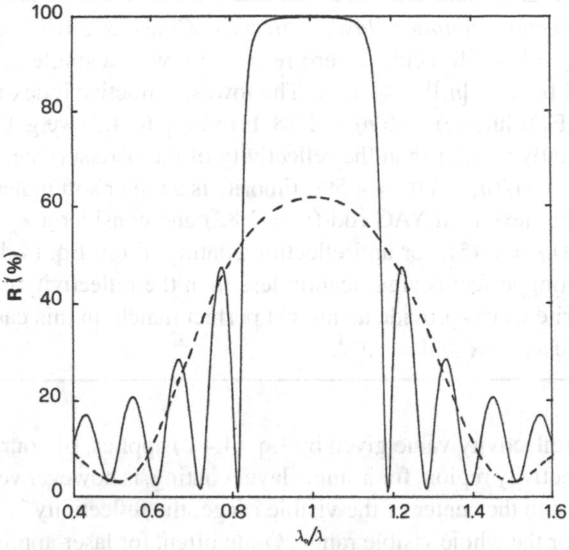

FIG. 4.9. Reflectivity versus wavelength curves of a $\lambda/4$ multilayer dielectric stack made of TiO_2 and SiO_2 for a total number of layer of 3, dashed curved, and 15, continuous curve (substrate material BK7-glass).

Example 4.1. *Peak reflectivity calculation in multilayer dielectric coatings.* We will consider TiO_2 and SiO_2 for the high and low index materials, respectively. At the Nd:YAG laser wavelength of $\lambda_0 = 1.06\ \mu m$, one has $n_H = 2.28$ and $n_L = 1.45$. Taking BK-7 glass as substrate material one has $n_s \cong 1.54$. From Eq. (4.4.1) we obtain $R \cong 61.8\%$ for $J = 3$ and $R = 99.8\%$ for $J = 15$. We also note that, the reflectance at a single interface, according to Eq. (4.3.1) is, in our example, $[(n_H - n_L)/(n_H + n_L)]^2 = 4.9\%$.

If the wavelength λ of the incident wave is different from λ_0, the reflectivity will of course be lower than the value given by Eq. (4.4.1). As representative examples, Fig. 4.17 shows curves of reflectivity versus wavelength for $J = 15$ and $J = 3$. One notices that the peak reflectivity value obviously increases with the number of layers and that the high reflectivity region gets wider and has steeper edges as the number of layers is increased. One can also observe that, for the high reflectivity curve, high reflectivity is maintained over a wavelength range $\Delta\lambda = \lambda - \lambda_0 \cong \pm(10\%)\lambda_0$.

To reduce the reflectivity of a given optical surface, a single layer coating of a material with refractive index lower than that of the substrate can be used. As one can easily see from Fig. 4.8b, since $n_L < n_s$, the first two reflections now have opposite phases if $n_L l_L = \lambda_0/4$. The overall reflectivity is thus reduced and, after taking account of all multiple reflections, one can show that the reflectivity at $\lambda = \lambda_0$ is given by

$$R = \left[\left(n_s - n_L^2\right) / \left(n_s + n_L^2\right)\right]^2 \qquad (4.4.2)$$

From this one notes that zero reflection would be obtained when $n_L = (n_s)^{1/2}$, a condition which is difficult to achieve, in practice, due to the limited number of available materials with low enough refractive index.

Example 4.2. *Single layer antireflection coating of laser materials.* Consider a BK-7 glass substrate for which, at $\lambda = 1.06\ \mu m$ one has $n_s = 1.54$. To achieve zero reflectivity with a single layer, the refractive index of the layer material should be $n_L = [n_s]^{1/2} \cong 1.24$. The lowest refractive index material available as a stable film is provided by MgF_2 (Fluorite) with $n_L = 1.38$. From Eq. (4.4.2) we get $R \cong 1.1\%$ which, although not zero, is still significantly smaller than the reflectivity of the bare surface, which, according to Eq. (4.3.1), is given by $R = [(n_s - 1)/(n_s + 1)]^2 = 4.5\%$. Fluorite is a rather soft material, however, and it can easily be scratched. Consider next a Nd:YAG rod ($n_s = 1.82$) and consider a $\lambda/4$ layer of SiO_2, a rather hard and durable material ($n_L = 1.45$), for antireflection coating. From Eq. (4.4.2) we get in this case $R = 3.4\%$ which while far from perfect is significantly less than the reflectivity of the bare surface ($R \cong 8.5\%$). One notes that Fluorite would provide an almost perfect match, in this case, the reflectivity according to Eq. (4.4.2) being reduced to $R \cong 4 \times 10^{-4}$.

The minimum reflectivity value given by Eq. (4.4.2) applies, of course, for $\lambda = \lambda_0$. The width of the low reflectivity region, for a single layer coating, is however very large. For example, if λ_0 corresponds to the center of the visible range, the reflectivity is reduced below that of the bare surface for the whole visible range. Quite often, for laser applications, even lower reflectivities than those considered in example 4.2 may be required (down to perhaps 0.1%). This can be achieved using more than one layer in the antireflection coating. A coating consisting of two, $\lambda/4$, layers of low and high refractive index material, with the sequence $n_s/n_L/n_H$, is often used for glass. A very hard and durable two-layer coating, which is often used is ZrO_2 ($n_H = 2.1$)-MgF_2 ($n_L = 1.38$). The region of low reflectivity is reduced, for this type of coating, with the reflectivity versus wavelength curve having a sharp, V-shaped, minimum. Such a coating is commonly referred to as a V-coating.

4.5. THE FABRY-PEROT INTERFEROMETER[5]

We now go on to consider a second example of multiple interference, the case of a Fabry-Perot (FP) interferometer. This interferometer, a common spectroscopic tool since its introduction in 1899, plays a very important role in laser physics for at least three different reasons: (1) On a fundamental level its physical behavior forms a basis to the behavior of optical resonators. (2) It is often used as a frequency selective element in a laser cavity. (3) It is often used as a spectrometer for analyzing the spectrum of the light emitted by a laser.

4.5.1. Properties of a Fabry-Perot Interferometer

The FP interferometer consists of two plane or spherical mirrors with power reflectivities R_1 and R_2, separated by a distance L and containing a medium of refractive index n_r. Although, for the ultimate performance, interferometers make use of spherical mirrors, we will, for simplicity, consider here the case of two plane and parallel mirrors. In this case, consider a plane wave of frequency ν incident on the interferometer in a direction making an angle θ' with the normal to the two mirrors (Fig. 4.10). This wave is indicated schematically by the ray

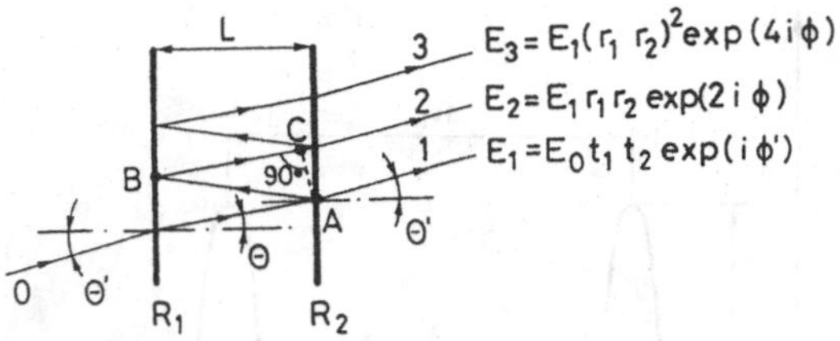

FIG. 4.10. Multiple-beam interference in a Fabry-Perot interferometer.

0 in Fig. 4.10. The output beam, leaving the interferometer, will consist of the superposition of the beam resulting from a single pass through the two mirrors (ray 1 in Fig. 4.10) with the beams arising from all multiple reflections, two of which are indicated by rays 2 and 3 in the figure. Thus, the electric field amplitude of the output beam E_t is obtained by summing the amplitudes E_l of all these beams, taking proper account of their corresponding phase-shifts. To illustrate this, the electric fields of the first three beams are also indicated in the figure. If all multiple reflections are taken into account, we get

$$E_t = \sum_{1}^{\infty}{}_l E_l = [E_0 t_1 t_2 \exp(j\phi')] \sum_{0}^{\infty}{}_m (r_1 r_2)^m \exp(2mj\phi) \tag{4.5.1}$$

In both Eq. (4.5.1) and Fig. 4.10, E_0 is the amplitude of the beam incident on the interferometer; t_1 and t_2 are the electric field transmissions of the two mirrors and r_1 and r_2 are the corresponding electric field reflectivities; ϕ' is the phase shift for a single pass and it also includes any phase shift due to passage through the two mirrors; 2ϕ is the phase difference between successive multiple reflections and is given by $2\phi = kL_s = 2kL\cos\theta = (4\pi n_r \nu/c)L\cos\theta$, where L_s is the sum of the lengths of the two segments AB and BC of Fig. 4.10, and where the angle θ is related to the incidence angle θ' by Snell's law ($n_r \sin\theta = \sin\theta'$). Note that the previous expression can, for simplicity, be transformed to

$$\phi = 2\pi L'\nu/c \tag{4.5.2}$$

where

$$L' = n_r L\cos\theta \tag{4.5.3}$$

The geometrical series appearing in Eq. (4.5.1) can be readily summed to give

$$E_t = E_0\, e^{j\phi'} \frac{t_1 t_2}{1 - (r_1 r_2)\exp(2j\phi)} \tag{4.5.4}$$

The power transmission T_{FP} of the Fabry-Perot interferometer is simply given by $T_{FP} = |E_t|^2/|E_0|^2$ and from Eq. (4.5.4) we get

$$T_{FP} = \frac{t_1^2\, t_2^2}{1 - 2r_1 r_2 \cos(2\phi) + r_1^2 r_2^2} \tag{4.5.5}$$

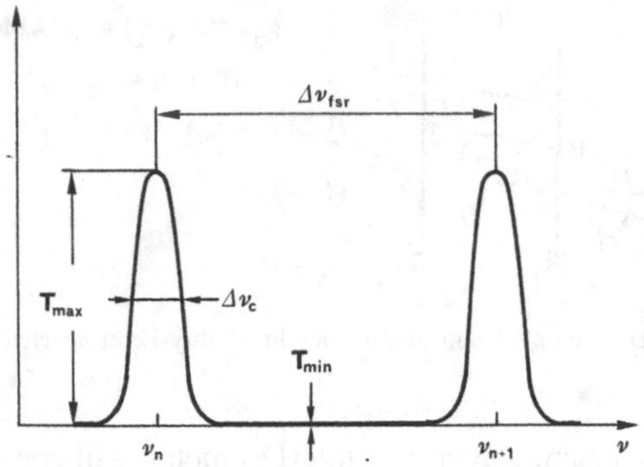

FIG. 4.11. Intensity transmission of a Fabry-Perot interferometer versus the frequency of the incident wave.

Since $R_1 = r_1^2$, $R_2 = r_2^2$, and, for a lossless mirror, $t_1^2 = 1 - r_1^2 = 1 - R_1$ and $t_2^2 = 1 - r_2^2 = 1 - R_2$, Eq. (4.5.5) transforms to

$$T_{FP} = \frac{(1 - R_1)(1 - R_2)}{\left[1 - (R_1R_2)^{1/2}\right]^2 + 4(R_1R_2)^{1/2}\sin^2\phi} \tag{4.5.6}$$

which is the final result of our calculation.

To illustrate the properties of the FP interferometer, Fig. 4.11 shows a plot of transmission T versus frequency of the incident wave for $R_1 = R_2 = 64\%$. This plot is obtained from Eq. (4.5.6), with Eq. (4.5.2) used for ϕ. One sees that the curve consists of a series of evenly spaced maxima. These maxima occur when $\sin^2\phi = 0$ in Eq. (4.5.2), i.e. when $\phi = m\pi$, where m is a positive integer. With the help of Eq. (4.5.2) the frequencies ν_n of these maxima are seen to be given by

$$\nu_n = mc/2L' \tag{4.5.7}$$

The frequency difference between two consecutive maxima, for reasons which will become clear at the end of this section, is called the free-spectral range of the interferometer, $\Delta\nu_{fsr}$. From Eq. (4.5.7) we immediately get

$$\Delta\nu_{fsr} = c/2L' \tag{4.5.8}$$

At a transmission maximum one has $\sin\phi = 0$ and the value of the transmission is seen from Eq. (4.5.6) to be

$$T_{max} = \frac{(1 - R_1)(1 - R_2)}{\left[1 - (R_1R_2)^{1/2}\right]^2} \tag{4.5.9}$$

Note that if $R_1 = R_2 = R$ then $T_{max} = 1$ irrespective of the value of the mirror reflectivity R. This result only holds if the mirrors have no absorption, as assumed in our analysis here.

The transmission minima occur when $\sin^2\phi = 1$, i.e., midway between maxima. The transmission at this minimum point is obtained from Eq. (4.5.6) as

$$T_{\min} = \frac{(1-R_1)(1-R_2)}{\left[1+(R_1R_2)^{1/2}\right]^2} \tag{4.5.10}$$

Note that, under usual circumstances, the value of $T_{\min}$ is very small (see example 4.3).

To calculate the width, $\Delta\nu_c$, of a transmission peak, we notice that, according to Eq. (4.5.6), the transmission will fall to $^1/_2$ of its maximum value for a displacement $\Delta\phi$ from the value $\phi = n\pi$ such that $4(R_1R_2)^{1/2}\sin^2\Delta\phi = [1-(R_1R_2)^{1/2}]^2$. Assuming that $\Delta\phi$ is much smaller than π we can make the approximation $\sin\Delta\phi \cong \Delta\phi$, which gives $\Delta\phi = \pm[1-(R_1R_2)^{1/2}]/2[R_1R_2]^{1/4}$. This last equation shows that the two "half-intensity" points, corresponding to $\Delta\phi_+$ and $\Delta\phi_-$, are symmetrically situated at either side of the maximum. If we let $\Delta\phi_c = \Delta\phi_+ - \Delta\phi_-$, then we get

$$\Delta\phi_c = \frac{1-(R_1R_2)^{1/2}}{(R_1R_2)^{1/4}} \tag{4.5.11}$$

and, from (4.5.2)

$$\Delta\nu_c = \frac{c}{2L'}\frac{1-(R_1R_2)^{1/2}}{\pi(R_1R_2)^{1/4}} \tag{4.5.12}$$

We now define the "*finesse*," F, of the interferometer as

$$F = \Delta\nu_{fsr}/\Delta\nu_c \tag{4.5.13}$$

From Eq. (4.5.8) and Eq. (4.5.12) we then get

$$F = \frac{\pi(R_1R_2)^{1/4}}{1-(R_1R_2)^{1/2}} \tag{4.5.14}$$

The finesse indicates how narrower the transmission peak is compared to the free spectral range: typically it is much greater than 1.

Example 4.3. *Free-spectral range, finesse and transmission of a Fabry-Perot etalon.* Consider a F–P interferometer made of a piece of glass with two plane-parallel surfaces coated for high reflectivity (often called a F–P etalon). If we assume $L = 1$ cm and $n_r = 1.54$, the free-spectral range for near normal incidence, i.e. for $\theta \cong 0$, is $\Delta\nu_{\text{fsr}} = c/2n_rL = 9.7$ GHz. If we now take $R_1 = R_2 = 0.98$, we get from Eq. (4.5.14) a finesse $F \cong 150$, so that $\Delta\nu_c = \Delta\nu_{\text{fsr}}/F = 65$ MHz. For a lossless coating, the peak transmission, according to Eq. (4.5.9) is $T_{\max} = 1$, while the minimum transmission, from Eq. (4.5.10), is $T_{\min} \cong 10^{-4}$. Note the very small value of $T_{\min}$.

The previous expressions and considerations hold for perfectly lossless mirrors. For finite mirror absorbance we will assume, for simplicity, the same reflectivity and the same transmission for the two mirrors, i.e. we will take $R_1 = R_2 = R$ and $t_1^2 = t_2^2 = T$, where T is the mirror transmission. From Eq. (4.5.5) one then readily gets that the transmission of the Fabry-Perot interferometer, T_{FP}, can now be written as

$$T_{FP} = \left[\frac{T}{1-R}\right]^2 \frac{(1-R)^2}{(1-R)^2 + 4R\sin^2\phi} \tag{4.5.14a}$$

Note tha T is now given by $T=1-R-A$ where A is the fraction of the incident power absorbed by the mirror (mirror absorption). For finite absorption, one therefore has $T<1-R$. The comparison of Eq. (4.5.14a) with Eq. (4.5.6), when $R_1=R_2=R$, then shows that mirror absorption reduces the overall transmission of the interferometer by a factor $[(T/(1-R)]^2$.

4.5.2. The Fabry-Perot Interferometer as a Spectrometer

After this general description of the properties of a FP interferometer we now describe its use as a spectrum analyzer. We consider the simplest case where the direction of the incident light is normal to the interferometer mirrors (i.e., $\cos\theta=1$) and the medium inside the interferometer is air ($n_r \cong 1$). We assume that the length L can be changed by a few wavelengths by, e.g., attaching one of the two FP plates to a piezoelectric transducer (scanning FP interferometer). To understand what happens in this case let us first consider a monochromatic wave at frequency ν (wavelength λ). According to the previous discussion, the transmitted light will exhibit peaks when $\phi=m\pi$, i.e., when the interferometer length is equal to $L=m\lambda/2$ (see Fig. 4.12a), where m is a positive integer. The change in L needed to shift from one transmission peak to the next one is then

$$\Delta L_{fsr} = \lambda/2 \tag{4.5.15}$$

The width of each transmission peak, ΔL_c, will be such that $(2\pi\nu/c_0)\,\Delta L_c=\Delta\phi_c$ where $\Delta\phi_c$ is given by Eq. (4.5.11). With the help of Eq. (4.5.14) we then get $\Delta L_c=\lambda/2F$. We therefore have

$$\Delta L_c = \Delta L_{fsr}/F \tag{4.5.16}$$

i.e. the analogous relation to Eq. (4.5.13).

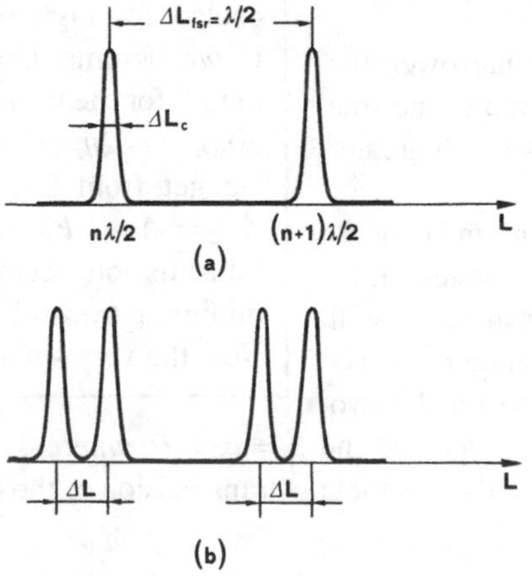

FIG. 4.12. Intensity transmission of a scanning Fabry-Perot interferometer when the incident wave is: (a) monochromatic, (b) made up of two, closely spaced, frequencies.

We now consider the case where two waves at frequencies ν and $\nu + \Delta\nu$ are incident on the interferometer. The wave at frequency $(\nu + \Delta\nu)$ will produce a set of transmission peaks displaced by a quantity ΔL from those corresponding to frequency ν (Fig. 4.12b). Since $2\pi L\nu/c = n\pi$, the displacement ΔL must be such that $2\pi(L + \Delta L)(\nu + \Delta\nu)/c = n\pi$, i.e. such that $\Delta L = -(\Delta\nu/\nu)L$. The two frequencies ν and $\nu + \Delta\nu$ will be resolved by the spectrometer if $|\Delta L| \geq \Delta L_c$. The equality sign in this expression corresponds to the minimum frequency interval $\Delta\nu_m$ which can be resolved, which gives $(\Delta\nu_m/\nu)L = \Delta L_c$. With the help of Eq. (4.5.16) and Eq. (4.5.15) we then get $(\Delta\nu_m/\nu)L = \lambda/2F$. Using Eq. (4.5.8) with $L' = L$ we obtain

$$\Delta\nu_m = \Delta\nu_{fsr}/F \tag{4.5.17}$$

Thus the finesse of the interferometer specifies its resolving power in terms of the free spectral range.

It must be noted that when $|\Delta L| = \Delta L_{fsr}$, i.e. when $\Delta\nu = \Delta\nu_{fsr} = c/2L$, the transmission peaks at frequencies $\nu + \Delta\nu$ and ν will be coincident, although shifted by one order relative to each other. Therefore, when $\Delta\nu > \Delta\nu_{fsr}$, an ambiguity by a multiple of $\Delta\nu_{fsr}$ occurs in the measurement of $\Delta\nu$. Thus, when using the interferometer to provide a measurement of frequency difference, a simple and unambiguous result is only obtained when $\Delta\nu < \Delta\nu_{fsr}$, which explains why $\Delta\nu_{fsr}$ is called the free spectral range of the interferometer. We can readily generalize the above result and say that if $\Delta\nu_{osc}$ is the spectral bandwidth of the incident light, then to avoid frequency ambiguity, we must have $\Delta\nu_{osc} \leq \Delta\nu_{fsr}$. If the equality is assumed to hold in this relation, then from Eq. (4.5.17) we get

$$\Delta\nu_m = \Delta\nu_{osc}/F \tag{4.5.18}$$

Thus the *finesse F* also provides a measure of how *finely* we can discriminate frequencies within the total spectral bandwidth $\Delta\nu_{osc}$.

Example 4.4. *Spectral measurement of an Ar^+-laser output beam.* We will consider an Ar-ion laser oscillating on its green line at $\lambda = 514.5$ nm wavelength. We will assume the laser to be oscillating on many longitudinal modes encompassing the full Doppler width of the laser line ($\Delta\nu_0^* = 3.5$ GHz). We will thus have $\Delta\nu_{osc} = \Delta\nu_0^* = 3.5$ GHz. To avoid frequency ambiguity, we must have $\Delta\nu_{fsr} = (c/2L) \geq 3.5$ GHz i.e. $L \leq 4.28$ cm. If we now assume a finesse $F = 150$ and take $L = 4.28$ cm, according to Eq. (4.5.18) we have, for the interferometer resolution, $\Delta\nu_m = \Delta\nu_{osc}/F \cong 23$ MHz. If, for example, the length of the laser cavity is $L_1 = 1.5$ m, consecutive longitudinal modes are separated (see next Chapter) by $\Delta\nu = c/2L_1 = 100$ MHz. Thus, since $\Delta\nu_m < \Delta\nu$, the *FP* interferometer is able to resolve these longitudinal modes. One can also observe that, since the frequency of the laser light is $\nu = c/\lambda \cong 5.83 \times 10^{14}$ Hz, the corresponding resolving power of the interferometer is $\nu/\Delta\nu_m = 2.54 \times 10^7$. This is a very high resolving power compared, e.g., to the best that can be obtained with a grating spectrometer ($\nu/\Delta\nu < 10^6$).

4.6. DIFFRACTION OPTICS IN THE PARAXIAL APPROXIMATION[6]

We shall consider a monochromatic wave under the so-called scalar approximation where the e.m. fields are uniformly (e.g. linearly or circularly) polarized. The electric field of the wave can then be described by a scalar quantity viz

$$E(x, y, z, t) = \tilde{E}(x, y, z)\exp(j\omega t) \tag{4.6.1}$$

where the complex amplitude $\tilde{E}$ must satisfy the wave equation in scalar form i.e.

$$\left(\nabla^2 + k^2\right)\tilde{E}(x, y, z) = 0 \tag{4.6.2}$$

with $k = \omega/c$. An integral solution for the field amplitude can be obtained using the Fresnel-Kirchoff integral. For a given field distribution $\tilde{E}(x_1, y_1, z_1)$ in the $z = z_1$ plane, the resulting field distribution $\tilde{E}(x, y, z)$, at a general plane at coordinate z along the propagation direction, turns out to be given by

$$\tilde{E}(x, y, z) = \frac{J}{\lambda} \iint_S \tilde{E}(x_1, y_1, z_1) \frac{\exp-(jkr)}{r} \cos\theta \, dx_1 \, dy_1 \tag{4.6.3}$$

In Eq. (4.6.3) r is the distance between point P_1, of coordinates (x_1, y_1), and point P, of coordinates (x, y), (see Fig. 4.13), θ is the angle that the segment P_1P makes with the normal to the plane $z = z_1$, the double integral is taken over the coordinates x_1, y_1 in the $z = z_1$ plane, and the limits are defined by some general aperture S located in the plane. One can see that Eq. (4.6.3) is really the expression of the Huygens principle in mathematical form. Indeed $\left[\tilde{E}(x_1, y_1, z_1)\, dx_1 dy_1\right]$ $[\exp-(jkr)]/r$ represents the Huygens' wavelet originating from the elemental area $dx_1 dy_1$ around P_1 and the field at point P is obtained by summing the wavelets coming from all points in the plane $z = z_1$. The term $\cos\theta$ is the so-called obliquity factor, the need for which was recognized by Fresnel. The (j/λ) term in front of the integral is a normalization factor which arises from a detailed treatment of the theory. It indicates that the Huygens wavelets have a $\pi/2$ phase-shift compared to the beam which is incident at $z = z_1$ plane.

We will now consider the E-field solutions either in differential [Eq. (4.6.2)] or integral forms [Eq. (4.6.3)], within the paraxial wave approximation where the wave is assumed to be propagating at a small angle, θ, to the z-direction. In this case we can write

$$\tilde{E}(x, y, z) = u(x, y, z)\exp-(jkz) \tag{4.6.4}$$

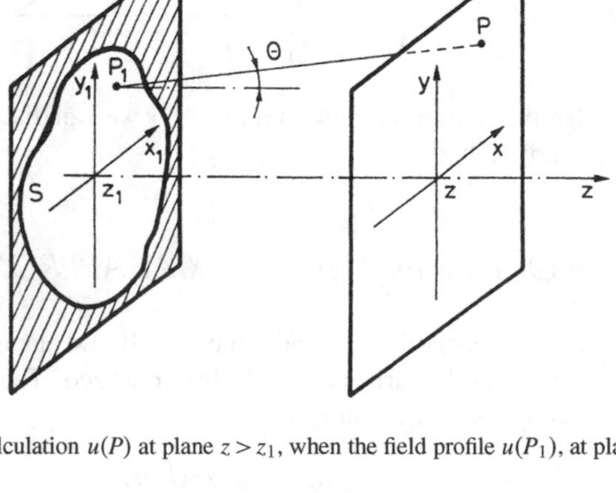

FIG. 4.13. Field calculation $u(P)$ at plane $z > z_1$, when the field profile $u(P_1)$, at plane $z = z_1$, is known.

where u is a slowly varying function, i.e., varying little on a wavelength scale, in z coordinate. Under the paraxial approximation, the substitution of Eq. (4.6.4) into Eq. (4.6.2) gives

$$\nabla_{\perp}^2 u - 2jk\frac{\partial u}{\partial z} = 0 \tag{4.6.5}$$

where $\nabla_{\perp}^2 = (\partial^2/\partial x^2) + (\partial^2/\partial y^2)$. Equation (4.6.5) is the paraxial wave equation.

To obtain an approximate form of Eq. (4.6.3), under the paraxial wave approximation, we write $\cos\theta \cong 1$ and $r = z - z_1$ in the amplitude factor of the spherical wavelet. In considering approximation of the phase factor, $-kr$, we must be more careful, however. In fact, consider, for instance, a distance $r \cong 1$ m and assume that this distance is evaluated with an accuracy of $\Delta r = 1$ μm. For the amplitude factor, this would give the very good relative accuracy of $\Delta r/r = 10^{-6}$. The phase accuracy, however, would be $\Delta\phi = k\Delta r = 2\pi\Delta r/\lambda$ and for $\lambda = 1$ μm, it would give $\Delta\phi = 2\pi$, an unacceptable level of accuracy as, for example, a phase change $\Delta\phi = \pi$ changes the sign of the entire phase term in the integral. Thus a better accuracy is needed for the phase term in Eq. (4.6.3). To this purpose we write the distance r between points P_1 and P of Fig. 4.13 as $r = \left[(z-z_1)^2 + (x-x_1)^2 + (y-y_1)^2\right]^{1/2}$. Under the paraxial wave approximation one has $[|x - x_1|, |y - y_1|] \ll |z - z_1|$. We can therefore write

$$\begin{aligned} r &= (z - z_1)\left[1 + \frac{(x-x_1)^2 + (y-y_1)^2}{(z-z_1)^2}\right]^{1/2} \\ &\cong (z - z_1) + \frac{(x-x_1)^2 + (y-y_1)^2}{2(z-z_1)} \end{aligned} \tag{4.6.6}$$

The substitution of Eq. (4.6.6) into the phase term of Eq. (4.6.3) then gives

$$\tilde{E}(x, y, z) = \frac{j\exp -jk(z - z_1)}{\lambda(z - z_1)} \iint \tilde{E}(x_1, y_1, z_1)\exp -jk\left[\frac{(x-x_1)^2 + (y-y_1)^2}{2(z-z_1)}\right] dx_1 dy_1 \tag{4.6.7}$$

which is the Huygens-Fresnel-Kirchoff integral in the so-called Fresnel approximation. The substitution of Eq. (4.6.4) into Eq. (4.6.7) then gives

$$u(x, y, z) = \frac{j}{\lambda L}\iint_s u(x_1, y_1, z_1)\exp -jk\left[\frac{(x-x_1)^2 + (y-y_1)^2}{2L}\right] dx_1 dy_1 \tag{4.6.8}$$

where we have put $L = z - z_1$. Equation (4.6.8) provides a solution for the E-field in integral form within the paraxial wave approximation while (4.6.5) gives the same solution in differential form. It can be shown, however, that the two forms are completely equivalent.

We next consider wave propagation, within the paraxial approximation, through a general optical system described by the $ABCD$ matrix of sect. 4.2. With reference to Fig. 4.14, we let $u(x_1, y_1, z_1)$ and $u(x, y, z)$ be the field amplitudes at planes $z = z_1$ and $z = z$ just before and after the optical system, respectively. We also assume that the Huygens principle applies to a general optical system of Fig. 4.14 provided that no field-limiting apertures are present in the optical system. This would for instance imply that any lens or mirror within the optical system

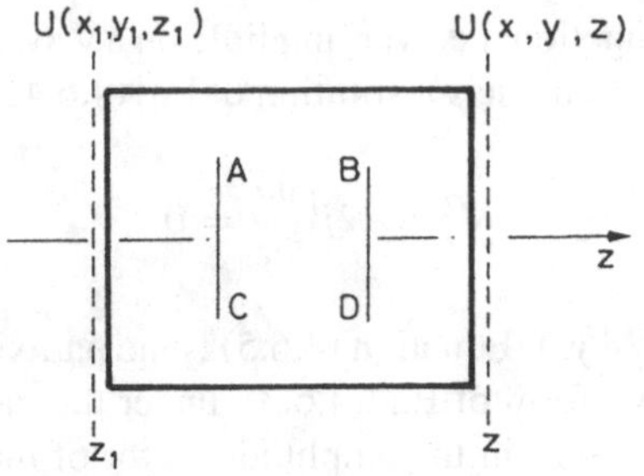

FIG. 4.14. Field calculation, $u(x, y, z)$, at plane z after an optical system described by the *ABCD* matrix, when the field profile, $u(x_1, y_1, z_1)$ at plane $z = z_1$ is known.

has an "infinite" aperture i.e. an aperture much wider than the transverse dimensions of the field*. According to this extension of the Huygens principle to a general optical system, the field $u(x, y, z)$ is obtained by the superposition of the individual wavelets emitted from plane $z = z_1$ and transmitted through this system. One then obtains[7]

$$u(x, y, z) = \frac{1}{B\lambda} \iint_S u(x_1, y_1, z_1) \exp -jk \left[\frac{A\left(x_1^2 + y_1^2\right) + D\left(x^2 + y^2\right) - 2x_1 x - 2yy_1}{2B} \right] dx_1 dy_1 \tag{4.6.9}$$

which constitutes a generalization of Eq. (4.6.8). Obviously, for free space we have (see Table 4.1) $A = D = 1$ and $B = L$ and Eq. (4.6.9) reduces to Eq. (4.6.8).

4.7. GAUSSIAN BEAMS

We now go on to discuss a very important class of E-field solutions, commonly called *Gaussian beams*. The properties of these beams, in the paraxial wave approximation, could be derived either via the paraxial wave equation Eq. (4.6.5) or via the Fresnel-Kirchoff integral in the Fresnel approximation [see Eqs. (4.6.8) and (4.6.9)]. We will follow the integral approach since it proves to be more useful also for describing the properties of optical resonators, to be discussed in the next Chapter.

4.7.1. Lowest-Order Mode

Consider a general optical system described by its corresponding (*ABCD*) matrix (see Fig. 4.14). We may ask the following question: is there any solution of Eq. (4.6.9) that retains its functional form as it propagates? In other words, is there any eigensolution of Eq. (4.6.9)? An answer is readily obtained if we assume that there is no limiting aperture in the $z = z_1$

* For "finite" apertures of the optical system, diffraction effects would be produced at these apertures thus sizeably changing the transmitted field.

plane so that the double integral of Eq. (4.6.9) can be taken between $-\infty$ and $+\infty$ for both x_1 and y_1 variables. In this case we can readily show by direct substitution in Eq. (4.6.9) that

$$u(x, y, z) \propto \exp -jk\left[\left(x^2+y^2\right)/2q\right] \tag{4.7.1}$$

where $q = q(z)$ is a complex parameter (often called the complex beam parameter of a Gaussian beam), is an eigensolution of Eq. (4.6.9). If we write, in fact,

$$u\left(x_1, y_1, z_1\right) \propto \exp -jk\left[\left(x_1^2+y_1^2\right)/2q_1\right] \tag{4.7.2}$$

we get from Eq. (4.6.9)

$$u(x, y, z) = \frac{1}{A+\left(B/q_1\right)} \exp -jk\, \frac{x^2+y^2}{2q} \tag{4.7.3}$$

where q is related to q_1 by the very simple law

$$q = \frac{A\, q_1+B}{C\, q_1+D} \tag{4.7.4}$$

Equation (4.7.4) is a very important relation, known as the *ABCD* law of Gaussian beam propagation. It bears an obvious similarity to Eq. (4.2.19) which shows how the radius of curvature of a spherical wave is transformed by an optical system. We will come back to this equation in sect. 4.7.3. for a deeper discussion.

We now go on to discuss a physical interpretation of the Gaussian solution of Eq. (4.7.1). For this we use Eqs. (4.7.1) and (4.6.4) and write

$$\tilde{E} \propto \exp -jk\left[z+\frac{x^2+y^2}{2q}\right] \tag{4.7.5}$$

Consider now a spherical wave with center at coordinates $x = y = z = 0$. Its field at point $P(x,\ y,\ z)$ can be written as $\tilde{E} \propto [\exp -jk\, R]\, /R$, where R is the wave's radius of curvature. Within the paraxial approximation, following a similar argument to that in Eq. (4.6.6), we write

$$R \cong z+\frac{x^2+y^2}{2\, R} \tag{4.7.6}$$

and the field of the spherical wave transforms to

$$\tilde{E} \propto \exp -jk\left[z+\frac{x^2+y^2}{2\, R}\right] \tag{4.7.7}$$

A comparison of Eq. (4.7.7) with Eq. (4.7.5) then shows that the Gaussian beam can be looked upon as a spherical wave of complex radius of curvature, q. To understand the meaning of this complex beam parameter we separate the real and imaginary part of $1/q$, i.e. we write

$$\frac{1}{q} = \frac{1}{R} - j\frac{\lambda}{\pi\, w^2} \tag{4.7.8}$$

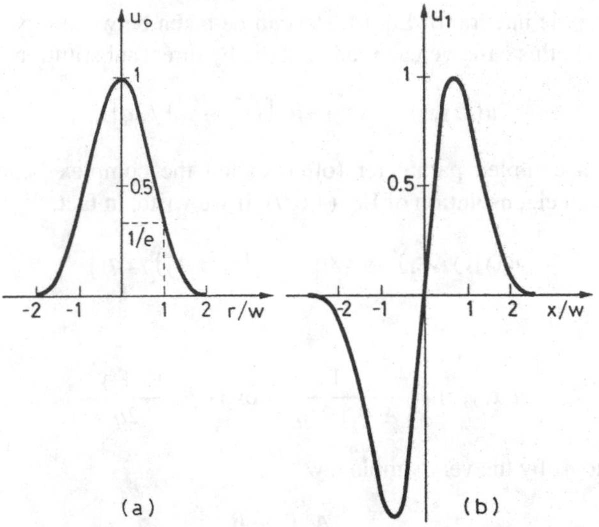

FIG. 4.15. Field profile of, (a), the lowest order and, (b), next order Gaussian mode.

The substitution of Eq. (4.7.8) into Eq. (4.7.5) then gives

$$\tilde{E}(x, y, z) \propto \exp -\frac{x^2 + y^2}{w^2} \times \exp -jk\left[z + \frac{x^2 + y^2}{2R}\right] \tag{4.7.9}$$

The amplitude factor on the right hand side of Eq. (4.7.9) i.e. $u_0 = \exp -[(x^2 + y^2)/w^2]$ is plotted in Fig. 4.15a vs r/w, where $r = [x^2 + y^2]^{1/2}$ is the radial beam coordinate. One sees that the maximum value is reached at $r = 0$ and that, for $r = w$, one has $u_0 = 1/e$. The quantity w therefore defines the transverse scale of the beam and is called the beam spot size (at the z-position considered). One can also notice that, since the beam intensity is given by $I \propto |\tilde{E}|^2$, we have $I = I_{\max} \exp -[2(x^2 + y^2)/w^2]$. If we define the spot size of the intensity profile, w_I as the value at which $I = I_{\max}/e$, we then have $w_I = w/\sqrt{2}$. Generally, when referring to a beam spot size, it is the field spot size, w, that is implied rather than the intensity spot size. Note that the intensity I reduces to $1/e^2$ of its maximum at a radial distance of one field's spot size. We now turn our attention to the phase factor in Eq. (4.7.9). A comparison with Eq. (4.7.7), which applies to a spherical wave, shows that the two expressions are identical. This leads us to identify R, in Eq. (4.7.8), as the radius of curvature of the spherical wavefront of the Gaussian beam. To see this better, consider the equiphase surface of the Gaussian beam which intercepts the z axis at a given position z'. The x, y, z coordinates of this surface must then satisfy the relation $kz + k(x^2 + y^2)/2R = kz'$, which gives

$$z = z' - \frac{x^2 + y^2}{2R} \tag{4.7.10}$$

Equation (4.7.10) thus shows that the equiphase surface is a paraboloid of revolution around the z-axis. It can be shown further that the radius of curvature of this paraboloid at $x = y = 0$, i.e. on the beam axis, is just equal to R. This demonstrates rather clearly why, within the paraxial wave approximation, the phase terms of the spherical wave, Eq. (4.7.7), and Gaussian beam, Eq. (4.7.9), are the same.

4.7.2. Free Space Propagation

Consider the propagation of the Gaussian beam of Eq. (4.7.1) along the positive z-direction without any restricting aperture in the x or y direction (i.e. in free-space). From Eq. (4.7.4) with $A = D = 1$ and $B = z$ we get

$$q = q_1 + z \tag{4.7.11}$$

Assume that at $z = 0$ one has $R = \infty$. We then write

$$(1/q_1) = -j\left(\lambda/\pi w_0^2\right) \tag{4.7.12}$$

where w_0 is the spot size at $z = 0$. We now write Eq. (4.7.11) as $(1/q) = 1/(q_1 + z)$, substitute $1/q$ from Eq. (4.7.8), and $1/q_1$ from Eq. (4.7.12), and separate the real and imaginary parts of the resulting equation. After some straightforward algebraic manipulation one arrives at the expressions for the spot size, w, and radius of curvature, R, of the equiphase surfaces, at z-coordinate, as

$$w^2(z) = w_0^2\left[1 + \left(\frac{\lambda z}{\pi w_0^2}\right)^2\right] \tag{4.7.13a}$$

$$R(z) = z\left[1 + \left(\frac{\pi w_0^2}{\lambda z}\right)^2\right] \tag{4.7.13b}$$

From Eqs. (4.7.3) and (4.7.12) we also write

$$u(x, y, z) = \left[\frac{1}{1 - j\left(\lambda z/\pi w_0^2\right)}\right] \exp -jk\frac{x^2 + y^2}{2q} \tag{4.7.14}$$

The complex factor in brackets in Eq. (4.7.14) can now be expressed in terms of its amplitude and phase. Using also the expression Eq. (4.7.8) for $(1/q)$, we get the expression for the field amplitude as

$$u(x, y, z) = \frac{w_0}{w} \exp -\frac{x^2 + y^2}{w^2} \exp -jk\frac{x^2 + y^2}{2R} \exp j\phi \tag{4.7.15}$$

where

$$\phi = \tan^{-1}\left(\frac{\lambda z}{\pi w_0^2}\right) \tag{4.7.15a}$$

Equation (4.7.15) together with the expression for $w(z)$, $R(z)$, and $\phi(z)$ given by Eqs. (4.7.13) and (4.7.15a), solve our problem completely. One can see from Eq. (4.7.13) that w, R, and ϕ (and hence the field distribution) depend only on w_0 (for given λ and z). This can be readily understood when we notice that, once w_0 is known, the field distribution at $z=0$ is known. In fact we know its amplitude, since the field distribution is a Gaussian function with spot size w_0, and its phase, since we have assumed $R=\infty$ for $z=0$. Once the field at $z=0$ is known, the corresponding field at $z>0$ is uniquely established as it can be calculated by means, for instance, of the Fresnel-Kirchoff integral Eq. (4.6.8). Again using Eq. (4.7.11), one can show that Eq. (4.7.13) holds also for negative z-values i.e. for forward propagation toward rather than from the $z=0$ plane. It should finally be noted that, if we define

$$z_R = \pi\, w_0^2 / \lambda \tag{4.7.16}$$

where z_R is called the *Rayleigh range* (whose significance will be discussed later), Eq. (4.7.13) can be put in the more suggestive form

$$w^2(z) = w_0^2\left[1 + (z/z_R)^2\right] \tag{4.7.17a}$$

$$R(z) = z\left[1 + (z_R/z)^2\right] \tag{4.7.17b}$$

$$\phi(z) = \tan^{-1}\,(z/z_R) \tag{4.7.17c}$$

Equation (4.7.15) together with Eq. (4.7.17) are the final results of our calculations. One sees that $u(x, y, z)$ is made up of a the product of an amplitude factor, $(w_0/w)\exp -\left[(x^2+y^2)/w^2\right]$, with a transverse phase factor, $\exp -jk\left[(x^2+y^2)/2R\right]$, and a longitudinal phase factor $\exp i\phi$. The physical meaning of these factors will now be discussed in some detail.

The amplitude factor in Eq. (4.7.15) shows that the beam, while propagating (both for $z>0$ and $z<0$), retains its Gaussian shape but its spot size changes according to Eq. (4.7.17a). One thus sees that $w^2(z)$ can be written as the sum of w_0^2 and $(\lambda z/\pi w_0)^2$, a term arising from beam diffraction. A plot of the normalized spot size, w/w_0, vs the normalized propagation length, z/z_R, is shown as a solid line in Fig. 4.16a for $z>0$. For $z<0$, the spot size is readily obtained from the same figure since $w(z)$ is a symmetric function of z. Thus the minimum spot size occurs at $z=0$ (hence referred to as the beam waist) and, for $z=z_R$, one has $w=\sqrt{2}w_0$. The Rayleigh range is thus the distance from the beam waist to where the spot has increased

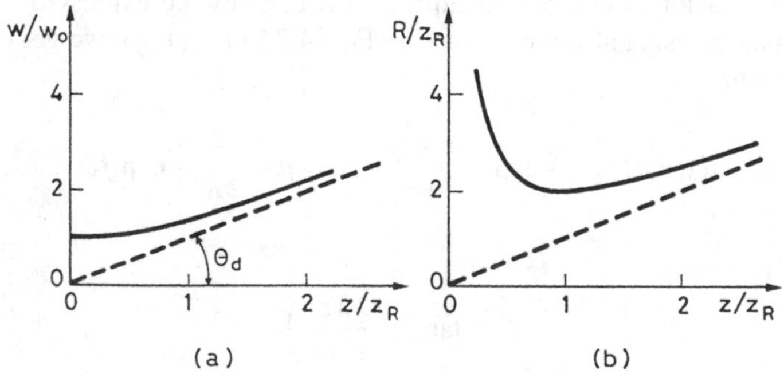

FIG. 4.16. Normalized values of the beam spot size, w, (a) and radius of curvature of the equiphase surface, R, (b), vs normalized values of the propagation length, z.

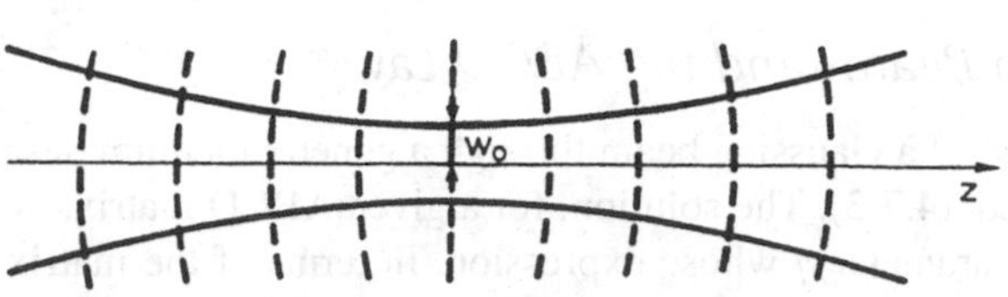

FIG. 4.17. Beam profile, continuous curves, and equiphase surfaces, dashead curves, for a TEM_{00} Gaussian mode.

by a factor $\sqrt{2}$. One can also observe that, for $z \to \infty$ (i.e., for $z \gg z_R$), one can write

$$w \approx w_0\, z/z_R = \lambda\, z/\pi\, w_0 \tag{4.7.18}$$

Equation (4.7.18) is also plotted as dashed line in Fig. 4.16a. At large distances, w increases linearly with z, hence one can define a beam divergence, due to diffraction, as $\theta_d = w/z$ and thus get

$$\theta_d = \lambda/\pi\, w_0 \tag{4.7.19}$$

The physical reason for the presence of the quantity w_0/w in the amplitude factor of Eq. (4.7.15) is also readily understood when one observes that, since the medium is assumed to be lossless, the total beam power must be the same at any plane z. This requires that $\int\int |u|^2 dxdy$ be independent of z. Now, it is just the presence of the quantity $w_0/w(z)$ that ensures that this condition holds. In fact, using Eq. (4.7.15), we can write

$$\iint |u|^2\, dxdy = \left(w_0^2/2\right) \int_{-\infty}^{+\infty} \exp\left(-\xi^2\right) d\xi \int_{-\infty}^{+\infty} \exp\left(-\eta^2\right) d\eta \tag{4.7.20}$$

where $\xi = \sqrt{2}x/w$ and $\eta = \sqrt{2}\, y/w$. By inspection, one can confirm that $\int\int |u|^2 dxdy$ is independent of z.

Let us now consider the transverse phase factor of Eq. (4.7.15). According to the discussion of the previous section, it indicates that for $z > 0$ the beam acquires, due to propagation, an approximately spherical wavefront with a radius of curvature R. A plot of the normalized radius of curvature, R/z_R vs the normalized variable, z/z_R, is shown in Fig. 4.16b for $z > 0$. For $z < 0$, the radius of curvature is readily obtained from the same figure since $R(z)$ is an antisymmetric function of z. One sees that $R \to \infty$ for $z = 0$, while R reaches its minimum value for $z = z_R$. For $z \gg z_R$ one has $R \approx z$ and the equation $R = z$ is also plotted as dashed line in Fig. 4.16b. Thus the wavefront is plane at $z = 0$ and, at large distances, increases linearly with z, just as for a spherical wave, being plane again at $z = \pm\infty$.

Lastly, we consider the longitudinal phase factor of Eq. (4.7.15). Using Eq. (4.6.4) one sees that the Gaussian beam has, besides the phase shift $-kz$ of a plane wave, an additional term $\phi(z)$ which changes from $-(\pi/2)$ to $(\pi/2)$ on going from $z \ll -z_R$ to $z \gg z_R$.

The results of Fig. 4.16 can be put together in a suggestive form as in Fig. 4.17, where the dimensions of the beam profile $2w(z)$, are shown as solid curves and the equiphase surfaces as dashed lines. The beam is seen to have a minimum dimension in the form of a "waist" at $z = 0$. Therefore, the corresponding spot size, w_0, is usually called the spot size at the beam waist or waist spot size. It should also be noted that, according to the convention used for the sign of wavefront curvature, since $R > 0$ for $z > 0$ and $R < 0$ for $z < 0$, the center of curvature is to the left of the wavefront for $z > 0$ and to the right of the wavefront for $z < 0$.

4.7.3. Gaussian Beams and the ABCD Law[8]

The propagation of a Gaussian beam through a general medium described by an ABCD matrix is given by Eq. (4.7.3). The solution, for a given ABCD matrix, then depends only on the complex beam parameter q whose expression, in terms of the matrix elements, is given by Eq. (4.7.4). This is a very important law of Gaussian beam propagation and it is often referred to as the ABCD – law of Gaussian beams. Its usefulness was already proved for the case of free-space propagation considered in the previous section. In this section we will further illustrate the importance of this law in some other examples which are somewhat more complex.

Example 4.5. *Gaussian beam propagation through a thin lens.* Consider a thin lens of focal length f. According to Eq. (4.7.4), the complex beam parameters just before, q_1, and just after the lens, q_2, are seen to be related by

$$\frac{1}{q_2} = \frac{C + (D/q_1)}{A + (B/q_1)} \tag{4.7.21}$$

With the help of the matrix elements of a lens given in Table 4.1, we then get

$$\frac{1}{q_2} = -\frac{1}{f} + \frac{1}{q_1} \tag{4.7.22}$$

Using Eq. (4.7.8) to express both $1/q_1$ and $1/q_2$, we can separately equate the real and imaginary parts of Eq. (4.7.22) to obtain the following relations between the spot sizes and the radii of curvature before and after the lens:

$$w_2 = w_1 \tag{4.7.23a}$$

$$\frac{1}{R_2} = \frac{1}{R_1} - \frac{1}{f} \tag{4.7.23b}$$

The physical relevance of Eq. (4.7.23) can now be discussed in connection with Fig. 4.18. Considering first Eq. (4.7.23a), one immediately see that its physical meaning is obvious since, for a thin lens, the beam amplitude distributions immediately before and after the lens must be the same i.e., there cannot be a discontinuous change of spot size (see Fig. 4.18a). To understand the meaning of Eq. (4.7.23b), consider first the propagation of a spherical wave through the same lens (Fig. 4.18b). Here a spherical wave originating from a point source P_1 is focused by the lens to the image point P_2. The radii of curvature R_1 and R_2 just before and after the lens will, in this case, be related by Eq. (4.2.20). A spherical lens can then be seen to transform the radius of curvature R_1 of an incoming wave to a radius R_2 of the outgoing wave according to Eq. (4.2.20). Since this is expected to occur irrespective of the transverse amplitude distribution, Eq. (4.2.20) is expected to hold also for a Gaussian beam, as indeed Eq. (4.7.23b) indicates.

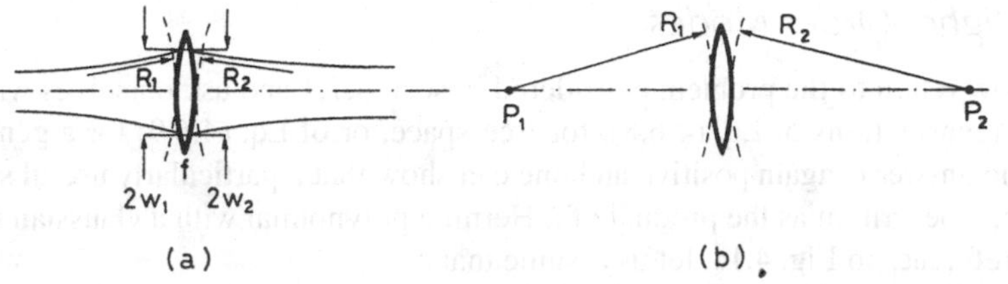

FIG. 4.18. Propagation through a lens of: (a) a Gaussian beam; (b) a spherical wave.

Example 4.6. *Gaussian beam focusing by a thin lens.* Consider now a Gaussian beam, with spot size w_{01} and with plane wavefront, entering a lens of focal length f (i.e., the beam waist is located at the lens). We are interested in calculating the beam waist position after the lens and its spot-size value w_{02}. According to Eq. (4.2.4) and Eq. (4.2.6), the transmission matrix for a lens of focal length, f, followed by a free-space length, z, is given by

$$\begin{vmatrix} 1 - z/f & z \\ -1/f & 1 \end{vmatrix} \tag{4.7.24}$$

The complex beam parameter, q_2, after this combination of lens plus free-space can again be obtained from Eq. (4.7.21) where the A,B,C,D elements are obtained from Eq. (4.7.24) and where $(1/q_1)$ is given by

$$(1/q_1) = -j\lambda/\pi w_{01}^2 = -j/z_{R_1} \tag{4.7.25}$$

with z_{R_1} being the Rayleigh range corresponding to the spot-size w_{01}. If now the coordinate z_m after the lens corresponds to the position where the beam waist occurs, then, according to Eq. (4.7.8), $1/q_2$ must also be purely imaginary. This means that the real part of the right hand side of Eq. (4.7.21) must be zero. With the help of Eq. (4.7.24) and Eq. (4.7.25) we then readily find that z_m is given by

$$z_m = f/\left[1 + (f/z_{R_1})^2\right] \tag{4.7.26}$$

Thus one sees, perhaps with some surprise, that the distance z_m from the lens, at which the minimum spot size occurs, is always smaller than the focal distance f. It should be noted, however, that, under typical conditions one usually has $z_{R_1} \gg f$, so that $z_m \approx f$. By equating the imaginary parts of both sides of Eq. (4.7.21) and using Eqs. (4.7.24) and (4.7.25) again, the spot size at the focal plane, w_{02}, is obtained as

$$w_{02} = \lambda f/\pi w_{01}\left[1 + (f/z_{R_1})^2\right]^{1/2} \tag{4.7.27}$$

Again for $z_{R_1} \gg f$ we obtain from (4.7.27)

$$w_{02} \approx \lambda f/\pi w_{01} \tag{4.7.28}$$

4.7.4. Higher-Order Modes

We now return to the problem considered in sect. 4.7.1 and ask ourselves whether there are other eigensolutions of Eq. (4.6.8), for free-space, or of Eq. (4.6.9) for a general optical system. The answer is again positive and one can show that a particularly useful set of eigen-solutions can be written as the product of a Hermite polynomial with a Gaussian function. In fact, with reference to Fig. 4.14, let us assume that

$$u(x_1, y_1, z_1) = H_l\left[\sqrt{2}x_1/w_1\right] H_m\left[\sqrt{2}y_1/w_1\right] \exp\left[-jk\left(x_1^2+y_1^2\right)/2q_1\right] \tag{4.7.29}$$

where H_l and H_m are Hermite polynomials of order l and m, q_1 is the complex beam parameter at $z=z_1$ and w_1 is the corresponding spot size. Substitution of Eq. (4.7.29) in the right hand side of Eq. (4.6.9) gives

$$u(x, y, z) = \left[\frac{1}{A+(B/q_1)}\right]^{1+l+m} H_l\left(\frac{\sqrt{2}x}{w}\right) H_m\left(\frac{\sqrt{2}y}{w}\right) \times \exp -jk\frac{(x^2+y^2)}{2q} \tag{4.7.30}$$

where q is the complex beam parameter after the optical system of Fig. 4.14 as given by Eq. (4.7.4), and w is the corresponding spot size.

For free space propagation, if we let the z_1 plane be the waist plane then one has $q_1 = j\pi w_0^2/\lambda$, where w_O is the spot size at the beam waist. On substituting the previous expression for q_1 into Eq. (4.7.30) and using Eq. (4.7.8) we obtain

$$\begin{aligned} u_{l,m}(x, y, z) &= (w/w_O)H_l\left[2^{1/2}x/w\right] H_m\left[2^{1/2}y/w\right] \exp\left[-\left(x^2+y^2\right)/w^2\right] \\ &\quad \times \exp\left\{-j\left[k\left(x^2+y^2\right)/2R\right] + j(1+l+m)\phi\right\} \end{aligned} \tag{4.7.31}$$

where ϕ is again given by Eq. (4.7.15a) and where, using Eq. (4.7.11) to obtain $q=q(z)$, it is seen that w and R are again given by Eqs. (4.7.13a) and (4.7.13b).

The lowest order mode is obtained from Eq. (4.7.31) on setting $l=m=0$. Since the Hermite polynomial of zeroth order is a constant, Eq. (4.7.31) reduces to the Gaussian solution already discussed in sect. 4.7.1. [see Eq. (4.7.15)]. This solution is called the TEM_{00} mode, where *TEM* stands for Transverse Electric and Magnetic (within the paraxial approximation both the electric and magnetic fields of the e.m. wave are approximately transverse to the z-direction) and where the indices 00 indicate zeroth order polynomials for both H_l and H_m in Eq. (4.7.31). The radial intensity profile of a TEM_{00} Gaussian mode, at any z-coordinate, will then be $I_{00}(x, y) \propto |u_{00}|^2 \propto \exp\left[-2(x^2+y^2)/w^2\right]$ and will only depend on the radial coordinate $r=(x^2+y^2)^{1/2}$. The mode thus corresponds to a circular spot [Fig. 4.19].

The next higher order mode is obtained from Eq. (4.7.31) by setting $l=1$ and $m=0$ (or $l=0$ and $m=1$). Since $H_1(x) \propto x$, the field amplitude is now given by $|u_{10}| \propto x \times \exp -\left[(x^2+y^2)/w^2\right]$. Thus, at a given x, the field profile will be described by a Gaussian function [see Fig. 4.15a] along the y-coordinate while, at a given y, it is described by the function $x \exp -\left(x^2/w^2\right)$ along the x-coordinate. This function, normalized to its peak value, is plotted vs x/w in Fig. 4.15b. This mode is called TEM_{10} and a picture of the corresponding intensity profile is shown in Fig. 4.19. The TEM_{01} [$l=0$ and $m=1$], is obtained simply by rotating the picture of the TEM_{10} mode in Fig. 4.19 by 90°.

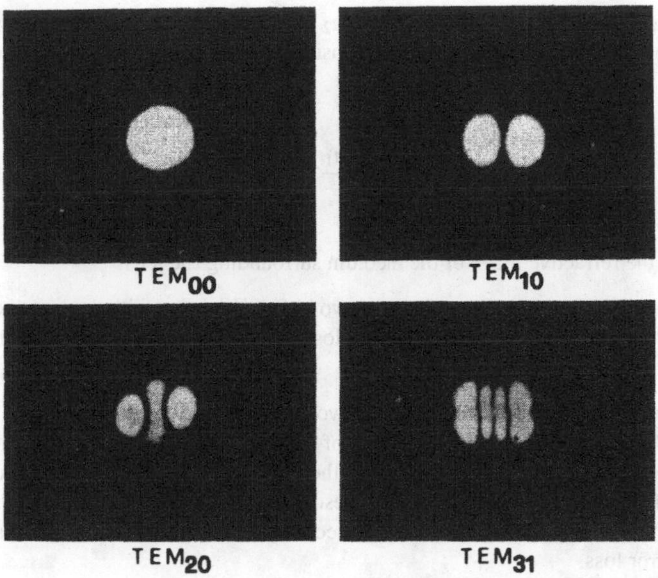

FIG. 4.19. Intensity patterns of some low-order Gaussian modes.

Two pictures of still higher order modes are also indicated in the figure. We note here the general result that the indices l and m give the number of zeros of the field (other than the zeros occurring at $x = \pm\infty$ and $y = \pm\infty$) along the x and y axes, respectively.

4.8. CONCLUSIONS

In this chapter a few topics from geometrical and wave optics which constitute a very useful background to the topics on optical resonators considered in next chapter, have been discussed. We have found, in particular, that the transformation effected on a ray of an optical element (such as an isotropic material, a thin lens, a spherical mirror etc.) can be described by a simple 2×2 matrix. The same matrix has also been found to describe the propagation of a Gaussian beam. A rather basic description of multilayer dielectric coatings and a somewhat more detailed discussion of a Fabry-Perot interferometer have also been presented.

PROBLEMS

4.1. Show that the ABCD matrix for a ray entering a spherical dielectric interface from a medium of refractive index n_1 to a medium of refractive index n_2 is

$$\begin{vmatrix} 1 & 0 \\ \dfrac{n_2 - n_1}{n_2}\dfrac{1}{R} & \dfrac{n_1}{n_2} \end{vmatrix}$$

where R is the radius of curvature of the spherical surface ($R > 0$ if the center is to the left of the surface).

4.2. Considering a thin lens of refractive index n_2 as a sequence of two, closely spaced, spherical dielectric interfaces of radii R_1 and R_2 and using the result of Problem 4.1, show that its focal length is given by the relation

$$\frac{1}{f} = \frac{n_2 - n_1}{n_1}\left(-\frac{1}{R_1} + \frac{1}{R_2}\right)$$

where n_1 is the refractive index of the medium surrounding the lens.

4.3. A Fabry-Perot interferometer is made up of two identical mirrors with the same power reflectivity, $R = 0.99$, and same fractional internal power loss, $A = 0.005$. Calculate the peak transmission and the finesse of the interferometer.

4.4. A Fabry-Perot interferometer, made up of two identical mirrors, air-spaced by a distance L, is illuminated by a monochromatic e.m. wave of tunable-frequency. From the measurement of the transmitted intensity versus the frequency of the input wave one finds that the free spectral range of the interferometer is 3×10^9 Hz and its resolution is 60 MHz. Calculate the spacing L of the interferometer, its finesse and the mirror reflectivity. If the peak transmission is 50%, calculate also the mirror loss.

4.5. A Fabry-Perot interferometer, made up of two identical mirrors air-spaced by a distance L, is illuminated by a 1-ps pulse from an external source at the wavelength $\lambda \cong 600$ nm. The output beam is observed to be made of a regular sequence of 1-ps pulses spaced by 10 ns. The energy of the pulses decreases exponentially with time with a time constant of 100 ns. Calculate the cavity length and the mirror reflectivity.

4.6. By direct substitution of Eq. (4.7.2) into the right hand side of Eq. (4.6.9) show that the double integral appearing in Eq. (4.6.9), when taken between $-\infty$ and $+\infty$, gives Eq. (4.7.3) where q is related to q_0 by Eq. (4.7.4).

4.7. A positive lens of focal length f is placed at a distance d from the waist of a TEM_{00} beam, of waist spot size w_0. Derive the expression for the focal length f (in terms of w_0 and d) that is required in order that the beam, leaving the lens, has a plane wavefront.

4.8. Show that the power contained in a TEM_{00} Gaussian beam of spot size w is given by $P = \left(\pi w^2/2\right) I_p$ where I_p is the peak ($r = 0$) intensity of the beam.

4.9. A given He-Ne laser, oscillating in a pure Gaussian TEM_{00} mode at $\lambda = 632.8$ nm with an output power of $P = 5$ mW is advertised as having a far-field divergence angle of 1 mrad. Calculate the spot size, the peak intensity and the peak electric field at the waist position.

4.10. The beam of an Ar laser, oscillating in a pure Gaussian TEM_{00} mode at $\lambda = 514.5$ nm with an output power of 1 W, is sent to a target at a distance of 100 m from the beam waist. If the spot size at the beam waist is $w_0 = 2$ mm, calculate, at the target position, the spot size, the radius of curvature of the phase front and the peak intensity.

4.11. Consider a TEM_{00} Gaussian beam of spot size w_1 entering a lens of diameter D and focal length f. To avoid excessive diffraction effects at the lens edge, due to truncation of the Gaussian field by the lens, one usually chooses the lens diameter according to the criterion $D \geq 2.25\, w_1$. Assume that: (i) the equality holds in the previous expression; (ii) the waist of the incident beam is located at the lens, i.e. $w_1 = w_{01}$; *(iii)* $f \ll z_{R_1} = \pi w_{01}^2/\lambda$; (iv) Eq. (4.7.27) is still valid. Under these conditions, express the minimum spot size after the lens as a function of the lens numerical aperture N.A. [N.A. $=\sin\theta$, where $\theta = \tan^{-1}(D/f)$, so that, for small θ, N.A. $\cong (D/f)$].

4.12. Suppose that a TEM_{00} Gaussian beam from a Ruby laser ($\lambda = 694.3$ nm) is transmitted through a 1 m diameter diffraction-limited telescope to illuminate a spot on the face of the moon. Assuming an earth-moon distance of $z \cong 348{,}000$ km and using the relation $D = 2.25\ w_{01}$ between telescope objective diameter and beam spot size (see previous problem) calculate the beam spot size on the moon (distortion effects from the atmosphere can be important, but are neglected here).

4.13. A Gaussian beam of waist spot size w_0 is passed through a solid plate of transparent material of length, L, and refractive index n. The plate is placed just in front of the beam waist. Using the *ABCD* law of Gaussian beam propagation, show that the spot size and radius of curvature of the phase front after the plate are the same as for propagation, in a vacuum, over a distance $L' = L/n$. According to this result, is the far-field divergence angle affected by the insertion of the plate?

4.14. From Eq. (4.7.26) show that a Gaussian beam of waist spot size w_{01} cannot be focused at a distance larger than $z_{R_1}/2$, where $z_{R_1} = \pi\ w_{01}^2/\lambda$. What is the focal length corresponding to this maximum focusing condition?

References

1. H. Kogelnik, Propagation of Laser Beams, in *Applied Optics and Optical Engineering*, ed. by R. Shannon and J.C. Wynant (Academic Press, New York, 1979), Vol. II, pp. 156–190.
2. M. Born and E. Wolf, *Principles of Optics*, 6th edition (Pergamon, Oxford, 1980) Sect. 1.5.
3. E. Ritter, Coatings and Thin-Film Techniques, in *Laser Handbook*, ed. by F.T. Arecchi and E.O. Schultz-Dubois (North-Holland, Amsterdam, 1972) Vol. 1, pp. 897–921.
4. *Thin Films for Optical Systems*, ed. by F. R. Flory (Marcel Dekker, New York 1995).
5. Ref. [2], sect. 7.6.
6. A.E. Siegman, *Lasers* (Oxford University Press, Oxford, U. K.), Chapt. 16.
7. Ref. [6], chapt. 20.
8. H. Kogelnik ant T. Li, Laser Beams and Resonators, *Appl. Opt.* 5, 1550 (1966).

5

Passive Optical Resonators

5.1. INTRODUCTION

This chapter deals with the theory of passive optical resonators i.e. where no active medium is present within the cavity. The most widely used laser resonators have either plane or spherical mirrors of rectangular (or, more often, circular) shape, separated by some distance L. Typically, L may range from a few centimeters to a few tens of centimeters, while the mirror dimensions range from a fraction of a centimeter to a few centimeters. Laser resonators thus differ from those used in the microwave field (see e.g. Sect. 2.2.1) in two main respects: (1) The resonator dimensions are much greater than the laser wavelength. (2) Resonators are usually open, i.e. no lateral surfaces are used. The resonator length is usually much greater than the laser wavelength because this wavelength usually ranges from a fraction of a micrometer to a few tens of micrometers. A laser cavity with length comparable to the wavelength would then generally have too low a gain to allow laser oscillation. Laser resonators are usually open because this drastically reduces the number of modes which can oscillate with low loss. In fact, with reference to example 5.1 to be considered below, it is seen that even a narrow linewidth laser such as a He-Ne laser would have a very large number of modes ($\approx 10^9$) if the resonator were closed. By contrast, on removing the lateral surfaces, the number of low-loss modes reduces to just a few (≈ 6 in the example). In these open resonators, in fact, only the very few modes corresponding to a superposition of waves traveling nearly parallel to the resonator axis will have low enough losses to allow laser oscillation.

According to the previous discussion, it is seen that open resonators have inevitably some losses due to diffraction of the e.m. field, which leads to some fraction of the energy leaving the sides of the cavity (*diffraction losses*). Strictly speaking, therefore, the mode definition given in Sect. 2.2.1 cannot be applied to an open resonator and true modes (i.e. stationary configurations) do not exist for such a resonator. In what follows, however, we shall see that standing-wave configurations having very small losses do exist in open resonators. We will

O. Svelto, *Principles of Lasers*,
DOI: 10.1007/978-1-4419-1302-9_5, © Springer Science+Business Media LLC 2010

therefore define, as a mode, an e.m. configuration whose electric field can be written as

$$\mathbf{E}(\mathbf{r}, t) = E_0\mathbf{u}(\mathbf{r}) \exp[(-t/2\tau_c) + j\omega t] \tag{5.1.1}$$

Here τ_c (the decay time of the square of the electric field amplitude) is called the cavity photon decay time.

Of the various possible resonators we make particular mention of the following types:

a. *Plane – Parallel (or Fabry–Perot) Resonator*. This consists of two plane mirrors set parallel to one another. To a first approximation the modes of this resonator can be thought of as the superposition of two plane e.m. waves propagating in opposite directions along the cavity axis, as shown schematically in Fig. 5.1a. Within this approximation, the resonant frequencies can be readily obtained by imposing the condition that the cavity length L must be an integral number of half-wavelengths, i.e. $L = n\lambda/2$, where n is a positive integer. This is a necessary condition for the electric field of the e.m. standing wave to be zero on the two mirrors. It then follows that the resonant frequencies are given by

$$\nu = n(c/2L) \tag{5.1.2}$$

It is interesting to note that the same expression Eq. (5.1.2) can also be obtained by imposing the condition that the phase shift of a plane wave due to one round-trip through the cavity must equal an integral number times 2π, i.e. $2kL = 2n\pi$. This condition is readily obtained by a self-consistency argument. If the frequency of the plane wave is equal to that of a cavity mode, the phase shift after one round trip must be zero (apart from an integral number of 2π) since only in this case will the amplitudes at any arbitrary point, due to successive reflections, add up in phase so as to give an appreciable total field. Note that, according to Eq. (5.1.2), the frequency difference between two consecutive modes, i.e. modes whose integers differ by one, is given by

$$\Delta\nu = c/2L \tag{5.1.3}$$

This difference is called the frequency difference between two consecutive longitudinal modes with the word longitudinal used because the number n indicates the number of half-wavelengths of the mode along the laser resonator, i.e. longitudinally.

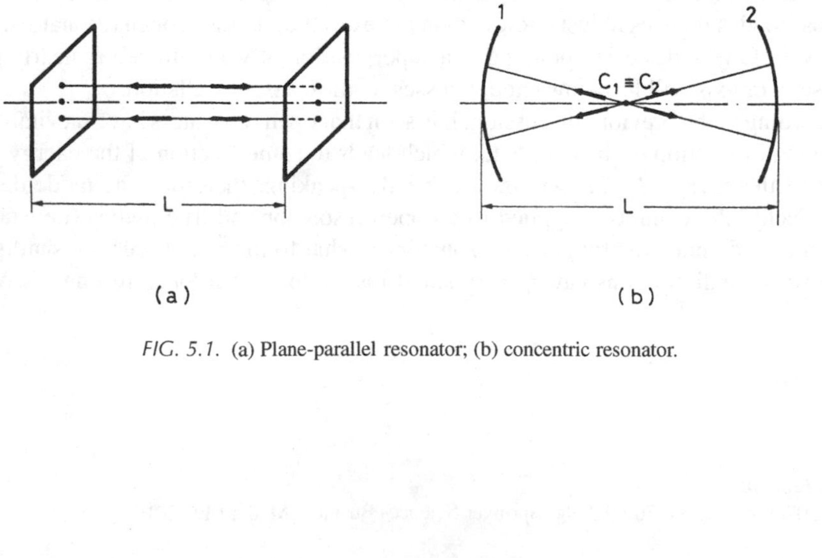

FIG. 5.1. (a) Plane-parallel resonator; (b) concentric resonator.

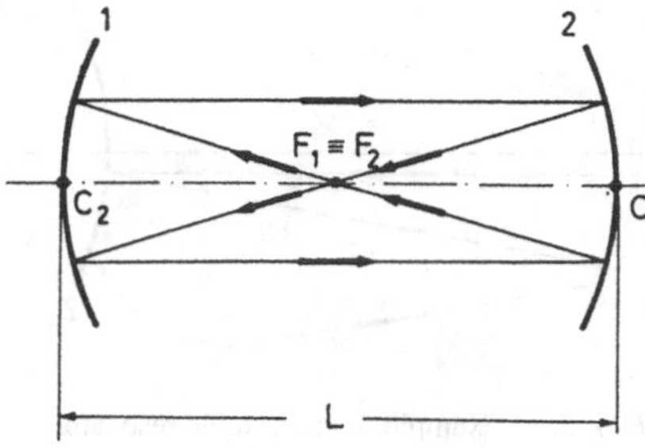

FIG. 5.2. Confocal resonator.

b. *Concentric (or Spherical) Resonator*. This consists of two spherical mirrors having the same radius R and separated by a distance L such that the mirror centers of curvature C_1 and C_2 are coincident (i.e. $L = 2R$) (Fig. 5.1b). The geometrical-optics picture of the modes of this resonator is also shown in the figure. In this case the modes are approximated by a superposition of two oppositly traveling spherical waves originating from the point C. The application of the above self-consistency argument again leads to Eq. (5.1.2) as the expression for the resonant frequencies and to Eq. (5.1.3) for the frequency difference between consecutive longitudinal modes.

c. *Confocal Resonator* (Fig. 5.2). This consists of two spherical mirrors of the same radius of curvature R and separated by a distance L such that the mirror foci F_1 and F_2 are coincident. It then follows that the center of curvature C of one mirror lies on the surface of the second mirror (i.e. $L = R$). From a geometrical-optics point of view, we can draw any number of closed optical paths of the type shown in Fig. 5.2 by changing the distance of the two parallel rays from the resonator axis C_1C_2. Note also that the direction of the rays can be reversed in Fig. 5.2. This geometrical optics description, however, does not give any indication of what the mode configuration will be, and we shall see that in fact this configuration cannot be described either by a purely plane or a purely spherical wave. For the same reason, the resonant frequencies cannot be readily obtained from geometrical-optics considerations.

Resonators formed by two spherical mirrors of the same radius of curvature R and separated by a distance L such that $R < L < 2R$ (i.e. somewhere between the confocal and concentric conditions) are also often used. In addition, we can have $L > R$. For these cases it is not generally possible to use a ray description in which a ray retraces itself after one or a few passes.

All of these resonators can be considered as particular examples of a general resonator consisting of two either concave ($R > 0$) or convex ($R < 0$) spherical mirrors, of different radius of curvature, spaced by some arbitrary distance L. These various resonators can be divided into two categories, namely, *stable* resonators and *unstable* resonators. A resonator will be described as unstable when an arbitrary ray, in bouncing back and forth between the two mirrors, will diverge indefinitely away from the resonator axis. An obvious example of an unstable resonator is shown in Fig. 5.3. Conversely, a resonator for which the ray remains bounded will be described as a stable resonator.

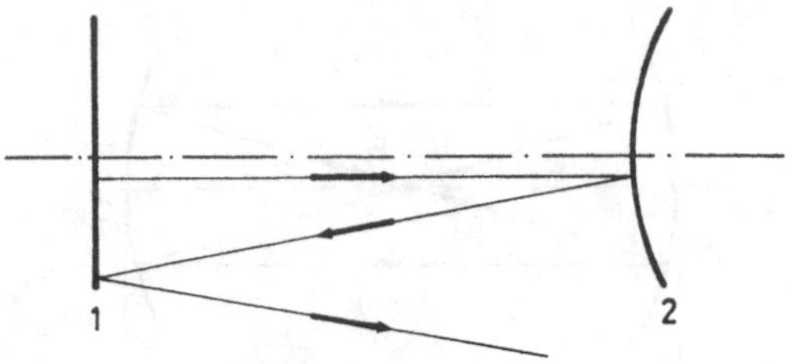

FIG. 5.3. Example of an unstable resonator.

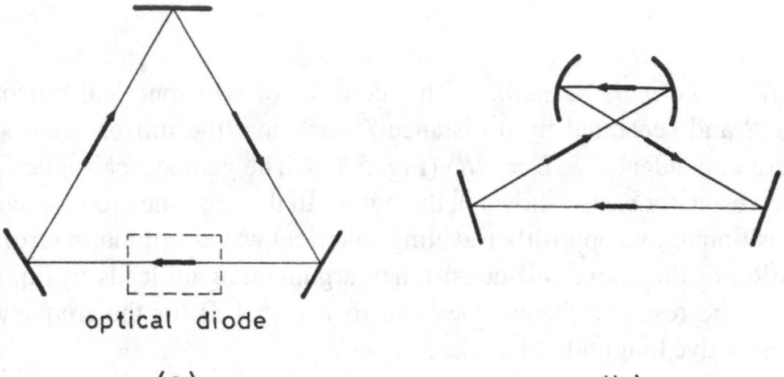

FIG. 5.4. (a) Simplest three-mirror ring resonator. (b) Folded ring resonator.

A particularly important class of laser resonator is the *ring resonator* where the path of the optical rays is arranged in a ring configuration (Fig. 5.4a) or in a more complicated configuration such as the folded configuration of Fig. 5.4b. In both cases the resonance frequencies can be obtained by imposing the condition that the total phase shift along the ring path of Fig. 5.4a or along the closed-loop path of Fig. 5.4b (continuous paths) be equal to an integral number of 2π. We then readily obtain the expression for the resonance frequencies as

$$\nu = nc/L_p \tag{5.1.4}$$

where L_p is the perimeter of the ring or the length of the closed-loop path of Fig. 5.4b, and n is an integer. Note that the arrows of the continuous paths of Fig. 5.4 can in general be reversed which means that e.g. in Fig. 5.4a the beam can propagate either clockwise or anticlockwise. Thus, in general, a standing wave pattern will be formed in a ring resonator. One can see, however, that, if a unidirectional device is used, allowing the passage of e.g. only the right to left beam in Fig. 5.4a (optical diode, see Sect. 7.8.2.2. for more details), then only the clockwise propagating beam can exist in the cavity. So the concepts of a cavity mode and cavity resonance frequency are not confined to standing-wave configurations. Note also that ring resonators can be either of the stable (such as in Fig. 5.4) or unstable configuration.

5.2. EIGENMODES AND EIGENVALUES[1]

> **Example 5.1.** *Number of modes in closed and open resonators.* Consider a He-Ne laser oscillating at the wavelength of $\lambda = 633$ nm, with a Doppler-broadened gain linewidth of $\Delta\nu_0^* = 1.7 \times 10^9$ Hz. Assume a resonator length $L = 50$ cm and consider first an open resonator. According to Eq. (5.1.3) the number of longitudinal modes which fall within the laser linewidth is $N_{open} = 2L\Delta\nu_0^*/c \cong 6$. Assume now that the resonator is closed by a cylindrical lateral surface with a cylinder diameter of $2a = 3$ mm. According to Eq. (2.2.16) the number of modes of this closed resonator which fall within the laser linewidth $\Delta\nu_0^*$ is $N_{closed} = 8\pi\nu^2 V\Delta\nu_0^*/c^3$, where $\nu = c/\lambda$ is the laser frequency and $V = \pi a^2 L$ is the resonator volume. From the previous expressions and data we readily obtain $N_{closed} = (2\pi a/\lambda)^2 N_{open} \cong 1.2 \times 10^9$ modes.

Consider a general two-mirror resonator, (Fig. 5.5a), consisting of two spherical mirrors of different radius of curvature (either positive or negative) spaced by a distance L and which may be either stable or unstable. Assume that a beam of general shape is launched in the cavity starting from e.g. mirror 1 and consider its propagation back and forth in the cavity. This propagation can be regarded as equivalent to that occurring in the periodic lens-guide structure of Fig. 5.5b with the same beam traveling in one direction, e.g. along the positive direction of the z-axis. One should note that the focal lengths f_1 and f_2 in Fig. 5.5b are related to the radii of curvature R_1 and R_2 of Fig. 5.5a by the well known relations $f_1 = R_1/2$ and $f_2 = R_2/2$. It should also be noted that the two diaphragms, 1 and 2, of diameter $2a_1$ and $2a_2$ situated after the corresponding lenses in Fig. 5.5b, simulate the apertures of the two mirrors of Fig. 5.5a. Now let $\tilde{E}(x_1, y_1, 0)$ be the complex field amplitude of the beam at some given point having transverse co-ordinates x_1 and y_1 at diaphragm 1, whose longitudinal co-ordinate is taken to be $z = 0$. The field amplitude $\tilde{E}(x, y, 2L)$ after one lens-guide period, i.e. at $z = 2L$, can be calculated, once $\tilde{E}(x_1, y_1, 0)$ and the lens-guide geometry (i.e. the quantities f_1, f_2, a_1, a_2 and L) are specified. For this calculation one can use e.g. the Huyghens–Fresnel propagation equation (see Sect. 4.6). The calculation can become somewhat involved for finite values of the apertures $2a_1$ and $2a_2$, as we shall see in Sect. 5.5.2. The calculation can be even more involved if one needs to consider the case of some additional optical elements (e.g. a lens or a sequence of lenses) being located within the cavity of Fig. 5.5a. In general, as a consequence of the linearity of the

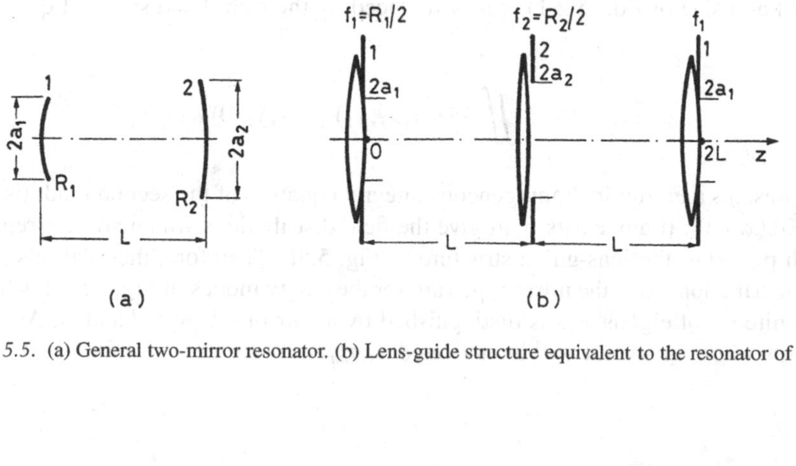

FIG. 5.5. (a) General two-mirror resonator. (b) Lens-guide structure equivalent to the resonator of (a).

Huyghens–Fresnel equation with respect to the field amplitudes, one can write

$$\tilde{E}(x, y, 2L) = (\exp -2jkL) \iint_1 K(x, y, x_1, y_1)\, \tilde{E}\,(x_1, y_1, 0)\, dx_1 dy_1 \tag{5.2.1}$$

where the double integral is taken over the aperture 1 at the input plane ($z = 0$) and where K is a function of the transverse co-ordinates of both input ($z = 0$) and output ($z = 2L$) planes, being known as the propagation kernel. A few examples of this kernel will be considered in Sect. 5.5.2. One can see however that, if $\tilde{E}(x_1,\ y_1,\ 0)$ were a bidimensional Dirac δ-function centered at co-ordinates x_1', y_1' i.e. if $\tilde{E}\,(x_1, y_1, 0) = \delta\left(x_1 - x_1', y_1 - y_1'\right)$, then from Eq. (5.2.1) one would have $\tilde{E}(x, y, 2L) = \exp(-2jkL)\ k\left(x, y; x_1', y'\right)$. Thus, apart from the phase factor $\exp(-2jkL)$, the kernel $K(x, y, x_1, y_1)$ represents the field at the output plane generated by a point-like source located at co-ordinates x_1, y_1 in the input plane.

Instead of considering a general beam propagating in the lens-guide structure of Fig. 5.5b let us now consider a beam whose transverse structure corresponds to that of a cavity mode of Fig. 5.5a. In this case, for self-consistency, the field must reproduce its shape after one lens-guide period. More precisely we require

$$\tilde{E}(x, y, 2L) = \tilde{\sigma}\ \exp\left(-2jkL\right) \tilde{E}\ (x, y, 0) \tag{5.2.2}$$

where the constant $\tilde{\sigma}$ is generally complex since the propagation kernel K is itself a complex function. We can therefore write

$$\tilde{\sigma} = |\tilde{\sigma}|\ \exp j\phi \tag{5.2.3}$$

where the amplitude $|\tilde{\sigma}|$ is expected to be smaller than one as a result of beam attenuation, due to diffraction losses. The phase ϕ then gives the additional contribution to the round trip (or single-period of the lens-guide) phase-shift besides the obvious one, i.e. $-2\,kL$, arising from the free-space propagation of a plane wave over the distance $2L$. According to Eqs. (5.2.2) and (5.2.3) the total single-period phase shift is

$$\Delta\phi = -2kL\ +\ \phi \tag{5.2.4}$$

If the left hand side of Eq. (5.2.1) is now replaced by the right hand side of Eq. (5.2.2) one obtains

$$\tilde{\sigma}\ \tilde{E}(x, y, 0) = \iint K(x, y, x_1, y_1)\ \tilde{E}(x_1, y_1, 0)\ dx_1\, dy_1 \tag{5.2.5}$$

which represents a Fredholm homogeneous integral equation of the second kind. Its eigensolutions, $\tilde{E}_{lm}(x, y, 0)$, if any exists, will give the field distributions which are self-reproducing after each period of the lens-guide structure of Fig. 5.5b. Therefore, they will also describe the field distributions over the mirror aperture for the cavity modes of Fig. 5.5a. Each solution in the infinite set of eigenstates is distinguished by a pair of integers, l and m. Accordingly, the corresponding eigenvalue will be indicated as $\tilde{\sigma}_{lm}$.

From the above discussion, the eigenvalues $\tilde{\sigma}_{lm}$ are seen to be such that $|\tilde{\sigma}_{lm}|^2$ gives the factor by which the beam intensity is changed as a result of one round trip. Since this change is due to diffraction losses, we must then have $|\tilde{\sigma}_{lm}|^2 < 1_j$ thus the quantity

$$\gamma_{lm} = 1 - |\tilde{\sigma}_{lm}|^2 \tag{5.2.6}$$

gives the round-trip fractional power loss due to diffraction. One can also see that, according to Eq. (5.2.4), $\Delta\phi_{lm} = -2\,kL + \phi_{lm}$ is the corresponding round-trip phase shift. For the field to be self-reproducing, we must then require that $\Delta\phi_{lm} = -2\pi\,n$, where n is an integer. We thus get $-2kL + \phi_{lm} = -2\pi n$, and, with the substitution $k = 2\pi\nu/c$, we obtain the cavity resonance frequencies as

$$\nu_{lmn} = \frac{c}{2L}\left[n + \frac{\phi_{lm}}{2\pi}\right] \tag{5.2.7}$$

Note that we have indicated explicitly that these frequencies are dependent on the values of the three numbers l, m, and n. The integers l and m represent the order of the eigensolution in Eq. (5.2.5) while the integer number n specifies the total phase shift of the beam, after one round trip, in units of 2π (i.e. $n = -\Delta\phi_{lm}/2\pi$).

As a conclusion of this section we can say that the eigenmodes and the eigenvalues of our problem can be obtained upon solving the integral equation Eq. (5.2.5). In fact, its eigensolutions, $\tilde{E}_{lm}$, give the field of the eigenmodes at all point in a given plane. For each mode $\tilde{E}_{lm}$, the corresponding eigenvalue $\tilde{\sigma}_{lm}$ then gives: (a) The round-trip diffraction loss, γ_{lm}, through its magnitude $|\tilde{\sigma}_{lm}|$ [see Eq. (5.2.6)]. (b) The resonance frequency, ν_{lmn}, through its phase, ϕ_{lm} [see Eq. (5.2.7)].

5.3. PHOTON LIFETIME AND CAVITY Q

Consider a given mode of a stable or unstable cavity and assume, for generality, that some losses other than diffraction losses are also present. For instance one may have mirror losses as a result of mirror reflectivity being smaller than unity. One may also have scattering losses in some optical element within the cavity. Under these conditions we want to calculate the rate of energy decay in the given cavity mode. To this purpose, let I_0 be the initial intensity corresponding to the field amplitude $\tilde{E}(x_1, y_1, 0)$ at a given transverse coordinate x_1, y_1. Let R_1 and R_2 be the (power) reflectivities of the two mirrors and T_i the fractional internal loss per pass due to diffraction and any other internal losses. The intensity $I(t_I)$ at the same point x_1, y_1 at a time $t_1 = 2L/c$, i.e. after one cavity round trip, will be

$$I(t_1) = R_1\,R_2(1 - T_i)^2\,I_0 \tag{5.3.1}$$

Note that, since T_i is defined here as the fractional internal loss per pass, the intensity is reduced by a factor $(1 - T_i)$ in a single pass and hence by a factor $(1 - T_i)^2$ in a double pass (round trip). The intensity, at the same transverse co-ordinate, after m round trips, i.e. at time

$$t_m = 2mL/c \tag{5.3.2}$$

is then

$$I(t_m) = [R_1 R_2 (1 - T_i)^2]^m I_0 \tag{5.3.3}$$

Let now $\phi(t)$ be the total number of photons in the given cavity mode at time t. Since the mode retains its shape after each round trip, we can set $\phi(t) \propto I(t)$. From Eq. (5.3.3) we can then write

$$\phi(t_m) = \left[R_1 R_2 (1 - T_i)^2\right]^m \phi_0 \tag{5.3.4}$$

where ϕ_0 is the number of photons initially present in the cavity. We can also set

$$\phi(t_m) = [\exp(-t/\tau_c)]\phi_0 \tag{5.3.5}$$

where τ_c is a suitable constant. In fact, a comparison of Eqs. (5.3.5) with (5.3.4) with the help of Eq. (5.3.2) shows that

$$\exp(-2mL/c\,\tau_c) = \left[R_1 R_2 (1 - T_i)^2\right]^m \tag{5.3.6}$$

from which one finds that τ_c is independent of the number of round trips, m, and is given by

$$\tau_c = -2L/c \ln\left[R_1 R_2 (1 - T_i)^2\right] \tag{5.3.7}$$

If we now assume that Eq. (5.3.5) holds, not only at times t_m, but also at any time t (>0), we can then write

$$\phi(t) \cong \exp(-t/\tau_c)\,\phi_0 \tag{5.3.8}$$

Example 5.2. *Calculation of the cavity photon lifetime.* We will assume $R_1 = R_2 = R = 0.98$ and $T_i \cong 0$. From Eq. (5.3.7) we obtain $\tau_c = \tau_T/[-\ln R] = 49.5\,\tau_T$, where τ_T is the transit time of the photons for a single-pass in the cavity. From this example we note that the photon lifetime is much longer than the transit time, a result which is typical of low loss cavities. If we now assume $L = 90$ cm, we get $\tau_T = 3$ ns and $\tau_c \cong 150$ ns.

In this way, we justify the assumption Eq. (5.1.1) for the mode field and identify Eq. (5.3.7) as the expression for the cavity photon lifetime. One can notice that Eq. (5.3.7), with the help of Eqs. (1.2.4) and (1.2.6), can readily be transformed to

$$\tau_c = L/c\gamma \tag{5.3.9}$$

We thus see that the cavity photon lifetime is just equal to the transit time $\tau_T = L/c$ of the beam in the laser cavity divided by the (logarithmic) cavity loss γ.

Having calculated the photon lifetime, the time behavior of the electric field, at any point inside the resonator can, according to Eq. (5.1.1) and within the scalar approximation, be written as $E(t) = \tilde{E}\exp[(-t/2\tau_c) + j\omega t]$, where ω is the angular resonance frequency of the mode. The same time behavior then applies for the field of the output wave leaving the cavity through one mirror as a result of finite mirror transmission. If we now take the Fourier transform of this field, we find that the power spectrum of the emitted light has a Lorentzian line shape with linewidth (FWHM) given by

$$\Delta\nu_c = 1/2\pi\,\tau_c \tag{5.3.10}$$

It should be noted that the spectrum of the emitted light, obtained in this way, does not exactly agree with the transmission spectrum shown for a Fabry–Perot interferometer in Sect. 4.5, whose shape is not Lorentzian [see Eq. (4.5.6)]. In particular, the expression for $\Delta\nu_c$ obtained here [see Eq. (5.3.10)], when combined with Eq. (5.3.7) with $T_i \cong 0$, does not coincide with that obtained in Sect. 4.5 [see Eq. (4.5.12) with $L' = L$]. This discrepancy can be traced back to the approximation made in writing Eq. (5.3.8). In numerical terms, however, the discrepancy between the two results is quite small, especially at high values of reflectivity, as can be seen from the following example. From now on we will therefore assume that the cavity line shape is Lorentzian with width given by Eq. (5.3.10) and that the cavity photon lifetime is given by Eq. (5.3.7).

Example 5.3. *Linewidth of a cavity resonance.* If we take again $R_1 = R_2 = 0.98$ and $T_i = 0$, from Eqs. (5.3.10) and (5.3.7) we get $\Delta\nu_c \cong 6.4307 \times 10^{-3} \times (c/2L)$, while from Eq. (4.5.12) we get $\Delta\nu_c \cong 6.4308 \times 10^{-3} \times (c/2L)$. For the particular case $L = 90$ cm, we then obtain $\Delta\nu_c \cong 1.1$ MHz. Even at the relatively low reflectivity values of $R_1 = R_2 = 0.5$, the discrepancy is not large. In fact from Eqs. (5.3.10) and (5.3.7) we get $\Delta\nu_c \cong 0.221 \times (c/2L)$, while from Eq. (4.5.12) $\Delta\nu_c \cong 0.225 \times (c/2L)$. Again for $L = 90$ cm we then obtain $\Delta\nu_c \cong 37.5$ MHz. Thus, in typical cases, $\Delta\nu_c$ may range from a few to a few tens of MHz.

Having discussed the cavity photon lifetime, we can now introduce the cavity quality factor, or Q factor, and derive its relation to the photon lifetime. For any resonant system, and in particular for a resonant optical cavity, one defines the cavity Q factor (usually abbreviated to cavity Q) as $Q = 2\pi$(energy stored)/(energy lost in one cycle of oscillation). Thus a high value of cavity Q implies low losses of the resonant system. Since, in our case, the energy stored is $\phi h\nu$ and the energy lost in one cycle is $h\nu(-d\phi/dt)(1/\nu) = -hd\phi/dt$, we have

$$Q = -2\pi\nu\phi/(d\phi/dt) \tag{5.3.11}$$

From Eq. (5.3.8) we then get

$$Q = 2\pi\nu\tau_c \tag{5.3.12}$$

which, with the help of Eq. (5.3.10), can be transformed to the more suggestive form

$$Q = \nu/\Delta\nu_c \tag{5.3.13}$$

Thus the cavity Q factor can be interpreted as the ratio between the resonance frequency, ν, of the given mode and its linewidth, $\Delta\nu_c$.

Example 5.4. *Q-factor of a laser cavity* According to example 5.2 we will again take $\tau_c \cong 150$ ns and assume $\nu \cong 5 \times 10^{14}$ Hz (i.e. $\lambda \cong 630$ nm). From Eq. (5.3.12) we obtain $Q = 4.7 \times 10^8$. Thus, very high Q-values can be achieved in a laser cavity and this means that a very small fraction of the energy is lost during one oscillation cycle

5.4. STABILITY CONDITION

Consider first a general two-mirror resonator (Fig. 5.6a) and a ray leaving point P_0 of a plane β inside the resonator e.g. just in front of mirror 1. This ray, after reflection from mirrors 2 and 1, will intersect the plane β at some point P_1. If we let r_0 and r_1 be the transverse

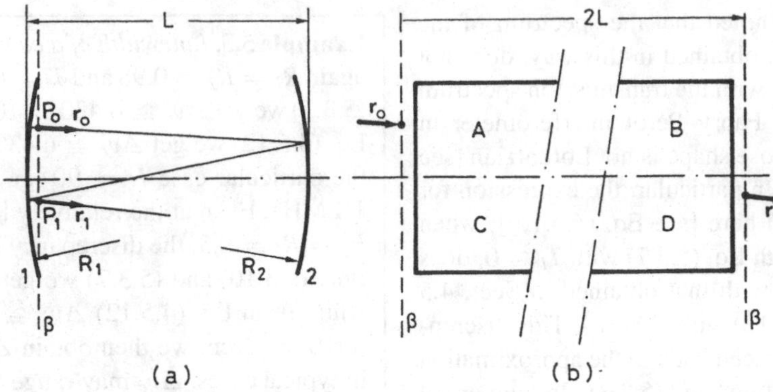

FIG. 5.6. (a) Stability analysis of a two-mirror resonator. (b) Stability analysis of a general resonator described by the *ABCD* matrix.

coordinates of P_0 and P_1 with respect the resonator axis and r_0' and r_1' the angles that the corresponding rays make with the axis, then according to Eq. (4.2.2) we can write

$$\begin{vmatrix} r_1 \\ r_1' \end{vmatrix} = \begin{vmatrix} A & B \\ C & D \end{vmatrix} \begin{vmatrix} r_0 \\ r_0' \end{vmatrix} \tag{5.4.1}$$

where the *ABCD* matrix is the cavity round trip matrix. The ray leaving point $P_1\left(r_1, r_1'\right)$ will, after one round trip, intersect the plane β at point $P_2\left(r_2, r_2'\right)$ given by

$$\begin{vmatrix} r_2 \\ r_2' \end{vmatrix} = \begin{vmatrix} A & B \\ C & D \end{vmatrix} \begin{vmatrix} r_1 \\ r_1' \end{vmatrix} = \begin{vmatrix} A & B \\ C & D \end{vmatrix}^2 \begin{vmatrix} r_0 \\ r_0' \end{vmatrix} \tag{5.4.2}$$

Therefore, after n round trips, the point $P_n(r_n, r_n')$ is given by

$$\begin{vmatrix} r_n \\ r_n' \end{vmatrix} = \begin{vmatrix} A & B \\ C & D \end{vmatrix}^n \begin{vmatrix} r_0 \\ r_0' \end{vmatrix} \tag{5.4.3}$$

For the resonator to be stable, we require that, for any initial point (r_0, r_0'), the point (r_n, r_n') should not diverge as n increases. This means that the matrix

$$\begin{vmatrix} A & B \\ C & D \end{vmatrix}^n$$

must not diverge as n increases.

The previous considerations can be readily extended to a general resonator whose round trip ray transformation is described by a general *ABCD* matrix e.g. a two-mirror resonator containing some other optical elements such as lenses, telescopes etc. (see Fig. 5.6b). In this case we again require that the n-th power of the *ABCD* matrix does not diverge as n increases.

For both resonators shown in Fig. 5.6a, b the ray starts from and arrives at the same plane β, which means that the refractive index is the same for both rays, input, $\mathbf{r}_0$, and output, $\mathbf{r}_1$. It then follows that the determinant of the matrix, *AD-BC*, has unit value. A theorem of matrix calculus,(2) sometimes referred to as Sylvester's theorem, then shows that, if we define an angle θ by the relation

$$\cos\theta = (A + D)/2 \tag{5.4.4}$$

one has

$$\begin{vmatrix} A & B \\ C & D \end{vmatrix}^n = \frac{1}{\sin\theta}\begin{vmatrix} A\sin n\theta - \sin(n-1)\theta & B\sin n\theta \\ C\sin n\theta & D\sin n\theta - \sin(n-1)\theta \end{vmatrix} \tag{5.4.5}$$

Equation (5.4.5) shows that the n-th power matrix does not diverge if θ is a real quantity. Indeed, if θ were complex, say $\theta = a + ib$, the terms proportional to e.g. $\sin n\theta$ in Eq. (5.4.5) could be written as $\sin n\theta = [\exp(jn\theta) + \exp(-jn\theta)]/2j = [\exp(jna - nb) + \exp(-jna + nb)]/2j$. The quantity $\sin n\theta$ would then contain a term growing exponentially with n, e.g. $[\exp(-jna + nb)]/2j$ for $b > 0$, and the overall n-th power matrix would thus diverge as n increases. So, for the resonator to be stable, we require θ to be real and, according to Eq. (5.4.4), this implies that

$$-1 < \left(\frac{A+D}{2}\right) < 1 \tag{5.4.6}$$

Equation (5.4.6) establishes the stability condition for the general resonator of Fig. 5.6b. In the case of the two-mirror resonator of Fig. 5.6a we can go one step further by explicitly calculating the corresponding *ABCD* matrix. We recall that a given overall matrix can be obtained by the product of matrices of the individual optical elements traversed by the beam, with the matrices written down in the reverse of the order in which the ray propagates through the corresponding elements. Thus, in this case, the *ABCD* matrix is given by the ordered product of the following four matrices: (1) Reflection from mirror 1, (2) free-space propagation from mirror 1 to 2, (3) reflection from mirror 2, (4) free-space propagation from mirror 2 to 1. With the help of Table 4.1 we then have

$$\begin{vmatrix} A & B \\ C & D \end{vmatrix} = \begin{vmatrix} 1 & 0 \\ -2/R_1 & 1 \end{vmatrix}\begin{vmatrix} 1 & L \\ 0 & 1 \end{vmatrix}\begin{vmatrix} 1 & 0 \\ -2/R_2 & 1 \end{vmatrix}\begin{vmatrix} 1 & L \\ 0 & 1 \end{vmatrix} \tag{5.4.7}$$

After performing the matrix multiplication of Eq. (5.4.7), we obtain

$$\frac{A+D}{2} = 1 - \frac{2L}{R_1} - \frac{2L}{R_2} + \frac{2L^2}{R_1 R_2} \tag{5.4.8}$$

Equation (5.4.8) can be readily transformed to

$$\frac{A+D}{2} = 2\left(1 - \frac{L}{R_1}\right)\left(1 - \frac{L}{R_2}\right) - 1 \tag{5.4.9}$$

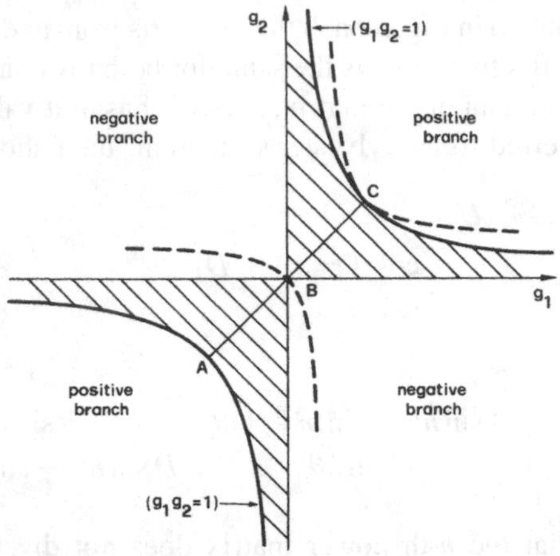

FIG. 5.7. g_1, g_2 stability diagram for a general spherical resonator. The stable region corresponds to the shaded parts of the figure. The dashed curves correspond to the possible confocal resonators.

It is now customary to define two dimensionless quantities for the cavity, called the g_1 and g_2 parameters, and defined as

$$g_1 = 1 - (L/R_1) \tag{5.4.10a}$$

$$g_2 = 1 - (L/R_2) \tag{5.4.10b}$$

In terms of these parameters, the stability condition of Eq. (5.4.6), with the help of Eq. (5.4.9), readily transforms to the very simple relation

$$0 < g_1g_2 < 1 \tag{5.4.11}$$

The stability condition Eq. (5.4.11) can be conveniently displayed in the g_1, g_2 plane [Fig. 5.7]. For this purpose we have plotted in Fig. 5.7, as heavy lines, the two branches of the hyperbola corresponding to the equation $g_1g_2 = 1$. Since the other limiting condition in Eq. (5.4.11), namely $g_1g_2 = 0$, implies either $g_1 = 0$ or $g_2 = 0$, one can readily see that the stable regions in the g_1, g_2 plane correspond to the shaded area of the figure. A particularly interesting class of two mirror resonators is that corresponding to points on the straight line *AC* making an angle of 45° with the g_1 and g_2 axes. This line corresponds to resonators having mirrors of the same radius of curvature (symmetric resonators). As particular examples of these symmetric resonators, we notice that those corresponding to points *A*, *B* and *C* of the figure are the concentric, confocal, and plane resonators, respectively. Therefore all three of these resonators lie on the boundary between the stable and unstable regions. For these resonators, only some particular rays, e.g. rays normal to the plane mirrors in Fig. 5.1a, do not diverge during propagation. For this reason, these resonators are also said to be marginally stable and, in general, the conditions $g_1g_2 = 0$ or $g_1g_2 = 1$ are described as being of marginal stability.

5.5. STABLE RESONATORS

To greatly simplify our analysis we will first consider a resonator with no limiting aperture. We will then briefly consider the effects of a finite aperture.

5.5.1. Resonators with Infinite Aperture[3,4]

With reference to Fig. 5.6b for a general resonator and to 5.5a for a two-mirror resonator we will assume no limiting aperture i.e. we will take $a_1 = a_2 = \infty$ in Fig. 5.5a. The field distribution $u(x, y, z)$ after one cavity round trip of Fig. 5.5a or after one period of the lens-guide system of Fig. 5.5b i.e. at $z = 2L$, can be obtained from Eq. (4.6.9) with $z_1 = 0$, where the *ABCD* matrix is the one-round-trip (or one-period) matrix. If we now take, at the $z_1 = 0$ plane, $\tilde{E}(x_1, y_1, 0) = u(x_1, y_1, 0)$, then, within the paraxial wave approximation and according to Eq. (4.6.4), we can write $\tilde{E}(x, y, 2L) = u(x, y, 2L)\exp(-2jkL)$. On inserting, into this relation, the expression for $u(x, y, z)$ given by Eq. (4.6.9) we get

$$\tilde{E}(x, y, 2L) = \exp(-2jkL) \int_{-\infty}^{+\infty}\int_{-\infty}^{+\infty} \left(\frac{i}{B\lambda}\right) \exp \left\{ -jK \left[\frac{A\,(x_1{}^2 + y_1{}^2) + D(x^2 + y^2) - 2x_1 x - 2y_1 y}{2B} \right] \right\} \times \tilde{E}(x_1, y_1, 0)dx_1dy_1 \tag{5.5.1}$$

A comparison of Eqs. (5.5.1) with (5.2.1) then shows that the propagation kernel $K(x, y; x_1, y_1)$, is given, in this case, by

$$K = \left(\frac{j}{B\lambda}\right) \exp -jk \left[\frac{A\,(x_1{}^2 + y_1{}^2) + D\,(x^2 + y^2) - 2x_1 x - 2y_1 y}{2B} \right] \tag{5.5.1a}$$

As explained in Sect. 4.7, the lowest order Gaussian solution, Eq. (4.7.1), and the general solution for higher order, Eq. (4.7.30), are eigensolutions of the propagation equation, Eq. (4.6.9), when no aperture is present within the optical system described by the given *ABCD* matrix. For these Hermite–Gaussian eigensolutions to describe the field distribution of the cavity eigenmodes, we must now require that the beam reproduces itself after one round trip. This means that if we let q_1 be the complex beam parameter of the Gaussian beam leaving plane β in front of e.g. mirror 1 of Fig. 5.8a, the complex beam parameter q after one round-trip must be equal to q_1. From the *ABCD* law of Gaussian beam propagation Eq. (4.7.4), if we set $q_1 = q$, we obtain

$$q = \frac{Aq + B}{Cq + D} \tag{5.5.2}$$

The q parameter must then satisfy the quadratic equation

$$Cq^2 + (D - A)q - B = 0 \tag{5.5.3}$$

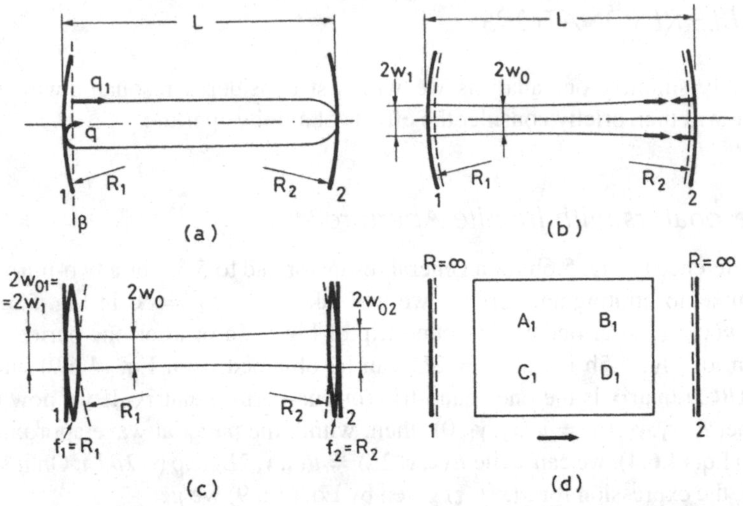

FIG. 5.8. (a) Calculation of the q-parameter for a two-mirror resonator. (b) Spot size and equiphase surfaces in a two mirror-resonator. (c) Transformation of a two-mirror resonator into a resonator with plane end-mirrors. (d) General resonator with two plane end-mirrors.

Since q must be a complex quantity one can see from the standard solution of a quadratic equation that the discriminant of Eq. (5.5.3) must be negative i.e.

$$(D - A)^2 + 4BC < 0 \tag{5.5.4}$$

Since however $AD - BC = 1$, Eq. (5.5.4) readily gives $(D + A)^2 < 4$, i.e. the same condition given by Eq. (5.4.6). This means that a Gaussian beam solution can only be found for stable resonators or, alternatively, that all stable resonators with infinite aperture have modes described by the general Hermite–Gaussian solution of Eq. (4.7.30).

5.5.1.1. Eigenmodes

Consider first the two-mirror resonator of Fig. 5.8b. To obtain an expression for the complex amplitude distribution, $u(x, y, z)$, at e.g. mirror 1 one just needs to calculate the complex beam parameter q, obtained as a solution of Eq. (5.5.3), for given values of the matrix elements A, B, C, and D. Having calculated the q-parameter, one obtains the real and imaginary parts of $1/q$ from which, in accordance with Eq. (4.7.8), the spot size, w and the radius of curvature of the wavefront, R, at the given position, are obtained. One can proceed in a similar way for calculating w and R at any position within the resonator including mirror 2 (Fig. 5.8b). For these calculations it is convenient to transform the resonator of Fig. 5.8b into that of Fig. 5.8c where e.g. the spherical mirror of radius R_1 is substituted by a combination of a plane mirror plus a thin lens of focal length $f_1 = R_1$.* The resonator of Fig. 5.8c is then seen to belong to a

* The fact that we are considering here an equivalent lens of focal length $f_1 = R_1$ while the focal length of the equivalent lens-guide structure was $f_1 = R_1/2$ (see Fig. 5.5b) may generate some confusion. One should note,

general class of resonators consisting of two plane mirrors and containing an optical element whose *single-pass* propagation matrix from mirror 1 to mirror 2 will be represented by the matrix elements A_1, B_1, C_1 and D_1 (Fig. 5.8d).

To obtain q from Eq. (5.5.3) we need to calculate the round-trip matrix for the general resonator of Fig. 5.8d. To this purpose we see that, according to Eq. (4.2.17), the matrix for a single-pass backwards, i.e. propagation from mirror 2 to mirror 1, is simply obtained from the $A_1B_1C_1D_1$ matrix by interchanging the elements A_1 and D_1. We also notice that the matrix of a plane mirror is readily obtained from that of a spherical mirror (see Table 4.1) by letting $R \to \infty$. One then sees that the matrix of a plane mirror is simply the unit matrix

$$\begin{vmatrix} 1 & 0 \\ 0 & 1 \end{vmatrix}$$

The round trip matrix, starting from mirror 1, is then simply given by

$$\begin{vmatrix} A & B \\ C & D \end{vmatrix} = \begin{vmatrix} D_1 & B_1 \\ C_1 & A_1 \end{vmatrix} \begin{vmatrix} A_1 & B_1 \\ C_1 & D_1 \end{vmatrix} = \begin{vmatrix} 2A_1D_1 - 1 & 2B_1D_1 \\ 2A_1C_1 & 2A_1D_1 - 1 \end{vmatrix} \tag{5.5.5}$$

From Eq. (5.5.5) one sees immediately that $A = D$ and from Eq. (5.5.3) one then gets

$$q = q_1 = j\sqrt{-\frac{B}{C}} = j\sqrt{-\frac{B_1D_1}{A_1C_1}} \tag{5.5.6a}$$

It should be noted that one can readily show that the stability condition Eq. (5.4.6) implies $B_1D_1/A_1C_1 < 0$. This means that q_1 is purely imaginary i.e. that the equiphase surface just in front of mirror 1 (see Fig. 5.8c, d) is plane. One could repeat the same argument starting from mirror 2 and shows that

$$q_2 = j\sqrt{-\frac{A_1B_1}{C_1D_1}} \tag{5.5.6b}$$

Since again $A_1B_1/C_1D_1 = (A_1/B_1)^2(B_1D_1/A_1C_1) < 0$, q_2 is also purely imaginary and the wavefront at mirror 2 is again plane. This means that the wavefront radius of curvature, after e.g. lens f_1 in Fig. 5.8c or in front of mirror 1 in Fig. 5.8b, is equal to R_1 and a similar argument applies for mirror 2. So, we reach the general conclusion that the equiphase surface on a cavity mirror always coincides with the mirror surface. This result can be understood from e.g. Fig. 5.8b where the field corresponding to the given eigenmode is considered in terms of a superposition of traveling waves. Then e.g. the right-traveling wave in Fig. 5.8b (indicated by left-to-right arrows) must transfer, upon reflection at mirror 2, into the left-traveling wave (indicated by right-to-left arrows). In geometrical optics terms, this implies that the propagating rays at mirror 2 must be orthogonal to the mirror surface. This means that the wavefront, being always orthogonal to these rays, must be coincident with the mirror surface at the mirror location.

however, that, due to the reflection at the plane mirror of Fig. 5.8c, the lens f_1 in the figure is traversed twice by the beam and its effect is thus equivalent to a single lens of overall focal length $f_1/2$.

The general results of Eqs. (5.5.6) can now be specialized to the two-mirror resonator. With reference to Fig. 5.8c we notice that, in propagating from mirror 1 to mirror 2, the beam passes through lens f_1, then through a free-space of length L and then a lens f_2. The A_1, B_1C_1, D_1 matrix is then simply obtained from the product of the corresponding three matrices with written order inverse to the propagation order. Using the matrices of Table 4.1, it is then a simple matter to show that

$$\begin{vmatrix} A_1 & B_1 \\ C_1 & D_1 \end{vmatrix} = \begin{vmatrix} g_1 & L \\ -(1-g_1g_2)/L & g_2 \end{vmatrix} \tag{5.5.7}$$

where g_1 and g_2 are given by Eqs. (5.4.10). From Eq. (5.5.6a) with the help of Eqs. (4.7.8) and (5.5.7) we then get

$$w_1 = \left(\frac{L\lambda}{\pi}\right)^{1/2} \left[\frac{g_2}{g_1(1-g_1g_2)}\right]^{1/4} \tag{5.5.8a}$$

Similarly, starting from Eq. (5.5.6b), we obtain

$$w_2 = \left(\frac{L\lambda}{\pi}\right)^{1/2} \left[\frac{g_1}{g_2(1-g_1g_2)}\right]^{1/4} \tag{5.5.8b}$$

which can be obtained straightforwardly from Eq. (5.5.8a) by interchanging the indices 1 and 2. Starting from the spot size $w_{01} = w_1$ of Fig. 5.8c, one can then calculate the spot size, w_0, at the beam waist using Eq. (4.7.27) with $f = f_1$ and $w_{02} = w_0$. We obtain

$$w_0 = \left(\frac{L\lambda}{\pi}\right)^{1/2} \left[\frac{g_1g_2(1-g_1g_2)}{(g_1+g_2-2g_1g_2)^2}\right]^{1/4} \tag{5.5.9}$$

Again, knowing the spot size, w_1, on mirror 1, one can obtain the waist distance from that mirror upon using the expression for z_m given by Eq. (4.7.26) with the substitutions $f = f_1 = R_1$ and $z_{R_1} = \pi w_{01}^2/\lambda$.

For a symmetric resonator one has $R_1 = R_2 = R$ and $g_1 = g_2 = g = 1 - (L/R)$ and both Eqs. (5.5.8a) and (5.5.8b) reduces to

$$w = \left(\frac{L\lambda}{\pi}\right)^{1/2} \left[\frac{1}{1-g^2}\right]^{1/4} \tag{5.5.10a}$$

while Eq. (5.5.9) gives

$$w_0 = \left(\frac{L\lambda}{\pi}\right)^{1/2} \left[\frac{1+g}{4(1-g)}\right]^{1/4} \tag{5.5.10b}$$

Example 5.5. *Spot sizes for symmetric resonators* The first case we shall consider is that of a confocal resonator, for which one has $g = 0$. From Eqs. (5.5.10 a and b) we get, respectively,

$$w_c = (L\lambda/\pi)^{1/2}, \quad w_{0c} = (L\lambda/2\pi)^{1/2} \tag{5.5.11}$$

where the suffix c stands for confocal. Equation (5.5.11) show that the spot size at the beam waist is, in this case, $\sqrt{2}$ smaller than that at the mirrors (Fig. 5.9a). For the case of a near-plane resonator, i.e. when $R \gg L$, we can write $g = 1 - \varepsilon$ where ε is a small, positive quantity. Neglecting higher order terms in ε, we get from Eq. (5.5.10)

$$\left(w_{np}/w_c\right) \cong \left(w_{0np}/w_c\right) \cong (1/2\varepsilon)^{1/4} \tag{5.5.12}$$

where the suffix np stands for near-plane and where the spot sizes have been normalized to the mirror spot size of a confocal resonator. Equation (5.5.12) show that, to first order, the two spot sizes are equal and thus the spot size is nearly constant over the length of the resonator (Fig. 5.9b). For the case of a near concentric resonator, i.e. when $L \cong 2R$, we can write $g = -1 + \varepsilon$ where again ε is a small, positive quantity. Neglecting terms of higher order in ε we get from Eq. (5.5.10)

$$(w_{nc}/w_c) = (1/2\varepsilon)^{1/4} \quad (w_{0nc}/w_c) = (\varepsilon/8)^{1/4} \tag{5.5.13}$$

where the suffix nc stands for near-concentric. Equation (5.5.13) show that the mirror spot size is given by the same expression, as a function of ε, as that for a near-plane resonator. The spot size at the beam waist, however, is now much smaller and it decreases with decreasing values of ε. The spot size behavior along the resonator is then as shown in Fig. 5.9c. Numerically, if we take $L = 1$ m and $\lambda = 514$ nm (an Argon laser wavelength) we get $w_c \cong 0.4$ mm for a confocal resonator. If we now consider a near plane resonator, still with $L = 1$ m and $\lambda = 514$ nm and with $R = 10$ m, we get $g = 0.9$ and, from Eq. (5.5.10) we obtain $w_o \cong 0.59$ mm and $w \cong 0.61$ mm. One should note the small values of beam spot size obtained in each case.

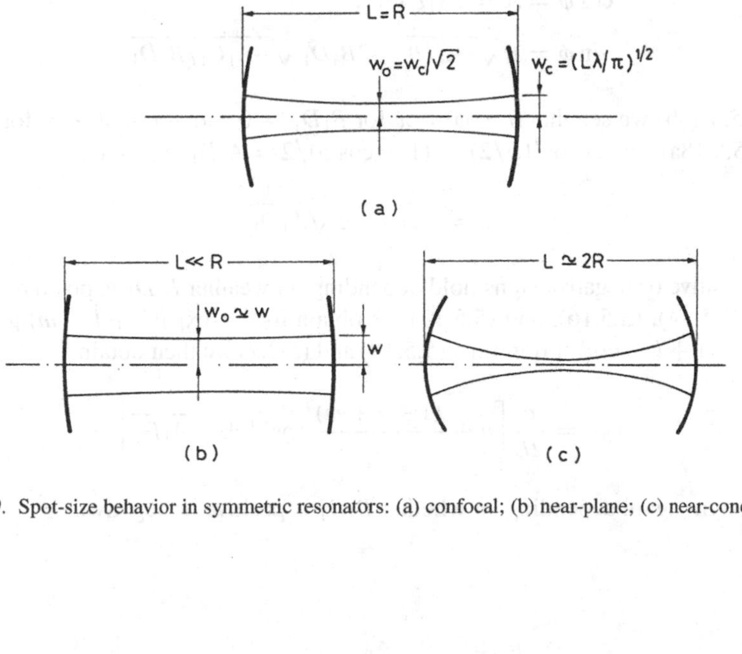

FIG. 5.9. Spot-size behavior in symmetric resonators: (a) confocal; (b) near-plane; (c) near-concentric.

5.5.1.2. Eigenvalues

A comparison of Eqs. (4.7.29) with (4.7.30) shows that, if the $ABCD$ matrix corresponds to the cavity round-trip matrix and if $q = q_1$, the field amplitude $u(x, y, 2L)$ after one round trip is equal to the initial field $u(x_1 = x, y_1 = y, z_1 = 0)$ except for the amplitude factor $1/[A + (B/q)]^{1+l+m}$. According to Eq. (5.2.2) it then follows that

$$\tilde{\sigma}_{lm} = \frac{1}{[A + (B/q)]^{1+l+m}} \tag{5.5.14}$$

From Eq. (5.5.3) we see that, since $A = D$, one has

$$q = j\sqrt{-B/C} \tag{5.5.15}$$

If we now write

$$\sigma = A + (B/q) \tag{5.5.16}$$

from Eq. (5.5.16), with the help of Eq. (5.5.15), we get $|\sigma|^2 = A^2 - BC = AD - BC = 1$. From Eq. (5.5.14) it then follows that the magnitude of $\tilde{\sigma}_{lm}$ is also unity and, according to Eq. (5.2.6), the diffraction loss, γ_{lm}, vanishes. This result is actually to be expected from our analysis since we stipulated at the outset that there were no limiting aperture (Fig. 5.8d) and considered, in particular, a two-mirror resonator with infinite mirror size (Fig. 5.8c).

To obtain an expression for the phase of the eigenvalue, $\tilde{\sigma}_{lm}$, we write

$$\sigma = \exp -j\phi \tag{5.5.17}$$

From Eqs. (5.5.17) and (5.5.16), with the help of Eqs. (5.5.15) and (5.5.5) we get

$$\cos\phi = A = 2A_1D_1 - 1 \tag{5.5.18a}$$

$$\sin\phi = B\sqrt{-C/B} = 2B_1D_1\sqrt{-A_1C_1/B_1D_1} \tag{5.5.18b}$$

From Eq. (5.5.18b) we see that $0 < \phi < \pi$ for $B_1D_1 > 0$ and $-\pi < \phi < 0$ for $B_1D_1 < 0$. From Eq. (5.5.18a) we get $\cos^2(\phi/2) = (1 + \cos\phi)/2 = A_1D_1$ and hence

$$\phi = 2\cos^{-1} \pm\sqrt{A_1D_1} \tag{5.5.19}$$

where the positive or negative signs hold depending on weather B_1D_1 is positive or negative. From Eqs. (5.5.14), (5.5.16), and (5.5.17) we obtain $\tilde{\sigma}_{lm} = \exp j(1 + l + m)\phi = \exp j\phi_{lm}$ where $\phi_{lm} = (1 + l + m)\phi$. From Eqs. (5.5.19) and (5.2.7) we then obtain

$$\nu_{lmn} = \frac{c}{2L}\left[n + \frac{(1 + l + m)}{\pi}\cos^{-1} \pm \sqrt{A_1D_1}\right] \tag{5.5.20}$$

the + or − signs again depending on whether B_1D_1 is positive or negative.

For the particular case of a two-mirror resonator, the matrix elements A_1 and D_1 are obtained from Eq. (5.5.7). Equation (5.5.20) then transforms to

$$\nu_{lmn} = \frac{c}{2L}\left[n + \frac{1+l+m}{\pi}\cos^{-1} \pm \sqrt{g_1 g_2}\right] \tag{5.5.21}$$

and, according to Eq. (5.5.7), the + or − sign is chosen according to whether g_2 (and hence g_1) is positive or negative.

Example 5.6. *Frequency spectrum of a confocal resonator* For a confocal resonator one has $g_1 = g_2 = 0$ and from Eq. (5.5.21) we get

$$\nu_{lmn} = \frac{c}{4L}[2n + (1 + l + m)] \tag{5.5.22}$$

The corresponding frequency spectrum is shown in Fig. 5.10a. One can observe that modes having the same value of $2n + l + m$ have the same resonance frequency although they correspond to different spatial configurations. These modes are said to be frequency-degenerate. It is also seen that, instead of the simple expression given by Eq. (5.1.2) for a plane parallel resonator, the frequency spacing between consecutive modes is now $c/4L$. The two consecutive modes, however, need to have different (l, m) values and $c/4L$ is seen to correspond to the frequency difference between two consecutive *transverse modes* [e.g. $(n, 0, 0) \rightarrow (n, 0, 1)$]. On the other hand, the frequency spacing between two modes with the same (l, m) values (e.g. TEM_{00}) and with n differing by 1 (i.e. the frequency spacing between adjacent *longitudinal modes*) is still $c/2L$, the same as for the plane parallel resonator.*

Example 5.7. *Frequency spectrum of a near-planar and symmetric resonator* In this case one has $g_1 = g_2 = g = 1 - (L/R)$, with $L/R \ll 1$. Thus g is positive and slightly less than unity. Accordingly one has $\cos^{-1} g = \cos^{-1}[1 - (L/R)] \cong (2L/R)^{1/2}$ and Eq. (5.5.21) becomes

$$\nu_{lmn} = \frac{c}{2L}\left[n + \frac{(1 + l + m)}{\pi}\left(\frac{2L}{R}\right)^{1/2}\right] \tag{5.5.23}$$

The corresponding frequency spectrum is shown in Fig. 5.10b. One can see that the frequency spacing between consecutive longitudinal modes is again $c/2L$, while the frequency difference between two consecutive transverse modes is $(c/2L)\,(2L/\pi^2 R)^{1/2}$.

* The usage of the terms "longitudinal mode" and "transverse mode" in the laser literature has sometimes been rather confusing, and can convey the (mistaken) impression that there are two distinct types of modes, viz., longitudinal modes (sometimes called axial modes) and transverse modes. In fact any mode is specified by three numbers, e.g., n, m, l of (5.5.24). The electric and magnetic fields of the modes are nearly perpendicular to the resonator axis. The variation of these fields in a transverse direction is specified by l, m while field variation in a longitudinal (i.e., axial) direction is specified by n. When one refers, rather loosely, to a (given) transverse mode, it means that one is considering a mode with given values for the transverse indices (l, m), regardless of the value of n. Accordingly a single transverse mode means a mode with a single value of the transverse indexes (l, m). A similar interpretation can be applied to the "longitudinal modes". Thus two consecutive longitudinal modes mean two modes with consecutive values of the longitudinal index n [i.e., n and $(n + 1)$ or $(n - 1)$.

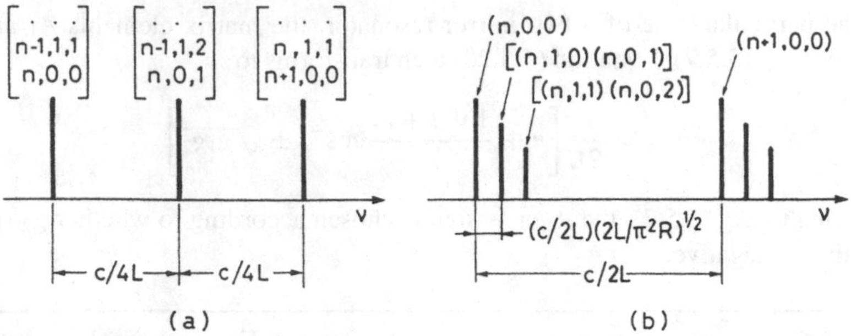

FIG. 5.10. (a): Mode spectrum of a confocal resonator. (b): Mode spectrum of a near-plane resonator.

5.5.1.3. Standing- and Traveling-Waves in a Two-Mirror Resonator

Following the discussion presented in the previous two sections about spot sizes and resonance frequencies in a general resonator, we are now ready to present a description of the corresponding behavior of the mode along the laser cavity. We will limit our discussion to a two-mirror resonator. The field inside this resonator, according to Eqs. (4.7.31) and (4.6.4), can be written as:

$$\tilde{E}_{lmn}(x, y, z) = \frac{w_o}{w} H_l \left[\frac{\sqrt{2}x}{w}\right] H_m \left[\frac{\sqrt{2}y}{w}\right] \exp\left[-\frac{x^2+y^2}{w^2}\right] \tag{5.5.24a}$$

$$\times \exp\left[-jkz + j(1+l+m)\phi\right] \tag{5.5.24b}$$

$$\times \exp\left[-jk(x^2+y^2)/2R\right] \tag{5.5.24c}$$

where $w(z)$, $R(z)$, and $\phi(z)$ are given by Eqs. (4.7.17) and can be calculated once the waist position and the corresponding spot size, w_0, are known. One should observe that the field eigenmode $\tilde{E}$ has been explicitly indicated to be dependent of the three subscripts l, m and n. The subscripts l and m come from the order of the Hermite polynomials involved in Eq. (5.5.24a). The subscript n is also explicitly indicated since $k = 2\pi\nu/c$ and the resonant frequency depends on the three indices l, m and n [see Eq. (5.5.21)]. An interesting interpretation of these indices is as follows: (1) The indices l and m give the field nulls along the x and y axis, respectively, as already pointed out in Sect. 4.7.4. (2) The index n, following the discussion in Sect. 5.1, gives the number of half-wavelengths of the standing wave mode along the resonator, i.e. it gives the number of field nulls along the z-direction.

To conclude this section, we consider the question of whether Eq. (5.5.24) represents a traveling or standing wave pattern for the field eigenmode. The answer depends on the form of the time behavior of the mode. If, according to Eq. (4.6.1) we write $E = \tilde{E}\exp(j\omega t)$ where $\omega = \omega_{lmn} = 2\pi\nu_{lmn}$ is the angular frequency of the mode resonance, then, from the longitudinal phase factor of Eq. (5.5.24b) we get, taking the example of a TEM_{00} mode, $E \propto \exp j[-kz + \phi + \omega t]$ which corresponds to a wave propagating in the positive z-direction. If, on the other hand, we write $E = \tilde{E}\exp(-j\omega t)$, we obtain a wave propagating in the negative z-direction. The standing-wave eigenmode is then obtained by the sum of these two waves, i.e. upon writing $E = \tilde{E}\cos\omega t$. Following the above argument one then realizes that, apart from a proportionality factor given by the mirror's transmission, $E = \tilde{E}\exp(j\omega t)$ also represents the wave escaping through mirror 2 and propagating in the positive z-direction.

5.5.2. Effects of a Finite Aperture

In Sect. 5.5.1.2 it was shown that, for a general resonator with no limiting aperture such as that of Fig. 5.8d, the diffraction loss vanishes. Indeed, to calculate these losses, one must take into account the actual size of any apertures present in the resonator (often a diaphragm is inserted in the resonator or the aperture is set by the transverse dimension of the active medium). The loss introduced by a finite aperture can, in fact, be appreciated with the help of Fig. 5.11 where a TEM_{00} mode is considered and where we indicate the transverse profile of this mode over the plane containing the aperture of diameter $2a$. The Gaussian TEM_{00} mode is seen to be truncated by this aperture and the dashed wings of the beam are therefore lost each time the beam passes through the aperture. This description is, however, an approximate one because, in fact, the introduction of a limiting aperture significantly modifies the field distribution, which would then no longer be precisely Gaussian.

To perform a correct and accurate calculation, we must return to the original integral equation, Eq. (5.2.5), and take into account there the finite size of the aperture. In the discussion that follows, we will limit ourselves to considering a two-mirror resonator, assuming the limiting aperture to be set by the finite mirror size.

Consider first a symmetric resonator ($R_1 = R_2 = R$ and $a_1 = a_2 = a$, see Fig. 5.12a) and its equivalent lens-guide structure (Fig. 5.12b). By virtue of the symmetry of the problem, we can limit our considerations to one period of length L and require that the field reproduces

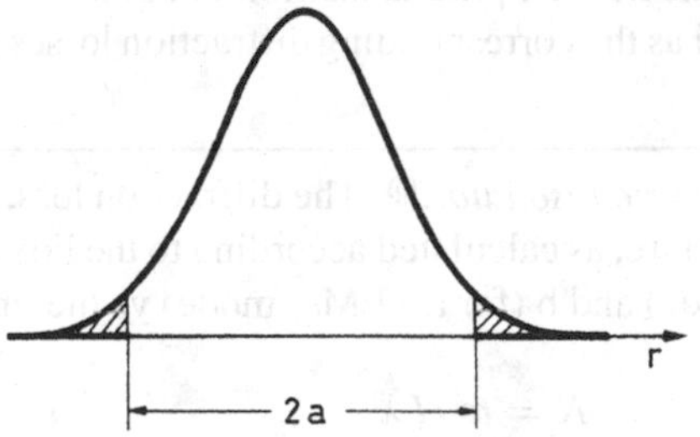

FIG. 5.11. Diffraction losses arising from beam truncation by an aperture of radius a.

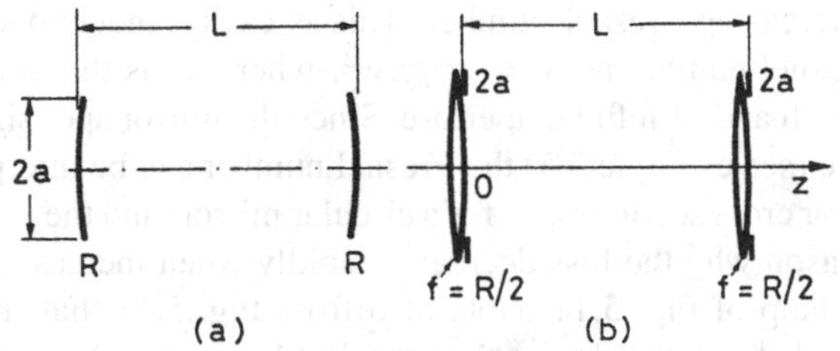

FIG. 5.12. (a) Mode and diffraction loss calculation in a symmetric resonator. (b) Equivalent lens-guide configuration.

its shape after this period. We then arrive at an integral equation similar to Eq. (5.2.5), namely

$$\tilde{\sigma}\tilde{E}(x, y, 0) = \int_{-a}^{+a}\int_{-a}^{+a} K(x, y, x_1, y_1)\tilde{E}(x_1, y_1, 0)dx_1dy_1 \tag{5.5.25}$$

where the double integral is taken over the limiting aperture and K is the single pass propagation kernel. Since the beam encounters no other limiting apertures when propagating from the diaphragm at $z = 0$ to that at $z = L$ (Fig. 5.12b), the kernel can be expressed as in Eq. (5.5.1a), where the *ABCD* matrix refers now to one period of length L. The matrix is then given by the product of the matrix of free-space propagation over a length L with the matrix of a lens of focal length $f = R/2$. One should note, however, that, since the double integral in Eq. (5.5.25) does not extend between $-\infty$ and $+\infty$, the eigensolutions no longer have the form of a product of a Hermite polynomial with a Gaussian function. To solve Eq. (5.5.25) one usually adopts some iterative procedure, generally with the help of a computer. An approach, often used, is the so-called Fox-Li iterative procedure, after Fox and Li[5] who first applied this procedure to obtain the eigenmodes of a plane-parallel resonator. One starts by assuming some field expression $\tilde{E}(x, y, 0)$ in the right hand side of Eq. (5.5.25) and then one calculates, by performing the double integral, the field $\tilde{E}(x, y, L)$ after one lens-guide period. This field is then inserted back into the right hand side of Eq. (5.5.25) and a new field $\tilde{E}(x, y, 2L)$ is then calculated by performing the double integration again, and so on. The procedure, although rather slow (it usually converges in a few hundred iterations), eventually leads to a field which does not change any more on each successive iteration, except for an overall amplitude reduction due to diffraction loss and a phase factor which accounts for the single-pass phase shift. In this way one can compute the field amplitude distribution of the lowest order mode and also of higher order modes, as well as the corresponding diffraction losses and resonance frequencies.

Example 5.8. *Diffraction loss of a symmetric resonator.*[6] The diffraction loss, per pass, for a symmetric two-mirror resonator of finite mirror aperture, as calculated according to the Fox and Li iterative procedure is plotted in Fig. 5.13a (for a TEM_{00} mode) and b (for a TEM_{01} mode) vs the Fresnel number

$$N = a^2/L\lambda \tag{5.5.26}$$

The calculation has been performed for a range of symmetric resonators, which are characterized by their corresponding g values. Note that, for a given g value and for a given mode (e.g. the TEM_{00} mode), the loss rapidly decreases with increasing Fresnel number. This is easily understood when, according to Eq. (5.5.11), one writes the Fresnel number as $N = a^2/\pi w_c^2$, where w_c is the mirror spot size for a confocal resonator of the same length and of infinite aperture. Since the mirror spot size does not change strongly upon changing the g-value (see example 5.5) the Fresnel number can be interpreted as a number proportional to the ratio of the mirror cross section (πa^2 for a circular mirror) and the mode cross-sectional areas (πw^2) on the mirror. The reason why the loss decreases rapidly when increasing the latter ratio is now readily appreciated with the help of Fig. 5.11. Note also from Fig. 5.13 that, for a given Fresnel number and g value, the TEM_{00} mode has lower losses than the TEM_{01} mode. The TEM_{00} mode actually turns out to have the loss lower than for any of the higher order modes. So the lowest order mode is identified as the lowest loss mode.

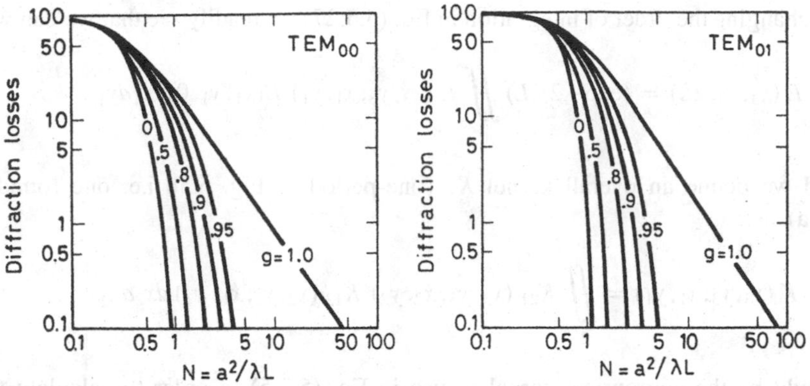

FIG. 5.13. Diffraction loss per transit versus Fresnel number for, (a), the TEM_{00} mode and, (b), for the TEM_{01} mode, for several symmetric resonators (after Li,[6] copyright 1965, American Telephone and Telegraph Company, reprinted with permission.)

Example 5.9. *Limitation on the Fresnel number and resonator aperture in stable resonators* To obtain oscillation on the TEM_{00} mode only, we must provide a sufficiently high value of diffraction losses, γ_{01}, for the TEM_{01} mode. On the other hand, to obtain a large value of the spot size we must design the resonator to operate near the instability boundary $g = 1$ or $g = -1$ (see example 5.5). Furthermore, if we consider e.g. a near-planar resonator, we cannot operate too close to the instability boundary or the resonator would become too sensitive to external perturbation (e.g. mirror tilt due to vibrations or temperature changes). We choose then, as an example, $\gamma_{01} = 10\%$ and $g < 0.95$ ($R < 20L$). From Fig. 5.13b we then get $N < 2$, which can be considered a typical result. Thus, for $L = 2$ m and $\lambda = 1.06\ \mu$m (a Nd:YAG laser wavelength) we obtain $a < 2$ mm while for $L = 2$ m and $\lambda = 10.6\ \mu$m (typical wavelength of a CO_2 laser) we get $a < 6.3$ mm.

Let us now consider the general two-mirror resonator of Fig. 5.5a and its equivalent lens-guide structure of Fig. 5.5b. If we let $\tilde{E}(x_1, y_1, 0)$ be the field at a general point (x_1, y_1) of the $z = 0$ plane in Fig. 5.5b, the field at point (x_2, y_2) of the $z = L$ plane is readily obtained as $\tilde{E}(x_2, y_2, L) = (\exp -jkL) \iint_1 K_{12}(x_2, y_2; x_1, y_1) \times \tilde{E}(x_1, y_1, 0)dx_1dy_1$ where K_{12} is the kernel for beam propagation from the $z = 0$ to the $z = L$ planes and where the double integral is taken over aperture 1. Similarly, the field at point (x_3, y_3) at $z = 2L$ plane is obtained as $\tilde{E}(x_3, y_3, 2L) = (\exp -jkL) \iint_2 K_{21}(x_3, y_3; x_2, y_2)\, \tilde{E}(x_2, y_2, L)dx_2dy_2$ where K_{21} is the kernel for beam propagation from the $z = L$ to $z = 2L$ plane and where the double integral is taken over aperture 2. The combination of the last two equations leads to

$$\tilde{E}(x_3, y_3, 2L) = (\exp -2jkL) \iint_2 K_{21}(x_3, y_3; x_2, y_2)dx_2dy_2 \times \iint_1 K_{12}(x_2, y_2; x_1, y_1)\tilde{E}(x_1, y_1, 0)dx_1dy_1 \qquad (5.5.27)$$

On interchanging the order of integration in Eq. (5.5.27) we readily see that we can write

$$\tilde{E}\,(x_3, y_3, 2L) = (\exp -2jkL) \iint_1 K\,(x_3, y_3; x_1, y_1)\ \tilde{E}(x_1, y_1, 0)dx_1dy_1 \tag{5.5.28}$$

provided we define an overall kernel K, (one-period in Fig. 5.5b i.e. one round trip in Fig. 5.5.a), as

$$K\,(x_3, y_3; x_1, y_1) = \iint_2 K_{21}\,(x_3, y_3; x_2, y_2)\ K_{12}\,(x_2, y_2, x_1, y_1)\,dx_2dy_2 \tag{5.5.29}$$

This would be the appropriate kernel to use in Eq. (5.2.5) in order to calculate the field eigenmodes and the corresponding eigenvalues.

5.5.3. Dynamically and Mechanically Stable Resonators

A very important problem which arises with stable resonators is to increase the beam spot size within the active medium to a size comparable to the transverse dimensions of the medium. In fact, considering for simplicity a symmetric two-mirror resonator, one can see from Eq. (5.5.10) that, to significantly increase the spot size within the laser cavity beyond the value established for a confocal cavity, one should choose a resonator much closer to the $g = \pm 1$ point (near-plane or near-concentric resonator). The cavity would then be too close to an instability boundary and would generally be very sensitive to any cavity perturbations such as those arising from variation of the pump power. We shall now consider a laser design which allows large spot sizes to be achieved within the active medium, the design being particularly insensitive to cavity perturbations arising either from changes of pump power or from mirror tilting (dynamically and mechanically stable resonator).[7]

We first consider a laser resonator consisting of two spherical mirrors of radii R_1 and R_2 and containing an active medium whose pump-induced thermal effects can be simulated by a thin lens whose dioptric power, $1/f$, is proportional to the pump power (Fig. 5.14a). This model corresponds well with the situation for solid-state lasers. In fact, some of the ideas which follow can also be applied to the more complex perturbations induced by the pump in a gas medium.

A first constraint for the design of the laser cavity of Fig. 5.14a can be obtained by the condition that the spot size in the active medium, w_a, be insensitive to the change of the lens diopric power. We thus write

$$dw_a/d(1/f) = 0 \tag{5.5.30}$$

A resonator for which this condition is satisfied is often referred to as *dynamically stable*. A second constraint can be obtained by the condition that the spot size, w_a, be comparable to the radius a of the active medium. So as not to introduce excessive diffraction losses due to beam truncation by this finite aperture, we can e.g. require that[8]

$$2a \cong \pi w_a \tag{5.5.31}$$

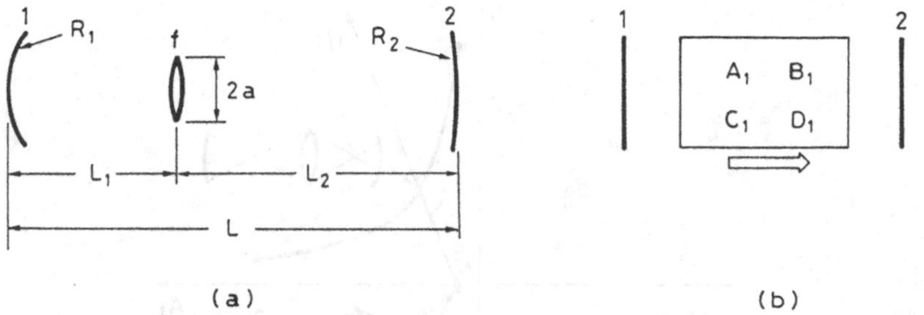

FIG. 5.14. (a) General two-mirror spherical resonator which includes a lens, of focal length f, simulating the thermal lens of the active medium. (b) Generalization of the resonator of (a), where the A_1, B_1, C_1, and D_1 elements of the one-way matrix include the matrix of the thermal lens.

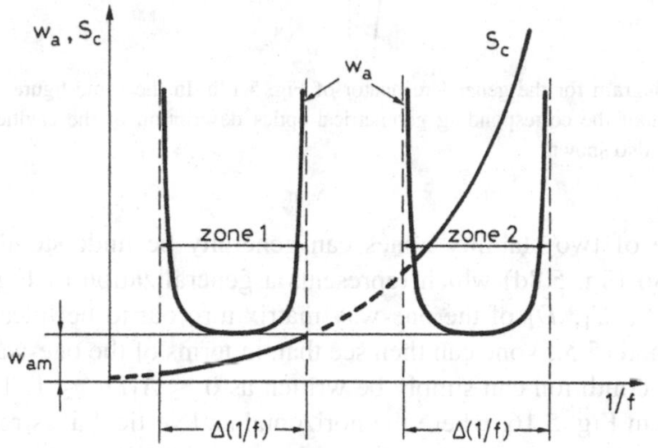

FIG. 5.15. Spot size in the active medium, w_a, and combined misalignment sensitivity, S_c, vs dioptric power, $1/f$, for the cavity of Fig. 5.14a.

For given values of a and $1/f$, Eqs. (5.5.30) and (5.5.31) provide a pair of equations for the cavity parameters R_1, R_2, L_1, and L.

One may now ask the question whether a dynamically stable point actually exists for the cavity of Fig. 5.14a. To answer this, we show in Fig. 5.15 the general behavior of w_a vs dioptric power $1/f$, for the above cavity for given values of the other cavity parameters. From this figure one notices the following general characteristic features: (1) Two dynamically stable points, i.e. satisfying Eq. (5.5.30), are found when the lens dioptric power is changed. (2) Both points correspond to a minimum of w_a, the minimum value, w_{am}, being the same for the two points. (3) The minima belong to two different stability zones, with the spot size actually diverging at each zone boundary. (4) The width, $\Delta(1/f)$, of the two zones is the same and satisfies a fundamental relationship with the minimum spot size given by the equation

$$\frac{\pi w_{am}^2}{\lambda}\Delta(1/f) = 2 \qquad (5.5.32)$$

independently of the values of the other cavity parameters.

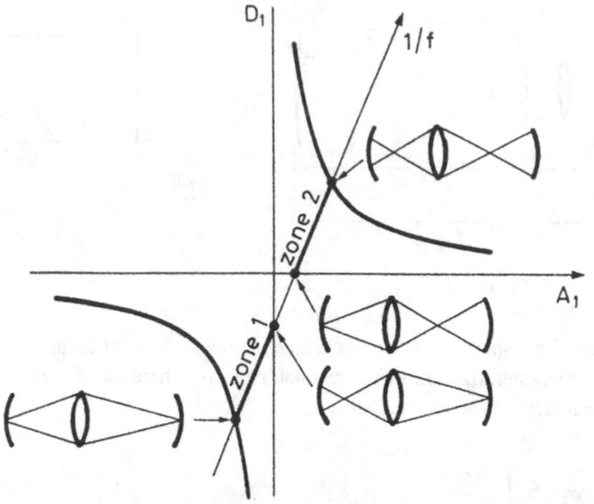

FIG. 5.16. Stability diagram for the general resonator of Fig. 5.14b. In the same figure, the two stability zones discussed in Fig. 5.15 and the corresponding geometrical-optics description of the cavities corresponding to the stability boundaries are also shown.

The existence of two stability zones can generally be understood with reference to Fig. 5.14b (see also Fig. 5.8d) which represents a generalization of Fig. 5.14a and where the elements A_1, B_1, C_1, D_1 of the one-way matrix turn out to be linear functions of $1/f$. From Eqs. (5.4.6) and (5.5.5) one can then see that, in terms of the one-way matrix elements, the cavity stability condition can simply be written as $0 \leq A_1D_1 \leq 1$. This stability condition is represented in Fig. 5.16, where the horizontal and vertical axes represent A_1 and D_1, respectively. Now, since A_1 and D_1 are linear functions of $1/f$, a plot of the values for A_1 vs the corresponding values of D_1, obtained by changing $1/f$, will show a linear relationship in the $A_1 - D_1$ plane (see Fig. 5.16). This straight line then generally intersects the stability boundaries at four distinct points which define two distinct stable zones. The laser beam configurations corresponding to these four limit points can be described by geometrical optics and are also shown in the same figure.

Having understood the origin of the two stability zones, it may be worth observing that the dioptric power of an optically pumped rod turns out to be given by[9]

$$\frac{1}{f} = \frac{k}{\pi a^2} P_a \tag{5.5.33}$$

where P_a is the pump power absorbed in the rod and k is a constant characteristic of the given material. If the expression for $1/f$ given by Eq. (5.5.33) is substituted in Eq. (5.5.32) and if, according to Eq. (5.5.31), one takes $(w_{am}/a) = (2/\pi)$ in the resulting expression, one can readily see that the range of acceptable absorbed power, ΔP_a, corresponding to each stability zone, is a constant for a given laser material (e.g. $\Delta P_a \cong 10$ W for a diode-pumped Nd:YAG).

From the above discussion, it would appear that the optical properties of the two stability zones are identical. A strong distinction between these zones is however revealed when one considers the misalignment properties of the laser cavity. We first define misalignment sensitivities, S_1 and S_2, for mirrors 1 and 2 according to the relations $S_1 = \delta r_{c1}/w_a\delta\theta_1$ and

$S_2 = \delta r_{c2}/w_a \delta\theta_2$, where, e.g. for mirror 1, δr_{c1} is the displacement of the beam center in the laser rod arising from a tilt $\delta\theta_1$ of mirror 1, and similarly for mirror 2. One can now define a combined misalignment sensitivity of the two mirrors as $S_c = \left[S_1^2 + S_2^2\right]^{1/2}$. A plot of this combined sensitivity vs lens dioptric power is also shown in Fig. 5.15. One can then see that one of the two zones, henceforth referred to as zone 1, is much less sensitive to mirror misalignment than the other, henceforth referred to as zone 2. The reason for the reduced sensitivity to misalignment for zone 1 can be understood by noting that the spot sizes at the mirrors are much smaller in zone 1 than in zone 2. Indeed, according to geometrical optics, one of the two stability boundaries of zone 1 corresponds to the beam being focused on both mirrors (see Fig. 5.16). Thus, when close to this boundary, the mirror spot size, w_m, is very small and the beam divergence, $\theta \approx \theta_d = \lambda/\pi w_m$, very large. Consequently, the mirror tilt, needed to produce a beam axis rotation comparable to the beam divergence, must also be large.

As a conclusion we can say that a dynamically and mechanically stable resonator can be designed for a general laser cavity describable as in Fig. 5.14b and comprising a variable element such as the thermally induced lens in the laser rod. The resonator should be chosen to belong to the more stable zone, zone 1, and must satisfy conditions Eqs. (5.5.30) and (5.5.31). In practice, instead of satisfying Eq. (5.5.30), the resonator can be designed to correspond to the center of zone 1. So, if the distance L_1, shown in Fig. 5.14a, is assumed to be the variable parameter, its value can be taken as the mean of its limiting values, L_1' and L_1'', in zone 1. From the geometrical optics description shown in Fig. 5.16, one then finds that L_1' and L_1'' must satisfy the conditions $L_1'^{-1} + L_2^{-1} = f^{-1}$ and $\left(L_1'' - R_1\right)^{-1} + L_2^{-1} = f^{-1}$, respectively, where L_2 is also shown in Fig. 5.14a. Once designed for a given focal length, f, and hence a given absorbed power, P_a, the resonator can then work for a range of absorbed pump power, ΔP_a, which is the same for a given active medium, independently of the cavity parameters.

5.6. UNSTABLE RESONATORS[10]

The stability condition for a generalized spherical resonator was discussed in Sect. 5.4 and the unstable regions were shown to correspond to the unshaded regions of the $g_1 - g_2$ plane in Fig. 5.7. Unstable resonators can be separated into two classes: (1) positive branch resonators, which correspond to the case $g_1 g_2 > 1$, and (2) negative branch resonators, which correspond to the case $g_1 g_2 < 0$.

Before going on to a quantitative discussion of unstable resonators, it is worth pointing out here the reasons why these resonators are of interest in the laser field. First we note that, according to the results obtained in example 5.5, for a stable resonator corresponding to a point in the $g_1 - g_2$ plane that is not close to an instability boundary, the spot size w is typically of the order of that given for the case of a confocal resonator and, for a wavelength of $\approx 1\,\mu$m, is usually smaller than 1 mm. We also note that, according to the discussion in example 5.9, a resonator aperture with radius $a < 2$ mm needs to be inserted in the laser resonator if oscillation is to be limited to the TEM_{00} mode. When oscillation is confined to a TEM_{00} mode of such a small cross section, the power (or energy) available in the output beam is necessarily limited. For unstable resonators, on the contrary, the field does not tend to be confined to the axis (see, for example, Fig. 5.3), and a large mode volume in a single transverse mode is possible. With unstable resonators, however, there is the problem that rays tend to walk off out of the cavity. The corresponding modes, therefore, have substantially greater (geometrical) losses than those of a stable cavity (where the losses are due to diffraction). This fact can, however, be used to advantage if these walk-off losses are turned into useful output coupling.

5.6.1. Geometrical-Optics Description

To establish the mode configurations of an unstable resonator, we can start by using a geometrical-optics approximation, as first done by Siegman.[11] To do this, we begin by recalling the two main results that were obtained for the eigensolutions of a stable resonator [see Eq. (5.5.24)]: (1)The amplitude is given by the product of a Hermite polynomial with a Gaussian function. (2)The phase distribution is such as to give a spherical wave front. The presence of the Gaussian function limits the transverse size of the beam and essentially arises from the focusing properties of a stable spherical resonator. The fact that the wave-front is spherical is, on the other hand, connected with the boundary conditions set by a spherical mirror. In the unstable case there are no Hermite–Gaussian solutions, as indeed discussed in connection with the solution of Eq. (5.5.4). Since the beam is no longer focused toward the resonator axis, but rather spread out over the whole resonator cross section, it is natural to assume, as a first approximation, that the solution has a constant amplitude over the resonator cross section while the wave front is still spherical, i.e. the solution is represented by a spherical wave. More precisely, since the mode can always be considered as being due to the superposition of two counter-propagating waves, we will assume that these consist of two counter-propagating *spherical waves*. It should be noted that one reaches the same conclusion by considering the solution of Eq. (5.5.3) in the unstable region. In this case, the discriminant of the quadratic equation in Eq. (5.5.3) is positive and one generally gets two real solutions for the parameter q and these will correspond to two spherical waves.

To calculate the mode field, we let P_1 and P_2 be the centers of curvature of the two spherical waves in the general two-mirror unstable resonator of Fig. 5.17a. By symmetry, P_1 and P_2 must lie on the resonator axis and their position are easily calculated by a self-consistency argument: the spherical wave originating from point P_1, after reflection at mirror 2, must give a spherical wave originating from P_2 and, vice-versa, the spherical wave originating from P_2, after reflection at mirror 1, must give a spherical wave originating from P_1. These two conditions lead to two equations, which can readily be established by a straightforward calculation based on geometrical optics, in the two unknowns, namely the positions of points P_1 and P_2. If these positions are expressed in terms of the dimensionless quantities r_1 and r_2 indicated in Fig. 5.17a, these last quantities turn out to be functions only of the resonator g_1, g_2 parameters. In fact, after some lengthy but straightforward calculations, one arrives at the relations

$$r_1^{-1} = g_1\left[1 - (g_1g_2)^{-1}\right]^{1/2} + g_1 - 1 \tag{5.6.1a}$$

$$r_2^{-1} = g_2\left[1 - (g_1g_2)^{-1}\right]^{1/2} + g_2 - 1 \tag{5.6.1b}$$

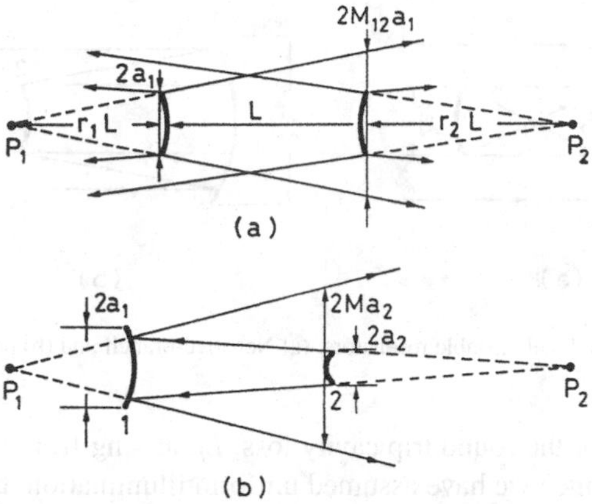

FIG. 5.17. (a) General, convex mirror, unstable resonator; (b) single-ended unstable resonator.

Having calculated r_1 and r_2 one can easily obtain from Fig. 5.17a the so-called single-pass magnification factor on going from mirror 1 to mirror 2, M_{12}, or from mirror 2 to mirror 1, M_{21}. For instance, M_{12} is defined as the increase in diameter of the spherical wave when propagating from mirror 1 to mirror 2. From simple geometrical considerations one gets from Fig. 5.17a

$$M_{12} = (1 + r_1)/r_1 \tag{5.6.2a}$$

Similarly one gets

$$M_{12} = (1 + r_2)/r_2 \tag{5.6.2b}$$

Usually, for laser applications, a single-ended resonator such as that of Fig. 5.17b, is of interest. In this case, the diameter of mirror 1, $2a_1$, must be larger than the transverse extent, at mirror 1, of the spherical wave originating from point P_2. We thus require $a_1 > M_{21}a_2$. With this condition, the only wave that emerges from the cavity is the spherical wave emitted by point P_1 escaping around mirror 2 (mirrors 1 and 2 are assumed to be 100% reflecting). This spherical wave starts from mirror 2 with a diameter $2a_2$ (see Fig. 5.17b) and returns to mirror 2, after one round trip, magnified by a factor M given by

$$M = M_{21}\, M_{12} = \left(1 + r_1^{-1}\right)\,\left(1 + r_2^{-1}\right) \tag{5.6.3}$$

where Eqs. (5.6.2) have been used. With the help of Eqs. (5.6.1), (5.6.3) readily gives

$$M = (2\,g_1 g_2 - 1) - 2\,g_1 g_2 \left[1 - (g_1 g_2)^{-1}\right]^{1/2} \tag{5.6.4}$$

which shows that M, the round trip magnification factor, depends only on the cavity g parameters. Note that, when $g_1 g_2 < 0$, M becomes negative and then it is the magnitude of this value that must be considered. Having calculated the round trip magnification factor, one can easily

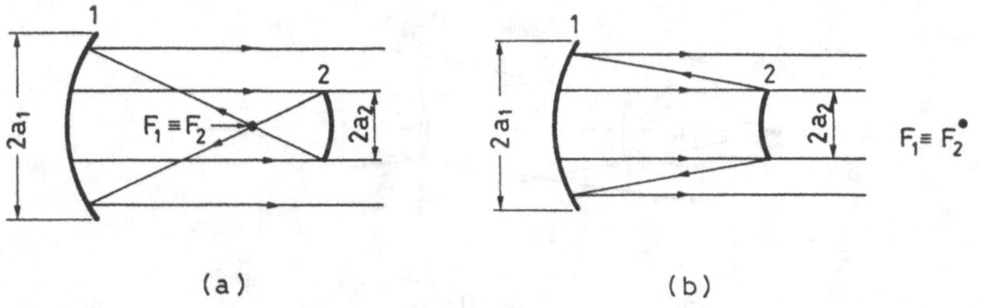

FIG. 5.18. Confocal unstable resonators: (a) Negative-branch and (b) positive-branch.

obtain the expression for the round trip cavity loss, L_i, arising from transmission around the output mirror. In fact, since we have assumed uniform illumination, the fraction of the beam power that is coupled out of mirror 2, after a round trip, is seen to be

$$L_i = \frac{S_2' - S_2}{S_2'} = \frac{M^2 - 1}{M^2} \tag{5.6.5}$$

where $S_2 = \pi a_2^2$ and $S_2' = \pi M^2 a_2^2$ are, respectively, the cross section for the beam originating from mirror 2 and that after one round trip. Note that, according to Eq. (1.2.4c) the round-trip logarithmic loss γ_i is given by $\gamma_i = -\ln(1 - L_i)$. Note also that γ_i, like M, is independent of mirror diameter $2a_2$.

Example 5.10. *Unstable confocal resonators* A particularly important class of unstable resonator is the confocal resonator, which can be of negative-branch or positive-branch. These are shown in Fig. 5.18a, b, respectively. In both cases, the two mirror foci F_1 and F_2 are coincident, and one can readily show that the resonators are represented, in the $g_1 - g_2$ plane, by the two branches of the hyperbola indicated as dashed curves in Fig. 5.7 [the equation of the hyperbola is $(2g_1 - 1)(2g_2 - 1) = 1$]. Of these various resonators, only the (symmetric) confocal one ($g_1 = g_2 = 0$) and the plane-parallel one ($g_1 = g_2 = 1$) lie on the boundary between the stable and unstable regions. All other confocal resonators are unstable and may either belong to the negative or positive branch of the instability region. As shown in Fig. 5.18 and as one can also show from Eq. (5.6.1), the mode consists of a superposition of a plane wave with a spherical wave originating from the common focus $F_1 = F_2$. The round-trip magnification factor M is simply given by $M = |R_1|/|R_2|$, where R_1 and R_2 are the two curvature radii of the two mirrors ($|R_1| > |R_2|$). If the aperture of diameter $2a_1$ at mirror 1 is made sufficiently large ($2a_1 > 2Ma_2$), only the plane beam will escape out of the cavity. Thus the beam escaping from a single-ended confocal resonator is a plane wave and this constitutes one of the main advantages of unstable confocal resonators. The round trip loss, or fractional output coupling, of this single-ended resonator is then given by Eq. (5.6.5).

5.6.2. Wave-Optics Description

The discussion so far has been based on a geometrical-optics approximation. To get a more realistic picture of the modes of an unstable resonator one must use a wave approach, e.g. use the integral equation Eq. (5.2.5), which arises from the Huyghens–Fresnel diffraction

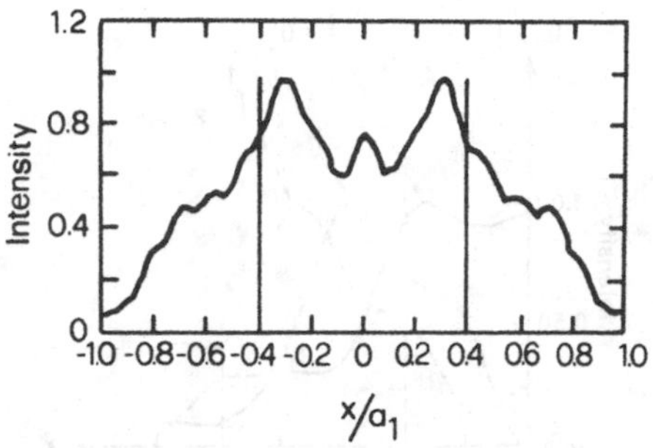

FIG. 5.19. Typical example of the radial behavior of mode intensity distribution in an unstable cavity obtained using a wave-optics calculation (after Rensch and Chester,[12] by permission).

equation Eq. (5.2.1). For unstable resonators, the limited aperture size of the output mirror constitutes an essential feature, since the beam must exit around this mirror. Consequently, the kernel K to be used in Eq. (5.2.5) can be obtained, in principle, by essentially the same procedures as those developed, for a stable cavity, in Sect. 5.5.2. Thus the solution of the integral equation can be obtained by an iterative approach such as the Fox-Li procedure discussed in that section. These calculations will not be discussed at any length here and we will limit ourselves to pointing out and commenting on a few relevant results.

A first important result is that the wave-optics description does indeed show that eigensolutions, i.e. field profiles which are self-reproducing after one round trip, do exist also for unstable resonators. To shows this in some detail, we will limit our discussion to a single-ended unstable confocal resonator and define an equivalent Fresnel number as $N_{eq} = [(M-1)/2] \times (a_2^2/L\lambda)$, for the positive branch, and as $N_{eq} = [(M+1)/2](a_2^2/L\lambda)$, for the negative-branch, with $2a_2$ being the diameter of the output mirror. A typical example of a computed plot of the radial intensity profile, which is self-reproducing after one round trip, is shown in Fig. 5.19. The calculation relates to a positive branch confocal resonator with $M = 2.5$ and $N_{eq} = 0.6$, and the intensity profile refers to the field just in front of mirror 2 (Fig. 5.18b) of a beam propagating to the right inside the resonator. The intensity profile in Fig. 5.19 is plotted vs the x (or y) transverse coordinate normalized to the radius, a_1, of mirror 1. To ensure a single-ended output, the condition $a_1 = 2.5\, a_2$ is assumed. Consequently, the vertical lines in the figure, occurring at $(x/a_1) = \pm 0.4$, mark the edge of the output mirror. Note the peculiar meaning of a round-trip self-reproducing profile for unstable resonators. Starting in fact from mirror 2, the left propagating spherical wave (see Fig. 5.18b), will arise only from that part of the beam of Fig. 5.19 for which $-0.4 \leq (x/a_1) \leq 0.4$. In fact, the remaining part of the beam escapes around mirror 2 to form the output beam. The part remaining in the resonator, after propagation over a round trip, will produce again, through the combined effect of spherical divergence and beam diffraction, the *whole intensity profile* of Fig. 5.19. The amplitude of the beam profile after one round trip will of course be smaller than the original value due to the loss represented by that part of the beam which has been coupled out of mirror 2. One should note that the beam intensity profile in Fig. 5.19 is quite different from the constant value assumed in the geometrical-optic theory, the difference being due to field diffraction, in particular from the edges of mirror 2. Indeed, one sees from Fig. 5.19 that, if x is interpreted as the radial distance from the mirror's center, several

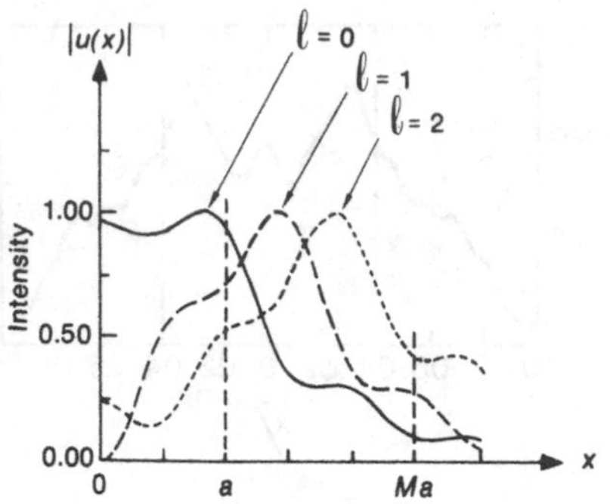

FIG. 5.20. Intensity profiles of the three lowest-order eigenmodes for a strip unstable resonator with $M = 25$ and $N_{eq} = 0.6$ (after Siegman,[10] by permission).

diffraction ring arising from the sharp edges of mirror 2 are present in the beam. Despite this significant difference between the intensity profile predicted by wave optics and that predicted by geometrical optics, the phase variation turns out to be remarkably similar in the two cases. Thus the wavefront turns out to be close to spherical, with radius almost equal to that predicted by geometrical optics (i.e. plane in this case).

A second relevant result of the wave-optics calculation is that for unstable resonator also, just as for stable, there exist different transverse modes, i.e. different self-reproducing spatial patterns. These modes generally differ from each other in the location and strength of the diffraction rings. An example of three such modes, again for a positive branch confocal unstable resonator, is shown in Fig. 5.20. Unlike the case of stable resonators, it is not possible, in this case, to make a clear distinction in terms of these field distribution between the lowest order and higher order modes. It should be noted, however, that the mode labeled $l = 0$ in the figure shows a field amplitude distribution which is more concentrated toward the beam axis. Thus, in this case, this mode will have the lowest loss i.e. it will be the "fundamental" mode.

A third characteristic result is found when one changes the equivalent Fresnel number, i.e. one changes either M or a_2, or L. In fact, at each integer value of the equivalent Fresnel number a different and distinct mode becomes the "lowest-order" i.e. the lowest-loss mode. This circumstance can be understood with the help of Fig. 5.21, where the magnitude of the eigenvalue σ is plotted vs N_{eq} for the three modes indicated in Fig. 5.20. One notes in particular that, since $\gamma = 1 - |\sigma|^2$, the $l = 1$ mode becomes the lowest order mode when N_{eq} becomes larger than one (and smaller than two). The reason for this circumstance arises from the fact that as N_{eq} increases, starting from e.g. the value $N_{eq} = 0.6$ of Fig. 5.20, the mode $l = 1$ contracts inwards while the mode $l = 0$ spreads outwards so that, at $N_{eq} \cong 1$, the role of the two modes is interchanged. One can also notice from Fig. 5.21 that, at each half-integer values of N_{eq}, there is a large difference between the losses of the "lowest order" mode and those of other modes. This might seem to suggest that a large transverse-mode discrimination can only be obtained under these conditions. It should be noted, however, that when the loss curves of two modes cross each other (i.e. for integer values of N_{eq} in Fig. 5.21), the intensity patterns of these two modes happen to become identical. Thus, at e.g. $N_{eq} = 1$, a large difference in loss exists between the $l = 2$ mode and the $l = 0$, $l = 1$ modes, which, in

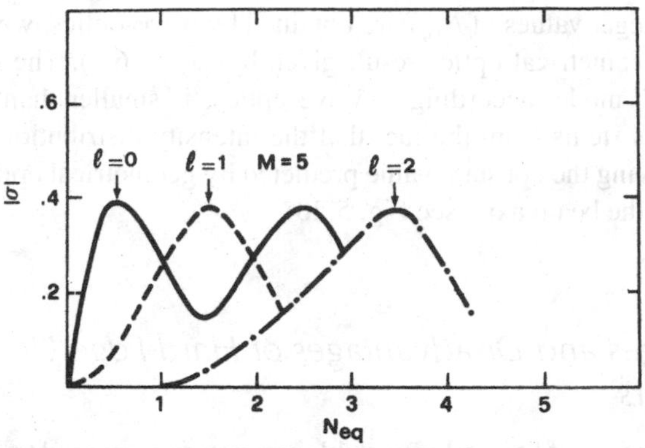

FIG. 5.21. Typical example of the oscillatory behavior of eigenvalue magnitude, σ, vs equivalent Fresnel number, N_{eq}, for the three consecutive modes of Fig. 5.20.

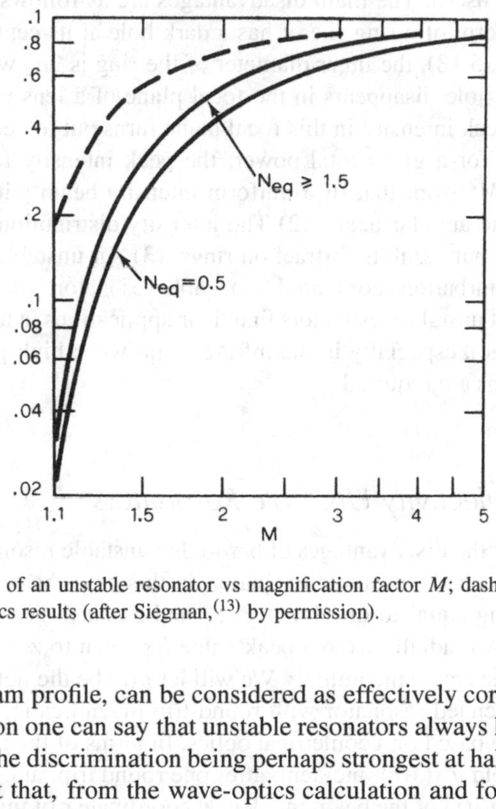

FIG. 5.22. Coupling losses of an unstable resonator vs magnification factor M; dashed curve: geometrical optics result; solid-lines: wave-optics results (after Siegman,[13] by permission).

terms of transverse beam profile, can be considered as effectively corresponding to the same mode.* As a conclusion one can say that unstable resonators always have a large transverse-mode discrimination, the discrimination being perhaps strongest at half-integer values of N_{eq}. One can also point out that, from the wave-optics calculation and for half-integer values of N_{eq}, one obtains a loss of the lowest order mode which is considerably smaller than the value predicted by geometrical optics. This result is apparent from Fig. 5.22 where the loss γ is plotted vs the round trip magnification factor M. In the figure the solid curves (which apply

* The two modes still differ with respect to the total round trip phase shift, i.e., they still differ in the field variation along the longitudinal z axis and thus in their resonance frequencies.

to successive half-integer values of N_{eq}) are obtained by wave-optics, while the dashed curve corresponds to the geometrical-optics result given by Eq. (5.6.5). The reason why the loss of the "lowest-order" mode, according to wave optics, is smaller than the value predicted by geometrical optics stems from the fact that the intensity distribution of the lowest order mode, rather than having the constant value predicted by geometrical optics, tends to be more concentrated toward the beam axis (see Fig. 5.20).

5.6.3. Advantages and Disadvantages of Hard-Edge Unstable Resonators

The main advantages of hard-edge unstable resonators compared to stable resonators can be summarized as follows: (1) Large, controllable mode volume; (2) good transverse-mode discrimination; (3) all reflective optics (which is particularly attractive in the infrared, where metallic mirrors can be used). The main disadvantages are as follows: (1) The output beam cross-section is in the form of a ring (i.e. it has a dark hole at its center). For example, in a confocal resonator (Fig. 5.18), the inner diameter of the ring is $2a_2$ while its outer diameter is $2Ma_2$. Although this hole disappears in the focal plane of a lens used to focus the beam (far-field pattern), the peak intensity in this focal plane turns out to decrease with decreasing ring thickness. In fact, for a given total power, the peak intensity for an annular beam is reduced by $(M^2 - 1)/M^2$ from that of a uniform-intensity beam with a diameter equal to the large diameter of the annular beam. (2) The intensity distribution in the beam does not follow a smooth curve, but exhibits diffraction rings. (3) An unstable resonator has greater sensitivity to cavity perturbations compared to a stable resonator. The above advantages and disadvantages mean that unstable resonators find their applications in high-gain lasers (so that M can be relatively large), especially in the infrared, and when high-power (or high-energy) diffraction-limited beams are required.

5.6.4. Variable-Reflectivity Unstable Resonators

Some, if not all, of the disadvantages of hard-edge unstable resonators can be overcome by using a variable reflectivity unstable resonator. In this case the reflectivity of the output mirror, rather than being equal to one for $r < a_2$ and equal to zero for $r > a_2$, as in the hard edge case, decreases radially from a peak value R_0 down to zero over a radial distance comparable to that of the active medium.[14] We will let $\rho(r)$ be the field reflectivity of mirror 2 and assume a single-ended resonator with round-trip magnification M. For simplicity we will follow an approach based on geometrical optics. In terms of the radial coordinate, r, we can then say that the field $u'_2(Mr)$, incident, after one round trip, at coordinate Mr of mirror 2, comes from the field $u_2(r)$ of the beam incident at coordinate r of mirror 2 at the start of the round trip. After taking into account the field reflectivity profile of mirror 2 and the round-trip magnification M, we can then write

$$u'_2(Mr) = \frac{\rho(r)\, u_2(r)}{M} \tag{5.6.6}$$

Note the quantity M appearing at the denominator of the right hand side of Eq. (5.6.6). This is a consequence of the fact that, after a magnification by a factor M, the beam area increases by a factor M^2. To conserve the power of the beam, the intensity must therefore decrease by a factor M^2 and the field by a factor M. If u_2 corresponds to a cavity mode, then it follows that $u'_2(r) = \sigma u_2(r)$ where σ is now a real quantity, with magnitude smaller than unity, in order to account for cavity losses. From Eq. (5.6.6) we then get

$$\sigma\, u_2(Mr) = \frac{\rho(r)\, u_2(r)}{M} \tag{5.6.7}$$

The eigensolutions $u_2(r) = u_{2l}(r)$ of Eq. (5.6.7) will give the field distributions inside the cavity, in front of mirror 2, while the eigenvalues of Eq. (5.6.7) will give the round-trip losses, due to the output coupling, according to the familiar relation (see Eq. (5.2.6))

$$\gamma = 1 - \sigma^2 \tag{5.6.8}$$

The first case that we shall consider is that of a Gaussian reflectivity profile.[11,12] We therefore write

$$\rho = \rho_0 \exp\left(-r^2/w_m^2\right) \tag{5.6.9}$$

where ρ_0 is the peak field reflectivity and w_m sets the transverse scale of the mirror reflectivity profile. One should note that, according to Eq. (5.6.9), the intensity reflectivity profile, which is the quantity usually measured experimentally, will be given by

$$R = R_0 \exp(-2r^2/w_m^2) \tag{5.6.10}$$

where $R_0 = \rho_0^2$ is the peak reflectivity. With the help of Eq. (5.6.9), the lowest order solution of Eq. (5.6.7) can be shown, by direct substitution, to be given by

$$u_{20}(r) = u_{20}(0) \exp(-r^2/w^2) \tag{5.6.11}$$

where

$$w^2 = (M^2 - 1)w_m^2 \tag{5.6.12}$$

The corresponding eigenvalue σ is

$$\sigma = \rho_0/M \tag{5.6.13}$$

so that, according to Eq. (5.6.8), the output coupling losses are given by

$$\gamma = 1 - (R_0/M^2) \tag{5.6.14}$$

The radial intensity distribution for the beam incident on mirror 2 is then given by

$$I_{in}(r) = I_{in}(0) \exp(-2r^2/w^2) \tag{5.6.15}$$

One notes that the radial profiles of both the field amplitude, u_{20}, and beam intensity, I_{in}, are described by Gaussian functions. On the other hand, the intensity of the output beam, I_{out}, is given by

$$\begin{aligned} I_{out}(r) &= I_{in}(r)[1 - R(r)] \\ &= I_{in}(0)[\exp(-2r^2/w^2) - R_0 \exp(-2M^2 r^2/w^2)] \end{aligned} \tag{5.6.16}$$

where Eqs. (5.6.15), (5.6.10) and (5.6.12) have been used. Note that I_{out} is not described by a Gaussian function and that, under appropriate conditions, one can expect an intensity profile that has a flat top for $r = 0$, a feature that is of interest for some applications. This circumstance occurs in fact when $(d^2 I_{out}/dr^2)_{r=0} = 0$. In this case, we find from Eq. (5.6.16) that the central reflectivity, R_0, and the cavity magnification, M, must satisfy the condition

$$R_0 M^2 = 1 \tag{5.6.17}$$

For this resonator, the round trip cavity losses will, according to Eqs. (5.6.14) and (5.6.17), be given by

$$\gamma = 1 - (1/M^4) \tag{5.6.18}$$

The above equations give the salient results for unstable resonators with mirrors of Gaussian reflectivity profile. Although these results are based on a simple geometrical optics approach, they are in good agreement with results based on a wave-optics approach for sufficiently large values of the equivalent Fresnel number ($N_{eq} \geq 5$).[15] For Gaussian reflectivity mirrors one can also use an elegant wave optics analysis based on a suitable *ABCD* matrix with complex matrix elements.[16]

Example 5.11. *Design of an unstable resonator with an output mirror having a Gaussian radial reflectivity profile* We will assume $\gamma = 0.5$ as the value which optimizes the output coupling of a given laser (see Chap. 7) and we will consider the case where the output beam has its flattest profile. From Eq. (5.6.18) we get $M^2 = \sqrt{2}$, from Eq. (5.6.17) $R_0 = 1/M^2 = 1/\sqrt{2} = 0.71$ and from Eq. (5.6.12) $w^2 = 0.41\, w_m^2$. The reflectivity profile and the corresponding intensity profiles inside and outside the resonator are all shown in Fig. 5.23. If we now let a be the radius of the active medium and if the medium is placed in front of mirror 2, the beam intensity profile within the medium will be given by $I_{in}(r)$. To avoid excessive beam truncation by the active medium aperture, i.e. to avoid excessively pronounced diffraction rings arising from this truncation, we can, e.g. impose the condition $I_{in}(a)/I_{in}(0) = 2 \times 10^{-2}$. We then obtain $a \cong 0.9\, w_m$ which, for a given aperture a, establishes the spot size w_m of the Gaussian reflectivity profile. As an example, if we take $a = 3.2$ mm we get $w_m = 3.5$ mm. Thus, to conclude, the Gaussian mirror must have a peak reflectivity of $R_0 \cong 71\%$, a spot size $w_m = 3.5$ mm and it must be used in an unstable cavity (e.g. a confocal cavity) with a round trip magnification of $M = [2]^{1/4} = 1.19$.

The second case that we shall consider is that of a super-Gaussian reflectivity profile.[17] Instead of Eqs. (5.6.9) and (5.6.10) we will now write

$$\rho = \rho_0 \exp\left(-r^n/w_m^n\right) \tag{5.6.19a}$$

$$R = R_0 \exp\left(-2r^n/w_m^n\right) \tag{5.6.19b}$$

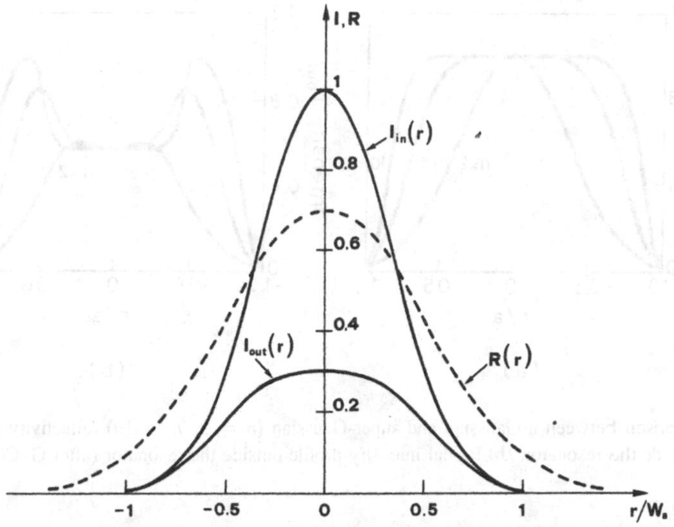

FIG. 5.23. Radial intensity profiles inside, I_{in}, and outside, I_{out}, an unstable cavity with a Gaussian reflectivity output coupler, $R(r)$ (case of the flattest profile for I_{out}).

and, for $n > 2$, Eqs. (5.6.19) describe curves with super-Gaussian reflectivity profile. The substitution Eqs. (5.6.19a) into (5.6.7) then gives

$$u_2(r) = u_2(0)\exp(-r^n/w^n) \tag{5.6.20}$$

where

$$w = w_m(M^n - 1)^{1/n} \tag{5.6.21}$$

Again we have $\sigma = \rho_0/M$ and $\gamma = 1 - \sigma^2 = 1 - (R_0/M^2)$. From Eq. (5.6.20) we now obtain

$$I_{in}(r) = I_{in}(0)\exp(-2r^n/w^n) \tag{5.6.22}$$

and the radial profiles of both u_2 and I_{in} are described by super-Gaussian functions of the same order, n, as that of the reflectivity profile. The intensity of the output beam, I_{out}, is readily obtained from $I_{out} = I_{in}(r)[1 - R(r)]$ and one notes that it is not described by a super-Gaussian function.

To make a comparison between the performance of unstable resonators with Gaussian and super-Gaussian reflectivity profiles, we show in Fig. 5.24a the intensity profiles, I_{in}, for $n = 2$ (Gaussian) and $n = 5, 10$ (super-Gaussian). The curves have all been normalized to their peak values and the corresponding spot size w in Eqs. (5.6.22) and (5.6.15) have been chosen so that exp-$(2a^n/w^n) = 2 \times 10^{-2}$, where a is the radius of the active medium. The comparison is therefore made for the same degree of beam truncation by the active medium. The main advantage of a super-Gaussian mirror compared to a Gaussian mirror is apparent from Fig. 5.24a: super-Gaussian mirrors of increasing super-Gaussian order, n, allow better exploitation of the active medium (i.e. the area of the mode, A_m, increases as n is increased).

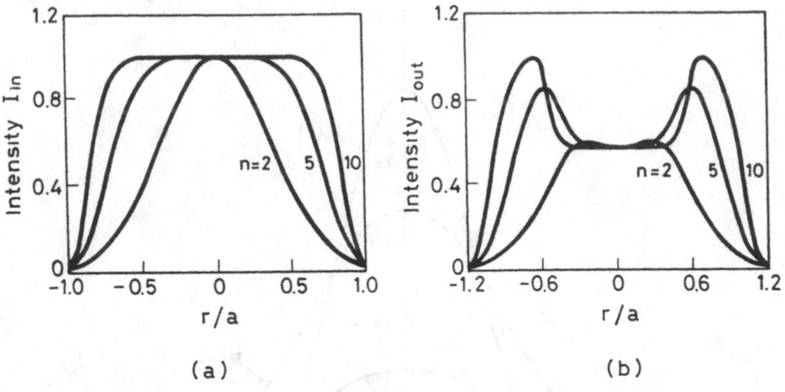

FIG. 5.24. Comparison between a Gaussian and super-Gaussian ($n = 5$, $n = 10$) reflectivity profile: (a) Radial intensity profile inside the resonator. (b) Radial intensity profile outside the resonator (after G. Cerullo et al.,[18] by permission).

On the other hand, the diffraction angle, θ_d, increases with increasing n, as can be understood from Fig. 5.24b. In this figure the corresponding radial intensity profiles, as predicted by the previous equations for $I_{out}(r)$ and for $R_0 = 0.45$ and $M = 1.8$, are shown. One sees that, as n increases, a hole of increasing depth appears in the output beam and this results in an increased beam divergence. As a consequence of these two conflicting tendencies, the beam brightness, which may be taken to be proportional to A_m/θ_d^2, has an optimum value as a function of n. It turns out that this optimum value depends on the cavity round trip magnification, M, and on peak mirror reflectivity R_0 but, for all practical cases, it ranges between 5 and 8.[18] Thus, in terms of beam brightness, super-Gaussian mirrors with super-Gaussian order $n = 5 \div 8$ provide the best choice for a variable-reflectivity unstable resonator.

5.7. CONCLUDING REMARKS

In this chapter a few of the most relevant features of stable and unstable resonators have been considered. It is shown, in particular, that, to obtain single transverse mode oscillation, one can use stable resonators provided that the Fresnel number is typically smaller than two. This usually means that the radius of the limiting aperture (e.g. the radius of the active medium) must typically be smaller than 2 mm at $\lambda = 1.06\,\mu$m and ~6.5 mm at $\lambda = 10.6\,\mu$m. For larger values of the active medium dimensions, unstable resonators need to be used. In this case, radially-variable reflectivity output mirrors of Gaussian or, better, super-Gaussian profile provide the best solution.

PROBLEMS

5.1. A two-mirror resonator is formed by a convex mirror of radius $R_1 = -1$ m and a concave mirror of radius $R_2 = 1.5$ m. What is the maximum possible mirror separation if this is to remain a stable resonator?

5.2. Consider a confocal resonator of length $L = 1$ m used for an Ar^+ laser at wavelength $\lambda = 514.5$ nm. Calculate: (a) the spot size at the resonator center and on the mirrors; (b) the frequency difference between consecutive longitudinal modes; (c) the number of non-degenerate modes falling within the Doppler-broadened width of the Ar^+ line ($\Delta\nu_0^* = 3.5$ GHz, see Table 2.2)

5.3. Consider a hemiconfocal resonator (plane-spherical resonator with $L = R/2$) of length $L = 2$ m used for a CO_2 laser at a wavelength of $\lambda = 10.6\,\mu$m. Calculate: (a) the location of the beam waist; (b) the spot size on each mirror; (c) the frequency difference between two consecutive TEM_{00} modes; (d) the number of TEM_{00} modes falling within the laser linewidth (consider a typical low pressure CO_2 laser and thus take $\Delta\nu \simeq 50$ MHz).

5.4. Consider a resonator consisting of two concave spherical mirrors both with radius of curvature 4 m and separated by a distance of 1 m. Calculate the spot size of the TEM_{00} mode at the resonator center and on the mirrors when the laser oscillation is at the Ar^+ laser wavelength $\lambda = 514.5$ nm.

5.5. How is the spot size modified at each mirror if one of the mirrors of the above problem is replaced by a plane mirror?

5.6. Using Eqs. (4.7.26) and (5.5.8a), show that the beam waist for the two-mirror resonator of Fig. 5.8b occurs at a distance, z_1, from mirror 1 given by $z_1 = (1 - g_1)g_2L/(g_1 + g_2 - 2g_1g_2)$.

5.7. One of the mirrors in the resonator of Problem 5.4 is replaced by a concave mirror of 1.5 m radius of curvature. Using the result of Problem 5.6, calculate: (1) the position of the beam waist; (2) the spot size at the beam waist and on each mirror.

5.8. A resonator consists of two plane mirrors with a positive lens inserted between the two mirrors. If the focal length of the lens is f, and L_1 and L_2 are the distances of the lens from the two mirrors, calculate: (1) the spot size at the lens position, and the spot sizes at each mirror; (2) the conditions under which the cavity is stable.

5.9. A triangular ring cavity is made up of three plane mirrors (Fig. 5.4a) with a positive lens inserted between two of the mirrors. If p is the length of the ring perimeter, calculate the position of minimum spot size, its value, and the spot size at the lens position. Also find the stability condition for this cavity.

5.10. A laser operating at $\lambda = 630$ nm has a power gain of 2×10^{-2} per pass and is provided with a symmetric resonator consisting of two mirrors each of radius $R = 10$ m and spaced by $L = 1$ m. Choose an appropriate size of mirror aperture in order to suppress TEM_{01} mode operation while allowing TEM_{00} mode operation.

5.11. On account of its relatively small sensitivity to mirror misalignment (see problem 5.16), a nearly hemispherical resonator (i.e. a plane-spherical resonator with $R = L + \Delta$ and $\Delta \ll L$) is often used for a He-Ne laser at $\lambda = 630$ nm wavelength. If the cavity length is $L = 30$ cm, calculate: (1) the radius of curvature of the spherical mirror so that the spot size at this mirror is $w_m = 0.5$ mm; (2) the location in the g_1–g_2 plane corresponding to this resonator; (3) the spot size at the plane mirror.

5.12. Consider the nearly hemispherical He-Ne resonator of the previous problem and assume that the aperturing effect produced by the bore of the capillary containing the He-Ne gas mixture (see Chap. 10) can be simulated by a diaphragm of radius, a, in front of the spherical mirror. If the power gain per pass of the He-Ne laser is taken to be 2×10^{-2}, calculate the diaphragm radius needed to suppress the TEM_{01} mode (Hint: Show that the round-trip loss of this resonator is the same as the single pass loss of a near concentric symmetric resonator of length $L_{nc} = 2L$ and $R_1 = R_2 = R$. To calculate diffraction loss, then use Fig. 5.13, assuming the loss for a resonator of negative g-value to be the same as that of the corresponding resonator with corresponding positive g-value).

5.13. Consider the $A_1B_1C_1D_1$ matrix of Fig. 5.8d and show that, for a stable cavity, one must have $0 < A_1D_1 < 1$ and $-1 < B_1C_1 < 0$. From these results, then show that $B_1D_1/A_1C_1 < 0$ so that q_1 in Eq. (5.5.6a) is purely imaginary.

5.14. For a stable two-mirror resonator one can define a misalignment sensitivity, δ, as the transverse shift of the intersection of the optical axis with a given mirror, normalized to the spot size on that mirror, for a unit angular tilt of one of the two mirrors. In particular, for mirror 1, one can define two misalignment sensitivity factors δ_{11} and δ_{12} as $\delta_{11} = (1/w_1)(dr_1/d\theta_1)$ and $\delta_{12} = (1/w_1)(dr_1/d\theta_2)$, where $dr_1/d\theta_i$ $(i = 1, 2)$ is the transverse change of beam center at mirror 1 for unit angular tilt of either mirror 1 or 2. Show that, for a confocal resonator, $(\delta_{11})_c = 0$ and $(\delta_{12})_c = (\pi w_s/\lambda)$.

5.15. Using the definitions given in the previous problem, show that, for a near-plane symmetric resonator, the misalignment sensitivity is such that $\delta_{11} = \delta_{12} = \delta_{21} = \delta_{22} = (\delta_{12})_c 4w^3/w_s^2$, where $(\delta_{12})_c$ is the misalignment sensitivity of the confocal resonator, w is the spot size on the mirror for the actual resonator and w_s is the mirror spot size of a confocal resonator of the same length. According to the above equation which of the two resonators is the less sensitive to mirror tilt?

5.16. Consider a nearly hemispherical resonator ($R = L + \Delta$ with $\Delta \ll L$) in which mirror 1 is the plane mirror. Show that we have in this case $\delta_{12} = (\delta_{12})_c(w_2/w_s)$ and $\delta_{21} = (\delta_{12})_c(w_s/w_2)$. Comparing this resonator with the long radius resonator of the previous problem, for the same value of mirror spot-size, i.e. $w = w_2$, what conclusion can be drawn with regard to the misalignment sensitivity of a nearly hemispherical resonator compared to that of a nearly flat resonator?

5.17. An unstable resonator consists of a plane mirror (mirror 1) and a convex mirror (mirror 2) of radius of curvature $R_2 = 2$ m, spaced by a distance $L = 50$ cm. Calculate: (1) The resonator location in the g_1, g_2 plane. (2) The location of points P_1 and P_2 of Fig. 5.17. (3) The condition under which the resonator is single-ended with beam output only occurring around mirror 2. (4) The round-trip magnification factor and the round trip losses.

5.18. A confocal unstable resonator is to be used for a CO_2 laser at a wavelength of $\lambda = 10.6\,\mu$m. The resonator length is chosen to be $L = 1$ m. Which branch would you choose for this resonator if the mode volume is to be maximized? Calculate the mirror apertures $2a_1$ and $2a_2$ so that: (1) $N_{eq} = 7.5$, (2) single-ended output is achieved, and (3) a 20% round-trip output coupling is obtained. Then find the radii of the two mirrors R_1 and R_2.

5.19. Using a geometrical-optics approach (and assuming lowest-order mode oscillation), calculate the round-trip loss of the resonator designed in the above problem. What are the shape and dimensions of the output beam?

5.20. Consider an unstable resonator consisting of a convex mirror, mirror 1, of radius R_1 and a plane mirror (mirror 2) spaced by a distance $L = 50$ cm. Assume that the plane mirror has a super-Gaussian reflectivity profile with a super-Gaussian order $n = 6$ and peak power reflectivity $R_0 = 0.5$. Assume also that the active medium consists of a cylindrical rod (e.g. a Nd:YAG rod) with radius $a \cong 3.2$ mm, placed just in front of mirror 2. To limit the round-trip losses to an acceptable value, assume also a round trip magnification $M = 1.4$. Calculate: (1) The spot size w of the field intensity, I_{in}, for a 2×10^{-2} intensity truncation by the active medium. (2) The corresponding mirror spot size w_m. (3) The cavity round trip losses. (4) The radius of curvature of the convex mirror.

References

1. A.E. Siegman, *Lasers* (University Science Books, Mill Valley, California, 1986) sect. 14.2
2. M. Born and E. Wolf, *Principles of Optics*, 6th ed. (Pergamon Press, London 1980) sect. 1.6.5
3. H. Kogelnik and T. Li, Laser Beams and Resonators, *Appl. Opt.* **5**, 1550 (1966)
4. Reference [1], Chap. 19
5. A.G. Fox and T. Li, Resonant Modes in a Maser Interferometer, *Bell. Syst. Tech. J.* **40**, 453 (1961)
6. T. Li, Diffraction Loss and Selection of Modes in Maser Resonators with Circular Mirrors, *Bell. Syst.Tech. J.* **44**, 917 (1965)
7. V. Magni, Resonators for Solid-State Lasers with Large-Volume Fundamental Mode and High Alignment Stability, *Appl. Opt.*, **25**, 107–117 (1986). See also *erratum Appl. Opt.*, **25**, 2039 (1986)
8. Reference [1], p. 666
9. W. Koechner, *Solid-State Laser Engineering*, Vol. 1, *Springer Series in Optical Sciences*, fourth edition (Springer, New York, 1996)
10. Reference [1], Chap. 22
11. A.E. Siegman, Unstable Optical Resonators for Laser Applications, *Proc. IEEE* **53**, 277–287 (1965)
12. D.B. Rensch and A.N. Chester, Iterative Diffraction Calculations of Transverse Mode Distributions in Confocal Unstable Laser Resonators, *Appl. Opt.* **12**, 997 (1973)
13. A.E. Siegman, Stabilizing Output with Unstable Resonators, *Laser Focus* **7**, 42 (1971)
14. H. Zucker, Optical Resonators with Variable Reflectivity Mirrors, *Bell. Syst. Tech. J.* **49**, 2349 (1970)
15. A.N. Chester, Mode Selectivity and Mirror Misalignment Effects in Unstable Laser Resonators, *Appl. Opt.* **11**, 2584 (1972)
16. Ref. [1], sect. 23.3
17. S. De Silvestri, P. Laporta, V. Magni and O. Svelto, Solid-State Unstable Resonators with Tapered Reflectivity Mirrors: The super-Gaussian Approach, *IEEE J. Quant. Electr.* **QE-24**, 1172 (1988)
18. G. Cerullo et al., Diffraction-Limited Solid State Lasers with Supergaussian Mirror, In: *OSA Proc. on Tunable Solid-State Lasers*, Vol. 5 ed. by M. Shand and H. Jenssen (Optical Society of America, Washington 1989) pp. 378–384

6 Pumping Processes

6.1. INTRODUCTION

We have seen in Chap. 1 that the process by which atoms are raised from level 1 to level 3 (for a three-level laser, Fig. 1.4a) or from level 0 to level 3 (for a four-level or a quasi-three-level laser, Fig. 1.4b) is called the pumping process. Usually it is performed in one of the following two ways: (i) *Optically*, i.e. by the cw or pulsed light emitted by a powerful lamp or by a laser beam. (ii) *Electrically*, i.e. by a cw, radio-frequency, or pulsed current flowing in a conductive medium such as an ionized gas or a semiconductor.

In *optical pumping* by an incoherent source, the light from a powerful lamp is absorbed by the active medium and the atoms are thereby pumped into the upper laser level. This method is particularly suited to solid-state or liquid lasers (i.e. dye lasers). The line-broadening mechanisms in solids and liquids produce in fact very considerable broadening, so that one is usually dealing with pump bands rather than sharp levels. These bands can, therefore, absorb a sizable fraction of the, usually broad-band, light emitted by the lamp. The availability of efficient and powerful, cw or pulsed, laser sources at many wavelengths has recently made *laser pumping* both attractive and practical. In this case, the narrow line emission from a suitable laser source is absorbed by the active medium. This requires that the laser wavelength fall within one of the absorption bands of the medium. It should be noted, however, that laser's monochromaticity implies that laser pumping needs not to be limited to just solid-state and liquid lasers but can also be applied to gas lasers, provided that one can ensure that the line emitted by the pumping laser coincides with an absorption line of the medium to be pumped. This situation occurs, for instance, in most far-infrared gas lasers (e.g., methyl alcohol or CH_3OH, in the vapor state) which are usually pumped by a suitable rotational-vibrational line of a CO_2 laser. For solid-state or liquid lasers, on the other hand, Argon ion lasers, for cw excitation, Nitrogen or Excimer lasers, for pulsed excitation, and Nd:YAG lasers and their second and third harmonics, either cw or pulsed, are often used. Whenever possible, however, semiconductor-diode lasers, due to the inherently high efficiency of these laser sources (overall optical to electrical efficiencies larger than 60% have been demonstrated),

O. Svelto, *Principles of Lasers*,
DOI: 10.1007/978-1-4419-1302-9_6,

are now commonly used (*diode-laser pumping*). Actually one can foresee that, in a not too far future, diode-laser pumping will become the dominant means of optical pumping, replacing even high power lamps.

Electrical pumping is usually accomplished by means of a sufficiently intense electrical discharge and it is particularly suited to gas and semiconductor lasers. Gas lasers, in particular, do not usually lend themselves so readily to lamp pumping because their absorption lines are typically much narrower than the usual broad-band emission of the pumping lamp. A notable exception that should be mentioned is the case of the optically pumped Cs laser, in which Cs vapor is pumped by a lamp containing low-pressure He. In this case the situation was quite favorable for optical pumping since the strong $\sim$ 390 nm He emission line (which is rather sharp owing to the low pressure used) happens to coincide with an absorption line of Cs. This laser, however, is no longer in use and its importance resides mostly in its historical significance as the most notable lamp-pumped gas laser and, particularly, as it was the earliest proposed laser scheme. Electrical pumping of gas lasers, on the other hand, can be a fairly efficient process (e.g. for pumping the CO_2 laser) because the linewidth of the excitation cross-section of a given transition by electron impact is usually quite large (from a few to a few tens of eV, see Figs. 6.25 and 6.27). This circumstance occurs because electron impact excitation, namely $e + A \rightarrow A^* + e$ where A is the species to be excited, is a non-resonant process. The surplus energy, above that needed to excite species A, is in fact left as kinetic energy of the scattered electron. By contrast, the process of optical excitation by an incoming photon of energy $h\nu$, namely $h\nu + A \rightarrow A^*$, is a resonant process because the photon energy must equal the excitation energy of species A. Actually, as discussed in Chap. 2, some line-broadening processes occur in this case on account of some energy, arising e.g. from thermal movement of species A (as in Doppler broadening), which can be added to the process. The resulting width of the absorption line, however, turns out to be quite small (e.g. $\approx 10^{-5}$ eV for Doppler broadening of Ne atoms) and this is the fundamental reason why optical pumping by a broad-band source would be so inefficient for a gas laser. In the case of semiconductor lasers, on the other hand, optical pumping could be used very effectively, since the semiconductor medium has a strong and broad absorption band. Indeed, a number of optically pumped semiconductor lasers (particularly by laser pumping) have been made to operate. Electrical pumping proves to be more convenient, however, since a sufficiently large current density can be made to flow through a semiconductor, usually in the form of a p-n or p-i-n diode.

The two pumping processes considered above, optical pumping and electrical pumping, are not the only ones available for pumping lasers. A form of pumping which is somewhat similar to optical pumping is involved when the medium is excited by a beam from an X-ray source (*X-ray pumping*). Likewise, a pumping process somewhat similar to electrical pumping is involved when the medium is excited by a beam of electrons from an electron-beam machine (*e-beam pumping*). Although both X-ray and e-beam pumping are able to deliver high pump powers or energies in a large volume of active medium (generally in gaseous form), these pumping mechanisms are not widely used, in practice, due to the complexity of the X-ray or e-beam apparatus. It should also be noted, in this contest, that possibly the shortest wavelength so far achieved in a laser ($\lambda \cong 1.4$ nm i.e. around the boundary between soft and hard X-ray region) has been achieved using the intense X-rays produced by a small nuclear detonation. The details of this laser are still classified but one can readily appreciate that this pumping configuration is not easily duplicated in the typical laboratory!

A conceptually different and rather interesting type of pumping is involved when the required inversion is produced as a direct result of an exothermic chemical reaction (*chemical pumping*). There are two general kinds of these reactions which can be used, namely: (i) Associative reactions, i.e. $A+B \rightarrow (AB)^*$, resulting in the molecule AB being left in an excited vibrational state. (ii) Dissociative reactions, e.g. where the dissociation is induced by a photon i.e. $AB + h\nu \rightarrow A + B^*$, resulting in species B (atom or molecule) being left in an excited state. Chemical pumping usually applies to materials in the gas phase and generally requires highly reactive and often explosive gas mixtures. On the other hand, the energy available in an exothermic reaction is often quite large and high powers, for cw operation, or energies, for pulsed operation, can be available for laser action if a good fraction of the available energy is converted into laser energy. These features have enabled chemical lasers to produce the largest cw laser powers so far available (2.2 MW for the so-called MIRACL laser, an acronym for Mid Infrared Advanced Chemical Laser). In view of the handling problems associated with reactive and hazardous materials, the use of these lasers has been confined to the military field, for use as directed energy weapons.

Another conceptually different type of pumping mechanism for gas molecules is by supersonic expansion of a gas mixture containing the particular molecule (*gas-dynamic pumping*). In this case, a suitable mixture, usually involving the CO_2 molecule as the active species (e.g. CO_2:N_2:H_2O in the 6:76:1 ratio), is used. The mixture is raised, in a suitable container, to a high pressure (e.g. $\approx$ 17 atm) and temperature (e.g. $\approx$ 1,400 K) by combustion of appropriate fuels (e.g. combustion of benzene, C_6H_6, and nitrous oxide, N_2O, thus automatically supplying hot CO_2 with a CO_2/H_2O ratio of 2:1). The CO_2 molecule in this mixture is, of course, not inverted but, due to the high temperature, a substantial fraction of molecules is found in the lower laser level ($\approx$ 25%) while a lower but still substantial fraction is found in the upper laser level ($\approx$ 10%). It should be noted, in fact, that the CO_2 laser is a roto-vibrational laser and the lower and upper laser levels of the ground electronic state can be significantly excited thermally, i.e., by having the mixture at a high temperature. The gas mixture is then made to expand, adiabatically, to a very low pressure (e.g. $\approx$ 0.09 atm) trough a row of expansion nozzles (an example of this expansion system can be found in the chemical laser section of Chap. 10). Due to expansion, the translational temperature of the mixture will be reduced to a much lower value (e.g. $\approx$ 300 K). Consequently, during the expansion process, upper and lower state populations will tend to relax to the, much lower, equilibrium values appropriate to this lower temperature. For a CO_2 laser, however, the lifetime of the upper state is appreciably longer than that of the lower state. This means that relaxation of the lower level will occur at an earlier stage, downstream in the expanding beam. Thus there will be a fairly extensive region, downstream from the expansion nozzle, where the population of the lower laser level has decayed, while that of the upper level has persisted at its initial value in the container. Thus a population inversion is created in this region via the expansion process. Gas-dynamic pumping has been mainly applied to CO_2 lasers and has yielded high cw powers ($\approx$ 100 kW). The complications of the system have been an obstacle to its use for civilian applications while its lower power, compared to chemical lasers, has put it at disadvantage for military applications.

As for the case of radiation-matter interaction, considered in Chaps. 2 and 3, where the ultimate goal was the calculation of both stimulated and spontaneous transition rates, so the ultimate goal here would be to calculate the pump rate per unit volume, R_p, as defined by Eq. (1.3.1). When pumping with a broad-band light source, i.e. a lamp, the calculation of R_p becomes rather involved.[1] This is also the case when pumping via electrons in a gas discharge, where a distribution of electron-velocities is involved.[2] So, we will limit ourselves here to a description of various pumping schemes with some discussion of the underlying physical mechanisms involved in the processes.

6.2. OPTICAL PUMPING BY AN INCOHERENT LIGHT SOURCE

In the case of optical pumping by a powerful incoherent source, i.e. a lamp, the pump light is emitted in all directions and, generally, over a broad spectrum. This light then needs to be transferred into the active medium. The object of the next section is to describe how this transfer can be achieved by a suitable optical system.

6.2.1. Pumping Systems[3]

The lamps used for laser pumping are, often, of cylindrical shape. Figure 6.1 shows two of the most commonly used pumping configurations when a single lamp is used. In both cases the active medium is taken to be in the form of a cylindrical rod with length and diameter about equal to those of the lamp. The diameter usually ranges from a few millimeters to some tens of millimeters and the length from a few centimeters to a few tens of centimeters. In Fig. 6.1a the lamp is placed along one of the two focal axes, F_1, of a specularly reflecting cylinder of elliptical cross-section (labeled 1 in the figure and usually referred to as the pumping chamber). The rod is placed along the second focal axis F_2. A well-known property of an ellipse is that a ray F_1P, leaving the first focus F_1, passes, after reflection by the elliptical surface, through the second focus F_2 (ray PF_2). This means that a large fraction of the light emitted by the lamp is conveyed, by the pumping chamber, to the active rod. High reflectivity of this chamber is achieved by vacuum deposition of a gold or silver layer on the inside surface of the cylinder. Figure 6.1b shows an example of what is known as a close-coupled configuration. The rod and the lamp are placed as close as possible and are closely surrounded by cylindrical reflector (labeled 1 in the figure). In this case, pumping chambers made of diffusely reflecting

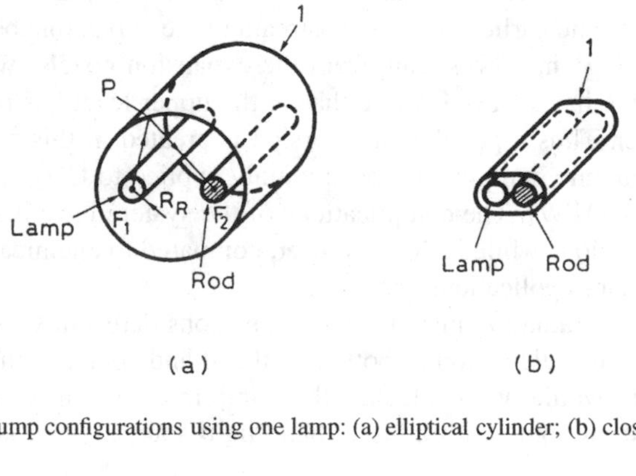

FIG. 6.1. Pump configurations using one lamp: (a) elliptical cylinder; (b) close-coupling.

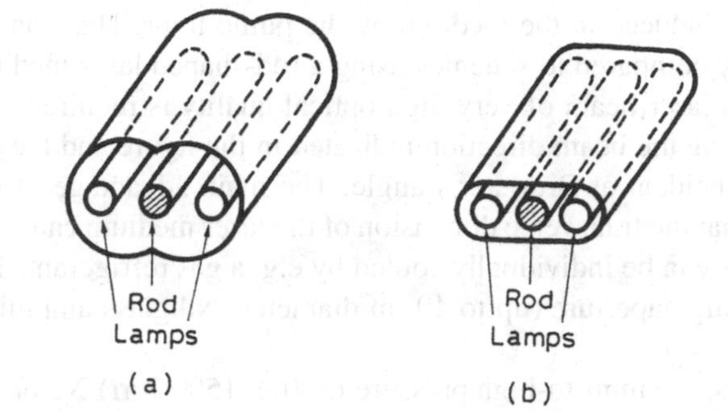

FIG. 6.2. Pump configurations using two lamps: (a) double-ellipse; (b) close-coupling.

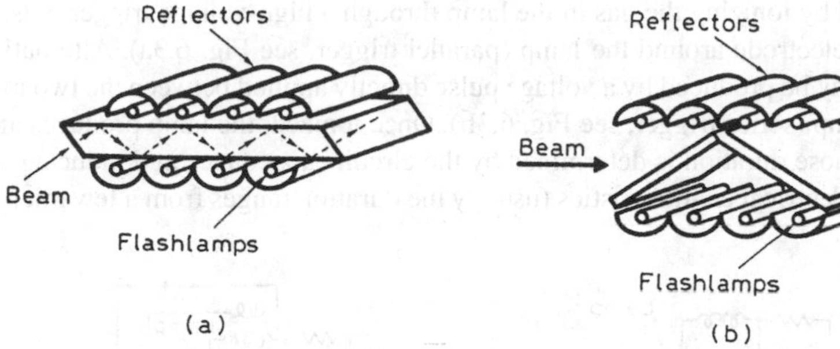

FIG. 6.3. Pumping configuration using many lamps: (a) Active medium in the form of a single slab with the laser beam traversing the slab in a zig-zag path. (b) Active medium made of many slabs inclined at Brewster's angle to the laser beam.

materials are often used instead of specular reflectors. For highly diffusing materials such as compressed $BaSO_4$ powders or white ceramic, which are very efficient scattering media, the efficiency for close-coupled configuration is usually not much less than that of specularly reflecting cylinders. The pump light distribution within the laser rod is much more uniform, however. Figure 6.2 shows two common examples of pumping chambers involving the use of two lamps. In Fig. 6.2a, the specularly reflecting cylinder consists of a double-ellipse sharing a common focal axis. The laser rod is placed along this axis while the two lamps lie on the other two focal axes of the ellipses. Figure 6.2b shows two lamps placed as close as possible to the laser rod (close-coupled configuration) the reflecting cylinder again being usually of diffusive material. The efficiencies of these two-lamp configurations are lower than for the corresponding single-lamp configurations of Fig. 6.1. The pump uniformity is however better and higher pump energies, for a given lamp loading, can be obtained from a two-lamp, compared to a single-lamp, configuration. For high-power or high-energy systems, multiple-lamp configurations have also been used. A widely-used configuration involves the active medium arranged in the form of a slab (Fig. 6.3a) or multiple slabs (Fig. 6.3b). In both cases each lamp is placed along e.g. the focal line of a parabolic reflecting cylinder so as to ensure uniform illumination of the slab(s). In Fig. 6.3a, laser action occurs by total internal reflections at the two slab faces. The advantage of this zig-zag beam path is that it averages out the stress-birefringence and thermal focusing induced in the medium by the pump light. This configuration, despite its greater complexity compared to schemes using a rod-shaped laser medium is particularly advantageous when a laser beam of very high optical quality is required. In Fig. 6.3b, laser action takes place along the beam direction indicated in the figure and the slabs are oriented so that the beam is incident at Brewster's angle. The main advantage of this configuration stems from the fact that the transverse dimension of the laser medium can be made very large. Furthermore the slabs can be individually cooled by e.g. a gas refrigerant. This configuration finds application in large aperture (up to 40 cm diameter) Nd:glass amplifiers used for laser fusion experiments.

For pulsed lasers, medium-to-high pressure ($500 \div 1500$ Torr) Xe or Kr flashlamps are used and the pump light pulse is produced by discharging, through the lamp, the electrical energy stored in a capacitor bank, charged by a suitable power supply (Fig. 6.4). A series inductance L is often used in the electrical circuit to limit the current rise-time. The discharge may be initiated by ionizing the gas in the lamp through a high-voltage trigger pulse applied to an auxiliary electrode around the lamp (parallel trigger, see Fig. 6.3a). Alternatively, the preionization may be produced by a voltage pulse directly applied between the two main electrodes of the lamp (series trigger, see Fig. 6.3b). Once ionized, the lamp produces an intense flash of light whose duration is determined by the circuit capacitance and inductance as well as by the lamp electrical characteristics (usually the duration ranges from a few microseconds

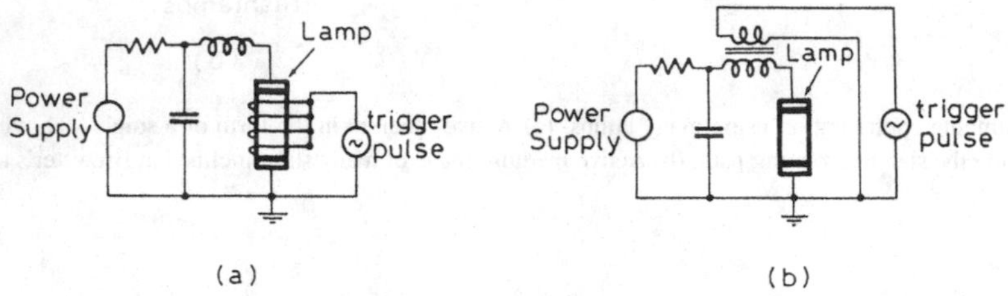

FIG. 6.4. Pulsed electrical excitation of a flashlamp using either an external trigger, (a), or series trigger configuration, (b).

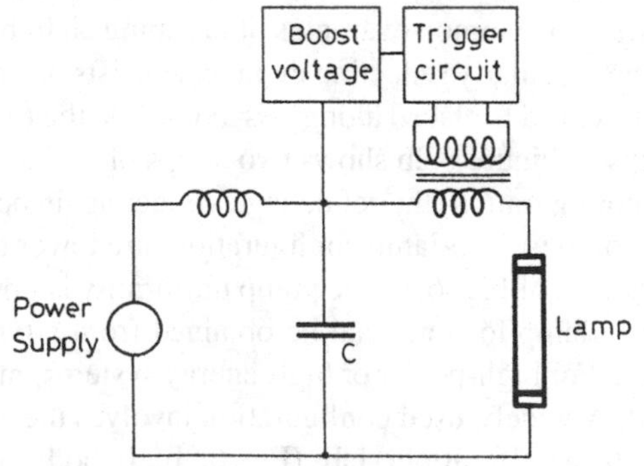

FIG. 6.5. Electrical excitation of a cw lamp.

to a few milliseconds). For cw lasers, high-pressure (1–8 atm) Kr lamps are most often used and the cw current may be delivered by a current regulated power supply, see Fig. 6.5 where the L/C filter network is used for ripple suppression. In this case, also, an electrical trigger pulse, usually from a series trigger, is needed to provide the required initial ionization. For reliable lamp starting, the voltage of the power supply must be boosted to a sufficiently high value and for a sufficiently long time, during the trigger phase, so as to ensure a high enough density of ions and electrons in the lamp to stabilize the discharge. This is conveniently done by impulsively exciting the trigger transformer through a low-current booster power supply.

6.2.2. Absorption of Pump Light

To understand the process of light emission by a lamp, we begin by showing in Fig. 6.6a the emission spectra, for pulsed excitation, of a Xe flashlamp at two typical current densities. For the case of cw excitation, Fig. 6.6b shows the emission spectrum of a cw Kr lamp at a current density of $J = 80\,\mathrm{A/cm^2}$. Actually the typical operating current density of a Kr lamp is somewhat higher than this, i.e. $J \cong 150\,\mathrm{A/cm^2}$, but this difference does not influence the discussion that follows. Note that, at the relatively low current density of a cw lamp, the emission is concentrated mostly in various Kr emission lines which are considerably broadened by the high gas pressure. By contrast, at the much higher current densities of a flashlamp, the spectrum also contains a broad continuous component arising from electron-ion recombination (*recombination radiation*) as well as from electrons deflected by ions during collisions (*bremsstralung radiation*). For both these phenomena, the emission arises from electron-ion interaction. Accordingly, the intensity of the emitted light is expected to be proportional to the product N_eN_i, where N_e and N_i are the electron and ion densities in the discharge. In a neutral gas discharge one has $N_e \cong N_i$ while the two densities are proportional to the discharge current density, J, by the well known relation $N_e = J/ev_{drift}$, where v_{drift} is the electron drift velocity. It then follows that, to a first approximation, the continuous component of the spectrum is expected to grow as J^2. By contrast, to a first approximation, the intensity of the line spectrum of Fig. 6.6b can be taken as proportional to N_e and hence to J. This is the reason

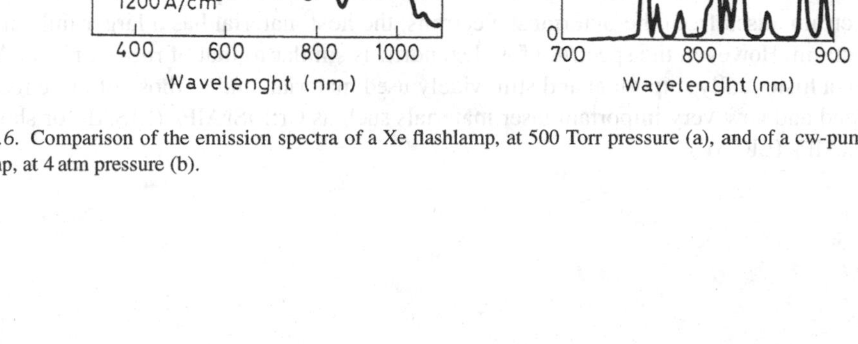

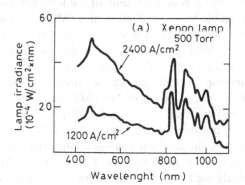

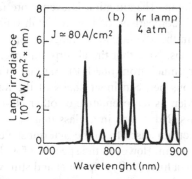

FIG. 6.6. Comparison of the emission spectra of a Xe flashlamp, at 500 Torr pressure (a), and of a cw-pumped Kr arc lamp, at 4 atm pressure (b).

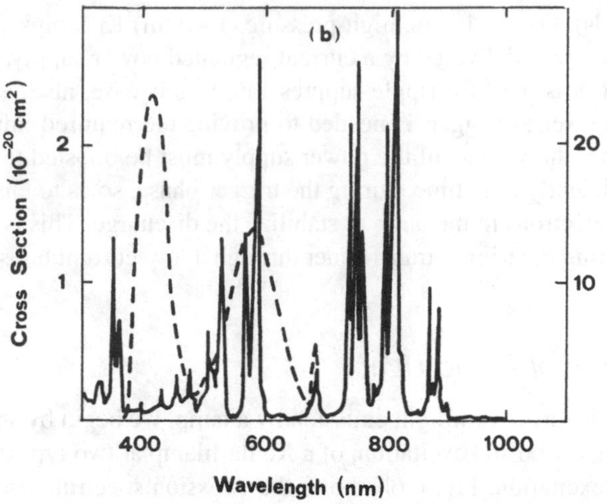

FIG. 6.7. Absorption cross section of Nd^{3+} ion in YAG (solid line) and of Cr^{3+} ion in Alexandrite (dashed line). The left-hand scale refers to the cross-section of Nd:YAG and the right-hand scale to alexandrite. For alexandrite, the average of the three values measured for polarization parallel to the *a*, b, and *c* axes has been taken.

why the continuous spectrum becomes dominant over the line spectrum at higher values of the current density (Fig. 6.6a) while it is not apparent at the much lower current densities of a cw lamp (Fig. 6.6b).

To understand the details of how the light emitted by the lamp is absorbed by the active medium, we begin by showing in Fig. 6.7, as a solid line, the absorption spectrum of Nd:YAG (Nd^{3+} in $Y_3Al_5O_{12}$ crystal) and, as a dashed line, the absorption spectrum of Alexandrite (Cr^{3+} in a $BeAl_2O_4$ crystal). In both cases, it is the dopant ion, present in the crystal as a trivalent ion impurity, which is responsible for the absorption and which also acts as the active element. A comparison of Fig. 6.7 with 6.6a indicates that the relatively broad spectra of both Nd^{3+} and Cr^{3+} ions allow a reasonably good utilization of the light emitted by a flashlamp. The situation is even more favorable for cw excitation of a Nd:YAG by a Kr lamp. A comparison of Fig. 6.7 with 6.6b shows, in fact, that some strong emission lines of Kr, in the $750 \div 900$ nm range, happen to coincide with the strongest absorption lines of Nd^{3+} ions. Note that the absorption spectrum of a rare-earth element, such as Nd^{3+}, does not vary much from one host material to another, since the absorption arises from electron transitions between inner shells of the ion. So the spectrum of Nd:YAG can be taken, to first order, as representative of other Nd-doped materials such as Nd:$YLiF_4$, Nd:YVO_4 and Nd:glass (Nd^{3+} ions in a glass matrix). For a transition metal dopant such as Cr^{3+}, where the spectrum arises from the outermost electrons, the host material has a larger influence on the spectrum. However, the spectrum for alexandrite is similar to that of ruby (Cr^{3+} in Al_2O_3 crystal), a historically important and still widely used material, and to those of more recently developed and now very important laser materials such as Cr:$LiSrAlF_6$ (LISAF for short) or Cr:$LiCaAlF_6$ (LICAF).

6.2.3. Pump Efficiency and Pump Rate

Consider first a cw laser pumped by a pump rate, R_p, which is assumed uniform troughout the volume of the pumped region, V. We can ask ourselves what would be the minimum pump power, P_m, needed to obtain a given pump rate R_p. With reference to Fig. 6.17, this would correspond to the case where the upper laser level were directly pumped from the ground state by, e.g. monochromatic pump photons of energy $h\nu_{mp}$, where ν_{mp} is the frequency difference between the ground level and the upper laser level. The minimum pump power, P_m, is then given by

$$P_m = (dN_2/dt)_p V h\nu_{mp} = R_p V h\nu_{mp} \tag{6.2.1a}$$

where V is the pumped volume of the active medium. We can now define a pump efficiency, η_p, as the ratio between this minimum pump power, P_m, and the actual electrical pump power, P_p, entering the lamp i.e.

$$\eta_p = P_m/P_p \tag{6.2.1}$$

For non-uniform pumping, we can then write

$$\begin{aligned} P_m &= h\,\nu_{mp} \int_a R_p dV = \\ &= h\,\nu_{mp} \langle R_p \rangle V \end{aligned} \tag{6.2.2}$$

where the integral is taken over the whole volume of the medium and $\langle R_p \rangle$ is the average of R_p in the medium. From Eqs. (6.2.1) and (6.2.2) we then get

$$\eta_p = (h\,\nu_{mp} \langle R_p \rangle V)/P_p \tag{6.2.3}$$

For a pulsed pumping system, we can, likewise, define η_p as

$$\eta_p = (h\nu_{mp} \int R_p dV dt)/E_p \tag{6.2.4}$$

where the integral is also taken over the whole duration of the pump pulse and E_p is the electrical pump energy given to the lamp.

To calculate or simply estimate the pumping efficiency, the pump process can be divided into four distinct steps: (i) the emission of radiation by the lamp; (ii) the transfer of this radiation to the active medium; (iii) the absorption in the medium; (iv) the transfer of the absorbed radiation to the upper laser level. Consequently, the pumping efficiency can be written as the product of four terms, namely,

$$\eta_p = \frac{P_r}{P_p}\frac{P_t}{P_r}\frac{P_a}{P_t}\frac{P_m}{P_a} = \eta_r \eta_t \eta_a \eta_{pq} \tag{6.2.5}$$

where: (i) $\eta_r = P_r/P_p$ is the ratio between the radiated power of the lamp in the wavelength range corresponding to the pump bands of the laser medium, P_r, and the total electrical pump power to the lamp, P_p. The efficiency η_r is referred to as the lamp *radiative efficiency* and is smaller than one because some of the electrical input power is emitted into not useful wavelength ranges or transformed as heath. (ii) $\eta_t = P_t/P_r$ is the ratio between the power actually transmitted to the the medium by the pumping system and the radiated power, P_r. The efficiency η_t is referred to as the *transfer efficiency* of the pump system and is smaller than one because not all radiative power emitted by the lamp, P_r, is conveyed into the active medium. (iii) $\eta_a = P_a/P_t$ is the ratio between the power actually absorbed by the medium, P_a, and the power entering into it, P_t. The efficiency η_a is referred to as the *absorption efficiency* and is smaller than one because not all power entering into the medium is there absorbed. (iv) $\eta_{pq} = P_m/P_a$ is the ratio between the minimum pump power considered above and the absorbed power P_a. The efficiency η_{pq} is referred to as the *power quantum efficiency* and is smaller than one because the absorbed power raises atoms to generally a few pump bands with energy larger than $h\nu_{mp}$, and because not all excited atoms then decay to the upper laser level.

Specific expressions for the above four efficiency terms can be obtained when the lamp spectral emission, pump geometry, medium absorption coefficient and geometry are known.[1] We will not undertake an in-depth consideration of this topic here, so we will limit ourselves to a discussion of a few typical results in the example that follows.

Example 6.1. *Pump efficiency in lamp-pumped solid state lasers* We will take as the active medium a cylindrical rod with 6.3 mm diameter pumped in a silvered elliptical pumping chamber with major axis of $2a = 34$ mm and minor axis $2b = 31.2$ mm. For each laser medium, the lamp current density is assumed to have the appropriate value for that laser configuration, ranging generally between 2,000 and $3,000\ A/cm^2$. Under these pumping conditions, the calculated values for the four efficiency terms $\eta_r, \eta_t, \eta_a, \eta_{pq}$ and for the overall pump efficiency, η_p, for ruby, alexandrite, Nd:YAG and Nd:glass are listed in Table 6.1. From this table we may notice, in particular, that: (i) The lamp radiative efficiency is typically less than 50% in all cases considered. (ii) In view of the larger Nd content in a glass and broader absorption bands of Nd:glass material, the overall efficiency of Nd:glass is almost twice that of Nd:YAG. (iii) The overall efficiency of Alexandrite is almost 3 times higher than for the other Cr^{3+} -doped materials, i.e. ruby. This is due mainly to the stronger absorption bands of Alexandrite owing to the higher Cr^{3+} content. Still higher pump efficiency, above the 10% level, are then expected for other Cr^{3+} -doped media such as Cr:LISAF and Cr:LICAF on account of the even higher (by more than an order of magnitude) Cr content. (iv) In all cases considered, the overall efficiency, being the product of four efficiency terms, turns out to be quite small ($3 \div 8\%$).

In concluding this section we note that, once the overall pump efficiency is calculated or, perhaps, simply estimated, the pump rate can be readily obtained from Eqs. (6.2.1a)

TABLE 6.1. Comparison between computed pumping efficiency terms for different laser materials[1]

Active Medium	η_r (%)	η_t (%)	η_a (%)	η_{pq} (%)	η_p (%)
Ruby	27	78	31	46	3.0
Alexandrite	36	65	52	66	8.0
Nd:YAG	43	82	17	59	3.5
Nd:Glass (Q-88)	43	82	28	59	5.8

and (6.2.1) as

$$R_p = \eta_p \left(\frac{P}{Al\,h\nu_{mp}} \right) \tag{6.2.6}$$

where A is the cross-sectional area of the pumped volume of the active medium and l is its length. This is the simple basic expression for the pump rate often used in the laser literature[4] and which will be used frequently in the following chapters. Note however that, to obtain R_p from Eq. (6.2.6), one needs to know η_p, implying that the detailed calculations, such as those discussed in,[1] need to have been performed by someone!

6.3. LASER PUMPING

Laser beams have often been used to pump other lasers since the early days of lasers, being used for example in the first demonstration of laser action in a dye medium.[5,6] In particular, Ar ion lasers are widely used to pump cw dye and Ti^{3+}:Al_2O_3 lasers, Excimer, Nitrogen and Copper Vapor lasers are used for pulsed pumping of dye lasers, Nd:YAG and its second harmonic beam are used as pumps for cw and pulsed dye and solid-state lasers (including color-center lasers). Laser pumping has become a very much more important pumping technique, however, since efficient and high power diode lasers have been developed and become widely available. A particularly interesting case is the use of diode lasers to pump other solid-state laser materials thus providing an all-solid-state laser. The most relevant examples include: (i) Nd:YAG, Nd:YLF, Nd:YVO_4 or Nd:glass pumped by GaAs/AlGaAs* Quantum Well(QW) lasers at $\sim$ 800 nm (typical oscillation wavelengths are around 1 μm, 1.3 μm and 0.95 μm). (ii) Yb:YAG, Er:glass or Yb:Er:glass pumped by InGaAs/GaAs strained QW lasers in the 950 ÷ 980 nm range (oscillation wavelength is around 1 μm for Yb and 1.54 μm for Er lasers). Note that, in the case of Er:Yb codoping, the pump light is absorbed by Yb^{3+} ions and then transferred to Er^{3+} lasing ions. (iii) Alexandrite, Cr:LISAF or Cr:LICAF pumped by GaInP/AlGaInP QW lasers in the 640 ÷ 680 nm range and oscillating in a $\sim$ 130 nm range at $\sim$ 840 nm. (iv) The Tm:Ho:YAG laser pumped by AlGaAs QW lasers at 785 nm and oscillating around 2.08 μm. Note that, in this case, the pump light is absorbed by Tm^{3+} ions and transferred to the Ho^{3+} lasing ions.

As a representative example of Nd ion lasers, we show in Fig. 6.8a the relevant plots of the absorption coefficient vs wavelength for both Nd:YAG, continuous line, and Nd:glass, dashed line. Note that Nd:YAG is most effectively pumped at a wavelength of $\lambda = 808$ nm and this is obtained by a $Ga_{0.91}Al_{0.09}As/Ga_{0.7}Al_{0.3}As$ QW laser whose emission bandwidth is typically 1 ÷ 2 nm wide. Nd:glass, on the other hand, due to its broader and featureless absorption profile, can be pumped over a broader range around the 800 nm peak. For the case of Yb-ion lasers, we show in Fig. 6.8b the relevant plots of the absorption coefficient vs wavelength for Yb:YAG (solid line) and Yb:glass (dashed line). Again the absorption coefficient for glass appears broader and featureless compared to that of YAG. The best pumping wavelength

* In all double-compound (A/B) semiconductor lasers, considered in this section, the first compound (A) refers to the active layer while the second one (B) is the so called cladding layer (see Chapt. 9, Sect. 9.4)

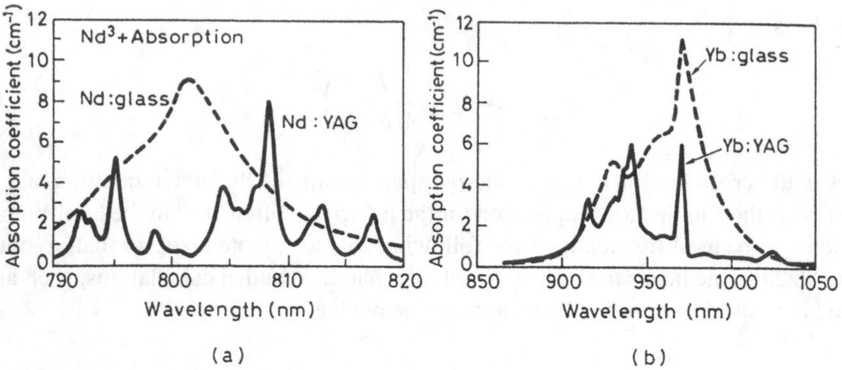

FIG. 6.8. Absorption coefficient vs wavelength in the wavelength range of interest for diode laser pumping: (a) Nd:YAG, solid line, and Nd:glass, dashed line. Neodymium concentration is $1.52 \times 10^{20}\ cm^{-3}$ for Nd:YAG (1.1 atomic % doping) and $3.2 \times 10^{20}\ cm^{-3}$ for Nd:glass (3.8% by weight of Nd_2O_3) (after ref.[16] by permission). (b) Yb:YAG, solid line, and Er:Yb:glass, dashed line. Ytterbium concentration is $8.98 \times 10^{20}\ cm^{-3}$ for Yb:YAG (6.5 atomic %) and $1 \times 10^{21}\ cm^{-3}$ for Yb:glass. The curves of Yb:YAG and Yb:glass are based on the corresponding plots of ref.[17] and,[18] respectively.

TABLE 6.2. Comparison between pumping parameters and laser wavelengths for different laser materials

	Nd:YAG	Yb:YAG	Yb:Er:glass	Cr:LISAF	Tm:Ho:YAG
Concentr.	1 at. %	6.5 at.%	1 mol.%	6.5 at.% Tm	0.36 at. %Ho
Pumping Diode	AlGaAs	InGaAs	InGaAs	GaInP	AlGaAs
Wav. (nm)	808	950	980	670	785
Active-ion conc. $[10^{20}\ cm^{-3}]$	1.38	9	10 [Yb]	0.9	8 [Tm]
			1 [Er]		0.5 [Ho]
Pump abs. coeff. (cm^{-1})	4	5	16	4.5	6
Oscillation Wav. (μm)	1.06	1.03	1.53	0.72 ÷ 0.84	2.08
	1.32, 1.34 0.947				

is 960 nm for Yb:YAG and 980 nm for glass and these wavelengths can be obtained from a InGaAs/GaAs QW laser (e.g. $In_{0.2}Ga_{0.8}As/GaAs$ for $\lambda = 980$ nm). The plots of the absorption coefficient vs wavelength for Cr^{3+} ion lasers (Alexandrite, Cr:LISAF, Cr:LICAF) show the general structureless shape of the dashed curve of Fig. 6.7. The peak absorption coefficient at 600 nm wavelength is $\sim 0.5\ cm^{-1}$ for Alexandrite and up to $50\ cm^{-1}\ cm^{-1}$ for Cr:LISAF. Note that the higher absorption coefficient in Cr:LISAF is due to the higher Cr concentration which can be used ($\sim$ 100 times higher than for Alexandrite) without incurring the problem of concentration quenching of the upper state lifetime. Due to the lack of suitable diode lasers at shorter wavelengths, pumping is achieved in the 640 ÷ 680 nm wavelength range, obtained from GaInP/AlGaInP QW lasers (e.g. $Ga_{0.5}In_{0.5}P/Al_{0.25}Ga_{0.25}In_{0.5}P$ for 670 nm wavelength) with GaInP being the active QW layer. Table 6.2 summarizes the most relevant pumping data for some of the active media considered above.

6.3.1. Laser Diode Pumps

There are essentially four types of pumping laser diodes, listed in order of increasing output power as: (i) Single-stripe; (ii) diode-array; (iii) diode-bar; (iv) stacked-bars.

At the lower end of the output power range ($P < 100\,\text{mW}$) one has the single-stripe semiconductor laser such as the index-guided laser of Fig. 6.9a. By means of a suitable insulating oxide layer, the diode current is confined to a $3 \div 5\,\mu\text{m}$ wide stripe which extends over the whole length of the diode. The emitted beam has an elliptical shape with a diameter in the direction perpendicular to the laser junction of $d_\perp = 1 \div 3\,\mu\text{m}$ and a diameter in the junction plane of $d_\parallel \cong 3 \div 6\,\mu\text{m}$. With such small spot-sizes, the beam is spatially coherent i.e. it is diffraction limited. In fact, in a typical situation, the divergence half-angle-cone at $1/e^2$ intensity point is $\theta_\perp = 20° = 0.35\,\text{rad}$, perpendicular to the junction. One then gets $\theta_\perp = 2\lambda/\pi d_\perp$ provided one takes, at $\lambda = 800\,\text{nm}$, $d_\perp \cong 1.4\,\mu\text{m}$. In the junction plane one typically has $\theta_\parallel = 5° = 0.09\,\text{rad}$ and again one gets $\theta_\parallel =\cong 2\lambda/\pi d_\parallel$ by taking $d_\parallel = 5.8\,\mu\text{m}$. [Gaussian distributions in the two planes, with spot sizes $w_{0\perp} \cong d_\perp/2$ and $w_{0\parallel} \cong d_\parallel/2$, are assumed so that beam divergence is calculated according to Eq. (4.7.19)]. Note that, in view of this strong difference between the beam divergences in the two directions, the beam has its major axis direction rotated by 90° after beam propagation just a few micrometers away from the diode exit face.

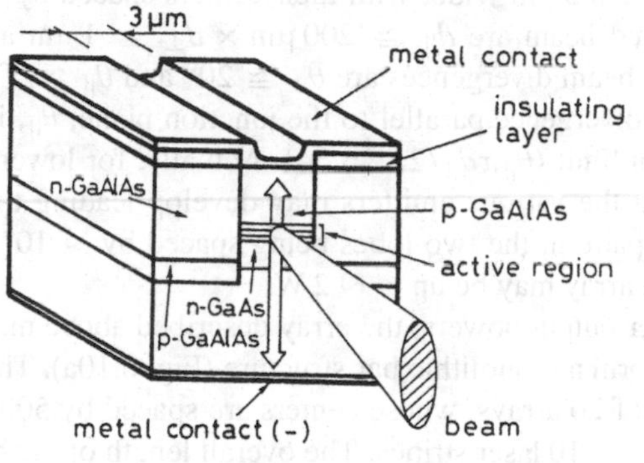

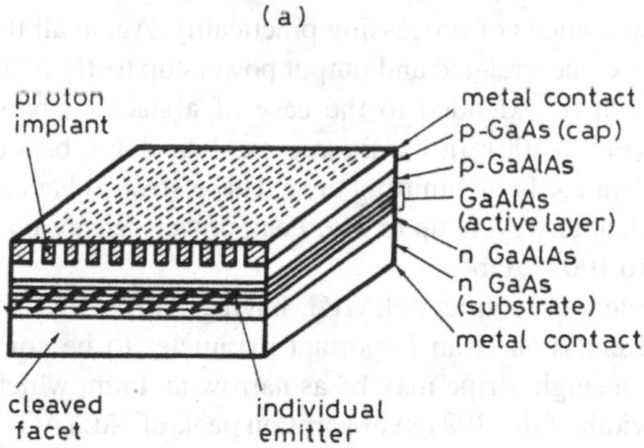

FIG. 6.9. (a) Single-stripe index-guided semiconductor diode laser. (b) Monolithic array of many stripes on a single semiconductors chip.

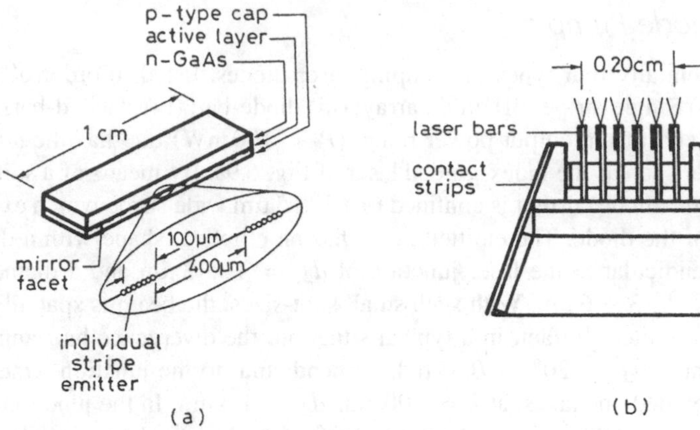

FIG. 6.10. (a) Monolithic 1-cm bar for cw operation. (b) Stacked bars for quasi cw operation.

To obtain larger output power values, one uses a monolithic array of diode-laser stripes, fabricated on the same semiconductor substrate (Fig. 6.9b). In typical cases the array may contain twenty stripes, each 5 μm wide, with their centers spaced by ~ 10 μm. The overall dimensions of the emitted beam are $d_{\parallel} \cong 200\ \mu m \times d_{\perp} \cong 1\ \mu m$ and, for arrays having uncorrelated phases, the beam divergences are $\theta_{\perp} \cong 20°$ and $\theta_{\parallel} \cong 5°$ i.e. the same as for a single stripe. The beam divergence parallel to the junction plane, $\theta_{\parallel}$, is now about 40 times more than the diffraction limit ($\theta_{\parallel}\pi d_{\parallel}/2\lambda \cong 34$). Actually, for lower power devices, some phase correlation among the various emitters may develop leading to a characteristic two-lobed angular emission pattern, the two lobes being spaced by ~ 10° and each ~ 1° wide. Output power from such array may be up to ~ 2 W.

To obtain still larger output powers, the array described above may be serially repeated in a single substrate to form a monolithic bar structure (Fig. 6.10a). The device shown in the figure is seen to consist of 20 arrays, whose centers are spaced by 500 μm, each array being 100 μm long and containing 10 laser stripes. The overall length of the bar is thus ≈ 1 cm, the limit being set by considerations of processing practicality. Again all the stripe emitters may be considered to be phase uncorrelated and output powers up to 10 ÷ 20 W are usual.

The bar concept can be extended to the case of a stack of bars which form a two-dimensional structure (Fig. 6.10b). In the figure, six, 1 cm long, bars are shown stacked so as to form an overall, 2 mm × 1 cm, emitting area. These stacked bars are so far intended for quasi-cw operation with a duty cycle up to 2%. Peak power density may be up to $1\ kW/cm^2$ and average power up to $100\ W/cm^2$.

To pump laser materials such as Nd:YAG, having narrow absorption lines, the width of the diode's spectral emission is an important parameter to be considered. The spectral emission bandwidth of a single stripe may be as narrow as 1 nm, which compares favorably with the ~ 2 nm bandwidth of the 808 nm absorption peak of Nd:YAG. For the case of arrays and, even more so, for bars or stacked bars, spectral emission may be substantially larger than this value due to compositional variation between stripes and temperature gradients, both leading to different stripe emission wavelength. Currently, the best results for a bar may be a spectral width as low as ~ 2 nm. To tune and to stabilize the emission wavelength, diode lasers are normally cooled by a thermoelectric cooler, for low power devices, and by liquid cooling for the highest powers. A temperature stability and accuracy of less than 1 C is usually required.

6.3.2. Pump Transfer Systems

For efficient pumping, the light emitted by the diode laser systems described above must be properly transferred to the active medium. There are, basically, two types of pump system geometry: (i) *Longitudinal (or axial) pumping*, where the pump beam enters the laser medium along the resonator axis. (ii) *Transverse pumping*, where the beam is conveyed to the active medium generally from one or more directions, transverse to the resonator axis. We shall consider the two cases separately because the diode lasers and pump transfer systems are somewhat different for the two cases.

6.3.2.1. Longitudinal Pumping

For longitudinal pumping, the beam emitted by the diode laser generally needs to be concentrated into a small (100 μm ÷ 1 mm diameter) and usually not so necessarily circular spot into the active medium. Three of the most common laser configurations are shown in Fig. 6.11a, b and c, respectively. In Fig. 6.11a, the laser rod is shown in a plane-concave resonator, the plane mirror being directly deposited to one rod face, and the pump beam focused on this face. In Fig. 6.11b and c two pump beams, from two different diode systems, are focused into the rod center from the two sides of the rod. The laser resonator may then consist either of a folded ring configuration (Fig. 6.11b) or a z-shaped folded linear cavity

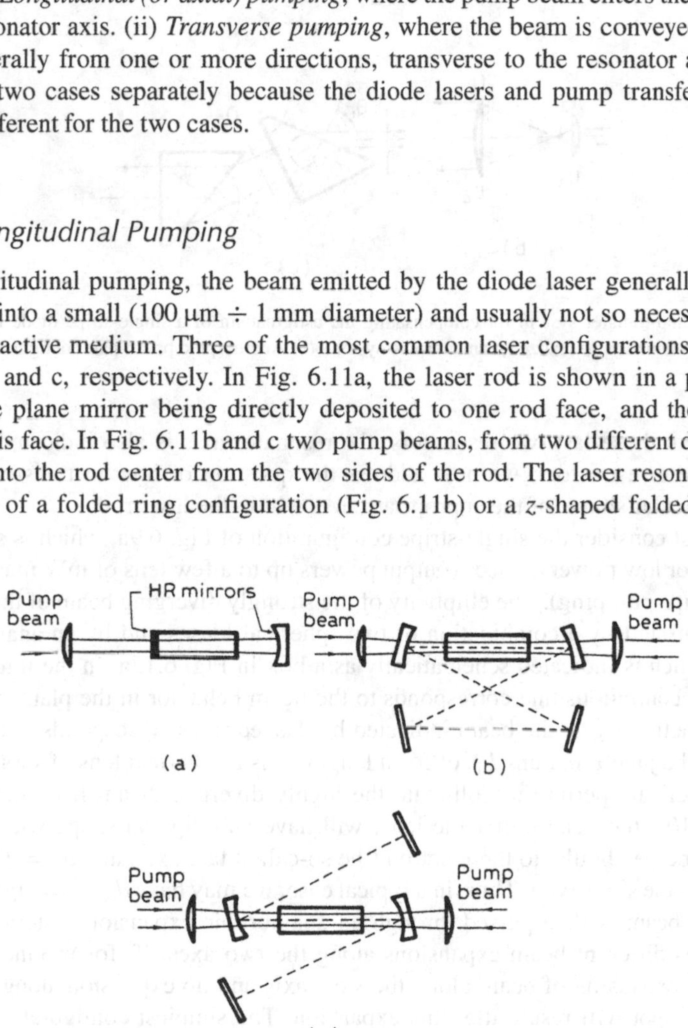

FIG. 6.11. Typical configurations for longitudinal diode laser pumping: (a) Single-ended pumping in a simple plane-concave resonator. (b) Double-ended pumping for a ring laser in a folded configuration. (c) Double-ended pumping for a z-shaped folded linear cavity.

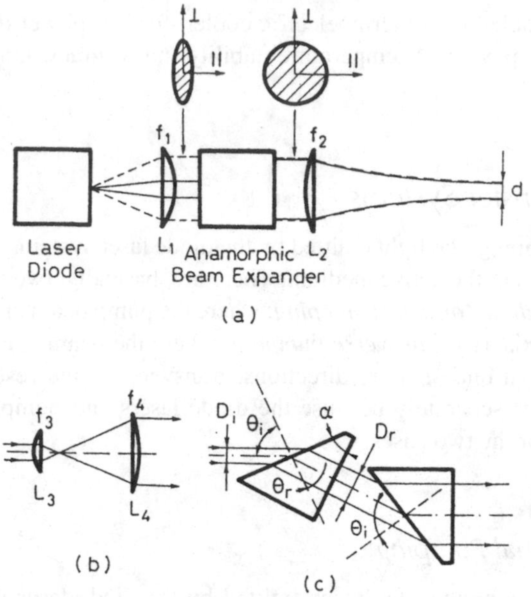

FIG. 6.12. (a) Pump-transfer system for compensating the astigmatism of a single-stripe diode laser. (b) Simple cylindrical lens combination to realize an anamorphic system. (c) Anamorphic prism-pair configuration.

(Fig. 6.11c). For these last two resonators, the resonator axis is also indicated by a dashed line. Given these resonators, we now address the question of how to transform the pump beam into a circular shape, of the appropriate size, within the laser rod.

Let us first consider the single-stripe configuration of Fig. 6.9a, which is still used as a pump source for low power devices (output powers up to a few tens of mW may be achieved with single stripe pumping). The ellipticity of the strongly diverging beam of the diode stripe can be compensated by a combination of two spherical lenses and by an anamorphic optical system, which is indicated schematically as a box in Fig. 6.12a. In the figure, the beam indicated by a continuous line corresponds to the beam behavior in the plane parallel to the laser diode junction while the beam indicated by dashed lines corresponds to the plane perpendicular to the junction. Lens L_1, of focal length f_1, is a spherical lens of short focal length and high numerical aperture to collimate the highly divergent beam from the laser diode. Since $\theta_\perp \cong 4\theta_{||}$, the beam, after the lens, will have an elliptical shape with a dimension $d_\perp = f_1 \,\mathrm{tg}\theta_\perp$ perpendicular to the junction (the so-called fast-axis) and $d_{||} = f_1 \,\mathrm{tg}\theta_{||}$ parallel to the junction (the slow axis). Thus, in a typical case, we may have $d_\perp/d_{||} = \mathrm{tg}\theta_\perp/\mathrm{tg}\theta_{||} \cong 4$. This elliptical beam is then passed through an anamorphic expansion system i.e. a system which provides different beam expansions along the two axes. If, for instance, the system provides a 4:1 expansion of beam along the slow axis and no expansion along the fast axis, then a circular spot will result after this expansion. The simplest configuration for such an anamorphic expander could perhaps be provided by the combination of two cylindrical lenses, L_3 and L_4, in a confocal (or telescopic) arrangement (Fig. 6.12b). If the two lenses have their focusing action in the plane containing the slow axis, there will be a beam expansion of f_4/f_3, where f_4 and f_3 are the focal lengths of the two lenses, for the beam in this plane (solid-line).

For the other plane, however, the two cylindrical lenses behave simply as plane parallel plates and the beam will thus be unaffected, in the fast-axis direction, by the beam expander. The anamorphic system of Fig. 6.12b is seldom used in practice, however, because, to save space, the system would require cylindrical lenses of short focal length and such lenses, if they are to be aberration free, are rather expensive. Thus, the anamorphic prism pair of Fig. 6.12c is usually employed.[7] In the figure we again consider the beam behavior only in the slow-axis plane (solid line). By simple geometrical considerations one can show that, after refraction at the front surface of the first prism, the incident beam, of diameter D_i, is enlarged to a diameter D_r such that $D_r/D_i = \cos\theta_r/\cos\theta_i$, where θ_i and θ_r are, respectively, the angles of incidence and refraction at the prism surface. Then, if the exit face of the first prism is made near normal to the beam direction, no refraction will occur at this face and the beam will pass through it unchanged. Under these conditions the beam magnification, M, after the first prism, will simply be given by

$$M = \frac{D_r}{D_i} = \frac{\cos\theta_r}{\cos\theta_i} \tag{6.3.1}$$

Let us now consider the passage of the beam through the second prism. If the prism is identical to the first one, is oriented as in Fig. 6.12c and if the angle of incidence at the entrance face is again equal to θ_i, then the beam will again be magnified by a factor M on traversing the second prism. The overall beam magnification is then equal to M^2 and the direction of the output beam is parallel to that of the input beam, although shifted laterally. In the fast axis plane, on the other hand, the two prisms behave as simple plates and so there is no beam magnification. Thus, for the example considered in Fig. 6.12a, if one chooses an anamorphic prism pair with appropriate values for θ_i and for the prism refraction index, n, one can readily arrange to have $M = 2$ i.e. an overall magnification of $M^2 = 4$. The beam, after the prism-pair, will thus have a circular shape. If the collimating lens L_1 in Fig. 6.12a has a sufficiently high numerical aperture to accept the highly diverging beam along the fast axis, and if the lens is, ideally, aberration free, the beam after this lens and, hence, after the prism-pair will still retain the diffraction limited quality of the original beam from the diode. Since the beam leaving the prism pair has a circular shape, the beam divergence will now be equal along the two axes. A spherical lens L_2, of appropriate focal length f_2, can then be used to focus the beam to a round spot of appropriate size in its focal plane (Fig. 6.11a) i.e. where the active medium is placed. If lens L_2 is also aberration free, the beam in the focal plane will again be of circular symmetry and diffraction limited.

Example 6.2. *Calculation of an anamorphic prism-pair system to focus the light of a single-stripe diode laser* We will consider the system configuration of Fig. 6.12 and a single-stripe laser with $\theta_\perp = 20°$ and $\theta_\| = 5°$, so that, assuming diffraction limited Gaussian distributions, we can take $d_\perp = 1.4\ \mu m$ and $d_\| = 5.8\ \mu m$. We will consider a collimating lens, L_1, of focal length $f_1 = 6.5$ mm. After lens L_1, the beam diameters along the fast and slow axes, will be respectively $D_\perp = 2f_1 \mathrm{tg}\theta_\perp = 4.73$ mm and $D_\| = 2f_1 \mathrm{tg}\theta_\| = 1.14$ mm. Each prism must then provide a magnification of $M = [D_\perp/D_\|]^{1/2} \cong 2$. Assuming the prisms to be made of fused silica, so that the refractive index at 800 nm wavelength is $n = 1.463$, then θ_i and θ_r are found from Eq. (6.3.1) and from Snell's law $\sin\theta_i = n \sin\theta_r$. The solution can be readily obtained either graphically or by a fast iterative procedure. For this procedure, we first assume a tentative value of θ_i and use Snell's law, with $n = 1.463$, to calculate a first value of θ_r. This value is then inserted into Eq. (6.3.1), with $M = 2$, to calculate a new value of θ_i corresponding to the first iteration, and so on. Starting from e.g. $\theta_i = 70°$, this iterative calculation rapidly converges, in a few iterations, to $\theta_i = 67.15°$ and $\theta_r \cong 39°$. Since the beam is assumed to exit normal to the second face of the prism, a simple geometrical argument shows that the apex angle of the prism must be $\alpha = \theta_r \cong 39°$. In this way, after the second prism, a circular beam with diameter $D_{\parallel} = D_{\perp} = 4.73$ mm is obtained. Let us now take the focal length of lens L_2 to be $f_2 = 26$ mm and assume that the beam is still diffraction limited after this lens. The beam spot size in the focal plane of this second lens will then be $d \cong 4\lambda f_2/\pi D \cong 5.52\ \mu$m [the expression which applies for Gaussian beam focusing is again used here, see Eq. (4.7.28)]. Note the very small value of the pump diameter which can, in principle, be achieved. Indeed one readily sees that the effect of the optical system in the fast axis plane (Fig. 6.12a) is to make a $f_2/f_1 \cong 4$ magnified image of the field distribution at the diode exit face. Since one has $d_{\perp} = 1.4\ \mu$m, we then expect $d = (f_2/f_1)\ d_{\perp} \cong 5.6\ \mu$m. To obtain such a small spot, however, lenses which are well corrected for spherical aberration must be used, in particular for the collimator lens, L_1. In a typical situation, account being taken of the finite resolving powers of lenses L_1 and L_2, the beam diameter in the focal plane of lens L_2 may be 5 ÷ 10 times larger In any case, the beam divergence in the focal plane of lens L_2 is given by $\theta \cong D/2f_2$, where D is the beam diameter at the lens position. If a rod of refractive index n_R is placed in the focal plane, then, due to beam refraction, the divergence is approximately reduced by a factor n_R. If we then take $n_R = 1.82$, as appropriate for YAG crystals, we then get $\theta_n \cong D/2f_2n_R = 0.05$ rad $\cong 3°$.

In the case of e.g. a 200 μm wide array, since the divergence angles $\theta_{\perp}$ and $\theta_{\parallel}$ are approximately the same as for a single-stripe, the configuration of Fig. 6.12a and c can still be used to produce a circular spot after the anamorphic prism-pair. Since the slow-axis beam divergence is however ∼ 40 times larger than the diffraction limit, the spot in the focal plane of lens L_2 would be elliptical with a 40:1 ratio between the two axes. Following the previous example and for a well corrected collimating lens L_1, the elliptical beam should have a 2.8 μm×112 μm dimensions. In practice, the aberrations of the optical system, which are more pronounced for the fast-axis direction, will tend to produce a more circular spot with a spot size of perhaps 150 μm. Another way, widely used for a diode arrays, for transferring the pump beam to the active medium is by means of a multimode optical fiber. For a 200 μm stripe, a fiber with a 200 μm core diameter can be used and the fiber may be butt-coupled to the diode. With this configuration, however, the fiber numerical aperture (N.A. $= \sin\theta_f$, where θ_f is the acceptance angle of the fiber) needs to have a sufficiently high value to accept the highly diverging beam of the diode, i.e. $\sin\theta_f > \sin\theta_{\perp} \cong 0.4$. The output beam after propagation in the fiber is circular but its divergence is established by the fiber N.A., i.e. $\theta_{out} = \theta_f$. In doing so, therefore, one worsen the slow axis divergence from $\theta_{in} = \theta_{\parallel}$ to $\theta_{out} = \theta_f \cong \theta_{\perp}$. To reduce the beam divergence, one can use a cylindrical lens of very short focal length to collimate the beam in the fast-axis direction to a diameter equal to the fiber diameter and then use a fiber of numerical aperture approximately equal to the slow-axis divergence, i.e. take $\theta_f \cong \theta_{\parallel}$. In this case, as shown in more details in the example which follows, the beam of a 200 μm wide array can be focused into a fiber of perhaps 250 ÷ 300 μm core diameter and N.A. of 0.1.

In the case of a 1 cm bar, a single 1 cm long cylindrical microlens can be used to focus each array of the bar into a single multimode fiber. Since each array is now typically 100 μm

Example 6.3. *Diode-array beam focusing into a multimode optical fiber* We will consider the simple configuration of Fig. 6.13, where a cylindrical lens of sufficiently short focal lens, f, is used to collimate the beam along the fast axis (dashed lines). The beam diameter, after the lens and along this axis, will then be given by $D_{\perp} = 2f\,\mathrm{tg}\theta_{\perp}$. Along the slow-axis, the cylindrical lens behaves like a plane-parallel plate and the beam (continuous line) will be essentially unaffected by the lens. (To draw attention to this in the figure, the cylindrical lens is drawn as a dashed line to indicate that it only focuses in the fast axis plane). The beam diameter in the slow axis plane, after the lens, will then be approximately $D_{||} \approx L_a + 2f\,\mathrm{tg}\theta_{||}$ where L_a is the length of the array. If one now sets the condition $D_{||} = D_{\perp}$ one obtains $f = L_a/2(\mathrm{tg}\theta_{\perp} - \mathrm{tg}\theta_{||})$. Taking $L_a = 200\ \mu\mathrm{m}$, $\theta_{\perp} = 20°$ and $\theta_{||} = 5°$, we get $f = 350\ \mu\mathrm{m}$, a focal length which can be obtained with fiber microlenses. With such a small value of focal length, the beam diameter after the lens will be $D = D_{||} = D_{\perp} = 2f\,\mathrm{tg}\theta_{\perp} = 254\ \mu\mathrm{m}$ which can easily be accepted into e.g. a 300 μm diameter, multimode fiber, but coupled to the microlens. For a well corrected fiber microlens, the beam divergence, after the lens, will mostly arise from the uncompensated divergence of the slow-axis beam. The fiber numerical aperture must then be N.A. $= \sin\theta_f \geq \sin\theta_{||} \cong 0.09$. The beam divergence of the light leaving the fiber, for a sufficiently long fiber, will then have circular symmetry and be equal to the fiber's N.A.

long (see Fig. 6.10a) fibers with 200 μm core diameter and 0.1 N.A. can be used for each array (see Fig. 6.13). In this way, one can convey the whole beam of the bar into 20 fibers, whose ends can be arranged into a circular fiber bundle of $1 \div 1.5$ mm diameter and overall divergence equal to the N.A., 0.1, of the fiber. The beam emitted by this bundle is then imaged into the rod along one (Fig. 6.11a) or two longitudinal directions (Fig. 6.11b, c). With this pump configuration, an overall transmission of the transfer system up to 85% has been demonstrated. Output powers up to $\sim$ 15 W in a TEM_{00} mode with an optical-to-optical efficiency of $\sim$ 50% have been obtained using a Nd:YVO_4 rod pumped by two such fiber-coupled diode bars.

A quite interesting and alternative approach has been demonstrated which allows the very asymmetric output beam from a diode bar or array to be reshaped so as to produce the same beam dimensions and divergences along the original fast-axis (vertical) and slow-axis (horizontal) directions. The technique involves sending the beam from a diode bar or array, after collimation in the fast direction by a fiber lens, to a tilted pair of parallel mirrors which, by multiple reflections of the beam, effectively chop it into several segments in the horizontal

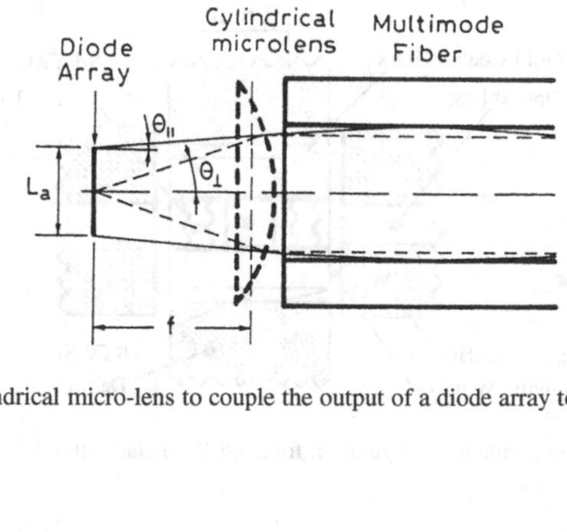

FIG. 6.13. Use of a cylindrical micro-lens to couple the output of a diode array to a multimode optical fiber.

direction and then stack these segments above each other, resulting in a rectangular shape.[(8)] In equalizing the beam parameters in the horizontal and vertical directions, the decrease in beam brightness in the vertical direction is compensated by increase in brightness in the horizontal direction so that overall brightness can be maintained. This shaped beam allows very intense longitudinal pumping, which is particularly effective for the otherwise difficult cases of low gain and quasi-three-level lasers.

6.3.2.2. Transverse Pumping

In the case of transverse pumping, active media in the shape of either slabs or rods can be used. Figure 6.14 shows a particularly interesting, transversely pumped, slab configuration.[(9)] Pumping is achieved through 25 individual laser arrays, each coupled to a 600 μm core diameter 0.4 N.A. fiber. The power of the beam exiting each fiber is ~ 9.5 W and the total power is 235 W. The fibers ends are spaced along the two sides of a 1.7 mm thick, 1.8 mm wide, miniature slab. The center line length of the slab is ~ 58.9 mm and this corresponds to 22 total internal reflections at the two slab faces (see Fig. 6.3a). Due to the averaging properties of the resulting zigzag pattern, the optical quality of the active medium, as seen by the beam, is excellent and an output power of 40 W in a TEM_{00} mode with an optical-to-optical efficiency of ~ 22% have been achieved. A particularly interesting configuration using a Nd:YAG rod is shown in Fig. 6.15.[(10)] The 4 mm diameter rod, cooled by water flowing in a surrounding

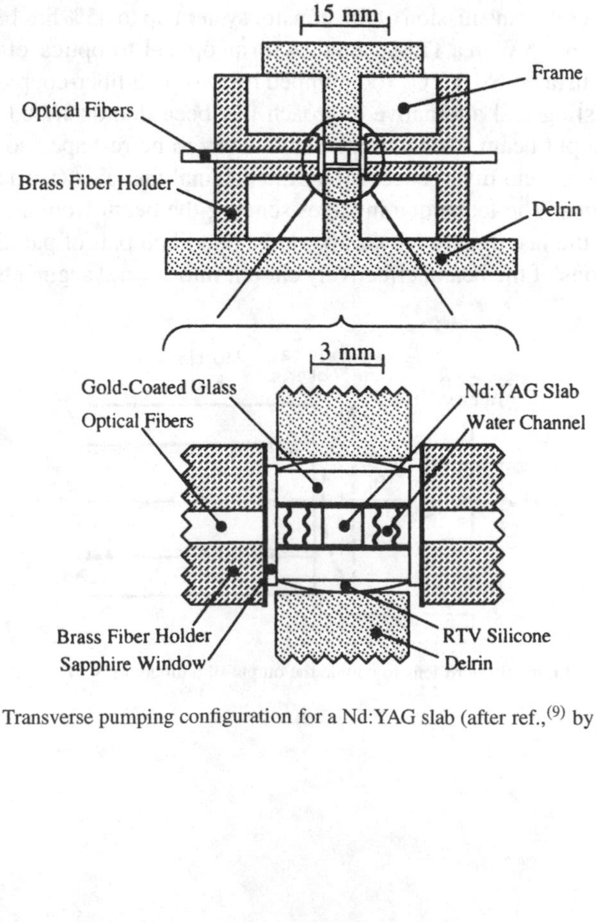

FIG. 6.14. Transverse pumping configuration for a Nd:YAG slab (after ref.,[(9)] by permission).

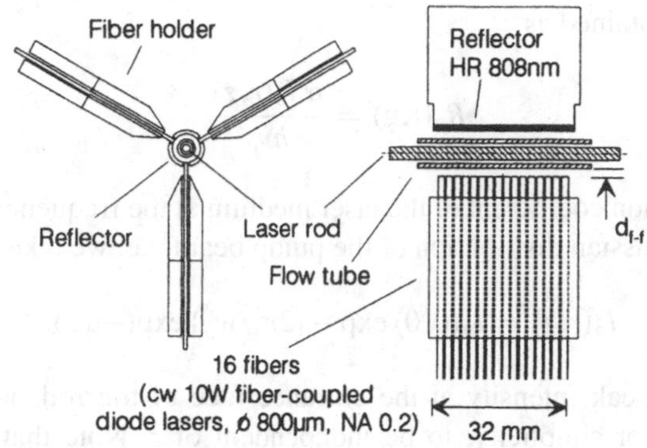

FIG. 6.15. Transverse pumping configuration for a Nd:YAG rod (after ref.,[10] by permission).

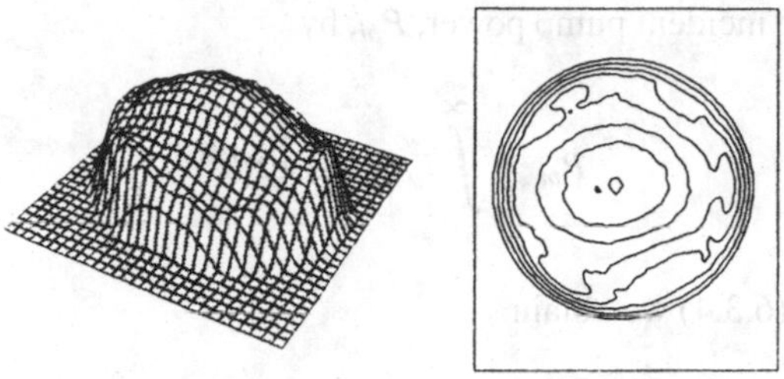

FIG. 6.16. Pump light distribution in the Nd:YAG rod for the transverse pump distribution of Fig. 6.15 (after ref.,[10] by permission).

tube, is radially pumped by either 3 or 5 pump modules placed in a circularly symmetric arrangement. Each pump module consists of sixteen, 800 μm core diameter 0.22 N.A., fibers, mounted side by side in a linear row with 2 mm center-to-center spacing. Into each fiber, the beam of a diode array with a nominal output power of 10 W is injected. The output beam from each fiber directly irradiates the laser rod without any additional focusing optics. A pump transfer efficiency of $\sim$ 80% is estimated for this transverse pump configuration. To help achieve sufficient absorption of the diode-laser radiation, pump light reflectors, facing each pump module, are mounted around the rod. For large enough fiber-to-rod distances the pump light distribution within the rod, achieved in this way, turns out to be rather uniform. As an example, Fig. 6.16 shows this pumping light distribution for a 13 mm fiber-to-flow-tube distance. Using both configurations of Fig. 6.15 an output power of $\sim$ 60 W in a TEM_{00} mode has been achieved with an optical-to-optical efficiency of 25%.

6.3.3. Pump Rate and Pump Efficiency

In the case of longitudinal pumping, if we let $I_p(r, z)$ be the pump intensity at the location inside the laser medium specified by radial coordinate r and longitudinal coordinate z, the pump rate is readily obtained as

$$R_p(r,z) = \frac{\alpha I_p(r,z)}{h\nu_p} \tag{6.3.2}$$

where α is the absorption coefficient of the laser medium at the frequency ν_p of the pump. We will now assume a Gaussian distribution of the pump beam, i.e. we take

$$I_p(r,z) = I_p(0,0)\exp - \left(2r^2/w_p^2\right)\exp(-\alpha z) \tag{6.3.3}$$

where $I_p(0,0)$ is the peak intensity at the entrance face of the rod, w_p is the pump spot-size which is taken, for simplicity, to be independent of z. Note that α is the absorption coefficient under the laser operating conditions and, to a good approximation, it coincides with the unpumped absorption coefficient since the population raised to the upper levels by the pumping process is usually only a small fraction of the total population. The intensity $I_p(0,0)$ is related to the incident pump power, P_{pi}, by

$$P_{pi} = \int_0^\infty I_p(r,0)2\pi\, r dr \tag{6.3.4}$$

From Eqs. (6.3.3) and (6.3.4) we obtain

$$I_p(0,0) = \frac{2P_{pi}}{\pi\, w_p^2} \tag{6.3.5}$$

The incident pump power, P_{pi}, is then related to the diode laser electrical power, P_p, by

$$P_{pi} = \eta_r \eta_t P_p \tag{6.3.6}$$

where η_r is the diode radiative efficiency and η_t is the efficiency of the pump transfer system. From Eq. (6.3.2), with the help of Eqs. (6.3.3), (6.3.5) and (6.3.6), we obtain

$$R_p(r,z) = \eta_r \eta_t \left(\frac{P_p}{h\nu_p}\right)\left(\frac{2\alpha}{\pi\, w_p^2}\right)\exp - \left(\frac{2r^2}{w_p^2}\right)\exp(-\alpha z) \tag{6.3.7}$$

It is shown in Appendix E that, as far as the threshold condition is concerned, the pump-rate which is effective for a given cavity mode is the average, $<R_p>$, of R_p taken over the field distribution of the mode. More precisely, if we let $u(r,z)$ be the complex field amplitude normalized to its peak value, $<R_p>$ is given by

$$<R_p> = \left(\int_a R_p|u|^2 dV\right) \Big/ \int_a |u|^2 dV \tag{6.3.8}$$

where the integrals are taken over the whole of the active medium. Let us consider a TEM_{00} single-longitudinal mode. If the spot size at the beam waist, w_0, is located in the laser rod and if the spot size is assumed constant along the rod, then, according to Eq. (5.5.24), with $R \to \infty$ and $\phi \cong 0$, one has

$$|u|^2 \propto \exp -\left(2r^2/w_0^2\right) \cos^2 kz \tag{6.3.9}$$

Equation (6.3.8) with the help of Eqs. (6.3.7) and (6.3.9) gives

$$<R_p> = \eta_r \eta_t \left(\frac{P_p}{h\nu_p}\right) \frac{2\{1 - \exp[-(\alpha l)]\}}{\pi\left(w_0^2 + w_p^2\right) l} \tag{6.3.10}$$

where l is the length of the laser rod. It should be noted that, in performing the integral along the z coordinate in Eq. (6.3.8), we have made the approximation $\int_0^l \exp -(\alpha z) \cos^2 kz dz \cong (1/2) \int_0^l \exp -(\alpha z) dz$, using the fact that, since $\cos^2(kz)$ changes much more rapidly with z then the $\exp -(\alpha z)$ term, we can substitute $\cos^2(kz)$ with its average value $<\cos^2(kz)> = (1/2)$. If we now define the absorption efficiency, η_a, as

$$\eta_a = 1 - \exp -(\alpha\, l) \tag{6.3.11}$$

equation (6.3.10) can then be put in the more suggestive form

$$<R_p> = \eta_p \left(\frac{P_p}{h\nu_p}\right) \frac{2}{\pi\left(w_0^2 + w_p^2\right) l} \tag{6.3.12}$$

where we have defined $\eta_p = \eta_r \eta_t \eta_a$. Equation (6.3.12) constitutes the final result of our calculation for the effective pump rate in the case of longitudinal pumping. Note that, for a given value of P_p, $<R_p>$ increases as w_p decreases so that the maximum value of $<R_p>$ would be attained for $w_p \to 0$. For very small values of pump spot-size, however, divergence of the pump beam in the active rod cannot be neglected with the result that the beam may actually become larger than the laser beam at the end of the rod. For this reason and to optimize the optical efficiency, the condition $w_p \cong w_0$ is often taken as a rough guide to the optimum case.

In the case of transverse pumping, we begin by writing the following obvious relation between pump rate and power, P_{pi}, incident on the rod

$$\int_a h\nu_p R_p \, dV = \eta_a P_{pi} \tag{6.3.13}$$

where η_a is the fraction of the incident power which is absorbed in the active medium. Note that, according to Eq. (6.3.11), the absorption efficiency η_a can be written as $\eta_a \cong (1 - \exp -\alpha D)$ where D is the relevant transverse dimension of the rod ($D \cong D_R$, where D_R is the rod diameter, for a single pass or $D \cong 2D_R$, for a double pass of the pump beam in the rod). Equation (6.3.13) allows the pump rate to be calculated once its spatial variation is known. If we take, as a simple case, $R_p = \text{const}$, we obviously obtain from Eq. (6.3.13) $R_p = \eta_r \eta_t \eta_a P_p / h\nu_p A l$ where A is the cross-sectional area of the rod and where Eq. (6.3.6) has been used. To calculate $<R_p>$ we consider a conceptually simpler model of the laser rod, where the active species is assumed to be confined to the central region of the rod, $0 \le r \le a$, while the rod is undoped, for $r > a$ (*cladded rod*). In this case, Eq. (6.3.9) can be taken to hold for any value of r while one has $R_p = \text{const}$ for $0 \le r \le a$ and $R_p = 0$ for $r > 0$. Then, from Eqs. (6.3.8) and (6.3.9), we obtain

$$<R_p> = \eta_p \left(\frac{P_p}{h\nu_p}\right) \frac{[1 - \exp - (2a^2/w_0^2)]}{\pi a^2 l} \tag{6.3.14}$$

where again we have written $\eta_p = \eta_r \eta_t \eta_a$. This equation constitutes the final result for our calculation of the effective pump rate in the case of transverse pumping.

For the comparison to be performed in sect. 6.3.5, it is appropriate to also calculate here the effective pump rate that applies for lamp pumping. Assuming the cladded rod model considered above and again taking R_p to be constant in the active medium, i.e., for $0 \le r \le a$, we obtain from Eqs. (6.2.6) and (6.3.8)

$$<R_p> = \eta_{pl} \left(\frac{P_p}{h\nu_{mp}}\right) \frac{[1 - \exp - (2a^2/w_0^2)]}{\pi a^2 l} \tag{6.3.15}$$

where η_{pl} is the pumping efficiency for lamp pumping given, according to Eq. (6.2.5), by $\eta_{pl} = \eta_r \eta_t \eta_a \eta_{pq}$.

6.3.4. Threshold Pump Power for Four-Level and Quasi-Three-Level Lasers

With the results obtained in the previous section for the effective pump rate, we can now go on to calculate the expected threshold pump rate and threshold pump power for a given laser. We will limit our considerations to two very important cases: (i) An ideal four-level laser, where pumped atoms are immediately transferred to the upper laser level, 2, and where the lower laser level, 1, is empty [see Fig. 1.4b]. (ii) An ideal quasi-three-level laser, where pumped atoms are again transferred immediately to the upper laser level, 2, and where the lower laser level is a sublevel of the ground level 1. The first case includes lasers such as Nd:YAG at $\lambda = 1.06\ \mu m$ or $\lambda = 1.32\ \mu m$, Ti:Al_2O_3, and Cr:LISAF or LICAF. The most important lasers belonging to the second category are Nd:YAG at $\lambda = 0.946\ \mu m$, Er:glass or Yb:Er:glass at $\lambda \cong 1.45\ \mu m$, Yb:YAG or Yb:glass, and Tm:Ho:YAG.

Let us first consider an ideal four-level laser and let us assume that the upper laser level actually consists of many strongly coupled sublevels whose *total* combined population will be called N_2. According to Eq. (1.2.5), the threshold value of this population, N_{2c}, can be written as $N_{2c} = \gamma/\sigma_e l$, where σ_e now indicates the effective stimulated emission cross-section [see sect. 2.7.2]. Actually, this previous expression only holds for a spatially uniform model i.e., when both R_p and the mode configuration, $|u|^2$, are considered to be spatially independent. When spatial dependency is taken into account, the previous expression for the threshold upper state population gets modified as follows (see Appendix *E*)

$$<N_2>_c = \gamma/\sigma_e l \tag{6.3.16}$$

where $<N_2>$ is the effective value of population, given by

$$<N_2> = \left(\int_a N_2|u|^2 dV\right) \Big/ \int_a |u|^2 dV \qquad (6.3.17)$$

The critical, or threshold, pump rate can then be obtained from the condition that the number of atoms raised by the pumping process must equal the number of atoms decaying spontaneously. Thus we get $R_p = N_{2c}/\tau$, where τ is the effective lifetime of the upper laser level, taking account of the decay of all the sublevels [see again sect. 2.7.2]. It then follows that

$$<R_p>_c = <N_2>_c/\tau \qquad (6.3.18)$$

From Eqs. (6.3.16) and (6.3.18) we get

$$<R_p>_c = \frac{\gamma}{\sigma_e l \tau} \qquad (6.3.19)$$

Once the threshold value of the pump rate is calculated, we can readily obtain the corresponding threshold pump power. Using Eq. (6.3.19) into Eqs. (6.3.12) and (6.3.14), we get in fact the following expressions

$$P_{th} = \left(\frac{\gamma}{\eta_p}\right)\left(\frac{h\nu_p}{\tau}\right)\left[\frac{\pi\left(w_0^2 + w_p^2\right)}{2\sigma_e}\right] \qquad (6.3.20)$$

$$P_{th} = \left(\frac{\gamma}{\eta_p}\right)\left(\frac{h\nu_p}{\tau}\right)\left\{\frac{\pi a^2}{\sigma_e\left[1 - \exp - \left(2a^2/w_0^2\right)\right]}\right\} \qquad (6.3.21)$$

which hold for longitudinal and transverse pumping, respectively. The expression of the threshold pump power for longitudinal pumping given by Eq. (6.3.20) agrees with that given by Kubodera, Otsuka, and Miyazawa.[19] Note that, again for longitudinal pumping, the threshold pump power increases as w_0 is increased because, as w_0 increases, the wings of the mode extend further into the less strongly pumped regions of the active medium. Likewise, for transverse pumping and for the cladded rod model considered above, the threshold pump power increases as w_0 is increased because, as w_0 increases, the wings of the mode extend further into the cladding, i.e., into the unpumped part of the medium. Similar considerations could be applied to the more realistic case of a rod without cladding. In this case, however, the calculation would be more involved because, in general, Eq. (6.3.9) would no longer apply and the true field distribution, account being taken of the aperturing effects established by the finite rod diameter, would have to be used. When w_0 is appreciably smaller than a (say $w_0 \leq 0.7a$), however, the field distribution is not greatly affected by the presence of this aperture and Eq. (6.3.21) can be assumed to hold also for a rod without cladding. In this case, of course, $\left[1 - \exp - \left(2a^2/w_0^2\right)\right]$ is very much closer to unity and, in calculating the threshold pump power, this term could even be omitted from Eq. (6.3.21). As we will show in the next Chapter, however, it is important to keep this term in Eq. (6.3.21) to obtain the correct expression for the slope efficiency.

For the comparison to be made in next section, it is also appropriate to calculate the threshold pump rate for lamp pumping. From Eqs. (6.3.15) and (6.3.19) we get

$$P_{th} = \left(\frac{\gamma}{\eta_{pl}}\right)\left(\frac{h\nu_{mp}}{\tau}\right)\left\{\frac{\pi\, a^2}{\sigma_e\left[1 - \exp - \left(2a^2/w_0^2\right)\right]}\right\} \qquad (6.3.22)$$

Let us now consider a quasi-three-level laser where the lower laser level, 1, is a sublevel of the ground level and assume that the population raised, from the pumping process, to the pump level(s) is immediately transferred to the upper laser level (ideal quasi-three-level laser). We will assume that all ground state sub-levels are strongly coupled and hence in thermal equilibrium and call N_1 the total combined population of level 1. We will also assume that the upper laser level, level 2, consists of a number of strongly coupled sublevels and call N_2 the total combined population of the upper level. The threshold values for the population of the two levels is again established by the condition that the total net gain equals the losses. For the space-dependent case and according to Eq. (6.3.16) we now obtain (see Appendix *E*)

$$[\sigma_e {<}N_2{>}_c - \sigma_a {<}N_1{>}_c]\, l = \gamma \qquad (6.3.23)$$

Where $<N_2>$ and $<N_1>$ again indicate spatially averaged values as in Eq. (6.3.17) and σ_e and σ_a are, respectively, the effective values of the stimulated emission and absorption cross sections. Since, for an ideal quasi-three-level laser, one has $N_1 + N_2 = N_t$ it follows that $<N_1> + <N_2> = N_t$ and using this expression in Eq. (6.3.23) we can readily calculate $<N_2>_c$. The effective value of the threshold pump rate must again satisfy Eq. (6.3.18) and, using the value of $<N_2>_c$ in this way calculated, we obtain

$$<R_p>_c = \frac{[\sigma_a N_t l + \gamma]}{(\sigma_e + \sigma_a)\, l\tau} \qquad (6.3.24)$$

Note that Eq. (6.3.24) obviously reduces to Eq. (6.3.19) if we let $\sigma_a \to 0$.

In our calculation of the corresponding threshold pump power, we limit our considerations to longitudinal pumping, since this is the only configuration which has allowed operation with a reasonably low threshold, in this case. From Eqs. (6.3.24) and (6.3.12) we obtain

$$P_{th} = \left(\frac{\sigma_a N_t l + \gamma}{\eta_p}\right)\left(\frac{h\nu_p}{\tau}\right)\left[\frac{\pi\left(w_0^2 + w_p^2\right)}{2(\sigma_e + \sigma_a)}\right] \qquad (6.3.25)$$

which agrees with the expression given by Fan and Byer.[20] Note again that Eq. (6.3.25) reduces to Eq. (6.3.20) if we let $\sigma_a \to 0$.

6.3.5. Comparison Between Diode-pumping and Lamp-pumping

Following the discussion presented in the previous sections, we are now ready to perform a general comparison between lamp pumping and diode pumping. The comparison can only be made for four-level lasers, since quasi-three-level lasers have mostly been operated by means of longitudinal pumping by diodes. To compare Eq. (6.3.22) with Eqs. (6.3.20)

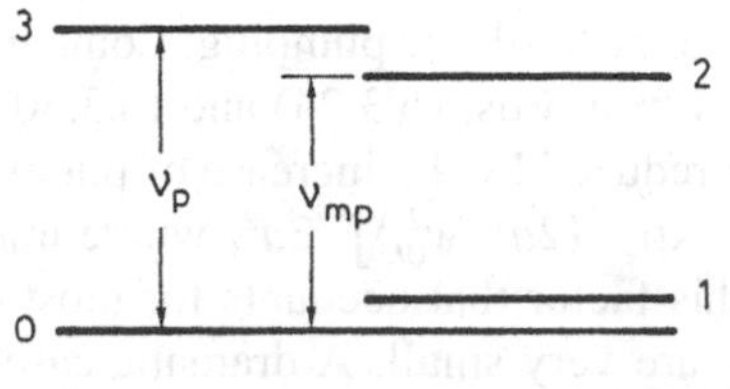

FIG. 6.17. Actual pump frequency, ν_p, and ideal minimum pump frequency, ν_{mp}, in a 4-level laser.

TABLE 6.3. Comparison between pumping efficiencies of lamp pumping and diode pumping

Pump Configuration	η_r (%)	η_t (%)	η_a (%)	η_{pq} (%)	η_p (%)
Lamp	43	82	17	59	3.5
Diode (longitudinal)	50	80	98	82	32
Diode (transverse)	50	80	90	82	30

and (6.3.21), it is convenient, for diode pumping, to define a pump quantum efficiency η_{pq} as $h\nu_{mp}/h\nu_p$, where ν_p is the actual pump frequency and ν_{mp} is the minimum pump frequency, i.e., the pump frequency that would have been required for direct pumping to the upper laser level (see Fig. 6.17). Equations (6.3.20) and (6.3.21) then readily transform to

$$P_{th} = \left(\frac{\gamma}{\eta_{pd}}\right)\left(\frac{h\nu_{mp}}{\tau}\right)\left[\frac{\pi\left(w_0^2 + w_p^2\right)}{2\sigma_e}\right] \tag{6.3.26}$$

$$P_{th} = \left(\frac{\gamma}{\eta_{pd}}\right)\left(\frac{h\nu_{mp}}{\tau}\right)\left\{\frac{\pi a^2}{\sigma_e\left[1 - \exp - \left(2a^2/w_0^2\right)\right]}\right\} \tag{6.3.27}$$

where we have defined $\eta_{pd} = \eta_p\eta_{pq} = \eta_r\eta_t\eta_a\eta_{pq}$ as the overall pump efficiency for diode pumping. Equations (6.3.23), (6.3.26) and (6.3.27) allow us now to make a general comparison between lamp pumping and diode pumping.

A first comparison can be made in terms of the four efficiency factors η_r, η_t, η_a, η_{pq} and hence of the overall pump efficiency $\eta_p = \eta_r\eta_t\eta_a\eta_{pq}$. Limiting ourselves to the case of Nd:YAG, Table 6.3 shows the estimated values of these efficiency factors where the values for lamp pumping have been taken from Table 6.1. In the case of longitudinal pumping by a diode laser, a 1 cm long crystal is considered while, for transverse pumping, a 4 mm diameter rod is assumed. Note that, despite the great diversity of the various pumping systems which have been considered so far, the comparison in terms of these four efficiency factors become very simple and instructive. One can see, in fact, that the radiative and transfer efficiencies are approximately the same for lamp and diode pumping and that the almost ten times increase in overall pump efficiency for diode pumping comes from the very large increase in absorption efficiency (by almost a factor 6) and a consistent increase of the pump quantum efficiency (by a factor of ~ 1.5). Note also that, in terms of pump efficiency, longitudinal and transverse pumping are roughly equivalent with a slightly smaller value of the absorption efficiency for transverse pumping.

A second comparison can be made with respect to threshold pump powers. According to Eqs. (6.3.24) and (6.3.31) and for the same value of rod cross-sectional area, the main difference in pump thresholds between lamp pumping and transverse pumping arises for the almost ten-fold increase in pump efficiency for diode pumping. Comparing longitudinal diode pumping to lamp pumping, one sees from Eqs. (6.3.24) and (6.3.30) that the pump threshold for diode-pumping, besides being reduced by the increase of pump efficiency, is further reduced by a factor $\left(w_0^2+w_p^2\right)\times\left[1-\exp-\left(2a^2/w_{0l}^2\right)\right]/2a^2$, where w_{0l} is the laser spot size for the case of lamp pumping. It is this factor that accounts for most of the reduction in threshold pump power when w_0 and w_p are very small. A dramatic case of this type occurs for fiber lasers where, for single mode fibers, the value of w_0 as well as that of w_p may be as small as $2\div 3\,\mu$m. If, for example, we take $w_0=w_p=2\,\mu$m, for the case of a fiber laser, and $a=2$ mm and $w_{0l}=0.5\,a$, in the case of lamp pumping, the expected reduction arising from the previous geometrical factor is by almost six orders of magnitude! This is the essential reason why fiber lasers exhibit such small pump thresholds. Comparing longitudinal and transverse pumping, we may note from Eqs. (6.3.22) and (6.3.23) that the pump threshold is lower for longitudinal compared to transverse pumping essentially by the ratio $\left(w_0^2+w_p^2\right)_l\left[1-\exp-\left(2a^2/w_{0t}^2\right)\right]/2a^2$, where the suffices l and t stand for longitudinal and transverse pumping, respectively. For the very small values of spot sizes, w_0 and w_p, that can be used in longitudinal pumping, this ratio may again have very small values. However, to achieve comparable outputs for the two cases, the TEM_{00} spot-size of the two cases will need to be more comparable. It is instructive, therefore, to make this comparison for the same value of spot size in each case i.e., for $(w_0)_l=(w_0)_t$. To avoid excessive diffraction effects arising from beam truncation at the aperture formed by the rod diameter, the spot size for transverse pumping must then be somewhat smaller than the rod radius, a. In practice, a value of $(w_0)_t\cong 0.7a$ may be chosen. Assuming best overlapping condition, i.e. $w_0=w_p$, for longitudinal pumping, it then follows that $\left(w_0^2+w_p^2\right)_l\times\left[1-\exp-\left(2a^2/w_{0t}^2\right)\right]/2a^2\cong 0.48$ and, under these conditions, the threshold pump power for longitudinal pumping may be only a factor ~ 0.5 smaller than that for transverse pumping.

Compared to lamp pumping, besides having much higher pump efficiency and very much lower pump threshold, diode pumping has the additional advantage of inducing a reduced thermal load in the active medium. In fact, for a given absorbed power P_a in the medium, the fraction $\eta_{pq}P_a$ is available in the upper laser level and, consequently, the fraction $\eta_{pq}(h\nu/h\nu_{mp})P_a$ is available as laser power, $h\nu$ being the energy of the laser photon. The power dissipated as heat is thus $[1\text{-}\eta_{pq}(h\nu/h\nu_{mp})]\ P_a$. From Table 6.2 one then sees that the thermal load for lamp pumping is ~ 2 larger than for diode pumping. This reduced thermal load has two beneficial effects: (i) Reduced thermal lensing and thermally-induced birefringence in the rod. (ii) Reduced thermal fluctuations of the refraction index of the medium for a given pump power fluctuation. Both these effects are important for obtaining solid-state laser operation on a single transverse and longitudinal mode of high quality.

6.4. ELECTRICAL PUMPING

We recall that this type of pumping is used for gas and semiconductor lasers. We will limit our considerations here to the case of gas lasers and defer discussion of the more straightforward case of semiconductor laser pumping to the semiconductor laser section of Chap. 9.

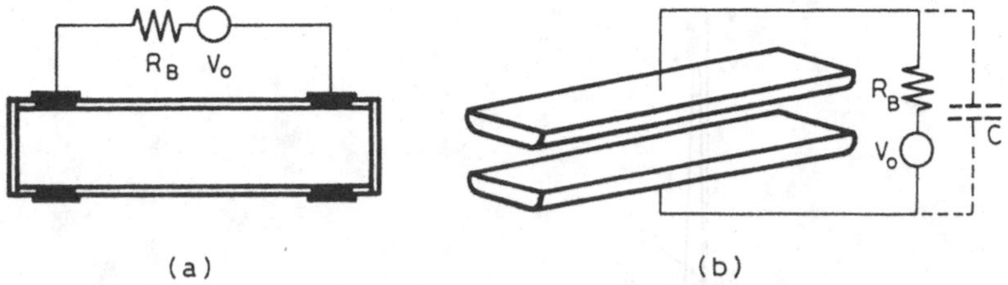

FIG. 6.18. Most frequently used pumping configurations for gas-discharge lasers: (a) Longitudinal discharge. (b) Transverse discharge.

Electrical pumping of a gas laser is achieved by allowing a current, which may be continuous (d.c. current) or at radio-frequency (r.f. current) or pulsed, to pass through the gas mixture. Generally, the current through the gas passes either along the laser axis direction (longitudinal discharge, Fig. 6.18a) or transversely to it (transverse discharge, Fig. 6.18b). Since the transverse dimension of a laser medium is usually much smaller than its longitudinal dimension, then, for the same gas mixture, the voltage needed in a transverse configuration is significantly less than for a longitudinal configuration. On the other hand a longitudinal discharge, when confined in a dielectric (e.g. a glass) tube, as in Fig. 6.18a, often provides a more uniform and stable pumping configuration. In fact. in the discussion that follows we will concentrate on the so-called *glow discharge*, where, due to the uniformity of the current density, a uniform bluish glow of light is observed from the discharge. The situation that needs to be avoided is that of an *arc discharge* where current is observed to flow in one or more streamers, emitting white light of high intensity (as in a lightning).

One requires the presence of a series resistance, R_B, often called the ballast resistance, as shown in both Fig. 6.18a and b, to stabilize the discharge at the desired operating point. To understand this feature, we show in Fig. 6.19, as a solid line, the voltage vs current characteristic of a gas discharge. Note that, in the operating region, the voltage across the discharge remains nearly constant as discharge current increases. A peak voltage, V_p, about an order of magnitude larger than this constant operating voltage is needed to induce gas break-down. Thus, the behavior of a discharge tube is very different from that of a simple resistor! In the same figure we also show, as a dashed line, the voltage vs current characteristic of a power supply giving a voltage, V_0, in series with a ballast resistance, R_B. One notes that the current will stabilize at either of the intersections A and C of the two curves (the intersection B corresponds to an unstable equilibrium situation). Thus, starting with a lamp that is initially unenergized and then applying the voltage from the power supply, the lamp will stabilize itself at point C with very little current flowing in the discharge. To reach the other stable point, A, the desired operating point, one can briefly raise the applied voltage so as to overcome the voltage barrier V_p. This is usually achieved by applying an over-voltage to the high voltage electrode for a long enough time to produce sufficient gas ionization (see also Fig. 6.4b and 6.5). Alternatively, a high voltage pulse may be applied to some auxiliary electrode (see also Fig. 6.4a).

Various different electrode structures are used for both longitudinal and transverse discharges. For a longitudinal discharge, the electrodes often have an annular structure with the cathode surface usually much larger than that of the anode to help reduce degradation due

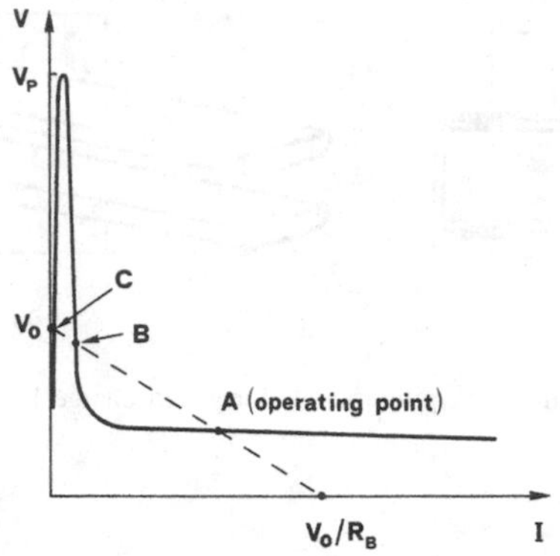

FIG. 6.19. Voltage, V, vs current, I, characteristic of a gas discharge (solid line) and of a power supply with a series resistance (dashed line).

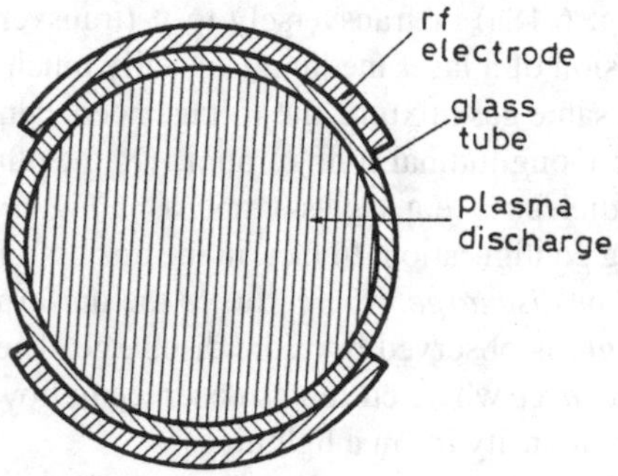

FIG. 6.20. Radio-frequency transverse excitation of a gas in a quartz tube.

to impact of the heavy ions. In a transverse discharge, the electrodes extend over the whole length of the laser material and the opposing surface of each electrode must have a very smooth curvature. In fact, if there is any sharp corner, the high electric field produced there may easily result in an arc formation rather than in a uniform discharge. Usually, longitudinal discharge arrangements are only used for cw lasers while transverse discharges are used with cw, pulsed, or rf lasers. A particularly interesting case of a transverse discharge, using rf excitation, is shown in Fig. 6.20, where the rf electrodes are applied to the outside of the discharge tube, usually made of glass. The presence of a finite thickness of the glass tube presents several advantages: (i) It acts as a series capacitor for the discharge whose impedance, at the frequency of the rf voltage, acts as an effective, capacitive, ballast for stabilizing the discharge. The loss of pump power in the resistive ballast, R_B, of Fig. 6.18 is thus avoided. (ii) Since the glass dielectric medium extends over the whole of the electrode structure, the problem of arc formation is greatly reduced. (iii) Since the gas mixture is not in contact with the electrodes, the plasma-chemical effects, occurring at the electrode surface and leading to dissociation of the mixture, are eliminated. When this configuration is applied to a CO_2 laser, for instance, an order of magnitude reduction in the electrode maintenance time can be gained and a factor of two decrease in the gas consumption rate.

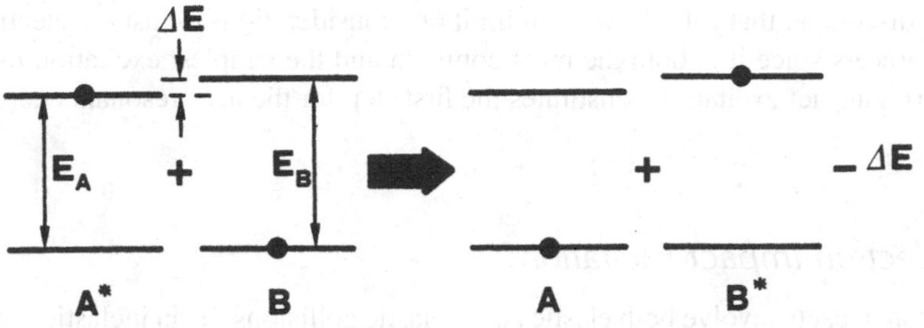

FIG. 6.21. Laser pumping by near-resonant energy transfer.

We now go on to present a general description of the physical phenomena leading to excitation in the gas. First we recall that, in an electrical discharge, both ions and free electrons are produced and, since these charged particles acquire additional kinetic energy from the applied electric field, they are able to excite a neutral atom by collision. The positive ions, owing to their much greater mass, are accelerated to much lower velocities than the electrons and therefore do not play any significant part in the excitation process. Therefore, electrical pumping of a gas usually occurs via one, or both, of the following two processes: (i) For a gas consisting of only one species, the excitation is only produced by electron impact, i.e., the process

$$e + \mathrm{X} \rightarrow \mathrm{X}^* + e \tag{6.4.1}$$

where X and X* represent the atom in the ground and excited state, respectively. Such a process is called a *collision of the first kind*. (ii) For a gas consisting of two species (say A and B), excitation can also occur as a result of collisions between atoms of different species through a process known as resonant energy transfer (see also Sect. 2.6.1). Referring to Fig. 6.21, let us assume that species B is in the ground state and species A is in the excited state, as a result of electron impact. We will also assume that the energy difference ΔE between the two transitions is less than kT. In this case, there is an appreciable probability that, after collision, species A will be found in its ground state and species B in its excited state. The process can be denoted by

$$\mathrm{A}^* + \mathrm{B} \rightarrow \mathrm{A} + \mathrm{B}^* - \Delta E \tag{6.4.2}$$

where the energy difference ΔE will be added to or subtracted from the translational energy of the colliding partners, depending on its sign. This is the reason why ΔE must be smaller than kT. This process provides a particularly effective way of pumping species B, if the upper state of A is metastable (forbidden transition). In this case, once A is excited to its upper level, it will remain there for a long time, thus constituting an energy reservoir for excitation of species B. A process of the type indicate in Eq. (6.4.2) is called a *collision of second kind**.

* Collisions of the *first kind* involve conversion of the kinetic energy of one species into potential energy of another species. In collisions of *the second kind*, potential energy is converted into some other form of energy (other than radiation) such as kinetic energy, or is transformed into potential energy (in the from of electronic, vibrational, or rotational energy) of another like or unlike species. Collisions of the second kind therefore include not only the reverse of collisions of the first kind (e.g., $e + \mathrm{X}^* \rightarrow e + \mathrm{X}$) but also, for instance, the conversion of excitation energy into chemical energy.

In the discussion that follows we will limit our considerations to just the electron impact excitation process since it is both the most common and the simplest excitation mechanism. Also, electron impact excitation constitutes the first step for the near-resonant energy transfer process.

6.4.1. Electron Impact Excitation

Electron impacts involve both elastic and inelastic collisions. In an inelastic collision, the atom may either be excited to a higher state or be ionized. Of the various possible excitations the one we are usually interested in is that which excites the atomic species to the desired upper laser level. In order to describe the above excitation phenomena by means of appropriate collision cross-sections, we will first consider the simple case of impact excitation by a collimated beam of mono-energetic electrons. If F_e is the electron flux (number of electrons per unit area per unit time), a total cross section σ_e can be defined in a similar way to the case of a photon flux [see Eq. (2.4.20)]. Thus, if we let dF_e be the change of flux that results from the beam traveling a distance dz in the material, we can write

$$dF_e = -\sigma_e N_t F_e dz \tag{6.4.3}$$

where N_t is the total population of the atomic species. Collisions that produce electronic excitation will only account for some fraction of the total cross section. In fact, the cross section for elastic collisions, σ_{el}, is usually the largest, its order of magnitude being $\sim 10^{-16}\,\mathrm{cm}^2$. If we now let σ_{e2} be the cross section for electronic excitation from the ground level to the upper laser level, then, according to Eq. (6.4.3), the rate of population of the upper state due to the pumping process is

$$(dN_2/dt)_p = \sigma_{e2} N_t F_e = N_t N_e \upsilon \sigma_{e2} \tag{6.4.4}$$

where υ is the electron velocity and N_e is the electron density. A calculation of the pump rate requires a knowledge of the value of σ_{e2}, which is expected to depend on the energy E of the incident electron, i.e. $\sigma_{e2} = \sigma_{e2}(E)$. In a gas discharge the electrons have a distribution of energies which can be described by the distribution $f(E)$; its meaning is that $dp = f(E)dE$ represents the elemental probability that the electron energy ranges between E and $E + dE$. In this case the rate of population of the upper state is obtained from Eq. (6.4.4) by averaging over this distribution, viz.

$$(dN_2/dt)_p = N_t N_e <\upsilon \sigma_{e2}> \tag{6.4.5}$$

where

$$<\upsilon\sigma> = \int \upsilon\sigma\,(E) f(E) dE \tag{6.4.6}$$

According to Eqs. (1.3.1) and (6.4.5) the pump rate is then given by

$$R_p = N_t N_e <\upsilon \sigma_{e2}> \tag{6.4.7}$$

Where $<\upsilon\sigma>$ is given by Eq. (6.4.6). The calculation of R_p thus requires the knowledge of the energy dependence of both σ and f. This dependence will be considered in the following sections.

6.4.1.1. Electron Impact Cross Section

The qualitative behavior of σ vs the electron energy E is indicated in Fig. 6.22 for the three cases: (1) optically allowed transition, (2) optically forbidden transition involving no change of multiplicity, (3) optically forbidden transition involving a change of multiplicity. In all three cases, the peak value of σ has been normalized to unity. Note that, in each case, there is a distinct threshold E_{th} for the cross section. As expected, the value of E_{th} turns out to be close to the energy of the transition involved. The cross section rises very sharply above threshold, reaches a maximum value, and thereafter decrease slowly. The peak value of σ and the width of the curve depend on the type of transition involved: (1) For an optically allowed transition, the peak value of σ can be typically 10^{-16} cm^2 and the width of the curve may be typically 10 times greater than the threshold energy (curve a of Fig. 6.22). (2) For an optically forbidden transition involving no change of multiplicity, the peak cross section is drastically reduced by nearly three orders of magnitude (to about 10^{-19} cm^2) and the width of the curve may be only $3 \div 4$ times the threshold energy (curve b of Fig. 6.22). (3) When a change of multiplicity is involved, the peak cross section may actually be larger than for an optically forbidden transition and the width of the curve may now be typically equal to or somewhat smaller than the threshold energy E_{th} (curve c of Fig. 6.22). It should be noted that, in any case, the width of the curve is roughly comparable to the threshold energy, i.e. to the transition energy. By contrast, the transition linewidths for photon absorption are much sharper (typically $10^{-4} \div 10^{-6}$ of the transition frequency). This very important circumstance

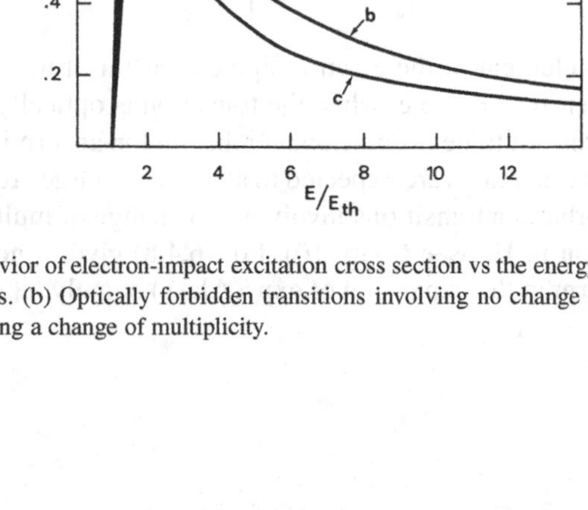

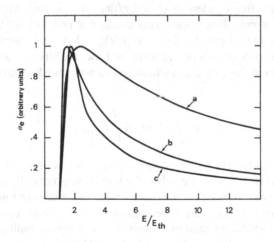

FIG. 6.22. Qualitative behavior of electron-impact excitation cross section vs the energy of the incident electron: (a) Optically allowed transitions. (b) Optically forbidden transitions involving no change of multiplicity. (c) Optically forbidden transitions involving a change of multiplicity.

arises from the fact that, as explained in Sect. 6.1, the electron impact excitation is basically a non-resonant phenomenon. This is the basic reason why excitation of a gaseous medium is performed much more effectively by a "polychromatic" source of electrons (such as in a gas discharge) than by a polychromatic light source (such as a lamp).

To provide a deeper insight into the mechanism involved in electron impact excitation, we now give a sketch of the procedure for a quantum mechanical calculation of the cross section σ. For optically allowed transitions or for optically forbidden transitions involving no change of multiplicity, the simplest, and often the most accurate, calculation uses the Born approximation. Before collision, the atom is described by the ground state wave-function u_1 and the incident electron by the plane wave function $\exp(j\mathbf{k}_0 \cdot \mathbf{r})$, where $\mathbf{k}_0$ is the electron wave vector and $\mathbf{r}$ is the vector describing the position of the incident electron with respect to a center situated e.g. at the nuclear position. After collision, the atom is described by the upper state wave-function u_2 and the scattered electron by the plane wave $\exp(j\mathbf{k}_n \cdot \mathbf{r})$, where $\mathbf{k}_n$ is the wave-vector of the scattered electron. For the discussion that follows, one needs to recall that $k = 2\pi/\lambda$, where λ is the deBroglie wavelength of the electron which can be expressed as $\lambda = (1.23/\sqrt{V})$ nm, where V is the electron energy in electron volts. The interaction has its origin in the electrostatic repulsion between the incident electron and the electrons of the atom. This interaction is assumed to be weak enough for there be only a very small probability of a transition occurring in the atom during the impact and for the chance of two such transitions to be negligible. In this case the Schrödinger equation for the problem can be linearized. It then turns out that the transition rate and hence the transition cross section can be expressed as

$$\sigma_e \propto \left| \int [u_2 \exp(j\mathbf{k}_n \cdot \mathbf{r})]^* [u_1 \exp(j\mathbf{k}_0 \cdot \mathbf{r})]\, dV \right|^2 \tag{6.4.8}$$

From the above expression for the deBroglie wavelength and assuming an electron energy of only a few eV, the wavelength $\lambda' = 2\pi/|\mathbf{k}_0 - \mathbf{k}_n| = 2\pi/|\Delta\mathbf{k}|$ is seen to be appreciably larger than the atomic dimensions. This means that $(\Delta\mathbf{k} \cdot \mathbf{r}) \ll 1$ for $|\mathbf{r}| \le a$, where a is the atomic radius. In this case the factor $\exp j[(\mathbf{k}_0 - \mathbf{k}_n) \cdot \mathbf{r}] = \exp j(\Delta\mathbf{k} \cdot \mathbf{r})$ appearing in Eq. (6.4.8) can be expanded as a power series of $(\Delta\mathbf{k} \cdot \mathbf{r})$. Since u_1 and u_2 are orthogonal functions, the first term in this expansion which gives a non-vanishing term for σ_e, is $j(\Delta\mathbf{k} \cdot \mathbf{r})$ and one gets

$$\sigma_e \propto \left| \int u_2^* \mathbf{r} u_1 dV \right|^2 \propto |\boldsymbol{\mu}_{21}|^2 \tag{6.4.9}$$

where $\boldsymbol{\mu}_{21}$ is the matrix element of the electric dipole moment of the atom [see Eq. (2.3.7)]. It then follows that, when $\boldsymbol{\mu}_{21} \neq 0$, i.e. when the transition is optically allowed, the electron impact cross section turns out to be proportional to the photon absorption cross section. Thus strong optically allowed transitions are expected to also show a large cross section for electron impact. For optically forbidden transitions involving no change of multiplicity ($\Delta S = 0$, e.g. the $1^1S \to 2^1S$ transition in He, see Chap. 10), Eq. (6.4.8) gives a non-vanishing value for the next-higher-order term in the expansion of $\exp j(\Delta\mathbf{k} \cdot \mathbf{r})$ namely $-(\Delta\mathbf{k} \cdot \mathbf{r})^2/2$. This means

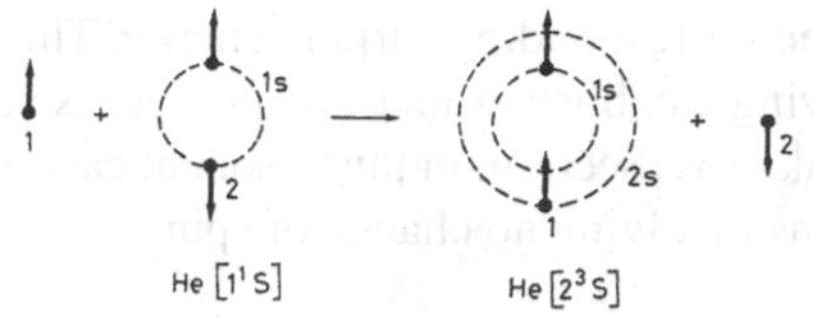

FIG. 6.23. Illustration of the phenomenon of electron exchange in the case of the $1\,^1S \rightarrow 2\,^3S$ transition in a He atom.

that σ_e can now be written as $\sigma_e \propto \left|\int u_2^*(\Delta\mathbf{k}\cdot\mathbf{r})^2 u_1 dV\right|^2$. This relation is completely different from the corresponding one that would apply in the case of a photon interaction, i.e., that due to a magnetic dipole interaction. It is therefore no surprise to find that the ratio between the two peak cross sections $\sigma_{\text{forbidden}}/\sigma_{\text{allowed}}$ is typically, in this case, about 10^{-3} while the same ratio was shown to be $\sim 10^{-5}$ for photon absorption [see (2.4.14)]. So one can make the assertion that optically forbidden transitions are relatively more easily excited by electron impact than by "photon impact," and this has some profound consequences for the operating principles of most gas lasers, since pumping is often achieved through optically forbidden transitions.

When a change of multiplicity is involved (e.g., the $1^1S \rightarrow 2^3S$ transition in He, see Capt. 10) the Born approximation gives a zero cross section in any order of the expansion of $\exp j(\Delta\mathbf{k}\cdot\mathbf{r})$. In fact, such a transition involves a spin change, while, within the Born approximation, the incoming electron, through its electrostatic interaction, can only couple to the orbital motion of the atom rather to its spin*. The theory, in this case is largely due to Wigner and its starting point is the observation that, in a collision, it is the total spin of the atom plus that of the incident electron that must be conserved, not necessarily that of the atom alone. Transitions may, therefore, occur via an electron *exchange collision*, where the incoming electron replaces the electron of the atom involved in the transition and this electron is in turn ejected by the atom. To conserve the total spin, the incoming electron must have its spin opposite to that of the ejected one. To clarify this exchange process we show in Fig. 6.23 the electron impact excitation of the $1^1S \rightarrow 2^3S$ transition in He. Note that the process can be visualized as the incident electron, labeled 1, is captured in the $2s$ state of He while the electron of the atom with opposite spin, labeled 2, is actually ejected. It should be pointed out, however, that this constitutes a very naive way of describing the phenomenon because, during the collision, the two electrons are quantum mechanically indistinguishable. From this simple description, however, one readily understands that this exchange mechanism must be of a more resonant nature than that considered in the Born approximation: there will be a high probability for this exchange to occur only if the energy of the incoming electron closely matches the transition energy. In this case, in fact, the electron energy is just what is needed to leave the electron 1, after collision, in the upper level, $2s$, while the second electron, electron 2, is ejected with zero velocity. For higher energies of the incident electron, the exchange process would leave electron 1 in the $2s$ orbital, while electron 2, ejected from

* This assumes a negligible spin-orbit coupling, which is true for light atoms (e.g., He, Ne), while it would not be true for heavy atoms like Hg.

the atom, would carry off the corresponding surplus energy. This would definitely be a less likely process to occur. Having established that this process is somewhat resonant, we can now understand why the peak cross section, in this resonant case, can be even higher than for optically forbidden transitions involving no change of spin.

6.4.2. Thermal and Drift Velocities

As already mentioned in previous sections, it is the electrons that are responsible for the phenomena occurring in a gas discharge. They acquire energy from the applied electric field and lose or exchange energy through three processes: (1) Inelastic collisions with the atoms, or molecules, of the gas mixture, which either raise the atom to one of its excited states or ionize it. These electron-impact excitation or ionization phenomena are perhaps the most important processes for laser pumping, hence the extended discussion in these sections. (2) Elastic collisions with the atoms. If we assume that the atoms are at rest before collision (the mean velocity for an atom is indeed much smaller than for an electron), the electron will lose energy upon collision. It can be shown by a straightforward analysis of the elastic collision process that, for random direction of the scattered electron, the electron loses, on average, a fraction $2(m/M)$ of its energy, where m is the mass of electron and M is the mass of the atom. Note that this loss is very small since m/M is small (e.g., $m/M = 1.3 \times 10^{-5}$ for Ar atoms). (3) Electron-electron collisions. For a gas which is ionized to a moderate degree, the frequency of such collisions is usually high since both particles are charged and exert forces on one another over a considerable distance. Moreover, since both colliding particles have the same mass, the energy exchange in the collision is considerable. As a result of the collision phenomena mentioned above and as a consequence of the electrons being accelerated by the electric field of the discharge, the electron "gas" in the plasma acquires a distribution of velocities. We can describe this by introducing the distribution $f(\upsilon_x, \upsilon_y, \upsilon_z)$ with the meaning that $f(\upsilon_x, \upsilon_y, \upsilon_z)d\upsilon_x d\upsilon_y d\upsilon_z$, gives the elemental probability that the electron is found with velocity components in a range $d\upsilon_x, d\upsilon_y, d\upsilon_z$, around $\upsilon_x, \upsilon_y, \upsilon_z$. Given this distribution we can define a *thermal velocity* υ_{th} so that

$$\upsilon_{th}^2 = <\upsilon^2> \tag{6.4.10}$$

where the average is taken over the velocity distribution. Similarly, we can define a *drift velocity*, υ_{drift}, as the average velocity along the field direction i.e.

$$\upsilon_{drift} = <\upsilon_z> \tag{6.4.11}$$

where the z-axis is taken along the field direction and where, again, the average is taken over the electron velocity distribution.

To make a rough calculation of both υ_{th} and υ_{drift} we make the simplifying assumption that, at each collision, some constant fraction δ of the kinetic energy of the electron is lost. A first equation can then be obtained from a power balance consideration: the average power lost by the electron must equal the average power delivered to the electron by the external field. To proceed with this, we note that the average kinetic energy of the electron is $mv_{th}^2/2$ while

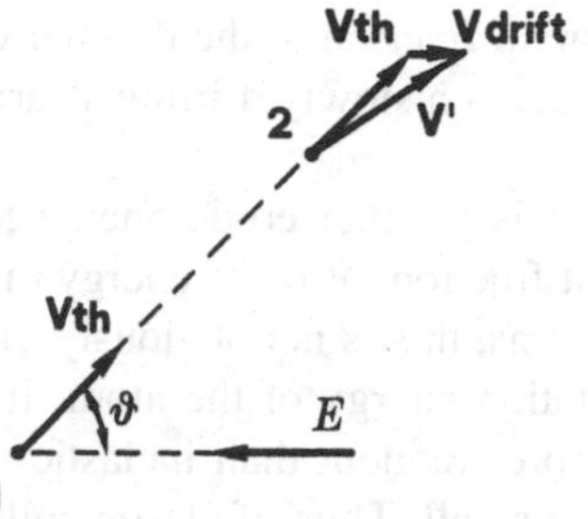

FIG. 6.24. Calculation of the drift velocity resulting from acceleration of an electron by the external electric field in between two consecutive collisions.

υ_{th}/l, where l is the electron mean free path, is the average collision rate. The average power lost by the electron is therefore $\delta(\upsilon_{th}/l)\left(m\upsilon_{th}^2/2\right)$, and this must equal the power supplied by the electric field, $\mathcal{E}$, namely, $e\mathcal{E}\upsilon_{drift}$. Hence

$$e\mathcal{E}\upsilon_{drift} = \delta(\upsilon_{th}/l)(m\upsilon_{th}^2/2) \tag{6.4.12}$$

The second equation is obtained from the requirement of an average momentum balance between two consecutive collisions. We assume that, after each collision, the electron is scattered in a random direction and hence it loses its preferential drift velocity. With reference to Fig. 6.24, the electron velocity at point 1, after the first collision, is thus assumed to have a magnitude equal to the thermal velocity υ_{th} and a direction making a general angle θ to the field direction. During its free flight between points 1 and 2, the electron will be accelerated by the electric field and, at point 2, just before the next collision, it will have acquired an additional velocity, υ_{drift}, along the field direction, with a direction opposite to the field. The impulse produced by the corresponding force will be $-e\mathcal{E}l/\upsilon_{th}$,, where l is the distance between points 1 and 2 (which is assumed, on the average, to be equal to the electron mean free path). This impulse can now be equated to the change of momentum, i.e., $(m\boldsymbol{\upsilon}' - m\boldsymbol{\upsilon}_{th}) = m\boldsymbol{\upsilon}_{drift}$. In terms of their magnitudes we can then write

$$e\mathcal{E}l = m\upsilon_{th}\upsilon_{drift} \tag{6.4.13}$$

which, together with Eq. (6.4.12), provides the two required equations. From these equations we get

$$\upsilon_{th} = (2/\delta)^{1/4}(e\mathcal{E}l/m)^{1/2} \tag{6.4.14}$$

and

$$\upsilon_{drift}(\delta/2)^{1/4}(e\mathcal{E}l/m)^{1/2} \tag{6.4.15}$$

Note that, on taking the ratio between Eqs. (6.4.15) and (6.4.14), we obtain

$$(\upsilon_{drift}/\upsilon_{th}) = (\delta/2)^{1/2} \tag{6.4.16}$$

We have already mentioned earlier that an electron, after undergoing elastic scattering with an atom, loses a fraction of its kinetic energy equal, on the average, to $2\,m/M$. If we then assume $\delta \cong 2\,m/M$, we get from Eq. (6.4.16), $(\upsilon_{drift}/\upsilon_{th}) \cong (m/M)^{1/2} \cong 10^{-2}$. This show that the drift velocity is a very small fraction of the thermal velocity, so that we can consider the movement of electrons in a gas as a slowly drifting swarm of randomly moving particles rather than a stream of particles.

The calculation given above is a rather crude one since it is based on the assumption that the electron loses a constant fraction, δ, of its energy in each collision. Although this is true for elastic collisions with atoms, this is not obviously true for inelastic collisions, where the energy lost equals the excitation energy of the atom. It should be noted that, although elastic collisions are actually more frequent than inelastic collisions, the energy lost in an elastic collision is, however, very small. Thus, if elastic collisions are the dominant process, the discharge would not provide a particularly efficient means for pumping a laser. Indeed, if elastic collisions were the predominant mechanism of electron cooling, most of the discharge energy would be used to heat up rather than to excite the atoms. It should also be noted that electron-electron collision does not play any role in the energy balance equation expressed by Eq. (6.4.12), since this process simply redistributes the electron velocities without changing their average energy.

6.4.3. *Electron Energy Distribution*

We now proceed to a consideration of the distribution of electron velocities or of electron energies in a gas discharge. If the energy redistribution due to electron-electron collisions is fast enough compared to the energy loss due to both elastic and inelastic collision with the atoms, then the prediction of statistical mechanics is that the distribution of electron velocities (or energies) is given by the Maxwell-Boltzmann (MB) distribution function. This can be described, for instance, by the energy distribution function $f(E)$, where $f(E)dE$ is the elemental probability for an electron to have its kinetic energy lying between E and $E + dE$. We then obtain

$$f(E) = \left(\frac{2}{\pi^{1/2}kT_e}\right)\left(\frac{E}{kT_e}\right)^{1/2} \exp -(E/kT_e) \tag{6.4.17}$$

where T_e is the electron temperature. One thus sees that, when the distribution can be described by the MB law, the electron temperature is the only parameter that needs to be specified for characterizing the distribution.

Once T_e is known, one can calculate υ_{th} from Eq. (4.10) using the electron energy distribution given by Eq. (6.4.17). Using the standard relation $\upsilon^2 = 2E/m$, we readily obtain from Eq. (6.4.10)

$$\upsilon_{th} = [3kT_e/m]^{1/2} \tag{6.4.18}$$

which relates υ_{th} to T_e. From Eqs. (6.4.18) and (6.4.14) we then obtain

$$T_e = \left[\left(\frac{2}{\delta}\right)^{1/2}\frac{e}{3k}\right](\mathcal{E}l) \tag{6.4.19}$$

Since the electron mean free path l is inversely proportional to the gas pressure p, Eq. (6.4.19) shows that, for a given gas mixture, T_e is proportional to the ratio $\mathcal{E}/p$. A more detailed treatment than the simple one leading to (6.4.14) shows that T_e is a function of $\mathcal{E}/p$ rather then being simply proportional to this ratio i.e.

$$T_e = f(\mathcal{E}/p) \tag{6.4.20}$$

The $\mathcal{E}/p$ ratio is thus the fundamental quantity involved in establishing a given electron temperature and it is often used in practice for specifying the discharge conditions.

We now address the question as to whether the electron energy distribution can actually be described by MB statistics. Indeed, one obvious reason for the distribution not being Maxwellian is that the MB distribution implies that the velocity distribution in space is isotropic. Actually, if this were the case, the drift velocity, as defined by Eq. (6.4.11), would be zero and hence there could be no current flowing in the discharge! We have seen, however, that the drift velocity is a very small fraction of the thermal velocity and, consequently, the effect of the drift velocity in altering the MB distribution may be considered to be negligible. An important case, however, where MB statistics constitutes only a crude approximation, occurs for a weakly ionized gas with high values for the electron impact cross sections e.g. for CO_2 or CO gas laser mixtures. In this case, in fact, due to the low electron concentration, the energy redistribution process arising from electron-electron collisions does not proceed at a sufficiently fast rate compared to that for inelastic collisions. As we shall discuss in more depth in the following example, one thus expects, in this case, to find dips in the energy distribution function at energies corresponding to specific transitions of the molecules. By contrast, for neutral atom or ion gas lasers, the electron density is much higher because these lasers are relatively inefficient, and, as discussed further in the second example that follows, the departure from a Maxwellian distribution is expected to be less significant.

Example 6.4. *Electron energy distribution in a CO_2 laser* We show in Fig. 6.25 the situation occurring for a CO_2:N_2:He gas mixture with a 1:1:8 ratio between the corresponding partial pressures. In the figure, the electron impact cross section for N_2 excitation up to the $\upsilon = 5$ vibrational level is shown[11] (the main pumping mechanism is, in fact, via energy transfer from an excited N_2 molecule to the lasing CO_2 molecule). As a result of the very high value of the peak cross section ($\approx 3 \times 10^{-16}\,\text{cm}^2$) for the N_2 molecule and, also, as a result of the low value of the current density required in a CO_2 laser (the CO_2 laser is one of the most efficient lasers), the assumption of a Maxwellian distribution is expected to be inadequate, in this case. To calculate the correct electron energy distribution, one then needs to perform an *ab initio* calculation using the appropriate electron transport equation (the Boltzmann transport equation) where all possible electron collision processes leading to excitation (or de-excitation) of the vibrational and electronic levels of all gas species are taken into account.[12] The electron distribution, $f(E)$, computed in this way for an $\mathcal{E}/p$ ratio of $\sim 8\,\text{V cm}^{-1}\,\text{Torr}^{-1}$ and corresponding to an average electron energy* of $\approx 1.7\,\text{eV}$ is indicated as a solid line in the same figure.[13] For comparison, the Maxwellian distribution, $f'(E)$, for the same average energy is also shown as a dashed line. One should note in the figure that the

* Although for a non-Maxwellian distribution the concept of temperature loses its meaning, one can still define an average electron energy and, as in the case of a Maxwellian distribution, this energy turns out to be a function of the $\mathcal{E}/p$ ratio.

depression of the $f(E)$ curve, compared to the Maxwellian curve, for $E > 2$eV is due to the very high value of the electron impact cross section for N_2. In fact, when accelerated by the electric field of the discharge, few electrons go beyond the $E = 2$ eV barrier since they would be immediately involved in N_2 excitation. Consequently, the electrons accumulate in the energy range below 2 eV.

Example 6.5. *Electron energy distribution in a He-Ne laser* In contrast to the results of the previous example, we show in Fig. 6.26 the situation that applies to a helium discharge under conditions appropriate to a He-Ne laser. In the figure, the two plots of the electron impact cross section to the 2^1S and 2^3S levels of He vs electron energy are shown. As in the previous case, in fact, the main pumping mechanism arises from energy transfer between an excited He atom to the Ne lasing atom. Note, however, that the peak values of the cross sections are, in this case, about two orders of magnitude smaller than for the N_2 molecule. Since the current density and hence the electron density are also much higher, the He-Ne laser being a rather inefficient laser, the Maxwellian distribution is expected to hold, in this case. Accordingly, we show in the same figure a Maxwellian distribution with a mean electron energy of 10 eV which is the average electron energy in a He-Ne laser corresponding to the optimum excitation condition [see sect. 6.4.5]. Note the much higher value of the average electron energy in this case compared to the previous case, a consequence of the fact that one needs to excite electronic energy levels rather than vibrational energy levels.

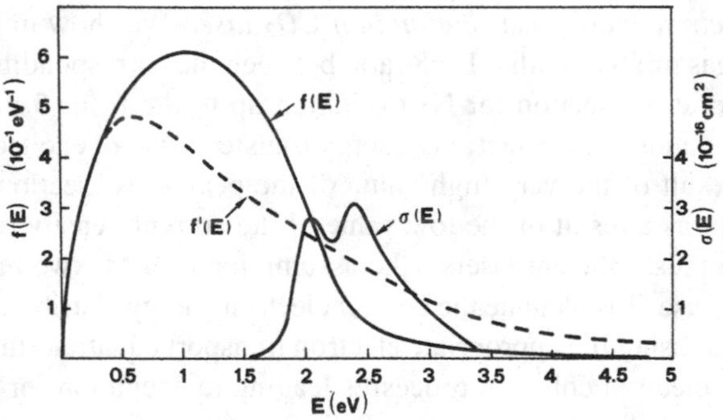

FIG. 6.25. Comparison of the electron energy distribution $f(E)$ for a 1:1:8 CO_2:N_2:He mixture (redrawn from Ref.[13]) with a Maxwellian distribution, $f'(E)$, of the same average energy: In the same figure, the electron impact cross section, $\sigma(E)$, for N_2 excitation up to the $\upsilon = 5$ vibrational level is also shown (redrawn from Ref.[11]). The redrawn curves are indicative of the physical situation rather than representing the actual original values shown in the cited references.

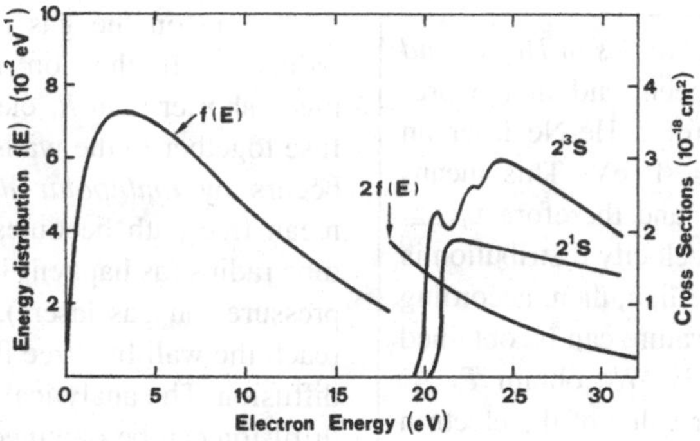

FIG. 6.26. Electron energy distribution, $f(E)$, and electron impact cross-section for the $1\,^1S \rightarrow 2\,^1S$ and $1\,^1S \rightarrow 2\,^3S$ transitions of He.

6.4.4. The Ionization Balance Equation

In an electrical discharge, electrons and ions are being continuously created in the discharge volume by electron impact. Ionization is produced by the hot electrons present in the discharge i.e., those whose energy is larger than the ionization energy of the atom. In the steady state, this ionization process must be counter-balanced by some electron-ion recombination process. Radiationless electron-ion recombination cannot occur within the discharge volume, however, because this process cannot conserve both the total momentum and total energy of the particles. To understand this statement, let us consider, for simplicity, head-on collisions. Upon invoking momentum conservation, the velocity υ of the recombined atom is obtained as $\upsilon = (m_1\upsilon_1 + m_2\upsilon_2)/(m_1 + m_2)$ where m_i $(i = 1, 2)$ are the masses and υ_i the velocities of the electron and ion before collision. On the other hand, for energy conservation, we must require $\left[\left(m_1\upsilon_1^2/2\right) + \left(m_2\upsilon_2^2\right)/2\right] = \left[(m_1 + m_2)\,\upsilon^2/2\right] + E_r$ where E_r is the energy released by the electron-ion recombination. For given values of m_1, m_2, υ_1,and υ_2, the momentum and energy conservation relations thus furnish us with two equations for the one unknown quantity, υ, the velocity of the recombined atom. Thus, in general, these two equations cannot both be satisfied. Radiative ion-electron recombination, on the other hand, is an unlikely process at the carrier concentrations holding for a gas laser. The recombination process can thus only occur in the presence of a third partner, M, since momentum and energy conservation can be conserved in a three-body collision process. In fact, again assuming head-on collisions, one now has a pair of equations in the two unknown υ, the velocity of the recombined atom, and υ_M, the velocity of the third partner, M, after collision. At the low pressures of a gas laser (a few Torr) and if the gas mixture is contained in a cylindrical tube, the necessary third partner M is simply provided by the tube walls. Thus, in a gas laser, electron-ion recombination only occurs at the tube walls.

One now needs to realize that, although the electron velocity is much larger than the ion velocity, the movement of electrons and ions to the walls must occur together. In fact, if electrons were arriving at the walls more rapidly than ions, a radial electric field would be established, which would accelerate the movement of the ions toward the wall and decelerate the electrons. For the usual electron and ion concentrations in a gas discharge, this space charge effect would be quite substantial, consequently electrons and ions move to the tube walls at the same rate. The movement can then occur by two different mechanisms,

Example 6.6. *Thermal and drift velocities in He-Ne and CO_2 lasers* Based on what has been said in the previous example, we will assume for a He-Ne laser an average electron energy $<E> \cong 10\,\text{eV}$. This means that $(mv_{th}^2/2) = <E> = 10\,\text{eV}$ and therefore $v_{th} \cong 1.9 \times 10^6\,\text{m/s}$. Since the electron velocity distribution is assumed, in this case, to be Maxwellian, then, according to Eq. (6.4.18), the electron temperature can be obtained from the relation $T_e = 2<E>/3k$. We obtain $T_e \cong 7.7 \times 10^4\,\text{K}$. Note the much higher value of the electron temperature compared to room temperature. To calculate the drift velocity, we make use of Eq. (6.4.16) and assume that the dominant cooling process for the electrons is via elastic collisions with the lighter He atoms. We then get $(v_{drift}/v_{th}) \approx (m/M_{\text{He}})^{1/2} \cong 1.16 \times 10^{-2}$, where M_{He} is the mass of the Helium, so that $v_{drift} \cong 2.2 \times 10^4\,\text{m/s}$. In the case of a CO_2 laser, based on the findings of example 6.4, we will assume an optimum electron energy value of $<E> \cong 1.7\,\text{eV}$. From the relation $(mv_{th}^2/2) = <E>$ we then get $v_{th} \cong 0.78 \times 10^6\,\text{m/s}$. The drift velocity can then be obtained from ref.[13] assuming an $\mathcal{E}/p$ ratio of $\sim 8\,\text{V cm}^{-1}\,\text{Torr}^{-1}$ and a 1:1:8 partial pressure ratio of the CO_2:N_2:He mixture. We get $v_{drift} \cong 6 \times 10^4\,\text{m/s}$. Note that, in this case, we cannot talk about an electron temperature since the electron energy distribution departs considerably from a Maxwellian distribution. Note also that, in both cases, the thermal velocity is $\sim 10^6\,\text{m/s}$ and the drift velocity is ~ 100 times smaller.

depending on the gas pressure p and tube radius R. If the ion mean free path is much shorter than R, electrons and ions diffuse together to the walls and recombination occurs by *ambipolar diffusion*. If the ion mean free path becomes comparable to the tube radius (as happens in the relatively low-pressure ion gas lasers), electrons and ions reach the wall by "free flight" rather than by diffusion. The analytical theory of ambipolar diffusion can be obtained[14] from the Schottky theory of a discharge in the so called positive column. In the low pressure limit, on the other hand, the *free fall* model of Tonks-Langmuir for the plasma discharge should be used.[15] The two theories are rather complicated and their description goes beyond the scope of this book. In both theories, however, a balance equation must always hold between the number of electron-ion pairs produced and the number of electron-ion pairs recombining at the walls (*ionization balance equation*). So, in the case of Schottky theory, the balance equation can be written, in our notations, as

$$<v\sigma_i>N_g = \frac{kT_e}{e}\mu_+\left(\frac{2.405}{R}\right)^2 \tag{6.4.21}$$

where σ_i is the ionization cross section, μ_+ is the ion mobility and R is the tube radius. One can now see that, for a given atomic species i.e., with a given expression for $\sigma_i = \sigma_i(E)$, the average value $<v\sigma_i>$ appearing in Eq. (6.4.21) will only depend on the electron temperature, T_e. Ionization is in fact produced by the most energetic electrons in the energy distribution and their number depends on the temperature T_e. One can also see that N_g is proportional to the gas pressure while the ion mobility is inversely proportional to it. Equation (6.4.21) can then be rearranged as

$$f(T_e) = C/(pR)^2 \tag{6.4.22}$$

where we have written $<v\sigma_i>/kT_e = f(T_e)$ and where C is a suitable constant. Thus, for a given atomic species, the ionization balance equation leads to a relation between T_e and pR in much the same way as the energy and momentum balance equation leads to a relation between T_e and $\mathcal{E}/p$ [see (6.4.20)]. The functional relation $f = f(T_e)$ in Eq. (6.4.22) is such that T_e increases as pR decreases. In fact, if for a given tube radius one decreases the gas pressure, the electron-ion recombination due to diffusion to the walls increases. The electron temperature has therefore to increase in order to maintain the balance between ionization and recombination. In the case of Tonks-Langmuir theory a similar functional relation again exists between T_e and the pR product.

6.4.5. Scaling Laws for Electrical Discharge Lasers

Equations (6.4.20) and (6.4.22) provide two fundamental relations which can be used to understand a number of aspects of the physical behavior of any gas discharge. For example, we can now explain why, in a stable glow discharge, the voltage across a discharge tube is essentially independent of the current which is flowing (see Fig. 6.19). In fact, if we consider some given gas tube i.e. with given values of tube radius and gas pressure, then, according to Eq. (6.4.22), the electron temperature is fixed. We then see from Eq. (6.4.20) that the electric field must also be fixed and thus independent of the discharge current.

Let us now see the consequences of Eqs. (6.4.20) and (6.4.22) for a gas laser discharge. First we should note that, for a given gas medium, an optimum value of electron temperature, T_{opt}, exists if we want to maximize the pump rate to the upper laser level. Too low a values of the electron temperature, in fact, would result in insufficient electron energy to excite the upper laser level. The electron energy will then be lost mostly through excitation of lower levels of the medium, including the lower laser level. Too high a value of the electron temperature, on the other hand, would lead to strong excitation of higher levels of the gas mixture (which may not be coupled to the upper laser level) or might produce excessive ionization of the gas mixture (which could result in a discharge instability, i.e., a transition from a glow discharge to an arc). If we then set $T_e = T_{opt}$ on the left hand side of both Eqs. (6.4.20) and (6.4.22), we obtain

$$(\mathcal{E}/p) = (\mathcal{E}/p)_{opt} \tag{6.4.23a}$$

$$(pD) = (pD)_{opt} \tag{6.4.23b}$$

Thus, for a given gas mixture, some optimum values exist for both pD and $\mathcal{E}/p$ if the mixture is to be used as the active medium of a gas laser. Equations (6.4.23) establish the *scaling laws* for any gas laser. As an example of applying these laws, let us assume that we start with the best operating conditions and that, for some reason, we want to decrease the tube diameter by e.g. a factor 2. Then, Eq. (6.4.23b) shows that we must increase the pressure of the gas mixture by the same factor if we want the laser to be still operating with optimum efficiency. If the pressure is doubled, then, according to Eq. (6.4.23a) the electric field, $\mathcal{E}$, in the gas discharge and hence the total voltage, V, across the laser tube, must also double. This means that the V versus I characteristic of the given laser tube (see solid line of Fig. 6.19) will scale up by a factor 2 in voltage, at any given current. The open circuit voltage of the power supply, V_0, and the ballast resistance, R_B, must then be designed so as to have the desired current flowing in the discharge tube.

6.4.6. Pump Rate and Pump Efficiency

To calculate the pump rate, we first recall the standard equation $J = e\upsilon_{drift}N_e$ which relates the current density, J, to the electron density, N_e, of a discharge. From Eq. (6.4.7) we then obtain

$$R_p = N_t \frac{J}{e}\left[\frac{<\upsilon\sigma>}{\upsilon_{drift}}\right] \tag{6.4.24}$$

If a Maxwellian electron energy distribution is taken, the term $<\upsilon\sigma>$ in Eq. (6.4.24) will depend only on the electron temperature, T_e. According to Eqs. (6.4.15) and (6.4.19) one readily sees that υ_{drift} also depends only on T_e. For given values of the gas pressure, p, and tube radius, R, the electron temperature remains constant, ideally at the optimum operating value. It then follows that the term in the square brackets of Eq. (6.4.24) is a constant, i.e. independent of the current density and one sees that R_p, in this simple model, increases linearly with the current density. Just as for optical pumping we can now define a pumping efficiency, η_p, as the ratio between the minimum pump power which would ideally be needed to achieve a given pump rate, R_p, and the actual electrical pump power, P_p, to the discharge. We thus write

$$\eta_p = \frac{R_p V_a h\nu_{mp}}{P_p} \tag{6.4.25}$$

where V_a is the volume of the active medium and ν_{mp} is the frequency difference between the ground level and the upper laser level. Note that, to a first approximation, η_p can be taken to be independent of the discharge current density since both R_p and P_p are proportional to J.

It should be noted that the expression for R_p given by Eq. (6.4.24) can only be taken as a qualitative guide to the complex phenomena occurring in gas laser pumping rather than as an accurate quantitative expression for the actual value of the pump rate. As already mentioned earlier, particularly for the most efficient gas lasers, the electron energy distribution is significantly different from a Maxwell-Boltzmann distribution and its calculation requires an *ab initio* treatment of the Boltzmann transport equation with knowledge of all possible electron collision processes leading to excitation (or de-excitation) of the rotational, vibrational, and electronic levels of all the gas species present in the discharge. Furthermore, the number of gas species may be much larger than the number of species initially introduced into the tube. For instance, for a CO_2:N_2:He mixture, various amounts are also found, in the discharge, of CO, O_2, N_2O etc. depending on the complicated plasma-chemical reactions occurring in the volume of the gas and at the electrodes. The calculation therefore gets quite involved, requiring the use of a computer and sometimes proving impractical due to lack of appropriate data on electron collision cross sections for all of the components in the gas mixture.[12] Detailed computer calculation have therefore only been performed for gas mixtures of particular importance such as the CO_2:N_2:He mixture used in high power CO_2 lasers. One (apparent) way to circumvent the problem is to assume that η_p is known or that it can be estimated. In this case we obviously obtain from Eq. (6.4.25)

$$R_p = \eta_p \frac{P_p}{Alh\nu_{mp}} \tag{6.4.26}$$

where A is the cross-sectional area of the active medium and l its length. This is the simple pump rate expression often used in the laser literature and which we ourselves shall use in the following chapters. As in the case of optical pumping, however, the usefulness of Eq. (6.4.26) relies on somebody having already performed the necessary calculations or on reliable estimates for the value of η_p being available.

Example 6.7. *Pumping efficiency in a CO_2 laser* As a particularly relevant example of the calculation of η_p, we show in Fig. 6.27 the computed results both for a 1:2:3 and 1:0.25:3 CO_2:N_2:He gas mixture.[13] The figure gives the percentage of total pump power going into the various excitation channels as a function of either the $\mathcal{E}/p$ ratio or the $\mathcal{E}/N$ ratio, where N is the total concentration of all species in the mixture. Curve I indicates the power going into elastic collisions, excitation of the ground state rotational levels of N_2 and CO_2, and excitation of the lower vibrational levels of the CO_2 molecule. Curves III and IV give, respectively, the power going into electronic excitation and ionization. Curve II gives the power going into excitation of the upper, (001), laser level of the CO_2 molecule and of the first five vibrational levels of N_2. Assuming a very efficient energy transfer between N_2 and CO_2, all this power will be available as useful pump power. Curve II therefore gives the pump efficiency of a CO_2 laser under the stated conditions. Note that, as discussed in sect. 6.4.5 for the electron temperature (which in this case is a meaningless concept, since the electron distribution is far from being Maxwellian), an optimum value of $\mathcal{E}/p$ exists. For too low a value of $\mathcal{E}/p$, the pump power is mostly lost in elastic collisions and excitation of the lower vibrational levels of CO_2. For too high a value of $\mathcal{E}/p$, electronic excitation becomes the dominant excitation channel. Note also that, at the optimum $\mathcal{E}/p$ value, a very high value of η_p can be obtained ($\sim$ 85% for the 1:2:3 mixture).

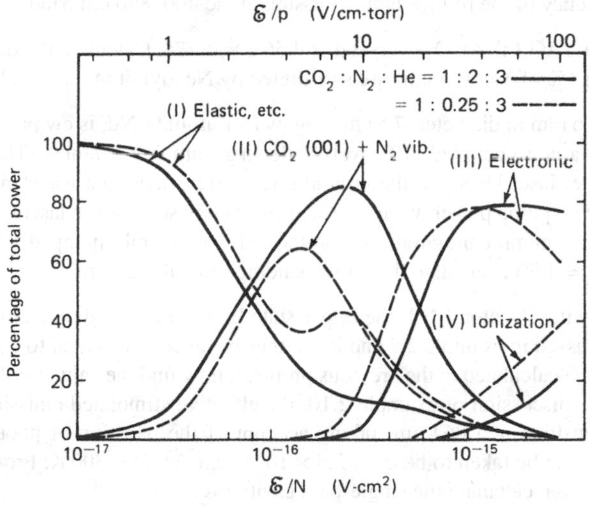

FIG. 6.27. Percentage of total pump powers that goes into the various excitation channels of a CO_2 laser (after ref.,[13] by permission).

6.5. CONCLUSIONS

In this chapter optical pumping and electrical pumping have been considered in particular detail. We have seen that, for both cases, the mechanisms underlying the pumping processes involve a variety of physical phenomena. This has given us the opportunity to acquire a reasonably in-depth knowledge of e.g. plasma emission from a lamp, coherent and incoherent emission of diode lasers used for laser pumping, and physical properties of electrical discharges. The system configurations used are also quite diverse and consideration of the analogies and similarities between these various configurations should help to devise other workable systems. Despite all this diversity, a unified treatment in term of pump efficiency allows effectiveness for laser pumping of the various configurations to be easily compared.

PROBLEMS

6.1. If pump light entering a laser rod is assumed to propagate in a radial direction within the rod, show that the absorption efficiency can be written as

$$\eta_a = \int [1 - \exp -(2\alpha R)] I_{e\lambda}\, d\lambda \Big/ \int I_{e\lambda}\, d\lambda$$

where R is the radius of the rod, α is the absorption coefficient, and $I_{e\lambda}$ is the spectral intensity of the light entering the rod.

6.2. Consider a 6 mm diameter Cr:LISAF laser rod pumped by a 500 Torr Xe flashlamp driven at 2400 A/cm^2 current density. Assume, for simplicity, that the absorption coefficient of Cr:LISAF vs wavelength can be considered to consist of two flat bands, each of which having a peak value of 4 cm^{-1}, the two bands being centered at 420 and 650 nm and having a 80 nm and 120 nm bandwidth respectively. Using the expression for η_a given in the previous problem, calculate the absorption efficiency of the rod for lamp emission in the 400–800 nm band.

6.3. The density of a YAG ($Y_3Al_5O_{12}$) crystal is 4.56 g × cm^{-3}. Calculate the density of Nd ions in the crystal when 1% of Yttrium ions are substituted by Neodymium ions (1 atom.% Nd).

6.4. A Nd:YAG rod, 6 mm in diameter, 7.5 cm long, with 1 atom.% Nd, is cw pumped by a high pressure Kr lamp in a close coupled diffusively reflecting pumping chamber. The energy separation between the upper laser level and the ground level corresponds to a wavelength of 940 nm. The measured threshold pump power when the rod is inserted in some given laser cavity is $P_{th} = 2$ kW. Assuming, for this pump configuration, that the rod is uniformly pumped with an overall pump efficiency of $\eta_p = 4.5\%$, calculate the corresponding critical pump rate.

6.5. For a Nd:YAG with 1% atom. Nd, the upper state lifetime is not significantly quenched by the mechanism discussed in example 2.8 and it can thus be taken to be equal to $\tau = 250$ μs. From the value of pump rate calculated in the previous problem, now find the value for the critical inversion. According to the discussion of example 2.10, the effective stimulated emission cross section for the Nd:YAG transition at 1.064 μm, taking account of the partition of population between the upper sublevels, can be taken to be $\sigma \cong 2.8 \times 10^{-19}$ cm^2 at T = 300 K. From the knowledge of the critical inversion, calculate the single-pass cavity loss.

6.6. The laser of problem 6.4 is to be pumped by sun light. The average day-time intensity of the sun, at the surface of the earth, may be taken to be ~ 1 kW/cm^2. Assume that a suitable optical system is used to allow transverse pumping of the laser rod. Assume also that: (i) 10% of the sun's spectrum is absorbed by the rod. (ii) The pump quantum efficiency is the same as for flashlamp pumping (see Table 6.1). (iii) The transmission of the light focusing optics is 90%. (iv) The pump light distribution within the rod is uniform. Given these assumptions, calculate the required area of the collecting optics to allow the laser to be pumped 2 times above threshold. The focusing system could be made by the, admittedly expensive, combination of two cylindrical lenses, with crossed axes, so as to make a 6 mm × 7, 5 cm image of the sun (i.e. suitable to transversely pump the rod). Knowing that the sun's disc as seen from the earth has a full angle of ~ 9.3 mrad, calculate the focal lengths of the two cylindrical lenses. Could you device a cheaper focusing scheme?

6.7. A Nd:YAG rod of 5 mm diameter, 5 cm long, with 1 atomic % Nd, is pumped by a Xe flashlamp in a close coupled diffusely reflecting pumping chamber. The measured threshold pump energy when the rod is placed in some given laser cavity is $E_{th} = 3.4$ J. Assume that: (i) The overall pumping efficiency is 3.5% (see Table 6.1). (ii) The emitted power from the flashlamp lasts for 100 μs and is constant during this time. Given these assumptions calculate the threshold pump rate R_{cp}. By solving the time-dependent rate equation which includes the effects of both pumping and spontaneous decay, calculate the threshold inversion. If the flash duration is increased to 300 μs, while still remaining constant in time, calculate the new pump rate and pump energy to reach threshold.

6.8. A 1 cm long Ti^{3+}:Al_2O_3 rod is longitudinally pumped by an Argon laser at 514.5 nm wavelength in a configuration similar to that of Fig. 6.11c. The absorption coefficient at the pump wavelength for the rod can, for this case, be taken to be $\alpha_p \cong 2\,cm^{-1}$. The transmission, at the pump wavelength of the cavity mirror, through which the pump beam enters the cavity, can be taken to be $\eta_t = 0.95$. The wavelength corresponding to the minimum pump frequency ν_{mp} (see Fig. 6.17) for Ti:sapphire is $\lambda_{mp} = 616$ nm. Calculate the overall pumping efficiency. If the pump beam is focused to a spot size of $w_p = 50$ μm in the laser rod, if the laser mode spot size is equal to pump spot size and if a single-pass cavity loss $\gamma = 5\%$ is assumed, calculate the optical pump power required from the Ar laser at threshold.

6.9. A 2 mm long Nd:glass rod, made of LHG-5 glass and with a Nd^{3+} concentration of $3.2 \times 10^{20}\,cm^{-3}$, is longitudinally pumped by a single-stripe AlGaAs Quantum Well laser at 803 nm wavelength in a configuration similar to that of Fig. 6.11a. The pump beam is made circular by e.g. the anamorphic system of Fig. 6.12 and focused into the rod to a spot size closely matching the laser mode spot size, $w_0 = 35$ μm. The pump transfer efficiency, including the transmission loss at the first cavity mirror of Fig. 6.11a can be taken to be 80%. Assuming an absorption coefficient at the pump wavelength of $9\,cm^{-1}$, an effective stimulated emission cross section $\sigma_e = 4.1 \times 10^{-20}\,cm^2$, an upper state lifetime of 290 μs and a total loss per pass of 0.35%, calculate the threshold pump power. Note the large difference in pump threshold between this case and that considered in the previous problem and explain the difference.

6.10. A Yb:YAG laser rod, 1.5 mm long, with 6.5 atomic % Yb doping, is longitudinally pumped in a laser configuration such as that of Fig. 6.11a by the output of an InGaAs/GaAs Quantum Well array at 940 nm wavelength focused to a spot size approximately matching the laser mode spot size, $w_0 = 45$ μm. The effective cross sections for stimulated emission and absorption, at the $\lambda = 1.03$ μm lasing wavelength and at room temperature, can be taken to be, respectively, $\sigma_e \cong 1.9 \times 10^{-20}\,cm^2$ and $\sigma_a \cong 0.11 \times 10^{-20}\,cm^2$ while the effective upper state lifetime is $\tau \cong 1.5$ ms. The transmission of the output coupling mirror is 3.5% so that, including other internal losses, the single pass loss may be estimated to be $\gamma \cong 2\%$. Calculate the threshold pump power under the stated conditions.

6.11. A Nd:YAG rod, 4 mm diameter, 6.5 cm long, with 1 atom. % Nd is transversely pumped, at 808 nm wavelength, in e.g. the pump configuration of Fig. 6.15. Assume that 90% of the optical power emitted from the fibers is absorbed, in a uniform way, in the rod. To obtain high power from the laser, an output mirror of 15% transmission is used. Including other internal losses, a loss per single pass of $\gamma = 10\%$ is estimated. If the effective stimulated emission cross section is taken to be $\sigma_e = 2.8 \times 10^{-19}\,\text{cm}^2$, calculate the optical power required from the fibers to reach laser threshold. Compare this value with that obtained in problem 6.9 and explain the difference.

6.12. Assuming a Maxwell-Boltzmann distribution for the electron energy, calculate the electron temperature, in eV, for a gas of electrons with average kinetic energy of 10 eV.

6.13. Suppose that an electron of mass m collides elastically with an atom of mass M. Assuming the atom to be at rest before the collision and that electrons are scattered isotropically, show that, as a result of the collision, the electron loses, on average, a fraction $2\,m/M$ of its energy.

6.14. A pulsed nitrogen laser operating in the ultraviolet ($\lambda = 337.1$ nm) requires an optimum electric field of $\sim 10\,\text{kV/cm}$ at its typical operating pressure of $p \cong 30$ Torr (for a tube cross section of 5×10 mm). A typical length for the nitrogen laser is ~ 1 m. Which of the two pumping configurations shown in Fig. 6.18 would you use for this laser?

6.15. Consider a 1-cm-radius discharge tube filled uniformly with both ions and electrons at a density $N_i = N_e = 10^{13}\,\text{cm}^{-3}$. If all electrons then disappeared leaving behind the positive ion charge, what would be the potential of the tube wall V relative to the tube center? Hence provide an explanation for the phenomenon of ambipolar diffusion.

6.16. Assume that the ionization cross section is a step function starting at an energy equal to the ionization energy E_i and having a constant value σ_i for higher energies. Assuming a Maxwellian distribution for the electron energy, show that the ionization rate can be written as

$$W_i = N_e \sigma_i \left(\frac{8kT_e}{\pi m}\right)^{1/2} \left(1 + \frac{E_i}{kT_e}\right) \exp\left(-\frac{E_i}{kT_e}\right)$$

6.17. The theory of ambipolar diffusion leads to the following relation between the electron temperature T_e and the product pD [compare with (6.4.22)]:

$$\frac{e^x}{x^{1/2}} = 1.2 \times 10^7 (CpD)^2$$

where C is a constant, for a given gas, and $x = (E_i/kT_e)$, where E_i is the ionization energy of the gas. Taking values appropriate for helium, $C = 3.2 \times 10^{-4} (\text{Torr} \times \text{mm})^{-1}$ and $E_i = 24.46$ eV, calculate the required value of pD for an electron temperature of $T_e = 80{,}000$ K.

6.18. The electron mean free path l can be obtained from the relation $l = 1/N\sigma$, where N is the atomic density and σ is the total electron-impact cross section of the atom. Assuming σ to be given by the elastic cross section σ_{el} and taking $\sigma_{el} = 5 \times 10^{-16}\,\text{cm}^2$ for He, calculate υ_{th} and υ_{drift} for an average electron energy of $E = 10$ eV, a He pressure of $p = 1.3$ Torr and temperature $T = 400$ K, and an applied electric field in the discharge of $\mathcal{E} = 30\,\text{V/cm}$.

6.19. A fluorescent lamp consists of a tube filled with about 3 Torr of Ar and a droplet of Hg, which provides a vapor pressure of ~ 3 mTorr at the normal operating temperature of $T = 300$ K. Thus, as far as the discharge parameters are concerned, the tube can be assumed to be filled only with Ar gas. The voltage required across the lamp for a tube length of 1 m is about 74 V. Assuming that the fraction lost by the electrons, per collision, is $\delta = 1.4 \times 10^{-4}$, assuming that the elastic collisions dominate all other collision processes and that $\sigma_{el} = 2 \times 10^{-16}\,\text{cm}^2$, calculate the electron temperature in the discharge.

References

1. P. Laporta, V. Magni, and O. Svelto, Comparative Study of the Optical Pump Efficiency in Solid-State Lasers, *IEEE J: Quantum Electron.* **QE-21**, 1211 (1985).
2. C.S. Willett, *An Introduction to Gas Lasers: Population Inversion Mechanisms*, (Pergamon Press, Oxford, 1974).
3. W. Koechner, *Solid-State Laser Engineering*, Vol. 1, *Springer Series in Optical Sciences*, fourth edition (Springer-Verlag, New York, 1996) Chapter 6, Sect. 6.3.
4. Ref. [3] Chapter 3, Sect. 3.4.
5. P.P. Sorokin and J.R. Lankard, Stimulated Emission Observed by an Organic Dye, Chloro-Aluminum Phtalocyanine, *IBM J. Res. Dev.* **10**, 162 (1966).
6. F.P. Schäfer, F.P.W. Schmidt, and J. Volze, Organic Dye Solution Laser, *Appl. Phys. Lett.* **9**, 306 (1966).
7. J. Berger, D.F. Welch, D.R. Scifres, W. Streifer, and P. S. Cross, High power, High Efficient Neodymium:Yttrium Aluminum Garnet Laser End Pumped by a Laser Diode Array, *Appl. Phys. Lett.* **51**, 1212 (1987).
8. W.A. Clarkson and D.C. Hanna, Two-mirror beam-shaping technique for high-power diode-bars, *Opt. Lett.* **21**, 375 (1996).
9. R.J. Shine, A.J. Alfrey, and R.L. Byer, 40-W cw, TEM_{00}-mode, Diode-Pumped, Nd:YAG Miniature-Slab Laser, *Opt. Lett.* **20**, 459 (1995).
10. D. Golla, M. Bode, S. Knoke, W. Schöne, and A. Tünnermann, 62-W cw TEM_{00} Nd:YAG Laser Side-pumped by Fiber-coupled Diode Lasers, *Opt. Lett.* **21**, 210 (1996).
11. G.J. Shultz, Vibrational Excitation of N_2, CO and H_2 by Electron Impact, *Phys. Rev.* **135A**, 988 (1964)
12. Ref. [2] Sect. 3.2.2.
13. J.J. Lowke, A.V. Phelps, and B.W. Irwin, Predicted Electron Transport Coefficients and Operating Characteristics of CO_2-N_2-He Laser Mixtures, *J. Appl. Phys.* **44**, 4664 (1973).
14. Ref. [2], Sect. 3.2.2.
15. C.C. Davis and T.A. King, Gaseous Ion Lasers, in *Advances in Quantum Electronics*, ed. by D.W. Goodwin (Academic Press, New York, 1975) Vol. 3, pp. 170–437.
16. T.Y. Fan and R.L. Byer, Diode Laser-Pumped Solid-State Lasers, *IEEE J. Quantum Electr.* **QE-24**, pp. 895–912 (1988)
17. T.Y. Fan, Diode-Pumped Solid-State Lasers, in *Laser Sources and Applications*, (SUSSP Publications and IOP Publications, 1996) pp. 163–193.
18. S.J. Hamlin, J.D. Myers, and M.J. Myers, *Proc. SPIE* **1419**,100 (1991).
19. K. Kubodera, K. Otsuka, and S. Miyazawa, Stable $LiNdP_4O_{12}$ Miniature Laser, *Appl. Opt.* **18**, pp. 884–890 (1979).
20. T.Y. Fan and R.L. Byer, Modeling and CW Operation of a Quasi-Three-Level 946 nm Nd:YAG Laser, *IEEE J. Quantum Electr.* **QE-23**, pp. 605–612 (1987).

7

Continuous Wave Laser Behavior

7.1. INTRODUCTION

In previous chapters, we have discussed several features of the components that make up a laser. These are the laser medium itself, whose interaction with an e.m. wave was considered in Chaps. 2 and 3, the passive optical resonator (Chap. 5) and the pumping system (Chap. 6). In this chapter we will make use of results from these earlier chapters to develop the theoretical background required to describe the continuous wave, c.w., laser behavior. The case of transient laser behavior will be considered in the next chapter. The theory developed here uses the so-called rate-equation approximation and the laser equations are derived on the basis of a simple notion that there should be a balance between the total atoms undergoing a transition and total number of photons which are being created or annihilated.[1,2] This theory has the advantage of providing a rather simple and intuitive picture of laser behavior. Furthermore, it gives sufficiently accurate results for most practical purposes. For a more refined treatment one should use either the semiclassical approach (in which the matter is quantized while the e.m. radiation is treated classically, i.e., through Maxwell's equations) or the full quantum electrodynamics approach (in which both matter and radiation are quantized). We refer the reader elsewhere for these more advanced treatments.[3]

7.2. RATE EQUATIONS

We will consider the rate equations for both a four-level laser (Fig 1.4b) and a quasi-three-level laser, where the lower laser level is a sublevel of the ground state. These two categories of laser include, in fact, the most important lasers currently in use. In the category of four level lasers one could mention, for instance: (i) Ionic crystal lasers, such as Neodymium lasers, in various hosts, for most of its many possible transitions, Chromium and Titanium doped lasers [with the exception of Ruby, Cr^{3+}:Al_2O_3, the first laser to operate,

O. Svelto, *Principles of Lasers*,
DOI: 10.1007/978-1-4419-1302-9_7,

which involves a pure three-level scheme]. (ii) Gas lasers, such as CO_2, Ar^+, He-Ne, He-Cd, Cu vapor, HF, N_2. In the category of quasi-three-level lasers one could mention many rare-earth ions in various crystal or glass hosts such as Yb, Er and Yb:Er, Ho, Tm and Tm:Ho, and, for its shortest wavelength transition, Nd again.

7.2.1. Four-Level Laser

We will consider an idealized four level scheme in which we assume that there is only one pump level or band (band 3 of Fig. 7.1) and that the relaxation from the pump band to the upper laser level, 2, as well as the relaxation from the lower laser level, 1, to the ground level proceed very rapidly. The following analysis remains unchanged, however, even if more than one pump band (or level) is involved provided that the decay from these bands to the upper laser level is still very fast. Under these conditions we can make the approximation $N_1 \cong N_3 \cong 0$ for the populations of the lower laser level and pump level(s). Thus we need only deal with two populations, namely the population N_2 of the upper laser level and the population N_g of the ground level. We will assume the laser to be oscillating on only one cavity mode and we let ϕ be the corresponding total number of photons in the cavity.

In a first treatment, we will consider the case of *space independent* rate equation i.e. we will assume that the laser is oscillating on a single mode and that pumping and mode energy densities are uniform within the laser material. As far as the mode energy density is concerned this means that the mode transverse profile must be uniform and that we are neglecting the effects of the standing wave character of the mode. Strictly speaking, the treatment that follows would then only apply for a unidirectional ring resonator with uniform transverse profile and where pumping is uniformly distributed in the active medium, clearly a rather special and simplified case. This case, although perhaps oversimplified, will help us to understand many basic properties of laser behavior. Features arising from space-dependency of both pump and mode patterns, will be discussed at some length later on in this chapter.

For the space-independent case, we can readily write the following three equations:

$$(dN_2/dt) = R_p - B\phi N_2 - (N_2/\tau) \tag{7.2.1a}$$

$$(d\phi/dt) = V_a B\phi N_2 - (\phi/\tau_c) \tag{7.2.1b}$$

In Eq. (7.2.1a) the pumping term R_p [see Eq. (1.3.1)] is based on the assumption of negligible depopulation of the ground state. Explicit expressions for the pumping rate, R_p, have already

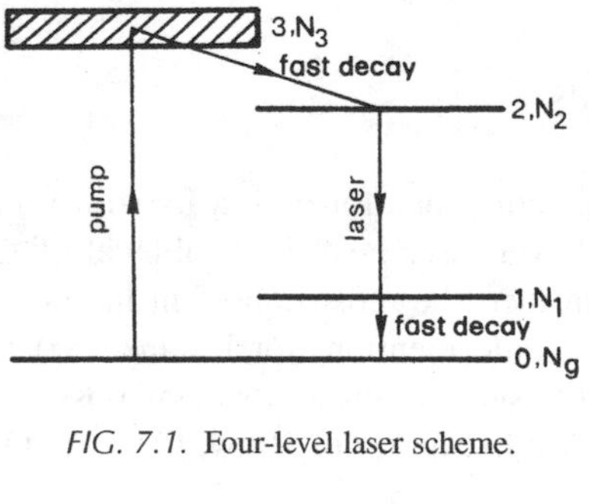

FIG. 7.1. Four-level laser scheme.

been derived in the previous Chapter both for optical and electrical pumping. The term $B\phi N_2$ in Eq. (7.2.1a) accounts for stimulated emission. It was shown in Chaps. 2 and 3 that the stimulated emission rate W is in fact proportional to the square of the magnitude of the electric field of the e.m. wave, and hence W can also be taken to be proportional to ϕ. The coefficient B will therefore be referred to as the stimulated transition rate per photon per mode. The quantity τ is the lifetime of the upper laser level, and, in general must take into account both radiative and non-radiative processes [see Eq. (2.6.18)]. It must also be noted that the upper laser level often consists of a combination of many strongly coupled sublevels. In this case the lifetime τ is intended to be the effective upper level lifetime, account being taken of the lifetimes of all upper state sublevels with a weight proportional to the corresponding sublevel population [see Eq. (2.7.19c)]. In Eq. (7.2.1b) the term $V_a B\phi N_2$ represents the growth rate of photon population due to stimulated emission and is obtained by a simple balance argument. In Eq. (7.2.1a), in fact, the term $B\phi N_2$ gives the rate of population decrease due to stimulated emission. Since each stimulated-emission process creates a photon, $B\phi N_2$ also represents the number of photons created in the unit time and in the unit volume of the medium. The photon growth rate must then be $V_a B\phi N_2$, where V_a is the volume of the mode in the active medium. Finally, the term ϕ/τ_c, where τ_c is the photon lifetime (see Sect. 5.3), accounts for the disappearance of photons due to cavity losses.

Before proceeding it is worth pointing out that, in Eq. (7.2.1b), a term accounting for spontaneous emission has not been included. Since, as already mentioned in Chap. 1, laser action is actually initiated by spontaneous emission, we would not expect Eq. (7.2.1) to account for the onset of laser oscillation. Indeed, if, at time $t = 0$, we put $\phi = 0$ on the right hand side of Eq. (7.2.1b), we get $(d\phi/dt) = 0$ implying that laser action does not start. To try to account for spontaneous emission one might be tempted to apply simple considerations of balance, starting with the term N_2/τ_r, τ_r being the radiative lifetime of level 2, which is included in the term N_2/τ of (7.2.1a). It might then be thought that the appropriate term in Eq. (7.2.1b), to account for spontaneous emission, would be $V_a(N_2/\tau_r)$. This is wrong, however. In fact, as seen in Chaps. 2 and 3, the spontaneously emitted light is distributed over the entire frequency range corresponding to the gain bandwidth and, furthermore, emission occurs into a 4π solid angle. The spontaneous emission term which is needed in Eq. (7.2.1b) must, however, only include the fraction of the spontaneously emitted light that contributes to the given mode (i.e., that is emitted in the same angular direction and in the same spectral bandwidth of the mode). The correct expression for this term can only be obtained by a quantized treatment of radiation-matter interaction. The result is particularly simple and instructive:[(4)] in a quantum electrodynamics treatment, Eq. (7.2.1b) transforms to

$$(d\phi/dt) = V_a B(\phi + 1)N_2 - (\phi/\tau_c) \tag{7.2.2}$$

Thus everything behaves as if there were an *extra photon* in the term describing stimulated emission. When the laser is oscillating, however, and unless very close to threshold, the number of photons in the laser cavity may easily range between $10^{10} \div 10^{16}$ for a cw laser, see example 7.1, and much more than this value for pulsed lasers. In the following analysis we will therefore not consider this extra term arising from spontaneous emission, and instead assume that an arbitrarily small number of photons ϕ_i, say $\phi_i = 1$, is initially present in the cavity just to allow laser action to start.

We are now interested in deriving an explicit expression for the quantity B, the stimulated emission coefficient per photon per mode, which is present in both Eqs. (7.2.1a) and (7.2.1b). So, we consider a resonator of length L in which an active medium of length l and refractive index n is inserted. Since, for the time being, we are considering a traveling wave beam, we let I be the intensity of this beam at a given cavity position and at time $t = 0$. Following the argument considered in Sect. 1.2, the intensity I' after one cavity round trip is $I' = I \times R_1 R_2 (1-L_i)^2 \exp(2\sigma N_2 l)$, where R_1 and R_2 are the power reflectivities of the two mirrors, L_i is the single pass internal loss of the cavity so that $(1-L_i)^2$ is the round trip cavity transmission, and $\exp(2\sigma N_2 l)$ is the round trip gain of the active medium. Note that, if the upper laser level is either degenerate or consists of many strongly coupled sublevels, one should use here the effective value of the cross section, as discussed in Sect. 2.7. We now write $R_1 = 1 - a_1 - T_1$ and $R_2 = 1 - a_2 - T_2$, where T_1 and T_2 are the power transmissions of the two mirrors and a_1 and a_2 are the corresponding fractional mirror losses. The change of intensity, $\Delta I = I' - I$, for a cavity round trip will then be

$$\Delta I = [(1 - a_1 - T_1)(1 - a_2 - T_2)(1 - L_i)^2 \exp(2\sigma N_2 l) - 1]I \qquad (7.2.3)$$

We will now assume that the mirror losses are equal ($a_1 = a_2 = a$) and so small that we can put $(1-a-T_1) \cong (1-a)(1-T_1)$ and $(1-a-T_2) \cong (1-a)(1-T_2)$. Then (7.2.3) obviously transforms to

$$\Delta I = [(1 - T_1)(1 - T_2)(1 - a^2)(1 - L_i)^2 \exp(2\sigma N_2 l) - 1]I \qquad (7.2.4)$$

Before proceeding it is convenient to introduce some new quantities, γ (see Sect. 1.2), which can be described as the logarithmic loss per pass, namely, [compare with (1.2.4)]:

$$\gamma_1 = -\ln(1 - T_1) \qquad (7.2.5)$$

$$\gamma_2 = -\ln(1 - T_2) \qquad (7.2.6)$$

$$\gamma_i = -[\ln(1 - a) + \ln(1 - L_i)] \qquad (7.2.7)$$

As already pointed out in Sect. 1.2, γ_1 and γ_2 are the logarithmic losses per pass due to the mirror transmission and γ_i is the logarithmic internal loss per pass. For brevity, however, we will simply call γ_1 and γ_2 the mirror losses and γ_i the internal loss. The logarithmic loss notation proves to be the most convenient way of representing laser losses, given the exponential character of the laser gain. It should be noted, however, that, for small transmission values, one has $\gamma = -\ln(1-T) \cong T$. Likewise, for very small values of a and L_i one has from Eq. (7.2.7) $\gamma_i \cong a + L_i$, so that the quantities γ really represent loss terms for the cavity. Obviously, one can see that the above approximations only hold for small values of cavity loss or mirror transmission. As an example, if we take $T = 0.1$ we get $\gamma = 0.104$, i.e., $\gamma \cong T$ while if we take $T = 0.5$ we get $\gamma = 0.695$.

With the help of this logarithmic loss notation, we can also define a total (logarithmic) loss per pass γ as

$$\gamma = \gamma_i + [(\gamma_1 + \gamma_2)/2] \qquad (7.2.8)$$

and proceed by substituting Eqs. (7.2.5)–(7.2.8) into Eq. (7.2.4). Upon making the additional assumption

$$[\sigma N_2 l - \gamma] \ll 1 \tag{7.2.9}$$

then the exponential function resulting from Eq. (7.2.4) can be expanded as a power series to yield

$$\Delta I = 2[\sigma N_2 l - \gamma] I \tag{7.2.10}$$

We now divide both sides of Eq. (7.2.10) by the time Δt taken for the light to make one cavity round trip, i.e., $\Delta t = 2L_e/c$, where L_e is the optical length of the resonator, given by

$$L_e = L + (n-1)l \tag{7.2.11}$$

If the approximation $\Delta I/\Delta t \cong dI/dt$ is used, we get

$$\frac{dI}{dt} = \left[\frac{\sigma\, lc}{L_e} N_2 - \frac{\gamma c}{L_e}\right] I \tag{7.2.12}$$

Since the number ϕ of photons in the cavity is proportional to I, a comparison of Eq. (7.2.12) with Eq. (7.2.1b) gives

$$B = \frac{\sigma\, lc}{V_a L_e} = \frac{\sigma c}{V} \tag{7.2.13}$$

$$\tau_c = \frac{L_e}{\gamma\, c} \tag{7.2.14}$$

where

$$V = (L_e/l) V_a \tag{7.2.15}$$

will be referred to as the mode volume within the laser cavity (the mode diameter is taken to be independent of the cavity longitudinal coordinate). Note that Eq. (7.2.14) generalizes the expression for photon lifetime given in Sect. 5.3. Note also that, if the upper laser level is actually made of several strongly coupled sub-levels and if N_2 is the total population of the upper laser level, then, according to the discussion in Sect. 2.7.2, the cross section σ to be used in Eq. (7.2.13) is the effective cross section, i.e., the true cross section times the fraction of the upper state population which is found in the sub-level from which laser action originates.

Once the explicit expressions for B and τ_c are obtained, then Eq. (7.2.1) provides, within the previous limitations and approximations, a description of both static and dynamic behavior of a four level laser. To simplify notations, we will write $N \equiv N_2 - N_1 \cong N_2$ as the population inversion. From Eq. (7.2.1) we then get

$$\frac{dN}{dt} = R_p - B\phi\, N - \frac{N}{\tau} \tag{7.2.16a}$$

$$\frac{d\phi}{dt} = \left[BV_a N - \frac{1}{\tau_c}\right]\phi \tag{7.2.16b}$$

These equations together with the expressions for B, τ_c and V_a, given by Eqs. (7.2.13), (7.2.14) and (7.2.15) respectively, describe, for a four level-laser, the laser behavior both in the c.w. and transient cases.

Before proceeding it is worth pointing out some caveats that apply to our considerations above. One such critical comment has already been made right at the beginning, namely that the results strictly hold only for a laser oscillating with uniform pump and mode energy distributions in the active medium. This important point would seem to indicate a severe limitation to the usefulness of the above equations. However, the results obtained below, using this simple model, will be seen to be very useful for understanding some basic aspects of laser behavior. Furthermore, at least for c.w. behavior, the much more complicated space-dependent case leads, as we shall see, to rather similar results, whose relevance can then better understood by comparison with the space-independent model. A second critical remark relates to the fact that these rate equations strictly apply only for single mode oscillation. For n oscillating modes, in fact, one generally needs to write $2n$ differential equations for the *amplitude and phase* of the field modes to properly account for the beating terms among the various modes. In fact, under appropriate conditions of locking between the phases of the modes, this leads to the phenomenon of mode-locking, to be described in next Chapter and which obviously cannot be described within a rate equation treatment. However, when many modes with random phases are oscillating, the overall beam intensity can, to a first approximation, be taken as the sum of the intensities of all modes. For a uniform transverse pump profile, the superposition of the mode transverse profiles will thus tend to produce a fairly uniform profile for the overall beam. Furthermore, with the oscillation of many modes having different longitudinal patterns, the corresponding total energy density will not show any pronounced standing wave pattern. In this case the picture can be greatly simplified by considering just one rate equation for the total number of photons, ϕ, i.e., summed over all modes, and Eqs. (7.2.16) can still be applied in an approximate fashion. A third critical remark relates to the fact that, in writing e.g. Eq. (7.2.3), we have implicitly assumed that the population inversion, during laser action, is independent of the longitudinal z coordinate. Actually, for large values of laser gain, both counter-propagating beams show strong dependence on the z coordinate and similarly for the inversion. Under such conditions, the laser behavior should be treated on a pass-by-pass basis as first done by Rigrod (the so-called Rigrod analysis[5]). At least for the c.w. case where Eq. (7.2.9) can be taken to hold, however, the expression for the output power as obtained by the Rigrod analysis coincides with that obtained via this far simpler treatment provided that one makes use of the γ notation, as indeed adopted here. For a pulsed laser, on the other hand, Eq. (7.2.9) will hold only when the laser is not driven far above threshold. Otherwise, Eqs. (7.2.16) can no longer be applied and the dynamic behavior of the laser must be analyzed on a pass-by-pass basis, as in the Rigrod analysis.[5] A fourth and perhaps more serious remark is that Eqs. (7.2.16) really do not apply to an inhomogeneously broadened line. To understand this point, let us consider, for simplicity, a non-Doppler inhomogeneously-broadened transition and assume the laser to be oscillating on a single frequency. The beam will then interact only with that fraction of population whose resonance frequency coincides with the laser frequency and, at sufficiently high intensity, the saturated gain profile, likewise what shown in Fig. 2.22 for an absorption profile, will show a hole located at this frequency. Clearly, in this case, the starting point of our analysis on beam amplification, namely Eq. (7.2.3) where N_2 is the *total* upper state population, is no longer valid. The situation is even more complicated for a Doppler-broadened inhomogeneous transition, where, if the laser is oscillating at a frequency sufficiently far from the central frequency of the transition, the right traveling beam and the left traveling beam interact with different sets of atoms or molecules. The c.w. behavior of an inhomogeneously broadened transition has been considered, notably by Casperson,[6] and shown to give results considerably different from those obtained via Eqs. (7.2.16).

Within the limitations discussed in the previous paragraph, we now take Eqs. (7.2.16) as valid for a first order description of laser behavior. Equations (7.2.16) then need to be solved under the appropriate conditions for the case under examination. So, to describe c.w. laser behavior, the case pertinent to this Chapter, we merely set the time derivative on the left hand side of both equations equal to zero. To describe transient laser behavior, we need $R_p = R_p(t)$ to be specified and we also need to know the initial conditions. For instance, if pumping is initiated at $t = 0$, the initial conditions will be $N(0) = 0$ and $\phi(0) = \phi_i$ where ϕ_i is a very small number of initial photons simulating the effect of spontaneous emission (e.g., $\phi_i = 1$). This will be discussed in the next Chapter. For both c.w. and transient laser behavior, however, once ϕ or $\phi(t)$ is known, the calculation of the output power through one of the cavity mirrors become straightforward. In fact, according to Eqs. (7.2.14) and (7.2.8) we can write

> **Example 7.1.** *Calculation of the number of cavity photons in typical c.w. lasers* We will first consider, as a low power example, a 50 cm long He-Ne laser, oscillating at $\lambda = 630$ nm with an output power of 10 mW. The transmission of the output mirror, for this low gain laser, may typically be $T_2 = 1\%$ so that $\gamma_2 = -\ln(1 - T_2) \cong 0.01$. From Eq. (7.2.18) we then get $\phi \cong 1.06 \times 10^{10}$ photons. As a high power example, we consider a 10 kW CO_2 laser oscillating at the wavelength of 10.6 μm. We take a cavity length of $L_e = 150$ cm and an output mirror transmission, for this higher gain laser, of $T_2 = 45\%$. We get $\gamma_2 = -\ln(1 - T_2) \cong 0.598$ and from Eq. (7.2.18) $\phi \cong 0.9 \times 10^{16}$ photons.

$$\frac{1}{\tau_c} = \frac{\gamma_i c}{L_e} + \frac{\gamma_1 c}{2L_e} + \frac{\gamma_2 c}{2L_e} \tag{7.2.17}$$

If we now substitute Eq. (7.2.17) into the right hand side of Eq. (7.2.16b) we recognize that e.g., the term $(\gamma_2 c/2L_e)\phi$ gives the rate of photon loss due to transmission through mirror 2. The output power through this mirror will therefore be given by

$$P_{out} = \left(\frac{\gamma_2 c}{2L_e}\right)(h\nu)\phi \tag{7.2.18}$$

Thus the solution of Eqs. (7.2.16) allows not only to calculate the internal laser behavior but also, through the simple relation Eq. (7.2.18), one of the most important laser parameter i.e., the output power. Viceversa, if the output power is known, Eq. (7.2.18) can be used to calculate the total number of cavity photons as shown in the following example.

7.2.2. Quasi-Three-Level Laser

In a quasi-three-level laser, the lower laser level, level 1 in Fig. 7.2, is a sub-level of the ground level and all ground state sub-levels are assumed to be strongly coupled and hence in thermal equilibrium. Likewise, the upper laser level, level 2 in Fig. 7.2, is a sublevel of a set of upper state sublevels which are also assumed to be in thermal equilibrium. In this case, we let

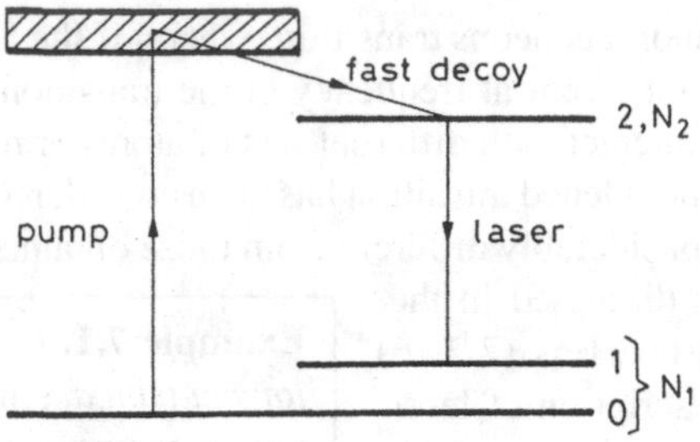

FIG. 7.2. Quasi-three-level laser scheme.

N_1 and N_2 be the total population of all ground state and all upper state sub-levels, respectively. We again assume a very fast decay from the pump level(s) to the upper state sub-levels, so that we will only be concerned with the populations N_1, and N_2 (ideal quasi-three-level case). Now let 0 represent the lowest sublevel of the ground state and assume that the energy separation between sub-levels 1 and 0 is comparable to kT. Then a non-negligible fraction of ground state population, N_1, will be present in the lower laser level (see Sect. 2.7.2), and this will result in absorption of laser photons. Following the discussion of Sect. 2.7.2, the rate equations for both upper and lower state laser sublevels can be written in terms of the total populations N_1 and N_2. The rate equations for a quasi-three-level laser can then be written in a similar way to that for the four-level case, taking account of the fact that absorption as well as stimulated emission of laser photons now occurs. We thus write

$$N_1 + N_2 = N_t \tag{7.2.19a}$$

$$(dN_2/dt) = R_p - \phi(B_e N_2 - B_a N_1) - (N_2/\tau) \tag{7.2.19b}$$

$$(d\phi/dt) = V_a\phi(B_e N_2 - B_a N_1) - (\phi/\tau_c) \tag{7.2.19c}$$

Where: N_t is the total population; τ is, again, the effective lifetime of level 2; B_e and B_a are now given by [compare with Eq. (7.2.13)]

$$B_e = \sigma_e c/V \tag{7.2.20a}$$

$$B_a = \sigma_a c/V \tag{7.2.20b}$$

where σ_e and σ_a are the effective cross sections for stimulated emission and absorption (see Sect. 2.7.2). The substitution of Eqs. (7.2.20) into Eqs. (7.2.19) gives

$$N_1 + N_2 = N_t \tag{7.2.21a}$$

$$\frac{dN_2}{dt} = R_p - \frac{\sigma_e c}{V}\phi(N_2 - f N_1) - \frac{N_2}{\tau} \tag{7.2.21b}$$

$$\frac{dq}{dt} = \left[\frac{V_a\sigma_e c}{V}(N_2 - f N_1) - \frac{1}{\tau_c}\right]\phi \tag{7.2.21c}$$

where we have put

$$f = \sigma_a/\sigma_e \tag{7.2.22}$$

Equations (7.2.21b) and (7.2.21c) suggest that we may now define a population inversion N as

$$N = N_2 - fN_1 \tag{7.2.23}$$

Using the pair of Eqs. (7.2.21a) and (7.2.23) one can obtain N_1 and N_2 in terms of N and N_t. The three Eqs. (7.2.21) can then be reduced to just two equations in the variables ϕ and N. After some straightforward manipulation one obtains

$$\frac{dN}{dt} = R_p(1+f) - \frac{(\sigma_e + \sigma_a)c}{V}\phi N - \frac{f N_t + N}{\tau} \tag{7.2.24a}$$

$$\frac{d\phi}{dt} = \left[\frac{V_a \sigma_e c}{V} N - \frac{1}{\tau_c}\right]\phi \tag{7.2.24b}$$

Within the limits discussed for the validity of Eqs. (7.2.16) and (7.2.24) describe the static and dynamic behavior of a quasi-three-level laser*. In the case of transient laser behavior and if pumping is initiated at $t = 0$, Eqs. (7.2.24) must be solved with the initial conditions $N(0) = -fN_1$ and $\phi(0) = \phi_i \cong 1$. Note that the photon rate equations for four-level, Eq. (7.2.16b), and quasi-three-level, Eq. (7.2.24b), lasers are the same. The rate equation for the population inversion is somewhat different, however. In particular, the stimulated term for a quasi-three-level laser is a factor $(\sigma_e + \sigma_a)/\sigma_e$ larger than that of a four-level laser. To understand this result, consider a unit volume, $V_a = 1$, and assume that, in a given time Δt, one photon has been created, by the stimulated processes, in this volume. According to Eq. (7.2.24b), this implies that $(\sigma_e c_0 N\phi/V)\Delta t = 1$. Using this result in Eq. (7.2.24a) one then sees that, correspondingly, the inversion N decreases by an amount equal to $\Delta N = (\sigma_e + \sigma_a)/\sigma_e$. One thus has $\Delta N > 1$, and this is understood when one notices that, due to this stimulated process, N_2 has decreased by 1 while N_1 has been increased by 1, so that, according to Eq. (7.2.23), the decrease of N must indeed be larger than 1. By contrast, for a four-level laser, the emission of a photon implies that N_2 decreases by 1 while N_1 remains essentially unchanged (i.e., zero) on account of the fast $1 \to 0$ decay. Thus, in this case, the decrease of N is simply equal to 1. Note also that, as expected, Eqs. (7.2.24) reduce to (7.2.16) when σ_a and hence f are set equal to zero.

Within the limits of a space-independent rate equation treatment, Eqs. (7.2.24) thus represent the final result of our calculation for a quasi-three-level level laser. For both c.w. and transient case, once N and ϕ are obtained by solving these equations with the appropriate boundary conditions, the output power through e.g. mirror 2 is again obtained from Eq. (7.2.18).

7.3. THRESHOLD CONDITIONS AND OUTPUT POWER: FOUR-LEVEL LASER

In this section we will investigate, for a four-level laser, the threshold conditions and the output power for a c.w. laser i.e. when R_p is time-independent. We first consider the behavior corresponding to the space-independent rate equations described previously. The results

* One should note that, as a quasi-3-level laser becomes progressively closer to a pure 3-level laser, the assumption that the ground state population is changed negligibly by the pumping process will eventually not be justified and the pump rate R_p could not be taken to be constant any more.

predicted from the space-dependent model will then be discussed and a comparison made between the two models.

7.3.1. Space-Independent Model

In the previous section, the rate equations of a four-level laser have been derived under the simplifying assumption of a very short lifetime of the lower laser level. Before going into a detailed calculation of the c.w. laser behavior under these conditions, it is worth deriving a necessary condition for c.w. oscillation when the lifetime of the lower laser level, τ_1, has a finite value. To do this we first note that the steady-state population of level 1 is of course established by a balance between populations entering and leaving that level. In the absence of oscillation, we thus write $(N_1/\tau_1) = (N_2/\tau_{21})$, where τ_{21} is the lifetime of the $2 \rightarrow 1$ transition. If, for simplicity, we consider the case where the two levels are actually single-levels having the same degeneracy, then, to get laser action, we require $N_2 > N_1$. From the expression above, this implies

$$\tau_1 < \tau_{21} \tag{7.3.1}$$

If this inequality is not satisfied, then laser action is only possible on a pulsed basis provided that the pumping pulse is shorter than or comparable to the lifetime of the upper laser level. Laser action, having begun, will continue until the number of atoms accumulated in the lower level, as a result of stimulated emission, is sufficient to wipe out the population inversion. These lasers are therefore referred to as *self-terminating*. If, on the other hand, Eq. (7.3.1) is satisfied and if R_p is sufficiently strong, then a steady-state oscillation condition will eventually be reached. In what follows we now examine this condition subject to the assumption that $\tau_1 \ll \tau_{21}$ so that Eqs. (7.2.16) can be considered to apply.

We begin by considering the *threshold condition* for laser action. Suppose that, at time $t = 0$, an arbitrarily small number ϕ_i of photons (e.g. $\phi_i = 1$) is present in the cavity due to spontaneous emission. From Eq. (7.2.16b) we then see that, to have $(d\phi/dt) > 0$, one must have $BV_aN > 1/\tau_c$. Laser action therefore initiates when the population inversion N reaches a critical value N_c given by (see also Sect. 1.2)

$$N_c = (1/BV_a\tau_c) = (\gamma/\sigma l) \tag{7.3.2}$$

where use has been made of Eqs. (7.2.13) and (7.2.14). The corresponding critical pump rate R_{cp} is then obtained from Eq. (7.2.16a) by letting $(dN/dt) = 0$, since we are at steady state, with $N = N_c$, and $\phi = 0$. The critical pump rate is then seen to correspond to the situation where the rate of pump transitions, R_{cp}, equals the spontaneous transition rate from level 2, N_c/τ. We thus get

$$R_{cp} = N_c/\tau = (\gamma/\sigma l\tau) \tag{7.3.3}$$

where use has been made of (7.3.2).

If $R_p > R_{cp}$, the photon number ϕ will grow from the initial value determined by spontaneous emission, and, if R_p is independent of time, ϕ will eventually reach some steady-state value ϕ_0. This value and the corresponding steady-state value, N_0, for the inversion are

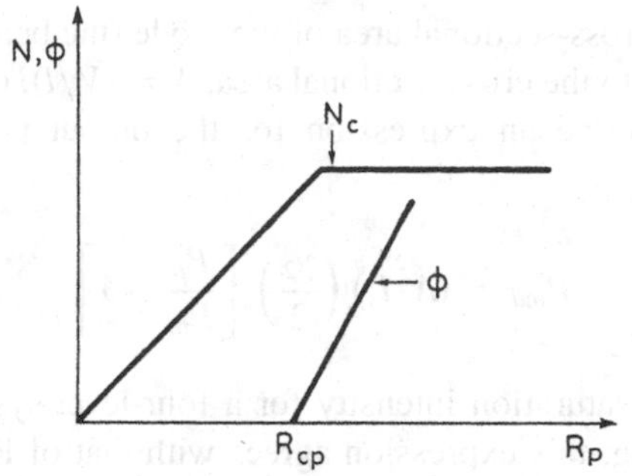

FIG. 7.3. Qualitative behavior of the population inversion N and total number of cavity photons ϕ as a function of the pump rate R_p.

obtained from Eqs. (7.2.16) by setting $(dN/dt) = (d\phi/dt) = 0$. This gives

$$N_0 = (1/BV_a\tau_c) = (\gamma/\sigma l) \quad (7.3.4a)$$

$$\phi_0 = V_a\tau_c[R_p - (N_0/\tau)] \quad (7.3.4b)$$

Note that, to obtain Eq. (7.3.4b), use has been made of Eqs. (7.3.4a), (7.2.15) and (7.2.14).

Equations (7.3.4) describe the c.w. behavior of a four level laser. We will now examine these equations in some detail. Comparing Eq. (7.3.4a) with Eq. (7.3.2) one can first observe that, even when $R_p > R_{cp}$, one has $N_0 = N_c$, i.e., the steady-state inversion always equals the critical or threshold inversion. To get a better understanding of this result, we show in Fig. 7.3 a plot of both N and ϕ vs the pump rate R_p. When $R_p < R_{cp}$, then $\phi_0 = 0$ and the inversion, N, increases linearly with R_p. When $R_p = R_{cp}$, one obviously has $N = N_c$ and $\phi_0 = 0$ still applies. If now R_p is increased above R_{cp}, Eqs. (7.3.4) show that, while N_0 remains fixed at the critical inversion, ϕ_0 increases linearly with R_p. In other words, the pump rate increases the inversion (i.e. the energy stored in the material), below threshold, while it increases the number of photons (i.e., the e.m energy stored in the cavity), above threshold.

We can now recast Eq. (7.3.4b) in a somewhat simpler form if we take the term $(N_0/\tau) = (N_c/\tau) = R_{cp}$ outside the square brackets. We obtain

$$\phi_0 = (V_a N_0)\frac{\tau_c}{\tau}(x - 1) \quad (7.3.5)$$

where

$$x = R_p/R_{cp} \quad (7.3.6)$$

is the amount by which the pump rate exceeds the threshold pump rate. One can now see that, both for optical and electrical pumping, one can write

$$x = P_p/P_{th} \quad (7.3.7)$$

where P_p is the pumping power and P_{th} is its threshold value. Using Eqs. (7.3.7) and (7.3.4a) in Eq. (7.3.5), this equation can be transformed to a somewhat more useful form:

$$\phi_0 = \frac{A_b\gamma}{\sigma}\frac{\tau_c}{\tau}\left[\frac{P_p}{P_{th}} - 1\right] \quad (7.3.8)$$

where $A_b = (V_a/l)$ is the cross-sectional area of the mode (the beam area) which is assumed to be smaller than or equal to the cross-sectional area, $A = (V/l)$, of the active medium.

We now go on to derive an expression for the output power. From Eqs. (7.2.18) and (7.3.8) we obtain

$$P_{out} = (A_b I_s)\left(\frac{\gamma_2}{2}\right)\left[\frac{P_p}{P_{th}} - 1\right] \tag{7.3.9}$$

where $I_s = h\nu/\sigma\tau$ is the saturation intensity for a four-level system [see Eq. (2.8.24)]. If mirror 1 is totally reflecting, this expression agrees with that of Rigrod,[5] obtained using a pass-by-pass analysis. Since a plot of P_{out} versus P_p yields a straight line intercepting the P_p axis at $P_p = P_{th}$, one can define the laser slope efficiency as

$$\eta_s = dP_{out}/dP_p \tag{7.3.10}$$

and η_s turns out to be constant for a given laser configuration. With the help of the previous expressions and of the equations derived in Chap. 6, we can get a very useful and instructive expression for η_s. Thus, inserting P_{out} from Eq. (7.3.9) into Eq. (7.3.10), we begin by writing

$$\eta_s = \frac{A_b h\nu}{\sigma\tau}\frac{\gamma_2}{2}\frac{1}{P_{th}} \tag{7.3.11}$$

For both lamp pumping and electrical pumping, using Eq. (6.2.6) or (6.4.26) in Eq. (7.3.3), we obtain

$$P_{th} = \frac{\gamma}{\eta_p}\left(\frac{h\nu_{mp}}{\tau}\right)\left(\frac{A}{\sigma}\right) \tag{7.3.12}$$

where we recall that ν_{mp} is the frequency difference between the upper laser level and the ground level and A is the area of the active medium. From Eqs. (7.3.11) and (7.3.12) one obtains

$$\eta_s = \eta_p\left(\frac{\gamma_2}{2\gamma}\right)\left(\frac{h\nu}{h\nu_{mp}}\right)\left(\frac{A_b}{A}\right) \tag{7.3.13}$$

One can then write

$$\eta_s = \eta_p\,\eta_c\,\eta_q\,\eta_t \tag{7.3.14}$$

where: (1) η_p is the pump efficiency. (2) $\eta_c = \gamma_2/2\gamma$ represents the fraction of generated photons coupled out of the cavity, which one can call the *output coupling efficiency*. Note that η_c is always smaller than 1 and it reaches the value 1 when $\gamma_1 = \gamma_i = 0$. (3) $\eta_q = h\nu/h\nu_{mp}$ gives the fraction of the minimum pump energy which is transformed into laser energy, referred to as the *laser quantum efficiency*. (4) $\eta_t = A_b/A$, gives the fraction of the active medium cross-section which is utilized by the beam cross section and may be called the transverse utilization factor of the active medium or the *transverse efficiency*. Note that the whole of the active medium is assumed to be pumped uniformly, in our case.

Example 7.2. *CW laser behavior of a lamp pumped high-power Nd:YAG laser* We consider the laser system of Fig. 7.4, where a 6.35 mm diameter, 7.5 cm long Nd:YAG rod, with 1% atomic concentration of the active Nd ions, is pumped in an elliptical pump chamber by a high-pressure Kr lamp. The laser cavity consists of two plane mirrors spaced by 50 cm. The reflectivity of one mirror is $R_1 = 100\%$ while that of the output coupling mirror is $R_2 = 85\%$. A typical curve of the output power, P_{out}, through mirror 2 (in multimode operation) vs electrical pump power, P_p, to the Kr lamp is shown in Fig. 7.5.[7] Note that one is dealing with a reasonably high power c.w. Nd:YAG laser with an output power exceeding 200 W. One may also observe that, since the laser is oscillating on many transverse and longitudinal modes then, according to the discussion in Sect. 7.2.1, it is reasonable to compare the experimental results with the theoretical predictions given by the preceding space-independent rate equation treatment. In fact, except for input powers just above threshold, the experimental points of Fig. 7.5 indeed show a linear relationship between output and input powers as predicted by Eq. (7.3.9). From the linear part of the curve, an extrapolated threshold of $P_{th} = 2.2$ kW is obtained. Above threshold, the output vs input power relation can be fitted by the equation

$$P_{out} = 53[(P_p/P_{th}) - 1] \tag{7.3.15}$$

where P_{out} is expressed in watts. The slope efficiency is then easily obtained from Eq. (7.3.15) as $\eta_s = (dP_{out}/dP_p) = 53/P_{th} = 2.4\%$. Equation (7.3.15) can be readily compared to Eq. (7.3.9) once we remember that, as discussed in example 2.10, the effective values of cross section and upper level lifetime for the $\lambda = 1.06\ \mu$m transition in Nd:YAG can be taken to be $\sigma = 2.8 \times 10^{-19}$ cm^2 and $\tau = 230\ \mu$s, respectively. The energy of the photon, at this wavelength, is obtained as $h\nu = 3.973 \times 10^{-19} \times (0.5/1.06) = 1.87 \times 10^{-19}$ J, where 3.973×10^{-19} J is the energy of a photon with a wavelength of 0.5 μm (see Appendix *H*). We then obtain the value of the saturation intensity as $I_s = h\nu/\sigma\tau = 2.9$ kW/cm^2. We now take $R_2 = (1 - a_2 - T_2) \cong (1 - T_2)$ since, for a good multilayer coating, mirror absorption, a_2, may be less than 0.1%. We then get $\gamma_2 = -\ln R_2 = 0.162$. Comparison of Eq. (7.3.15) with Eq. (7.3.9) then gives $A_b \cong 0.23$ cm^2, to be compared with the cross-sectional area of the rod $A \cong 0.317$ cm^2.

To compare the measured slope efficiency and the extrapolated threshold with the values predicted by calculation, we need to know γ, i.e., γ_i. Now, since $\gamma_1 = 0$, (7.3.12) with the help of (7.2.8) can be rearranged as

$$\frac{-\ln R_2}{2} + \gamma_i = \eta_p \left(\frac{\sigma}{A}\right) \left(\frac{P_{th}\tau}{h\nu_{mp}}\right) \tag{7.3.16}$$

Thus, if several measurements are made of the threshold pump power at different mirror reflectivities R_2, a plot of $\gamma_2 = -\ln R_2$ vs P_{th} should yield a straight line. In fact, this is what is found experimentally, as shown in Fig. 7.6. The intercept of this straight line with the γ_2 axis gives, according to Eq. (7.3.16), the value of the internal losses (Findlay and Clay analysis[9]). From Fig. 7.6 we get $\gamma_i \cong 0.038$, which gives a total loss $\gamma = (\gamma_2/2) + \gamma_i \cong 0.12$.

Once the total losses are known, we can use Eq. (7.3.14) to compare the measured slope efficiency, $\eta_s = 2.4\%$, with the theoretical predictions. We take, in fact, $\eta_c = \gamma_2/2\gamma \cong 0.68$. We also take $\eta_q = \lambda_{mp}/\lambda = 0.89$, where $\lambda_{mp} = 0.94\ \mu$m is the wavelength corresponding to the transition from the upper laser level to the ground level (see Fig. 2.15) in Nd:YAG, and, according to the previous calculation, $\eta_t = A_b/A. \cong 0.72$. From Eq. (7.3.14) we get $\eta_p = 5.5\%$, which appears to be a reasonable value of pump efficiency for Kr pumping (see also Table 6.1). The predicted value of P_{th} can now be readily obtained from Eq. (7.3.12) once we take into account that $h\nu_{mp} \cong 2.11 \times 10^{-19}$ J. We obtain $P_{th} \cong 2.26$ kW, in good agreement with the experimental result. A knowledge of the total losses also allows one to calculate the threshold inversion. From Eq. (7.3.2) one finds $N_c \cong 5.7 \times 10^{16}$ ions/cm^3. For a 1% atomic doping, the total Nd concentration is $N_t = 1.38 \times 10^{20}$ ions/cm^3. Thus $N_c/N_t = 4.1 \times 10^{-4}$, which shows that the population inversion is a very small fraction of the total population.

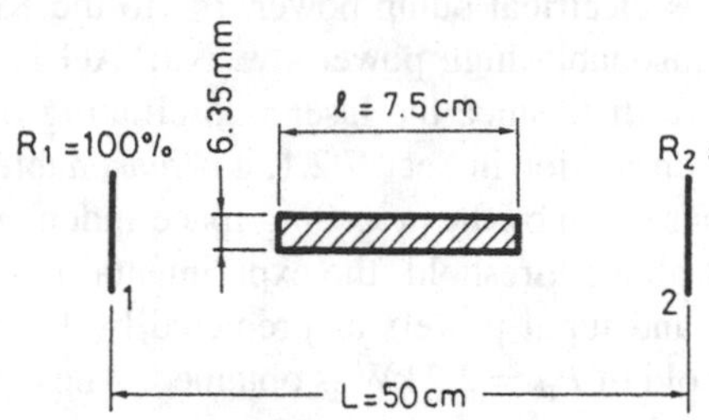

FIG. 7.4. Possible cavity configuration for a, lamp-pumped, cw Nd:YAG Laser.

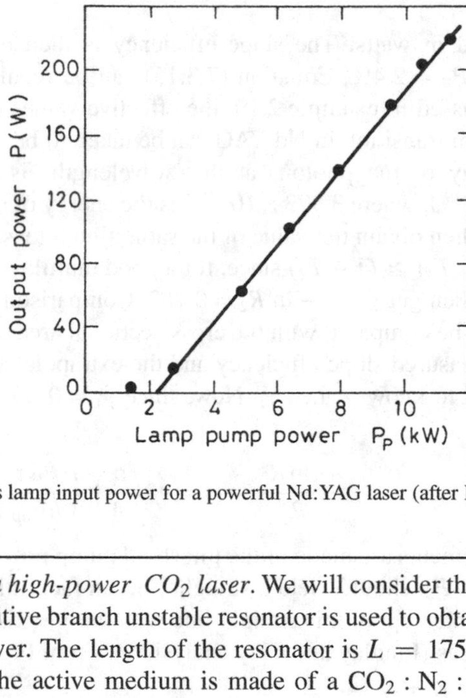

FIG. 7.5. Output power vs lamp input power for a powerful Nd:YAG laser (after Koechner,[7] by permission).

Example 7.3. *CW laser behavior of a high-power CO_2 laser*. We will consider the laser system indicated schematically in Fig. 7.7 where a positive branch unstable resonator is used to obtain a large mode volume and hence high values of output power. The length of the resonator is $L = 175$ cm while the length of the laser medium is $l = 140$ cm. The active medium is made of a $CO_2 : N_2$: He gas mixture with a 1:1:8 partial pressure ratio and with a total pressure of 100 Torr. For cooling reasons, the mixture flows transversely to the resonator axis. Gas excitation is provided by a d.c. electric discharge between two electrodes as indicated schematically in the figure (transverse discharge, see also Fig. 10.16). Typical performance data for the output power, P_{out}, vs input electrical pump power, P_p, are shown in Fig. 7.8.[10] The data points can be fitted by the equation

$$P_{out} = 6.66[(P_p/P_{th}) - 1] \quad (7.3.17)$$

where P_{out} is given in kW and P_{th} is the extrapolated threshold input power ($P_{th} \cong 44$ kW). Note that we are dealing here with a high power CO_2 laser giving an output power which exceeds 10 kW.

At 100 Torr pressure, the CO_2 laser line is predominantly broadened by collisions. From example 3.3, in fact, assuming a gas temperature of $T = 400$ K, we find for this case $\Delta\nu_c \cong 430$ MHz, while Doppler broadening only amounts to $\sim$ 50 MHz (see example 3.2). For the given cavity length, the frequency separation between consecutive longitudinal modes is $\Delta\nu = c/2L = 107$ MHz and, sufficiently far above threshold, a few longitudinal modes are expected to oscillate. Furthermore a few transverse modes are also expected to oscillate. In fact, the equivalent Fresnel number (see Sect. 5.6.2) is rather large in our case ($N_{eq} = 7.4$), so that a few transverse modes are expected to have comparable losses (see also Fig. 5.21). Consequently, the transverse beam profile within the laser cavity is expected to be rather uniform. We are thus dealing with conditions where the previous rate-equation treatment should provide a reasonable approximation and, since the CO_2 laser operates on a four-level scheme, Eq. (7.3.9) can be used for the comparison with Eq. (7.3.17). For this purpose we need to know the transmission T_2 of the output mirror. Since the transverse beam profile is assumed to be rather uniform, we will use the geometrical-optics approximation. One thus finds that T_2, which is equal to the round trip cavity loss of the unstable resonator, is given by [see Eq. (5.6.5)] $T_2 = (M^2 - 1)/M^2 = 0.45$. In the previous expression M is the round-trip magnification factor of the resonator and is given by $M = R_1/|R_2| = 1.35$, where R_1 and R_2 are the radii of the two mirrors ($R_2 < 0$, since mirror 2 is a concave mirror). A comparison of Eq. (7.3.17) with Eq. (7.3.9), using $\gamma_2 = -\ln(1 - T_2) \cong 0.6$, then yields $A_b I_s = 22.3$ kW. The beam diameter in the laser cavity is (see also Fig. 5.18b) $D = 2Ma_2 = 7.6$ cm, where $2a_2 = 5.7$ cm is the diameter of the output coupling mirror (see Fig. 7.7). One thus gets $A_b = \pi D^2/4 \cong 45\ \text{cm}^2$ and hence $I_s \cong 500\ \text{W/cm}^2$. This value is in agreement with the best theoretical estimates of the saturation intensity for a CO_2 laser of this kind.[11]

From the data of Fig. 7.8 we can now go on to evaluate the unsaturated (i.e. when laser action is prevented) gain coefficient g expected for the laser medium at an input power $P \cong 140$ kW. In fact we have

$$g = N_2\,\sigma = \frac{P_p}{P_{th}} N_{20}\,\sigma = \frac{P_p}{P_{th}}\frac{\gamma}{l} \tag{7.3.18}$$

where N_2 and N_{20} are the, unsaturated, upper state populations at $P_p = 140$ kW and $P_p = P_{th}$, respectively. Note that the expression Eq. (7.3.4) has been used for N_{20} so that the saturated gain coefficient, $g_0 = N_{20}\sigma$, turns out to be simply given by $g_0 = \gamma/l$. To calculate either g or g_0, we thus need to know the single-pass cavity loss γ. So, we assume mirror absorption and scattering losses of 2%. In fact, for this high power laser oscillating at the 10.6 μm wavelength, polished, water cooled Copper mirrors are used, which have substantially higher losses than multilayer dielectric mirrors. We then have $\gamma_i \cong 0.02$ and, since $\gamma_1 = 0$ and $\gamma_2 = 0.6$, we obtain $\gamma \cong 0.32$. Substitution of this last value into Eq. (7.3.18) gives $g = 6.3 \times 10^{-3}\ \text{cm}^{-1}$. The unsaturated gain coefficient can easily be obtained experimentally by measuring the gain coefficient of the laser medium with both mirrors removed. The measured values of gain coefficient, for this type of laser, are in fairly good agreement with the values calculated here.[12]

We now compare the experimental value of slope efficiency of Fig. 7.8 with the theoretical predictions. We will assume $\eta_p \cong 0.8$ (see Fig. 6.28), $\eta_c = \gamma_2/2\gamma = 0.94$, $\eta_t \cong 1$ and $\eta_q = h\nu/h\nu_{mp} = 0.4$ (see the CO_2 laser energy levels in Chap. 10). From Eq. (7.3.14) we then obtain $\eta_s = 0.3$, which is appreciably higher than the experimental value obtained from Fig. 7.8 ($\eta_s \cong 0.21$). This discrepancy can be attributed to at least two separate causes: (1) The transverse utilization factor η_t may be appreciably smaller than 1. Perhaps by coincidence, if we were using the same value of η_t found in the previous problem, $\eta_t = 0.73$, the theoretical result would be in almost exact agreement with the experimental one. (2) The data of Fig. 7.7 refer to a partially closed-cycle system and, in this case, the products of the electric discharge (mostly CO and O_2) are likely to accumulate in the gas mixture, thus reducing the pumping efficiency below the theoretical value of 80%. Actually, it is a matter of fact that slope efficiencies larger than $\sim$ 20% are seldom found in practice for any CO_2 laser. The discussion presented above thus helps ones understanding of how the slope efficiency is further reduced from the already reduced value established by the quantum efficiency ($\eta_q = 40\%$).

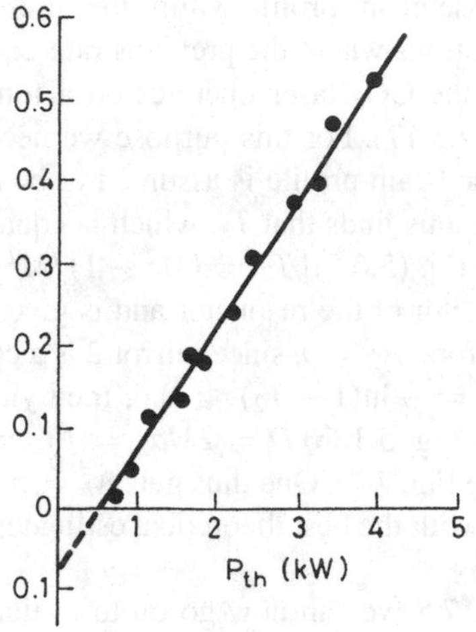

FIG. 7.6. Threshold pump power as a function of mirror reflectivity (after Koechner,[8] by permission).

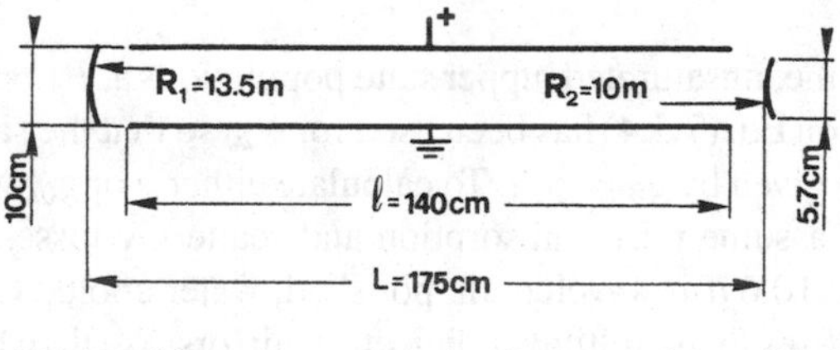

FIG. 7.7. Possible cavity configuration for a powerful cw CO_2 laser.

7.3.2. Space-Dependent Model

We shall consider now the case where the mode distribution and the pump rate are spatially dependent. In this case the inversion will also be spatially dependent and the rate equation treatment becomes more complicated. We will therefore limit ourselves, here, to a discussion of the most relevant results and refer to Appendix E for a detailed treatment. We assume a cylindrical symmetry and let u be the field amplitude of the given mode,

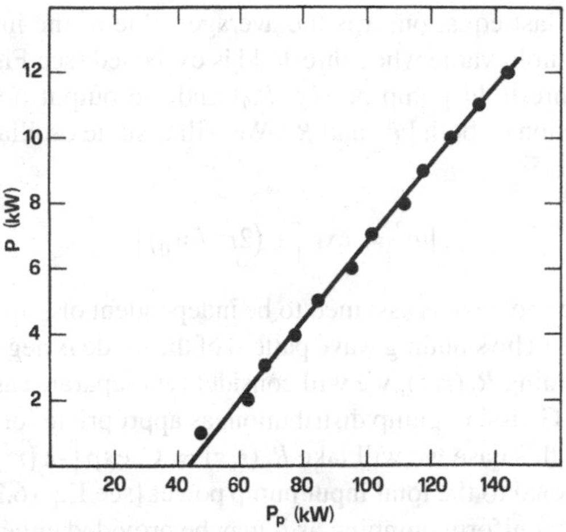

FIG. 7.8. Output power, P, versus electrical discharge power, P_p, for a powerful cw CO_2 laser.

normalized to its peak value. For simplicity, we will take u to be independent of the longitudinal coordinate, z, along the resonator while we will generally take the pump rate to be dependent on both radial and longitudinal coordinates, i.e. $R_p = R_p(r, z)$.

As far as the threshold conditions are concerned, it will be shown in the Appendix E that Eq. (7.3.2) still holds for the average value of N i.e.,

$$\langle N \rangle_c = (\gamma/\sigma l) \tag{7.3.19}$$

where the average is taken over the squared amplitude of the field distribution, viz [see also Eq. (6.3.17)]

$$\langle N \rangle = \left(\int_a N|u|^2 dV \right) \Big/ \int_a |u|^2 dV \tag{7.3.20}$$

and the integrals are taken over the volume of the active medium. At each point of the active medium, below or at threshold, an equilibrium must exist between the number of atoms raised by the pumping process and those decaying spontaneously i.e. $R_p(r, z) = N(r, z)/\tau$. At threshold, we then have

$$\langle R_p \rangle_c = \frac{\langle N \rangle_c}{\tau} = \frac{\gamma}{\sigma l \tau} \tag{7.3.21}$$

where $\langle R_p \rangle$ is the average of $R_p(r, z)$ over the squared amplitude of the field distribution [see Eq. (6.3.8)] and where Eq. (7.3.19) has been used.

Above threshold, from the condition $d\phi/dt = 0$, one now finds that the average gain must equal total losses i.e.

$$\sigma l \langle N \rangle_0 = \gamma = \sigma l \langle N \rangle_c \tag{7.3.22}$$

Thus, according to this last equation, it is the average value of the inversion, $<N>_0$, which gets clamped at its threshold value when threshold is exceeded (see Fig. 7.3).

To calculate the threshold pump power, P_{th}, and the output power, P_{out}, we need to specify the spatial variation of both $|u|^2$ and R_p. We will assume oscillation on a TEM_{00} mode and take

$$|u|^2 = \exp\left[-\left(2r^2 / w_0^2\right)\right] \tag{7.3.23}$$

This means that: (i) The spot size is assumed to be independent of z and equal to the spot size, w_0, at the beam waist. (ii) The standing wave pattern of the mode is neglected [see, for comparison, Eq. (6.3.9)]. Regarding $R_p(r, z)$, we will consider two separate cases: (i) Uniform pumping, i.e. $R_p = \text{cons.}$ (ii) Gaussian pump distribution, as appropriate for longitudinal pumping, e.g., by diode-lasers. In this case we will take $R_p(r, z) = \mathrm{C}\ \exp\left[-2\left(r^2/w_p^2\right)\right]\exp\text{-}(\alpha z)$, where C is a constant proportional to the total input pump power [see Eq. (6.3.7)].

Let us first consider uniform pumping as it may be provided either by electrical or lamp-pumping. Then, from either Eq. (6.2.6) or (6.4.26) we obtain

$$R_p = \eta_p \frac{P_p}{\pi\ a^2 l\ h\nu_{mp}} \tag{7.3.24}$$

where a cylindrical medium of radius a has been considered. We will now consider the cladded rod geometry (see Sect. 6.3.3), where the active species is assumed to be confined to the central region of the rod, $0 \le r \le a$, while the rod is undoped for $r > a$. In this case we don't have to be concerned with the effects of beam truncation due to the finite aperture of the medium. Thus Eq. (7.3.23) can be taken to hold for $0 \le r \le \infty$ while one has $R_p = \text{cons.}$ for $0 \le r \le a$ and $R_p = 0$ for $r > 0$. One can now substitute Eq. (7.3.24) into Eq. (7.3.21) and then use Eq. (7.3.23) to calculate the average value of R_p. One obtains, in this way, an expression for P_{th}, which is the same as that in (6.3.22) provided one replaces η_{pl} there by η_p. The calculation of the output power then proceeds as discussed in Appendix *E* and we limit ourselves to quoting and discussing the final result. So, we define a normalized pump power, x, as

$$x = P_p/P_{mth} \tag{7.3.25}$$

where P_{mth} is the minimum threshold which occurs when $w_0 \ll a$ and, according to (6.3.22), is given by

$$P_{mth} = \left(\frac{\gamma}{\eta_p}\right)\left(\frac{h\nu_{mp}}{\tau}\right)\left(\frac{\pi a^2}{\sigma_e}\right) \tag{7.3.26}$$

We also define a normalized value of the output power y as

$$y = P_{out}/P_s \tag{7.3.27}$$

where P_s is a saturation power given by

$$P_s = \frac{\gamma_2}{2}\frac{\pi w_0^2}{2} I_s \tag{7.3.28}$$

The resulting relation between x and y is then

$$x = \frac{y}{\ln\left(\frac{1+y}{1+\beta y}\right)} \tag{7.3.29}$$

where

$$\beta = \exp\left[-\left(2a^2 / w_0^2\right)\right] \tag{7.3.30}$$

We see that the relation between normalized output power, y, and amount by which threshold is exceeded, x, is a little complicated and quite different from the simple one predicted by the space-independent rate-equations [see Eq. (7.3.9)]*. For comparison, we have plotted as solid lines in Fig. 7.9 the normalized output power, y, vs the normalized pump power, x, for $w_0 \ll a$, $w_0 = 0.7a$, and $w_0 = \sqrt{2}a$. One can see that, particularly when $w_0 \ll a$, the relation between y and x is no longer linear, with the derivative, dy/dx, increasing with x. To understand this behavior it is appropriate to calculate the slope efficiency $\eta_s = dP_{out}/dP_p = (P_s/P_{mth})(dy/dx)$, where Eqs. (7.3.25) and (7.3.27) have been used. With the help of Eqs. (7.3.26) and (7.3.28) one readily sees that η_s can be expressed again as in Eq. (7.3.14), with $\eta_q = h\nu/h\nu_{mp}$, provided that the transverse efficiency is now defined as

$$\eta_t = \left[\frac{(\pi w_0^2/2)}{\pi a^2}\frac{dy}{dx}\right] \tag{7.3.31}$$

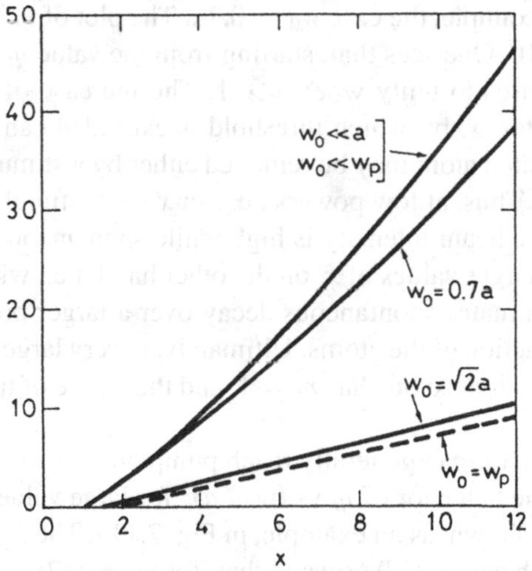

FIG. 7.9. Normalized output power, y, vs normalized pump power, x, for a laser oscillating on a TEM_{00} mode. The *continuous curves* refer to the case of uniform pumping, in a rod of radius a, at several values of the mode spot size w_0. The *dashed curve* refers to the case of a Gaussian distribution of pump light with a spot size w_p such that $w_0 = w_p$.

* If we take $P_s = A_b I_s \gamma_2/2$ in (7.3.9), this equation simply gives $y = (x-1)$, where y and x are again given by (7.3.27) and (7.3.25).

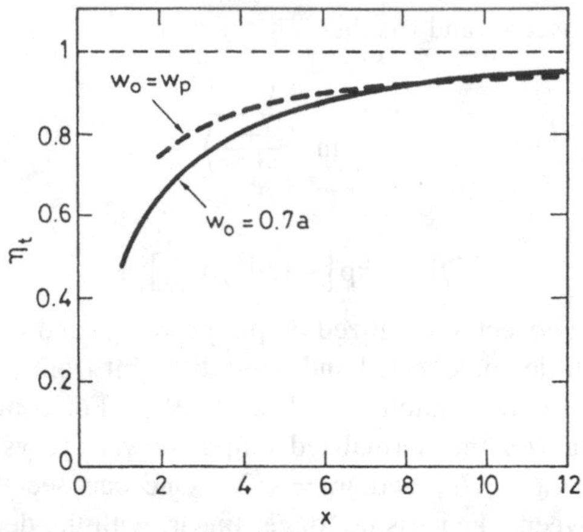

FIG. 7.10. Plot of the transverse efficiency, η_t, versus normalized pump power, x. The continuous and dashed curves refer to the case of uniform pumping and Gaussian-beam pumping, respectively.

In the case of transverse pumping by diode lasers and again for uniform pumping, one obtains the same expression for the slope efficiency with the only difference that now $\eta_q = h\nu/h\nu_p$. Note that when $\beta \to 0$ i.e. when $w_0 \ll a$, one has $(dy/dx) = 2$ for $y \to 0 (x \to 1)$, and the transverse efficiency becomes $\eta_t = \left(\pi w_0^2/\pi a^2\right)$. It should also be noted that, since (dy/dx) increases with x, the transverse efficiency will also increase with x. To understand this point we will consider, as an example, the case $w_0 = 0.7a$. The plot of η_t vs x for this case is shown as a solid line in Fig. 7.10. One sees that, starting from the value $\eta_t \cong 0.97(w_0/a)^2 \cong 0.473$, at low powers, η_t increases to unity when $x \gg 1$. The increase of transverse efficiency, η_t, with increase in the factor, x, by which threshold is exceeded can be understood by noting that the energy of an excited atom may be removed either by a stimulated emission process or by a spontaneous decay. Thus, at low powers i.e., for $x \to 1$, stimulated emission will prevail near the beam axis where beam intensity is high while spontaneous decay will prevail in the wings of the beam. For larger values of x, on the other hand, i.e., with increased beam power, stimulated emission dominates spontaneous decay over a larger portion of the pump profile and hence for a larger fraction of the atoms. Ultimately, at very large powers, all excited atoms decay by stimulated emission, so one has $\eta_t = 1$, and the whole of the pump profile is utilized by the laser beam.[6]

For diode pumping, one can generally reach pump powers well above threshold and it is instructive to consider the behavior of η_t vs $(w_0 / a)^2$ for large values of x. The behavior of η_t vs $(w_0 / a)^2$ at $x = 10$ is shown, as an example, in Fig. 7.11a. The figure shows that, to get e.g. $\eta_t > 90\%$, one needs to have $w_0 > 0.66a$ and that, for $w_0 = 0.7a$, one obtains the rather high value of 94%. However, for the usual case of a medium without cladding, it is generally not beneficial to increase the spot size even further since this would lead to excessive diffraction losses arising from the rod aperture.

For a Gaussian distribution of the pump beam, the calculation, also presented in Appendix *E*, proceeds in a similar way. We again define a normalized output power, y, as in Eq. (7.3.27), where P_s is again given by Eq. (7.3.28), and a normalized pump power, x, as in Eq. (7.3.25) where now P_{mth} is the minimum threshold for Gaussian-beam pumping and is

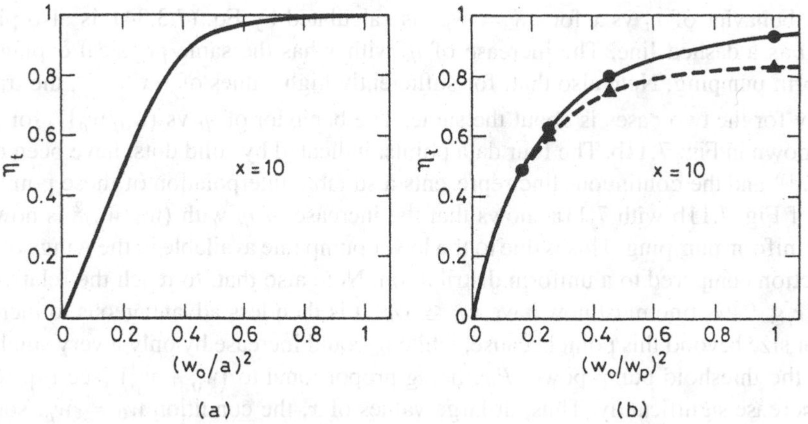

FIG. 7.11. (a) Plot of the transverse efficiency, η_t, versus $(w_0/a)^2$, at a normalized pump power $x = 10$, in the case of uniform pumping (w_0 is the mode spot-size and a is the rod radius). (b) Plot of the transverse efficiency, η_t, versus $(w_0/w_p)^2$, at a normalized pump power $x = 10$, in the case of Gaussian-beam pumping (w_0 is the mode spot-size and w_p is the spot-size of the pump distribution). The *closed circles* refer to a four-level laser while the *closed triangles* refer to a quasi-three-level laser with $B = \sigma_a N_t l/\gamma = 1$.

obtained from Eq. (6.3.20) when $w_0 \ll w_p$. We get

$$P_{mth} = \left(\frac{\gamma}{\eta_p}\right)\left(\frac{h\nu_p}{\tau}\right)\left(\frac{\pi w_p^2}{2\sigma_e}\right) \tag{7.3.32}$$

The relation between y and x then turns out to be given by the following expression

$$\frac{1}{x} = \int_0^1 \frac{t^\delta\, dt}{1 + yt} \tag{7.3.33}$$

where $\delta = (w_0/w_p)^2$. This equation, while differing in notation, agrees with that originally given by Moulton.[13] For $w_0 \ll w_p$, one has $\delta \to 0$ and Eq. (7.3.33) gives the same result as that obtainable from Eq. (7.3.29) when $w_0 \ll a (\beta \to 0)$. Thus the plot of y vs x for this case is the same as that for uniform pumping (see Fig. 7.9). For very small values of spot size, in fact, the beam sees no distinction between uniform or Gaussian pump distributions. For $w_0 = w_p$, one has $\delta = 1$ and the integration of Eq. (7.3.33) gives

$$x = \frac{y}{\left[1 - \frac{\ln(1+y)}{y}\right]} \tag{7.3.34}$$

This expression is also plotted in Fig. 7.9 as a dashed line. We again can calculate the slope efficiency as $\eta_s = dP_{out}/dP_p = (P_s/P_{mth})(dy/dx)$ and we again find that η_s can be expressed as in Eq. (7.3.14) where, as for transverse diode pumping $\eta_q = h\nu/h\nu_p$, and where now

$$\eta_t = \frac{\pi w_0^2}{\pi w_p^2}\frac{dy}{dx} \tag{7.3.35}$$

The behavior of η_t vs x for $w_0 = w_p$, as calculated by Eq. (7.3.34), is also plotted in Fig. 7.10 as a dashed line. The increase of η_t with x has the same physical explanation as for uniform pumping. Note also that, for sufficiently high values of $x (x > 7)$, the transverse efficiency for the two cases is about the same. The behavior of η_t vs $(w_0/w_p)^2$, for $x = 10$, is then shown in Fig. 7.11b. The four data points, indicated by solid dots, have been obtained from ref.[13] and the continuous line represents a suitable interpolation of these points. Comparison of Fig. 7.11b with 7.11a shows that the increase of η_t with $(w_0/w_p)^2$ is now slower than for uniform pumping. This is due to the lower pump rate available in the wings of a Gaussian function compared to a uniform distribution. Note also that, to reach the relatively high value of e.g. 94%, one must now have $w_0 \cong w_p$. It is then less advantageous to increase the laser spot size beyond this point because, while η_t could increase by only a very small further amount, the threshold pump power P_{th}, being proportional to $(w_0^2 + w_p^2)$ [see Eq. (6.3.20)], would increase significantly. Thus, at large values of x, the condition $w_0 = w_p$, sometimes also called the mode-matching condition, may be taken as more or less the optimum situation. The dashed line in Fig. 7.9 then gives the output power, at a given pump power, for this case. It is perhaps worth to noticing that, by comparing this curve with that shown in the same figure for $w_0 = 0.7a$, one may get the mistaken impression that, for a given value of x and for the same spot size w_0, the output power available with Gaussian-beam pumping is smaller than that with uniform pumping. The two cases of Gaussian and uniform pumping should be compared, however, not only with the same beam spot size but also with the same pumped area i.e., with $(\pi w_p^2/2) = \pi a^2$. If we now consider, for Gaussian-beam pumping, the case $w_0 = w_p$ and if we further assume equal values of mode spot-size for the two cases, one then has $(\pi w_0^2/2) = \pi a^2$. It follows that, for uniform pumping, the curve to be considered for the comparison is that for which $w_0 = \sqrt{2}a$. This curve is also shown in Fig. 7.9 and one sees that the dashed (Gaussian pumping) and continuous (uniform pumping) curves are now almost coincident. Note that the curve $w_0 = \sqrt{2}a$ is only shown for the sake of comparison because, for an actual active medium without any cladding, this situation would produce excessive diffraction losses due to the aperture of the active medium.

So far, the standing wave character of a mode has been neglected i.e., $|u|^2$ has been written as in Eq. (7.3.23) rather than as in (6.3.9). This would be correct for a unidirectional ring resonator (see Fig. 5.4a) while, for most other cases e.g., using a two-mirror resonator, a well defined standing wave pattern is formed when the laser is oscillating on a single longitudinal mode*. The effect on the output power of a standing wave pattern, for a mode with uniform profile, has been considered by Casperson.[6] As far as the slope efficiency is concerned, the results obtained can be represented in terms of a fifth efficiency factor, to be introduced in the right hand side of Eq. (7.3.14), which one can refer to as the longitudinal utilization factor of the pump distribution, η_l, or *longitudinal efficiency*. The value of η_l is $\eta_l = (2/3) = 0.666$ at threshold and it increases to e.g. $\eta_l = (8/9) = 0.89$ when ten times above threshold. The physical origin of η_l is similar to that discussed for η_t, namely that, at threshold, only atoms around the peaks of the standing wave decay predominantly by stimulated emission, while, for atoms near the zeros of the pattern, spontaneous decay prevails. At increasing values of

* A notable exception occurs for the twisted-mode technique,[14] where the two oppositely traveling beams in the active medium consist of a right and left circularly polarized waves, respectively, and no standing wave pattern is thus produced.

x i.e., at increasing energy densities, more atoms around the field nodes will then undergo stimulated rather than spontaneous decay and the longitudinal efficiency therefore increases.

As a conclusion to this section we can say that, when space dependence is taken into account, the problem becomes somewhat more complicated. The expressions for threshold inversion and threshold pump rate, however, remain identical to those obtained in a space-independent treatment provided that appropriate average values $<N>_c$ and $<R_p>_c$ are used. Note that, as shown in Sect. 6.3.3, this result also holds when the standing wave pattern of the mode is taken into account. The expression for the output power as a function of the amount by which threshold is exceeded, becomes more complicated. In terms of slope efficiency, however, the results are very simple and suggestive and can be directly related to those obtained for the space-independent case.

Example 7.4. *Threshold and Output Powers in a Longitudinally Diode-Pumped Nd:YAG Laser* As a representative example of longitudinal diode pumping we will consider the laser configuration of Fig. 7.12 where a 1 cm long Nd:YAG rod is pumped by a 100 μm wide laser array at 805–808 nm wavelength.[15] The coupling optics consists of a 6.5 mm focal length, 0.615 *N.A.*, collecting lens, a $\times 4$ anamorphic prism pair, and a 25 mm lens to focus the pump light into the rod (see Fig. 6.12). The Nd:YAG resonant cavity is formed by a plane mirror directly coated on one face of the rod and a 10 cm-radius, 95% reflecting, mirror spaced by approximately 5.5 cm from the plane mirror. About 93% of the pump power is transmitted into the rod through the plane mirror. In this geometry, the TEM_{00} mode waist occurs at the planar reflector and its spot size can be calculated to be $w_0 \cong 130\,\mu$m (thermally induced lensing in the rod is neglected). The spot size of the pump beam provides good mode matching with this TEM_{00} laser mode. The laser operating characteristics are indicated in Fig. 7.13. The threshold pump power is $P_{th} \cong 75$ mW and, at an optical pump power of $P_p = 1.14$ W, an output power of $P_{out} = 370$ mW is obtained. At this output power, the measured optical to optical slope efficiency is $\eta_s \cong 40\%$.

To compare the threshold pump power with the expected value, we assume that the transverse pump beam distribution can be approximated by a Gaussian function and take $w_p \cong w_0 = 130\,\mu$m. From (6.3.20), with $h\nu_p = 2.45 \times 10^{-19}$ J, $\sigma_e = 2.8 \times 10^{-19}$ cm^2 and $\tau = 230\,\mu$s, we obtain $(\gamma/\eta_p) \cong 3.7 \times 10^{-2}$. For a 5% transmission of the output mirror we have $\gamma_2 \cong 5 \times 10^{-2}$ and, assuming an internal loss per pass $\gamma_i - 0.5 \times 10^{-2}$, we obtain $\gamma = \gamma_i + (\gamma_2/2) = 3 \times 10^{-2}$. From the previously obtained value of γ/η_p we then get $\eta_p \cong 81\%$, which includes the overall transmission of the coupling optics and the transmission of the plane mirror at the pump wavelength. Note that the absorption efficiency of the pump radiation in the laser rod, $\eta_a = [1\text{-}\exp\text{-}(\alpha l)]$ in a single pass, can be taken to be unity for an average absorption coefficient of $\sim 6\,\text{cm}^{-1}$ in the 805–808 nm band (see Fig. 6.8a) and for a rod length of $l = 1$ cm. We can now compare the measured slope efficiency with the expected value. Since the threshold power for $w_0 \cong w_p$ is 75 mW, the minimum threshold power, which occurs when $w_0 \to 0$, is expected to be half this value i.e. $P_{mth} \cong 38$ mW. Thus, at 1.14 W input pump power one has $x \cong 30$. At this value of x, from Eq. (7.3.34) one obtains $y = 26$ and, from Eq. (7.3.35), one gets $\eta_t \cong 0.97$. We then have $\eta_c = (\gamma_2/\gamma) = 0.83$ and $\eta_q = (807/1060) = 0.76$. The expected overall optical to optical slope efficiency is thus $\eta_s = \eta_p\eta_c\eta_t\eta_q = 0.49$ in fair agreement with the measured one. Note that the longitudinal efficiency has not been taken into account because the laser is oscillating on many longitudinal modes whose different standing wave patterns add to produce a fairly uniform energy density distribution along the laser rod. According to Eqs. (7.3.27) and (7.3.28) the expected output power at 1.14 W input power is

$P_{out} = yP_s = 500$ mW, i.e., somewhat larger than the experimental one. Some of the discrepancy can perhaps be attributed to thermal effects in the laser rod, which, at the highest pump powers, increase the losses and decrease the spot size w_0.

It should be noted that the quoted 40% efficiency refers to the optical-to-optical efficiency. To obtain the overall electrical-to-optical slope efficiency we need to multiply the optical efficiency by the radiative efficiency, η_r, of the array. Again from Fig. 7.13 one obtains $\eta_r \cong 29\%$ so that the overall electrical to optical slope efficiency is about 11.6%.

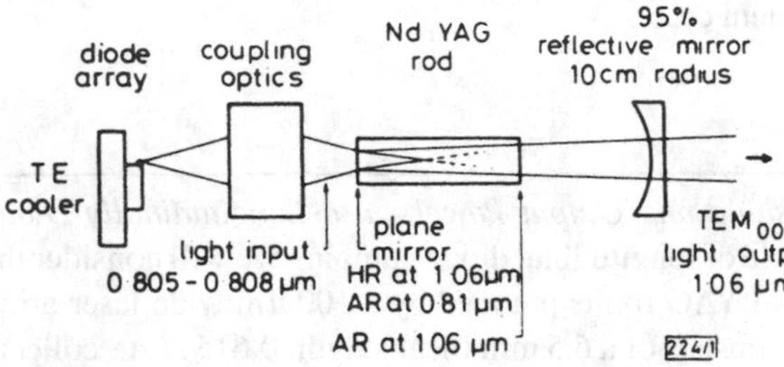

FIG. 7.12. Schematic illustration of the experimental set-up of a Nd:YAG laser, longitudinally pumped by a diode array (after ref.,[15] by permission).

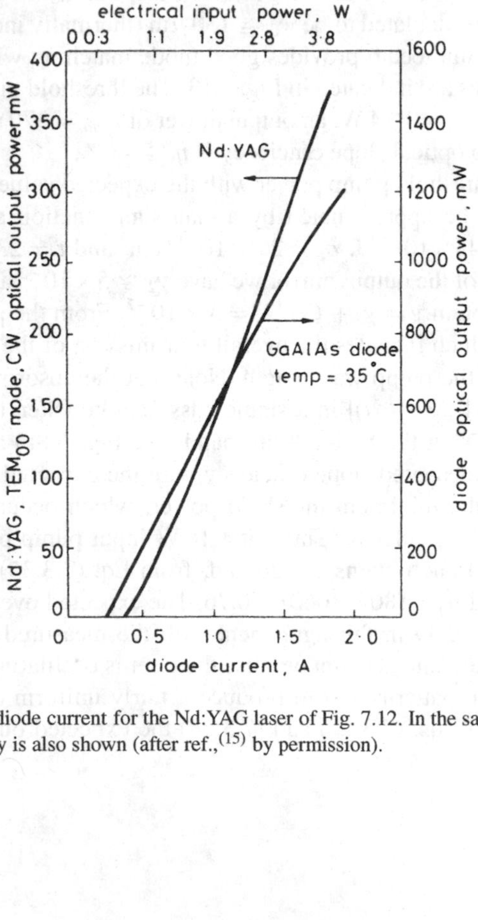

FIG. 7.13. Output power vs diode current for the Nd:YAG laser of Fig. 7.12. In the same figure the output power vs current of the laser diode array is also shown (after ref.,[15] by permission).

7.4. THRESHOLD CONDITION AND OUTPUT POWER: QUASI-THREE-LEVEL LASER

We shall now investigate the threshold condition and output power for a quasi-three-level laser. The laser behavior will first be considered within the space-independent rate-equation model of Sect. 7.2.2. The results predicted from the space-dependent model will then be discussed and a comparison made between the two models.

7.4.1. Space-Independent Model

The analysis for a quasi-three-level laser proceeds in a similar way to that for a four-level case, starting now from Eqs. (7.2.24).

The threshold inversion is obtained by putting $(d\phi/dt) = 0$ in Eq. (7.2.24b), thus giving

$$N_c = \frac{V}{V_a \sigma_e c \tau_c} = \frac{\gamma}{\sigma_e l} \tag{7.4.1}$$

i.e., the same expression as for a four-level laser. The critical pump rate is then obtained from Eq. (7.2.24a) by setting $(dN/dt) = 0$, $\phi = 0$, and $N = N_c$. We obtain

$$R_{cp} = \frac{fN_t + N_c}{(1+f)\,\tau} \tag{7.4.2}$$

Since in most cases one has $f \ll 1$, one can write $(1+f) \cong 1$ in the denominator of Eq. (7.4.2). Comparison of Eq. (7.4.2) with Eq. (7.3.3) then shows that the critical pump rate for a quasi-three-level is increased over that for a four-level laser by the presence of the additional term fN_t in the numerator of Eq. (7.4.2). In typical situations, this term may perhaps be ~ 5 times larger than N_c. For the case of uniform pumping by a diode laser, according to the discussion in Sect. 6.3.3, we can write $R_p = \eta_p P_p / h\nu_p A l$ where A is the cross sectional area of the active medium and l its length. Using this expression in Eq. (7.4.2) we obtain the threshold pump power as

$$P_{th} = \frac{h\nu_p}{\eta_p \tau} \frac{(f\,N_t + N_c)Al}{(1+f)} \tag{7.4.3}$$

With the help of Eqs. (7.2.22) and (7.4.1), one can put Eq. (7.4.3) in the more suggestive form [compare with Eq. (7.3.12)]

$$P_{th} = \frac{\gamma(1+B)}{\eta_p} \left(\frac{h\nu_p}{\tau}\right) \left(\frac{A}{\sigma_e + \sigma_a}\right) \tag{7.4.4}$$

where we have set

$$B = \sigma_a N_t l / \gamma \tag{7.4.5}$$

Above threshold, the c.w. inversion, N_0, and the cw photon number, ϕ_0, are obtained from Eq. (7.2.24) by letting $(dN/dt) = (d\phi/dt) = 0$. Just as for the four-level laser, N_0 is again seen to be equal to N_c while ϕ_0, as obtained from Eq. (7.2.24a) with the help of Eq. (7.4.2), is given by

$$\phi_0 = \frac{V}{N_0(\sigma_e + \sigma_a)c} \frac{f N_t + N_0}{\tau}(x - 1) \tag{7.4.6}$$

where $x = R_p/R_{cp} = P_p/P_{th}$ and again represents the amount by which threshold is exceeded. Equation (7.4.6), using Eq. (7.4.1) for N_0 and with the help of Eqs. (7.2.14), (7.2.15) and (7.2.22), transforms to

$$\phi_0 = \left[\frac{A_b \gamma(1 + B)}{\sigma_e + \sigma_a}\right] \left(\frac{\tau_c}{\tau}\right) (x - 1) \tag{7.4.7}$$

where we have set $V_a = A_b l$, A_b being the beam area in the medium, which is assumed to be smaller than or equal to the cross-sectional area of the active medium.

The output power through e.g. mirror 2, is obtained from Eq. (7.2.18), using Eqs. (7.4.7) and (7.2.14), as [compare with Eq. (7.3.9)]

$$P_{out} = \left[\frac{A_b(1 + B)}{\sigma_e + \sigma_a}\right] \left(\frac{h\nu}{\tau}\right) \left(\frac{\gamma_2}{2}\right) \left(\frac{P_p}{P_{th}} - 1\right) \tag{7.4.8}$$

The laser slope efficiency, $\eta_s = (dP_{out}/dP_p)$, is then readily obtained from Eq. (7.4.8) as

$$\eta_s = \left[\frac{A_b(1 + B)}{\sigma_e + \sigma_a}\right] \left(\frac{h\nu}{\tau}\right) \left(\frac{\gamma_2}{2}\right) \left(\frac{1}{P_{th}}\right) \tag{7.4.9}$$

With the help of Eqs. (7.4.4), (7.4.9) gives

$$\eta_s = \eta_p \left(\frac{\gamma_2}{2\gamma}\right) \left(\frac{h\nu}{h\nu_p}\right) \left(\frac{A_b}{A}\right) \tag{7.4.10}$$

Thus, given the same parameters, the slope efficiency of a quasi-three-level laser is predicted to be the same as that of a four level laser. At first sight this result might seem unexpected owing to the increased loss, in a quasi-three-level laser, arising from ground state absorption. However, the energy removed via ground-state absorption actually raises atoms to the upper laser level and these atoms are then available for producing stimulated emission.

7.4.2. Space-Dependent Model[16,17]

We now consider the case where the mode distribution and pump rate are assumed to be spatially dependent. We limit ourselves to a discussion of the most relevant results corresponding to the active medium being longitudinally pumped by a beam with Gaussian transverse profile. We again refer to Appendix *E* for a more detailed treatment of this case as well as of the case of uniform pumping. We assume that the field intensity profile is again described by Eq. (7.3.23) and the spatial distribution of the pump rate by (6.3.7) which is repeated here for convenience

$$R_p(r, z) = \eta_r \eta_t \left(\frac{P_p}{h\nu_p}\right) \left(\frac{2\alpha}{\pi\, w_p^2}\right) \exp\left[-\left(\frac{2r^2}{w_p^2}\right)\right] \exp(-\alpha\, z) \tag{7.4.11}$$

As far as the threshold condition is concerned, one again finds that Eq. (7.4.1) still holds for the average value of N i.e.,

$$<N>_c = (\gamma/\sigma_e l) \tag{7.4.12}$$

where the average value is calculated according to Eq. (7.3.20). To obtain the threshold pump rate we notice that, at each point of the active medium, below or at threshold, an equilibrium must exist between the number of atoms raised by the pumping process and the number of atoms decaying spontaneously. Thus, from Eq. (7.2.24a), we obtain $R_p = [fN_t + N(r,z)]/(1+f)\tau$. Upon averaging this expression over the mode intensity distribution and using the expression for threshold inversion given by Eq. (7.4.12) we obtain

$$<R_p>_c = \frac{\sigma_a N_t l + \gamma}{(\sigma_e + \sigma_a) l \tau} \tag{7.4.13}$$

If the pump rate expression Eq. (7.4.11) is substituted into the left hand side of Eq. (7.4.13) and if the average of R_p over the field intensity profile is calculated, we end up with the threshold pump power expression of given by Eq. (6.3.25) which is repeated here in a slightly different form [compare with Eq. (7.4.4)]

$$P_{th} = \frac{\gamma(1+B)}{\eta_p}\left(\frac{h\nu_p}{\tau}\right)\left[\frac{\pi\,(w_0^2 + w_p^2)}{2(\sigma_e + \sigma_a)}\right] \tag{7.4.14}$$

where B is again given by Eq. (7.4.5).

Above threshold, using the condition $(d\phi/dt) = 0$, one again finds that the average gain must equal losses, thus giving

$$<N>_0 = <N>_c = \gamma/\sigma_e l \tag{7.4.15}$$

The calculation of output power is considered in some detail in Appendix E and we limit ourselves here to quoting and discussing the final result. First, we define x as the factor by which threshold is exceeded as in Eq. (7.3.25) and a minimum threshold power P_{mth} as the threshold value which holds for $w_0 \ll w_p$ and $B \ll 1$ $[\sigma_a N_t l \ll \gamma]$. From Eq. (7.4.14) we get

$$P_{mth} = \frac{\gamma}{\eta_p}\left(\frac{h\nu_p}{\tau}\right)\left[\frac{\pi\,w_p^2}{2(\sigma_e + \sigma_a)}\right] \tag{7.4.16}$$

We also define a normalized output power as in Eq. (7.3.27) where now

$$P_s = \frac{\gamma_2}{2}\left[\frac{\pi\,w_0^2}{2(\sigma_e + \sigma_a)}\right]\left(\frac{h\nu}{\tau}\right) \tag{7.4.17}$$

The relation between y and x then turns out to be

$$x = \frac{1 + B\frac{\ln(1+y)}{y}}{\int_0^1 \frac{t^{\alpha}dt}{1+yt}} \tag{7.4.18}$$

where $\alpha = (w_0 / w_p)^2$. Note that Eq. (7.4.18) reduces to Eq. (7.3.33) when $B \Rightarrow 0$ $[\sigma_a N_t l \ll \gamma]$ i.e., for negligible absorption arising from the lower laser level. Apart from the difference in notations, Eq. (7.4.18) is the same as Eq. (25) of ref.[16]

We can again calculate the slope efficiency as $\eta_s = (dP_{out}/dP_p) = (P_s/P_{mth})(dy/dx)$. With the help of Eqs. (7.4.16) and (7.4.17) we obtain

$$\eta_s = \eta_p \left(\frac{\gamma_2}{2\gamma}\right)\left(\frac{h\nu}{h\nu_p}\right)\left(\frac{\pi w_0^2}{\pi w_p^2}\frac{dy}{dx}\right) \tag{7.4.19}$$

i.e. the same expression as for a 4-level laser. The behavior of the transverse efficiency, $\eta_t = \left(\pi w_0^2 / \pi w_p^2\right)(dy / dx)$, for $x = 10$ and $B = 1$ is also shown in Fig. 7.11b. The three data points indicated by closed triangles have been obtained from the computed results of ref.[16] while the dashed line is a suitable interpolation. Note that, when $(w_0 / w_p)^2 \ll 1$, the value of η_t tends to coincide with the corresponding value for a four-level laser. In this case, in fact, the population raised to the upper laser level via ground state absorption is mostly useful for a stimulated emission process and ground state absorption does not degrade the laser efficiency. At higher values of $(w_0 / w_p)^2$, however, e.g. $(w_0 / w_p)^2 = 1$, ground state absorption in the wings of the inversion profile leads predominantly to spontaneous rather than stimulated decay. Accordingly, the value of η_t becomes smaller than the corresponding value for a four level laser. It can be shown, however, that, on further increasing the pump power, i.e., the value of x, η_t tends to equality with the corresponding value for a four-level laser. At sufficiently high values of x, in fact, the normalized output power y becomes large enough that $B[\ln(1+y)]/y \ll 1$. Under this condition Eq. (7.4.18) becomes identical to Eq. (7.3.33).

Example 7.5. *Threshold and Output Powers in a Longitudinally Pumped Yb:YAG Laser* As a representative example we will consider a $l = 2.5$ mm thick 6.5 atomic % Yb:YAG laser disc longitudinally pumped, at $\lambda_p = 941$ nm, by a $Ti^{3+}:Al_2O_3$ laser.[18] One face of the disc is plane and coated for high reflectivity at the laser wavelength ($\lambda = 1.03\ \mu m$). The other face is concave with a 1 cm radius of curvature and coated to give a power reflectivity, $R_2 = 90\%$, at the laser wavelength. Under these conditions, the calculated spot size at the beam waist, i.e., at the plane mirror, is $w_0 = 28\ \mu m$ and the spot size can be taken as approximately constant along the resonator. The measured pump spot size in the laser disc is $w_p = 31\ \mu m$. The measured values of output power vs absorbed pump power, at $T = 300$ K, are indicated by triangles in Fig. 7.14. To compare these results with theoretical predictions we note that the total Yb concentration is $N_t \cong 9 \times 10^{20}\ cm^{-3}$ and that, from the measured absorption coefficient at $\lambda = 1.03\ \mu m$ [Fig. 6.8b], one gets $\sigma_a \cong 1.2 \times 10^{-21}\ cm^2$. The effective value of the stimulated emission cross section is then evaluated in ref.[18] to be $\sigma_e \cong 18 \times 10^{-21}\ cm^2$. Assuming that there are no other losses except the output coupling loss, we get $\gamma = \gamma_2/2 \cong (1 - R_2)/2 = 5 \times 10^{-2}$. From Eq. (7.4.14) we then obtain $P_{th} \cong 83$ mW to be compared with the experimental value of 70 mW in Fig. 7.14. We now calculate the predicted slope efficiency at the maximum pump power of 180 mW. We first note that one has $\gamma = \gamma_2/2$ and that, since the data in Fig. 7.14 are expresses with respect to absorbed pump power, we must set $\eta_p = 1$ in Eq. (7.4.19). The expression for the slope efficiency then reduces to $\eta_s = (h\nu/h\nu_p) \times \left(w_0^2 dy/w_p^2 dx\right)$. To calculate the transverse efficiency $\eta_t = \left(w_0^2 dy/w_p^2 dx\right)$, we first observe that, according to Eq. (7.4.16), one has $P_{mth} = 6.5$ mW so that, at $P_p = 180$ mW, one gets $x = (P_p/P_{mth}) = 27.7$. The transverse efficiency can then be estimated from Fig. 4e of ref.[18] by taking into account that $\eta_t = dS/dF$, $B = \sigma_a N_t l/\gamma \cong 5$, $a = w_p/w_0 \cong 1.1$, $(F/F_{th}) = xa^2/(1+a^2)(1+B) \cong 2.16$, where the quantities S, F, and F_{th} are defined in the cited reference. We obtain $\eta_t = dS/dF \cong 70\%$ so that $\eta_s = (h\nu/h\nu_p)\eta_t \cong 63\%$. A more exact calculation of the output power vs pump power, as obtained directly from Eq. (7.4.18), is also plotted as a solid line in Fig. 7.14. From this calculation one gets a more exact value of the predicted slope.efficiency of $\eta_s = 59\%$ at $P_p = 180\,\text{mW}$, to be compared with the value of 56% obtained from the four points in the figure corresponding to the highest experimental powers.

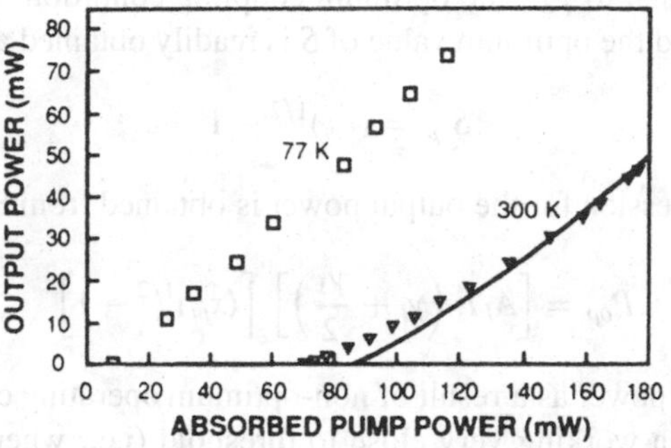

FIG. 7.14. Plots of the output power vs absorbed pump power for a Ti:sapphire-pumped Yb:YAG laser, for liquid nitrogen cooled operation (77 K) and at room temperature operation (300 K) (after ref.,[18] by permission).

7.5. OPTIMUM OUTPUT COUPLING

For a fixed pump rate, there is some value for the transmission, T_2, of the output mirror that maximizes the output power.[19] Physically, the reason for this optimum arises from the fact that, as T_2 is increased, we have the following two contrasting circumstances: (1) The output power tends to increase due to the increased mirror transmission. (2) The output power tends to decrease since the increased cavity losses cause the number of cavity photons, ϕ_0, to decrease

To find the optimum output coupling condition, we will limit ourselves to a four-level laser and consider the space-independent model. The optimum transmission is then obtained from Eq. (7.3.9) by imposing the condition $dP_{out}/d\gamma_2 = 0$ for a fixed value of pump power P_p. We must obviously take into account the fact that, according to Eq. (7.3.12), P_{th} is also a function of γ_2. From Eq. (7.3.12) one can then write

$$P_{th} = P_{mth} \frac{\gamma}{\gamma_i + (\gamma_1/2)} \tag{7.5.1}$$

where P_{mth}, the minimum threshold pump power, is the threshold pump power for zero output coupling, $\gamma_2 = 0$. Equation (7.3.9) can then be transformed to

$$P_{out} = \left[A_b I_s \left(\gamma_i + \frac{\gamma_1}{2}\right)\right] S \left(\frac{x_m}{S+1} - 1\right) \tag{7.5.2}$$

where

$$S = \frac{(\gamma_2/2)}{\gamma_i + (\gamma_1/2)} \tag{7.5.3}$$

and

$$x_m = P_p/P_{mth} \tag{7.5.4}$$

The only term in Eq. (7.5.2) that depends on γ_2 is the quantity S, which, according to Eq. (7.5.3), is proportional to γ_2. The optimum coupling condition can then be obtained by setting $dP_{out}/dS = 0$ and the optimum value of S is readily obtained as

$$S_{op} = (x_m)^{1/2} - 1 \tag{7.5.5}$$

The corresponding expression for the output power is obtained from Eq. (7.5.2) as

$$P_{op} = \left[A_b I_s \left(\gamma_i + \frac{\gamma_1}{2}\right)\right] \left[(x_m)^{1/2} - 1\right]^2 \tag{7.5.6}$$

The reduction in output power as a result of non-optimum operating conditions becomes particularly important when working very close to threshold (i.e., when $x_m \cong 1$). Well above threshold, however, the output power becomes rather insensitive to a change of output power around the optimum value. As an example, Fig. 7.15 shows a normalized plot of P_{out} vs S, for $x_m = 10$. According to Eq. (7.5.2), one has $P_{out} = 0$ for $S = 0$ (i.e., $\gamma_2 = 0$) and $S = 9$, while, according to Eq. (7.5.5) one gets $S_{op} \cong 2.16$. From Fig. 7.15 one can now see that changes of, S, i.e., of the output coupling, around the optimum value by as much as 50% only result in ~ 5% reduction of the output power.

In the case of a space-dependent model, similar considerations could be developed starting from Eq. (7.3.29) (uniform pumping) or from Eq. (7.3.34) (Gaussian pumping). However, at pump powers well above threshold, the relation between output power and pump

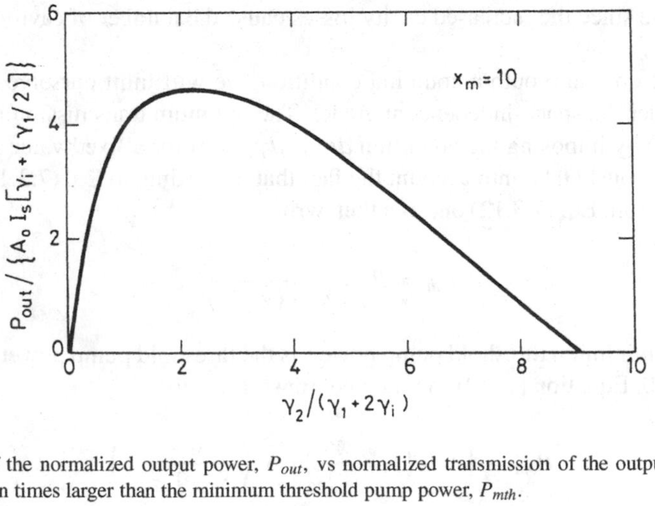

FIG. 7.15. Plot of the normalized output power, P_{out}, vs normalized transmission of the output mirror, γ_2, for a pump power, P_p, ten times larger than the minimum threshold pump power, P_{mth}.

power tends to become linear (see Fig. 7.9), as predicted by Eq. (7.3.9). Upon also taking into account the relative insensitivity of P_{out} to variation around the optimum output coupling, one can then make the approximation of using the optimum coupling expression given by Eq. (7.5.5) for this case also, where P_{mth} is obtained from Eq. (7.3.26) (uniform pumping) or Eq. (7.3.32) (Gaussian pumping) by letting $\gamma = \gamma_i + (\gamma_1/2)$.

Example 7.6. *Optimum output coupling for a lamp-pumped Nd:YAG laser* We consider the laser configuration discussed in example 7.2 (see Fig. 7.4 and 7.5) and calculate the optimum transmission of the output mirror when the laser is pumped by a lamp input power of $P_p = 7\,\text{kW}$. Since the threshold power, P_{th}, in Fig. 7.5 was measured to be 2.2 kW, then, according to Eq. (7.5.1) with $\gamma_1 = 0$, we obtain $P_{mth} = P_{th}(\gamma_i/\gamma) \cong 697\,\text{W}$, where the values $\gamma_i = 0.038$ and $\gamma = 0.12$, as obtained in example 7.2, have been used for the internal loss and the total loss, respectively. We then obtain $x_m = P_p/P_{mth} \cong 10$ so that, from Eq. (7.5.5), we find $S_{op} \cong 2.17$. From Eq. (7.5.3) we finally get $(\gamma_2)_{op} \cong 0.165$, which corresponds to an optimum transmission of $(T_2)_{op} = 1 - \exp\left[-(\gamma_2)_{op}\right] \cong 15\%$ i.e., agreeing with the value actually used in Fig. 7.4.

7.6. LASER TUNING

The gain linewidth of some lasers (e.g., dye lasers or vibronic solid-state lasers) is very wide and for various applications one has a requirement of tuning the laser output wavelength away from line center and across the entire available linewidth. In other cases, lasers may exhibit gain on more than one transition (e.g., CO_2 laser or Ar laser), the strongest of which would usually oscillate, whereas one may need to tune the laser wavelength away from the strongest line. In both the above circumstances one usually employs of a wavelength selective element within the laser cavity.

In the middle infrared, such as for the CO_2 laser, one generally uses, as one of the cavity mirrors, a diffraction grating aligned in the so-called Littrow configuration (Fig. 7.16a). In this configuration, for a given angular setting of the grating there is a particular wavelength (labeled λ_1 in the figure) that is reflected exactly back into the resonator and wavelength tuning is thus achieved by grating rotation.

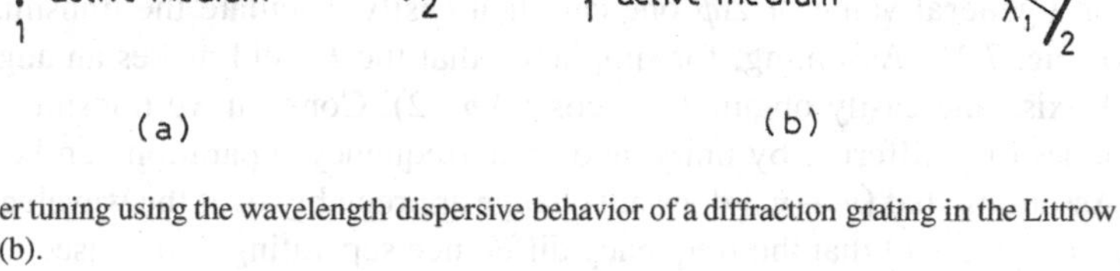

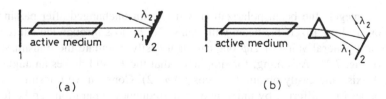

FIG. 7.16. Laser tuning using the wavelength dispersive behavior of a diffraction grating in the Littrow configuration, (a), or a prism, (b).

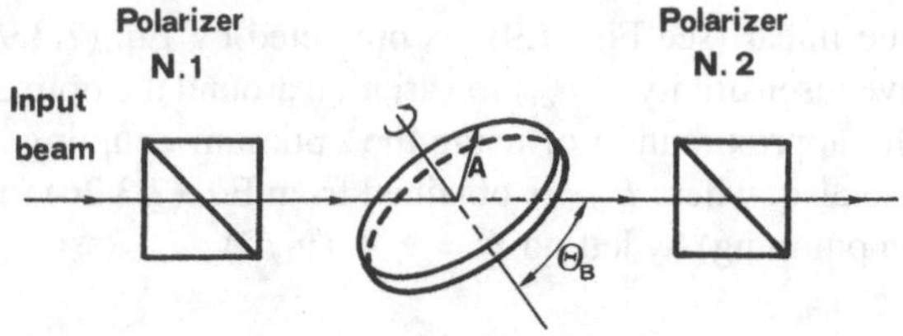

FIG. 7.17. Use of a birefringent filter as a wavelength-selective element.

In the visible or near IR spectral region, it is more common to use a dispersive prism with faces close to Brewster's angle with respect to the laser beam (Fig. 7.16b). Again, for a given angular setting of the prism, a particular wavelength (labeled λ_1 in the figure) is reflected from mirror 2 exactly back into the resonator. Tuning is thus achieved by prism rotation.

A third wavelength-selective element, which is becoming increasingly popular in the visible or near IR spectral region, makes use of a birefringent filter inserted within the laser cavity. The filter simply consists of a plate of a suitable birefringent crystal (e.g., quartz or KDP) inclined at Brewster's angle, θ_B, to the beam direction (Fig. 7.17). The optical axis, A, of the crystal is assumed to be in a plane parallel to the surface of the plate. Let us first suppose that the birefringent plate is placed between two polarizers with parallel orientation. This orientation is assumed to be such as to transmit the E-field in the plane of incidence of the plate. Then, the input beam will not suffer any reflection loss upon entering the plate, since this is inclined at Brewster's angle. Provided the optic axis is neither perpendicular nor parallel to the plane of incidence, the input beam will contain both ordinary and extraordinary components. These components will experience a difference in phase shifts $\Delta\phi = 2\pi(n_e - n_o)L_e$, where n_o and n_e are the refractive indices for the ordinary and extraordinary beams, respectively, and L_e is the plate thickness along the beam direction within the plate. After passing through the plates, unless $\Delta\phi$ is an integer number of 2π, the two components will combine to form a resultant beam with elliptical polarization and the presence of second polarizer will then lead to loss for this elliptically polarized light. On the other hand, if $\Delta\phi$ is an integral number of 2π, i.e., if

$$\frac{2\pi}{\lambda}(n_e - n_o)L_e = 2l\pi \tag{7.6.1}$$

where l is an integer, the beam polarization will remain unchanged after passing through the plate. For ideal polarizers the beam will then suffer no loss through the entire system of Fig. 7.17. For a general value of $\Delta\phi$ one can then easily calculate the transmission, T, of the system of Fig. 7.17. Assuming, for simplicity, that the E-field makes an angle of 45° to the optical A-axis, one easily obtain $T = \cos^2(\Delta\phi / 2)$. Consecutive transmission maxima have their values of l differing by unity, and their frequency separation can be found from Eq. (7.6.1). Assuming that $(n_e - n_o)$ does not change appreciably over the wavelength range of interest, one has the result that the frequency difference separating two consecutive maxima, i.e., the free spectral range, $\Delta\nu_{fsr}$, isgiven by

$$\Delta\nu_{fsr} = \frac{c_0}{(n_e - n_o)L_e} \tag{7.6.2}$$

Accordingly, the plate thickness, which usually ranges between 0.3 and 1.5 mm, determines the width of the tuning curve and thus the resolving power. The thinner the plate, the greater the available tuning range and the lower the resolving power. Tuning of one transmission peak can then be achieved upon rotating the plate around the normal to the surface. By doing so, in fact, one changes the value of n_e, which depends on the angle between the optical axis and the electric field vector, and hence changes the plate birefringence $\Delta n = n_e - n_o$. Note finally that, in low gain lasers such as cw gas or dye lasers, one can dispense with the two polarizers if other polarizing element, such as the active medium or indeed the Brewster's angle surfaces of the birefringent plate itself, provide sufficient loss discrimination between the two polarizations.

Example 7.7. *Free spectral range and resolving power of a birefringent filter* Consider a dye laser operating at the wavelength $\lambda = 600$ nm in which a birefringent filter consisting of a $L = 1.5$ mm thick potassium dihydrogen phosphate (KDP) crystal is inserted for laser tuning. The ordinary and extraordinary refractive indices, at the laser wavelength, are $n_o = 1.47$ and $n_e = 1.51$. The Brewster angle for this plate is $\theta_B \cong \tan^{-1} n \cong 56.13°$, where n has been taken as the average of n_o and n_e. The Brewster angle inside the crystal is then given by Snell's law as $\theta'_B = 33.9°$ so that $L_e = L/\cos\theta'_B \cong 1.81$ mm. If the orientation of the optic axis A in Fig. 7.17 is near orthogonal to the beam direction, the refractive index of the extraordinary beam to be used in Eq. (7.6.2) is just $n_e = 1.51$, and from Eq. (7.6.2) one gets $\Delta\nu_{fsr} \cong 4.14 \times 10^{12}$ Hz. The corresponding wavelength interval between two consecutive peaks is $\Delta\lambda = \lambda(\Delta\nu_{fsr}/\nu) \cong 5$ nm, where $\nu = c/\lambda = 5\times10^{14}$ Hz is the frequency of the radiation. Since the transmission of the birefringent filter of Fig. 7.17 is equal to $T = \cos^2(\Delta\phi/2)$, one can readily show that the width of the transmission curve (full width between half-maximum points) is just equal to $\Delta\lambda/2$ i.e. it is equal to ~ 2.5 nm.

7.7. REASONS FOR MULTIMODE OSCILLATION

Lasers generally tend to oscillate on many modes. The reason for this behavior arises essentially from the fact that the frequency separation of the modes is usually smaller, and often very much smaller, than the width of the gain profile. If, for example, we take $L = 1$ m, the frequency difference between two consecutive longitudinal modes will be $\Delta\nu = c/2L = 150$ MHz. The laser linewidth, on the other hand, may range from ~ 1 GHz, for a Doppler broadened transition of a visible or near IR gas laser, up to 300 GHz or more for a transition of a crystal ion in a solid state material. The number of modes within the laser linewidth may thus range from a few to a few thousands and the gain difference between these modes, particularly when a few thousand modes are considered, becomes very small. At first sight one would therefore expect a significant fraction of these modes to be excited at a sufficiently high pump rate.

The above, seemingly straightforward, conclusion needs to be examined more carefully, however. In fact, in the early days of laser development, it was argued that in principle lasers should always tend to oscillate on a single mode, provided the gain line was homogeneously broadened. The argument can be followed with the help of Fig. 7.18, in which the laser gain profile is plotted vs frequency for increasing values of the pump rate. For simplicity,

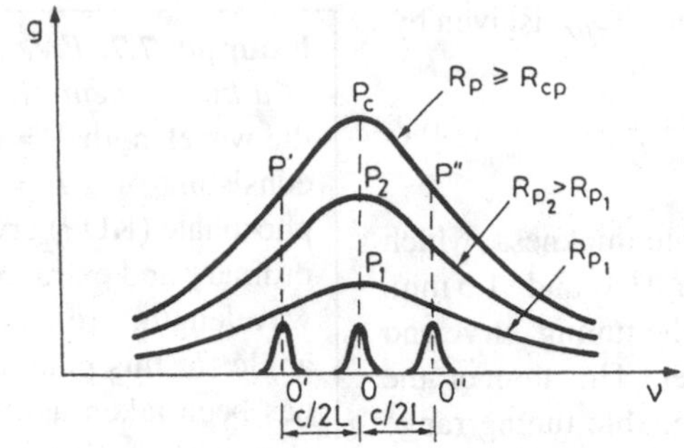

FIG. 7.18. Frequency dependence of laser gain coefficient versus pump rate, R_p, under saturation condition (*homogeneous line*).

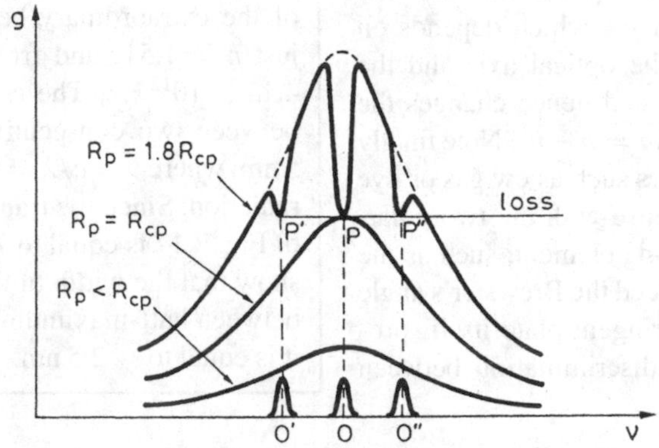

FIG. 7.19. Frequency dependence of laser gain coefficient versus pump rate, R_p, under saturation conditions (*inhomogeneous line*): frequency hole-burning behavior.

one cavity mode is assumed to be coincident with the peak of the gain curve. It is further assumed that oscillation occurs on the TEM_{00} mode, so that the mode frequencies are all separated by $c/2L$ (see Fig. 5.10). The laser gain coefficient is given by (2.4.35), where, for a homogeneous line, the cross section is given by (2.4.18). Oscillation will start on the central mode when the inversion $N = N_2 - N_1$, or the average inversion for the space-dependent model, reaches a critical value N_c giving a gain equal to the cavity losses [see Eq. (7.3.2), or (7.3.18)]. However, even when R_p is increased above the threshold value, in the steady state the inversion remains fixed at the critical value N_c. The peak gain, represented by the length OP in Fig. 7.18, will therefore remain fixed at the value OP_c when $R_p \geq R_{cp}$. Since the line is homogeneously broadened, its shape cannot change and the whole gain curve will remain the same for $R_p \geq R_{cp}$, as indicated in Fig. 7.18. The gain of other modes, represented by the lengths O′P′, O″P″, etc., will always remain smaller than the value, OP_c, for the central mode. If all modes have the same losses, then, in the steady state, only the central mode should oscillate. The situation is quite different for an inhomogeneous line (Fig. 7.19). In this case the cross section to be used into (2.4.35) is given by (2.4.23), i.e., it is given by the superposition of the cross sections for the individual atoms, whose transition frequencies are distributed in a given spectrum described by the function $g^*(\nu_0' - \nu_0)$. Accordingly, it is possible to "burn holes" in the gain curve as was discussed, for an absorption curve, in Sect. 2.8.3 [see Fig. 2.22]. Therefore, when R_p is increased above R_{cp}, the gain of the central mode remain fixed at the critical value OP_c, while the gain of the other modes O′P′, O″P″, etc., can keep on increasing up to the corresponding threshold value. In this case, if the laser is operating somewhat above threshold, then more than one mode can be expected to oscillate.

Shortly after the invention of the laser it was actually observed experimentally that multimode oscillation occurred both for inhomogeneous (e.g., gas laser) and homogeneous (e.g., ruby laser) lines. This last result appeared to be in conflict with the argument given above. This inconsistency was later removed[20] by taking into account the fact that each mode has a well-defined standing-wave pattern in the active medium. For the sake of simplicity, we will consider two modes whose standing-wave patterns are shifted by $\lambda/4$ in the active medium (Fig. 7.20a)*. We will assume that mode 1 in Fig. 7.20 is the center mode of Fig. 7.18, so that it is the first to reach threshold. However, when oscillation on mode 1 sets in, the inversion around those points where the electric field is zero (points A, B, etc.) will mostly be left undepleted and the inversion can continue growing there even when the laser is above threshold. This situation is illustrated in Fig. 7.20b, where the spatial distribution of the population inversion in the laser medium is indicated†. Accordingly, mode 2, which initially had a lower gain, will experience a gain growing with pump rate since it uses inversion from those regions that have not been depleted by mode 1. Therefore, sufficiently far above threshold, mode 2

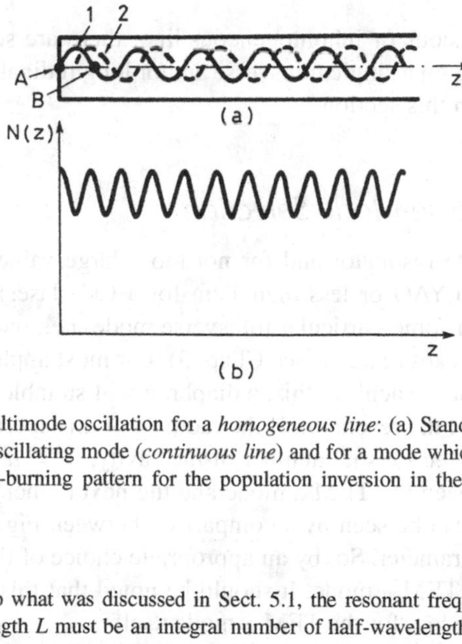

FIG. 7.20. Explanation of multimode oscillation for a *homogeneous line*: (a) Standing-wave mode-field configurations, in the medium, for the oscillating mode (*continuous line*) and for a mode which may oscillate above threshold (*dashed line*). (b) Spatial-hole-burning pattern for the population inversion in the laser medium produced by the oscillating mode.

* We recall that, according to what was discussed in Sect. 5.1, the resonant frequencies can be obtained via the condition that the cavity length L must be an integral number of half-wavelengths, i.e., $L = n(\lambda/2)$, where n is a positive integer. Two consecutive longitudinal modes, having their number n differing by 1, are thus shifted by $(\lambda/2)$ upon going from one mirror to the other. If the active medium is just placed at the resonator center, these two modes have their spatial patterns shifted by just $(\lambda/4)$ in the medium, corresponding to the two modes of Fig. 7.20.

† According to the discussion presented in Sections 7.3.2 and 7.4.2, it is the average value, $< N >$, of the inversion, as defined by (7.3.20), which, above threshold, remains clamped to the threshold value.

can also be set into oscillation and this will obviously occur when its gain equals its losses. Thus, for a homogeneous line, the multimode oscillation is not due to holes burned in the gain curve (*spectral hole burning*) but to holes burned in the spatial distribution of inversion within the active medium (*spatial hole burning*, Fig. 7.20b). It should also be noticed that the phenomenon of spatial hole burning does not play any significant role for an inhomogeneous line, because, in this case, different modes (with large enough frequency separation) interact with different sets of atoms and the hole-burning pattern of one set of atoms is ineffective for the other mode.

The conclusion of this section is that a laser always tend to oscillate on many modes. For a homogeneous line this is due to spatial hole burning while, for an inhomogeneous line, this is due to spectral hole burning. It should be noted, however, that, in the case of a homogeneous line, when a few modes are oscillating with frequencies around the center of the gain line, the spatial variation of inversion will be essentially smeared out due to the presence of the corresponding, spatially-shifted, standing-wave patterns of these modes. In this case, the homogeneous character of the line prevents other modes, further away from the center of the gain line, from oscillating. So, compared with an inhomogeneous line, a homogeneous line restricts oscillation to a smaller number of modes centered around the peak of the gain line.

7.8. SINGLE-MODE SELECTION

For either a homogeneous or inhomogeneous line, there are several methods for constraining a laser to oscillate on a single transverse and/or longitudinal mode, and these will be discussed at some length in this section.

7.8.1. Single-Transverse-Mode Selection

In the case of a stable resonator and for not too a large value of laser spot size (e.g. less then 0.5 mm for a Nd:YAG or less than 1 cm for a CO_2 laser), it is relatively easy to make the laser oscillate on some particular transverse mode, i.e., one with prescribed values of the transverse mode indexes l and m (see Chap. 5). For most applications, oscillation on a TEM_{00} mode is desired and, to achieve this, a diaphragm of suitable aperture size is inserted at some point on the axis of the resonator. If the radius a of this aperture is sufficiently small, it will dictate the value of the Fresnel number of the cavity, $N = a^2/L\lambda$. As a is decreased, the difference in loss between the TEM_{00} mode and the next higher order modes (TEM_{01} or TEM_{10}) will increase, as can be seen by a comparison between Fig. 5.13a and Fig. 5.13b at the same value of the g parameter. So, by an appropriate choice of the aperture size, one can enforce oscillation on the TEM_{00} mode. It should be noted that this mode-selecting scheme inevitably introduces some loss for the TEM_{00} mode itself.

For large diameters of the active medium, as discussed in Sect. 5.5.2, it is not possible to obtain a mode spot size comparable to this diameter without incurring serious problems of instability in the size of the transverse mode profile. With reference to example 5.9, one can in fact show that the Fresnel number, $N = a^2/L\lambda$ where a in this case would be the radius of the active medium, should not exceed a value of about 2. For larger values of this radius one then need to resort to unstable resonators. In particular, as discussed in Sect. 5.6.2, if the equivalent Fresnel number is chosen to have a half-integer value, a large loss discrimination will occur between the lowest-order and the higher-order modes (see Fig. 5.20). In this case, the output beam is in the form of a ring, a shape which is not always convenient. The best way to obtain oscillation on the lowest-order mode, for this case, would then be to use an unstable cavity with a radially variable output coupler of Gaussian or, even better, super-Gaussian profile (Sect. 5.6.4).

7.8.2. Single-Longitudinal-Mode Selection

Even when a laser is oscillating on a single transverse mode, it can still oscillate on several longitudinal modes (i.e., modes differing in their value of the longitudinal mode index n). These modes are separated in frequency by $\Delta\nu = c/2L$. Isolation of a single longitudinal mode can be achieved, in some cases, by using such a short cavity length that $\Delta\nu > \Delta\nu_0/2$, where $\Delta\nu$ is the width of the gain curve*. In this case, if a mode is tuned to coincide with the center of the gain curve, the two adjacent longitudinal modes are far enough away from line center that, for a laser not too far above threshold, they cannot oscillate. The requirement for this mode-selecting scheme can then be written as

$$L \leq c/\Delta\nu_0 \tag{7.8.1}$$

So, for example, if the equality applies in Eq. (7.8.1) and if one mode is coincident with the peak of the gain profile, the two adjacent modes will see an unsaturated gain coefficient that, both for a Gaussian or Lorentzian line, is half that of the peak gain. In particular, for a Gaussian line, one can easily understand from Fig. 7.19 that single longitudinal mode operation is achieved for $R_{cp} \leq R_p \leq 2R_{cp}$. It should be noted that, to tune one mode to coincidence with the line center, one needs to mount one cavity mirror on e.g. a piezo-electric transducer. Upon applying a voltage to this transducer, one can thereby produce a small controllable change of the cavity length [one can show that the cavity length needs to be changed by $\lambda/2$ to shift the comb (Fig. 5.10 with $l = m = 0$) of the longitudinal modes by one mode spacing].

The method discussed above can be used effectively with a gas laser, notably with a He-Ne laser, where gain linewidths are relatively small (a few GHz or smaller). For instance, in the case of a He-Ne laser oscillating on its red transition one has $\Delta\nu_0^* \cong 1.7$ GHz and from Eq. (7.8.1) we obtain $L \leq 17.5$ cm. For solid-state-lasers, on the other hand, the gain linewidth is usually much larger (a few hundreds GHz) and, to fulfill Eq. (7.8.1), the equivalent cavity length must typically be appreciably smaller than 1 mm (*microchip-lasers*). For lasers with much larger bandwidths (e.g., dye lasers or tunable solid-state lasers) the cavity length to fulfill Eq. (7.8.1) becomes too small to be practical. In this case and also when longer lengths of the active medium are needed (e.g. for high power lasers), longitudinal mode selection can be achieved by a variety of other techniques which are the subject of next two sections.

* If a tuning element, such as those in Figs. 7.16 and 7.17, is inserted in the laser cavity and if the corresponding linewidth, ranging in actual cases between 0.1 nm 1 nm, is smaller than that of the gain medium, the linewidth $\Delta\nu_0$ to be considered in this section is that of the tuning element rather than that of the active medium. Notably, this case occurs for dye lasers or for tunable solid-state lasers.

7.8.2.1. Fabry-Perot Etalons as Mode-Selective Elements

A common way to achieve single-longitudinal-mode oscillation for both homogeneous or inhomogeneous lines is to insert, within the cavity, one or more Fabry-Perot (FP) etalons. These consist of a plane-parallel plate of transparent material (fused quartz or glass for visible or near IR wavelengths) whose two plane surfaces are coated to a suitably high reflectivity value R.

We first consider the case where a single FP etalon is used, inclined at an angle θ to the resonator axis (Fig. 7.21). According to the discussion in Sect. 4.5.1, the transmission maxima of the etalon will occur at frequencies ν_n given by

$$\nu_n = \frac{nc_0}{2n_r L' \cos\theta'} \tag{7.8.2}$$

where n is an integer, θ' is the refraction angle of the beam within the etalon, n_r is the etalon refractive index, and L' is its length. Since L' is much smaller than the cavity length L, only a very small tilt of the angle θ (and hence of θ'), away from the $\theta = \theta' = 0$ position, is needed to tune a transmission maximum of the etalon to coincide with the mode nearest the peak of the laser gain profile (Fig. 7.22). If now the frequency separation, $\Delta\nu = c/2L$, between two adjacent longitudinal modes is $\geq \Delta\nu_c/2$, where $\Delta\nu_c$ is the linewidth of an etalon transmission peak, the etalon will select the mode nearest to line center from its neighbors*. According to (4.5.13), the discrimination between adjacent longitudinal modes requires that

$$\frac{\Delta\nu_c}{2} = \frac{\Delta\nu'_{fsr}}{2F} \leq \Delta\nu \tag{7.8.3}$$

where $\Delta\nu'_{fsr}$ is the free-spectral range and F is the finesse of the etalon. To ensure single longitudinal mode operation we also require the etalon free-spectral range $\Delta\nu'_{fsr}$ to be larger than or equal to half of the gain linewidth $\Delta\nu_0$, otherwise the two neighboring transmission peaks of the etalon will allow the corresponding cavity modes to oscillate. The discrimination between adjacent transmission maxima of the etalon then requires that

$$\Delta\nu'_{fsr} \geq \Delta\nu_0/2 \tag{7.8.4}$$

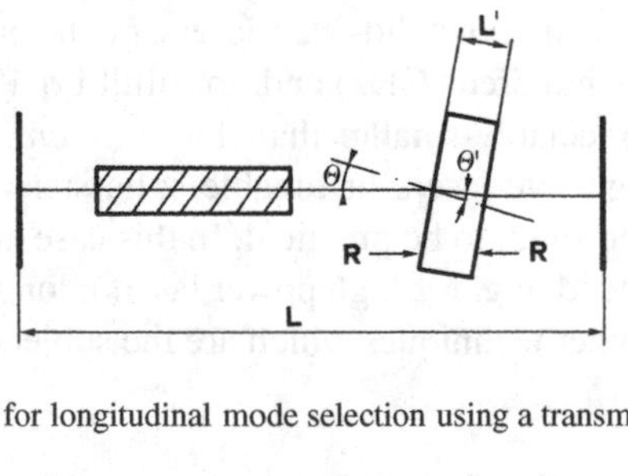

FIG. 7.21. Configuration for longitudinal mode selection using a transmission Fabry-Perot etalon.

* More precisely, the single-pass etalon transmission losses of the two neighboring modes will, in this case, be $\geq$ 50%.

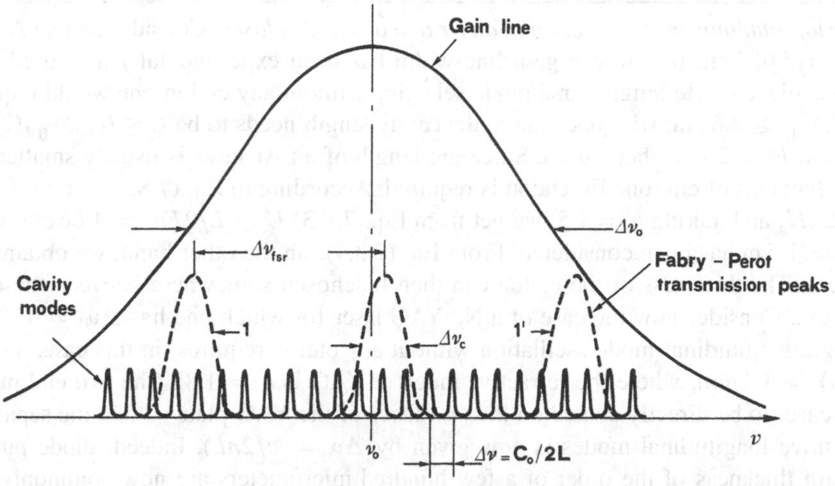

FIG. 7.22. Longitudinal mode selection using a transmission Fabry-Perot etalon.

From Eqs. (7.8.3) and (7.8.4) we then find that $(\Delta\nu_0/2) \leq \Delta\nu'_{fsr} \leq 2F\Delta\nu$, which requires, as a necessary condition, that $(\Delta\nu_0/2) \leq 2F\Delta\nu$ i.e., that

$$L \leq (c/\Delta\nu_0)2F \tag{7.8.5}$$

Comparing this equation with Eq. (7.8.1) we can see that, compared to a resonator without any etalon, the cavity length can now be increased by a factor $2F$. Assuming, for example, $F = 30$ (there are various factors, such as flatness of the etalon surfaces and beam walk-off in the etalon, which limit the value of the finesse achievable in this case), one then sees that, the use of a Fabry-Perot etalon allows a substantial increase in cavity length, while still ensuring single-longitudinal-mode operation.

If the cavity length does not satisfy condition Eq. (7.8.5), then single-longitudinal-mode operation cannot be achieved using one FP etalon and two or more etalons are then needed. In the case of two etalons, the thicker etalon is required to discriminate against adjacent longitudinal modes of the cavity. Its free spectral range, $\Delta\nu'_{fsr}$, must then satisfy condition Eq. (7.8.3). A second thinner etalon must then discriminate against the adjacent transmission maxima of the first etalon and, at the same time, its free spectral range, $\Delta\nu''_{fsr}$, must be larger than or equal to the half-width of the gain curve (i.e., $\Delta\nu''_{fsr} \geq \Delta\nu_0/2$). To achieve both these conditions, it can be shown that the cavity length needs now to satisfy the following relation

$$L \leq (c/\Delta\nu_0)(2F)^2 \tag{7.8.6}$$

A comparison between Eqs. (7.8.1), (7.8.5), and (7.8.6) then shows that, to achieve single-longitudinal mode operation without an etalon, with one etalon, or with two etalons the cavity length must satisfy the respective conditions $L \leq c/\Delta\nu_0$, $c/\Delta\nu_0 \leq L \leq (c/\Delta\nu_0)2F$, or $(c/\Delta\nu_0)2F \leq L \leq (c/\Delta\nu_0)(2F)^2$.

Example 7.8. *Single-longitudinal-mode selection in an Ar and a Nd:YAG laser* Consider first an Ar laser oscillating on its $\lambda = 514.5$ nm line whose gain line-width has been experimentally measured to be $\Delta\nu_0^* = 3.5$ GHz. To achieve single longitudinal mode selection without any etalon one would require a cavity length $L \leq c/\Delta\nu_0^* \cong 8.6$ cm; with one etalon, the cavity length needs to be $L \leq (c/\Delta\nu_0^*)(2F) \cong 5.14$ m, where a finesse $F = 30$ has been used. Since the length of an Ar laser is usually smaller than 2 m but larger than a few tens of cm, one FP etalon is required. According to Eq. (7.8.2), for $\cos\theta' \cong 1$ one has $\Delta\nu'_{fsr} \cong c/2n_rL'$, and, taking $n_r = 1.5$, we get from Eq. (7.8.3) $L' \geq L/2Fn_r = 1.66$ cm, where a cavity length of $L = 1.5$ m has been considered. From Eq. (7.8.4), on the other hand, we obtain $L' \leq (c/\Delta\nu_0^* n_r) = 5.71$ cm. The thickness of the etalon can then be chosen somewhere between these two values e.g. $L' = 3.7$ cm. Consider now the case of a Nd:YAG laser for which one has $\Delta\nu_0 = 120$ GHz (at $T = 300$ K). Single-longitudinal-mode oscillation without any etalon requires, in this case, a cavity length $L \leq (c/\Delta\nu_0 n) = 1.4$ mm, where the refractive index of YAG is $n = 1.82$ (the two end mirrors are assumed, in this case, to be directly coated on the two faces of the YAG plate so that the separation between two consecutive longitudinal modes is now given by $\Delta\nu = c/2nL$). Indeed, diode pumped Nd:YAG platelets with thickness of the order of a few hundred micrometers are now commonly used and even commercially available (*microchip lasers*). When a single FP etalon is used, then, according to Eq. (7.8.5) the cavity length can be increased up to $L \cong 9.5$ cm (assuming again $F = 30$).

7.8.2.2. Single Mode Selection via Unidirectional Ring Resonators

For a homogeneously broadened transition, single-longitudinal mode operation can automatically be achieved, or at least greatly facilitated, if the laser cavity is in the form of a ring and oscillation is constrained to be unidirectional (see Fig. 5.4a). In this case, in fact, the phenomenon of spatial hole burning within the active medium does not occur and, as discussed in Sect. 7.7, the laser tends to oscillate on a single mode. Actually, if the transition is only partly homogeneously broadened and particularly when the gain profile is very broad, some further bandwidth selecting elements such as birefringent filters and/or Fabry-Perot etalons may also be needed. An additional advantage of this unidirectional ring configuration is that higher output power is available since the whole of the active material rather than just those regions around the maxima of the standing-wave pattern contributes to the laser output.

To achieve unidirectional ring operation, a unidirectional device or *optical diode*, giving preferential transmission for one direction of beam propagation, needs to be inserted within the cavity. In principle the device can be made as shown in Fig. 7.23. Here the wave going in one direction, e.g., from left to right, passes first through an input polarizer (polarizer 1), then through a rod of suitable transparent material (e.g., glass) to which a longitudinal dc magnetic field is applied (*Faraday rotator*), and then through an output polarizer (polarizer 2) which has the same orientation as the first polarizer*. When a linearly polarized optical beam passes through the Faraday rotator with beam axis along the magnetic field direction, the output beam still consists of a linearly polarized wave whose plane of polarization is however rotated about the beam axis. The sense of rotation, seen by an observer facing the oncoming beam, depends on the relative direction of magnetic field and beam propagation direction.

* To avoid confusion and have a consistent way of describing the sense of a rotation of polarization, we will always assume that the observer is facing the oncoming light beam.

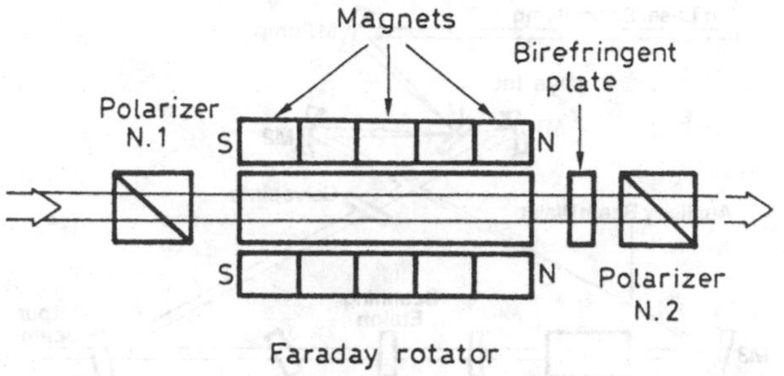

FIG. 7.23. Unidirectional device using a Faraday rotator (*optical-diode*).

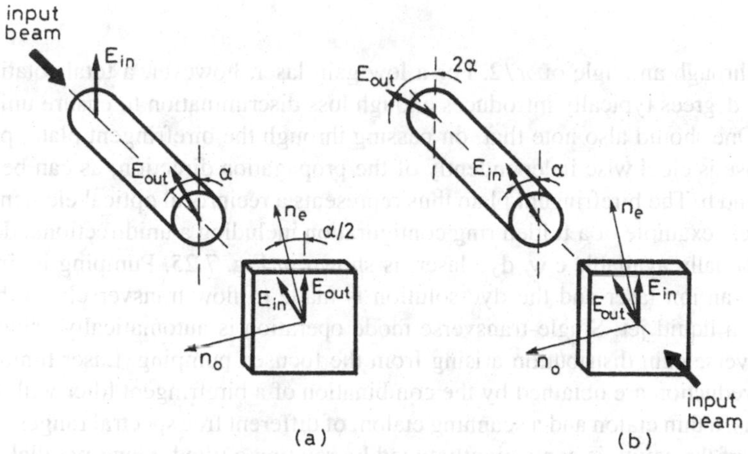

FIG. 7.24. Polarization rotation for a combination of a Faraday-rotator and a $\lambda/2$ birefringent plate for beam propagation from *left-to right*, (a), and from *right-to-left*, (b).

This means that, if polarization rotation, is seen by an observer facing the incoming beam to occur e.g. anti-clockwise for left to right propagating beam (Fig. 7.24a), the polarization rotation will be seen (the observer again facing the oncoming beam) to occur clockwise for right to left propagating beam (Fig. 7.24b). For this reason, the Faraday rotator is said to represent a non-reciprocal element. The beam is then passed through a birefringent plate having a $\lambda/2$ optical path difference between the two polarizations. The phase shift between the two polarizations will then be equal to π, i.e. $2\pi(n_o - n_e)l/\lambda = \pi$, where l is the plate length, and, if the polarization of the input beam makes an angle $\alpha/2$ to the extraordinary axis, the plate will rotate the polarization, clockwise (seen by the observer facing the beam) by an angle α (Fig. 7.24a). Thus, if the Faraday rotator rotates the polarization, anti-clockwise, by an angle α, the two rotations exactly cancel and no attenuation is suffered by the beam on passing through the output polarizer (polarizer 2 of Fig. 7.23). In the case of a beam traveling in the opposite direction right to left as in Fig. 7.24b, however, polarization rotation will again have a clockwise sense on passing through the birefringent plate (Fig. 7.24b), the two rotations will add to each other and the beam experiences a loss in passing through the second polarizer (polarizer 1 in Fig. 7.23). Note that this loss may even reach 100% if the total

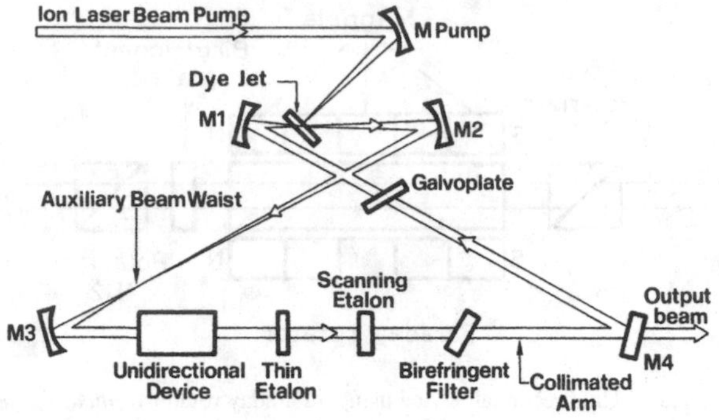

FIG. 7.25. High-power single-longitudinal mode dye laser using a unidirectional ring cavity.

rotation is through an angle of $\pi/2$. For a low gain laser, however, a total rotation of only about a few degrees typically introduces enough loss discrimination to ensure unidirectional operation. One should also note that, on passing through the birefringent plate, polarization rotation sense is clockwise independently of the propagation direction, as can be seen from Fig. 7.24a and b. The birefringent plate thus represents a reciprocal optical element.

A typical example of a folded ring configuration including a unidirectional device, used in a commercially available c.w. dye laser, is shown in Fig. 7.25. Pumping is, in this case, provided by an ion laser and the dye solution is made to flow transversely to the beam in the form of a liquid jet. Single-transverse mode operation is automatically achieved owing to the transverse gain distribution arising from the focused pumping. Laser tuning and gain bandwidth reduction are obtained by the combination of a birefringent filter with two Fabry-Perot etalons, a thin etalon and a scanning etalon, of different free spectral ranges. The optical path length of the cavity is conveniently tuned by rotating a tilted, plane-parallel, glass plate inside the resonator (the galvoplate). Single longitudinal mode operation is then ensured by an unidirectional device consisting of a Faraday rotator and a birefringent plate. Note that no polarizers are used because enough polarization loss is provided, in this case, by the faces of various optical elements inclined at Brewster's angle.

A more recent and rather interesting example of a unidirectional ring resonator using a non-planar cavity, used in a commercially available Nd:YAG laser, is shown in Fig 7.26.[21] The resonator is made from a small slab (e.g. $3 \times 6 \times 8$ mm) of Nd:YAG, whose faces B and D are cut at such an angle that the beam follows the non-planar path, BCD, shown in the figure, where point C is on the upper surface of the slab. Permanent magnets provide a magnetic field in the direction shown in the figure. The beam undergoes total internal reflection at surfaces B, C and D, and is reflected at surface A by a multilayer dielectric coating that acts as the output coupler. The Nd:YAG slab provides both the active material and Faraday rotator and is longitudinally pumped by the beam from a semiconductor laser (not shown in the figure). The non-planarity of the laser path creates an effect that is analogous to rotation by a half-wave plate. Assume that the planes of incidence at the corner faces B (plane ABC) and D (plane CDA) are at an angle β to the plane of incidence at the front face A (plane DAB). Assume also that the plane of incidence at the top face C (plane BCD) is perpendicular to those at the corner faces. The rotations of the planes of incidence then result in a net polarization-rotation, and image-rotation of 2β after the three reflections at points B, C, and D. Finally, the polarization

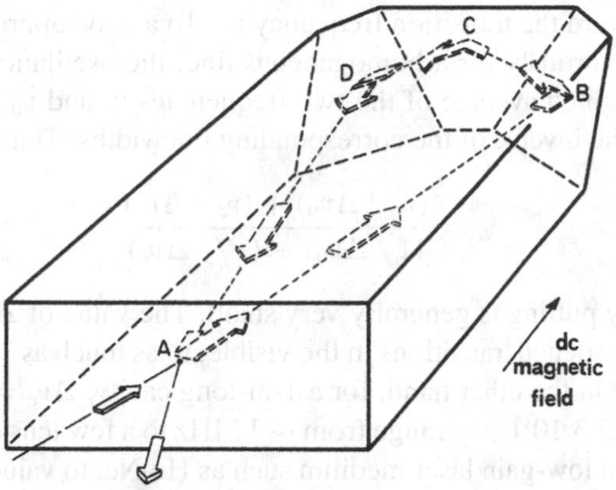

FIG. 7.26. Single-longitudinal-mode Nd:YAG laser using a unidirectional and non-planar ring cavity (non-planar ring oscillator, NPRO, after Kane and Byer,[21] by permission).

selective element is simply the multilayer dielectric coating at surface A, whose reflectivity depends on beam polarization. Since the homogeneously broadened linewidth of Nd:YAG is much smaller than that of a dye laser and since the frequency separation between longitudinal modes in Fig. 7.26, owing to the small cavity dimensions, is much larger than that of Fig. 7.25, no further frequency selective elements (such as birefringent filters or Fabry-Perot etalons) are needed in this case. Single-transverse-mode operation is again automatically achieved owing to the transverse gain distribution arising from the focused pumping. A compact and monolithic single-mode device is thus achieved.

7.9. FREQUENCY-PULLING AND LIMIT TO MONOCHROMATICITY

Let us assume that oscillation occurs on a cavity mode of frequency ν_c which is different from the center frequency ν_0 of the transition. We let $\Delta\nu_c$ and $\Delta\nu_0$ be the widths of the cavity mode resonance and of the laser transition, respectively, and address ourselves to the question of finding the laser frequency ν_L and the width of the output spectrum $\Delta\nu_L$ (Fig. 7.27).

The calculation of ν_L can be carried out within the semiclassical approximation. It can be shown[3,22] that ν_L will be in some intermediate position between ν_0 and ν_c i.e., the oscillation

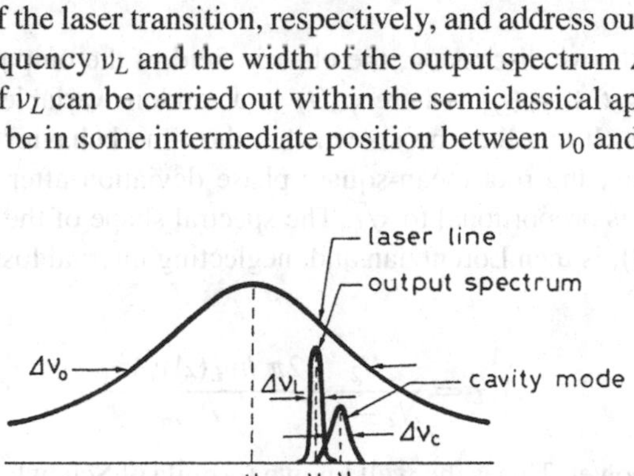

FIG. 7.27. Frequency pulling and spectral output of a single mode laser.

frequency is pulled toward the transition frequency ν_0. To a good approximation for an inhomogeneous line and rigorously for a homogeneous line, the oscillation frequency turns out to be given by the weighted average of the two frequencies ν_c and ν_0, the weighting factors being proportional to the inverse of the corresponding linewidths. Thus we have

$$\nu_L = \frac{(\nu_0 / \Delta\nu_0) + (\nu_c / \Delta\nu_c)}{(1 / \Delta\nu_0) + (1 // \Delta\nu_c)} \tag{7.9.1}$$

The effect of frequency pulling is generally very small. The value of $\Delta\nu_0$ may range from $\sim$ 1 GHz, for Doppler-broadened transitions in the visible, to as much as 300 GHz for solid-state lasers (see Table 2.1). On the other hand, for a 1-m-long cavity, $\Delta\nu_c = 1/2\pi\tau_c = \gamma c/2\pi L_e$ [see Eqs. (7.2.14) and (5.3.10)] may range from $\sim$ 1 MHz to a few tens of MHz (for γ ranging from $\sim$ 1%, typical of a low-gain laser medium such as He-Ne, to values of the order of 50% for high-gain materials). We thus see that the weighting factor $(1/\Delta\nu_c)$ is more than three order of magnitude larger than $(1/\Delta\nu_0)$.

We now turn our attention to the calculation of the width $\Delta\nu_L$ of the laser output spectrum when oscillation occurs on this single mode. Its ultimate limit is established by spontaneous emission noise or, more precisely, by the zero-point fluctuations of the laser mode field. Since these fluctuations can only be treated correctly via a full quantum electrodynamics approach to the problem (see Sect. 2.3.2), derivation of the expression for this limit is beyond the scope of our present treatment. It can be shown that, although both the amplitude and phase of the zero-point field fluctuate randomly, the spectral broadening of the output arises predominantly from random phase fluctuations while very small field fluctuations are induced by the amplitude fluctuations of the zero-point field. This can be traced back to the fact that, as discussed earlier in this chapter, the laser cavity photon population, and hence the output power, is quite insensitive to the number q_i of photons considered to be initially present in the cavity to simulate the effect of spontaneous emission. To be more precise, we can note that, according to Eq. (7.2.2), the rate of increase of cavity photons due to the "extra-photon" arising from spontaneous emission, is given, in the steady state, by $(d\phi/dt)_{se} = V_a B N_0$ and this term must be compared with the stimulated one which, again according to Eq. (7.2.2), is given by $V_a B N_0 \phi_0$. Since ϕ_0 may range from 10^{10} to 10^{16} (see example 7.1), it is apparent that the spontaneous emission term has a negligible effect on the number of cavity photons, i.e., to the field amplitude.

According to the above discussion, the electric field of the output beam can be written as $E(t) = E_0 \sin[2\pi\nu_L t + \varphi_n(t)]$, where $\varphi_n(t)$ is a random variable accounting for zero-point field fluctuations. It can then be shown that the time behavior of $\varphi_n(t)$ is typical of a diffusion process, i.e., the root-mean-square phase deviation after a time t, $\Delta\varphi(t) =< [\varphi_n(t) - \varphi_n(0)]^2 >^{1/2}$, is proportional to $\sqrt{t}$. The spectral shape of the emitted light, i.e., the power spectrum of $E(t)$, is then Lorentzian and, neglecting internal losses, the spectral width (FWHM) is given by[(23)]

$$\Delta\nu_L = \frac{N_2}{N_2 - N_1} \frac{2\pi\, h\nu_L (\Delta\nu_c)^2}{P} \tag{7.9.2}$$

where P is the output power. This is the well known formula of Schawlow-Townes, introduced by these two authors in their original proposal of the laser,[(24)] which establishes the *quantum limit* to laser linewidth.

Typically, the linewidth predicted by Eq. (7.9.2) turns out to be negligibly small compared to that produced by various other cavity disturbances, to be discussed later on, except for the very important case of a semiconductor laser. The reasons for this exception are that, as shown in the following example, $\Delta\nu_c$ for a semiconductor laser is typically about 5 orders of magnitude larger than that of e.g. a He-Ne laser. In fact, careful experiments done with GaAs lasers showed the actual linewidth to be about 50–100 times larger than the value predicted by Eq. (7.9.2). This observation was later on understood in terms of a new phenomenon peculiar of a semiconductor laser. The fluctuations of electron-hole density caused by spontaneous emission produce, in fact, a measurable fluctuation of the refractive index of the laser medium. The resulting fluctuation in optical cavity length thereby produces a fluctuation of the cavity frequency and hence of the oscillation frequency. Thus, in a semiconductor laser, the right hand side of Eq. (7.9.2) must be multiplied by a factor that we shall denote as α^2 and which is considerably larger than 1. The factor α is called the Henry-factor, named after the scientist who first provided an explanation of this phenomenon.[25]

Example 7.9. *Limit to laser linewidth in He-Ne and GaAs semiconductor lasers* Consider first the case of a single-mode He-Ne laser oscillating on its red transition ($\lambda = 632.8$ nm, $\nu_0 \cong 4.7 \times 10^{14}$ Hz). We will take $L_e = 1$ m, $\gamma = 1\%$ and we also assume an output power of $P = 1$ mW. From Eqs. (7.2.14) and (5.3.10) we get respectively $\tau_c = 3.3 \times 10^{-7}$ s and $\Delta\nu_c \cong 4.7 \times 10^5$ Hz. From Eq. (7.9.2), taking $N_2/(N_2 - N_1) \cong 1$, we obtain $\Delta\nu_L \cong 0.43$ mHz. Consider next the case of a single-mode GaAs semiconductor laser ($\lambda = 850$ nm) with a cavity length $L = l = 300\,\mu$m and with a power reflectivity at the two end faces of $R = 0.3$ (i.e. equal to the Fresnel losses of the uncoated semiconductor surfaces). Neglecting all other cavity losses we get $\gamma = -\ln(R) \cong 1.03$ and hence $\tau_c = nL/c\gamma = 3.4$ ps, where $n = 3.5$ is the refractive index of GaAs. We then get $\Delta\nu_c = 1/2\pi\tau_c \cong 4.7 \times 10^{10}$ Hz and, from Eq. (7.9.2) assuming $N_2/(N_2 - N_1) = 3$ and $P = 3$ mW, we obtain $\Delta\nu_L \cong 3.2$ MHz. Note that the laser linewidth is, in this case, almost 10 orders of magnitude larger than in the case of a He-Ne laser on account of the much shorter cavity decay time and hence of the much larger cavity linewidth.

According to the above example, the linewidth of a typical semiconductor laser arises from *quantum noise* and, in practice, is difficult to reducing below 1 MHz. In the case of a He-Ne laser and for all other lasers of relevance for obtaining small oscillation linewidths (such as Nd:YAG, CO_2 or Ar lasers), the linewidth determined by the Schawlow-Townes formula, even for modest powers of a few mW, is always well below 1 Hz and down to the mHz. Since $\nu_L = c/\lambda \cong 4.7 \times 10^{14}$ Hz, the relative monochromaticity of this laser, set by zero-point fluctuations, would be $(\Delta\nu_L/\nu_L) \cong 2.7 \times 10^{-18}$. Now, let us examine the cavity length stability requirement to keep the resonator frequency stable to the same degree. From Eq. (5.1.2) with $n = $ const., we find $(\Delta L/L) \cong -(\Delta\nu_c/\nu_c) \cong 2.7 \times 10^{-18}$ so that, e.g., with $L = 1$ m, we have $|\Delta L| \cong 2.7 \times 10^{-9}$ nm. This means that a cavity length variation by a quantity $\approx 10^{-8}$ smaller than a typical atomic dimension is already enough to induce a shift of the cavity frequency ν_c and hence of the oscillation frequency ν_L that is comparable to the oscillation linewidth given by Eq. (7.9.2). Thus, the limit to monochromaticity is, in practice, set by changes of cavity length induced by vibrations or thermal drifts, as we shall see in next section. These changes, arising from noise disturbances of perhaps less fundamental nature than those considered above, are often said to be due to *technical noise*.

7.10. LASER FREQUENCY FLUCTUATIONS AND FREQUENCY STABILIZATION

To consider laser frequency fluctuations, let us consider an active medium of refractive index n_m and length l in a laser cavity of length L in air. The effective cavity length is then given by $L_e = n_a(L - l) + n_m l$, where n_a is the refractive index of air. We can then separate the mode frequency changes into two parts: (i) *Long-term drifts*, i.e. occurring in a time longer than, say, 1 s, of either L or n_a which are mainly caused by temperature drifts or slow pressure changes of the ambient air surrounding the laser. (ii) *Short-term fluctuations* caused e.g. by acoustic vibrations of mirrors leading to cavity-length changes, by acoustic pressure waves which modulate n_a, or by short-term fluctuations of n_m, due e.g., to fluctuations of the discharge current in a gas laser or to air bubbles in the jet flow of a dye laser. In optically pumped solid-state lasers, fluctuations in pump power cause temperature fluctuations which in turn change the refractive index and therefore the optical length of the cavity.

To illustrate the influence of long-term drifts on cavity length, let α be the thermal expansion coefficient of the elements (e.g. invar rods) which determine the mirror separation. We then have $|\Delta\nu_L/\nu_L| \cong |\Delta L/L| = \alpha\Delta T$, where ΔT is the temperature change of the laser environment. Slow pressure changes, on the other hand, contribute to the frequency drift by an amount $|\Delta\nu_L| = \nu_L|\Delta n_a|(L - l)/L_e = \nu_L(n_a - 1)|\Delta p/p|(L-l)/L_e$, where Δn_a is the change of refractive index of the air as arising from the the slow change, Δp, of the ambient pressure, p.

Example 7.10. *Long term drift of a laser cavity* Taking, for invar, $\alpha = 1 \times 10^{-6}\,\mathrm{K}^{-1}$ and considering a frequency in the central part of the visible spectrum, i.e. $\nu_L \cong 5 \times 10^{14}$ Hz, from the previous expressions we obtain that the frequency drift, $\Delta\nu_L$, due to a thermal change, ΔT, amounts to $|\Delta\nu_L| = \nu_L\alpha\Delta T = (5 \times 10^8\ \Delta T)$ Hz. This shows that a change of ΔT of, e.g., only 0.1 K would produce a frequency drift of $\sim$ 50 MHz. To calculate the frequency drift due to a slow pressure change, we note that, for a gas laser, one typically has $(L - l)/L_e \cong 0.2$ while for air one has $n_a \cong 1.00027$. Again for $\nu_L \cong 5 \times 10^{14}$ Hz, we can then write $|\Delta\nu_L| \cong 2.7 \times 10^{10}|\Delta p/p|$Hz. Thus, for a relative pressure change $|\Delta p/p| \cong 3 \times 10^{-3}$, which can readily occur during one hour time, one has $|\Delta\nu_L| \cong 80$ MHz.

According to the previous example, to reduce long-term frequency drifts below e.g., 1 MHz one needs to use very low expansion materials for the spacer elements, down to an expansion coefficient perhaps smaller than $1 \times 10^{-7}\,\mathrm{K}^{-1}$ and to stabilize the ambient temperature below 0.01 °C. One also needs to enclose the laser in a pressure stabilized chamber. The reduction of short-term frequency fluctuations is an even more difficult problem and one needs to rely on good vibration isolation optical tables and efficient covers of the whole laser path. It is therefore generally difficult to reduce short term frequency fluctuations below the 1 MHz level, except for monolithic and compact solid-state lasers, such as the NPRO laser considered in Fig. 7.26, where short-term frequency fluctuations of about 10 kHz have been measured.

To characterize the spectrum of laser frequency fluctuations, let us write the electric field of the output beam as $E(t) = E_0 \sin[2\pi\nu_L t + \varphi_n(t)]$ where ν_L is the central laser frequency and $\varphi_n(t)$ describes the noise phase fluctuations. The instantaneous frequency can then be written as $\nu(t) = \nu_L + \mathrm{d}\varphi_n(t)/2\pi\mathrm{d}t = \nu_L + \nu_n(t)$, where $\nu_n(t)$ is the frequency noise. It is this frequency noise that is fundamentally related to linewidth or frequency stability, and, consequently, it is the term that needs to be characterized. The measure that is used to characterize

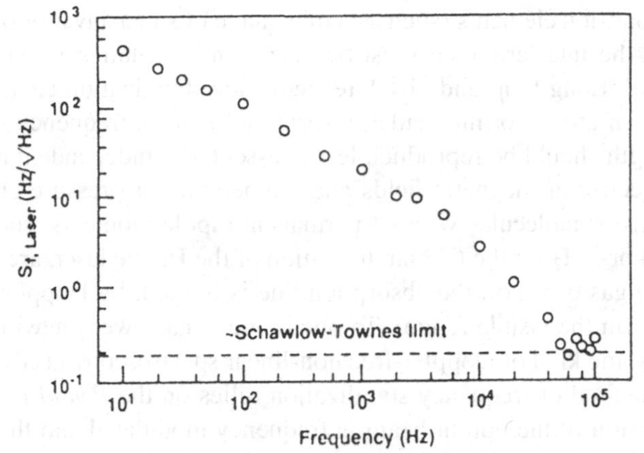

FIG. 7.28. Frequency noise spectra from a free-running diode-pumped Nd:YAG laser (after ref.,[26] by permission).

frequency fluctuations is the spectral power density of frequency noise (see Appendix G), represented by $S_\nu(\nu_m)$ and having units of Hz^2/Hz. Here, ν_m is called the *offset frequency*, which can be thought of as the frequency at which the phase $\varphi_n(t)$ is being modulated by the noise. The way $S_\nu(\nu_m)$ is measured in practice is to transform $\nu_n(t)$ into e.g., a voltage signal $V_n(t)$ by a frequency to voltage converter and then measure the power spectrum of $V_n(t)$ by an electronic spectrum analyzer. As a representative example Fig. 7.28 shows the square root of the frequency noise spectrum, $\sqrt{S_\nu(\nu_m)}$, from a free-running diode-pumped monolithic Nd:YAG laser.[26] In the same figure the Schawlow-Townes limit for $\sqrt{S_\nu(\nu_m)}$ is also indicated. In fact, for a Lorentzian line as that predicted by the Schawlow-Townes theory, it can be shown that the spectral power density of frequency fluctuations is white, i.e., $S_\nu(\nu_m)$ is a constant given by[27]

$$S_\nu(\nu_m) = \Delta\nu_L/\pi \tag{7.10.1}$$

where $\Delta\nu_L$ the linewidth [given by Eq. (7.9.2) for the Schawlow-Townes theory]. Note the large increase of the noise spectrum for offset frequencies smaller than 100 kHz as due, in this case, to acoustic disturbances and pump power fluctuations.

For the most sophisticated applications, e.g. for gravitational wave detection, the laser noise spectrum needs to be strongly reduced and, to this purpose, one needs to use techniques for active stabilization of the cavity length. To achieve this, one of the cavity mirrors is mounted on a piezoelectric transducer and frequency stabilization is achieved by applying a feedback voltage, via a suitable electronic circuit, to the transducer. By sending a fraction of the laser radiation to a frequency discriminator of sufficient high resolution and stability, the voltage fluctuations at the output yield the required error signal. The sharp transmission (or reflection) lines of a high finesse Fabry-Perot (FP) interferometer or a sharp absorption line of an atomic or molecular gas kept in a low pressure cell are often used as frequency discriminators. Fabry-Perot interferometers with finesse larger than 10^5, using mirrors with absorption and scattering losses of a few parts per million, have been used. For a 1 m long FP interferometer, sharp transmission lines with width in the range of a few kHz have been obtained. To reduce the frequency fluctuations of the FP cavity, the mirrors must be mounted using very low expansion elements (such as a tube made of super-invar or of very-low expansion ceramic) and the interferometer must be placed in a container providing pressure and temperature control. Long term and absolute frequency stabilization can, however, only be achieved by using an atomic or molecular absorption line as a frequency reference. A good reference wavelength should be reproducible and essentially independent of external perturbations such as electric or magnetic fields and temperature or pressure changes. Therefore transitions in atoms or molecules without permanent dipole moments, such as CH_4 for the 3.39 μm transition or $^{129}I_2$ for the 633 nm transition of the He-Ne laser, are the most suitable. For a low pressure gas or vapor, the absorption line is limited, by Doppler broadening, to a width of $\sim$ 1 GHz (in the visible range). To obtain much narrower linewidths down perhaps to the kHz range, some kind of Doppler-free non-linear spectroscopy needs to be used.[28]

A common method of frequency stabilization relies on the *Pound-Drever technique*[29] where a small fraction of the output beam is frequency modulated and then passed through the frequency discriminator, i.e. the FP interferometer or the absorption cell. To understand the principle of operation of this technique, we need first to point out that, any element exhibiting a transmission which changes with frequency will also induce, on the incident wave, a phase-shift which depends on the wave frequency. For a FP interferometer used in transmission, the phase shift can be obtained from (4.5.4). For an absorption line, on the other hand, the phase shift can be written as $\phi = 2\pi nl/\lambda$, where l is the length of the absorption cell and n is the refractive index of the medium. For a Lorentzian line, the refractive index n can be related to the absorption coefficient, α, of the medium by the dispersion relation

$$n(\nu - \nu_0) = 1 + \frac{c}{2\pi\nu}\frac{\nu_0 - \nu}{\Delta\nu_0}\alpha(\nu - \nu_0) \tag{7.10.2}$$

where n_0 is the refractive index sufficiently far away from the resonance line, ν is the frequency of the e.m. wave, ν_0 is the transition frequency, and $\Delta\nu_0$ is the transition width. Note that, for $\nu = \nu_0$, one has $n = n_0$, and the transition makes no contribution to the refractive index. For an inhomogeneous line, one must add, at frequency ν, the phase shifts induced from all atoms, with their transition frequencies, ν_0', being now distributed according to the function $g^*\left(\nu_0' - \nu_0\right)$. The refractive index of the medium is then obtained from (7.10.2), averaging over the frequency distribution $g^*\left(\nu_0' - \nu_0\right)$. We get

$$n_{eff} = n_0 + \frac{cN_t}{2\pi\nu}\int \frac{\nu_0' - \nu}{\Delta\nu_0}\sigma_h\left(\nu - \nu_0'\right)g^*\left(\nu_0' - \nu_0\right)d\nu_0' \tag{7.10.3}$$

where N_t is the total ground-state population and σ_h is the homogeneous cross-section. According to Eq. (7.10.2) or (7.10.3), for a given profile of the absorption coefficient $\alpha = \alpha(\omega - \omega_0)$ (see Fig. 7.29a), the corresponding frequency shift generally takes the form shown in Fig. 7.29b. For simplicity, the phase shift at the line central frequency, $\phi_0 = 2\pi n_0 l/\lambda$, is taken to be zero. A rather similar curve applies for a FP interferometer and, so, Fig. 7.29 provides a general representation of the phase-shift of our frequency discriminator (i.e., the absorption cell or the FP interferometer).

We now consider a frequency-modulated beam and write its electric field as $E(t) = E_0 \exp\left[j\omega t + j\Gamma \sin\left(\omega_m t\right)\right]$, where Γ is the phase modulation index and ω_m is the modulation frequency. We can then expand this beam in terms of Bessel functions and write

$$E(t) = E_0 e^{j\omega t} \sum_{-\infty}^{+\infty} {}_n J_n(\Gamma) e^{jn\omega_n t} \tag{7.10.4}$$

where J_n is the Bessel function of order n. If we limit our considerations to the first two side-bands at frequency $\pm\omega_m$, we obtain from (7.10.4)

$$E(t) = E_0 e^{j\omega t}\left[-J_1(\Gamma)e^{-j\omega_m t} + J_0(\Gamma) + J_1(\Gamma)e^{j\omega_m t}\right] \tag{7.10.5}$$

where we have used the property $J_{-1}(\Gamma) = -J_1(\Gamma)$. The electric field of the wave after leaving the frequency discriminator will be given by

$$E(t) = E_0 e^{j\omega t}\left[-J_1(\Gamma)e^{-j\omega_m t - j\phi_{-1}} + J_0(\Gamma)e^{-j\phi_0} + J_1(\Gamma)e^{j\omega_m t - j\phi_1}\right] \tag{7.10.6}$$

where ϕ_0, ϕ_{-1}, and ϕ_1 are the phase shifts associated with the carrier wave and with the two sidebands, respectively (Fig. 7.29b). If the beam transmitted by the discriminator is sent to a quadratic detector, the detected photocurrent will be proportional to EE^*, where E^* is the complex conjugate of E. The component of the photocurrent at frequency ω_m will then be proportional to

$$(EE^*)_{\omega_m} = 2|E_0|^2 \mathrm{Re}\left\{J_0 J_1 e^{j\omega_m t}\left[e^{j(\phi_0 - \phi_1)} - e^{j(\phi_{-1} - \phi_0)}\right]\right\} \tag{7.10.7}$$

where Re stands for real part. If now the carrier frequency ω of the wave coincides with the discriminator central frequency $\omega_0 = 2\pi\nu_0$, one has $\phi_0 = 0$ and $\phi_{-1} = -\phi_1$ (see Fig. 7.29).

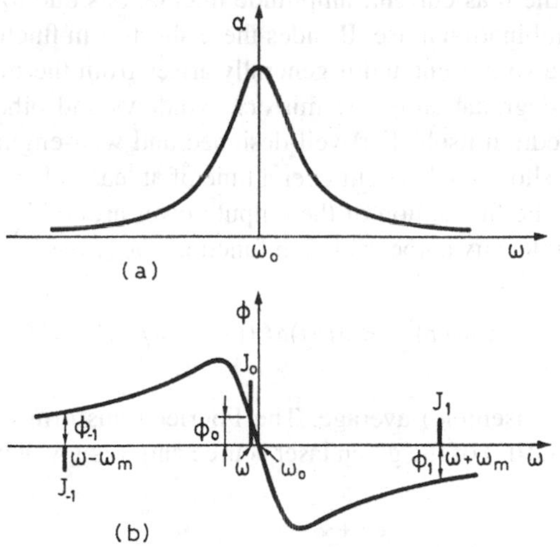

FIG. 7.29. Pound-Drever technique for frequency stabilization to the transmission minimum of an absorption cell (or to the transmission peak of a Fabry-Perot interferometer).

One then gets from (7.10.7) $(EE^*)_{\omega_m} = 0$. If, on the other hand, $\omega \neq \omega_0$ and if we assume $\omega_m \gg 2\pi\Delta\nu_0$, then we can write $\phi_{-1} \cong -\phi_1$ and from (7.10.7) we get

$$(EE^*)_{\omega_m} = -4|E_0|^2 J_0 J_1 \sin(\phi_0)\sin(\omega_m t - \phi_1) \tag{7.10.8}$$

The sign of the component of the photocurrent at frequency ω_m will therefore depend on the sign of ϕ_0 i.e., on weather ω is above or below ω_0. This component can then be used as the error signal for the electronic feed-back loop, to force the carrier frequency of the wave to coincide with the central frequency of the discriminator. The precision by which this can be achieved depends on the gain of the feedback loop and on its bandwidth. With very sharply defined frequency discriminators ($\Delta\nu_0 \cong 30$ kHz) short-term frequency drifts in the 100 mHz range have been achieved, in this way.[(30)]

7.11. INTENSITY NOISE AND INTENSITY NOISE REDUCTION

In the previous sections, we have seen that spontaneous emission and cavity length fluctuations induce only a frequency noise and, under these conditions, the field amplitude of the output beam can be taken to be independent of time. There are, however, other perturbations of a laser which result in amplitude fluctuations i.e., produce an *intensity noise*. The most common perturbations of this type can be listed as follows: (i) For a gas laser: fluctuations of the power supply current, instability of the discharge process, and mirror misalignments owing to resonator vibrations. (ii) For dye lasers: density fluctuations of the dye jet solution and the presence of air bubbles in the solution. (iii) For solid-state lasers: pump fluctuations (both for lamp pumping and diode pumping), and cavity misalignments. (iv) For semiconductor lasers: fluctuations of the bias current, amplitude fluctuations due to spontaneous emission and electron-hole recombination noise. Besides these short-term fluctuations, long term drift of the output power is also present and it generally arises from thermal misalignment of the laser cavity and from degradation of the mirrors, windows and other optical components, including the active medium itself. For well-designed and well-engineered lasers, however, this power degradation should only occur over a time of at least a few thousands hours.

If we let $\delta P(t)$ be the fluctuation of the output power around its average value $< P >$, one can first define an intensity autocorrelation function $C_{PP}(\tau)$ as

$$C_{PP}(\tau) = <\delta P(t)\delta P(t+\tau)>/<P>^2 \tag{7.11.1}$$

where $<>$ stands for (ensemble) average. The Fourier transform of $C_{PP}(\tau)$ is called the *relative intensity noise* (RIN) of the given laser source and it is given by

$$\text{RIN}(\omega) = \int_{-\infty}^{+\infty} C_{PP}(\tau)\exp(j\omega\tau)d\tau \tag{7.11.2}$$

Obviously, $C_{PP}(\tau)$ is obtained from (7.11.2) by taking the inverse Fourier transform, i.e.,

$$C_{PP}(\tau) = \frac{1}{2\pi}\int_{-\infty}^{+\infty} \mathrm{RIN}(\omega)\exp(-j\omega\tau)d\omega \tag{7.11.3}$$

A typical RIN spectrum as obtained from a single mode, diode-pumped, NPRO oscillator is shown in Fig. 7.30a (curve 1). Note that the vertical scale is expressed in dB/Hz, a notation which may create some confusion and whose meaning is as follows

$$\mathrm{RIN(dB/Hz)} \times \Delta\nu = 10\log[\mathrm{RIN}(\nu) \times \Delta\nu] \tag{7.11.4}$$

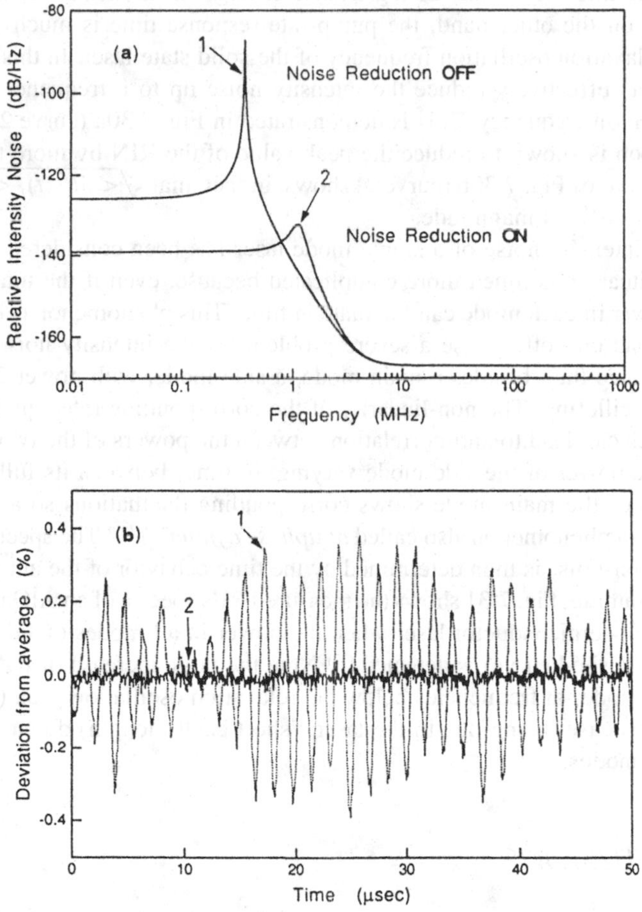

FIG. 7.30. (a): Typical relative intensity noise spectra for a diode-pumped Nd:YAG laser using the NPRO configuration (see Fig. 7.26) without (*curve* 1) and with (*curve* 2) active noise reduction. (b) Corresponding plots of typical relative fluctuations of the output power (after ref.[31] by permission).

where $RIN(\nu) = 2\pi RIN(\omega)$ and $\Delta\nu = 1$ Hz. One can see that the RIN spectrum is strongly peaked at a frequency ($\nu \cong 300$ kHz in the figure) corresponding to the laser relaxation oscillation frequency (see next Chapter). The corresponding relative fluctuation of the output power, $\delta P(t)/<P>$, is shown in Fig. 7.30b (curve 1). From this figure one can observe that the root-mean-square variation of $\delta P(t)$, i.e., $\sqrt{<\delta P^2(t)/<P>^2}$, is $\sim 2 \times 10^{-3}$. The same result can be obtained from Fig. 7.30a if one takes into account that the (3 dB) width, $\Delta\nu$, of the relaxation oscillation peak is roughly 2 kHz while the RIN value at the peak is ~ -85 dB/Hz. In fact, according to (7.11.4), we have $RIN(\nu) = 10^{-8.5} \cong 3.16 \times 10^{-9}$ Hz^{-1}. From Eqs. (7.11.1) and (7.11.3) with $\tau = 0$, we then obtain $C_{PP}(0) =< \delta P^2(t)/<P>^2 \cong RIN(\nu) \times \Delta\nu \cong 6.32 \times 10^{-6}$ and so $\sqrt{<\delta P^2(t)/<P>^2} \cong 2.5 \times 10^{-3}$.

To reduce the intensity noise, a negative feedback is often applied to the pump power supply. The time needed to establish this feedback is limited by the discharge response time (pump-rate response time). Accordingly, for a gas laser, this feedback scheme cannot be used to e.g. reduce the intensity noise arising from discharge instabilities. For a diode-pumped solid-state laser, on the other hand, the pump-rate response time is much shorter than the inverse of the relaxation oscillation frequency of the solid state laser. In this case a negative feedback loop can effectively reduce the intensity noise up to a frequency larger than the oscillation relaxation frequency. This is demonstrated in Fig. 7.30a (curve 2) where the use of a feedback loop is shown to reduce the peak value of the RIN by more than 35 dB. The corresponding curve of Fig. 7.30b (curve 2) shows, in fact, that $\sqrt{<\delta P^2(t)/<P>^2}$ is reduced by more than one order of magnitude.

So far, the intensity noise of a single mode laser has been considered. For multimode oscillation the situation is much more complicated because, even if the total power is kept constant, the power in each mode can fluctuate in time. This phenomenon is known as *mode-partition-noise* and can often pose a severe problem for the intensity noise in each mode. Assume for instance that, besides a main mode, a side-mode, with power 20 dB below the main mode, is oscillating. The non-linearity of the corresponding rate equations provides a mechanism which can lead to anti-correlation between the powers of the two modes.[(20)] This may result in the power of the side mode varying, in time, between its full value and zero while the power of the main mode shows corresponding fluctuations so as to keep a constant total power (a phenomenon also called *antiphase dynamic*).[(32)] The spectral frequency of this mode-partition-noise is then determined by the time behavior of the antiphase-dynamics process. As an example, Fig. 7.31 shows the measured RIN spectra of an AlGaAs Fabry-Perot-type (see Chap. 9) semiconductor laser when the power in all modes (solid curve) or in the dominant mode (dashed curve) is detected.[(33)] Note the large increase of RIN which occurs, in the dominant mode, at frequencies below the relaxation oscillation peak ($\cong 2.5$ GHz, the high value, in this case, being due to the laser's short cavity length) due to the presence of other oscillating modes.

7.12. CONCLUSIONS

In this chapter, a few topics related to the cw behavior of both a four-level and quasi-three-level laser has been considered in some detail. The space-independent rate-equations, under the simplest decay-rate conditions, have been developed first (ideal four-level and quasi-three-level laser) and the cw behavior predicted by these equations, including optimum

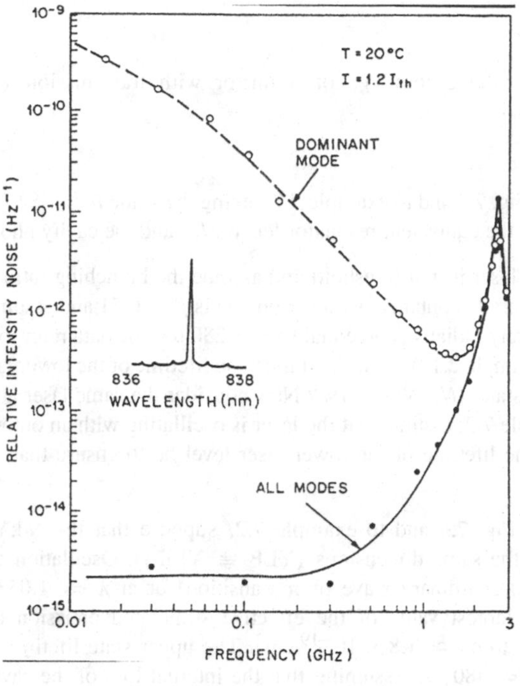

FIG. 7.31. Experimental observation of mode partition noise in a multimode semiconductor laser (after ref.[33] by permission).

coupling conditions, has been considered. The results obtained from the space-dependent equations have then been discussed at some length. It should be recalled that the rate-equation formulation represents the simplest way to describe cw, and transient, laser behavior. In order of increasing accuracy, and complexity, one should consider the semiclassical and the quantum electrodynamics treatments. For the cw case, however, the equations for the semiclassical treatment reduce to the rate equations. The full quantum treatment, on the other hand, is required to correctly describe the start of laser oscillation as well as the fundamental limit to laser frequency noise. When, however, the number of photons in a given cavity mode is much larger than one, the (average) results of the quantum treatment coincide with those of the semiclassical treatment. It should also be noted that the rate equations, in their simplest form as given here, only apply to relatively few cases. In most cases, there are more than just four levels involved, and the rate equation treatment becomes correspondingly more complicated. In fact it could be said that, in general, each laser has its own particular set of rate equations. The equations considered in this chapter, however, provide a model that can be readily extended to handle more complicated situations.

Besides topics which can be directly discussed in terms of rate equations, other subjects of fundamental importance for cw laser behavior have also been discussed, namely: (i) Reasons for multimode oscillation, methods of single mode selection, and laser tuning. (ii) Limit to monochromaticity, for single mode lasers, as well as field fluctuations of the output beam, both in frequency and amplitude. (iii) Methods to actively reduce both frequency and amplitude fluctuations. The ensemble of these topics constitute the minimum set of knowledge required for a balanced and up-to-date understanding of cw laser behavior.

PROBLEMS

7.1. Calculate the logarithmic loss, γ, of a mirror with transmission $T = 80\%$ and negligible internal loss.

7.2. Prove Eq. (7.2.11)

7.3. With reference to Fig. 7.4 and to example 7.2, taking the value $n = 1.82$ for the refractive index of Nd:YAG, calculate the equivalent resonator length, L_e, and the cavity photon decay time.

7.4. Consider a 4-level laser just at threshold and assume the branching ratio of the $2 \to 1$ transition compared to the overall spontaneous transition rate is $\beta = 0.51$ and assume that the overall upper-state lifetime is purely radiative and equal to $\tau = 230\ \mu s$ (the data refer to the 1.064 μm transition of Nd:YAG, see example 2.13). How short must the lifetime of the lower laser level, 1, be to ensure that, in the steady state, $(N_1/N_2) < 1\%$? Now consider the same laser above threshold and, with reference to example 7.2, assume that the laser is oscillating with an output power $P_{out} = 200$ W. How short must the lifetime of the lower laser level be to ensure that, under these conditions, $(N_1/N_2) < 1\%$?

7.5. With reference to Fig. 7.4 and to example 7.2, suppose that the Nd:YAG rod is replaced by a Nd:YLF rod of the same dimensions (YLF $\equiv$ $YLiF_4$). Oscillation can then occur at either $\lambda = 1.047\ \mu$m (extraordinary wave or π-transition) or at $\lambda = 1.053\ \mu$m (ordinary wave or σ-transition). The largest value of the effective stimulated emission cross section is for the π-transition, equal to $\sigma_e \cong 1.8 \times 10^{-19}$ cm^2. The upper state lifetime is the same for the two transitions, i.e., $\tau = 480\ \mu s$. Assuming that the internal loss of the cavity and the lamp pumping efficiency remain the same as for Nd:YAG, calculate the threshold inversion and the threshold pump power and compare the results with those of Nd:YAG. Assuming that the energy separation between the $^4F_{3/2}$ upper laser level and the ground level remains the same as for Nd:YAG and taking the same value of the beam area A_b, calculate the slope efficiency.

7.6. In the case of the previous problem, calculate the optimum output mirror transmission and the corresponding optimum output power when the laser is pumped by a lamp input power $P_p = 7$ kW.

7.7. For the high power CO_2 laser of Figs 7.7 and 7.8, and with reference to the data considered in example 7.3, calculate the optimum output coupling and the optimum power for an input pump power $P_p = 140$ kW. The resulting optimum value of γ_2 turns out to be substantially smaller than that considered in example 7.3., i.e., the unstable resonator of Fig. 7.7 is substantially overcoupled. This overcoupling is intentional, to increase the peak intensity of the output beam when focused by a lens. With reference to the focusing properties of the annular beam of an unstable resonator discussed in Sect. 5.6.3., show in fact that the ~ 12 kW beam of Fig. 7.8 (at $P_p = 140$ kW) produce a higher intensity at the focus of a lens compared to that of the optimum beam considered in this problem.

7.8. Consider the lamp-pumped Nd:YAG laser of Fig. 7.4. The pumping beam induces a thermal lens of focal length f in the rod whose dioptric power $1/f$ is proportional to the pump power, P_p. At a pump power of $P_p = 7$ kW and for the rod dimensions of Fig. 7.4, the thermal induced focal length is $f \cong 25$ cm.[34] Now assume that, to calculate the cavity mode, the rod of Fig. 7.4 can be simulated by a this lens, of focal length $f \cong 25$ cm, placed at the resonator center. In this case, calculate the TEM_{00} mode spot size at the lens and at the mirror location.

7.9. One can show that the radial extension of a TEM_{lm} mode of higher order (i.e., for $l \cong m \gg 1$) is approximately given by $w_{lm} \cong \sqrt{l}\, w$, where w is the spot size of the corresponding TEM_{00} mode.

The maximum number of Hermite-Gaussian modes that can fit within a rod of radius a will be such that $w_{lm} \cong a$.[35] Using this argument and the results obtained in the previous problem, calculate the approximate number of transverse modes which oscillate in the configuration of Fig. 7.4. The beam divergence of a TEM_{lm} mode ($l \cong m \gg 1$), θ_{lm}, is then approximately given by $\theta_{lm} \cong \sqrt{l}\lambda/\pi w_0$, where w_0 is the spot size of the TEM_{00} mode at the beam waist.[35] Calculate then the beam divergence for the case of Fig. 7.4 assuming that it is equal to the divergence of the highest order mode which is oscillating.

7.10. Consider the Nd:YAG laser of example 7.4 and assume that the optimum output coupling can be calculated by the formula established via the space-dependent case of Sect. 7.5. Calculate the optimum output coupling and, using this value of γ_2, calculate, with the help of Eq. (7.3.34), the expected value of the output power at a diode-laser pump power of $P_p = 1.14$ W.

7.11. Consider a 4 mm diameter 56 mm long 0.9 at. % Nd:YAG rod side-pumped at 807 nm wavelength by fiber-coupled diode lasers.[36] Assume that the laser is oscillating on a TEM_{00} mode with a constant spot-size within the active medium of $w_a \cong 1.4$ mm. Let the transmission of the output mirror be $T_2 = 15\%$ and assume a total internal loss of $\gamma_i = 3.8\%$ (see example 7.2). Calculate the output power at a laser-diode optical pump power of $P_p = 370$ W, corresponding to an absorbed power of $P_{ap} = 340$ W, and the corresponding slope efficiency. Compare the calculated values with the experimental values of ref.[36] and try to explain the discrepancy.

7.12. Consider again the diode-pumped laser of the previous problem, assume that, at the stated pump power, the thermally-induced lens in the rod has focal length $f = 21$ cm, and take a symmetric flat-flat resonator as in Fig. 7.4. Under the simplifying assumption that the rod can be simulated by a thin lens of focal length f, calculate the distance of the two plane mirrors from this lens to obtain a spot-size, at the lens position, of $w_a = 1.4$ mm. Calculate also the corresponding spot-size at the two mirrors.

7.13. Prove Eq. (7.4.7)

7.14. Consider a He-Ne laser oscillating on its red, $\lambda = 632.8$ nm in air, transition and assume a length of the gas tube of $l = 20$ cm, a tube radius of 1 mm, a 0.1 Torr partial pressure of Ne atoms, an output coupler transmission of 1% and a single pass internal loss of 0.5%. According to example 2.12, the effective cross section and the overall lifetime of the laser transition can be taken to be $\sigma_e = 3 \times 10^{-13}$ cm^2 and $\tau = 150$ ns, respectively. Assume, for simplicity, that the lifetime of the lower state is much shorter than that of the upper state. Calculate the threshold inversion, the ratio of this inversion to the total Ne population, and the critical pump rate. The upper laser level is predominantly pumped through the 2^1S state of He which lies ~ 20.5 eV above the ground state. Assuming unit quantum efficiency in this near-resonant energy-transfer process, calculate the minimum threshold pump power. For an output power of 3 mW, calculate also the ratio of the number of photons emitted by stimulated emission to the number of atoms that decay spontaneously.

7.15. An Ar ion laser oscillating on its green, $\lambda = 514.5$ nm, transition has a 10% unsaturated gain per pass. The resonator consists of two concave spherical mirrors both of radius of curvature $R = 5$ m and spaced by $L = 100$ cm. The output mirror has a $T_2 = 5\%$ transmission while the other mirror is nominally 100% reflecting. Identical apertures are inserted at both ends of the resonator to obtain TEM_{00} mode operation. Neglecting all other types of losses, calculate the required aperture diameter.

7.16. The linewidth, $\Delta\nu_0^* = 50$ MHz, of a low-pressure CO_2 laser is predominantly established by Doppler broadening. The laser is operating with a pump power twice the threshold value. Assuming that one mode coincides with the transition peak and equal losses for all modes, calculate the maximum mirror spacing that would still allow single longitudinal mode operation.

7.17. Consider an Ar ion laser oscillating on its green, $\lambda = 514.5$ nm, transition and assume this transition to be Doppler broadened to a width $\Delta\nu_0^* = 3.5$ GHz. Assume a cavity length $L_e = 120$ cm, a length of the Ar tube of 100 cm, and take a single pass cavity loss of $\gamma = 10\%$. According to example 2.12., we can take an effective value of the stimulated emission cross section and upper state lifetime of $\sigma_e = 2.5 \times 10^{-13}$ cm^2 and $\tau = 5$ ns, respectively. Assume the lifetime of the lower laser level much shorter than that of the upper laser level and that one cavity mode coincides with the transition peak. Calculate the threshold inversion for this central mode and the threshold pump rate. At how great a pump rate above threshold does oscillation start on the two adjacent longitudinal modes?

7.18. A He-Ne laser is oscillating on three adjacent longitudinal modes, the central one being coincident with the center of the laser transition. The cavity length is 50 cm and the output coupling 2%. Knowing that the laser linewidth is $\Delta\nu_0^* = 1.7$ GHz, calculate the mode spacing.

7.19. Assume that one cavity mirror is mounted on a piezoelectric transducer. Show that the comb of longitudinal modes shifts by approximately one comb spacing for a $\lambda/2$ translation of the transducer.

7.20. Consider a single longitudinal mode He-Ne laser and assume that the oscillating frequency is made to coincide with the frequency of the transition peak by the use of a piezoelectric transducer attached to a cavity mirror. How far can the mirror be translated before a mode-hop (i.e., a switch of oscillation to the next mode) occurs?

7.21. An Ar ion laser oscillating on its green, $\lambda = 514.5$ nm, transition has a total loss per pass of 4%, an unsaturated peak gain, $G_p = \exp(\sigma_p Nl)$, of 1.3, and a cavity length of 100 cm. To select a single longitudinal mode, a tilted and coated quartz ($n_r = 1.45$) Fabry-Perot etalon with a 2 cm thickness is used inside the resonator. Assuming, for simplicity, that one cavity mode is coincident with the peak of the transition (whose linewidth is $\Delta\nu_0^* = 3.5$ GHz), calculate the etalon finesse and the reflectivity of the two etalon faces to ensure single longitudinal mode operation.

7.22. With reference to Fig. 7.31, evaluate the relative root-mean-square fluctuation of the output power of the semiconductor laser for the dominant mode and for all modes.

References

1. H. Statz and G. de Mars, Transients and Oscillation Pulses in Masers, in *Quantum Electronics*, ed. by C.H. Townes (Columbia University Press, New York, 1960), pp. 530–537
2. R. Dunsmuir, Theory of Relaxation Oscillations in Optical Masers, *J. Electron. Control* **10**, 453–458 (1961)
3. M. Sargent, M.O. Scully, and W.E. Lamb, *Laser Physics* (Addison-Wesley, London, 1974)
4. R. H. Pantell and H.E. Puthoff, *Fundamentals of Quantum Electronics*, (Wiley, New York, 1969), Chap. 6, Sect. 6.4.2
5. W. W. Rigrod, Saturation Effects in High-Gain Lasers, *J. Appl. Phys.* **36**, 2487–2490 (1965)
6. L. W. Casperson, Laser Power Calculations: Sources of Error, *Appl. Opt.* **19**, 422–431 (1980)
7. W. Koechner, *Solid-State Laser Engineering*, Vol. 1, Springer Series in Optical Sciences, fourth edition (Springer-Verlag, Berlin, 1996), Chap. 3, adapted from Fig. 3.21
8. Reference,[7] Chap. 3, Fig. 3.22
9. D. Findlay and R. A. Clay, The Measurement of Internal Losses in 4-Level Lasers, *Phys. Lett.* **20**, 277–278 (1966)
10. Private communication, Istituto di Ricerca per le Tecnologie Meccaniche, Vico Canavese, Torino
11. M. C. Fowler, Quantitative Analysis of the Dependence of CO_2 Laser Performance on Electrical Discharge Properties, *Appl. Phys. Lett.* **18**, 175 (1971)
12. E. Hoag *et al.*, Performance Characteristics of a 10 kW Industrial CO_2 Laser System, *Appl. Opt.* **13**, 1959 (1974)

13. P. F. Moulton, An Investigation of the Co : MgF_2 Laser System, *IEEE J. Quant. Electr.* **QE-21**, 1582–1588 (1985)
14. V. Evtuhov and A. E. Siegman, A Twisted-Mode Technique for Obtaining Axially Uniform Energy Density in a Laser Cavity, *Appl. Opt.* **4**, 142–143 (1965)
15. J. Berger *et al.*, 370 mW, 1.06 μm, cw TEM_{00} Output from a Nd:YAG Laser Rod End-Pumped by a Monolithic Diode Array, *Electr. Lett.* **23**, 669–670 (1987)
16. W. P. Risk, Modeling of Longitudinally Pumped Solid-State-Lasers Exhibiting Reabsorption Losses, *J. Opt. Soc. Am. B* **5**, 1412–1423 (1988)
17. T. Y. Fan and R. L. Byer, Modeling and CW Operation of a Quasi-Three-Level 946 nm Nd:YAG Laser, *IEEE J. Quant. Electr.* **QE-23**, 605–612 (1987)
18. P. Lacovara *et al.*, Room-Temperature Diode-Pumped Yb:YAG Laser, *Opt. Lett.* **16**, 1089–1091 (1991)
19. A. Yariv, Energy and Power Considerations in Injection and Optically Pumped Lasers, *Proc. IEEE* **51**, 1723–1731 (1963)
20. C. L. Tang, H. Statz, and G. de Mars, Spectral Output and Spiking Behavior of Solid-State Lasers, *J. Appl. Phys.* **34**, 2289–2295 (1963)
21. T. J. Kane and R. L. Byer, Monolithic, Unidirectional Single-Mode Nd:YAG Ring Laser, *Opt. Lett.* **10**, 65 (1985)
22. A. E. Siegman, *Lasers* (University Science Books, Hill Valley, California, 1986), Chap. 12, Sect. 12.2
23. A. Yariv, *Optical Electronics* (Saunders College Publishing, Forth Worth, 1991), Sect. 10.7
24. A. L. Schawlow and C. H. Townes, Infrared and Optical Masers, *Phys. Rev.* **112**, 1940–1949 (1958)
25. C. H. Henry, Theory of Linewidth of Semiconductor Lasers, *IEEE J. Quant. Electr.* **QE-18**, 259 (1982)
26. T. Day, E. K. Gustafson, and R. L. Byer, Sub-Hertz Relative Frequency Stabilization of Two Diode Laser Pumped Nd:YAG Lasers Locked to a Fabry-Perot Interferometer, *IEEE J. Quant. Electr.* **QE-28**, 1106 (1992)
27. D. K. Owens and R. Weiss, Measurement of the Phase Fluctuation in a He-Ne Zeeman Laser, *Rev. Sci. Instrum.* **45**, 1060 (1974)
28. W. Demtröder, *Laser Spectroscopy*, Second Edition (Springer-Verlag, Berlin, 1996) Chap. 7
29. R. W. T. Drever, *et al.*, Laser Phase and Frequency Stabilization using an Optical Resonator, *Appl. Phys. B* **31**, 97–105 (1983)
30. N. Uekara and K. Ueda, 193-mHz Beat Linewidth of Frequency Stabilized Laser-Diode-Pumped Nd:YAG Ring Lasers, *Opt. Lett.* **18**, 505 (1993)
31. Introduction to Diode-Pumped Solid-State Lasers, LIGHTWAVE Electronics Corp. Techn. Information N. 1 (1993)
32. K. Otsuka, Winner-Takes-All and Antiphase States in Multimode Lasers, *Phys. Rev. Lett.* **67**, 1090–1093 (1991)
33. G. P. Agrawal and N. K. Dutta, *Long-Wavelength Semiconductor Lasers*, (Van Nostrand Reinhold, New York, 1986), Fig. 6.11, by permission.
34. Reference(7) Sect. 7.1.1. and Fig. 7.5
35. Reference,(22) Chapt. 17, Sects. 17.5 and 17.6
36. D. Golla *et al.*, 62-W CW TEM_{00} Mode Nd:YAG Laser Side-Pumped by Fiber-Coupled Diode-Lasers, *Opt. Lett.* **21**, 210–212 (1996)

8

Transient Laser Behavior

8.1. INTRODUCTION

In this chapter, we will consider a few cases where the pump rate and/or cavity losses are time dependent. We will also consider situations in which a nonlinear optical element, such as a saturable absorber, is inserted in the laser cavity, where the non-linearity leads to the laser departing from stable cw operation. For these various cases we are thus dealing with transient laser behavior. The transient cases to be considered can be separated into two categories: (i) Cases, such as relaxation oscillations, Q-switching, gain switching and cavity dumping, where, ideally, a single mode laser is involved and which can be described by a rate equation treatment. (ii) Cases where many modes are involved, e.g. mode-locking, and for which a different treatment needs to be considered. This requires a description in terms of either the fields of all oscillating modes (frequency domain description) or in terms of a self-consistent circulating pulse within the cavity (time domain description).

8.2. RELAXATION OSCILLATIONS

We will first consider the case of a step-function pump rate. Thus we assume that $R_p = 0$ for $t < 0$ and $R_p(t) = R_p$ (independent of time) for $t > 0$. We will also assume the laser to be oscillating in a single mode so that a simple rate-equation treatment can be properly applied. As seen in the previous chapter, the rate equations are nonlinear in the variables $N(t)$ and $\phi(t)$ since they involve products of the form ϕN. Consequently, analytical solutions for this case or for other cases about to be considered, are generally not possible and one often needs to resort to numerical computation.[1,2]

As a representative example, Fig. 8.1 shows one of the first computed plots of $N(t)$ and $\phi(t)$ as carried out for a three level laser such as a ruby laser.[2] In this case the initial condition for the population inversion is $N(0) = -N_t$, where N_t is the total population, because, at

O. Svelto, *Principles of Lasers*,
DOI: 10.1007/978-1-4419-1302-9_8,

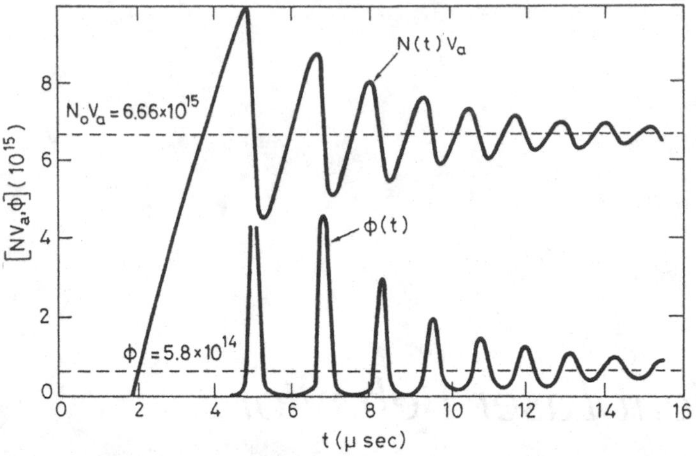

FIG. 8.1. Example of the temporal behavior of the total inversion, $V_aN(t)$, and photon number, $\phi(t)$, for a three-level laser (after reference,[2] by permission).

time $t = 0$, the entire population is in the lower laser level, 1 (see Fig. 1.4a). The initial condition for the total number of cavity photons is then $\phi(0) = \phi_i$, where ϕ_i can be taken to be some small integer (e.g. $\phi_i = 1$) which must just needed to allow laser action start. It should also be noted that a similar behavior to that of Fig. 8.1 is also expected for a four-level laser, one of the main differences being that, in this case, the initial condition is $N(0) = 0$. Thus, if the time origin of Fig. 8.1 is shifted approximately to the time $t = 2$ μs, where the population inversion become zero in the figure, the curves of Fig. 8.1 can be used to provide a qualitative description for the case of a four level laser, as well. Several features of this figure are now worth pointing out: (i) After a time $t = 2$ μs, the population inversion keeps growing due to the pumping process while the photon number remains at its initial low value, as determined by quantum field fluctuations, until the inversion crosses the threshold value ($N_0V_a = 6.66 \times 10^{15}$ in the figure). From this time on, roughly for $t > 4$ μs, the population exceeds the threshold value and the number of cavity photons can begin to grow. From either Eq. (7.2.16b) (four level laser) or Eq. (7.2.24b) (quasi-three level laser) one finds in fact that $d\phi/dt > 0$ when $N > N_c$, where N_c is the critical or threshold inversion. (ii) After threshold is exceeded, the photon number requires some time to grow, from its initial value $\phi_i = 1$, to a value e.g. equal to the steady state value ($\phi_0 = 5.8 \times 10^{14}$ in the figure) and, meanwhile, the population can continue growing due to the pumping process. (iii) When the photon number becomes large enough (roughly when $\phi > \phi_0$), the stimulated emission process becomes dominant over the pumping process. The population then begins to decrease and, at the time corresponding to the maximum of $\phi(t)$, $N(t)$ has dropped back to N_c. This can be readily shown from either Eq. (7.2.16b) or Eq. (7.2.24b) since, when $d\phi/dt = 0$, one has $N = N_c$. (iv) After this photon peak, the population inversion is then driven below N_c by the continuing high rate of stimulated emission. Thus the laser goes below threshold and the photon number also decreases. (v) When this photon number decreases to a sufficiently low value (roughly when $\phi < \phi_0$) the pumping process again becomes dominant over the stimulated emission process. The population inversion can now begin growing again and the whole series of events considered at points (i)–(iv) repeats itself. The photon number, $\phi(t)$, is then seen to display a regular sequence of peaks (or "laser spikes") of decreasing amplitude with consecutive peaks being, approximately, equally spaced in time. The output power would therefore show a similar time behavior. This aspect of regular oscillation for the output power is usually referred to as a *damped relaxation oscillation*. The time behavior of the population inversion then undergoes a similar oscillatory behavior, the oscillation of $N(t)$ leading that of $\phi(t)$ by about half the oscillation period since one must first produce a population rise of $N(t)$ to then have a corresponding rise of the photon number, $\phi(t)$. It should also be noted that, since a steady state solution is eventually reached, this solution corresponding to that given by Eqs. (7.3.4a) and (7.3.4b) for a four level-laser or by Eqs. (7.4.1) and (7.4.6) for a quasi-three-level laser, the computer calculation confirms that these solutions correspond to a stable operating condition.

8.2.1. Linearized Analysis

For small oscillations about the steady-state values (e.g., roughly for $t > 14\ \mu s$ in Fig. 8.1), the dynamical behavior can be described analytically. In fact, if we write

$$N(t) = N_0 + \delta N(t) \tag{8.2.1}$$

$$\phi(t) = \phi_0 + \delta\phi(t) \tag{8.2.2}$$

and assume $\delta N \ll N_0$ and $\delta\phi \ll \phi_0$, we can neglect the product $\delta N \delta\phi$ in the expression $N\phi$ appearing in the rate equations, and these equations become linear in the variables δN and $\delta\phi$. Limiting ourselves to the case of a four-level laser, we can substitute Eqs. (8.2.1) and (8.2.2) into Eqs. (7.2.16a) and (7.2.16b). Since N_0 and ϕ_0 must satisfy the same equations with the time derivatives being set to zero, we readily obtain from Eq. (7.2.16)

$$(\mathrm{d}\delta N/\mathrm{d}t) = -\delta N[B\phi_0 + (1/\tau)] - BN_0\delta\phi \tag{8.2.3}$$

$$(\mathrm{d}\delta\phi/\mathrm{d}t) = BV_a\phi_0\delta N \tag{8.2.4}$$

Note in particular that Eq. (8.2.4) has been obtained from Eq. (7.2.16b) using the fact that $BV_aN_0 - (1/\tau_c) = 0$. Substitution of Eq. (8.2.4) into Eq. (8.2.3) gives

$$\frac{d^2\delta\phi}{dt^2} + [B\phi_0 + (1/\tau)]\frac{d\delta\phi}{dt} + (B^2V_aN_0\phi_0)\,\delta\phi = 0 \tag{8.2.5}$$

We now look for a solution of the form

$$\delta\phi = \delta\phi_0 \exp(pt) \tag{8.2.6}$$

The substitution of Eq. (8.2.6) into Eq. (8.2.5) then shows that p must obey the equation

$$p^2 + \frac{2}{t_0}p + \omega^2 = 0 \tag{8.2.7}$$

where we have put

$$(2/t_0) = [B\phi_0 + (1/\tau)] \tag{8.2.8}$$

and

$$\omega^2 = B^2 V_a N_0 \phi_0 \tag{8.2.9}$$

The solution of Eq. (8.2.7) is obviously given by

$$p = -\frac{1}{t_0} \pm \left[\frac{1}{t_0^2} - \omega^2\right]^{1/2} \tag{8.2.10}$$

The first case we consider is where $(1/t_0) < \omega$. In this case the square root in Eq. (8.2.10) gives an imaginary number so that we can write $p = -(1/t_0) \pm j\omega'$ where

$$\omega' = \left[\omega^2 - (1/t_0)^2\right]^{1/2} \tag{8.2.11}$$

In this case, from Eq. (8.2.6) $\delta\phi$ is seen to correspond to a damped sinusoidal oscillation, i.e.,

$$\delta\phi = C \exp(-t/t_0) \sin(\omega' t + \beta) \tag{8.2.12}$$

where C and β are established by the initial conditions. If Eq. (8.2.12) is then substituted into Eq. (8.2.4) we find that δN is also described by a damped sinusoidal oscillation. Assuming $(1/t_0) \ll \omega'$, we get

$$\delta N \cong \frac{\omega' C}{B V_a \phi_0} \exp(-t/t_0) \cos(\omega' t + \beta) \tag{8.2.13}$$

Note that $\delta N(t)$ leads $\delta\phi(t)$ by half an oscillation period as already discussed previously since one must first have a growth of inversion $\delta N(t)$ before one can have a growth of $\delta\phi(t)$.

Equations (8.2.8) and (8.2.9) can be recast in a form more amenable to calculation if the explicit expressions for N_0 and ϕ_0 given by Eqs. (7.3.4a) and (7.3.4b) are used. We readily obtain

$$t_0 = 2\tau/x \tag{8.2.14}$$

$$\omega = [(x-1)/\tau_c \tau]^{1/2} \tag{8.2.15}$$

where $x = R_p/R_{cp}$, is the amount by which threshold is exceeded. Note that, while the damping time t_0 of the oscillation is determined by the upper state lifetime, the oscillation period $T = 2\pi/\omega' \cong 2\pi/\omega$ is determined by the geometrical mean of τ and the photon lifetime τ_c.

Example 8.1. *Damped oscillation in a Nd:YAG and a GaAs laser* We first consider the single mode Nd:YAG laser of Fig. 7.26 and assume that the above space-independent relaxation oscillation theory can be applied to this diode-pumped NPRO laser. Assuming the laser to be $x = 5$ times above threshold, we obtain from Eq. (8.2.14) $t_0 = 92$ μs, where we have taken $\tau = 230$ μs. We will also take $l = 11.5$ mm as the round trip path length of the NPRO resonator, assume a $T = 0.4\%$ output coupling transmission of the laser and a round trip cavity loss of $L = 0.5\%$. The round trip loss will then be $\gamma \cong (T + L) = 0.9\%$ and the cavity photon decay time $\tau_c = nl/c\gamma \cong 7.8$ ns, where $n = 1.82$ is the refractive index of the YAG material. From Eq. (8.2.15) we then obtain $\nu = \omega/2\pi \cong 238\,\text{kHz}$ for the frequency of the relaxation oscillation. Note that, in this case, we have $t_0 \gg 1/\omega$, thus the approximation $\omega' \cong \omega$ is justified. Note also that the spectrum of this damped oscillation is Lorentzian with a width $\Delta\nu_0 = 1/2\pi\, t_0 = 1.73\,\text{kHz}$, and this will also be the 3-dB width of the relaxation oscillation peak of the RIN spectrum for this laser (see Fig. 7.30a). Consider next a typical GaAs injection laser with a cavity length $L = l = 300\,\mu\text{m}$, in which the two end faces are cleaved and act as the cavity mirrors. According to Eq. (4.3.1) the power reflectivity of both mirrors will, in this case, be equal to $R = [(n-1)/(n+1)]^2 \cong 0.3$, where $n = 3.35$ is the refractive index of GaAs. We thus have $\gamma_1 = \gamma_2 = -\ln R = 1.2$. We will also assume a distributed loss coefficient of $\alpha_0 = 60\,\text{cm}^{-1}$ along the semiconductor length, so that we can write $\gamma_i = \alpha_0 L = 1.8$. We thus get $\gamma = \gamma_i + [(\gamma_1 + \gamma_2)/2] = 3$ and $\tau_c = L_e/c\gamma = nL/c\gamma = 1.1\,\text{ps}$. The upper state lifetime may be taken to be $\tau \cong 3\,\text{ns}$. Assuming $x = 1.5$ we get from Eq. (8.2.14), $t_0 = 4\,\text{ns}$, and from Eq. (8.2.15), $\nu = \omega/2\pi \cong 2\,\text{GHz}$. In this case also we have $t_0 \gg 1/\omega$ and the approximation $\omega' \cong \omega$ is again justified. Note also that, according to this calculation, the relaxation oscillation peak of the RIN spectrum of this laser is expected to be in the range of some GHz (see Fig. 7.31).

If the condition $t_0 > 1/\omega$ is not satisfied, the two solutions for p given by Eq. (8.2.10) are both real and negative. In this case the time behavior of $\delta\phi(t)$ consists of a superposition of two exponentially damped decays. To have $t_0 < 1/\omega$ we must, according to Eqs. (8.2.14) and (8.2.15), have

$$(\tau_c/\tau) > 4(x-1)/x^2 \tag{8.2.16}$$

The right hand side of this equation has a maximum value of 1 when $x = 2$. This means that, if $\tau_c > \tau$, Eq. (8.2.16) is satisfied for any value of x. This situation usually occurs in gas lasers, which therefore generally do not exhibit spiking behavior.

Example 8.2. *Transient behavior of a He-Ne laser* Consider a He-Ne laser oscillating on its red transition ($\lambda = 632.8\,\text{nm}$). In this case one has $\tau = 50\,\text{ns}$. Assuming a cavity length of $L = 50\,\text{cm}$, an output coupling of 1% and neglecting all other losses, we obtain $\gamma = \gamma_2/2 = 5\times10^{-3}$, and $\tau_c = L/c\gamma = 322\,\text{ns}$. We thus get $\tau_c > \tau$ and Eq. (8.2.16) is satisfied for any value of x. From Eqs. (8.2.14) and (8.2.15) with $x = 1.5$, we obtain $t_0 = 66.6\,\text{ns}$ and $\omega \cong 5.6\times10^6\,\text{Hz}$. From Eq. (8.2.10) we then see that the two lifetimes describing the decay are 1 μs and 33.3 ns.

Before ending this section it is worth noting that the linearized analysis we have just considered also applies to a slightly different case, i.e., when one needs to test the stability of a given steady-state solution by a *linear stability analysis*. We assume, in this case, that the laser is already operating in the steady state and that a small step-perturbation is applied (i.e., $\delta N = \delta N_0$ and $\delta\phi = \delta\phi_0$ at $t = 0$, where δN_0 and $\delta\phi_0$ are two known quantities). According to the discussion given above the perturbation introduced at time $t = 0$ will decay with time either by a damped sinusoidal oscillation or by a biexponential law. The steady state solutions N_0 and ϕ_0, which were discussed in the previous Chapter, therefore correspond to a stable equilibrium.

8.3. DYNAMICAL INSTABILITIES AND PULSATIONS IN LASERS

The simple results obtained in the previous section appear to conflict with many experimental results, observed from the earliest days of lasers, which indicated that, many lasers, even when operating cw, tend to exhibit a continuous pulsating behavior, sometimes irregular and sometimes regular in character. A classical example of this type is shown in Fig. 8.2, where pulsations observed in the first cw-excited ruby laser are indicated.[3] It can be seen that the output consists of a train of pulses irregularly spaced in time and of random amplitude (*irregular spiking*). Furthermore, these pulsations do not tend to a steady state value as in Fig. 8.1. This kind of unstable behavior has been the subject of more than 25 years of theoretical and experimental investigations, revealing that this behavior can be attributed to a variety of reasons that are briefly summarized below.[4]

In single mode lasers, one of the main causes of instability arises from external and usually accidental modulation of laser parameters such as pump rate or cavity losses. For random modulation, this simply leads to the laser intensity noise already discussed in Sect. 7.11. For sinusoidal modulation, the time behavior can be described in terms of rate equations by writing, e.g. for pump modulation, $R_p = R_{p0} + \delta R_p \exp(j\omega t)$, with $\delta R_p \ll R_{p0}$. According to Eqs. (8.2.1) and (8.2.2), we can then write $N(t) = N_0 + \delta N_0 \exp(j\omega t)$ and $\phi(t) = \phi_0 + \delta\phi_0 \exp(j\omega t)$, with $\delta N_0 \ll N_0$ and $\delta\phi_0 \ll \phi_0$, and solve the corresponding linearized equations. As seen in the previous section, a single mode laser presents a natural resonance at its relaxation oscillation frequency ω_R, given by e.g. Eq. (8.2.11) for a 4-level laser. A sinusoidal pump modulation thus forces the laser to exhibit small oscillations at the modulation frequency, ω, and the oscillation amplitude will be a maximum when ω coincides with ω_R. For a white spectrum of pump modulation, an intensity noise spectrum peaking at ω_R will therefore be observed (see Fig. 7.30a). Besides this instability of a technical origin, single mode lasers, under special circumstances, can also show a natural dynamical instability leading to pulsations and even chaotic behavior. For instance, in the case of a homogeneously broadened transition, the laser must be driven sufficiently far above threshold (typically more than ten times) and the cavity linewidth, $\Delta\nu_c$, must be sufficiently *larger* than the transition linewidth $\Delta\nu_0$ (so-called *bad-cavity case*). Conditions of this sort have been experimentally realized in, specially prepared, optically pumped far infrared lasers. Dynamical instabilities of this type can only be accounted for by a semiclassical treatment of laser behavior, i.e. by means of the Maxwell-Bloch equations.[5]

In multimode lasers, a new type of instability may easily set in, due to e.g. a pump-modulation-induced switching in time between one mode and another or from one set of modes to another set.[6] This instability leads to a kind of antiphase motion (antiphase dynamics) among the modes and it can be adequately described by a rate equation treatment in

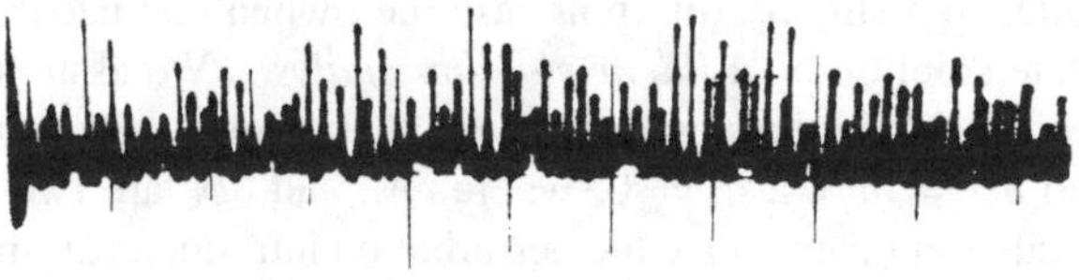

FIG. 8.2. Typical time behavior of early cw-pumped solid-state lasers. Time scale is 50 μs/div. (after reference,[3] by permission).

which, for a homogeneous line, the cross saturation effect due to spatial hole burning is taken into account.[7]

To summarize, we can say that single mode lasers do not usually exhibit dynamical instability but instead show some, possibly quite pronounced, intensity noise due to unavoidable perturbations of laser parameters. On the other hand, multimode lasers may also present additional instabilities due to a kind of antiphase motion among the oscillating modes. Depending on the amplitude of modulation of the laser parameters, the type of laser, and whether the line is homogeneously or inhomogeneously broadened, this instability may lead either to a mode partition noise (see Sect. 7.11) or, even, to strong laser pulsations.

8.4. Q-SWITCHING

We have seen in the previous chapter that, under cw operation, the population inversion gets clamped to its threshold value when oscillation starts. Even under the pulsed operating conditions considered in Sect. 8.2, the population inversion can only exceed the threshold value by a relatively small amount (see Fig. 8.1) due to the onset of stimulated emission. Suppose now that a shutter is introduced into the laser cavity. If the shutter is closed, laser action is prevented and the population inversion can then reach a value far in excess of the threshold population for the case where the shutter is absent. If the shutter is now opened suddenly, the laser will exhibit a gain that greatly exceeds losses and the stored energy will be released in the form of a short and intense light pulse.[8] Since this operation involves switching the cavity Q factor from a low to a high value, the technique is usually called *Q-switching*. The technique allows the generation of laser pulses of duration comparable to the photon decay time (i.e. from a few nanoseconds to a few tens of nanoseconds) and high peak power (in the megawatt range).

8.4.1. Dynamics of the Q-Switching Process

To describe the Q-switching dynamical behavior, we assume that a step pump pulse is applied to the laser starting at time $t = 0$, i.e., $R_p(t) = 0$ for $t < 0$ and $R_p(t) = R_p =$ const. for $0 < t < t_P$ and that, meanwhile, the shutter is closed (Fig. 8.3a). For $0 < t < t_P$, the time behavior of the population inversion can then be calculated from Eq. (7.2.16a), for a 4-level laser, or Eq. (7.2.24a), for a quasi-3-level laser, with ϕ set to zero. For instance, for a 4-level laser, we obtain

$$N(t) = N_\infty[1 - \exp(-t/\tau)] \tag{8.4.1}$$

where the asymptotic value N_∞ is given by

$$N_\infty = R_p\tau \tag{8.4.2}$$

as one can readily obtain from Eq. (7.2.16a) by putting $dN/dt = 0$. The time behavior of $N(t)$ is also shown in Fig. 8.3a. From Eq. (8.4.1) and from Fig. 8.3a we see that the duration t_P of the pump pulse should ideally be comparable to or shorter than the upper state

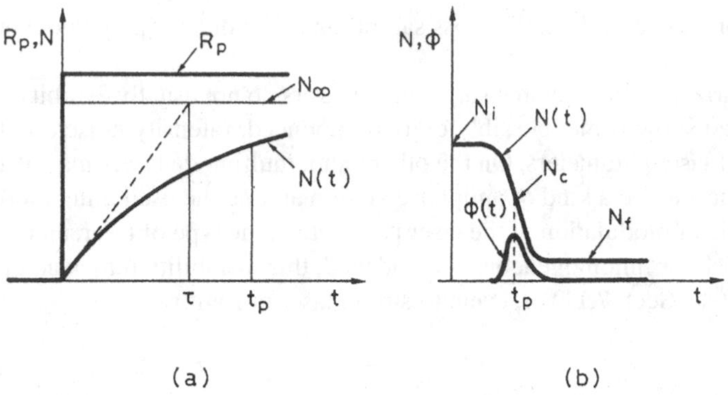

FIG. 8.3. Sequence of events in a Q-switched laser: (a) Idealized time behavior of the pump rate, R_p, and of the population inversion, N, before Q-switching. (b) Time behavior of population inversion, N, and photon number, ϕ, after Q-switching (fast switching case).

lifetime τ. In fact, for $t_P \gg \tau$, $N(t)$ would not undergo any appreciable increase and the pump power, rather than being accumulated as inversion energy, would be wasted through spontaneous decay. From Eq. (8.4.2) we also see that, to achieve a sufficiently large inversion, one needs a long lifetime τ. Thus Q-switching can be used effectively with electric-dipole-forbidden laser transitions where τ generally falls in the millisecond range. This is the case of most solid-state lasers (e.g., Nd, Yb, Er, Ho in different host materials, Cr doped materials such as alexandrite, Cr:LISAF, and ruby) and some gas lasers (e.g. CO_2 or iodine). On the other hand, for semiconductor lasers, dye lasers, and a number of important gas lasers (e.g., He-Ne, Ar, Excimers) the laser transition is electric-dipole allowed and the lifetime is of the order of a few to a few tens of nanoseconds. In this case, with the usual values of pump rates, R_p, available, the achievable inversion N_∞ is too low to be of interest for Q-switching.

Let us now assume that the shutter is suddenly opened at time $t = t_P$ so that the cavity loss, $\gamma(t)$, is switched from a very high value, corresponding to the shutter closed, to the value, γ, of the same cavity with the shutter open (*fast-switching*). We will now take the time origin at the instant when switching occurs (Fig. 8.3b). The time behavior of the population inversion, $N(t)$, and of the number of photons, $\phi(t)$, can be obtained through the rate equations with the simplifying assumption that, during the short time of the Q-switching process, the effect of the decay term N/τ can be neglected. The qualitative behavior of $N(t)$ and $\phi(t)$ will then be as depicted in Fig. 8.3b. The population inversion starts from the initial value N_i, which can be obtained from Eq. (8.4.1) for $t = t_P$, then remains constant for some time and finally begins to be depleted when the cavity photon number reaches a sufficiently high value. When $N(t)$ eventually falls to the threshold inversion N_c, the photon number reaches its peak value, as discussed earlier, for the case of relaxation oscillations, in Sect. 8.2. From this time on, the laser exhibits net loss rather than net gain and, as a consequence, the photon number decreases to zero. During the same time, the population inversion decreases to a final value N_f, which is left in the active medium, its value being established by the dynamics of the Q-switching process (see Sect. 8.4.4). Note that the time scales in Fig. 8.3a, b are very different. In fact, the time scale of Fig. 8.3a is set by the value of the upper state lifetime and

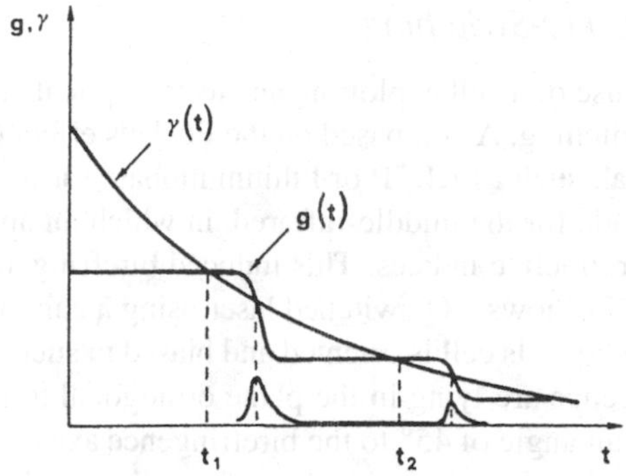

FIG. 8.4. Sequence of events in the slow switching case demonstrating the occurrence of multiple pulses. In the figure, $g(t) = \sigma N(t)l$, where l is the length of the active medium, represents the laser gain.

thus corresponds to the ms range (usually, 100 μs ÷ 1 ms). The time scale of Fig. 8.3b turns out to be of the order of the cavity photon decay time (see Sect. 8.4.4) and hence falls in the ns range (usually, 5 ÷ 50 ns).

So far we have been considering the dynamical behavior corresponding to fast switching, where the switching of cavity losses is treated as instantaneous. In practice fast switching requires that the switching time be appreciably shorter than the time taken for photons to build up to their peak value (several photon decay times i.e., typically, from a few tens to a few hundredths of a nanosecond). In the case of *slow-switching*, the dynamical behavior is somewhat more complicated and multiple pulses may result. This behavior is depicted in Fig. 8.4 where the cavity loss $\gamma(t)$ is assumed to decay from its high value to its final value in a relatively long time. In the same figure we also show the time behavior of the single pass gain, $g(t) = \sigma Nl$, and of the cavity photon number $\phi(t)$. We see that the first pulse starts at time t_1, at the instant when the decreasing loss $\gamma(t)$ becomes equal to the instantaneous gain $g(t)$. The pulse then reaches its peak value at the time when the gain, due to saturation, becomes equal to the loss. After this first pulse, the gain is driven below the loss and further oscillation cannot occur until the switch opens further, thus decreasing the loss below the gain. A second pulse can then be produced (occurring at time t_2 in the figure) whose peak again occurs at a time where gain saturation makes the gain equal to loss.

8.4.2. Methods of Q-Switching[9]

There are several methods that have been developed to achieve switching of the cavity Q and, in this section, we will limit ourselves to a discussion of the most commonly used, namely: (i) Electro-optical shutters. (ii) Rotating prisms. (iii) Acousto-optical switches. (iv) Saturable absorbers. These devices are generally grouped into two categories, *active* and *passive Q-switches*. In an active Q-switching device, one must apply some external active operation to this device (e.g. change the voltage applied to the electro-optical shutter) to produce Q-switching. In a passive Q-switch, the switching operation is automatically produced by the optical nonlinearity of the element used (e.g. saturable absorber).

8.4.2.1. Electro-Optical Q-Switching

These devices make use of a cell exploiting an electro-optical effect, usually the Pockels effect, to induce the Q-switching. A cell based on the Pockels effect (*Pockels cell*) consists of a suitable nonlinear crystal, such as KD*P or lithium niobate for the visible-to-near-infrared region, or cadmium telluride for the middle-infrared, in which an applied dc voltage induces a change in the crystal's refractive indices. This induced birefringence is proportional to the applied voltage. Figure 8.5a shows a Q-switched laser using a suitable combination of polarizer and Pockels cell. The Pockels cell is oriented and biased in such a way that the axes x and y of the induced birefringence are lying in the plane orthogonal to the axis of the resonator. The polarizer axis makes an angle of 45° to the birefringence axes.

Consider now a laser beam propagating from the active medium toward the polarizer-Pockels-cell combination with a polarization parallel to the polarizer axis. Ideally, this beam will be totally transmitted by the polarizer and then incident on the Pockels cell. The E-field of the incoming wave will thus be at 45° to the birefringence axes x and y of the Pockels cell and can be resolved into components E_x and E_y (Fig. 8.5b) with their oscillations in phase. After passing through the Pockels cell, these two components will have experienced different phase shifts, giving rise to a phase difference

$$\Delta\varphi = k\Delta nL' \tag{8.4.3}$$

where $k = 2\pi/\lambda$, $\Delta n = n_x - n_y$ is the value of the induced birefringence, and L' is the crystal length. If the voltage applied to the Pockels cell is such that $\Delta\varphi = \pi/2$, then the two field components leaving the Pockels cell will differ in phase by $\pi/2$. This means that, when E_x is maximum, E_y will be zero and vice versa, i.e., the wave becomes circularly polarized (Fig. 8.5c). After reflection at the mirror, the wave passes once more through the Pockels cell and its x and y components acquire an additional, $\Delta\varphi = \pi/2$, phase difference. So the total

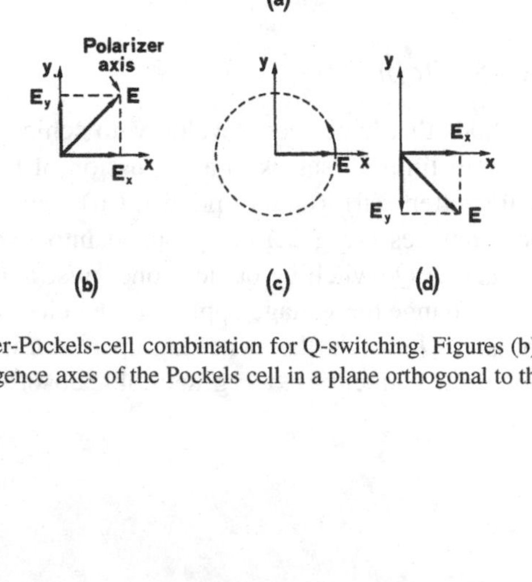

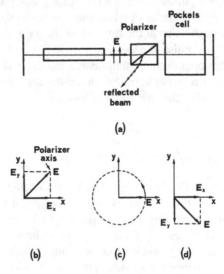

FIG. 8.5. (a) Possible polarizer-Pockels-cell combination for Q-switching. Figures (b), (c), and (d) show the E-field components along the birefringence axes of the Pockels cell in a plane orthogonal to the resonator axis.

phase difference now becomes π, so that, when e.g., E_x is at its maximum (positive) value, E_y will be at its maximum (negative) value as shown in Fig. 8.5d. As a result, the overall field, **E**, is again linearly polarized but with a polarization axis at 90° to that of the original wave of Fig. 8.5b. This beam is therefore not transmitted by the polarizer and is instead reflected out of the cavity (see Fig. 8.5a). This condition corresponds to the Q-switch being closed. The switch is then opened by removing the bias voltage to the Pockels cell. In this case the induced birefringence disappears and the incoming light is transmitted without change of polarization. Note that the required voltage for operation in this arrangement is called the $\lambda/4$ voltage or the "quarter-wave voltage," since the quantity $\Delta nL'$ i.e., the difference in optical path lengths for the two polarizations, is $\lambda/4$, as can be seen from Eq. (8.4.3).

Pockels cell Q-switches are very widely used. Depending upon the particular nonlinear crystal used in the cell, the particular arrangement of applied field, the crystal dimensions, and the value of the wavelength involved, the $\lambda/4$ voltage may range between 1 and 5 kV. This voltage must then be switched off in a time, t_s, smaller than the build-up time of the Q-switched pulse (typically $t_s < 20$ ns).

8.4.2.2. Rotating Prisms

The most common mechanical means of Q-switching involves rotating one of the end mirrors of the laser resonator about an axis perpendicular to the resonator axis. In this case, the high-Q condition is reached when the rotating mirror passes through a position parallel to the other cavity mirror. To simplify the alignment requirements, a 90° roof-top prism with roof edge perpendicular to the rotation axis is often used instead of an ordinary mirror (Fig. 8.6). Such a prism has the property that, for light propagating orthogonal to the roof edge (see Fig. 8.6), the reflected beam is always parallel to the incident beam regardless of any rotation of the prism about its roof edge. This ensures that the alignment between the prism and the other cavity mirror is always achieved in the plane orthogonal to the roof. The effect of rotation is then to bring the prism into alignment in the other direction.

Rotating-prism Q-switches are simple and inexpensive devices and can be made for use at any wavelength. They are rather noisy, however, and, due to the limited speed of the rotating motor, they generally result in slow Q-switching. For a typical multi-transverse-mode solid

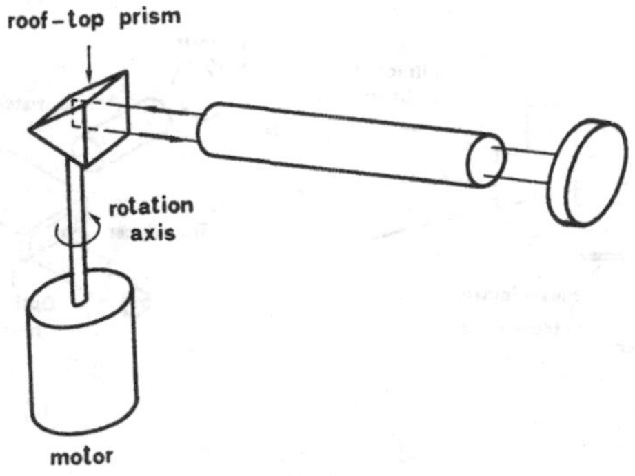

FIG. 8.6. Mechanical Q-switching system using a rotating 90° roof-top prism.

state laser, for instance, the beam divergence is around a few mrad. The high-Q situation then corresponds to an angular range of ~1 mrad around the perfect alignment condition. Thus, even for a motor rotating at the fast speed of 24,000 rpm (400 Hz), the duration of the high Q-switching condition is about 400 ns. This slow switching time can sometimes result in the production of multiple pulses.

8.4.2.3. Acousto-Optic Q-Switches

An acousto-optic modulator consists of a block of transparent optical material (e.g., fused quartz in the visible to near infrared and germanium or cadmium selenide in the middle-far infrared) in which an ultrasonic wave is launched by a piezoelectric transducer bonded to one side of the block and driven by a radiofrequency oscillator (Fig. 8.7a). The side of the block opposite to the transducer side is cut at an angle and has an absorber for the acoustic wave placed on its surface (see Fig. 8.7b). With back reflection of the acoustic wave thus suppressed, only a traveling acoustic wave is present in the medium. The strain induced by the ultrasonic wave results in local changes of the material refractive index through the photoelastic effect. This periodic change of refractive index acts then as a phase grating with period equal to the acoustic wavelength, amplitude proportional to the sound amplitude, and which is traveling at the sound velocity in the medium (traveling-wave phase grating). Its effect is to diffract a fraction of the incident beam out of the incident beam direction.[10] Thus, if an acousto-optic cell is inserted in a laser cavity (Fig. 8.7b), an additional loss will be present, due to beam diffraction, while the driving voltage to the transducer is applied. If the driving voltage is high enough, this additional loss will be sufficient to prevent the laser from oscillating. The laser is then returned to its high-Q condition by switching off the transducer voltage.

To gain a more detailed understanding of the operation of an acousto-optic modulator, we now consider the case where the length L' of the optical medium is sufficiently large that the grating acts as a thick phase grating. For this to be the case, the following condition must be satisfied

$$\frac{2\pi\lambda L'}{n\lambda_a^2} \gg 1 \qquad (8.4.4)$$

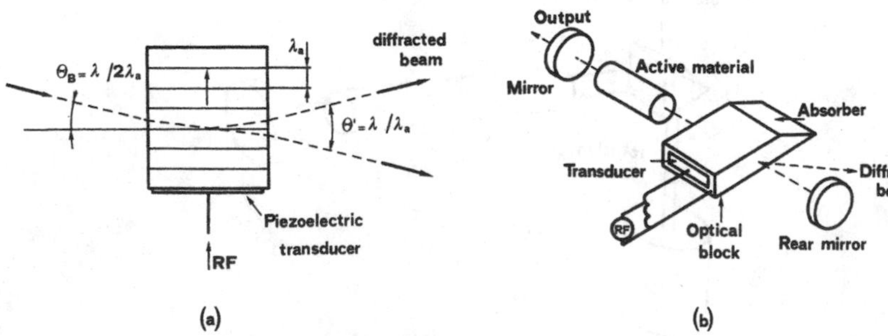

FIG. 8.7. (a) Incident, transmitted, and diffracted beams in an acousto-optic modulator (Bragg regime). (b) Q-switched laser arrangement incorporating an acousto-optic modulator.

where λ is the wavelength of the incident beam, n is the material refractive index, and λ_a is the wavelength of the acoustic wave. In typical cases, condition given by Eq. (8.4.4) requires that L' be larger than 1 cm. In this case, known as the Bragg regime, a single beam is diffracted out of the cavity at an angle $\theta' = \lambda/\lambda_a$. Note that this angle is equal to the divergence angle of a beam of wavelength λ diffracting out of an "aperture" of size λ_a. Maximum diffraction efficiency is then achieved when the angle of the incident light, θ_B, satisfies the condition $\theta_B = \lambda/2\lambda_a$ (Fig. 8.7a), originally derived by Bragg for X-ray diffraction from crystallographic planes. In such a case, the diffracted beam can be thought of as arising from specular reflection of the incident beam at the phase planes produced by the acoustic wave. For sufficiently high values of the rf drive power to the piezoelectric transducer, a relatively large fraction η of the incident beam can be diffracted out of the cavity (typical diffraction efficiencies are about $1 \div 2\%$ per watt of rf power). Note that condition given by Eq. (8.4.4) can be approximated as $(\lambda L'/n\lambda_a) \gg \lambda_a$ which can be interpreted as saying that a "wavelet" diffracted by an aperture λ_a, at the crystal entrance, spreads out, at the crystal exit, by an amount $\lambda L'/n\lambda_a$ which must be much larger than each aperture λ_a. Under this condition, in fact, each wavelet diffracted at the crystal entrance will be summed, before exiting the crystal, with wavelets produced by other apertures λ_a of the crystal, thus resulting in volume diffraction*.

> **Example 8.3.** *Condition for Bragg regime in a quartz acousto-optic modulator* We will consider an acousto-optic cell driven at a frequency of $\nu_a = 50$ MHz and take $\upsilon = 3.76 \times 10^5$ cm/s as the shear-wave velocity in quartz. The acoustic wavelength is then $\lambda_a = \upsilon/\nu_a = 75$ μm. Taking $n = 1.45$ for the refractive index of quartz at $\lambda = 1.06$ μm, from Eq. (8.4.4) we obtain $L' \gg 1.3$ mm. Thus, for a crystal length of about 5 cm, the condition to be in the Bragg diffraction regime is amply satisfied. Note that, in this example, the beam is diffracted at an angle $\theta' = \lambda/\lambda_a \cong 0.8°$ to the incident beam direction and the angle of incidence at the modulator must be $\theta_B = \lambda/2\lambda_a \cong 0.4°$.

Acousto-optic modulators have the advantage of low optical insertion losses, and, for repetitive Q-switching, they can readily be driven at high repetition rates (kHz). The loss introduced in the low-Q situation is rather limited, however, and the Q-switching time is rather long (being mainly established by the time taken for an acoustic wavefront to traverse the laser beam). These modulators are therefore used, primarily, for repetitive Q-switching of low-gain cw-pumped lasers (e.g. Nd:YAG).

8.4.2.4. Saturable-Absorber Q-Switch

The three Q-switching devices considered so far fall in the category of active Q-switches since they must be driven by an appropriate driving source (Pockel cell voltage power supply, rotating motor or rf oscillator). We now consider a case of passive Q-switching exploiting the non-linearity of a saturable absorber, this being by far the most common passive Q-switch in use so far.

A saturable absorber consists of a material which absorbs at the laser wavelength and which has a low value of saturation intensity. It is often in the form of a cell containing a

* When $(2\pi\lambda L') \ll n\lambda_a^2$, the acoustic grating behaves like a thin phase-grating and the cell is said to be operating in the Raman-Nath regime. This regime is seldom used for acousto-optic Q-switching owing to the higher requirement for rf power per unit volume of the cell.

solution of a saturable dye in an appropriate solvent (e.g., the dye known as BDN, bis 4-dimethyl-aminodithiobenzil-nickel, dissolved in 1,2-dichloroethane for the case of Nd:YAG). Solid state (e.g., BDN in a cellulose acetate, F_2:LiF, or Cr^{4+}:YAG, again for a Nd:YAG laser) or gaseous saturable absorbers (e.g., SF_6 for CO_2 lasers) are also used. To a first approximation, a saturable absorber can be treated as a two-level system with a very large peak cross section (10^{-16} cm^2 is typical for a saturable dye). It then follows from Eq. (2.8.11) that the corresponding saturation intensity I_s is comparatively small ($1 \div 10\,\text{MW/cm}^2$), and the absorber becomes almost transparent, due to saturation, for a comparatively low incident-light intensity.

To understand the dynamical behavior of a saturable-absorber Q-switch, let us assume that a cell containing this absorber, having a peak absorption wavelength coincident with the laser wavelength, is introduced in the laser cavity. As a typical case, assume that the initial, i.e., unsaturated, absorption of the cell is 50%. Then, laser action will start only when the gain of the active medium compensates the loss of the saturable absorber plus the unsaturable cavity losses. Owing to the large value of cell absorption, the required critical population inversion is thus very high. When laser action eventually starts, the beam intensity inside the laser cavity, $I(t)$, will build up from the starting noise, I_n, arising from spontaneous emission. To appreciate the full time evolution of $I(t)$ we show in Fig. 8.8a a logarithmic plot of $I(t)/I_n$ versus time in a typical situation (see example 8.5).When the laser intensity becomes equal to I_s, which occurs at time $t = t_s$ in the figure, the absorber begins to bleach owing to saturation. The rate of growth of laser intensity is thus increased, this in turn results in an increased rate of absorber bleaching, and so on. The overall result is a very rapid bleaching of the saturable absorber. Since I_s is comparatively small, the inversion still left in the laser medium, after bleaching of the absorber, is essentially the same as the initial inversion, i.e., very large. After the absorber has bleached, the laser will thus have a gain well in excess of the losses, and

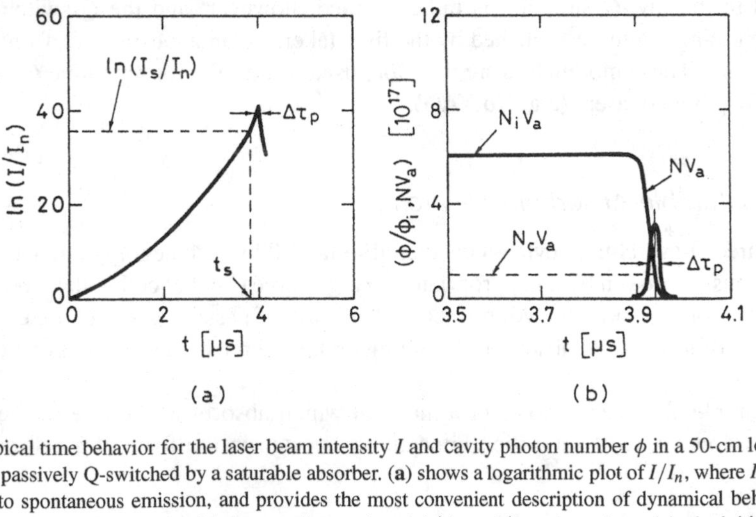

FIG. 8.8. Typical time behavior for the laser beam intensity I and cavity photon number ϕ in a 50-cm long Nd:YAG laser which is passively Q-switched by a saturable absorber. (**a**) shows a logarithmic plot of I/I_n, where I_n is the noise intensity due to spontaneous emission, and provides the most convenient description of dynamical behavior before saturation of the saturable absorber. (**b**) shows a linear plot of $\phi/\phi_i = I/I_n$, where $\phi_i \cong 1$ is the initial number of photons due to spontaneous emission, this providing the most convenient description of time evolution around the peak of the pulse.

a giant pulse will be produced. The dynamical behavior during this last phase is illustrated in more detail in Fig. 8.8b, where the quantity $\phi(t)/\phi_i = I(t)/I_n$ [$\phi(t)$ is the cavity photon number and $\phi_i \cong 1$ is the initial value due to spontaneous emission] is now plotted vs time, in a linear scale and in a restricted time interval. The time behavior of the total inversion NV_a, where V_a is the volume of the mode in the active medium, is also shown in the same figure (see example 8.5). One sees that, as in any other Q-switching case, the photon number now keeps increasing rapidly until saturation of the inversion and hence of the gain sets in. The pulse thus reaches a maximum value when the inversion becomes equal to the critical inversion, N_c, of the laser without saturable absorber and the pulse thereafter decreases.

An important feature to be noticed from Fig. 8.8 is that the time taken for the pulse to increase from the noise level to the peak value is very long ($t_p \cong 3.94$ μs). This is essentially due to the fact that, during the unbleached phase of the saturable absorber, i.e. for $t < t_s$, laser gain barely exceeds the high threshold value established by the presence of the, as yet unsaturated, absorber. The growth of laser intensity is thus very slow and a large number of passes is required for the beam to reach its peak value (~2,370 in the example considered in Fig. 8.8). This results in a natural selection of cavity modes.[11] Suppose in fact that two modes have single pass unsaturated gains g_1 and g_2 ($g = \sigma Nl$) and single pass losses γ_1 and γ_2. Since the two modes start from the same intensity as established by spontaneous emission, the ratio of the two intensities, at time $t = t_s$, i.e., before saturable absorber saturation, will be given by

$$\frac{I_1}{I_2} = \left[\frac{e^{(g_1-\gamma_1)}}{e^{(g_2-\gamma_2)}}\right]^n \tag{8.4.5}$$

where n is the number of round trips within the cavity up to a time $t = t_s$ ($n \cong 2,310$ in the example of Fig. 8.8). If we now let $\delta = (g_1-\gamma_1)-(g_2-\gamma_2)$ be the difference between the two net gains, from Eq. (8.4.5) we can write $(I_1/I_2) = \exp n\delta$. We thus see that, even assuming the very modest value for δ of 0.001, for $n = 2,310$ we get $(I_1/I_2) = \exp 2.3 \cong 10$. Thus even a very modest discrimination of either gain or loss between the two modes results in a large discrimination between their intensities at time $t = t_s$, and hence also at the peak of the pulse, which occurs shortly afterward (~100 ns in the example). As a result, single-mode operation can be achieved rather easily in the case of a saturable absorber Q-switch. Note that, in the case of active Q-switching, this mode selection mechanism is much less effective since the laser build-up from noise is much faster and the total number of transits may now be of the order of only 10 or 20*.

Passive Q-switching by saturable absorbers provides the simplest method of Q-switching. Photochemical degradation of the absorber, particularly for dye saturable absorbers, was the main drawback to this type of Q-switch. This situation is now changing with the advent of solid-state absorbers which do not degrade. The use of passive Q-switching has therefore mainly restricted to low average-power devices.

* It should be noted, however, that one can operate an active Q-switch in an analogous fashion to the saturable absorber by setting the low Q (high loss) condition to a value that permits lasing to start (known as "prelasing") before gradually switching to high Q after a long prelase has allowed mode selection to occur.[38]

8.4.3. Operating Regimes

Q-switched lasers can operate in either one of the following two ways: (1) Pulsed operation (Fig. 8.9). In this case the pump rate $R_p(t)$ is generally in the form of a pulse of duration comparable to the upper state lifetime τ (Fig. 8.9a). Without Q-switching, the population inversion, $N(t)$, would reach a maximum value and decrease thereafter. The cavity Q is switched at the time when $N(t)$ reaches its maximum value ($t = 0$ in Fig. 8.9b). Then, for $t > 0$, the number of photons begins to grow, leading to a pulse whose peak occurs at some time τ_d after switching. As a result of the growth of the photon number, the population inversion $N(t)$ will decrease from its initial value N_i (at $t = 0$) to the final value N_f left after the pulse is over. It should be noted that, according to a comment already made in connection with Fig. 8.3, the time scales for $t < 0$ and $t > 0$ are completely different. In fact, the time scale of events for $t < 0$ falls in the ms range while the time scale of events for $t > 0$ falls in the ns range. Q-switched lasers with a pulsed pump can obviously be operated repetitively, and typical repetition rates are from a few to a few tens of Hz. (2) Continuously pumped, repetitively Q-switched operation (Fig. 8.10). In this case a cw pump, R_p, is applied to the laser (Fig. 8.10a) and the cavity losses are periodically switched from a high to a low value (Fig. 8.10b). The laser output then consists of a continuous train of Q-switched pulses (Fig. 8.10c). During each pulse, the inversion will fall from its initial value N_i (before Q-switching) to a final value N_f (after the Q-switched pulse) (Fig. 8.10d). The population inversion is then restored to its initial value N_i by the pumping process before the next Q-switching event. Since the time taken to restore the inversion is roughly equal to the upper state lifetime τ, the time τ_p between two consecutive pulses must be equal to or shorter than τ. In fact, if τ_p were much longer than τ, most of the available inversion would be lost by spontaneous decay. Therefore, repetition rates of cw pumped Q-switched lasers are typically from a few kHz to a few tens of kHz.

Electro-optical and mechanical shutters as well as saturable absorbers are commonly used for pulsed operation. For repetitive Q-switching of continuously pumped lasers (which have lower gain than pulsed lasers) acousto-optic Q-switches and, sometimes, mechanical

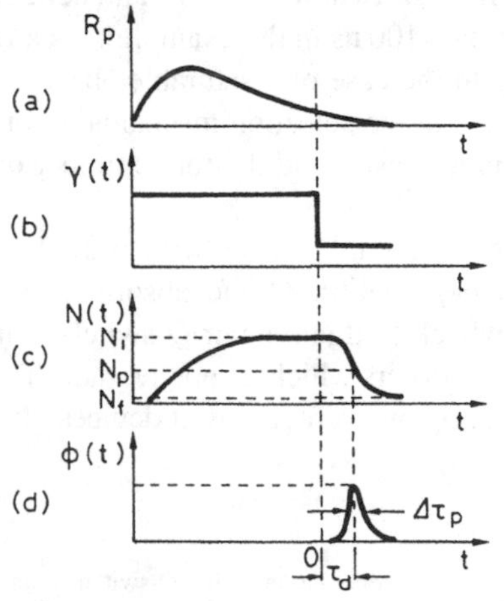

FIG. 8.9. Development of a Q-switched laser pulse in pulsed operation. The figure shows the time behavior of: (a) the pump rate R_p; (b) the resonator loss γ, (c) the population inversion N; (d) the number of photons ϕ.

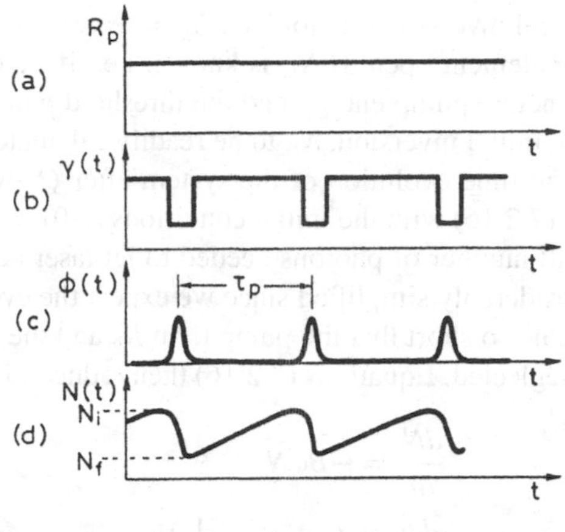

FIG. 8.10. Development of Q-switched laser pulses in a repetitively Q-switched, cw pumped, laser. The figure shows the time behavior of: (a) the pump rate R_p; (b) the resonator losses γ; (c) the number of photons ϕ; (d) the population inversion N.

shutters are commonly used. With low-power cw lasers, a saturable absorber within the cavity can, under appropriate conditions, lead to repetitive Q-switching operation. In this case, the repetition rate of the Q-switched pulses is established by the non-linear dynamics of the absorber rather than by an external control.

8.4.4. Theory of Active Q-Switching

For the sake of simplicity we will only consider the case of active Q-switching and we will further assume the switching to be instantaneous (fast switching case).[12] The dynamical behavior of the laser can again be obtained from Eqs. (7.2.16) and (7.2.24) for four- and quasi-three-level lasers, respectively.

We will first consider a four-level laser operating in a pulsed regime (Fig. 8.9) and assume that, for $t < 0$, the losses are large enough for the laser to be below threshold. If Q-switching is performed when $N(t)$ has attained its maximum value, the corresponding initial inversion can be obtained from Eq. (7.2.16a) by setting $(dN/dt) = 0$. We thus get

$$N_i = \tau R_p(0) \tag{8.4.6}$$

where $R_p(0)$ is the pump rate value when Q-switching occurs (i.e., at $t = 0$). We now assume that the time behavior of $R_p(t)$ is always the same whatever the value of $\int R_p dt$, i.e., of the pump energy. We can then put $R_p(0) \propto \int R_p dt$ so that, for example, if $\int R_p dt$ is doubled then $R_p(0)$ will also double. Thus, if we let E_p be the pump energy corresponding to the given pump rate, since $E_p \propto \int R_p dt$, we will then have $E_p \propto R_p(0)$ and, according to Eq. (8.4.6), $E_p \propto N_i$. Therefore, if we let N_{ic} and E_{pc} be the initial inversion and the corresponding pump energy, respectively, when the laser is operated just at threshold, we can write

$$(N_i/N_{ic}) = (E_p/E_{pc}) = x \tag{8.4.7}$$

where $x = (E_p/E_{pc})$ is the amount by which threshold is exceeded. Since N_{ic} is the critical inversion for normal laser action (i.e., when the Q-switching element is open), its value can be obtained by the usual critical inversion relation i.e., $N_{ic} = N_c = \gamma/\sigma l$, where γ is the cavity loss with the Q-switching element open. If N_{ic} is known, i.e., if γ, σ, and l are known, and if the ratio x between the actual pump energy and the threshold pump energy is also known, then Eq. (8.4.7) allows the initial inversion, N_i, to be readily calculated.

Once N_i is known, the time evolution of the system after Q-switching, i.e., for $t > 0$, can be obtained from Eq. (7.2.16) with the initial conditions $N(0) = N_i$ and $\phi(0) = \phi_i$. Here again ϕ_i is just some small number of photons needed to let laser action start [$\phi_i \cong 1$]. The equations can now be considerably simplified since we expect the evolution of both $N(t)$ and $\phi(t)$ to occur on a time scale so short that the pump term R_p and the spontaneous decay term N/τ in (7.2.16a) can be neglected. Equations (7.2.16) then reduce to

$$\frac{dN}{dt} = -B\phi N \tag{8.4.8a}$$

$$\frac{d\phi}{dt} = \left(V_a BN - \frac{1}{\tau_c}\right)\phi \tag{8.4.8b}$$

Before proceeding it is worth remembering that, according to Eq. (8.2.12), the population N_p corresponding to the peak of the photon pulse (see Fig. 8.9c), i.e., when $(d\phi/dt) = 0$, is

$$N_p = 1/V_a B\tau_c = \gamma/\sigma l \tag{8.4.9}$$

which is the same as the critical inversion N_c. This result, with the help of Eq. (8.4.7), allows us to express the ratio N_i/N_p in a form that will be useful for the discussion that follows, viz.,

$$(N_i/N_p) = x \tag{8.4.10}$$

After these preliminary considerations, we are ready to proceed with a calculation of the peak power of the laser output pulse, P_p, through, e.g., mirror 2. According to Eq. (7.2.18) we have

$$P_p = \left(\frac{\gamma_2 c}{2L_e}\right) h\nu\phi_p \tag{8.4.11}$$

where ϕ_p is the number of photons in the cavity at the peak of the laser pulse. To calculate ϕ_p we take the ratio between Eqs. (8.4.8a) and (8.4.8b). Using Eq. (8.4.9) also, we get

$$\frac{d\phi}{dN} = -V_a\left(1 - \frac{N_p}{N}\right) \tag{8.4.12}$$

which can be readily integrated to give

$$\phi = V_a[N_i - N - N_p \ln(N_i/N)] \tag{8.4.13}$$

where, for simplicity, the small number ϕ_i has been neglected. At the peak of the pulse we then get

$$\phi_p = V_a N_p\left(\frac{N_i}{N_p} - \ln\frac{N_i}{N_p} - 1\right) \tag{8.4.14}$$

which readily gives ϕ_p once N_p and the ratio (N_i/N_p) are known through Eqs. (8.4.9) and (8.4.10) respectively. The peakpower is then obtained from Eqs. (8.4.11), (8.4.14) and (8.4.9) as

$$P_p = \frac{\gamma_2}{2}\left(\frac{A_b}{\sigma}\right)\left(\frac{h\nu}{\tau_c}\right)\left(\frac{N_i}{N_p} - \ln\frac{N_i}{N_p} - 1\right) \tag{8.4.15}$$

where $A_b = V_a/l$ is the beam area and where the expression for τ_c, given by Eq. (7.2.14), has also been used.

To calculate the output energy, E, we begin by noticing that

$$E = \int_0^\infty P(t)dt = \left(\frac{\gamma_2 c}{2L_e}\right) h\nu \int_0^\infty \phi dt \tag{8.4.16}$$

where $P(t)$ is the time behavior of the output power and where Eq. (7.2.18) has been used again. The integration in Eq. (8.4.16) can be carried out easily by integrating both sides of Eq. (8.4.8b) and by noting that $\phi(0) = \phi(\infty) \cong 0$. We then get $\int_0^\infty \phi dt = V_a\tau_c \int_0^\infty B\phi N dt$. The quantity $\int_0^\infty B\phi N dt$ can then be obtained by integrating both sides of Eq. (8.4.8a), to give $\int_0^\infty B\phi N dt = (N_i - N_f)$ where N_f is the final inversion (see Fig. 8.3b). We thus get $\int_0^\infty \phi dt = V_a\tau_c(N_i - N_f)$ so that Eq. (8.4.16) becomes

$$E = (\gamma_2/2\gamma)(N_i - N_f)(V_a h\nu) \tag{8.4.17}$$

Note that Eq. (8.4.17) can be readily understood when we notice that $(N_i - N_f)$ is the available inversion and this inversion produces a number of photons $(N_i - N_f)V_a$. Out of this number of photons emitted by the medium, only the fraction $(\gamma_2/2\gamma)$ is available as output energy. To calculate E from Eq. (8.4.17) one needs to know N_f. This can be obtained from Eq. (8.4.13) by letting $t \to \infty$. Since $\phi(\infty) \cong 0$, we get

$$\frac{N_i - N_f}{N_i} = \frac{N_p}{N_i} \ln\frac{N_i}{N_f} \tag{8.4.18}$$

which gives N_f/N_i as a function of N_p/N_i. We can now define the quantity $\eta_E = (N_i - N_f)/N_i$ appearing in Eq. (8.4.18) as the inversion (or energy) utilization factor. In fact, out of the initial inversion N_i, the inversion which has been actually used is $(N_i - N_f)$. In terms of η_E Eq. (8.4.18) can be recast in the form

$$\eta_E(N_i/N_p) = -\ln(1 - \eta_E) \tag{8.4.19}$$

Figure 8.11 shows a plot of the energy utilization factor η_E versus (N_i/N_p) as obtained from Eq. (8.4.19). Note that, for large values of (N_i/N_p), i.e., for pump energy far exceeding the threshold pump energy, the energy utilization factor tends to unity. Note also that, in terms of η_E, Eq. (8.4.17), with the help of Eq. (8.4.9) can be put in the simpler and more suggestive form

$$E = \left(\frac{\gamma_2}{2}\frac{N_i}{N_p}\eta_E\right)\left(\frac{A_b}{\sigma}\right) h\nu \tag{8.4.20}$$

where $A_b = V_a/l$ is again the beam area.

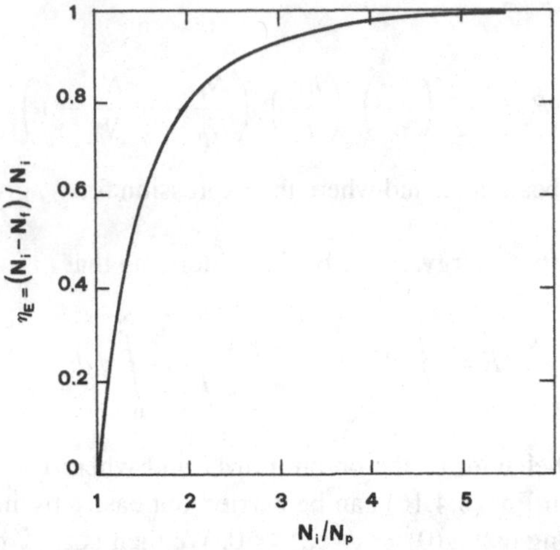

FIG. 8.11. Energy utilization factor η_E versus the ratio, N_i/N_p, between the initial inversion and the peak inversion.

Once the output energy and peak power are known, we can get an approximate value, $\Delta\tau_p$, for the width of the output pulse by defining it as $\Delta\tau_p = E/P_p$. From Eqs. (8.4.20) and (8.4.15) we get

$$\Delta\tau_p = \tau_c \frac{(N_i/N_p)\eta_E}{[(N_i/N_p) - \ln(N_i/N_p) - 1]} \tag{8.4.21}$$

Note that $\Delta\tau_p/\tau_c$ only depends upon the value of $(N_i/N_p) = x$, and, for (N_i/N_p) ranging e.g. from 2 to 10, $\Delta\tau_p$ turns out to range between 5.25 and 1.49 times the photon decay time τ_c. In particular, if we take $(N_i/N_p) = x = 2.5$ we find from Fig. 8.11 that $\eta_E = 0.89$ and from Eq. (8.4.21) $\Delta\tau_p \cong 3.81\ \tau_c$. It should be noted, however, that the expression Eq. (8.4.21) provides only an approximate value of $\Delta\tau_p$. The emitted pulse is in fact somewhat asymmetric and one can, more precisely, define a pulse risetime τ_r and a pulse fall time τ_f as the time intervals from the peak of the pulse to its half-power points. For the example $(N_i/N_p) = x = 2.5$ just considered, the computer values for τ_r and τ_f are $\tau_r = 1.45\tau_c$ and $\tau_f = 2.06\tau_c$. We see that, in this example, the approximate value for $\Delta\tau_p$ calculated from Eq. (8.4.21) is about 9% higher than the actual computed value $\tau_r + \tau_f$, a result that holds approximately for any value of (N_i/N_p).

We can now proceed to calculate the time delay τ_d between the peak of the pulse and the time of Q-switching (see Fig. 8.9). This delay can be approximated by the time required for the photon number to reach some given fraction of its peak value. If, for instance, we choose this fraction to be (1/10), no appreciable saturation of the inversion is expected to occur up to this point, and we can make the approximation $N(t) \cong N_i$ in Eq. (8.4.8b). With the help of Eqs. (8.4.9) and (8.4.10), then Eq. (8.4.8b) transforms to $(d\phi/dt) = (x - 1)\phi/\tau_c$, which upon integration gives

$$\phi = \phi_i \exp[(x - 1)t/\tau_c] \tag{8.4.22}$$

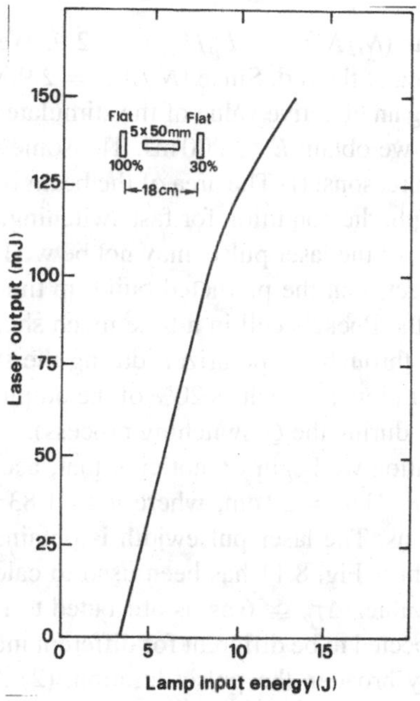

FIG. 8.12. Laser output energy versus input energy to the flashlamp for a Q-switched Nd:YAG laser, whose geometrical dimensions are shown in the inset (after Koechner,[13] by permission).

The time delay τ_d is obtained from Eq. (8.4.22) by putting $\phi = \phi_p/10$. Setting $\phi_i = 1$, we get

$$\tau_d = \frac{\tau_c}{x-1} \ln\left(\frac{\phi_p}{10}\right) \tag{8.4.23}$$

where ϕ_p is given by Eq. (8.4.14). Note that, since ϕ_p is a very large number ($\approx 10^{17}$ or more, see next example) and since it appears in the logarithmic term of Eq. (8.4.23), τ_d does not change much if we choose a different fraction of ϕ_p in this logarithmic term, e.g., $(\phi_p/20)$.

Example 8.4. *Output energy, pulse duration, and pulse build-up time in a typical Q-switched Nd:YAG laser* Figure 8.12 shows a typical plot of laser output energy, E, versus input energy, E_p, to the flashlamp for a Q-switched Nd:YAG laser. The rod and cavity dimensions are also indicated in the inset of the figure.[13] The laser is operated in a pulsed regime and is Q-switched by a KD*P (deuterated potassium dihydrogen phosphate, i.e., KD_2PO_4) Pockels cell. From the figure we observe that the laser has a threshold energy $E_{cp} \cong 3.4$ J and gives, e.g., an output energy $E \cong 120$ mJ for $E_p \cong 10$ J. At this value of pump energy the laser pulsewidth is found experimentally to be ~6 ns.

We can now proceed to a comparison of these experimental results with those predicted from the previous equations. We will neglect mirror absorption and so put $\gamma_2 \cong -\ln R_2 = 1.2$ and $\gamma_1 \cong 0$. Internal losses of the polarizer-Pockels cell combination are estimated to be $L_i \cong 15\%$, while, in comparison, the internal losses of the rod can be neglected. We thus get $\gamma_i = -\ln(1 - L_i) \cong 0.162$ and $\gamma = [(\gamma_1 + \gamma_2)/2] + \gamma_i = 0.762$. The predicted value of laser energy, at $E_p = 10$ J, can be obtained from Eq. (8.4.20) once we notice that, for our case, one has $(N_i/N_p) = (E_p/E_{cp}) = 2.9$. We now assume $A_b \cong A = 0.19\,\text{cm}^2$, where A is the cross-sectional area of the rod. Since $(N_i/N_p) = 2.9$, we find from Fig. 8.11 that $\eta_E \cong 0.94$ and from Eq. (8.4.20), assuming an effective value of the stimulated emission cross section of $\sigma = 2.8 \times 10^{-19}\,\text{cm}^2$ (see example 2.10), we obtain $E \cong 200\,\text{mJ}$. The somewhat larger value predicted by the theory can be attributed to two main reasons: (i) The area of the beam is certainly smaller than that of the rod. (ii) Due to the short cavity length, the condition for fast switching, namely that the switching time is much shorter than the build-up time of the laser pulse, may not be well satisfied in our case. Later on in this example, it will be shown, in fact, that the predicted build-up time of the Q-switched pulse, τ_d, is about 20 ns. It is difficult to switch the Pockels cell in a time much shorter than this value and, as a consequence, some energy will be lost, through the polarizer, during the switching process (in some typical cases, with pulses of this short a duration, as much as 20% of the output energy can be found to be switched out of the cavity by the polarizer, during the Q-switching process).

To calculate the predicted pulse duration we begin by noticing that, according to Eq. (7.2.11), the effective resonator length is $L_e = L + (n-1)l \cong 22\,\text{cm}$, where $n \cong 1.83$ for Nd:YAG, so that from Eq. (7.2.14) we obtain $\tau_c = L_e/c\gamma \cong 1\,\text{ns}$. The laser pulsewidth is obtained through Eq. (8.4.21) as $\Delta\tau_p = \tau_c \eta_E x/(x - \ln x - 1) \cong 3.3\,\text{ns}$, where Fig. 8.11 has been used to calculate η_E. The discrepancy between this value and the experimental value, $\Delta\tau_p \cong 6\,\text{ns}$, is attributed to two factors: (1) Multimode oscillation. In fact, the build up time is expected to be different for different modes owing to their slightly different gain and this should appreciably broaden the pulse duration. (2) As already mentioned, the condition for fast switching may not be completely satisfied in our case, and the pulsewidth is expected to be somewhat broadened by slow switching.

The build-up time of the Q-switched pulse can be obtained from Eq. (8.4.23) once ϕ_p is known. If we take $N_p = \gamma/\sigma l \cong 5.44 \times 10^{17}\,\text{cm}^{-3}$ and assume $V_a = A_b\, l \cong Al \cong 1\,\text{cm}^3$, from Eq. (8.4.14) we obtain $\phi_p \cong 4.54 \times 10^{17}$ photons, so that from Eq. (8.4.23), with $\tau_c = 1\,\text{ns}$ and $x = 2.9$, we get $\tau_d \cong 20\,\text{ns}$.

Example 8.5. *Dynamical behavior of a passively Q-switched Nd:YAG laser* We consider a laser cavity with equivalent length $L_e = 50\,\text{cm}$ in which the active medium is a Nd:YAG rod of diameter $D = 5\,\text{mm}$. We assume the laser to be passively Q-switched by a saturable absorber with saturation intensity $I_s = 1\,\text{MW/cm}^2$ and we also assume that the cell containing the saturable absorber solution has an unsaturated loss of $L = 50\%$. We take the output mirror reflectivity to be $R_2 = 74\%$ and we neglect all other cavity losses. We thus have $\gamma_2 = -\ln R_2 \cong 0.3$ for the loss of the output-coupling mirror and $\gamma_a = -\ln(1-L) = 0.693$ for the unsaturated loss of the saturable absorber. The total unsaturated loss will then be $\gamma_t = \gamma_a + (\gamma_2/2) = 0.843$. We now assume that the pump rate provides a square pulse lasting for a time $t_P = 100\,\mu\text{s}$ (see Fig. 8.13). According to Eq. (8.4.1) the population inversion, at the end of the pumping pulse and in the absence of laser action, is given by

$$N(t_P) = N_\infty[1 - \exp(-t_P/\tau)] = 0.35N_\infty \tag{8.4.24}$$

where we have taken $\tau = 230\,\mu\text{s}$. Oscillation threshold will be reached at time t_{th} (see Fig. 8.13) such that

$$\sigma N(t_{th})l = \gamma_t \tag{8.4.25}$$

We now set the pumping rate to exceed the laser threshold by 10% so that

$$N(t_P) = 1.1N(t_{th}) \tag{8.4.26}$$

where $N(t_P)$ is the inversion which would be present at $t = t_P$ in the absence of laser action (see Fig. 8.13). From Eqs. (8.4.26) and (8.4.24) we obtain $N(t_{th}) \cong 0.32\,N_\infty$ so that, from Eq. (8.4.1), we get $t_{th} \cong 88$ µs and from Eq. (8.4.25) $\sigma N_\infty l = \gamma_t/0.32 = 2.64$.

For $t > t_{th}$ the laser will show a net gain, $g_{net}(t')$, which, before appreciable absorber saturation occurs, can be written as $g_{net}(t) = \sigma Nl - \gamma_t \cong \sigma l(dN/dt)_{th}t'$, where we have changed to a new reference time axis t' whose origin is at $t = t_{th}$. From Eq. (8.4.1) and using the previously calculated value of t_{th}, one has $(dN/dt)_{th} = (N_\infty/\tau)\exp(-t_{th}/\tau) = 0.68\,(N_\infty/\tau)$. The net gain will then be given by $g_{net}(t') = 0.68(\sigma N_\infty l)(t'/\tau)$ and, using the previously calculated value of $(\sigma N_\infty l)$, we get

$$g_{net}(t') \cong 1.8(t'/\tau) \tag{8.4.27}$$

Once the expression for the net gain has been calculated, the growth of the cavity photons in the laser cavity can be obtained from the equation

$$(d\phi/dt) = (g_{net}/t_T)\phi \tag{8.4.28}$$

where $t_T = L_e/c \cong 1.66$ ns is the single-pass transit time for the laser cavity. Equation (8.4.28) could be readily derived by an argument very similar to that used to obtain Eq. (7.2.12). If Eq. (8.4.27) is substituted into Eq. (8.4.28) and the resulting equation integrated, we obtain

$$\phi(t') = \phi_i \exp\left[0.9\left(\frac{t_T}{\tau}\right)\left(\frac{t'}{t_T}\right)^2\right] \tag{8.4.29}$$

Note that, since the net gain increases linearly with time [see Eq. (8.4.27)], $\phi(t')$ increases exponentially with t'^2 [see Eq. (8.4.29) and see also Fig. 8.8a, where time t coincides with time t' of this example]. To calculate from Eq. (8.4.29) the time t'_s at which saturation occurs, we need to relate $\phi(t')$ to the circulating beam intensity $I(t')$. To arrive at this situation we first observe that, if two beams of the same intensity, I, are traveling in opposite directions within a laser cavity, the spatially averaged value of the energy density within the cavity will be $\rho = 2I/c$ [compare with Eq. (2.4.10)]. The relation between the photon number and the circulating laser intensity within the cavity will then be $\phi = \rho A_b L_e/h\nu = 2IA_bL_e/ch\nu$, where A_b is the beam area. From the previous expression, taking $A_b = \pi D^2/4 \cong 0.196\,\text{cm}^2$, we find that the photon number ϕ_s corresponding to the saturation intensity, $I_s = 1\,\text{MW/cm}^2$, is $\phi_s \cong 3.49 \times 10^{15}$. From Eq. (8.4.29), taking $\phi_i = 1$, we then obtain $(t'_s/t_T) \cong 2{,}347$ i.e., $t'_s \cong 3.89$ µs and hence $t_s \cong 92$ µs (see Fig. 8.13). Therefore, starting from the noise level, it takes ~2,350 transits for the light to reach an intensity equal to the absorber saturation intensity. From this point on the saturable absorber bleaches very rapidly.

For $t > t_s$, the dynamical behavior of the laser can be calculated approximately by assuming that the saturable absorber is completely bleached. The time evolution can then be obtained from the equations of this section by assuming, as initial inversion, $N_i = N(t_s)$. We then have for the initial gain $g_i = \sigma N_i l = \sigma N_\infty l[1 - \exp-(t_s/\tau)] = 0.87$, while the gain at the peak of the pulse is $g_p = \sigma N_p l = \gamma = 0.15$. We thus get $N_i/N_p = g_i/g_p = 5.8$. The total initial inversion in the mode volume V_a is then given by

$N_iV_a = N_iA_bl = (A_b/\sigma)g_i$. Taking $\sigma = 2.8 \times 10^{-19}\,\text{cm}^2$ (see previous example) and the previously calculated value of A_b and g_i, we obtain $N_iV_a = 6.09 \times 10^{17}$ ions (see Fig. 8.8b). Likewise we have $N_pV_a = N_iV_a/5.8 = 1.05 \times 10^{17}$ ions. The calculation of the peak photon number is then readily obtained from Eq. (8.4.14) as $\phi_p = 3.19 \times 10^{17}$ photons. The calculation of the peak pulse delay time τ_d and pulsewidth $\Delta\tau_p$ are also readily obtained from Eqs. (8.4.23) and (8.4.21), respectively, once the cavity photon decay time is calculated as $\tau_c = t_T/\gamma \cong 11$ ns. We get $\tau_d \cong 88$ ns and $\Delta\tau_p \cong 21$ ns, so that pulse peak occurs approximately at a time $t'_p \cong t'_s + \tau_d + (\Delta\tau_p/2) = 3.99\,\mu$s (see Fig. 8.8b).

We now go on to consider the case of a continuously pumped, repetitively Q-switched, laser (Fig. 8.10). We first note that, after switching and during the evolution of the Q-switched pulse, Eq. (8.4.8) still apply. The expressions for peak power, output energy, and pulse duration are therefore still given by Eqs. (8.4.15), (8.4.20), and (8.4.21) respectively. What does change, however, is the expression for (N_i/N_p), which is no longer given by Eq. (8.4.10) since it is determined by a different pump dynamics. In fact, we now require that, in the time τ_p between two consecutive pulses, the pump rate must reestablish the initial inversion, starting from the population, N_f, which was left after the preceding Q-switching event. From Eq. (7.2.16a), putting $\phi = 0$, we get, upon integration,

$$N_i = (R_p\tau) - (R_p\tau - N_f)\exp(-\tau_p/\tau) \tag{8.4.30}$$

From Eqs. (7.3.6), (7.3.3) and (8.4.9) we have $R_p\tau = xN_c = xN_p$ and Eq. (8.4.30) then gives

$$x\frac{N_p}{N_i}[1 - \exp(-1/f^*)] = 1 - \frac{N_f}{N_i}\exp(-1/f^*) \tag{8.4.31}$$

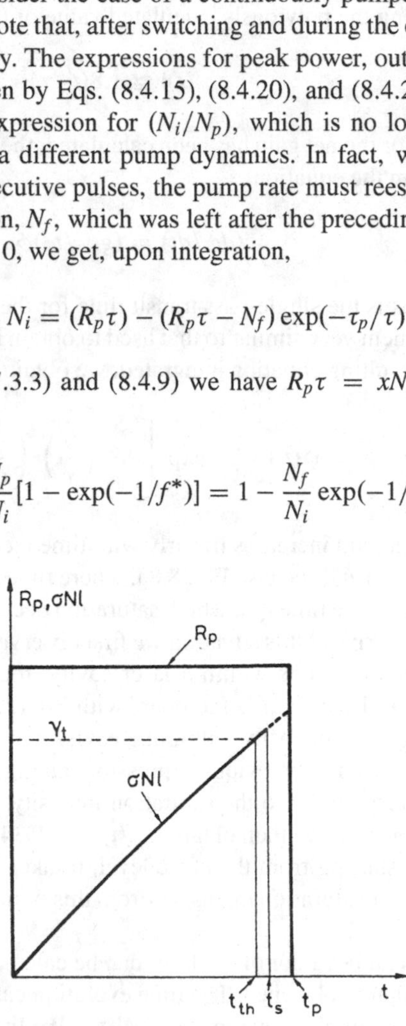

FIG. 8.13. Time evolution of the pump rate and of the laser gain σNL for a square pump of duration $t_P = 100\,\mu$s and for a medium with relaxation time $\tau = 230\,\mu$s.

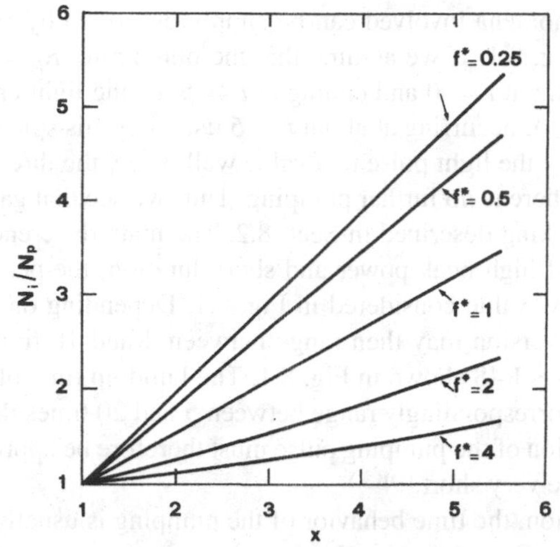

FIG. 8.14. Case of a cw-pumped repetitively Q-switched laser. Plot of (N_i/N_p) versus the amount x by which threshold is exceeded, for several values of the normalized pulse repetition rate f^*.

where x is the amount by which the cw pump exceeds its threshold value and $f^* = \tau f$, where $f = 1/\tau_p$ is the laser repetition rate. Equation (8.4.31) together with Eq. (8.4.18), which still holds, provide a pair of equations that can be solved for both (N_i/N_p) and (N_i/N_f), once x and f^* are known. Fig. 8.14 shows the solution for (N_i/N_p) versus the amount x by which threshold is exceeded, plotted for several values of the normalized frequency f^*. For given values of x and f^*, Fig. 8.14 gives the corresponding value of (N_i/N_p). Once (N_i/N_p) is known, the quantity (N_i/N_f) or, equivalently, the energy utilization factor η_E, can be obtained from Fig. 8.11. When (N_i/N_p) and η_E have been calculated, P_p, E, and $\Delta\tau_p$ are readily obtained from Eqs. (8.4.15), (8.4.20), and (8.4.21), respectively. Note that, within the range we have been considering for the variables x and f^*, the relation between (N_i/N_p) and x is close to linear.

The calculations for a quasi-three-level laser would proceed in a similar way starting from Eq. (7.2.24). Because of space limitations, these calculations are not presented here.

8.5. GAIN SWITCHING

Gain switching, like Q-switching, is a technique that allows the generation of a laser pulse of short duration (generally from a few tens to a few hundreds of nanoseconds) and high peak power. Unlike Q switching, however, where the losses are rapidly switched to a low value, in the case of gain switching, it is the laser gain that is rapidly switched to a high value. Gain switching is achieved by using a pumping pulse that is so fast that the population inversion and hence the laser gain reach a value considerably above threshold before the number of cavity photons has had time to build up to a sufficiently high level to deplete the inversion.

The physical phenomena involved can be simply described by referring to the spiking situation depicted in Fig. 8.1. If we assume that the pump rate, $R_p = R_p(t)$, is in the form of a square pulse starting at $t = 0$ and ending at $t \cong 5$ μs, the light emission will consist of just the first spike of $\phi(t)$, occurring at about $t = 5$ μs. After this spike, in fact, the inversion will have been driven by the light pulse to a value well below the threshold value and it does not grow thereafter as there is no further pumping. Thus we see that gain switching is similar in character to laser spiking described in Sect. 8.2. The main difference arises from the fact that, to obtain pulses of high peak power and short duration, the peak value of R_p must be much larger than the cw value considered in Fig. 8.1. Depending on the peak value of this pump rate, the peak inversion may then range between 4 and 10 times the threshold value rather than the value of ~1.48 shown in Fig. 8.1. The build up time of the laser radiation up to its peak value may correspondingly range between 5 and 20 times the cavity photon decay time τ_c. The time duration of the pumping pulse must therefore be approximately equal to this build-up time and hence very short.[14]

In an actual situation, the time behavior of the pumping is usually in the form of a bell-shaped pulse rather than a square pulse. In this case we require that the peak of the photon spike occurs at an appropriate time in the trailing edge of the pumping pulse. In fact, if this peak were to occur at e.g., the peak of the pumping pulse, there could be enough pump input left after the laser pulse to allow the inversion to exceed threshold again and thus produce a second, although weaker, laser pulse. If, on the other hand, the photon peak were to occur much later in the tail of the pulse, this would imply that there was insufficient pumping time for the inversion to grow to a sufficiently high value. The above discussion implies that, for a given duration of pump pulse, there is some optimum value for the peak pump rate. By decreasing the pump duration, the optimum value of the peak pump rate increases and a more intense and narrower laser pulse is produced.

It should be noted that, given a sufficiently fast and intense pump pulse, any laser can in principle be gain switched even if the spontaneous decay of its upper laser level is allowed by electric dipole interaction and the lifetime correspondingly falls in the nanosecond range. In this case the pumping pulse and the cavity photon decay time must be appreciably shorter than this lifetime and, correspondingly, very short gain switched pulses, i.e., with duration shorter than ~1 ns, can be obtained.

Example 8.6. *Typical cases of gain switched lasers* The most common example of a gain-switched laser is the electrically pulsed TEA (Transversely Excited at Atmospheric pressure; see Chap. 10) CO_2 laser. Taking a typical cavity length of $L = 1$ m, a 20% transmission of the output mirror, and assuming that the internal losses arise only from this mirror's transmission, we get $\gamma \cong 0.1$ and $\tau_c = L/c\gamma \cong 30$ ns. Assuming that the laser build up time is ten times longer, we see that the duration of the pumping pulse should last ~300 ns, in agreement with experimental findings.

As typical examples of gain switched lasers using active media with upper state lifetime, τ, in the nanosecond range, we mention the case of a short-cavity dye laser (e.g., Rhodamine 6G dye laser, $\tau \cong 5$ ns) pumped by the fast (~0.5 ns) pulse of an atmospheric pressure N_2 laser or the case of a semiconductor laser (e.g., GaAs, $\tau \cong 3$ ns) pumped by a very short (~0.5 ns) current pulse. In both cases, gain switched laser pulses of ~100 ps duration can be obtained.

8.6. MODE-LOCKING

Let us now consider a laser which is oscillating on a rather large number of longitudinal modes. Under ordinary circumstances, the phases of these modes will have random values and, for cw oscillation, the beam intensity will show a random time behavior. As an example, Fig. 8.15 shows the time behavior of the square of the electric field amplitude, $|A(t)|^2$, of the output beam for the case of $N = 31$ oscillating modes, each with the same amplitude E_0, and evenly separated in frequency by the frequency difference $\Delta\nu$ between consecutive longitudinal modes. One sees that the output beam consists of a random sequence of light pulses. Despite this randomness, since these pulses arise from the sum of N frequency components which are evenly spaced in frequency, the pulse waveform of Fig. 8.15 has the following general properties which are a characteristic of a Fourier series: (i) The waveform is periodic with a period $\tau_p = 1/\Delta\nu$. (ii) Each light pulse of the random waveform has a duration $\Delta\tau_p$ roughly equal to $1/\Delta\nu_L$ where, $\Delta\nu_L = N\Delta\nu$, is the total oscillating bandwidth. Thus, for lasers with relatively large gain bandwidths, such as solid-state, dye or semiconductor lasers, $\Delta\nu_L$ may be comparable to this gain bandwidth and hence short noise pulses, with durations of picoseconds or less, can be produced. Note that, since the response time of a conventional photodetector is usually much longer than a few picoseconds, one does not resolve this complex time behavior in the detected output of a, random phase, multimode laser, and instead its average value is monitored. This value is simply the sum of powers in the modes and hence is proportional to NE_0^2.

Let us now suppose that the oscillating modes, while still having equal or comparable amplitudes, are somehow made to oscillate with some definite relation between their phases. Such a laser is referred to as mode locked, and the process by which the modes are made to adopt a definite phase relation is referred to as *mode locking*.(15) Mode-locked lasers will be considered at some length in this section.

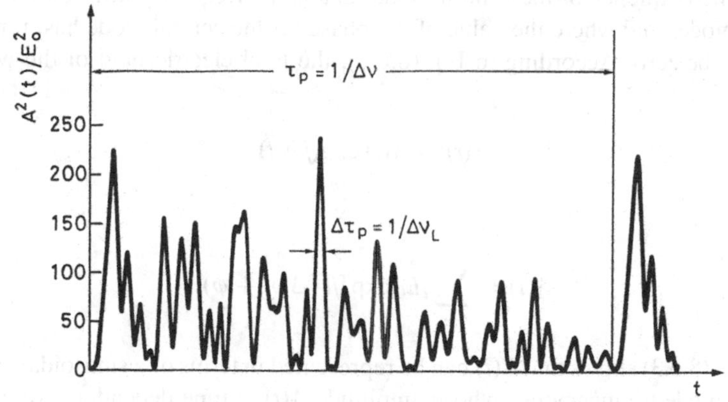

FIG. 8.15. Example of time behavior of the squared amplitude of the total electric field, $|A(t)|^2$, for the case of 31 oscillating modes, all with the same amplitude E_0 and with random phases.

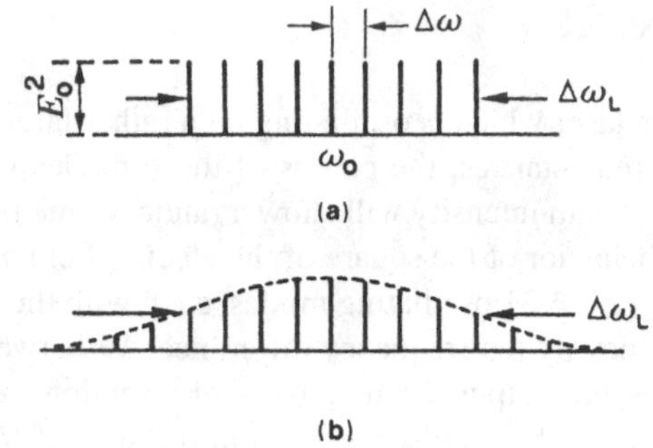

FIG. 8.16. Mode amplitudes (represented by vertical lines) versus frequency for a mode-locked laser. (a) Uniform amplitude. (b) Gaussian amplitude distribution over a bandwidth (FWHM) $\Delta\omega_L$.

8.6.1. Frequency-Domain Description

We will first describe mode-locking in the frequency domain and consider, as a first example, the case of $2n+1$ longitudinal modes oscillating with the same amplitude E_0 (Fig. 8.16a). We will assume the phases φ_l of the modes in the output beam to be locked according to the relation

$$\varphi_l - \varphi_{l-1} = \varphi \tag{8.6.1}$$

where φ is a constant. The total electric field $E(t)$ of the e.m. wave, at any given point in the output beam, can be written as

$$E(t) = \sum_{-n}^{+n} {}_l E_0 \exp\{j\,[(\omega_0 + l\Delta\omega)\,t + l\varphi]\} \tag{8.6.2}$$

where ω_0 is the frequency of the central mode, $\Delta\omega$ is the frequency difference between two consecutive modes and where the value of the phase for the central mode has, for simplicity, been taken to be zero. According to Eq. (8.6.2), the total electric field of the wave can be written as

$$E(t) = A(t)\exp(j\omega_0 t) \tag{8.6.3}$$

where

$$A(t) = \sum_{-n}^{+n} {}_l E_0 \exp\,[jl\,(\Delta\omega t + \varphi)] \tag{8.6.4}$$

Equation (8.6.3) shows that $E(t)$ can be represented in terms of a sinusoidal carrier wave, at the center-mode frequency ω_0, whose amplitude $A(t)$ is time dependent. To calculate the time behavior of $A(t)$, we now change to a new time reference t' such that $\Delta\omega t' = \Delta\omega t + \varphi$. In terms of the new variable t', Eq. (8.6.4) transforms to

$$A(t') = \sum_{-n}^{+n} {}_l E_0 \exp jl(\Delta\omega\, t') \tag{8.6.5}$$

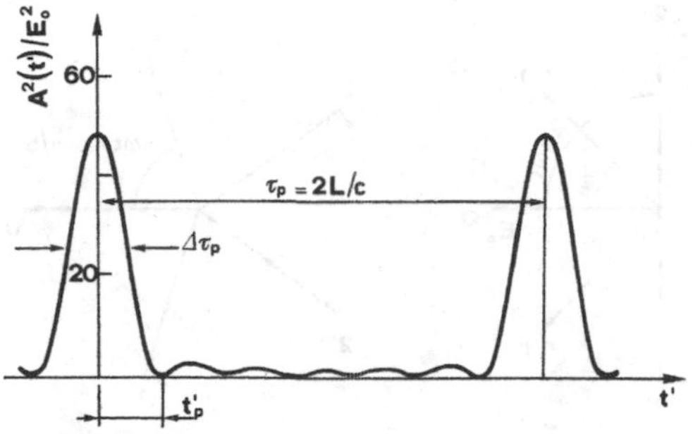

FIG. 8.17. Time behavior of the squared amplitude of the electric field for the case of seven oscillating modes with locked phases and equal amplitudes, E_0.

and the sum appearing in the right-hand side can be easily recognized as a geometric progression with a ratio $\exp j(\Delta\omega t')$ between consecutive terms. Summation of this progression can then be easily performed and we obtain

$$A(t') = E_0 \frac{\sin[(2n+1)\,\Delta\omega\, t'/2]}{\sin[\Delta\omega\, t'/2]} \tag{8.6.6}$$

To help understanding the physical significance of this expression, Fig. 8.17 shows the quantity $A^2(t')/E_0^2$, $A^2(t')$ being proportional to the beam intensity, versus time t', for $2n+1 = 7$ oscillating modes. It is seen that, as a result of the phase-locking condition Eq. (8.6.1), the oscillating modes interfere so as to produce a train of evenly spaced light pulses. The pulse maxima occur at those times for which the denominator of Eq. (8.6.6) vanishes. In the new time reference t', the first maximum occurs for $t' = 0$. Note that, at this time, the numerator of Eq. (8.6.6) also vanishes and, upon making the approximation $\sin\alpha \cong \alpha$, which holds for small values of α, we readily see from Eq. (8.6.6) that $A^2(0) = (2n+1)^2 E_0^2$. The next pulse will occur when the denominator of Eq. (8.6.6) again vanishes and this will happen at a time t' such that $(\Delta\omega t'/2) = \pi$. Two successive pulses are therefore separated by a time

$$\tau_p = 2\pi/\Delta\omega = 1/\Delta\nu \tag{8.6.7}$$

where $\Delta\nu$ is the frequency separation between two consecutive oscillating modes. For $t' > 0$, the first zero for $A^2(t')$ in Fig. 8.17 occurs when the numerator of Eq. (8.6.6) again vanishes. This occurs at a time t'_p such that $\left[(2n+1)\,\Delta\omega t'_p/2\right] = \pi$. Since the width $\Delta\tau_p$ (FWHM) of $A^2(t')$, i.e. of each laser pulse, is approximately equal to t'_p, we thus have

$$\Delta\tau_p \cong 2\pi/(2n+1)\Delta\omega = 1/\Delta\nu_L \tag{8.6.8}$$

where $\Delta\nu_L = (2n+1)\Delta\omega/2\pi$ is the total oscillating bandwidth (see Fig. 8.16a).

The mode-locking behavior of Fig. 8.17 can be readily understood if we represent the field components of Eq. (8.6.5) by vectors in the complex plane. The l-th amplitude component would thus correspond to a complex vector of amplitude E_0 and rotating at the angular velocity $l\Delta\omega$. At time $t' = 0$, all these vectors are seen from Eq. (8.6.5) to have zero phase

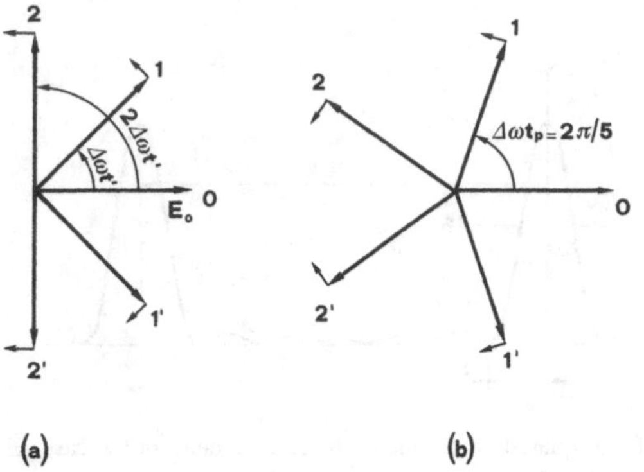

FIG. 8.18. Representation of the cavity mode amplitudes in the complex plane for the case of five modes. Figure (a) show the situation at a general time $t' > 0$, while FIG. (b) depicts the time instant at which the sum of the five mode amplitudes is zero.

and, accordingly, they lie in the same direction in Fig. 8.18, which we assume to be the horizontal direction. The total field will, in this case, be $(2n+1)E_0$. For $t' > 0$, the vector corresponding to the central mode remains fixed, the vectors of the modes with $l > 0$ i.e., with $\omega > \omega_0$, will rotate in one direction (e.g., counterclockwise) while the vectors of the modes with $\omega < \omega_0$ will rotate in the opposite (clockwise) sense. Thus, for the case of e.g., five modes, the situation at some later time t', will be as indicated in Fig. 8.18a. If now the time t' is such that mode 1 has made a 2π rotation (which occurs when $\Delta\omega t' = 2\pi$), mode -1 will also have rotated (clockwise) by 2π, while modes 2 and -2 will have rotated by 4π. All these vectors will therefore be aligned again with that at frequency ω_0, and the total field amplitude will again be $(2n+1)E_0$. Thus the time interval τ_p between two consecutive pulses must be such that $\Delta\omega\tau_p = 2\pi$, as indeed shown by Eq. (8.6.7). Note that, in this picture, the time instant t'_p at which $A(t')$ vanishes (see Fig. 8.17) correspond to the situation where all vectors are evenly spaced around the 2π angle (Fig. 8.18b). To achieve this condition, mode 1 must have made a rotation of only $2\pi/5$, or, more generally for $(2n+1)$ modes, of $2\pi/(2n+1)$. The time duration t'_p and hence the pulse duration $\Delta\tau_p$ thus turn out to be given by Eq. (8.6.8).

Before proceeding further it is worth summarizing and commenting on the main results that have been obtained so far. We have found that, under the mode locking condition given by Eq. (8.6.1), the output beam consists of a train of mode-locked pulses, the duration of each pulse, $\Delta\tau_p$, being about equal to the inverse of the oscillating bandwidth $\Delta\nu_L$. Again this result comes about again from a general property of a Fourier series. Now, since $\Delta\nu_L$ can be of the order of the width of the gain line $\Delta\nu_0$, very short pulses (down to a few picoseconds) can be expected to result from mode-locking of solid-state or semiconductor lasers. For dye or tunable solid-state lasers, the gain linewidth can be at least a factor 100 times larger, so that very much shorter pulsewidths are possible and indeed have been obtained (e.g., ~25 fs for Rhodamine 6G dye laser and ~7 fs for Ti:sapphire laser). In the case of gas lasers, on the other hand, the gain linewidth is much narrower (up to a few GHz) and relatively long pulses are generated (down to ~100 ps). Note also that the peak power of the pulse is proportional to $(2n+1)^2E_0^2$, while for modes with random phases the average power is simply the sum of powers in the modes and hence is proportional to $(2n+1)E_0^2$. Therefore, for the same number of oscillating modes and their field amplitudes E_0, the ratio between the peak pulse power in the mode-locked case and the average power in the non-mode-locked case is equal to the number, $(2n + 1)$, of oscillating modes, which, for solid state or liquid lasers, can be rather high ($10^3 \div 10^4$). We thus see that mode-locking is useful not only for producing pulses of very short duration but also for producing high peak power.

So far we have restricted our considerations to the rather unrealistic case of an equal-amplitude mode-spectrum (Fig. 8.16a). In general the spectral envelope is expected to have a bell-shaped form and, as a characteristic example, we will consider an envelope with a Gaussian distribution (Fig. 8.16b). We can therefore write the amplitude E_l of the l-th mode as

$$E_l^2 = E_0^2 \exp\left[-\left(\frac{2l\Delta\omega}{\Delta\omega_L}\right)^2 \ln 2\right] \tag{8.6.9}$$

where $\Delta\omega_L$ represents the bandwidth (FWHM) of the spectral intensity. If we again assume that the phases are locked according to Eq. (8.6.1) and that the phase of the central mode is equal to zero, we again find that $E(t)$ can be expressed as in Eq. (8.6.3) where the amplitude $A(t)$, in the time reference t', is given by

$$A(t') = \sum_{-\infty}^{+\infty} {}_l E_l \exp j(l\Delta\omega\, t') \tag{8.6.10}$$

If the sum is approximated by an integral [i.e., $A(t) \cong \int E_l \exp j(l\Delta\omega t) dl$], the field amplitude $A(t)$ is seen to be proportional to the Fourier transform of the spectral amplitude E_l. We then find that $A^2(t)$, i.e., the pulse intensity, is a Gaussian function of time, which can be written as

$$A^2(t) \propto \exp\left[-\left(\frac{2t}{\Delta\tau_p}\right)^2 \ln 2\right] \tag{8.6.11}$$

where

$$\Delta\tau_p = 2\ln 2/\pi\Delta\nu_L = 0.441/\Delta\nu_L \tag{8.6.12}$$

Here, the term $\Delta\tau_p$, appearing in Eq. (8.6.11), represents the width (FWHM) of the pulse intensity.

As a conclusion to our discussion of the two examples given above, we can say that, when the mode-locking condition Eq. (8.6.1) holds, the field amplitude turns out to be given by the Fourier transform of the magnitude of the spectral amplitude. In such a case, the pulsewidth $\Delta\tau_p$ is related to the width of the laser spectrum $\Delta\nu_L$ by the relation $\Delta\tau_p = \beta/\Delta\nu_L$, where β is a numerical factor (of the order of unity), which depends on the particular shape of the spectral intensity distribution. A pulse of this sort is said to be *transform-limited*.

Under locking conditions different from Eq. (8.6.1), the output pulse may be far from being transform limited. As an example, instead of Eq. (8.6.1), which can be written as $\varphi_l = l\varphi$, we consider the situation where

$$\varphi_l = l\varphi_1 + l^2\varphi_2 \tag{8.6.13}$$

where φ_1 and φ_2 are two constants. If we again assume a Gaussian amplitude distribution such as in Eq. (8.6.9), the Fourier transform of the spectrum can again be analytically calculated and $E(t)$ can be written as

$$E(t) \propto \exp\left[-\alpha t^2\right] \exp\left[j\left(\omega_0 t + \beta t^2\right)\right] \tag{8.6.14}$$

where the two constants α and β are related to $\Delta\omega_L$ and φ_2. For brevity, these relations are not given here since they are not needed for what follows. What is important to notice here, however, are the following three points: (i) The beam intensity, being proportional to $|E(t)|^2$, is still described by a Gaussian function whose pulsewidth $\Delta\tau_p$ (FWHM), in terms of the parameter α is equal to

$$\Delta\tau_p = (2\ \ln 2/\alpha)^{1/2} \tag{8.6.15}$$

(ii) The presence of a phase term $l^2\varphi_2$ in Eq. (8.6.13), which is quadratic in the mode index l, results in $E(t)$ having a phase term, βt^2, which is quadratic in time. This means that the instantaneous carrier frequency of the wave, $\omega(t) = d(\omega_0 t + \beta t^2)/dt = \omega_0 + 2\beta t$, now has a linear frequency sweep (or *frequency chirp*). (iii) Depending of the value of φ_2, the product $\Delta\tau_p\Delta\nu_L$ can be much larger than the minimum value, 0.441, given by Eq. (8.6.12). To understand this point, it is easiest to go back to a calculation of the spectrum of a field such as in Eq. (8.6.14). The spectral intensity again turns out to be given by a Gaussian function whose bandwidth $\Delta\nu_L$ (FWHM) is given by

$$\Delta\nu_L = \frac{0.441}{\Delta\tau_p}\left[1 + \left(\frac{\beta\Delta\tau_p^2}{2\ \ln 2}\right)\right]^{1/2} \tag{8.6.16}$$

where the expression for α given by Eq. (8.6.15) has been used. Equation (8.6.16) shows that, if $\beta \neq 0$, one has $\Delta\tau_p\Delta\nu_L > 0.441$ and that, for $\beta\Delta\tau_p^2 \gg 1$, i.e., for sufficiently large values of the frequency chirp, the product $\Delta\tau_p\Delta\nu_L$ becomes much larger than 1. The physical basis for this result can be understood by noting that the spectral broadening now arises both from the pulsed behavior of $|E(t)|^2$, i.e., from the amplitude modulation of $E(t)$ [which accounts for the first term on the right-hand side of Eq. (8.6.16)] and from the frequency chirp term $2\beta t$ of $E(t)$ [which accounts for the second term on the right-hand side of Eq. (8.6.16)].

8.6.2. Time-Domain Picture

We recall that, with the mode locking condition given by Eq. (8.6.1), two consecutive pulses of the output beam were found to be separated by a time τ_p given by Eq. (8.6.7). Since $\Delta\nu = c/2L$, where L is the cavity length, τ_p turns out to be equal to $2L/c$, which is just the cavity round trip time. It should be noted at this point that the spatial extent Δz of a typical mode-locked pulse is usually much shorter than the cavity length [e.g. for a pulse of duration $\Delta\tau_p = 1$ ps, one has $\Delta z = c\Delta\tau_p = 0.3$ mm while a laser cavity is typically several tens of centimeter long]. The oscillating behavior inside the laser cavity can therefore be visualized as consisting of a single ultrashort pulse, of duration $\Delta\tau_p$ given by Eq. (8.6.8), which propagates back and forth within the cavity. In such a case, in fact, the output beam would obviously

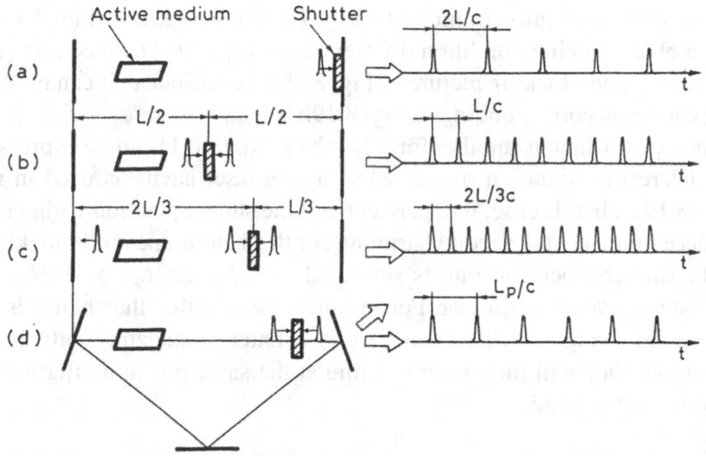

FIG. 8.19. Use of a fast cavity shutter to achieve mode-locking: (a) Shutter placed at one cavity end thus leading to an output pulse train with a repetition frequency of $\Delta\nu = c/2L$. (b) Shutter placed at a distance $L/2$, from one cavity mirror. (c) Shutter placed at a distance, $L/3$, from one cavity mirror. The output pulse repetition rates for cases (b) and (c) are $2\Delta\nu$ and $3\Delta\nu$, respectively, these being examples of harmonic mode-locking. Figure (d) represents the case of ring-laser mode-locking where the shutter position required for establishing mode-locking becomes irrelevant.

consist of a train of pulses with time separation between consecutive pulses equal to the cavity round trip time. This is the so-called *time-domain picture* of mode-locking. According to this picture, we readily understand that the mode-locking condition given by Eq. (8.6.1) can be achieved by placing a suitably fast shutter at one end of the cavity (Fig. 8.19a). In fact, if an initially non-mode-locked beam is present within the cavity, its spatial amplitude distribution can be represented as in Fig. 8.15 with time t being replaced by z/c, where z is the longitudinal coordinate along the laser cavity. Let us now assume that the shutter is periodically opened with a period $T = 2L/c$, possibly at the time where the most intense noise pulse of Fig. 8.15 reaches the shutter. If the opening time of the shutter is comparable with the duration of this noise pulse, then only this pulse will survive in the laser cavity and thus produce the mode-locking situation of Fig. 8.19a. Following a similar argument, one can now realize that, if the shutter is placed at the cavity center and if the shutter is periodically opened with a period $T = L/c$, the mode locking situation described in Fig. 8.19b will develop. In this case, two ultrashort pulses are present in the cavity and they are located and traveling in such a way as to cross each other at the shutter position when the shutter is opened. By the same argument, if the shutter is placed at a distance $L/3$ from one cavity mirror and if the shutter is periodically opened with a period $T = 2L/3c$, the mode-locking situation described in Fig. 8.19c will develop. In this case, three ultrashort pulses are present in the cavity and they are located and traveling in such a way that two pulses always cross at the shutter position when the shutter is open. Note that, for the cases of Fig. 8.19b and 8.19c, the repetition rate for the train of output pulses is $2\Delta\nu$ and $3\Delta\nu$, respectively, where $\Delta\nu = c/2L$ is the pulse repetition rate of the mode-locking case considered in Fig. 8.19a. For this reason, the mode-locking situations represented in Fig. 8.19b and c are referred to as cases of *harmonic mode-locking*. By contrast to these cases, the mode-locking situation of Fig. 8.19a is sometimes referred to as mode-locking at the fundamental frequency or *fundamental mode-locking*. Note also that the frequency domain description of mode-locking for the two cases of Fig. 8.19b and c must correspond to a phase locking condition different from Eq. (8.6.1), since this condition was shown to lead to the mode-locking picture of Fig. 8.19a. For instance, it can be shown that the phase locking condition corresponding to Fig. 8.19b is $\varphi_{l+1} - \varphi_l = \varphi_l - \varphi_{l-1} + \pi$, instead of $\varphi_{l+1} - \varphi_l = \varphi_l - \varphi_{l-1}$ which is another form in which Eq. (8.6.1) can be expressed.

A rather interesting situation occurs when a ring laser cavity is used in mode-locked operation (Fig. 8.19d). In this case, independently of the shutter position within the cavity, the laser will produce fundamental, second harmonic, or third harmonic mode-locking depending on whether the shutter repetition rate is set equal to c/L_p, $2c/L_p$, or $3c/L_p$, where L_p is the length of the ring cavity perimeter. For instance, assume that the shutter is opened with a repetition rate c/L_p (Fig. 8.19d). Then, if two counter-propagating pulses meet once at the shutter position, they will then keep meeting at the same position after each round trip, independently of shutter position.

8.6.3. Methods of Mode-Locking

The methods of mode-locking, like those of Q-switching, can be divided into two categories: (1) Active-mode-locking, in which the mode-locking element is driven by an external source. (2) Passive mode-locking, in which the element which induces mode-locking is not driven externally and instead exploits some non-linear optical effect such as saturation of a saturable absorber or non-linear refractive index change of a suitable material.

8.6.3.1. Active Mode-Locking

There are three main types of active mode-locking (ML), namely: (1) Mode-locking induced by an amplitude modulator (*AM mode-locking*). (2) Mode-locking induced by a phase modulator (*FM mode-locking*). (3) Mode-locking induced by a periodic modulation of the laser gain at a repetition rate equal to the fundamental cavity frequency $\Delta\nu = c/2L$ (*ML by synchronous pumping*). We will discuss AM mode-locking in most details, this type being the most popular, and then give a briefer discussion of the FM mode-locking. Mode-locking by synchronous pumping will not be discussed here since it is now less widely used. In fact, it only applies to active media with nanosecond relaxation time, notably dye media, and, to obtain the shortest pulses, requires that the modulation rate of the pump be equal, to within high precision, to the fundamental frequency of the laser cavity. For this reason, pulse durations shorter than 1 ps are difficult to achieve from a synchronously pumped dye laser.

To describe AM-mode-locking, we suppose a modulator to be inserted in the cavity, which produces a time-varying loss at frequency ω_m. If $\omega_m \neq \Delta\omega$, where $\Delta\omega = 2\pi\Delta\nu$, $\Delta\nu$ being the frequency difference between longitudinal modes, this loss will simply amplitude modulate the electric field, $E_l(t)$, of each cavity mode to give

$$E_l(t) = E_0[1 - (\delta/2)(1 - \cos\omega_m t)]\cos(\omega_l t + \phi_l) \tag{8.6.17}$$

where ω_l is the mode frequency, ϕ_l its phase and where δ is the depth of amplitude modulation, which means that the field amplitude is modulated from E_0 to $E_0(1 - \delta)$. Note that the term $E_0(\delta/2)\cos\omega_m t \times \cos(\omega_l t + \phi_l)$ in Eq. (8.6.17) can be written as $(E_0\delta/4)\{\cos[(\omega_l + \omega_m)t$

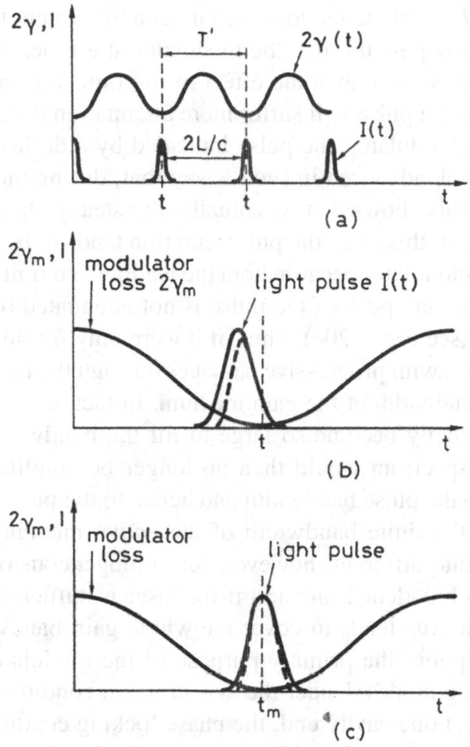

FIG. 8.20. Time-domain description of AM mode-locking: (a) steady state condition; (b) light pulse arriving before the time t_m of minimum loss; (c) pulse-shortening occurring when the pulse arrives at time t_m.

$+\phi_l]+\cos[(\omega_l-\omega_m)t+\phi_l]\}$. Thus $E_l(t)$ actually contains two terms oscillating at the frequencies $\omega_l \pm \omega_m$ (modulation side-bands). If now $\omega_m = \Delta\omega$, these modulation side-bands will coincide with the adjacent mode frequencies of the resonator. These two side-bands will thus give contributions to the field equations of the two adjacent cavity modes. So, the equations for cavity modes become coupled i.e., the field equation of a given cavity mode will contain two contributions arising from the modulation of the adjacent modes.[16] If the modulator is placed very close to one cavity mirror, this mode-coupling mechanism can then be shown to lock the mode phases according to Eq. (8.6.1).

The details of the operation of AM mode-locking can be more readily understood in the time domain rather than in the frequency domain. Thus, Fig. 8.20a shows the time behavior of the cavity round-trip power losses* 2γ which are modulated with a modulation period $T = 2\pi/\omega_m$. We will assume the modulator to be placed at one end of the cavity (see Fig. 8.19a). If now $\omega_m = \Delta\omega$, the modulation period T will be equal to the cavity round-trip time and the stable steady-state condition will correspond to light pulses passing through the modulator at the times t_m when a minimum loss of the modulator occurs (Fig. 8.20a). Indeed, if a pulse is assumed to pass through the modulator at a time of minimum loss, it will return to the

* In a mode-locked linear cavity, it proves to be simpler to talk in terms of round-trip loss and round trip gain, rather than in terms of the corresponding single-pass values.

modulator after a time, $2L/c$, where the loss is again at a minimum. If, on the other hand, the pulse is assumed to initially pass through the modulator at a time, e.g., slightly shorter than t_m (solid-line pulse in Fig. 8.20b), then the effect of the time-varying loss of the modulator is that the leading edge of the pulse will suffer more attenuation than the trailing edge. Thus, after passing through the modulator, the pulse indicated by a dashed line in Fig. 8.20b will result, having its pulse peak advanced in such a way that, during the next passage, the peak will arrive closer to t_m. This shows that, eventually, the steady state situation of Fig. 8.20a will be reached. Actually, in this case, the pulse-duration tends to be shortened each time the pulse passes through the modulator because both the leading and trailing edge of the pulse are somewhat attenuated while the peak of the pulse is not attenuated by the time varying loss, $2\gamma_m(t)$, of the modulator (see Fig. 8.20c). Thus, if it were only for this mechanism, the pulse-duration would tend to zero with progressive passages through the modulator. This is however prevented by the finite bandwidth of the gain medium. In fact, as the pulse becomes shorter, its spectrum would eventually become so large to fill the bandwidth of the laser medium. The wings of the pulse spectrum would then no longer be amplified. This constitutes the fundamental limitation to the pulse bandwidth and hence to the pulse duration.

The way in which the finite bandwidth of the active medium influences the steady-state pulse duration is quite different, however, for homogeneous or inhomogeneous lines. For an inhomogeneously broadened line and if the laser is sufficiently far above threshold, the oscillating bandwidth $\Delta\nu_L$ tends to cover the whole gain bandwidth $\Delta\nu_0^*$. In fact, in a frequency-domain description, the primary purpose of the modulator is to lock the phases of these already oscillating modes. Under the synchronism condition, $\omega_m = \Delta\omega$, and if the AM modulator is placed at one cavity end, the phase locking condition given by Eq. (8.6.1) develops and, assuming for simplicity a Gaussian distribution for the mode amplitudes, we get from Eq. (8.6.12).

$$\Delta\tau_p \cong 0.441/\Delta\nu_0^* \tag{8.6.18}$$

By contrast to this situation, for a homogeneous line, the phenomenon of spatial hole burning tends to concentrate the width of the oscillating spectrum in a narrow region around the central frequency ν_0 (see Sect. 7.7). Thus, assuming the laser to be originally unlocked, the noise light pulses (see Fig. 8.15) would tend to be appreciably longer than $1/\Delta\nu_0$, where $\Delta\nu_0$ is the width of the gain line. In this case, the mechanism described in Fig. 8.20c is really effective in shortening the pulse duration, i.e., in broadening its spectrum. This pulse narrowing is however counteracted by pulse broadening occurring when the pulse passes through the active medium and so undergoes spectral narrowing. The theory of active mode-locking, for a homogeneously broadened gain medium, has been given a detailed and elegant treatment by Kuizenga and Siegman[17] and, later, presented in a more general framework by Haus.[18] We will limit ourselves here to just quoting the most relevant results and refer to Appendix F for a more detailed treatment. The intensity profile turns out to be well described by a Gaussian function whose width $\Delta\tau_p$ (FWHM) is approximately given by

$$\Delta\tau_p \cong 0.45/(\nu_m\Delta\nu_0)^{1/2} \tag{8.6.19}$$

where ν_m is the frequency of the modulator ($\nu_m = \omega_m/2\pi = c/2L$, for fundamental harmonic mode-locking). If the pulsewidth expressions for inhomogeneous, Eq. (8.6.18), and homogeneous, Eq. (8.6.19), lines are compared at the same value of the laser linewidth (i.e., for $\Delta\nu_0^* = \Delta\nu_0$), we see that, since $(\nu_m/\Delta\nu_0) \ll 1$, one has $(\Delta\tau_p)_{\text{hom}} \gg (\Delta\tau_p)_{\text{inhom}}$. One should also note that the pulse-narrowing mechanism depicted in Fig. 8.20c does not play any appreciable role in the case of a inhomogeneous line, although it is still obviously present. In this case, in fact, short noise pulses, with duration about equal to the inverse of the gain bandwidth, are already present even under non-mode-locking conditions. The main role of the modulator is then to establish a synchronism between the oscillating modes so that, out of the noise pulses of Fig. 8.15, only one pulse survives, the pulse passing through the modulator at the time of minimum loss (Fig. 8.20a).

Example 8.7. *AM mode-locking for a cw Ar and Nd:YAG laser* We first consider a mode-locked Ar-ion laser oscillating on its $\lambda = 514.5$ nm green transition, this transition being Doppler-broadened to a width of $\Delta\nu_0^* = 3.5$ GHz. From Eq. (8.6.18) we then get $\Delta\tau_p \cong 126$ ps. We consider next a mode-locked Nd:YAG laser oscillating on its $\lambda = 1.064$ μm transition whose width is phonon broadened to a value $\Delta\nu_0 \cong 4.3\ \text{cm}^{-1} = 129$ GHz at $T = 300$ K. We take a laser cavity with an optical length $L_e = 1.5$ m and consider the case where the AM modulator is located at one cavity end (Fig. 8.19a). We then get $\nu_m = c/2L_e = 100$ MHz, and, from Eq. (8.6.19), $\Delta\tau_p \cong 125$ ps. Note that, on account of the different expressions of $\Delta\tau_p$ for a homogeneous or inhomogeneous line, the pulsewidths for the two cases are almost the same despite the linewidth of Nd:YAG being almost 30 times wider than that of Ar-ion.

To describe FM mode-locking, we suppose a modulator, whose refractive index n is sinusoidally modulated at frequency ω_m, to be inserted at one end of the cavity. Any given mode of the cavity will therefore be subjected to a time varying phase shift given by $\varphi = (2\pi L'/\lambda)\times n(t)$, where L' is the modulator length. These phase modulated modes will show sidebands [see Eq. (7.10.5)] whose frequencies, for $\omega_m = \Delta\omega$, coincide with those of the neighboring modes. Thus, the cavity modes become coupled again and their phases locked,(16) although the locking condition turns out to be different from that given by Eq. (8.6.1). In the time domain, this FM mode-locking produces pulses as indicated in Fig. 8.21. In this case, two stable mode-locking states can occur, i.e., such that the light pulse passes through the modulator either at each minimum of $n(t)$ (solid-line pulses) or at each maximum (dotted-line pulses). To get some physical understanding of what happens in this case, we first observe that, since the optical length of the modulator is $L'_e = n(t)L'$, this type of modulation actually results in a modulation of the overall optical length, L_e, of the cavity. In its effect, the cavity is thus equivalent to one without a modulator but where the position of one cavity mirror is oscillating at frequency ω_m. Either one of the two stationary situations of Fig. 8.21 thus corresponds to mode-locked pulses striking this moving mirror when it is at either of its extreme positions (i.e., when the mirror is stationary). Note that, after reflection by this moving mirror, the pulse

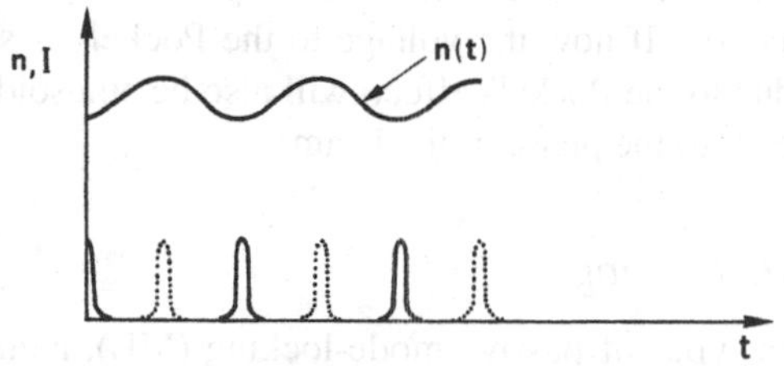

FIG. 8.21. FM mode-locking. Time behavior of modulator refractive index n and of output intensity I.

will acquire a nearly parabolic phase variation, of either positive sign (for the solid-line pulses) or negative sign (for the dotted-line pulses), and its spectrum will be slightly broadened. The overall phase modulation of the pulse and the corresponding pulse duration will then be established by the condition that the spectral broadening produced at each reflection from the moving mirror must be compensated by the spectral narrowing resulting from the passage through the amplifier. A stability analysis could also be performed to understand what happens when the pulse strikes the mirror not exactly at a stationary point. However, this analysis turns out to be rather complicated and, due to the somewhat limited importance of this type of mode-locking, it will not be considered here. In fact, this type of locking is much less frequently used in practice for two main reasons: (i) The pulses are frequency modulated. (ii) Mode-locking tends to be somewhat unstable in the sense that switching between the two states of Fig. 8.21 often occurs in practice.

For a pulsed and generally high gain laser, AM mode-locking is commonly achieved by a Pockels cell amplitude modulator. A possible configuration may be that shown in Fig. 8.5a with the Pockels cell voltage being sinusoidally modulated from zero to some fraction of the $\lambda/4$ voltage. For a cw pumped and generally low gain laser, AM mode-locking is more commonly achieved with an acousto-optic modulator, owing to its lower insertion loss compared to a Pockels cell modulator. However, the acousto-optic modulator used for mode-locking differs significantly from that used for Q-switching and discussed in Fig. 8.7. In fact, the face to which the piezoelectric transducer is bonded and the opposite face of the optical material are now cut parallel each other. The sound wave launched into the material by the transducer is then reflected back by the opposite face of the material. So, if the length of the optical block is equal to an integral number of half-wavelengths of the sound wave, an acoustic standing-wave pattern is produced. Since the amplitude of the standing-wave is sinusoidally modulated in time, the same will happen for the diffraction losses. It can be shown, however, that, if the sound wave is oscillating at frequency ω, the diffraction loss will be modulated at frequency 2ω. Consider, in fact, an acoustic standing-wave of the form $S = S_0 (\cos \omega t)(\sin kz)$. The modulator loss will reaches a maximum whenever a maximum *amplitude* of the standing-wave pattern occurs and this maximum is reached twice in an oscillation period (i.e., at $t = 0$ and $t = \pi/\omega$). The modulator loss is thus modulated at frequency 2ω and, for fundamental ML (see Fig. 8.19a), ML is achieved when: (i) the modulator is put as near as possible to one cavity mirror; (ii) the modulation frequency 2ν is set equal to $(c/2L)$ and, accordingly, the transducer is driven at a frequency equal to $c/4L$ (e.g., $\nu = 50\,\text{MHz}$ for $L = 1.5\,\text{m}$). In the case of FM mode-locking (both for pulsed or cw lasers), a Pockels cell electro-optic phase modulator is commonly used. In this case one of the two axes, e.g., x, of induced birefringence (see Fig. 8.5b) is oriented along the polarizer axis. So, the beam does not rotate its polarization when passing through the Pockels cell but, rather, acquires a phase shift given by $\phi = (2\pi L'/\lambda)n_x$, where L' is the Pockels cell length and n_x is its refractive index for polarization along x-direction. If now the voltage to the Pockels is sinusoidally modulated, the refractive index n_x, due to the Pockels effect, will also be sinusoidally modulated and the same modulation will occur to the phase of the beam.

8.6.3.2. Passive Mode Locking

There are four main types of passive mode-locking (ML), namely: (1) *Fast saturable absorber* ML, which makes use of the saturation properties of a suitable absorber (e.g. a dye molecule or a semiconductor) with very short upper state lifetime. (2) *Kerr Lens*

Mode-locking (KLM), which exploits the self-focusing property of a suitable transparent non-linear optical material. (3) *Slow saturable absorber* ML, which exploits the dynamic saturation of the gain medium. (4) *Additive Pulse* ML (APM), in which one exploits the self-phase-modulation induced in a suitable non-linear optical element inserted in an auxiliary cavity, coupled to the main cavity and of identical length. In this case, the pulse shortening mechanism arises from the interference of the main pulse in the laser cavity with the pulse coupled back from the auxiliary cavity and which has been phase-modulated by the non-linear material. APM locking requires however that the optical lengths of the two cavities be equal with an accuracy typically of a fraction of the laser wavelength. For this reason, this type of ML is not so widely used as the other techniques and will not be discussed further here.

To illustrate mode-locking by a fast saturable absorber, we consider an absorber with low saturation intensity and with relaxation time much shorter than the duration of the mode-locked pulses. The theory of mode-locking by a fast saturable absorber, for a homogeneously broadened gain medium, has been treated in detail, notably by Haus.[(19)] We will only quote here the most relevant results and refer to Appendix F for a more detailed treatment.

For low values of intracavity beam intensity, I, compared to the absorber's saturation intensity, I_s, the cavity round-trip power-loss* can be written as

$$2\gamma_t = 2\gamma - 2\gamma'(I / I_s) \tag{8.6.20}$$

where γ is the unsaturable single-pass loss and γ' is the low-intensity single-pass loss of the saturable absorber†. Suppose now that the absorber is very thin and placed in contact with one cavity mirror (Fig. 8.19a) and that the laser is initially oscillating with mode phases un-locked. The intensity of each of the two traveling waves will be made up of a random sequence of light bursts (see Fig. 8.15) and, for the initially low peak intensity of these bursts, the saturated round-trip power gain, $2g_0$, will be roughly equal to the cavity unsaturable loss. However, the most intense pulse of Fig. 8.15, as a result of absorber saturation, will suffer the least attenuation in the saturable absorber. If certain special conditions are met,[(19)] this pulse can then grow faster than the others and, after many round trips, the situation described in Fig. 8.22 will eventually be established, where, for simplicity, the gain medium and saturable absorber are assumed to be together at one end of the cavity. In this case a single, intense, ML pulse survives in the cavity and, due to the reduced loss arising from the more pronounced absorber saturation, the average power will increase compared to the unlocked case and, correspondingly, the round-trip saturated gain, $2g_0'$, will now become smaller than the round-trip unsaturable loss of the cavity. Accordingly, a time window of net gain is established during the passage of the pulse, i.e., between times t_1 and t_2 in the figure, the pulse tails seeing a net loss and the pulse peak a net gain. If it were only for this mechanism, the pulse would be progressively shortened after each pass through the absorber-amplifier combination. However, a steady state condition is again established by the balance between this pulse shortening

* In a mode-locked linear cavity it is preferable to think in terms of round-trip loss and gain rather than in terms of the corresponding single-pass values

† According to (2.8.12) and for $I \ll I_s$, the absorbance, γ_a, of an absorber of length l_a can be written as $\gamma_a = \alpha l_a = \alpha_0\, l_a[1 - (I/I_s)]$ and the intensity independent term $\alpha_0\, l_a$ can be included in the total unsaturable loss γ.

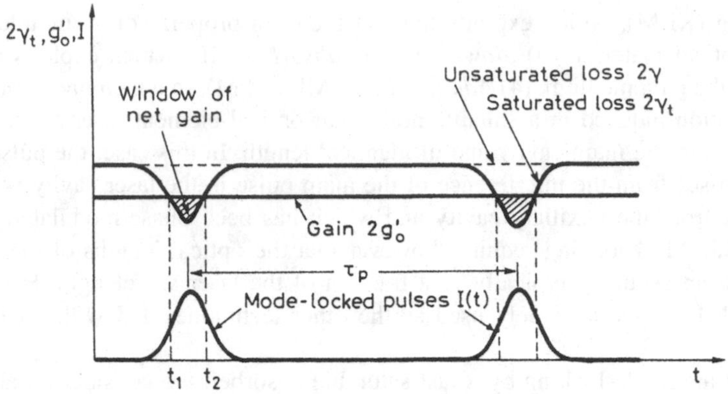

FIG. 8.22. Passive ML by a fast saturable absorber.

mechanism and pulse broadening arising from the finite gain bandwidth. The steady state pulse amplitude turns out, in this case, to be described by a hyperbolic secant function, viz.,

$$E(t) \propto \operatorname{sech}(t/\tau_p) \tag{8.6.21}$$

The duration $\Delta\tau_p$ of the pulse intensity (FWHM) is related to τ_p by $\Delta\tau_p \simeq 1.76\tau_p$ and is given by

$$\Delta\tau_p \cong \frac{0.79}{\Delta\nu_0}\left[\frac{g_0'}{\gamma'}\right]^{1/2}\left[\frac{I_s}{I_p}\right]^{1/2} \tag{8.6.22}$$

where $\Delta\nu_0$ is the gain bandwidth (FWHM) and I_p is the peak intensity of the pulse. Note that the physical picture described in Fig. 8.22 actually applies to long lifetime (hundreds of μs) gain media such as crystalline or glass solid state media. In this case, in fact, no appreciable variation of gain occurs during the passage of the pulse and the saturated gain, g_0', is established by the average intracavity laser power.

For a simple two-level system, the absorber's saturation intensity is given by $I_s = h\nu/2\sigma\tau$ [see Eq. (2.8.11)], and since τ must be very short (~ a few ps or shorter), the required low value of saturation intensity calls for very large values of the absorption cross section σ (~10^{-16} cm^2 or larger). It thus follows that the most commonly used saturable absorbers are either solutions of fast dye molecules or semiconductors. In the case of dye solutions, cyanine dyes consisting of a long chain of the form (–CH = CH–)$_n$, where n is an integer and terminated by two aromatic end groups, are often used. The upper state relaxation time of the cyanine dyes used for mode-locking is typically some tens of picoseconds and is established by non-radiative decay arising both from internal conversion (see Fig. 3.6) and from rotation of the aromatic rings. Thus, the absorber remains saturated for a time roughly equal to this relaxation time and ML pulses shorter than a few picoseconds cannot be obtained. For a semiconductor saturable absorber, the absorber's recovery typically shows a multi-component decay, namely: (i) A fast decay ($\tau \approx 100$ fs), due to intraband thermalization of electrons within the conduction band, arising from electron-electron collisions. (ii) A slower relaxation (≈1 ps) due to intraband thermalization of the conduction-band electrons with the lattice, arising from electron-phonon collisions. (iii) A still slower relaxation (from a few picoseconds to some nanoseconds) due to interband radiative and non-radiative decay. The longest relaxation time will result in the lowest saturation intensity, and this is helpful for starting the ML process. The fastest relaxation time then provides the fast saturable absorber mechanism needed to produce short pulses. A particularly interesting solution consists in integrating a multiple-quantum-well saturable absorber between two mirrors whose spacing is such that the resulting Fabry-Perot etalon operates at anti-resonance, i.e., at a point where a minimum of transmission or a maximum of reflection occurs (see Fig. 4.11). If the etalon is used as one cavity mirror, the laser intensity within the etalon may be substantially reduced compared to the value of the laser cavity. This offers the considerable advantages of increasing, in a controlled manner, the effective value of the saturation intensity, decreasing the effective unsaturable losses, and increasing the damage threshold.[20] The effectiveness of this simple-to-use anti-resonance-Fabry-Perot-saturable absorber (A-FPSA) has been widely proven for generating both picosecond and femtosecond laser pulses from several wide-bandwidth solid state lasers.

Another fast passive mode-locking technique relies on the lens-effect induced in a suitable material by a Kerr-type non-linearity and is thus referred to as *Kerr-Lens-Mode-Locking* (KLM).[21,22] Consider first an optical material, such as quartz or sapphire, traversed by a light beam of uniform intensity I. At sufficiently high intensity, the refractive index of the medium will be influenced to a readily observable extent by the field intensity, i.e., one can generally write $n = n(I)$. The first term of a Taylor expansion of n vs I will be proportional to I and one can thus write

$$n = n_0 + n_2 I \quad (8.6.23)$$

where n_2 is a positive coefficient which depends on the material (e.g. $n_2 \cong 4.5 \times 10^{-16}\,\mathrm{cm}^2/\mathrm{W}$ for fused quartz and $n_2 \cong 3.45 \times 10^{-16}\,\mathrm{cm}^2/\mathrm{W}$ for sapphire). This phenomenon is known as the *optical Kerr effect* and is generally due to a hyper-polarizability of the medium occurring at high electric fields and arising from either a deformation of the electronic orbitals of the atoms or molecules or from a reorientation of the molecules (for a gas or liquid). For a solid, only deformation of the atom's electron cloud can occur and the optical Kerr effect is very fast, the response time being of the order of a rotation period of the outermost electrons of the atom (a few femtoseconds). Assume now that the beam intensity, in a medium exhibiting the optical Kerr effect (a *Kerr medium*), has a given transverse profile, e.g., Gaussian. The intensity at the beam center will then be larger than in

Example 8.8. *Passive mode-locking of a Nd:YAG and Nd:YLF laser by a fast saturable absorber* We consider a cw Nd:YAG laser passively mode-locked by a $\sim$0.6 μm thick multiple-quantum-well ($\sim$50 wells) InGaAs/GaAs A-FPSA.[20] We take $g_0' = 2\%$, $\gamma' = 1\%$, $\Delta\nu_0 = 4.5\,\mathrm{cm}^{-1} \cong 135\,\mathrm{GHz}$ at $T = 300\,\mathrm{K}$, and $I_p = 0.3\,I_s$. From Eq. (8.6.22) we then get $\Delta\tau_p \cong 15\,\mathrm{ps}$. Note that, in this case, the absorber is heavily saturated and Eq. (8.6.22) can only be taken as a first order approximation to the calculation of the predicted pulse-width. For the case of a Nd:YLF laser, we will assume the same value of the unsaturable and saturable losses as for Nd:YAG. We therefore assume the same value of g_0'. The gain linewidth of Nd:YLF, $\Delta\nu_0'$, is taken to be about three-times larger than that of Nd:YAG (i.e., $\Delta\nu_0' \cong 13\,\mathrm{cm}^{-1}$) and the comparison then made at the same value of output power i.e., at the same value of the pulse energy $E \cong I_p\,\Delta\tau_p$. From Eq. (8.6.22) one readily finds that the pulsewidth $\Delta\tau_p'$ for this case is related to the pulsewidth of the previous case by the relation $\Delta\tau_p' = \left(\Delta\nu_0/\Delta\nu_0'\right)^2\,\Delta\tau_p$. For $\Delta\nu_0' = 2.89\Delta\nu_0$ we then get $\Delta\tau_p' \cong 1.8\,\mathrm{ps}$. Note the strong dependence of laser pulsewidth on gain linewidth under these conditions.

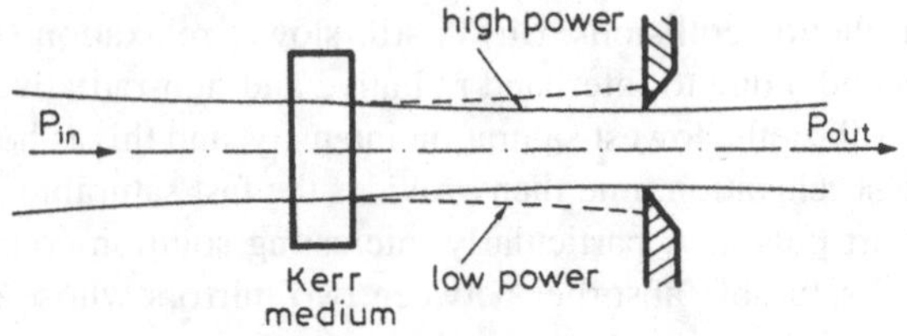

FIG. 8.23. Non-linear loss element exploiting the optical Kerr effect in a suitable non-linear material.

the wings and, according to Eq. (8.6.23), a non-linear refractive index change, $\delta n = n_2 I$, will be induced which is positive at the beam center and goes to zero in the wings of the beam. For a Gaussian beam profile, i.e., $I = I_p \exp -2(r/w)^2$, where I_p is the peak intensity and w is the (field) spot size, the non-linear phase shift acquired by the beam in traversing a length l of the medium will be $\delta\phi = 2\pi\delta n l/\lambda = (2\pi n_2 I_p l/\lambda)\exp -2(r/w)^2 \cong (2\pi n_2 I_p l/\lambda) \times [1-2(r/w)^2]$. Thus, to first order in $(r/w)^2$, $\delta\phi$ can be taken as a parabolic function of (r/w), which is equivalent to saying that a spherical lens is induced in the medium by the optical Kerr effect. In fact this induced lens may lead to beam focusing when the beam power exceeds a critical value, a phenomenon known as *self-focusing*. A non-linear loss element providing a loss of the general form of Eq. (8.6.20) can then be realized as shown schematically in Fig. 8.23. In fact, at higher beam intensities, the beam will be focused more strongly at the aperture and, therefore, less loss will be experienced at this aperture. If now the non-linear loss element of Fig. 8.23 is correctly located within a laser cavity, passive ML may be achieved according to the mechanism described in Fig. 8.22 for a fast saturable absorber. In fact, the time response of KLM is very short so that, for all practical purposes, it can be taken to be instantaneous. By appropriate control of cavity dispersion, the fastest ML pulses have been achieved by this technique, using ultra-broadband gain media (bandwidths of $\approx$100 THz).

Although many passively mode-locked lasers make use of fast saturable absorbers, slow saturable absorbers, under special circumstances, can also lead to mode locking and this type of ML is often referred to as *slow-saturable-absorber ML*. The special circumstances which are required can be summarized as follows: (i) The relaxation time of both absorber and amplifier must be comparable to the cavity round trip time. (ii) The saturation fluence of both gain medium [$\Gamma_{sg} = h\nu/\sigma_g$, see Eq. (2.8.29)] and saturable absorber [$\Gamma_{sa} = h\nu/2\sigma_a$, see Eq. (2.8.17)] must be sufficiently low to allow both media to be saturated by the intracavity laser fluence. (iii) The saturation fluence of the gain medium must be comparable to, although somewhat larger than that of the saturable absorber. The physical phenomena that lead to mode-locking, in this case, are rather subtle[(23)] and will be described with the help of Fig. 8.24, where, for simplicity, it is supposed that both saturable absorber and active medium are together at one end of the cavity. Before the arrival of the mode-locked pulse, the gain is assumed to be smaller than the losses, so that the early part of the leading edge of the pulse will suffer a net loss. If the total energy fluence of the pulse has a suitable value, the accumulated energy fluence of the pulse may become comparable to the saturation fluence of the absorber during the leading edge of the pulse. Saturation of the saturable absorber will begin to occur so that, at some time during the pulse leading edge (time t_1 in Fig. 8.24), the absorber loss becomes equal to the laser gain. For $t > t_1$ the pulse will then see a net gain rather than a net loss. However, if the saturation energy fluence of the gain medium has a

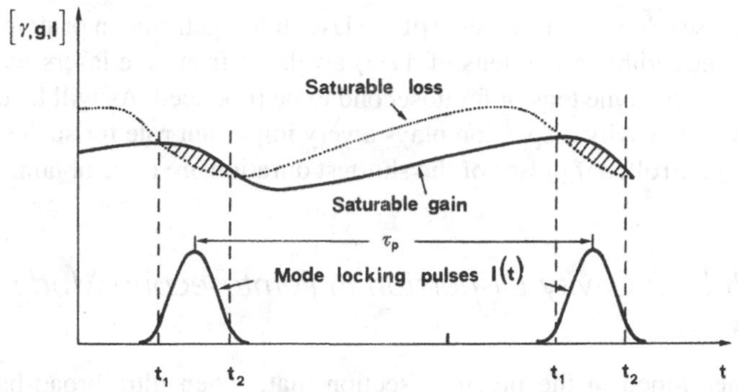

FIG. 8.24. Time-domain picture of slow-saturable-absorber mode locking. Note that the figure is not to scale since the time duration of a mode-locked pulse is typically in the hundreds femtosecond range while the time interval τ_p between two consecutive pulses, i.e., the cavity round trip time is typically a few nanoseconds.

suitable value (typically ~2 times higher than that of the absorber), gain saturation will be produced so that, at some time during the trailing edge of the pulse (time t_2 in Fig. 8.24), the saturated gain becomes equal to the saturated loss. For $t > t_2$ the pulse will then see a net loss again rather than a net gain and a time window of net gain will thus be established for $t_1 < t < t_2$. Thus, after each pass through the absorber-amplifier combination, the pulse is shortened and a steady state condition is again established by the balance between this pulse shortening mechanism and pulse broadening arising from the finite gain bandwidth. One thus expects a pulse duration again comparable to the inverse of the gain bandwidth $\Delta\nu_0$.

The evolution toward mode-locking, for this slow-saturable-absorber plus dynamic-gain-saturation mechanism, can be described by assuming the laser to be initially oscillating with unlocked phases. The saturated gain will then be equal to the unsaturated loss and, under appropriate circumstances, the most energetic pulse within its noisy time-pattern (see Fig. 8.15) will begin to produce the time-window net-gain mechanism described in Fig. 8.24. This process will then continue to occur after each passage through the laser cavity until only one laser pulse survives and the situation described in Fig. 8.24 occurs. Note that, after the mode locked pulse has passed through the absorber-amplifier combination and before the arrival of the next one, the saturable loss must recover to its unsaturated value by spontaneous (i.e., radiative plus non-radiative) decay. The corresponding decay time must then be appreciably shorter than the cavity round trip time. During the same time interval, the gain medium must partially but not completely recover to the steady state value established by the pumping process [see Fig. 8.3a] so as to leave a saturated gain smaller than the loss. This means that the lifetime of the gain medium must be somewhat longer than the cavity round trip time. We reiterate that the saturation fluences of both amplifier and absorber must be sufficiently low to allow the two media to be saturated by the laser pulse. So, this type of mode-locking can be made to occur with short-lifetime (~ a few nanoseconds) high cross section (~10^{-16} cm^2) gain media such as dyes or semiconductors. As saturable absorbers, saturable dyes with lifetime of a few nanoseconds (determined by spontaneous emission) are often used. By contrast to this situation, this type of mode-locking cannot occur with long lifetime (hundreds of μs) gain media such as crystalline or glass solid-state media, where dynamic gain saturation cannot occur. When the delicate conditions for this type of mode-locking can be met, however, very short light pulses down to the inverse of the laser linewidth can, in principle, be obtained. The large gain bandwidths (a few tens of THz) available from dye lasers would then allow pulses with duration of some tens of femtosecond to be produced. As will be discussed in the next section, however, cavity dispersion plays a very important role for such short pulses and its value must be controlled if pulses of the shortest duration are to be obtained.

8.6.4. The Role of Cavity Dispersion in Femtosecond Mode-Locked Lasers

We have mentioned in the previous section that, when ultra-broad-band gain-media (bandwidths as large as 100 THz) are involved, cavity dispersion plays an important role in establishing the shortest pulse duration that can be achieved in ML operation. We will consider this point here by first making a short digression to provide a reminder of the concepts of phase velocity, group velocity, and group delay dispersion in a dispersive medium.

8.6.4.1. Phase-Velocity, Group-Velocity and Group-Delay-Dispersion

Consider first a plane, linearly polarized, monochromatic e.m. wave, at frequency ω, propagating along the z direction of a transparent medium. The electric field $E(t, z)$ of the wave can then be written as $E = A_0 \exp j(\omega t - \beta z)$, where A_0 is a constant and where the propagation constant, β, will generally be a function of the angular frequency ω. The relation $\beta = \beta(\omega)$ is a characteristic of the given medium and is referred to as the *dispersion relation* of the medium (see Fig. 8.25). Since now the total phase of the wave is $\phi_t = \omega t - \beta z$, the velocity of a given phase front will be such that the elemental changes dt and dz, of the temporal and spatial coordinates, must satisfy the condition $d\phi_t = \omega dt - \beta dz = 0$. This

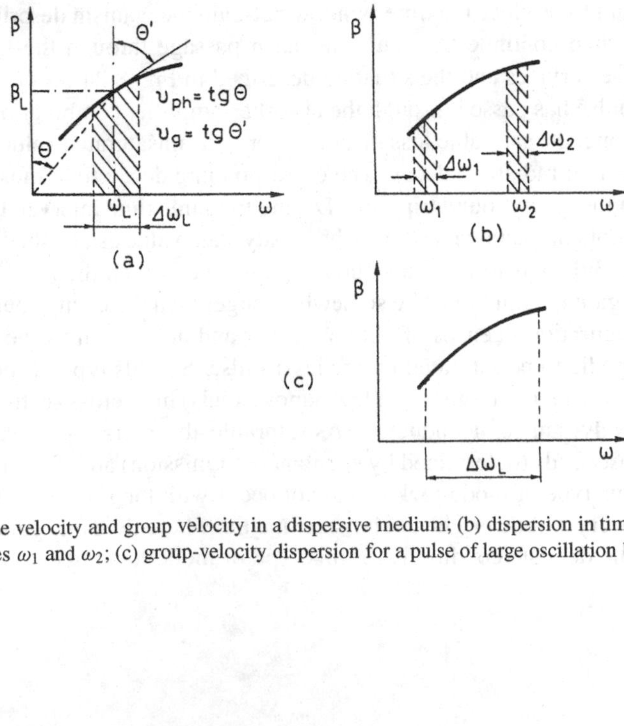

FIG. 8.25. (a) Phase velocity and group velocity in a dispersive medium; (b) dispersion in time delay for two pulses of carrier frequencies ω_1 and ω_2; (c) group-velocity dispersion for a pulse of large oscillation bandwidth $\Delta\omega_L$.

shows that the phase front moves at a velocity

$$v_{ph} = \frac{dz}{dt} = \frac{\omega}{\beta} \tag{8.6.24}$$

referred to as the *phase velocity* of the wave.

Consider next a light pulse traveling in the medium and let ω_L and $\Delta\omega_L$ be, respectively, the center frequency and the width of the corresponding spectrum (Fig. 8.25a). Assume also that the dispersion relation, over the bandwidth $\Delta\omega_L$, can be approximated by a linear relation, *viz.* $\beta = \beta_L + (d\beta/d\omega)_{\omega=\omega_L}(\omega - \omega_L)$, where β_L is the propagation constant corresponding to the frequency ω_L. In this case, upon considering a Fourier expansion of wave, one can show that the electric field of the wave can be expressed as (see Appendix G)

$$E(t, z) = A[t - (z/v_g)] \exp[j(\omega_L t - \beta_L z)] \tag{8.6.25}$$

where A is the pulse amplitude, $\exp[j(\omega_L t - \beta_L z)]$ is the carrier wave, and v_g is given by

$$v_g = \left(\frac{d\omega}{d\beta}\right)_{\beta=\beta_L} \tag{8.6.26}$$

The fact that the pulse amplitude is a function of the variable $t - (z/v_g)$ means that the pulse propagates without changing its shape and at a speed v_g. This velocity is referred to as the *group velocity* of the pulse and, according to Eq. (8.6.26), is given by the slope of the ω vs β relation at $\omega = \omega_L$ (i.e. $v_g = \text{tg}\,\theta'$, see Fig. 8.25a). Note also that, for a general dispersion relation such as that of Fig. 8.25a, the phase velocity of the carrier wave ($v_{ph} = \text{tg}\,\theta$, see Fig. 8.25a) will be different from the group velocity.

According to the previous considerations, the pulse, after traversing the length l of the medium, will be subjected to a time delay

$$\tau_g = \frac{l}{v_g} = l\left(\frac{d\beta}{d\omega}\right)_{\omega_L} = \phi'(\omega_L) \tag{8.6.27}$$

For the previous equation, we have defined a phase ϕ, dependent on the frequency ω, such that

$$\phi(\omega - \omega_L) = \beta(\omega - \omega_L)l \tag{8.6.28}$$

and we have called $\phi'(\omega_L) = [d\phi(\omega - \omega_L)/d\omega]_{\omega_L}$. For obvious reasons, the quantity $\tau_g = \phi'(\omega_L)$ is referred to as the *group delay* of the medium at frequency ω_L.

Let us now see what happens when two pulses, with bandwidths $\Delta\omega_1$ and $\Delta\omega_2$ centered at ω_1 and ω_2, respectively, are traveling in the medium ($\omega_2 > \omega_1$, see Fig. 8.25b). If the slope of the dispersion relation is different at the two frequencies, the two pulses will travel at different group velocities v_{g_1} and v_{g_2}. Thus, if the peaks of the two pulses enter the medium at the same time, then, after traversing the length l of the medium, they will become separated in time by a delay

$$\Delta\tau_g = \phi'(\omega_2) - \phi'(\omega_1) \cong \phi''(\omega_1) \times (\omega_2 - \omega_1) \tag{8.6.29}$$

where we have used the symbol $\phi''(\omega_1) = [d^2\phi/d\omega^2]_{\omega_1}$. Note that the last equation holds exactly if the relation between ϕ and ω, in the frequency range between ω_1 and ω_2, can be approximated by a parabolic law viz.

$$\phi = \phi_L + \left(\frac{d\phi}{d\omega}\right)_{\omega_L} (\omega - \omega_L) + \frac{1}{2}\left(\frac{d^2\phi}{d\omega^2}\right)_{\omega_L} (\omega - \omega_L)^2 \qquad (8.6.30)$$

Consider next the case of a light pulse with a bandwidth $\Delta\omega_L$ so large that it is no longer a good approximation to describe the dispersion relation by a linear law (Fig. 8.25c). In this case, different spectral regions of the pulse will travel with different group velocities and, consequently, the pulse will broaden as it propagates. Again assuming that the dispersion relation, within the bandwidth $\Delta\omega_L$, can be approximated by a parabolic law, then, according to Eq. (8.6.29), the broadening of the pulse due to dispersion, $\Delta\tau_d$, will be given approximately by the difference in group delay between the fastest spectral component and the slowest one. According to Eq. (8.6.29) we then have

$$\Delta\tau_d \cong |\phi''(\omega_L)|\, \Delta\omega_L \qquad (8.6.31)$$

The quantity $\phi''(\omega_L)$ is referred to as the *group delay dispersion* (GDD) of the medium at frequency ω_L. Its magnitude gives the pulse broadening per unit bandwidth of the pulse. From Eqs. (8.6.28) and (8.6.31) one then sees that $\Delta\tau_d$ can also be written as

$$\Delta\tau_d \cong l\left|\left(\frac{d^2\beta}{d\omega^2}\right)_{\omega_L}\right| \Delta\omega_L \qquad (8.6.32)$$

The quantity, GVD, expressed by

$$\text{GVD} = (d^2\beta\, /\, d^2\omega)_{\omega_L} = [d(1/v_g)\, /\, d\omega]_{\omega_L} \qquad (8.6.33)$$

is usually referred to as the *group velocity dispersion* at frequency ω_L. Its magnitude gives the pulse broadening per unit length of the medium and per unit bandwidth of the pulse. It should be observed that the concept of group velocity dispersion is straightforward, in application, only for a homogeneous medium. For an inhomogeneous or multi-component medium, such as the two prism pairs of Fig. 8.26 or a multilayer dielectric mirror, the concept of group delay dispersion is more easy to consider.

8.6.4.2. Limitation on Pulse Duration due to Group-Delay Dispersion

When a dispersive medium is present within a ML laser cavity, an approximated value of the steady state pulse duration can be obtained from the condition that the relative time shortening, $(\delta\tau_p/\tau_p)_s$, due to the net gain time window (see Fig. 8.22 or 8.24), must equal the pulse broadening due to both the gain medium, $(\delta\tau_p/\tau_p)_g$, and the dispersive medium, $(\delta\tau_p/\tau_p)_D$. For the sake of simplicity, we will consider here a ring cavity in which the pulse is sequentially passing through the gain medium, the dispersive medium, and through whatever element provides the self-amplitude-modulation (e.g., a fast saturable absorber). We will assume that the light pulse has a Gaussian intensity profile with pulsewidth (FWHM) $\Delta\tau_p$, and that the dispersive medium can be described, at a general frequency ω, by its group-delay dispersion $\phi'' = \phi''(\omega)$. The gain medium is assumed to be homogeneously broadened and will be described by its saturated single pass gain, $g_0 = N_0\sigma l$, and its linewidth (FWHM) $\Delta\omega_0$.

For small changes of pulse duration, the relative pulse broadenings, $(\delta\tau_p/\tau_p)_g$ and $(\delta\tau_p/\tau_p)_D$, after passing through the gain medium and the dispersive medium, are shown in Appendix G to be given, respectively, by

$$\left(\frac{\delta\tau_p}{\tau_p}\right)_g = \left(\frac{2\ln 2}{\pi^2}\right)\left(\frac{1}{\Delta\tau_p^2\Delta\nu_0^2}\right)g_0 \tag{8.6.34}$$

where $\Delta\nu_0 = \Delta\omega_0/2\pi$, and by

$$\left(\frac{\delta\tau_p}{\tau_p}\right)_D = (8\ln^2 2)\frac{\phi''^2}{\Delta\tau_p^4} \tag{8.6.35}$$

where ϕ'' is calculated at the central frequency, ω_L, of the laser pulse.

The steady state pulse duration can now be obtained from the condition that the relative time shortening, $(\delta\tau_p/\tau_p)_s$, due to the net gain time window must be equal to the pulse broadening due to both the gain medium and the dispersive medium. From Eqs. (8.6.34) and (8.6.35) we then get

$$\left(\frac{\delta\tau_p}{\tau_p}\right)_s = 0.14\frac{g_0}{\Delta\tau_p^2\Delta\nu_0^2} + 3.84\frac{\phi''^2}{\Delta\tau_p^4} \tag{8.6.36}$$

Note that the two terms on the right hand side of Eq. (8.6.36) are inversely proportional to $\Delta\tau_p^2$ and $\Delta\tau_p^4$, respectively. This means that the importance of GDD in establishing the pulse duration becomes more important as $\Delta\tau_p$ decreases. To get an estimate of the pulse-width at which GDD begins to become important we equate the two terms on the right hand side of Eq. (8.6.36). We get

$$\Delta\tau_p = \left[\frac{27.4}{g_0}\right]^{1/2} |\phi''|\,\Delta\nu_0 \tag{8.6.37}$$

Assuming, as an example, $\Delta\nu_0 \cong 100$ THz (as appropriate for a Ti:sapphire gain medium) and $\phi'' = 100\ \text{fs}^2$ (equivalent, at $\lambda \cong 800$ nm, to the presence of $\sim$2 mm of quartz material in the cavity) and $g_0 = 0.1$, we get from (8.6.37) $\Delta\tau_p \cong 162$ fs. This means that, to get pulses shorter than $\sim$150 fs, down to perhaps the inverse of the gain linewidth $\Delta\nu_0$ ($\Delta\tau_p' = 1/\Delta\nu_0 \cong 4$ fs), one needs to reduce GDD by about an order of magnitude.

To obtain the shortest pulses, when second order dispersion, GDD, is suitably compensated, one also needs to compensate higher order dispersion terms in the power expansion Eq. (8.6.30). Of course, the next term to be considered would be third order dispersion, TOD, defined as TOD $= \phi''' = \beta''' l$, where the third derivatives are taken at the laser's center frequency, ω_L. We shall not consider, at any length, the effects produced by TOD on an incident pulse, and, for a discussion of this topic, we refer to the literature.[24] We will just point out that, in the case of e.g., a Ti:sapphire laser ($\Delta\nu_0 \cong 100$ THz), TOD begins to play a pulse-limiting role for pulses shorter than $\sim$30 fs.

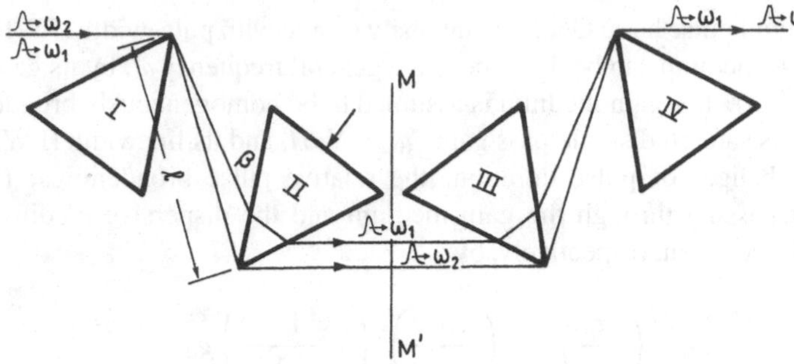

FIG. 8.26. Four-prism sequence having a negative and controllable second order group delay dispersion.

8.6.4.3. Dispersion Compensation

According to the previous example, to achieve pulses shorter than ~150 fs, the GDD of the laser cavity, contributed by its various optical elements (e.g., the active medium itself and the cavity mirrors), needs to be controlled in broad-band ML lasers. We have also seen in the previous example that fused quartz, at $\lambda \cong 800$ nm and more generally in the visible range, provides a positive value of ϕ'' and this is also the case for all media which are said to exhibit normal dispersion, i.e., for most common optical media. So, to compensate for cavity GDD, one needs a suitable element providing a negative ϕ'' i.e., showing anomalous dispersion.

The now classical solution to providing a negative and controllable second order GDD in a laser cavity utilizes the four prism sequence shown in Fig. 8.26.[25] The prisms are generally used at minimum deviation (i.e., with incidence angle equal to the refraction angle) and cut at such an apex angle that the rays enter and leave each prism at Brewster's angle. The entrance face of prism II is made parallel to the exit face of prism I, and the exit face of prism II is parallel to the entrance face of prism I and so on. The plane *MM'*, normal to the rays between prisms II and III and midway between the two prisms, is a plane of symmetry for the ray-paths. To understand the principle of operation we first observe that, according to Eq. (8.6.29), to obtain a negative GDD one must have $\Delta\tau_g < 0$ i.e., $\tau_g(\omega_2) < \tau_g(\omega_1)$ for $\omega_2 > \omega_1$. This is exactly what the four-prism sequence does. In fact, the angular dispersion of the prisms is such that two pulses at ω_2 and ω_1, entering the prism sequence at the same time and in the same direction, will propagate along the two different paths indicated in the figure. Owing to the longer path lengths, in prisms II and III, for the pulse at ω_1 compared to that at ω_2, the overall path length for pulse at ω_1 results to be longer than that at ω_2. This means that $\tau_g(\omega_1) > \tau_g(\omega_2)$, i.e. GDD $= \phi'' < 0$. For simplicity, we will not present the resulting expression for ϕ'' here. We merely limit ourselves to observing that ϕ'' depends linearly on the distance l between the two prism couples and, as an example, for quartz prisms, a length $l = 250$ mm gives a negative dispersion which can compensate the positive GDD, at $\lambda \cong 800$ nm, of a quartz element with a thickness of 6.6 mm (i.e., $\phi'' \cong -360\,\text{fs}^2$).

The four prism sequence of Fig. 8.26 proves to be a convenient way of introducing a negative GDD in a laser cavity for the following main reasons: (i) Since all faces are at Brewster's angle to the beam path, losses introduced by the system are low. (ii) The negative value of GDD can be coarsely changed by changing the separation l of the two prism pairs. (iii) By translating any one of the prisms along an axis normal to its base (e.g., prism II), one changes the total length of the optical medium traversed by the beam. This motion thus introduces, in a finely controlled way, a positive (material) dispersion, of adjustable size, without altering the ray directions and hence the negative dispersion due to the geometry of the ray paths. (iv) The transmitted beam is collinear with the incident one and this facilitates the insertion of the four-prism sequence in an already aligned cavity. Finally, it should be noted that, since MM' is a symmetry plane, one can use just the first two prisms, in a two-mirror resonator, provided that one cavity mirror is plane and located at the MM' position. In this case, the GDD per pass is of course half that of the four-prism sequence.

It should be noted that the four-prism sequence of Fig. 8.26 introduces not only a second order, ϕ'', but also a third order dispersion, ϕ''', and this term turns out to be the dominant contribution to the overall TOD of a typical femtosecond laser cavity. As with second order dispersion, the third order dispersion depends on the ray path geometry in Fig. 8.26, hence its value is proportional to the prism separation. The ratio ϕ'''/ϕ'' is therefore a characteristic of only the prism material and laser wavelength. In this respect, fused quartz proves to be one the best optical materials, with the ratio ϕ'''/ϕ'' having the lowest value [e.g., $\phi'''(\omega_L)/\phi''(\omega_L) = 1.19\,\text{fs}$ at the frequency of the Ti:sapphire laser, i.e., at $\lambda \cong 800\,\text{nm}$]. Thus, to achieve the smallest value of TOD, one must start with a cavity with the smallest value of positive ϕ'' so as to require the smallest values of both ϕ'' and ϕ''' from the four-prism sequence.

An alternative way of compensating cavity dispersion is to use, instead of the two-prism couple of Fig. 8.26, a dispersive element which introduces a negative GDD which is wavelength independent (i.e., such that $\phi''' \cong 0$). A very interesting solution, in this regard, is the use of chirped multilayer dielectric mirrors in the laser cavity.(26) The mirror consists of a large number (~40) of alternating low- and high-refractive-index layers whose thickness progressively increases, in a suitable manner, in going toward the substrate. In this way, the high frequency components of the laser pulse spectrum are reflected first and the low frequency components are reflected further on in the multilayer thickness. Hence the group delay of the reflected beam increases with decreasing values of ω thus giving $\phi'' < 0$. For the appropriate, computer-optimized, design of the spatial frequency chirp of the layers, one can also obtain a value of ϕ'' which, within the bandwidth of interest, is approximately constant with frequency i.e., so that $\phi''' \cong 0$. Alternatively, again by computer optimization, the GDD can be required to exhibit a slight linear variation with frequency with a slope suitable for compensating the TOD of other cavity components (e.g., the gain medium). The main limitation of chirped mirrors stems from the fact that the amount of negative GDD which is typically obtainable is rather small ($\sim -50\,\text{fs}^2$). To achieve the required GDD one must then arrange for the laser beam to undergo many bounces on the mirror.

8.6.4.4. Soliton-type of Mode-Locking

Consider a medium exhibiting the optical Kerr effect so that its refractive index can be described as in Eq. (8.6.23), and assume that a light pulse, of uniform transverse intensity profile, is traveling through the medium in the z-direction. After a length z, the carrier wave of the pulse acquires a phase term $\varphi = \omega_L t - \beta_L z$ given by

$$\varphi = \omega_L t - \frac{\omega_L(n_0 + n_2 I)}{c} z \tag{8.6.38}$$

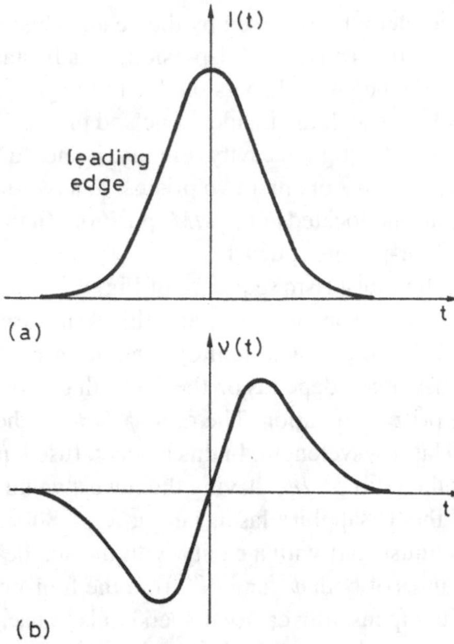

FIG. 8.27. Self-phase-modulation phenomenon. Time behavior of the pulse frequency, Fig. (b), when a bell-shaped pulse, Fig. (a), is traversing a medium exhibiting the optical Kerr effect.

where I is the light pulse intensity. Now, since $I = I(t)$, the instantaneous carrier frequency of the pulse will be given by

$$\omega = \frac{\partial\,(\omega_L t - \beta_L z)}{\partial\, t} = \omega_L - \frac{\omega_L n_2 z}{c}\frac{\partial\, I}{\partial\, t} \tag{8.6.39}$$

Thus, the carrier frequency, $\omega = \omega(t)$, is linearly dependent on the negative time derivative of the corresponding light intensity. So, for a bell shaped pulse as in Fig. 8.27a, the phase $\varphi = \varphi(t)$ will be time-modulated by the beam intensity and the carrier frequency will vary with time as indicated in Fig. 8.27b. This phenomenon is called *self-phase-modulation* (SPM).

We note that, around the peak of the pulse, i.e., around the region where the time behavior of the pulse can be described by a parabolic law, the frequency chirp induced by SPM increases linearly with time. Suppose now that the medium has a negative GDD. In this case, it is shown in Appendix G that, during propagation in this medium, the pulse tends to acquire an instantaneous frequency chirp which decreases linearly with time*. The two effects thus tend to cancel each other and one could expect that, under appropriate conditions, the effect of SPM can exactly cancel that due to dispersion for the whole pulse. The above intuitive picture is confirmed by a detailed calculation.[27] In fact, if a pulse is propagating in a medium showing an optical Kerr effect and a negative GVD (such as a silica optical

* One may observe that the pulse spectrum must remain unchanged while the pulse propagates through a passive medium such as the dipersive medium considered here. In such a medium, however, the pulse broadens upon propagation [see (8.6.32)] and the spectral contribution arising from the finite pulse duration decreases. It then follows that the pulse must also acquire an appropriate frequency modulation to keep the spectrum unchanged.

fiber at $\lambda > 1300\,\text{nm}$), the pulse propagation, due to the presence of SPM, is described by a nonlinear wave equation which admits pulsed solutions that propagate without distortion. The time behavior of the corresponding electric field, for the lowest order solution, turns out to be described by a hyperbolic secant function [i.e., $A(t') \propto \text{sech}\,(t'/\tau_p)$], i.e., the pulse is unchirped and the *whole* pulse propagates without distortion as a result of the mutual compensation of SPM and GVD, as discussed above. These solutions are called solitary solutions of the nonlinear wave equation or *solitons*. One of the most interesting properties of the lowest order soliton is that its pulsewidth $\Delta\tau_p$ (FWHM) is related to the pulse peak power, P_p, by the equation[27]

$$\Delta\tau_p^2 = \frac{3.11|b_2|}{\gamma P_p} \tag{8.6.40}$$

where $b_2 = (d^2\beta/d\omega^2)_{\omega_L}$, $\gamma = n_2\omega_L/cA_{eff}$, and A_{eff} is the effective area of the beam ($A_{eff} = \pi w^2$ for a Gaussian beam of spot size w). One can also show that $\gamma/2$ is simply the nonlinear phase shift per unit length per unit peak power [see also Eq. (8.6.38)].

One can now ask the question as to whether solitary pulses can be produced in a ML laser cavity containing a medium exhibiting the optical Kerr effect and with overall negative GDD. The answer to this question is that solitary solutions, alone, are unstable in a ML laser cavity[28] but can become stabilized by some nonlinear loss mechanism producing a self amplitude modulation (see Fig. 8.22 or Fig. 8.24). In this case, if one neglects pulse broadening arising from the finite gain bandwidth and higher order dispersion, one gets an approximate steady-state solution that is equivalent to the fundamental soliton propagation in optical fibers with anomalous dispersion. In particular, according to Eq. (8.6.40), the pulse duration turns out to be proportional to the inverse of the pulse energy E ($E \cong 2.27\,P_p\Delta\tau_p$) according to the relation[29]

$$\Delta\tau_p = \frac{3.53|\phi''|}{\delta E} \tag{8.6.41}$$

where ϕ'' is the GDD for the round trip in the laser cavity and δ is the nonlinear round-trip phase shift per unit power in the Kerr medium and thus given by $\delta = \gamma l_K$, where l_K is the length of the Kerr medium. From Eq. (8.6.41), taking $\phi'' = -200\,\text{fs}^2$, $\delta \approx 10^{-6}\,\text{W}^{-1}$, and $E \approx 50\,\text{nJ}$ (as appropriate for a ML Ti:sapphire laser) one obtains $\Delta\tau_p \cong 14\,\text{fs}$. Indeed, solitary solutions have been observed in both ML Ti:sapphire and ML dye lasers by carefully adjusting the laser parameters.[29,30] In both these cases, ML was exploiting a mechanism of self amplitude modulation, such as that occurring by KLM or by a slow saturable absorber combined with dynamic gain saturation. In both cases, very short (10–20 fs) pulses can be generated either by the soliton mechanism alone or by the combined action of pulse broadening due to the finite gain bandwidth. It is also important to note that very short ML light pulses (~100 fs) have been produced by this soliton-type mechanism even using ML elements providing a much slower time window of net gain (e.g., semiconductor saturable absorbers with picosecond relaxation time).[31] In this case, the ML element just helps to stabilizing the soliton solution while the pulse duration is essentially determined by the soliton equation given by Eq. (8.6.41).

8.6.5. Mode-Locking Regimes and Mode-Locking Systems

Mode-locked lasers can be operated either with a pulsed or cw pump, and depending on the type of mode-locking element and type of gain medium used, the ML regimes can be rather different. Some examples will be briefly discussed in this section.

In the pulsed case (Fig. 8.28a), the time duration $\Delta\tau_p'$ of the mode-locked train envelope has a finite value which is established by the mode-locking method being used. As already discussed in Sect. 8.6.3.1, active AM- and FM-mode-locking are commonly achieved by means of a Pockels cell providing electro-optic amplitude or phase modulation, respectively, and, in this case, $\Delta\tau_p'$ is generally determined by the duration of the pump pulse. This occurs, for instance, for gain media with fast recovery time (τ of around a few nanoseconds e.g., dye lasers) which cannot operate Q-switched. In this case $\Delta\tau_p'$ may typically be a few tens of microseconds. Passive ML is usually achieved by fast (a few tens of ps) dye-solution saturable absorbers, and, for gain media with a slow recovery time (τ of around a few hundreds of microseconds, as applies e.g., to solid-state lasers), the presence of the saturable absorber will result not only in mode-locked but also in Q-switched operation. In this case the duration $\Delta\tau_p'$ of the mode-locked train will be established by the same consideration that determine the duration $\Delta\tau_p$ of Q-switched pulse behavior (generally a few tens of nanoseconds, see Sect. 8.4.4). Note that, when a slow saturable absorber (τ of around a few nanoseconds) is used with a slow gain medium*, passive Q-switching with single-mode-selection rather than ML will tend to occur, due to the mode-selecting mechanism discussed in Sect. 8.4.2.4.

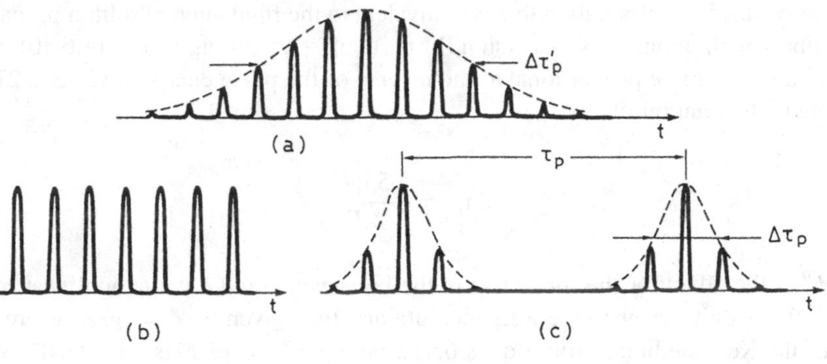

FIG. 8.28. Different mode-locking regimes: (a) ML with a pulsed pump. (b) cw ML with a cw pump. (c) ML with a cw pump and fast saturable absorber, showing the simultaneous occurrence of ML and repetitive Q-switching.

* It must be noted that we have been using the terms "fast" and "slow" in regard to recovery time in different ways for the cases of absorber and gain medium. The recovery time of a saturable absorber is considered to be slow when its value (typically a few nanoseconds) is comparable to a typical cavity round trip time. This lifetime is typical for absorbers whose decay is determined by spontaneous emission via an electric-dipole-allowed transition. The recovery time is considered to be fast (a few picoseconds or shorter) when it is comparable to a typical duration of a mode-locked pulse. By contrast, the lifetime of a gain medium is considered to be fast when comparable to the cavity round trip time. This occurs for an electric-dipole-allowed laser transition. The lifetime of a gain medium is considered to be slow when it corresponds to an electric-dipole-forbidden transition.

In the case of active ML and a cw pump, the output beam consists of a continuous train of mode-locked pulses (Fig. 8.28b), whose repetition rate will depend on whether fundamental or harmonic ML is involved (see Fig. 8.19). As already discussed in Sect. 8.6.3.1., active ML is, in this case, usually achieved by an acousto-optic modulator, on account of its lower insertion loss compared to a Pockels cell modulator. Continuous wave passive ML by slow-saturable-absorber-ML can be achieved using slow saturable absorbers combined with fast gain media (notably dye lasers). Continuous wave passive ML can also be achieved with non-linear elements providing a fast non-linear loss (such as a fast saturable absorber or Kerr-lens nonlinear element). When the latter is used with a slow gain medium (a solid-state medium), however, care must be exercised to avoid the simultaneous occurrence of repetitive Q-switching.[32,33] If this situation is not avoided, the system may operate either in repetitive Q-switching with mode-locking (Fig. 8.28c) or in repetitive Q-switching without mode-locking. In both of these cases, the time duration of the Q-switched pulse, $\Delta\tau_p$, and the Q-switching repetition rate, $1/\tau_p$, (see Fig. 8.28c) are established by the dynamics of the passive Q-switching process.

A large number of different lasers have been made to operate mode-locked, both actively and passively, including many gas lasers (e.g., He-Ne, Ar ion and CO_2 lasers), all of the commonly used solid-state lasers, many semiconductor lasers and many dye lasers. As representative examples, Table 8.1. shows the most common media providing picosecond and femtosecond laser pulses, in cw ML, together with the corresponding values of gain linewidth $\Delta\nu_0$, peak stimulated emission cross section σ, and upper state lifetime τ. In the same table, the shortest pulse duration, $\Delta\tau_p$, so far achieved and the minimum pulse duration, $\Delta\tau_{mp} \cong 0.44/\Delta\nu_0$, achievable from that particular laser is also shown. One should remember that, according to Eq. (7.3.12), the threshold pump power is inversely proportional to $\sigma\tau$. Thus, for a given gain medium, $1/\sigma\tau$ can be taken as a figure of merit for achieving the lowest threshold while, of course, $1/\Delta\nu_0$ represents a figure of merit to produce the shortest pulses.

TABLE 8.1. Most common media providing picosecond and femtosecond laser pulses together with the corresponding values of: (a) gain linewidth, $\Delta\nu_0$; (b) peak stimulated emission cross-section, σ; (c) upper state lifetime, τ; (d) shortest pulse duration so far reported, $\Delta\tau_p$; (e) shortest pulse duration, $\Delta\tau_{mp}$, achievable from the same laser

Laser medium	$\Delta\nu_0$	$\sigma[10^{-20}\ \text{cm}^2]$	$\tau[\mu\text{s}]$	$\Delta\tau_p$	$\Delta\tau_{mp}$
Nd:YAG $\lambda = 1.064\ \mu\text{m}$	135 GHz	28	230	5 ps	3.3 ps
Nd:YLF $\lambda = 1.047\ \mu\text{m}$	390 GHz	19	450	2 ps	1.1 ps
Nd:YVO_4 $\lambda = 1.064\ \mu\text{m}$	338 GHz	76	98	<10 ps	1.3 ps
Nd:glass $\lambda = 1.054\ \mu\text{m}$	8 THz	4.1	350	60 fs	55 fs
Rhodamine 6G $\lambda = 570\ \text{nm}$	45 THz	2×10^4	5×10^{-3}	27 fs	10 fs
Cr:LISAF $\lambda = 850\ \text{nm}$	57 THz	4.8	67	18 fs	8 fs
Ti:sapphire	100 THz	38	3.9	6–8 fs	4.4 fs

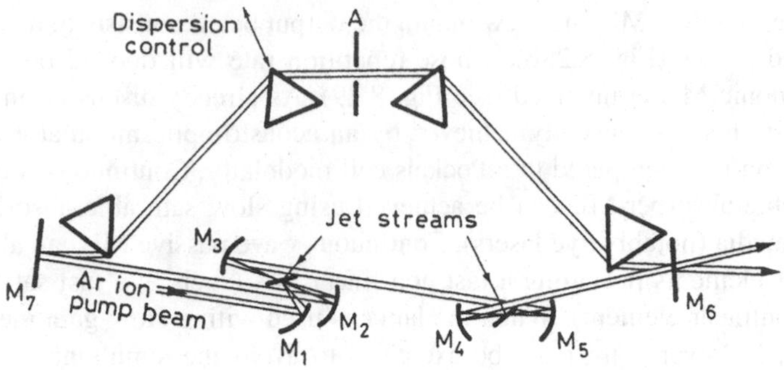

FIG. 8.29. Arrangement for a colliding-pulse mode-locked (CPM) ring dye laser including two-prism pairs for dispersion control.

Note that, since $\sigma \propto 1/\Delta\nu_0$, the lasers capable of the shortest pulses tend to have the highest threshold. Mode-locked lasers are also quite diversified in terms of their configuration, and it is beyond the scope of this book to provide detailed descriptions of these many and varied systems. We will therefore limit ourselves to describing just two cases of cw femtosecond lasers, representing particularly relevant and up-to-date examples: (i) The colliding-pulse mode-locked (CPM) Rhodamine 6G laser; (ii) the Ti:sapphire KLM laser.

In the CPM Rhodamine 6G dye laser (Fig. 8.29), a ring laser cavity is used and the dye active medium, which is located at the beam waist between the two focusing mirrors M_2 and M_3, consists of a solution of Rhodamine 6G in ethylene glycol, flowing in a jet stream orthogonal to the plane of the figure. The laser is passively mode-locked by a slow saturable absorber which is located at the beam waist between the two focusing mirrors M_4 and M_5 and consists of a solution of DODCI in ethylene glycol, again flowing as a jet stream orthogonal to the plane of the figure.[34] The active medium is quasi-longitudinally pumped by a cw Ar ion laser beam which is focused in the jet stream by mirror M_1. The ring configuration leads to the generation of two oppositely traveling femtosecond laser pulses that meet each time (i.e., collide) at the saturable absorber jet position. Due to the associated formation of a standing wave pattern in the saturable absorber, absorber saturation is enhanced and this increases the peak net gain generated (Fig. 8.24). The two pulses meet at time intervals separated by L_p/c, where L_p is the length of the ring perimeter. The Rhodamine 6G dye jet is positioned at a distance $L_p/4$ from the saturable absorber. As can be readily seen from Fig. 8.30, this ensures that the single pulses that pass through the Rhodamine 6G are all equally spaced in time by $L_p/2c$. This time symmetry allows the saturated gain of Rhodamine 6G to be the same for the two pulses, thus providing the best ML performances. To control the cavity GDD, the four-prism sequence of Fig. 8.26 is also introduced in the ring. In this way, with the overall GDD minimized, pulses of ~50 fs can be generated. Under special operating conditions, ML can be further enhanced by a soliton-type mechanism, via the phenomenon of SPM within the jet streams and with the required amount of negative GDD being provided by the four-prism system.[30] In this operating regime, pulses down to 27 fs have been generated (the shortest for a ML dye laser).

In the Ti:sapphire KLM laser, a 10 mm thick plate of Ti:sapphire is longitudinally pumped, in a z-folded linear laser cavity, again by a focused beam of an Ar ion laser

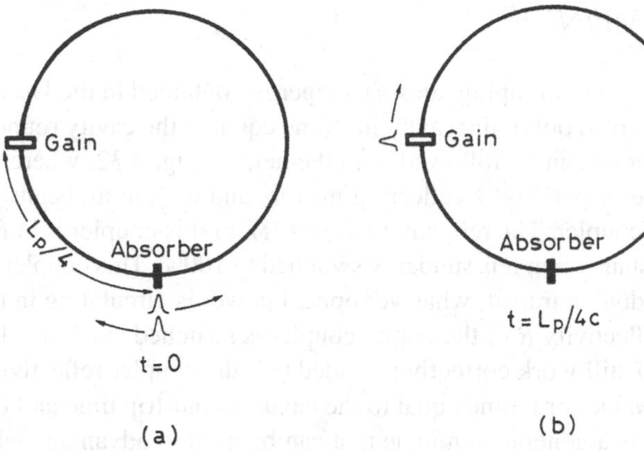

FIG. 8.30. Schematic representation of a CPM ring dye laser: (a) time $t = 0$ when the two counter-propagating pulses meet (collide) at the location of the saturable absorber; (b) time $t = L_p/4c$, where L_p is the length of the ring perimeter, when one pulse passes through the dye gain medium.

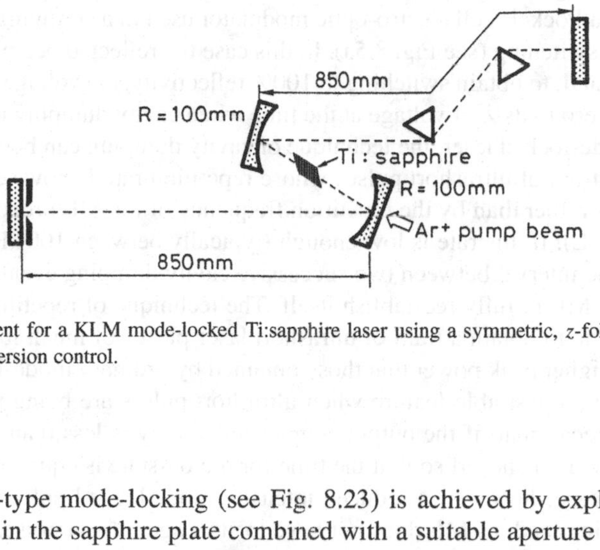

FIG. 8.31. Arrangement for a KLM mode-locked Ti:sapphire laser using a symmetric, *z*-folded, linear cavity and one prism pair for dispersion control.

(Fig. 8.31). KLM-type mode-locking (see Fig. 8.23) is achieved by exploiting the optical Kerr lens induced in the sapphire plate combined with a suitable aperture placed at one end of the cavity. To obtain the largest value of Kerr-lens nonlinearity, so that the laser can be self-starting, the two arms of the z-shaped cavity must be made of equal lengths.[35] A two-prism sequence (i.e., just the first half of the four prism sequence of Fig. 8.26) is used for GDD control. Under optimum operating conditions, when the two-prism sequence compensates the positive GDD of sapphire, pulses as short as 30 fs can be obtained from such a laser. By decreasing the sapphire thickness down to ~2 mm, one can reduce the positive GDD of the active medium to a sufficiently low value to be compensated by the negative GDD of intracavity chirped mirrors in a multipass configuration.[36] In this way, the problem of cavity third order dispersion is greatly reduced and pulses down to the record value of 6–8 fs have been obtained.[36,37]

8.7. CAVITY DUMPING[38]

The technique of cavity dumping allows the energy contained in the laser, in the form of laser photons, to be coupled out of the cavity in a time equal to the cavity round-trip time. The principle of this technique can be followed with the help of Fig. 8.32, where the laser cavity is assumed to be made of two 100% reflecting mirrors and the output beam is taken from a special kind of output coupler. The reflectivity, $R = R(t)$, of this coupler is in fact held at zero until a desired time instant when it is suddenly switched to 100%. This coupler will thus dump out of the cavity, in a double transit, whatever optical power is circulating in the laser cavity. Alternatively, if the reflectivity R of the output coupler is switched to a value less than 100%, the cavity dumper will still work correctly provided that the coupler reflectivity is held, after switching, at its high value for a time equal to the cavity round-trip time and then returned to zero. Cavity dumping is a general technique that can be used to advantage whether the laser be mode-locked, or cw or Q-switched. In the discussion that follows we will limit ourselves to considering the case of cavity dumping of a mode-locked laser, this being one of the cases where cavity dumping is used most often in practice.

For pulsed ML lasers, cavity dumping is usually carried out at the time when the intracavity mode-locked pulse reaches its maximum value (see Fig. 8.28a). In this way a single ultrashort pulse of high intensity is coupled out of the laser cavity. This type of dumping is often obtained by a Pockels cell electro-optic modulator used in a configuration that is similar to that used for Q-switching (see Fig. 8.5a). In this case the reflected beam from the polarizer is taken as output and, to obtain switching to 100% reflectivity, the voltage to the Pockels cell is switched from zero to its $\lambda/4$ voltage at the time when cavity dumping is required.

For a cw mode-locked laser, the technique of cavity dumping can be used in a repetitive way to produce a train of ultrashort pulses whose repetition rate is now set by the repetition rate of the dumper rather than by the repetition frequency of the ML process (e.g., $\nu = c/2L$ for fundamental ML). If this rate is low enough (typically between 100 kHz and 1 MHz) the corresponding time interval between two successive cavity dumping events (1–10 μs) allows sufficient time for ML to fully reestablish itself. The technique of repetitive cavity dumping therefore allows one to obtain a train of ultrashort laser pulses of much lower repetition rate and hence much higher peak power that those obtained by ordinary mode-locking. These two properties are often a desirable feature when ultrashort pulses are being put to use in some applications. We recall that, if the output coupler reflectivity is less than 100%, the coupler needs to be switched on and off so that the time for the on-state is equal to the cavity round-trip time. In this case, however, a reduced-intensity mode-locked pulse will remain in the cavity after dumping, and, as ML does not have to start again from noise, the system works

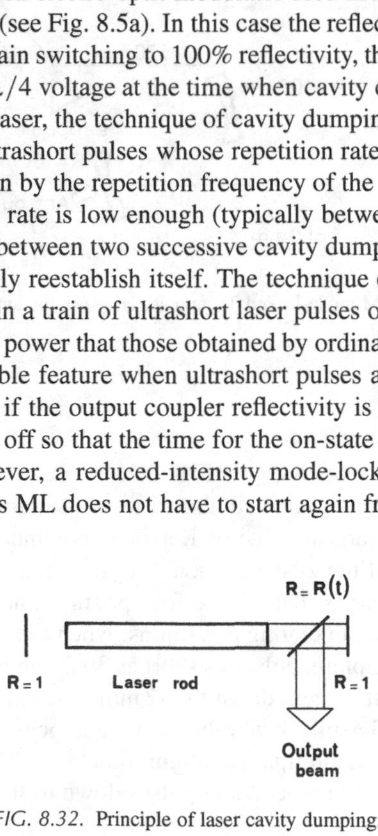

FIG. 8.32. Principle of laser cavity dumping.

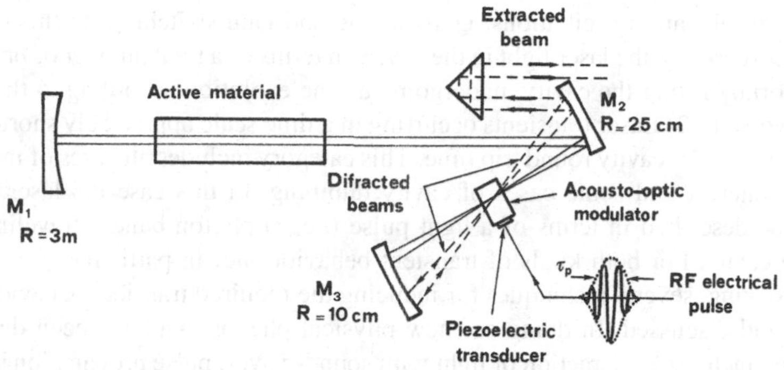

FIG. 8.33. Common arrangement for cavity dumping of a cw-pumped laser (e.g., Nd:YAG or Ar ion). Mirrors M_1–M_3 are nominally 100% reflecting at the laser wavelength. The broken lines indicate the beam which is diffracted by the modulator. For cavity dumping of a mode-locked laser, a mode-locker is also inserted at one end of the cavity (e.g., near mirror M_1).

in a more reliable way. In view of its lower insertion loss, the dumping device often used for this application is the acousto-optic cavity dumper. It consists of an acousto-optic modulator working in the Bragg regime and in the traveling-wave mode where the diffracted beam is taken as the output coupled beam. Its configuration is shown in Fig. 8.33 and it differs from that shown, for *Q*-switching, in Fig. 8.7a in three main aspects: (1) The rf oscillator, which drives the piezoelectric transducer, is now oscillating at a much higher frequency (e.g., $\nu =$ 380 MHz). Its output is gated in such a way that the rf envelope is a pulse of duration τ_p equal to the cavity round trip time (e.g., $\tau_p = 10$ ns). Cavity dumping thus occurs when the resulting acoustic pulse interacts with the cavity beam. This pulse must therefore be synchronized with the circulating mode-locked pulse so that the two pulses meet in the modulator. Note that the high carrier frequency serves the double purpose of allowing amplitude modulation by short pulses ($\tau_p = 10$ ns) and of producing a higher diffraction angle θ_d. In fact, since $\theta_d = \lambda/\lambda_a$, where λ_a is the acoustic wavelength, the diffraction angle increases linearly with the carrier frequency. (2) The beam is focused to a very small spot-size in the modulator. The time duration of the optical coupling is in fact established not only by the duration of the acoustic pulse but also by the pulse transit time through the focused laser beam. As an example, taking a spot diameter of $d = 50\ \mu$m and a sound velocity of $\upsilon = 3.76 \times 10^5$ cm / s (shear-wave velocity in fused quartz) we get $t = d/\upsilon = 13.3$ ns. (3) The circulating and the diffracted laser pulses are made to interact twice with the acoustic pulse within the modulator. This is achieved with the help of mirror M_3, which also refocuses the scattered beam back into the modulator. In this way, higher diffraction efficiency (~70%) can be obtained.

8.8. CONCLUDING REMARKS

In this chapter, several cases of transient laser behavior have been considered in some detail. Generally speaking, these cases can be grouped into two categories: (1) Laser transients occurring on a time scale appreciably longer than the cavity round trip time. This includes the phenomena of relaxation oscillations, Q-switching and gain switching. In this case, to first order, one can describe the laser light in the cavity in terms of a total number of photons, more or less uniformly filling the cavity, undergoing a time evolution according to the dynamical situation involved. (2) Laser transients occurring in a time scale appreciably shorter and often much shorter than the cavity round trip time. This category includes all cases of mode-locking of practical interest and some cases of cavity dumping. In this case the laser light in the cavity can be described in terms of a light pulse (i.e., a photon bunch) traveling back and forth in the cavity. For both kinds of transient behavior and, in particular, for Q-switching and mode-locking, several techniques for inducing the required transient behavior have been introduced and discussed. In doing so, new physical phenomena have been described and characterized including interaction of light with sound-waves, pulse propagation in dispersive media and some nonlinear optical phenomena such as self-phase-modulation, self-focusing and the formation of solitary waves. We have thus acquired some new understanding of light-matter interaction particularly under transient conditions.

The level of treatment has been subjected to several limitations. All but the simplest analytical treatments have been avoided, aiming rather to develop a deeper physical understanding of the complex phenomena involved. In particular, complications arising from any spatial variation of the laser beam and of the pump rate (e.g., variations of the transverse beam profile) have been ignored here. So, the material introduced in this chapter represents the minimum base-knowledge which, in the author's opinion, is required for a comprehensive and balanced understanding of transient laser behavior.

PROBLEMS

8.1. For the Nd:YAG Q-switched laser considered in Fig. 8.12, calculate the expected threshold energy, output energy, and pulse duration, for $E_{in} = 10$ J, when the output coupling is reduced to 20%.

8.2. Consider a Pockels cell in the so-called longitudinal configuration i.e., with the dc field applied along the direction of the beam passing through the nonlinear crystal. In this case, the induced birefringence, $\Delta n = n_x - n_y$, is given by $\Delta n = n_o^3 r_{63}\, V/L'$, where n_o is the ordinary refractive index, r_{63} is the appropriate nonlinear coefficient of the material, V is the applied voltage and L' is the crystal length. Derive an expression for the voltage required to keep the polarizer-Pockels-cell combination of Fig. 8.5a in the closed position.

8.3. For a Pockels cell made of KD_2PO_4 (deuterated potassium dihydrogen phosphate, also known as KD*P) the value of the r_{63} coefficient at $\lambda = 1.06\ \mu$m is $r_{63} = 26.4 \times 10^{-12}$ m/V, while $n_o = 1.51$. From the result of Problem 8.2 calculate the voltage that needs to be applied for the closed position.

8.4. The Nd:YAG laser of Figs. 7.4 and 7.5 is pumped at a level of $P_{in} = 10$ kW and repetitively Q-switched, at 10 kHz repetition rate, by an acousto-optic modulator (whose insertion losses are assumed negligible). Calculate the energy, and duration of the output pulses, as well as the peak and average powers expected for this case.

8.5. Derive the expressions for output energy and pulse duration which apply for a Q-switched quasi-three level laser.

8.6. A He-Ne laser beam with wavelength (in air) of $\lambda = 632.8$ nm is deflected by a $LiNbO_3$ acousto-optic deflector operating, in the Bragg regime, at the acoustic frequency of 1 GHz (the highest acoustic frequency which is possible in $LiNbO_3$ without introducing excessive losses). Assuming a sound velocity in $LiNbO_3$ of 7.4×10^5 cm/s and a refractive index $n = 2.3$, calculate the angle through which the beam is deflected.

8.7. Consider a Rhodamine 6G dye laser, transversely pumped by an atmospheric pressure nitrogen laser (see Chap. 9) with a pulse duration t_p sufficiently short to make the laser operate gain-switched. Assume: (i) The dye dissolved in ethanol ($n = 1.36$). (ii) A length $l = 5$ mm of the dye cell, the whole length being pumped by the focused beam of the nitrogen laser. (iii) A circular cross section of the pumped region with a diameter $D = 50\,\mu$m. (iv) The two end mirrors directly attached to the ends of the dye cell (i.e., $L \cong l$) one mirror being 100% reflecting and the other having a power transmission of 50%. (v) All other cavity losses negligible. (vi) An effective stimulated emission cross section, at the $\lambda = 570$ nm lasing wavelength, of $\sigma_e = 2 \times 10^{-16}$ cm^2. (vii) A square pump pulse of duration t_p much shorter than the upper state lifetime of the Rhodamine 6G solution ($\tau \cong 5$ ns). Assume also that the pump amplitude is such as to produce a peak inversion 4 times larger than the threshold inversion.
Calculate: (i) The time at which threshold is reached. (ii) The time behavior of the net gain. (iv) The time behavior of the photon number, neglecting gain saturation. (v) The time duration of the pump pulse, t_p, such that, at the end of the pump pulse, the photon number has reached the value $\phi_p/20$, where the peak value ϕ_p is calculated according to the theory of fast switching.

8.8. Derive Eq. (8.6.6)

8.9. Assuming a phase relation as in Eq. (8.6.13), show that the resulting electric field of the pulse can be written as in Eq. (8.6.14) and calculate the values of α and β as a function of $\Delta\omega_L$ and φ_2.

8.10. The oscillation bandwidth (FWHM) of a mode-locked He-Ne laser is 1 GHz, the spacing between consecutive modes is 150 MHz, and the spectral envelope can be approximately described by a Gaussian function. For fundamental mode-locking, calculate the corresponding duration of the output pulses and the pulse repetition rate.

8.11. Assume that the phase relation between consecutive modes is such that $\varphi_{l+1} - \varphi_l = \varphi_l - \varphi_{l-1} + \pi$ and that the spectral amplitude is constant over $2n$ modes. Show that the pulse repetition rate is now equal to $2\Delta\nu$, where $\Delta\nu = c/2L$ (case of second-harmonic mode-locking). [Hint: label the modes with the indices $l = 0, 1, 2, \ldots$ starting from the mode with the lowest frequency and show that one can write $\varphi_l = l\varphi$, where φ is a constant, for even l, and $\varphi_l = l\varphi + (\pi/2)$ for odd l. Then consider, separately, the even-mode sum and the odd-mode sum and show that they produce two, time-intercalated, mode-locked trains......].

8.12. Prove Eq. (8.6.16)

8.13. Assuming a uniform laser output spectrum as in Fig. 8.16a, calculate the ratio between the peak power for fundamental mode-locked operation and the average power when the modes have random phases.

8.14. For a random-noise intensity pattern as in Fig. 8.15, the probability density p_I (i.e., $p_I\, dI$ gives the elemental probability that the beam intensity is measured to be between I and $I + dI$) is given by $p_I \propto \exp -(I/I_0)$ (see also Fig. 11.11b). Calculate the average beam intensity and the probability that, in an intensity measurement, one finds a value exceeding $2I_0$.

8.15. By approximating the sum over all modes in Eq. (8.6.10) with an integral, an important characteristic of the output behavior is lost. What is it?

8.16. Consider a KLM-mode-locked Ti:sapphire laser and, according to Eq. (8.6.20), assume that the total round trip cavity losses can be written as $2\gamma_t = 2\gamma - kP$, where P is the peak intracavity laser power and where the nonlinear loss coefficient k, due to the KLM mechanism, can be taken to be $\approx 5\times 10^{-8}\,W^{-1}$. Assume a saturated round-trip gain $2g_0' \cong 0.1$, a gain bandwidth of 100 THz, and an intracavity laser energy of $W = 40$ nJ. Calculate the pulse duration achievable in the limiting case where the effects of cavity dispersion and self-phase modulation can be neglected.

8.17. Assuming a GVD for quartz at $\lambda \cong 800$ nm of 50 fs^2/mm, calculate the maximum thickness of a quartz plate that an initially unchirped 10 fs pulse, of Gaussian intensity profile, can traverse if the output pulse duration is not to exceed the input pulse duration by more than 20%.

8.18. Consider a Nd:phosphate glass laser, whose linewidth can be taken to be homogeneously broadened to $\Delta\nu_0 \cong 6$ THz. Assuming a saturated round-trip gain of 5% and a peak round-trip loss change, due to a fast self-amplitude-modulation mechanism, of 2%, calculate the expected width of the mode-locked pulses.

References

1. H. Statz and G. de Mars, Transients and Oscillation Pulses in Masers, in *Quantum Electronics*, ed. C. H. Townes (Columbia University Press, New York 1960) p. 530.
2. R. Dunsmuir, Theory of Relaxation Oscillations in Optical Masers, *J. Electron. Control*, **10**, 453–458 (1961).
3. D. F. Nelson and W. S. Boyle, A Continuously Operating Ruby Optical Maser, *Appl. Optics*, **1**, 181 (1962).
4. N. B. Abraham, P. Mandel, and L. M. Narducci, Dynamical Instabilities and Pulsations in Lasers, in *Progress in Optics, Vol. XXV*, ed. by E. Wolf (North-Holland, Amsterdam, 1988), pp 3–167.
5. M. Sargent, M. O. Scully, and W. E. Lamb, *Laser Physics* (Addison-Wesley, London, 1974).
6. C. L. Tang, H. Statz, and G. de Mars, Spectral Output and Spiking Behavior of Solid-State Lasers, *J. Appl. Phys.* **34**, 2289–2295 (1963).
7. K. Otsuka *et al.*, Alternate Time Scale in Multimode Lasers, *Phys. Rev.*, **46**, 1692–1695 (1992).
8. R. W. Hellwarth, Control of Fluorescent Pulsations, in *Advances in Quantum Electronics* ed. by J. R. Singer (Columbia University Press, New York, 1961), pp 334–341.
9. W. Koechner, Solid-State Laser Engineering, Fourth Edition, Springer Series in Optical Sciences (Springer-Verlag, Berlin, 1996) Chap. 8.
10. A. Yariv, *Optical Electronics*, fourth ed. (Saunders College Publ., Fort Worth, 1991), Chap. 12.
11. W. R. Sooy, The Natural Selection of Modes in a Passive Q-Switched Laser, *Appl. Phys. Lett.* **7**, 36–37 (1965).
12. W. G. Wagner and B. A. Lengyel, Evolution of the Giant Pulse in a Laser, *J. Appl. Phys.* **34**, 2040–2046 (1963).
13. W. Koechner, Solid-State Laser Engineering, Vol. 1, Springer Series in Optical Sciences (Springer-Verlag, Berlin, 1976) Chap. 11, adapted from Fig. 11.23.
14. L. W. Casperson, Analytical Modeling of Gain-Switched Lasers. I. Laser Oscillators, *J. Appl. Phys.*, **47**, 4555–4562 (1976).
15. A. E. Siegman, *Lasers*, (Oxford University Press, Oxford, 1986) Chap. 27 and 28.
16. Reference 15, Section 27.5.
17. D.J. Kuizenga and A.E. Siegman, FM and AM Mode Locking of the Homogenous Laser-Part I: Theory, *IEEE J. Quantum Electr.* **QE-6**, 694–708 (1970).
18. H. Haus, A Theory of Forced Mode-Locking, *IEEE J. Quantum Electr.*, **QE-11**, 323–330 (1975).
19. H. Haus, Theory of Mode Locking with a Fast Saturable Absorber, *J. Appl. Phys.*, **46**, 3049–3058 (1975).
20. U. Keller, Ulrafast All-Solid-State Laser Technology, *Appl. Phys. B*, **58**, 347–363 (1994).
21. D. E. Spence, P. N. Kean, W. Sibbett, 60-fs Pulse Generation from a Self-mode-Locked Ti:Sapphire Laser, *Opt. Letters*, **16**, 42–44 (1991).
22. M. Piché, Beam Reshaping and Self-Mode-Locking in Nonlinear Laser Resonators, *Opt. Commun.*, **86**, 156–158 (1991).
23. G. H. C. New, Pulse Evolution in Mode-Locked Quasi-Continuous Lasers, *IEEE J. Quantum Electron.*, **QE-10**, 115–124 (1974).
24. G. P. Agrawaal, *Nonlinear Fiber Optics*, second edition (Academic Press, San Diego 1995) Section 3.3.

25. R. L. Fork, O. E. Martinez, and J. P. Gordon, Negative Dispersion Using Pairs of Prisms, *Opt. Letters*, **9**, 150–152 (1984).
26. R. Szipöcs, K. Ferencz, C. Spielmann, F. Krausz, Chirped Multilayer Coatings for Broadband Dispersion Control in Femtosecond Lasers, *Opt. Letters*, **19**, 201–203 (1994).
27. Reference 26, Chapter 5.
28. H. A. Haus, J. G. Fujimoto, E. P. Ippen, Structures for Additive Pulse Mode Locking, *J. Opt. Soc. Am. B*, **8**, 2068–2076 (1991).
29. C. Spielmann, P. F. Curley, T. Brabec, F. Krausz, Ultrabroadband Femtosecond Lasers, *IEEE J. Quantum Electron.*, **QE-30**, 1100–1114 (1994).
30. J. A. Valdmanis, R. L. Fork, J. P. Gordon, Generation of Optical Pulses as Short as 27 Femtosecond Directly from a Laser Balancing Self-Phase Modulation, Group-Velocity Dispersion, Saturable Absorption, and Gain Saturation, *Opt. Letters*, **10**, 131–133 (1985).
31. U. Keller *et al.*, Semiconductor Saturable Absorber Mirrors (SESAMs) for Femtosecond to Nanosecond Pulse Generation in Solid-State Lasers, *J. Select. Topics Quantum Electronics*, Dec 1996.
32. H. A. Haus, Parameter Ranges for cw Passive Modelocking, *IEEE J. Quantum Electron.*, **QE-12**, 169–176 (1976).
33. F. X. Kärtner *et al.*, Control of Solid-State Laser Dynamics by Semiconductor Devices, *Opt. Engineering*, **34**, 2024–2036 (1995).
34. R. L. Fork, I. Greene, and C. V. Shank, Generation of Optical Pulses Shorter than 0.1 ps by Colliding Pulse Mode Locking, *Appl. Phys. Letters*, **38**, 671–672, (1991).
35. G. Cerullo, S. De Silvestri, V. Magni, Self-Starting Kerr Lens Mode-Locking of a Ti:Sapphire Laser, *Opt. Letters*, **19**, 1040–1042 (1994).
36. A. Stingl, M. Lenzner, Ch. Spielmann, F. Krausz, R. Szipöcs, Sub-10-fs Mirror-Controlled Ti:Sapphire Laser, *Opt. Letters*, **20**, 602–604 (1995).
37. I. D. Jung *et al.*, Self-Starting 6.5-fs Pulses from a Ti:Sapphire Laser, *Opt. Letters*, **22**, 1009–1011 (1997).
38. Reference [9], Section 8.6.

9
Solid-State, Dye, and Semiconductor Lasers

9.1. INTRODUCTION

In this chapter, the most important types of lasers involving high density active media are considered, namely solid-state, dye and semiconductor lasers. The chapter concentrates on those examples that are in widest use and whose characteristics are representative of a whole class of lasers. The main emphasis here is on stressing the physical behavior of the laser and relating this behavior to the general concepts developed in the previous chapters. Some engineering details are also provided with the main aim again of helping to provide for a better physical insight into the behavior of the particular laser. To complete the picture, some data relating to laser performances (e.g., oscillating wavelength(s), output power or energy, wavelength tunability, etc.) are also included to help providing some indication of the laser's applicability. For each laser, after some introductory comments, the following items are generally covered: (1) Relevant energy levels; (2) excitation mechanisms; (3) characteristics of the laser transition(s); (4) engineering details of the laser structure(s); (5) characteristics of the output beam; (6) applications.

9.2. SOLID-STATE LASERS

The use of the term solid-state laser is generally reserved for those lasers having as their active species ions introduced as an impurity in an otherwise transparent host material (in crystalline or glass form). Thus, semiconductor lasers are not usually included in this category, the mechanisms for pumping and for laser action being in fact quite different. These will be considered in a separate section.

O. Svelto, *Principles of Lasers*,
DOI: 10.1007/978-1-4419-1302-9_9, © Springer Science+Business Media LLC 2010

Ions belonging to one of the series of transition elements of the Periodic Table, in particular rare-earth (RE) or transition metal ions, are generally used as the active impurities. On the other hand, for host crystals, either oxides, e.g. Al_2O_3, or fluorides, e.g. $YLiF_4$ [abbreviated as YLF], are most often utilized.(1) The Al^{3+} site is too small to accommodate rare earth ions and it is generally used for transition metal ions. A suitable combination of oxides, to form synthetic garnets such as $Y_3Al_5O_{12} = (1/2)(3\,Y_2O_3 + 5\,Al_2O_3)$, are often used and the Al^{3+} site can accommodate transition metal ions while the Y^{3+} site can be used for RE ions. Other oxides include YVO_4 for Nd^{3+} ions and alexandrite for Cr^{3+} ions. Among the fluorides, YLF is used as host for rare earths while $LiSrAlF_6$ (abbreviated as LISAF) or $LiCaAlF_6$ (abbreviated as LICAF) are used for transition metals, most notably for Cr^{3+} ions. A comparison between oxides and fluorides shows that oxides have the advantages of being harder, with better mechanical and thermo-mechanical (i.e. higher thermal fracture limit) properties. Fluorides, on the other hand, show better thermo-optical properties (i.e. lower thermally-induced lensing and birefringence). Glasses of either the silicate (i.e. based on SiO_2) or phosphate (i.e. based on P_2O_5) family have so far been used only for RE ions. Compared to crystals, many glasses have much lower melting temperatures and, therefore, are easier and cheaper to fabricate even in much larger dimensions. On the other hand, glasses have much lower (by ~ an order of magnitude) thermal conductivity which leads to much worse thermo-mechanical and thermo-optical properties. A comparison between various glasses reveals silicates to have better thermal and mechanical properties while phosphates to show better thermo-optical and non-linear optical properties.

The general electronic structure of a RE is $4f^N\,5s^2\,5p^6\,5d^0\,6s^2$ as shown for Nd, Er, Yb, Tm, and Ho in Table 9.1 where, for comparison, the structure of Xe is also indicated. When the RE is inserted in a host material, the two $6s$ electrons and one of the $4f$ electrons are used for the ionic binding so that the RE presents itself as a triply ionized ion (e.g. $N - 1 = 3$ for Nd^{3+}). The remaining $N-1$ electrons can then arrange themselves in different states of the $4f$ shell, resulting in a large number of energy levels. In fact, these states are split by three types of interaction, namely Coulomb interaction among the $4f^{N-1}$ electrons, spin-orbit coupling, and crystal field interaction. The Coulomb interaction is the strongest of these three and splits

TABLE 9.1. Electronic configurations of some rare earth and transition metals of interest as laser active impurities. For reference, the fundamental configuration of Xe is also shown

Xenon, Xe	$(Kr)4d^{10}5s^25p^6$
Neodymium, Nd	$(Xe)4f^45d^06s^2$
Holmium, Ho	$(Xe)4f^{11}5d^06s^2$
Erbium, Er	$(Xe)4f^{12}5d^06s^2$
Thulium, Tm	$(Xe)4f^{13}5d^06s^2$
Ytterbium, Yb	$(Xe)4f^{14}5d^06s^2$
Chromium, Cr	$(Ar)3d^54s^1$
Titanium, Ti	$(Ar)3d^24s^2$
Cobalt, Co	$(Ar)3d^74s^2$
Nickel, Ni	$(Ar)3d^84s^2$

the $4f$ states into sublevels which are typically separated by $\sim 10{,}000\,\mathrm{cm}^{-1}$. Spin-orbit coupling then splits each term into manifolds typically separated by $\sim 3{,}000\,\mathrm{cm}^{-1}$. Crystal field interaction produces the weakest perturbation (weakened by the screening effect of the $5s^2$ and $5p^6$ orbitals), thus further splitting each sub-level into a manifold with energy separation typically of $200\,\mathrm{cm}^{-1}$. All relevant absorption and emission features are due to transitions between these $4f$ states ($4f$–$4f$ transitions). Electric dipole transitions within the $4f$ shell are parity-forbidden and it needs a mixture of wavefunctions with opposite parity, brought about by the crystal-field interaction, to create non-zero, although still weak, transition probabilities. Thus, one generally finds long (hundreds of μs) radiative lifetimes. Furthermore, due to the screening from the $5s^2$ and $5p^6$ orbitals, electron-phonon coupling turns out to be very weak. One thus has sharp transition lines and weak non-radiative decay channels for low ion-doping (ion-ion interaction can lead to non-radiative decay at high RE ion concentrations, see Fig. 2.13). From the above considerations one expects large values of the overall lifetime, τ, and of the product $\sigma\tau$, where σ is the peak cross-section. This implies a low threshold pump power for laser action since, e.g. for a four level laser, the threshold pump rate is proportional to $1/\sigma\tau$ [see Eq. (7.3.3)].

The electronic configurations for those transition metals, of interest for laser action, are also shown in Table 9.1. Note that the electronic configuration of the most important active species, i.e., Cr, is given by $(\mathrm{Ar})3d^5 4s^1$, while those of Ti, Co, and Ni can be written in the general form $(\mathrm{Ar})3d^N 4s^2$ (with $N=2$ for Ti, 7 for Co and 8 for Ni). When in an ionic crystal, the $4s^1$ electron and two $3d$ electrons of Cr are used for the ionic binding and Cr is present as a triply ionized ion with 3 electrons left in the $3d$ shell. For Titanium, the two $4s$ electrons and one $3d$ electron are used for the ionic binding and Ti is present as triply ionized ion with only one electron left in the $3d$ shell. For both Co and Ni the two $4s$ electrons are used for the binding and these elements are present as doubly ionized ions. In all cases, the remaining electrons in the $3d$ orbital can arrange themselves in a large number of states (e.g. 24 for Cr^{3+}) and all the absorption and emission features of transition metal ions arise from $3d-3d$ transitions. Lacking the screening which occurs for RE ions, the $3d$ states interact strongly with the crystal field of the host and, as we shall see later, this is the fundamental reason for the vibronic character, leading to wide absorption and emission bands, for most of the corresponding transitions. Again electric dipole transitions within the $3d$ shell are parity forbidden but, due to the stronger crystal field compared to the RE case, the $3d-3d$ transitions are more allowed and thus the lifetimes are significantly shorter (a few μs) than those of the $4f-4f$ transitions of RE ions. Compared to e.g. Nd:YAG, the transition cross sections are somewhat smaller so that the product $\sigma\tau$ is now typically one order of magnitude smaller.

As a conclusion to this section, it is worth noting that ions belonging to the actinide series, notably U^{3+}, were also used in the early days of laser development (actually the U^{3+} laser was the second solid state laser to be developed, i.e. immediately after the ruby laser). These ions are essentially no longer used but they deserve a mention here for historical reasons.

9.2.1. The Ruby Laser

This type of laser was the first to be made to operate (T. H. Maiman, June 1960[2,3]) and still continues to be used in some applications.[4] As a naturally occurring precious stone, ruby

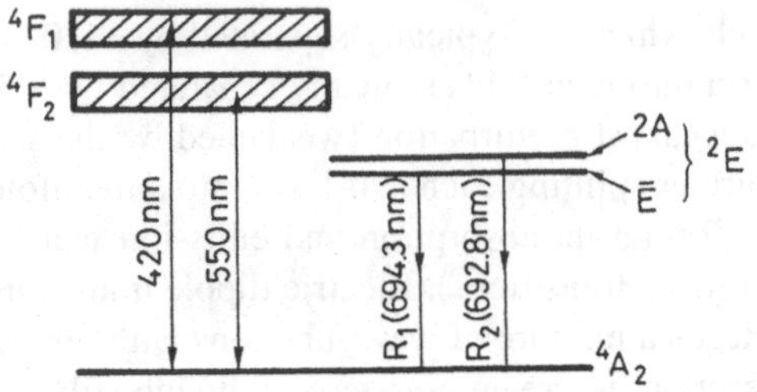

FIG. 9.1. Simplified energy levels of ruby.

has been known for at least 2,000 years. It consists of a natural crystal of Al_2O_3 (corundum) in which some of the Al^{3+} ions happen to have been replaced by Cr^{3+} ions. For the laser material, one uses artificial crystals obtained by crystal growth from a molten mixture of Al_2O_3 with a small percentage of Cr_2O_3 (0.05% by weight).[(4)] Without the addition of Cr_2O_3, the crystal that forms is colorless and it is known as sapphire. Due to the strong green and violet absorption bands of Cr^{3+} ions, it only needs the small addition of Cr_2O_3 to give the crystal a pink color (pink ruby). In the case of gem stones, the Cr^{3+} concentration is about an order of magnitude larger, giving them a strong red color (red ruby).

The energy levels of ruby are due to one of the three electrons of the Cr^{3+} ion, in the $3d$ inner shell, under the influence of the octahedral field at the Al site in the Al_2O_3 lattice. The corresponding levels, of interest for laser action, are shown in Fig. 9.1. The notation used to label the levels is derived from group theory and is not discussed at any length here. We merely limit ourselves to pointing out that the superscript to the left of each letter indicates the multiplicity of the state while the letter indicates the particular rotational symmetry of the state. Thus, as an example, the ground ${}^4A_2$ state has a multiplicity $(2S + 1) = 4$, i.e., $S = 3/2$ where S is the total spin quantum number of the three $3d$ electrons. This means that the spins of these electrons are all parallel in this case.

Ruby has two main pump bands ${}^4F_1$ and ${}^4F_2$ and the peaks of the transitions from the ground ${}^4A_2$ level to these bands occur at the wavelengths of 550 nm (green) and 420 nm (violet), respectively (see also Fig. 6.7). The two bands are connected by a very fast (ps) nonradiative decay to both $2\bar{A}$ and $\bar{E}$ states, which together form the ${}^2E$ state. The $2\bar{A}$ and $\bar{E}$ states are themselves connected to each other by a very fast nonradiative decay, which leads to a fast thermalization of their populations, thus resulting in the $\bar{E}$ level being the most heavily populated. Since the total spin of the ${}^2E$ state is 1/2, the ${}^2E \to {}^4A_2$ transition is spin-forbidden. The relaxation time of both $2\bar{A}$ and $\bar{E}$ levels to the ground state is thus very long ($\tau \cong 3$ ms), actually one of the longest among all solid-state laser materials.

From the discussion above it is now apparent that the level $\bar{E}$ accumulates the largest fraction of the pump energy, and is thus a good candidate as the upper laser level. In fact, laser action usually occurs on the $\bar{E} \to {}^4A_2$ transition (R_1 line) at the wavelength $\lambda_1 = 694.3$ nm (red). It should be noted, however, that the frequency separation between $2\bar{A}$ and $\bar{E}$ levels ($\sim$29 cm^{-1}) is small compared to kT/h ($\sim$209 cm^{-1} at $T = 300$ K) so that the $2\bar{A}$ population is comparable to, although slightly smaller than, the $\bar{E}$ level population. It then follows that it is also possible to obtain laser action on the $2\bar{A} \to {}^4A_2$ transition (R_2 line, $\lambda_1 = 692.8$ nm). It is anyhow apparent that ruby operates as a three-level laser (actually, together with Er lasers, it represents the most notewhorty example of a three-level laser). As already discussed in

TABLE 9.2. Optical and spectroscopic parameters of ruby for room temperature operation

Property	Values and units
Cr_2O_3 doping	0.05 wt. %
Cr^{3+} concentration	1.58×10^{19} ions/cm^3
Output wavelengths	694.3 nm (R_1 line)
	692.9 nm (R_2 line)
Upper laser level lifetime	3 ms
Linewidth of R_1 laser transition	11 cm^{-1}
Stim. emission cross-section σ_e	2.5×10^{-20} cm^2
Absorption cross section σ_a	1.22×10^{-20} cm^2
Refractive index ($\lambda = 694.3$ nm)	$n = 1.763$ ($E \perp c$)
	$n = 1.755$ ($E \| c$)

connection with Fig. 2.10, the R_1 transition is, at room temperature, predominantly homogeneously broadened, the broadening arising from the interaction of the Cr^{3+} ions with lattice phonons. The width of the transition (FWHM) is $\Delta\nu_0 \cong 11\,\text{cm}^{-1}$ (330 GHz) at $T = 300$ K. As a summary, Table 9.2 shows some relevant optical and spectroscopic parameters of ruby at room temperature.

Ruby lasers are usually operated in a pulsed regime. For this, the pump configuration of Fig. 6.1 using a medium-pressure (~500 Torr) xenon flashtube, is generally utilized. Typical rod diameters range between 5 and 10 mm with a length between 5 and 20 cm. It should be noted that a helical flashtube surrounding the active rod was used in the earliest ruby lasers. Since this laser operates on a three-level scheme, the threshold pump energy is typically an order of magnitude higher than that of other solid-state lasers operating with four level schemes (e.g. Neodimium lasers). Due to the long upper state lifetime, ruby lasers lend themselves readily to Q-switched operation and, due to the relatively broad laser linewidth, they can also produce short pulses ($\sim 5 \div 10$ ps) in mode-locked operation. Both active and passive methods can be used for Q-switching and mode-locking. When slow saturable absorbers are used for Q-switching, the laser tends to operate on a single transverse and longitudinal mode due to the mode selecting mechanism discussed in Sect. 8.4.2.4. With fast saturable absorbers (usually solutions of cyanine dyes), simultaneous Q-switched and mode-locked operation occurs (see Fig. 8.28a). Peak powers of a few tens of MW, for Q-switching, and a few GW, when also mode-locked, are typical. Since the gain of the R_2 line is somewhat smaller than for the R_1 line, laser action on the R_2 line can be selected by using, for instance, the dispersive system of Fig. 7.16b. Ruby lasers can also run cw, transversely pumped by a high-pressure mercury lamp or longitudinally pumped by an Ar ion laser.

Ruby lasers, once very popular, are now less widely used since, on account of their higher threshold, they have been superseded by competitors, such as Nd:YAG or Nd:glass lasers. In fact, ruby lasers were extensively used in the past for the first mass production of military rangefinders, an application in which this laser is now completely replaced by other solid-state lasers (Nd:YAG, Nd:glass, Yb:Er:glass). Ruby lasers are, however, still sometimes used for a number of scientific and technical applications where their shorter wavelength, compared to e.g., Nd:YAG, represents an important advantage. This is for instance the case of pulsed holography, where Nd:YAG lasers cannot be used owing to the lack of response, in the infrared, of the high-resolution photographic materials which are used.

9.2.2. Neodymium Lasers

Neodymium lasers are the most popular type of solid-state laser. The host medium is often a crystal of $Y_3Al_5O_{12}$ (commonly called YAG, an acronym for yttrium aluminum garnet) in which some of the Y^{3+} ions are replaced by Nd^{3+} ions. Besides this oxide medium, other host media include some fluoride (e.g. $YLiF_4$) or vanadate (e.g. YVO_4) materials as well as some phosphate or silicate glasses. Typical doping levels in e.g. Nd:YAG are ~1 atomic %. Higher doping generally leads to quenching of fluorescence and also results in strained crystals since the radius of the Nd^{3+} ion is somewhat larger (by ~14%) than that of the Y^{3+} ion. The doping levels used in Nd:glass (~4% of Nd_2O_3 by weight) are somewhat higher than the value for Nd:YAG. The undoped host materials are usually transparent and, when doped, generally become pale purple in color because of the Nd^{3+} absorption bands in the red.

9.2.2.1. Nd:YAG

A simplified energy-level scheme for Nd:YAG is shown in Fig. 9.2. As discussed above, these levels arise from the three inner shell $4f$ electrons of the Nd^{3+} ion which are effectively screened by 8 outer electrons ($5s^2$ and $5p^6$). The energy levels are only weakly influenced by the crystal field of YAG and the Russell-Saunders coupling scheme of atomic physics can then be used. Level notation is accordingly based on this scheme and the symbol characterizing each level is in the form ${}^{2S+1}L_J$, where S is the total spin quantum number, J is the total angular momentum quantum number, and L is the orbital quantum number. Note that the allowed values of L, namely $L = 0, 1, 2, 3, 4, 5, 6, \ldots$ are expressed, for historical reasons, by the capital letters $S, P, D, F, G, H, I, \ldots\ldots$, respectively. Thus the ${}^4I_{9/2}$ ground level correspond to a state in which $2S + 1 = 4$ (i.e., $S = 3/2$), $L = 6$, and $J = L - S = 9/2$. Each level is $(2J + 1)$-fold degenerate corresponding to the quantum number m_J running from $-J$ to $+J$ in unit steps. In the octahedral symmetry of the YAG crystal field, states with the same value of $|m_J|$ have the same energy in the presence of the Stark effect and each ${}^{2S+1}L_J$ level is split into $(2J+1)/2$ doubly degenerate sublevels. Thus the ${}^4I_{11/2}$ and ${}^4F_{3/2}$ levels are split into 6 and 2 sublevels, respectively (see Fig. 9.2). Note that, since the degeneracy of all sublevels

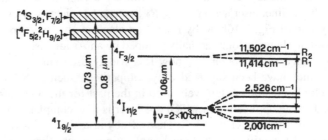

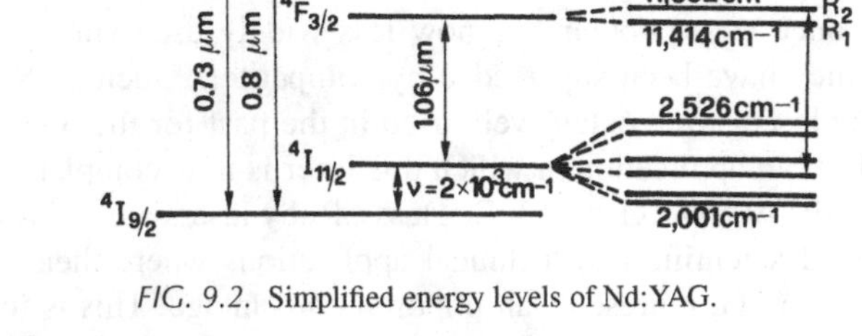

FIG. 9.2. Simplified energy levels of Nd:YAG.

is always the same (i.e., $g = 2$), we can disregard this degeneracy and consider each sublevel of Fig. 9.2 as if it were a single non-degenerate level.

The two main pump bands of Nd:YAG occur at ~730 and ~800 nm, respectively, although higher-lying absorption bands (see Fig. 6.7) also play an important role, notably for flash-lamp pumping. These bands are coupled by a fast nonradiative decay to the $^4F_{3/2}$ level from where the decay to the lower I levels occurs (i.e., to $^4I_{9/2}$, $^4I_{11/2}$, $^4I_{13/2}$, etc. levels, see Figs. 9.2 and 2.15). However, the rate of this decay is much slower ($\tau \cong 230\,\mu s$) than that of the nonradiative decay, because the transition, in the isolated ion, is forbidden via the electric-dipole interaction (the selection rule for electric-dipole allowed transitions is $\Delta J = 0$ or ± 1) but becomes weakly allowed due to the crystal field interaction. Note that nonradiative decay is not so important because the decay due to ion-ion interactions (see Fig. 2.13b) does not play an important role at the stated Nd ion concentrations and because multiphonon decay is also not so effective due both to the screening of the $5s^2$ and $5p^6$ states and the large energy gap between $^4F_{3/2}$ and the nearest level below it. This means that level $^4F_{3/2}$ accumulates a large fraction of the pump power and is therefore a good candidate as the upper level for laser action.

From the above discussion it is now apparent that several laser transitions are possible between $^4F_{3/2}$ and several lower-lying I levels. Out of these transitions, it turns out that the $^4F_{3/2} \rightarrow {}^4I_{11/2}$ is the strongest one. Level $^4I_{11/2}$ is then coupled by a fast (hundreds of ps) nonradiative decay to the $^4I_{9/2}$ ground level so that thermal equilibrium between these two levels is very rapidly established. Then, since the energy difference between the $^4I_{11/2}$ and $^4I_{9/2}$ levels is almost an order of magnitude larger than kT, according to Boltzmann statistics, level $^4I_{11/2}$ may, to a good approximation, be considered to be empty at all times. Thus laser operation on the $^4F_{3/2} \rightarrow {}^4I_{11/2}$ transition corresponds to a four-level scheme. Actually, according to the above discussion, the $^4F_{3/2}$ level is split by the Stark effect into two sublevels (R_1 and R_2) while the $^4I_{11/2}$ level is split into six sublevels. It then turns out that laser action usually occurs from the upper, R_2, sublevel to a particular sublevel of the $^4I_{11/2}$ level, this transition having the highest value of stimulated-emission cross section. The transition occurs at $\lambda = 1.064\,\mu m$ (near infrared) this being the most widely used lasing wavelength for Nd:YAG lasers. It should be noted that laser action can also be obtained on the $^4F_{3/2} \rightarrow {}^4I_{13/2}$ transition (see Fig. 2.15, $\lambda = 1.319\,\mu m$ being the wavelength of the strongest transition in this case) provided the multilayer dielectric coatings of the cavity mirrors have high reflectivity at $\lambda = 1.319\,\mu m$ and sufficiently low reflectivity at $\lambda = 1.064\,\mu m$ wavelength (see Fig. 4.9). With diode-laser pumping, laser action has also been made to occur effectively on the $^4F_{3/2} \rightarrow {}^4I_{9/2}$ transition. In this case the transition, at a wavelength $\lambda = 946$ nm (see Fig. 2.15), is to a sub-level of the $^4I_{9/2}$ state which, despite being a high lying sublevel, is still appreciably populated according to Boltzmann statistics, and the system operates as a quasi-three-level laser. In the case of the usual $\lambda = 1.064\,\mu m$ transition, and probably for all the other cases, the laser transition is homogeneously broadened at room temperature owing to interaction with lattice phonons. The corresponding width is $\Delta\nu \cong 4.2\,cm^{-1} = 126$ GHz at $T = 300$ K. This makes Nd:YAG a good candidate for mode-locked operation and pulses as short as 5 ps have indeed been obtained by passive mode-locking (see Example 8.8). The long lifetime of the upper laser level ($\tau \cong 230\,\mu s$) also makes Nd:YAG very suitable for Q-switched operation. Table 9.3 gives a summary of relevant optical and spectroscopic parameters for Nd:YAG at room temperature.

TABLE 9.3. Optical and spectroscopic parameters of Nd:YAG, Nd : YVO_4, Nd:YLF, and Nd:glass (phosphate). In the table, N_t is the concentration of the active ions, τ is the fluorescence lifetime, $\Delta\nu_0$ is the transition linewidth (FWHM), σ_e is the effective stimulated emission cross section. Data refer to room temperature operation

	Nd:YAG $\lambda = 1.064\,\mu m$	Nd : YVO_4 $\lambda = 1.064\,\mu m$	Nd:YLF $\lambda = 1.053\,\mu m$	Nd:glass $\lambda = 1.054\,\mu m$ (Phosphate)
Nd doping [at. %]	1 at. %	1 at. %	1 at. %	3.8% Nd_2O_3 by weight
N_t [10^{20} ions/cm^3]	1.38	1.5	1.3	3.2
$\tau[\mu s]$	230	98	450	300
$\Delta\nu_0$ [cm^{-1}]	4.5	11.3	13	180
σ_e [10^{-19} cm^2]	2.8	7.6	1.9	0.4
Refractive index	$n = 1.82$	$n_o = 1.958$ $n_e = 2.168$	$n_O = 1.4481$ $n_e = 1.4704$	$n = 1.54$

Nd:YAG lasers can operate either cw or pulsed and can be pumped either by a lamp or by an AlGaAs semiconductor laser.[(5)] For lamp pumping, linear lamps in single-ellipse (Fig. 6.1a), close-coupling (Fig. 6.1b) or multiple-ellipse configurations (Fig. 6.2) are commonly used. Medium pressure (500–1,500 Torr) Xe lamps and high-pressure (4–6 atm) Kr lamps are used for the pulsed and cw cases, respectively. If a rod is used as the active medium, the rod diameter ranges typically between 3 and 6 mm with a length between 5 and 15 cm. To reduce pump-induced thermal lensing and thermal birefringence, a slab configuration (Fig. 6.3a) is also sometimes used. The slope efficiency is about 3% for both cw and pulsed operation and average output powers up to a few kW (1–3 kW) are common. Longitudinally diode-pumped lasers (Fig. 6.11) with cw output powers up to ~15 W and transversely diode-pumped lasers (Figs. 6.14 and 6.15) with cw output powers well above 100 W are now available. The slope efficiency, for diode-pumping, is much higher than for lamp pumping and may exceed 10%.

Nd:YAG lasers are widely used in a variety of applications, among which we mention: (1) Material processing such as drilling and welding. For drilling applications the beam of a repetitively pulsed laser is focused on the material (average powers of 50–100 W are commonly used with $E = 5$–10 J pulse energy, $\Delta\tau_p = 1$–10 ms pulse duration, and $f = 10$–100 Hz repetition rate). For welding applications, the repetitively pulsed laser beam is conveyed to the working region through a, 0.5–2 mm diameter, optical fiber (average powers up to 2 kW are now commonly handled in this way). In this application, high power Nd:YAG lasers are superseding their direct competitors (high power CO_2 lasers) on account of the system flexibility offered by optical fiber delivery. (2) Medical applications. For coagulation and tissue disruption, cw Nd:YAG lasers with power up to 50 W are used and the beam is delivered through an optical fiber, inserted into a conventional endoscope, to the internal organs (lungs, stomach, bladder) of the human body. Repetitively Q-switched Nd:YAG lasers are used for photodisruption of transparent membranes of pathological origin which can appear in the anterior chamber of the eye (e.g., secondary cataract) or for iridoctomy. (3) Laser ranging, in particular for laser rangefinders and target designators used in a military contest. In this case Q-switched lasers are used ($E \approx 100$ mJ, $\Delta\tau_p = 5$–20 ns, $f = 1$–20 Hz). (4) Scientific applications: in this case, Q-switched lasers, with their second harmonic ($\lambda = 532$ nm), third harmonic ($\lambda \cong 355$ nm) and fourth-harmonic beams ($\lambda = 266$ nm), as well as mode-locked lasers are commonly used in a variety of applications. Lastly, it should be noted that diode-pumped Nd:YAG lasers, with intracavity second harmonic generation, giving a green ($\lambda = 532$ nm) cw output power up to ~10 W have become available, providing an all-solid-state alternative to the Ar laser for many of its applications.

9.2.2.2. Nd:Glass[6]

As already mentioned, the relevant transitions of the Nd^{3+} ion involve the three electrons of the $4f$ shell, and these are screened by 8 outer electrons in the $5s$ and $5p$ configuration. Accordingly, the energy levels of Nd:glass are approximately the same as those of Nd:YAG. Thus the strongest laser transition again occurs at about the same wavelength ($\lambda \cong 1.054\ \mu m$ for phosphate glass, see Table 9.3). The linewidths of the laser transitions are, however, much larger as a result of the inhomogeneous broadening arising from the local field inhomogeneities typical of a glass medium. In particular the main $\lambda \cong 1.054\ \mu m$ laser transition is much broader (by ~40 times) while the peak cross section is somewhat smaller (by ~7 times) than that of Nd:YAG. The larger bandwidth is of course a desirable feature for mode-locked operation and, indeed, diode-pumped passively-mode-locked Nd:glass lasers have produced ultrashort pulses (~100 fs). The smaller cross-section, on the other hand, is a desirable feature for pulsed high-energy systems since the "threshold" inversion for the parasitic process of amplified spontaneous emission, ASE [see Eq. (2.9.4)], is correspondingly increased. Thus more energy per unit volume can be stored in Nd:glass compared to Nd:YAG before the onset of ASE [see Example 2.13]. It should also be noted that glass, due to its lower melting temperature, can be fabricated more easily than YAG and active media of much larger dimensions can therefore be produced. Finally, since the pump absorption bands of Nd:glass are also much broader than those of Nd:YAG and Nd^{3+} concentrations are typically twice as great, the pumping efficiency of a lamp-pumped Nd:glass rod is somewhat larger (~1.6 times) than that of a Nd:YAG rod of the same dimensions (see Table 6.1). Against these advantages of Nd:glass compared to Nd:YAG, we must set the disadvantage arising from its much lower thermal conductivity (thermal conductivity of glass is about ten times smaller than that of Nd:YAG). This has limited applications of Nd:glass lasers mainly to pulsed laser systems of rather low repetition rate (<5 Hz) so that thermal problems in the active medium (rod or slab) can be avoided*.

As discussed above, Nd:glass lasers are often used in applications where a pulsed laser of low repetition rate is required. This is for instance the case for some military rangefinders and some scientific Nd lasers. A very important application of Nd:glass is in the form of laser amplifiers in the very high energy systems used in laser-driven fusion experiments. Systems based on Nd:glass amplifiers have indeed been built in several countries, the largest one being in the USA (Nova laser, Lawrence Livermore National Laboratory) and delivering pulses with energy of ~100 kJ and peak power of 100 TW ($\Delta\tau_p = 1$ ns). The laser makes use of a chain

* An exception to this is provided by glass fiber lasers, where long length and small transverse dimension eliminate the thermal problem and have allowed cw outputs in excess of 100 W.

of several Nd:glass amplifiers, the largest of which consists of Nd:glass disks (see Fig. 6.3b) of ~4 cm in thickness and ~75 cm in diameter. A national ignition facility delivering pulses of much higher energy (~10 MJ, Lawrence Livermore National Laboratory) and a similar system delivering about the same output energy (~2 MJ, Limeil Center) are presently being built in the USA and in France, respectively.

9.2.2.3. Other Crystalline Hosts

Many other crystal materials have been used as hosts for the Nd^{3+} ion and we limit ourselves here to mentioning $YLiF_4$ [YLF] and YVO_4.

Compared to YAG, YLF has better thermo-optical properties[7] (pump induced thermal lensing and thermal birefringence) and lamp pumped Nd:YLF lasers are used to obtain TEM_{00}-mode cw-beams of better quality and higher output power. The larger linewidth of Nd:YLF compared to Nd:YAG (~3 times, see Table 9.3) makes Nd:YLF lasers particularly attractive for mode-locked operation both with lamp and diode pumping (see Example 8.8). The mechanical and thermo-mechanical properties of YLF are however worse than those of YAG and this make YLF rods more difficult to handle and easier to break. It should also be noted that the 1,053 nm emission wavelength of Nd:YLF provides a good match with the peak gain wavelength of Nd:glass:phopsphate lasers (see Table 9.3). Mode-locked Nd:YLF lasers are accordingly used as the first stage in the large energy systems used for laser fusion experiments.

Compared to Nd:YAG, Nd:YVO_4 has much larger peak cross section ($\sigma_e \cong 7.6 \times 10^{-19}\ \mathrm{cm}^2$) and much shorter fluorescence lifetime ($\tau = 98\ \mu s$). The product $\sigma\tau$ is about the same for the two cases and one thus expects approximately the same threshold. For a given inversion, however, the gain coefficient of Nd:YVO_4 is about 3 times larger than that of Nd:YAG and this makes a Nd:YVO_4 laser less sensitive to cavity losses. Longitudinally-diode-pumped Nd:YVO_4 of high cw power (~15 W) are now commercially available and, for this application, Nd:YVO_4 seems to be preferred to Nd:YAG.

9.2.3. Yb:YAG

The Yb:YAG laser is the most noteworthy example of a quasi-three level laser. It oscillates at ~1.03 μm wavelength and thus presents itself as direct competitor to Nd:YAG laser. Since it operates on a quasi-three-level laser scheme, it is usually pumped by semiconductor laser diodes which can provide the intense pumping required.[8]

A simplified scheme of the energy level diagram of Yb:YAG is shown in Fig. 9.3. The level structure is particularly simple here, only one excited manifold, $^2F_{5/2}$, being present, because Yb^{3+} is one electron short to a full 4f shell (see Table 9.1), this shell thus acting as if it contained one electron hole. The two main absorption lines occur at 968 and 941 nm, respectively, the two lines have approximately the same value of the peak absorption cross section, the line at 941 nm being usually preferred for diode-pumping due to its larger width. The main gain line occurs at 1.03 μm (quasi-three-level laser). Table 9.4 shows some relevant optical and spectroscopic parameters of Yb:YAG at room temperature. Note, in particular, the long lifetime, $\tau = 1.16$ ms, of essentially radiative origin, which is indicative of a good storage medium.

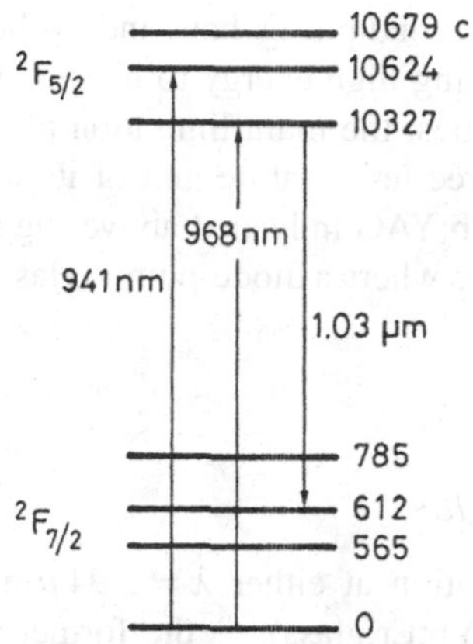

FIG. 9.3. Energy level diagram of Yb:YAG.

TABLE 9.4. Optical and spectroscopic parameters, at room temperature, of the most important quasi-three-level laser materials. Note that, for Yb:Er:glass, the effective value of the stimulated emission and absorption cross sections are about the same and the laser can be considered to operate on (almost) a pure three-level scheme

	Yb:YAG $\lambda = 1.03\,\mu m$	Nd:YAG $\lambda = 946\,nm$	Tm:Ho:YAG $\lambda = 2.091\,\mu m$	Yb:Er:glass $\lambda = 1.54\,\mu m$ (Phosphate)
doping [at. %]	6.5 at. %	1.1 at. %		
N_t [10^{20} ions/cm^3]	8.97	1.5	8 [Tm] 0.5 [Ho]	10 [Yb] 1 [Er]
τ[ms]	1.16	0.23	8.5	8
$\Delta\nu_0$ [cm^{-1}]	86	9.5	42	120
σ_e [10^{-20} cm^2]	1.8	2.4	0.9	0.8
σ_a [10^{-20} cm^2]	0.12	0.296	0.153	0.8
Refractive index	$n = 1.82$	$n = 1.82$	$n = 1.82$	$n = 1.531$

Yb:YAG lasers are pumped, in a longitudinal pumping configuration using a pump at $\lambda_p = 943$ nm wavelength, usually by InGaAs/GaAs strained quantum well, QW, lasers although they can be also pumped by a Ti:sapphire laser. The optical-to-optical efficiency turns out to be rather high (~60%), a result mainly of the high pump quantum efficiency ($\eta_q = h\nu/h\nu_p = \lambda_p/\lambda = 91.5\%$).[9] Average output powers well in excess of 50 W have so far been achieved.[10] Compared to Nd:YAG, the Yb:YAG laser offers the following main favorable properties: (1) Very low quantum defect [$(h\nu_p - h\nu)/h\nu_p \cong 9\%$] and hence very low fractional heating. (2) Long radiative lifetime of the upper state making Yb:YAG a good medium for Q-switching. (3) Due to the simple energy level structure, one can use high doping levels (6.5 at. % are usually used) without incurring in fluorescence quenching phenomena due to ion-ion interaction. (4) Broad emission bandwidth (~86 cm^{-1}) indicating suitability for mode-locked operation (sub-picosecond pulses have indeed been obtained). (5) Low stimulated emission cross-section allowing high energy to be stored before the onset of ASE. By contrast to these favorable properties, the main limitation of Yb:YAG comes from the high threshold, a result of its quasi-three-level nature and of its low stimulated-emission cross-section. The various features of Yb:YAG indicated above suggest that it can be better suited than Nd:YAG for many applications where a diode-pumped laser at a wavelength around 1 μm will be needed.

9.2.4. Er:YAG and Yb:Er:glass

Erbium-lasers can emit radiation at either $\lambda = 2.94\,\mu$m wavelength (for Er:YAG) or at $\lambda = 1.54\,\mu$m wavelength (for Yb:Er:glass),[11] the former wavelength being particularly interesting for biomedical applications and the latter wavelength being attractive for application situations where eye-safety is important and for optical communications in the third transparency-window of optical fibers.

In the case of Er:YAG, the Er^{3+} ion occupies Y^{3+} ion sites in the lattice and the relevant energy levels of the laser are shown in Fig. 9.4a. Laser oscillation can take place on either the ${}^4I_{11/2} \rightarrow {}^4I_{13/2}$ transition ($\lambda = 2.94\,\mu$m)[12] or on the ${}^4I_{13/2} \rightarrow {}^4I_{15/2}$ transition ($\lambda \cong 1.64\,\mu$m). Due to their interest for biomedical applications, Er:YAG lasers, oscillating on the $\lambda = 2.94\,\mu$m transition, have been subject to much development. The spectroscopy of Er:YAG indicates that, for flash-lamp excitation, the ${}^4I_{11/2}$ upper laser level is pumped via light absorbed from transitions at wavelengths shorter than 600 nm. For diode laser pumping, diode lasers oscillating at $\lambda = 970$ nm (InGaAs/GaAs strained QW) are used. The lifetime of the upper state (~0.1 ms) is much shorter than that of the lower state (~2 ms) and the laser is therefore usually operated in a pulsed regime. Despite this unfavorable lifetime ratio, the laser, particularly when diode pumped, can also operate cw under the usual operating conditions. This possibility arises from the fact that, due to the high Er concentrations used (10–50 at. %), a strong $Er^{3+} - Er^{3+}$ interaction occurs thus leading to an efficient ${}^4I_{13/2} \rightarrow {}^4I_{9/2}$ upconversion transition (see Fig. 2.13c). This process eventually leads to energy being recycled from the ${}^4I_{13/2}$ lower level to the ${}^4I_{11/2}$ upper laser level.

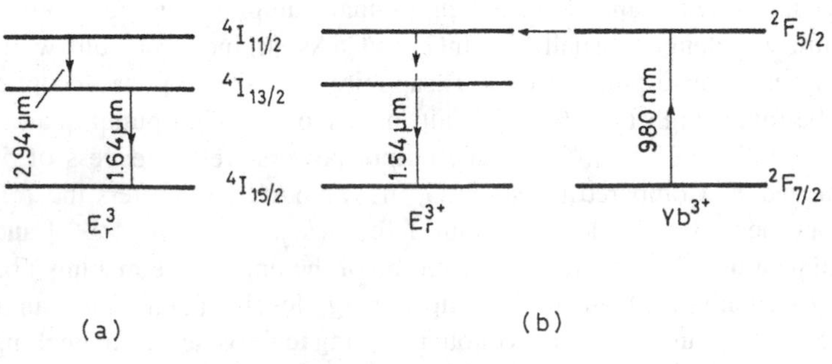

FIG. 9.4. Relevant energy level diagram of: (a) Er:YAG; (b) Yb:Er:phosphate-glass.

In a flash-lamp pumped configuration, the Er:YAG rod has typical dimensions comparable to those of a flash-lamp pumped Nd:YAG laser (e.g. 6 mm diameter × 7.5 cm length) and it is usually pumped using either an elliptical-cylinder or a close-coupling pumping chamber (Fig. 6.1). Flash-lamp pumped Er:YAG lasers giving output energies up to 1 J with repetition rates up to 10 Hz are commercially available. Due to the very strong water absorption spectrum around $\lambda = 2.94\,\mu m$, flashlamp-pumped Er:YAG lasers are particularly interesting for biomedical applications and, in particular, for plastic surgery. The human body in fact consists of ~70% water content and the skin penetration depth of a $2.94\,\mu m$ wavelength Er:YAG laser is around $5\,\mu m$. More recently, diode-pumped Er:YLF lasers, oscillating at $\lambda = 2.8\,\mu m$, have produced cw operation with good optical-to-optical slope efficiency (~35%) and sufficiently high output power (>1 W).

The relevant energy levels of the Yb:Er:phosphate-glass laser are shown in Fig. 9.4b. The laser can be pumped either by a flashlamp[13] or by a cw diode[14] and oscillates on the $^4I_{13/2} \rightarrow {}^4I_{15/2}$ transition ($\lambda = 1.54\,\mu m$). For $1.54\,\mu m$ Er lasers, the Er concentration must be kept low to avoid the detrimental effect, in this case, of the up-conversion mechanism mentioned above. So, for both flash-lamp and diode-laser pumping, the Er absorption coefficients turn out to be too small for efficient laser operation and, to increase the pump absorption, codoping with Yb^{3+} ions (and Cr^{3+} ions for flash-lamp pumping) is used. With diode-pumping around the 980 nm wavelength, the pump power is mainly absorbed by the Yb^{3+} ions ($^2F_{7/2} \rightarrow {}^2F_{5/2}$ transition) and excitation is then efficiently transferred to the $^4I_{11/2}$ Er level by a Förster-type dipole-dipole interaction (see Sect. 2.6.1). The $^4I_{11/2}$ Er level then decays relatively quickly ($\tau \cong 0.1$ ms), by multiphonon relaxation, to the $^4I_{13/2}$ upper laser level. The lifetime of this level in phosphate glass is particularly long ($\tau \cong 8$ ms) thus making it very suitable for laser action. It should be noted that the peak of the gain spectrum of Er:glass is only slightly Stokes-shifted to longer wavelengths compared to the peak of the absorption spectrum, both spectra arising from several transitions between the $^4I_{13/2}$ and the $^4I_{15/2}$ manifolds. Thus the Yb:Er:glass laser behaves almost like a pure three-level laser. Other relevant spectroscopic and optical properties of the Yb:Er:glass laser are shown in Table 9.4.

Q-switched flash-lamp pumped Cr:Yb:Er:glass lasers are used as eye-safe rangefinders. The $1.5\,\mu m$ wavelength is in fact particularly safe for the eye.[15] Diode-pumped cw Yb:Er:glass lasers have notable potential applications in optical communications and for free space optical measurements where eye safety is of concern.

9.2.5. Tm:Ho:YAG[16]

The relevant energy levels of the Tm:Ho:YAG laser are shown in Fig. 9.5. Both Tm^{3+} and Ho^{3+} ions occupy Y^{3+}-ion sites in the lattice. Typical Tm concentrations are rather high (4–10 at. %) while the concentration of Ho ions is an order of magnitude smaller. For flash-lamp pumping, the active medium is also sensitized by Cr^{3+} ions, which substitute for Al^{3+} ions in the YAG crystal. In this case the pump energy is absorbed mainly via the $^4A_2 \rightarrow {}^4T_2$ and $^4A_2 \rightarrow {}^4T_1$ transitions* of the Cr^{3+} ions and then efficiently transferred to the $^3F_4$ level of the

* Note that, from a group-theory view-point, the $^4T_2$ and $^4T_1$ states of Cr^{3+} ion considered here are equivalent to the $^4F_2$ and $^4F_1$ states of Ruby (see Fig. 9.1), the last notation of states thus being an old notation.

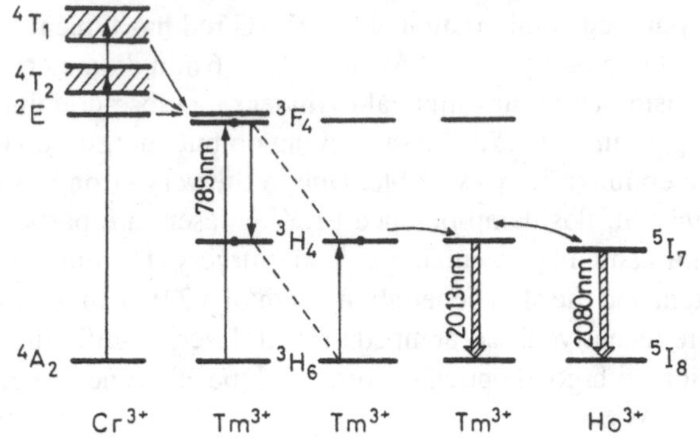

FIG. 9.5. Relevant energy level diagram of Cr:Tm:Ho:YAG system.

Tm^{3+} ion by a Förster-type ion-ion interaction. For cw diode-pumping, the $^3F_4$ level of Tm^{3+} is directly pumped by AlGaAs semiconductor lasers at 785 nm wavelength, so that codoping with Cr^{3+} ions is not needed. For both flash-lamp and diode-laser pumping, excitation to the $^3F_4$ level of Tm^{3+} ion is then followed by a cross-relaxation process, between adjacent ions, of the form $\mathrm{Tm}(^3F_4) + \mathrm{Tm}(^3H_6) \rightarrow 2\mathrm{Tm}(^3H_4)$. This process converts one excited Tm ion in the $^3F_4$ state into two excited Tm ions both in the $^3H_4$ state. For the high Tm concentrations being used, this cross-relaxation process becomes dominant over the $^3F_4$ radiative decay and leads to an overall pump quantum efficiency of nearly 2. A fast spatial migration of the excited energy between Tm ions, again due to Förster-type ion-ion interaction, then occurs until the excitation reaches a Tm ion very near to a Ho ion. In this case, energy transfer to the $^5I_7$ level of Ho occurs followed by laser action on the Ho^{3+} $^5I_7 \rightarrow {}^5I_8$ transition. Laser action actually occurs between the lowest sub-level of the $^5I_7$ manifold to a sublevel, $\sim 462\,\mathrm{cm}^{-1}$ above the ground sub-level, of the $^5I_8$ manifold at $\lambda = 2.08\,\mu\mathrm{m}$ wavelength (quasi-three-level laser). Without Ho-doping, the crystal can lase on the $^3H_4 \rightarrow {}^3H_6$ Tm transition at $\lambda = 2.02\,\mu\mathrm{m}$ wavelength.

When flash-lamp pumped, the active medium is in the form of a rod of the same typical dimensions as those of the Er:YAG rod considered in the previous section, and again pumped in an elliptical cylinder or close-coupling configuration (see Fig. 6.1). Output energies up to 1 J in a $\sim 200\,\mu\mathrm{s}$ long pulse, and a slope efficiency of up to 4% with a repetition rate below 10 Hz, are typical laser operating figures. This laser may find interesting applications in the biomedical field since biological tissue also has a strong absorption around $2\,\mu\mathrm{m}$ (although less strong than at the $2.94\,\mu\mathrm{m}$ wavelength of the Er-laser). When diode-pumped, a longitudinal pumping configuration such as that of Fig. 6.11a is often used. Given the strong absorption coefficient of Tm^{3+} ions at the pump wavelength ($\alpha_p \cong 6\,\mathrm{cm}^{-1}$), the thickness of the active medium is now typically 2–3 mm and the medium is generally cooled to low temperatures (−10 to −40 °C) to reduce the thermal population of the lower laser level.

Eye-safe coherent laser radar systems using Tm:Ho:YAG lasers are used for remote measurements of wind velocity in the atmosphere. This involves a single frequency diode-pumped Tm:Ho laser, used to injection-seed a flash-lamp-pumped, Pockels cell *Q*-switched, Cr:Tm:Ho:YAG slave oscillator.

9.2.6. Fiber Lasers[17]

In a fiber laser the active medium is the core of the fiber, doped with a rare earth. Most commonly, this fiber is of a single mode type and is made of silica. The pump beam is launched longitudinally along the fiber length and may be guided either by the core itself as occurs for the laser mode (conventional single mode fiber laser) or by an inner cladding around this core (double-clad fiber-laser). It should be noted that, although fiber lasers were first demonstrated in the early days of laser development,[18] they have become of practical interest only in recent years after the advent of suitable diode-lasers allowing efficient pumping and of the techniques for fabricating doped single-mode silica fibers.

In a conventional single-mode fiber, the transverse dimensions of both pump, w_p, and laser, w_0, beams are comparable to the core radius a (typically $a \cong 2.5\,\mu$m). Thus both w_p and w_0 are 10–50 times smaller than the corresponding typical values of a bulk device (see Examples 7.4 and 7.5). From either Eq. (6.3.20) or (6.3.25), holding for a 4-level and a quasi-3-level laser respectively, one sees that the threshold pump power, P_{th}, is proportional to $(w_0^2 + w_p^2)$. Hence, for the same values of laser parameters (e.g. γ, σ_e, η_p, τ for a 4-level laser), P_{th} is expected to be smaller in a fiber laser compared to a bulk device by two to three orders of magnitude. Thus, again according to examples 7.4 and 7.5, threshold pump powers well below 1 mW are expected and indeed achieved in fiber lasers. This argument also shows that laser action can be obtained for active media of very low radiative quantum efficiency and hence of very short lifetime τ. On the other hand, the expression for the laser slope efficiency for a 4-level and for a quasi-three-level laser [the two are identical, see Eqs. (7.3.13) and (7.4.10)], turns out to be independent of the upper state lifetime and only dependent on the pump efficiency η_p. Thus, even for a transition with a low radiative quantum efficiency, a high laser slope efficiency can be obtained provided that most of the pump power is absorbed (i.e., $\eta_p \cong 1$). So, transitions which look unpromising in bulk media can still show a low enough threshold in a fiber to be diode-pumped and show a high slope efficiency. It should be noted that an interesting effect occurring at the high pump powers (up to ~100 mW, see Sect. 6.3.1) available from single-transverse-mode diode lasers is ground state depletion. Consider for instance a 4-level laser such as Nd:glass and let F_p be the pump photon flux (assumed, for simplicity, to be uniform in the core) and N_g and N_2 the populations of the ground level and upper laser level, respectively. In the absence of laser action and under cw conditions one can then simply write the following balance equation

$$\sigma_p F_p N_g = N_2/\tau \tag{9.2.1}$$

where σ_p is the pump absorption cross section and τ is the upper state lifetime. Thus, to have $N_g = N_2$ one must have $F_p = (I_p/h\nu_p) = (1/\sigma_p\tau)$, where I_p is the pump intensity and $h\nu_p$ is the energy of a pump photon. According to Fig. 6.8a and Table 9.3, we now take, for a Nd:silica fiber, $\sigma_p = 2.8 \times 10^{-20}\,\text{cm}^2$ and $\tau = 300\,\mu\text{s}$. From the previous expression we get $I_p = (h\nu_p/\sigma_p\tau) \cong 25\,\text{kW/cm}^2$ so that $P_p = I_p A_{\text{core}} \cong 0.25\,\text{mW}$, where A_{core} is the area of the core, taken to be $\approx 10^{-7}\,\text{cm}^2$. Thus, in the example considered, more than half of the ground-state population is raised to the upper laser level at pump powers below 1 mW. Given the ease with which pump-induced depletion of the ground-state population occurs, it follows that typical pump powers can deplete absorption over lengths very much exceeding the small-signal extinction length ($l = 1/\alpha_p$, where α_p is the small-signal absorption coefficient at the pump wavelength). It can be shown in fact that, if the pump power exceeds this saturation

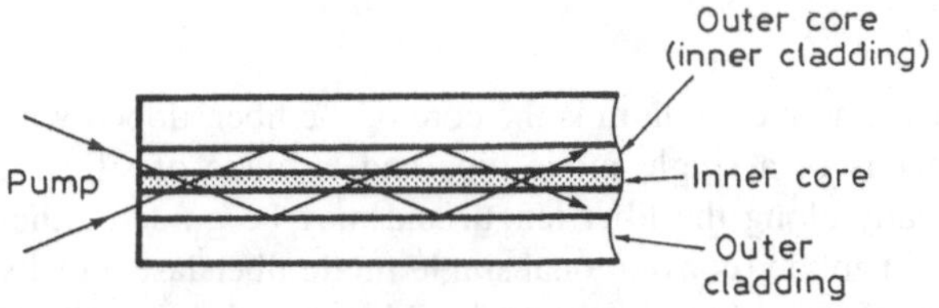

FIG. 9.6. Scheme of cladding pumping.

power P_p by a factor x, the pump power will penetrate the fiber to a distance roughly x times the extinction length. This circumstance must be taken into account in choosing the optimum fiber length.

In conventional end-pumped single-mode fibers, one needs a pump source which is itself diffraction limited. So, for diode pumping, only the single-stripe device of Fig. 6.9a can be used, consequently the pump power is limited to $\sim$100 mW (and $\sim$1 W in the case of MOPA diode lasers) and the output power is limited correspondingly. On the other hand, high-power diode lasers (see Figs. 6.9b and 6.10) show a rather poor beam quality, unsuitable for direct launch into the end of the fiber core. The solution to this problem is provided by the technique referred to as *cladding-pumping* and illustrated in Fig. 9.6. The core, which may be monomode, lies within a lower index inner-cladding which in turn lies within an outer-cladding of yet lower index. Pump light can be end-launched into the inner cladding, with a much less stringent beam-quality requirement compared with launching into the core. This pump light, while propagating in the inner cladding, is then progressively absorbed into the core with an effective absorption coefficient which is smaller than the true absorption coefficient of the core by a factor of the order of the ratio of the inner-cladding area to the core area. Thus, for a given doping of the core, the length of the fiber must be correspondingly increased to allow efficient absorption of the pump power. Provided propagation losses of the pump, in the inner cladding, and of the lasing mode, in the core, are not excessively increased by this increase of fiber length, then efficient pumping by a multimode diode and efficient monomode lasing can be achieved. Thus, the cladding-pumping scheme can provide a very simple means of enhancing the brightness of a (diode) pump source by efficiently converting it to a monomode laser output. Cladding pumped Nd-doped fibers and Yb-doped fibers with output powers of several watts ($4 \div 10$ W) are commercially available and power levels in excess of 30 W have been demonstrated.

As discussed above, the high values of the pump power available via laser pumping enable a considerable fraction of the ground state population, in a conventional monomode fiber, to be raised to some upper level of the active ion. Under this condition, a second pump photon of the same or of different wavelength, can raise this population to a still higher level. From this level laser action can then take place to a lower level so that the energy of the emitted photon is actually higher than pump photon energy (see Fig. 9.7). A laser working on such a scheme, where two or more than two pump photons, of equal or different wavelength, are used, is referred to as an *upconversion laser*. While such schemes have worked with bulk media, they have become much more practical with the advent of fiber lasers exploiting fiber materials of a special kind. In silica fibers, in fact, the main limitations to achieving this upconversion scheme stem from non-radiative decays of the levels involved, usually occurring via

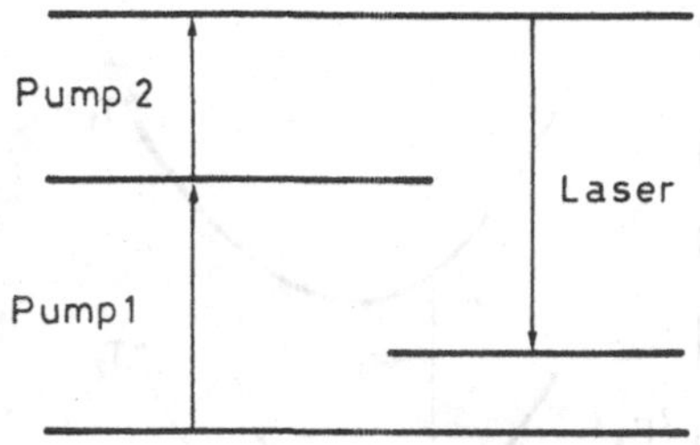

FIG. 9.7. Scheme of an up-conversion laser.

multi-phonon deactivation. As explained in Sect. 2.6.1, the probability of such a decay is a very strong function of the number of phonons that need to be emitted in the process. In this case, the relevant phonon energy is the maximum energy in the phonon spectrum of the host material, since the non-radiative decay rate increases strongly, for a given transition, with increasing values of this energy. For fused silica, this energy correspond to $\sim$1,150 cm^{-1}, which results in rapid nonradiative decay for energy gaps of less then $\sim$4,500 cm^{-1}. A substantial reduction of the rate for nonradiative decay can be obtained by using host materials with lower phonon energy. Among such materials which can be fabricated into fiber, the most widely used one consists of a mixture of heavy metal fluorides, referred to as ZBLAN [an acronym for Zirconium, Barium, Lanthanum, Aluminum and Sodium (Na)], which has a maximum phonon energy of only 590 cm^{-1}. A few years ago, the undoped fiber was already available at an advanced stage of development since, on account of the correspondingly reduced infrared absorption of heavy metal fluorides, it was developed as a possible route to ultra-low-loss fibers. As an example of performance capability, ZBLAN fibers doped with Tm^{3+}, when pumped by three-photons of the same wavelength ($\lambda = 1{,}120$–$1{,}150$ nm), have produced a very efficient upconversion laser in the blue ($\lambda = 480$ nm) giving output powers in excess of 200 mW. ZBLAN fibers doped with Pr^{3+}, when pumped by two photons at $\sim$1,010 nm and $\sim$835 nm, have produced laser action on several transitions from blue to red ($\lambda = 491$, 520, 605, 635 nm) giving e.g. up to $\sim$20 mW of output power in the blue. These figures hold promise for a practical, all-solid-state blue upconversion laser source.

9.2.7. Alexandrite Laser[19]

Alexandrite, chromium-doped chrysoberyl, is a crystal of $BeAl_2O_4$ in which Cr^{3+} ions replace some of the Al^{3+} ions (0.04–0.12 at. %). This laser may be considered as the archetype of what is now a large class of solid-state lasers usually referred to as *tunable solid-state lasers*. The emission wavelength of these lasers can in fact be tuned over a wide spectral bandwidth (e.g. $\Delta\lambda \cong 100$ nm around $\lambda = 760$ nm for alexandrite). Tunable solid-state lasers include, among others, Ti:sapphire and Cr:LISAF, to be considered in next sections, as well as Co:MgF_2 ($\Delta\lambda \cong 800$ nm around $\lambda = 1.9\ \mu$m), Cr^{4+}:YAG ($\Delta\lambda \cong 150$ nm around $\lambda = 1.45\ \mu$m) and Cr^{4+}:Forsterite ($Cr^{4+}: Mg_2SiO_4$, $\Delta\lambda \cong 250$ nm around $\lambda = 1.25\ \mu$m). In this category one could also include color center lasers,[20] which are broadly tunable in the near infrared (at wavelengths ranging between 0.8 to 4 μm). Color center lasers, once rather popular, have declined in popularity and importance due to problems associated with handling

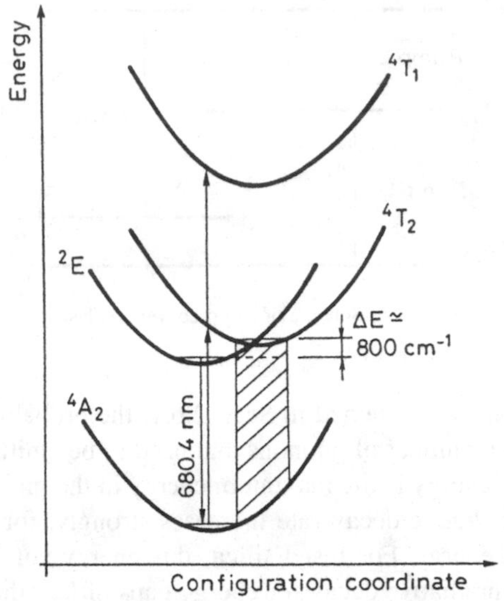

FIG. 9.8. Energy level diagram of alexandrite laser in a configuration-coordinate model.

and storage of the active medium and due to the advent of new competitors in the same wavelength range (i.e., other tunable solid-state lasers or parametric oscillators, to be considered in Chap. 12). For these reasons they will not be considered any further here.

The energy states of the Cr^{3+} ion in $BeAl_2O_4$ are qualitatively similar to those of Cr^{3+} in other hosts with octahedral crystalline field such as, e.g., ruby, which we have already considered. It is therefore important to understand why the alexandrite laser is tunable while ruby laser is not, at least not to anything like the same extent. To help explaining this point, Fig. 9.8 shows a simplified scheme for the energy states as a function of a configuration coordinate of the Cr^{3+} ion (i.e., the distance between this ion and the 6 surrounding O^{2-} anions of the octahedron, see Figs. 3.3 and 9.9a). One sees from Fig. 9.8 that the equilibrium coordinate for both the $^4T_2$ and $^4T_1$ states, due to their symmetry, is shifted to a larger value than that of $^4A_2$ and ^2E states*. As in other Cr^{3+}-doped hosts, the decay between the $^4T_2$ and ^2E states is via a fast internal conversion (decay-time of less than 1 ps) probably due to the level-crossing which occurs between the two states. These two states can therefore be considered to be in thermal equilibrium at all times, and, since the energy difference between the bottom vibrational levels of the $^4T_2$ and ^2E states in alexandrite ($\Delta E \cong 800\,\text{cm}^{-1}$) is only a few kT, an appreciable population will be present in the vibrational manifold of the $^4T_2$ state when the ^2E state has been populated. Invoking the Franck-Condon principle, one sees that the vibronic transitions from the $^4T_2$ state end in empty vibrational levels of the $^4A_2$ state, thus becoming the preferred laser transition. Excitation will then be terminated by phonon decay to the lowest vibrational level of the $^4A_2$ state. Because there is a very large number of vibrational levels involved, the resulting emission is in the form of a broad continuous band ($\lambda = 700$–$800\,\text{nm}$). In keeping with the physical description given above, this type of laser is also referred to as a *phonon terminated laser* or a *vibronic laser*. Note that, in the ruby laser,

* Note that, as already pointed out, the $^4T_2$ and $^4T_1$ states of Cr^{3+} ion, being considered here, are equivalent to the $^4F_2$ and $^4F_1$ states of Ruby (see Fig. 9.1).

laser action takes place between the ^2E and $^4A_2$ states while the phonon terminated transition does not occur because the energy difference between the $^4T_2$ (old $^4F_2$) and ^2E states is much larger ($\Delta E \cong 2,300\,\text{cm}^{-1}$) and hence there is no appreciable population in the $^4F_2$ level. Note also that, in alexandrite, laser action can occur, as for ruby, on the $^2E \to {}^4A_2$ transition (compare Fig. 9.8 with Fig. 9.1). In this case, however, alexandrite operates on a three-level scheme, the threshold is much higher, and the emission wavelength occurs at a somewhat different value ($\lambda = 680.4$ nm).

Pumping of alexandrite takes place mostly through its green and blue absorption bands ($^4A_2 \to {}^4T_2$ and $^4A_2 \to {}^4T_1$ transitions, see Fig. 9.8) which are very similar to those of ruby. The effective values of the lifetime and stimulated emission cross section of the $^4T_2$ upper laser state can be calculated roughly by assuming that the upper level consists of two strongly coupled levels with energy spacing of $\Delta E \cong 800\,\text{cm}^{-1}$, these levels being the lowest vibrational levels of the $^4T_2$ and ^2E states (see Fig. 2.16 and Example 2.11). The upper state lifetime, at $T = 300$ K, then turns out to be $\tau \cong 200\,\mu$s, which is almost the same as that of Nd:YAG. Note that, although the true lifetime of the $^4T_2$ state is much shorter ($\tau_T \cong 6.6\,\mu$s), the effective lifetime is considerably increased by the presence of the long-lived ^2E state ($\tau_E \cong 1.5$ ms, the $^2E \to {}^4A_2$ being spin forbidden), which thus acts as a reservoir for the $^4T_2$ state. Due to the coupling of these two states, the effective cross section of the laser transition ($\sigma_e \cong 0.8 \times 10^{-20}\,\text{cm}^2$) turns out to be considerably smaller than the true value. One should also note that both τ and σ_e are temperature dependent because the relative population of the two states depends on temperature. Table 9.5 provides a summary of some optical and spectroscopic data relevant to the tunable laser transition of alexandrite.

From an engineering viewpoint, alexandrite lasers are similar to Nd:YAG lasers. In fact, alexandrite lasers are usually lamp-pumped in a pumping chamber as in Fig. 6.1 or 6.2 and, although they can operate cw, the much smaller cross-section compared to e.g. Nd:YAG, makes pulsed operation more practical. The laser can operate either in the free-running regime (output pulse duration ~200 μs) or *Q*-switched regime (output pulse duration ~50 ns) and is usually pulsed at relatively high repetition rate (10–100 Hz). Due to the strong increase of effective emission cross section with temperature, the laser rod is usually held at an elevated temperature (50–70 °C). The performances of a pulsed alexandrite laser in terms of output

TABLE 9.5. Optical and spectroscopic parameters, at room temperature, of the most important tunable solid-state laser materials. Note that the density of active ions, N_t, for both Cr:LISAF and Cr:LICAF has been given at ~1% molar concentration of CrF_3 in the melt

	Alexandrite	Ti:sapphire	Cr:LISAF	Cr:LICAF
doping [at. %]	0.04–0.12 at. %	0.1 at. %	up to 15 at. %	up to 15 at. %
N_t [10^{19} ions/cm^3]	1.8–5.4	3.3	10 [~1 at. %]	10 [1 at. %]
Peak wavelength [nm]	760	790	850	780
Tuning range [nm]	700–820	660–1,180	780–1,010	720–840
σ_e [10^{-20} cm^2]	0.8	28	4.8	1.3
$\tau[\mu s]$	260	3.2	67	170
$\Delta\nu_0$ [THz]	53	100	83	64
Refractive indices	$n_a = 1.7367$	$n_o = 1.763$		
	$n_b = 1.7421$	$n_e = 1.755$	$n_e = 1.4$	$n_e = 1.39$
	$n_c = 1.7346$			

vs input energy and slope efficiency are similar to those of a Nd:YAG laser using a rod of the same dimensions. Average powers up to 100 W at pulse repetition rates of ~250 Hz have been demonstrated. Flashlamp-pumped alexandrite lasers have proved to be useful when high average power at $\lambda \cong 700$ nm wavelength is needed (such as in laser annealing of silicon wafers) or where tunable radiation is required (as in the case of pollution monitoring).

9.2.8. Titanium Sapphire Laser[21–23]

The titanium sapphire, Ti:Al_2O_3, laser is the most widely used tunable solid state laser. It can be operated over a broad tuning range ($\Delta\lambda \cong 400$ nm, corresponding to $\Delta\nu_0 \cong 100$ THz), thus providing the largest bandwidth of any laser.

To make Ti:sapphire, Ti_2O_3 is doped into a crystal of Al_2O_3 (typical concentrations range between 0.1 and 0.5% by weight) and Ti^{3+} ions then occupy some of the Al^{3+}-ion sites in the lattice. The Ti^{3+} ion possesses the simplest electronic configuration among the transition ions, only one electron being left in the 3*d* shell. The second 3*d* electron and the two 4*s* electrons of the Ti atom (see Table 9.1) are in fact used for the ionic bonding to the oxygen anions. When Ti^{3+} substitutes for an Al^{3+} ion, the Ti ion is situated at the center of an octahedral site whose 6 apexes are occupied by O^{2-} ions (Fig. 9.9a). Assuming, for simplicity, a field of perfect octahedral symmetry*, the fivefold degenerate (neglecting spin) *d*-electron levels, of an isolated Ti^{3+} ion, are split by the crystal field of the 6 nearest-neighbor oxygen anions into a triply degenerate $^2T_2$ ground state and a doubly degenerate ^2E upper state (Fig. 9.9b). As usual, the notation for these crystals incorporating a transition metal is derived

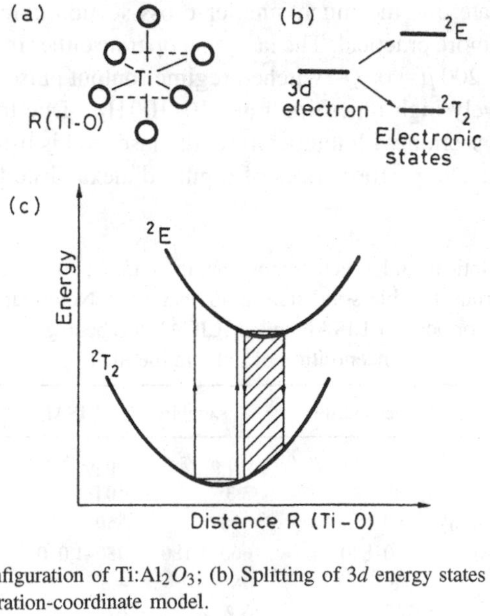

FIG. 9.9. (a) Octahedral configuration of Ti:Al_2O_3; (b) Splitting of 3*d* energy states in an octahedral crystal field. (c) Energy states in a configuration-coordinate model.

* For a more exact treatment, see[21,22]

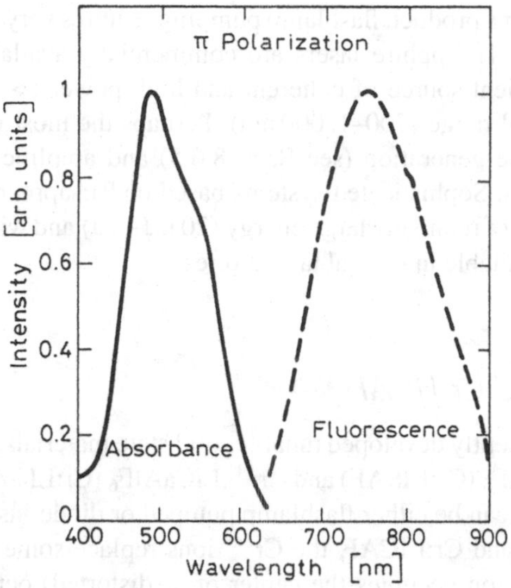

FIG. 9.10. Absorption and fluorescence bands of Ti:sapphire (after reference,[54] Copyright 1983 PennWell Publishing Co., by permission).

from group theory. When spin is also taken into account ($S = 1/2$ for this essentially one electron system), the two states acquire a multiplicity of $2S + 1 = 2$, as denoted by the upper-left suffix attached to each letter. In a configuration coordinate model, where this coordinate is just the Ti-O separation, the two states can be represented as in Fig. 9.9c. Note that the rather strong interaction of the $3d$ electron with the crystal field results in the equilibrium distance being appreciably larger for the upper state compared to the lower state. This circumstance is particularly relevant because it produces absorption and fluorescence bands which are wide and widely separated, as shown in Fig. 9.10. Finally, it should be noted that a particularly relevant feature of the Ti^{3+} ion in an octahedral site is that it only has one excited state (i.e., the ^2E state). This eliminates the possibility of excited state absorption (e.g. as arising from the $^4T_2 \rightarrow {}^4T_1$ transition in alexandrite), an effect that limits the tuning range and reduces the efficiency of many other transition metal lasers.

In accordance with the discussion above and again involving the Franck-Condon principle it follows that laser action takes place from the lowest vibrational level of the ^2E state to some vibrational level of the ground $^2T_2$ state. Some relevant optical and spectroscopic properties of this phonon terminated transition are listed in Table 9.5. Note that the upper-state lifetime ($\tau \cong 3.2\,\mu$s at $T = 300$ K, the radiative lifetime being $\tau_r \cong 3.85\,\mu$s) is much shorter than for alexandrite because there is no lengthening effect arising from a reservoir of population in another excited state, as for alexandrite. The stimulated emission cross section, on the other hand, is much (~40 times) larger than in alexandrite and comparable to that of Nd:YAG. Note the large bandwidth of the laser transition, the largest amongst commonly used solid-state lasers.

CW Ti:sapphire lasers are pumped by the green output of an Ar laser while, in pulsed operation, frequency doubled Nd:YAG or Nd:YLF lasers as well as flashlamps are used. Due to the small value of the $\sigma\tau$ product, flashlamp pumping requires very intense lamps. Nonetheless, flashlamp pumped Ti:sapphire lasers are commercially available. Argon pumped cw lasers provide a convenient source of coherent and high power (>1 W) light which is tunable over a wide spectral range (700–1,000 nm). Perhaps the most important application of a Ti:sapphire laser is the generation (see Sect. 8.6.5) and amplification (see Chap. 12) of femtosecond laser pulses. Sophisticated systems based on Ti:sapphire lasers and Ti:sapphire amplifiers, giving pulses of relatively large energy (20 mJ–1 J) and with femtosecond duration (20–100 fs), are now available in several laboratories.

9.2.9. Cr:LISAF and Cr:LICAF[24,25]

Two of the more recently developed tunable solid state materials, based on Cr^{3+} as active species, are Cr^{3+}:$LiSrAlF_6$ (Cr:LISAF) and Cr^{3+}:$LiCaAlF_6$ (Cr:LICAF). Both materials offer a wide tuning range and can be either flashlamp pumped or diode-laser pumped.

In both Cr:LISAF and Cr:LICAF, the Cr^{3+} ions replace some of the Al^{3+} ions in the lattice and the impurity ion occupies the center of a (distorted) octahedral site surrounded by 6 fluorine ions. Thus, to a first approximation, the general energy level picture, in the configuration-coordinate representation, as presented for Alexandrite also holds for this case (see Fig. 9.8). The corresponding absorption and fluorescence spectra, for the electric field parallel or perpendicular to the *c*-axis of the crystal (LISAF and LICAF are uniaxial crystals), are shown in Fig. 9.11 for Cr:LISAF. Note that the two main absorption bands centered at 650 and 440 nm, respectively, arise from ${}^4A_2 \rightarrow {}^4T_2$ and ${}^4A_2 \rightarrow {}^4T_1$ transitions. Note also that the sharp features superimposed on the ${}^4T_2$ band arise from absorption to the ${}^2E$ and ${}^2T_1$ states (the latter state is not shown in Fig. 9.8). Thus the ${}^2E$ state is now located within

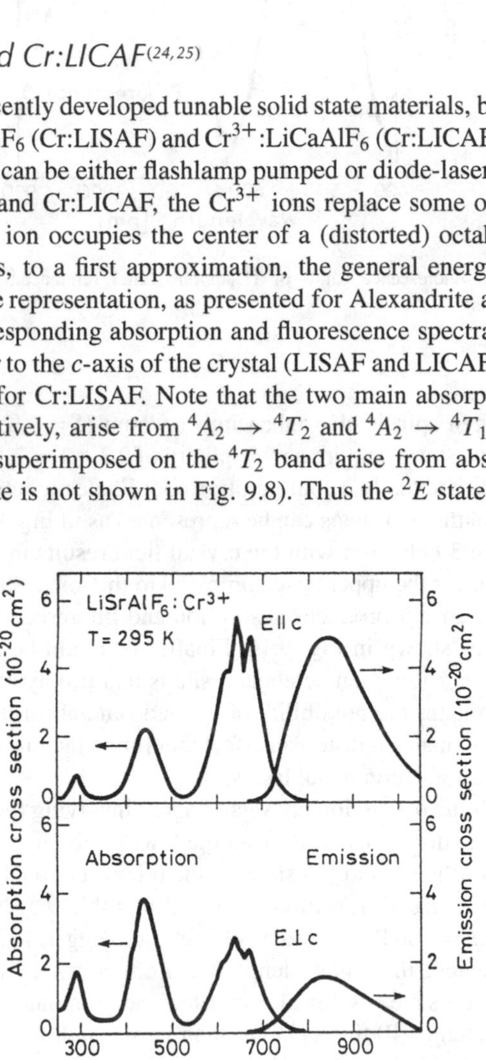

FIG. 9.11. Absorption and fluorescence bands of Cr:LISAF for polarization parallel and perpendicular to the optical *c*-axis of the crystal (after[24] by permission).

the $^4A_2 \rightarrow {}^4T_2$ absorption band which implies that the lowest vibrational level of $^4T_2$ must now be located appreciably below the ^2E state. Due to the fast relaxation between the two states, it then follows that the most heavily populated state is now $^4T_2$, so that the ^2E state does not play any role as an energy reservoir, as for alexandrite. This is also evidenced by the fact that the lifetime of the $^4T_2$ state has been measured to be roughly independent of temperature. Other relevant optical and spectroscopic parameters of the two laser materials are listed in Table 9.5. Note that, amongst the tunable solid state laser materials shown in the Table, Cr:LISAF exhibits the largest value of the $\sigma\tau$ product. Thus, due to its larger values of both the cross section and the $\sigma\tau$ product and its wider tuning range (the tuning range of Cr:LICAF is limited by excited state absorption), Cr:LISAF is generally preferred to Cr:LICAF.

Cr:LISAF has found applications as a flashlamp or diode-pumped laser source providing tunability around 850 nm. The large gain linewidth makes this medium attractive for the generation of femtosecond pulses. For this application, Kerr-lens mode locked Cr:LISAF lasers, end-pumped by GaInP/AlGaInP *QW* laser diodes at 670 nm wavelength, in a configuration such as that shown in Fig. 8.31, have been developed. Large, flashlamp pumped, Cr:LISAF amplifier systems, to amplify femtosecond pulses from either a Ti:sapphire or a Cr:LISAF mode-locked laser, have also been developed. Other potential applications of Cr:LISAF are in tunable systems for pollution monitoring and for spectroscopy.

9.3. DYE LASERS[26]

Dye lasers make use of an active medium consisting of a solution of an organic dye in a liquid solvent such as ethyl or methyl alcohol, glycerol, or water. Organic dyes constitute a large class of polyatomic molecules containing long chains of conjugated double bonds [e.g., $(\text{–C} =)_n$]. Laser dyes usually belong to one of the following classes: (1) Polymethine dyes, which provide laser oscillation in the red or near infrared (0.7–1.5 μm). As an example Fig. 9.12a shows the chemical structure of the dye 3, 3′ diethyl thiatricarbocyanine iodide, which oscillates in the infrared (at a peak wavelength, $\lambda_p = 810$ nm). (2) Xanthene dyes, whose laser operation is in the visible. As an example Fig. 9.12b shows the chemical structure of the widely used rhodamine 6G dye ($\lambda_p = 590$ nm). (3) Coumarin dyes, which oscillate in the blue-green region (400–500 nm). As an example Fig. 9.12c shows the chemical structure of coumarin 2, which oscillates in the blue ($\lambda_p = 450$ nm).

9.3.1. Photophysical Properties of Organic Dyes

Organic dyes usually show wide absorption and fluorescence bands without sharp features, the fluorescence being Stokes-shifted to longer wavelengths than the absorption, a feature reminiscent of the tunable solid state laser materials considered in the previous sections. As an example, Fig 9.13 shows the relevant absorption and emission characteristics of rhodamine 6G in ethanol solution.

To understand the origin of the features shown in Fig. 9.13, we first need to consider the energy levels of a dye molecule. A simple understanding of these levels can be obtained using the so-called free-electron model,[27] which is illustrated here by considering the case

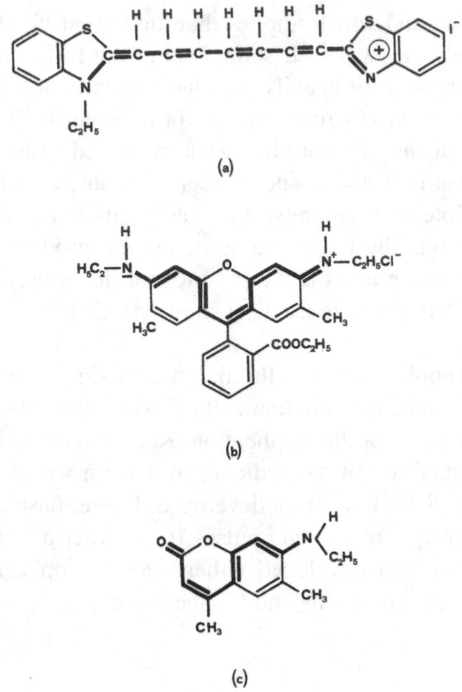

FIG. 9.12. Chemical structure of some common dyes: (a) 3, 3′ diethyl thiatricarbocyanine iodide; (b) rhodamine 6G; (c) coumarin 2. In each case the chromophoric region of the dye is indicated by heavier lines.

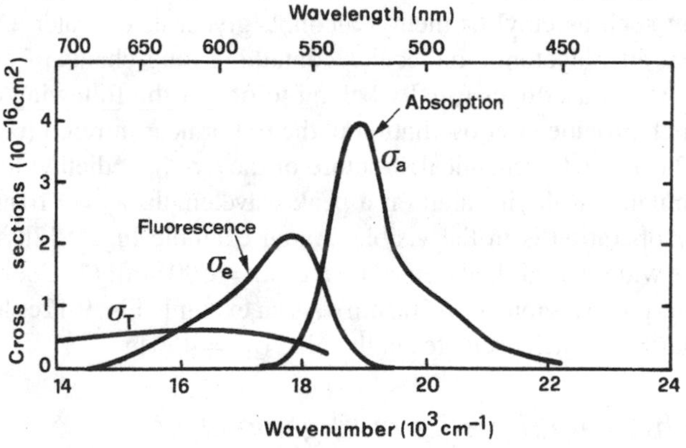

FIG. 9.13. Absorption cross section, σ_a, singlet-singlet stimulated-emission cross section, σ_e, and triplet-triplet absorption cross section, σ_T, for an ethanol solution of rhodamine 6 G.

of the cyanine dye shown in Fig. 9.14a. The π-electrons of the carbon atoms are then seen to form two planar distributions, one above and one below the plane of the molecule (dotted regions in both Fig. 9.14a, b). The π-electrons are assumed to move freely within their planar distributions, limited only by the repulsive potential of the methyl groups at the end of the dye

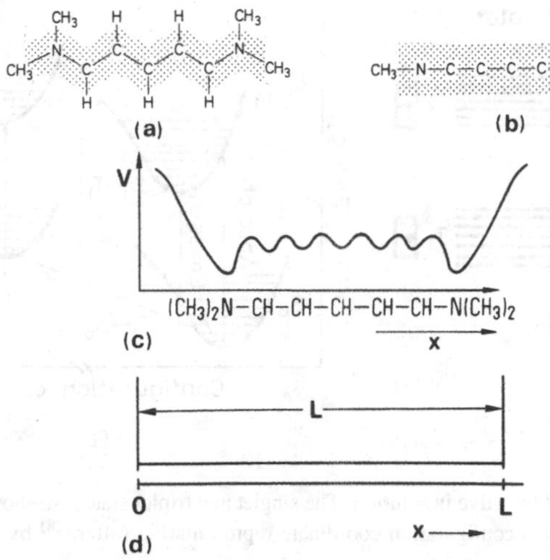

FIG. 9.14. Free-electron model for the electronic energy states of a dye molecule (after,[27] by permission).

chain, and the electronic states of the molecule originate from these electrons. To first order, the energy levels of the electrons are then simply those of a free electron in a potential well of the form shown in Fig. 9.14c. If this well is approximated by a rectangular one (Fig. 9.14d), the energy levels are then known to be given by

$$E_n = \frac{h^2 n^2}{8mL^2} \tag{9.3.1}$$

where n is an integer, m is the electron mass, and L is the length of the well. It is important to remark, at this point, that dye molecules have an even number of electrons in the π-electron cloud*. If we let the number of these electrons be $2N$, the lowest energy state of the molecule will correspond to the situation where these electrons are occupying the lowest N energy levels, each level being occupied by two electrons with opposite spin. This molecular state will thus have a total spin equal to zero and will thus be a singlet state, labeled S_0 in Fig. 9.15. An approximate value for the energy of the uppermost electrons of this state, E_N, is obtained from Eq. (9.3.1) by letting $n = N$. In Fig. 9.15a, the highest occupied level and the next one above it are indicated by two squares one above the other, and the S_0 state thus corresponds to the situation where the lower box is full, having two electrons, and the upper one is empty. The first excited singlet state (labeled S_1 in the figure) corresponds to one of the two highest-lying electron having been promoted, without flipping its spin, to the next level up. The energy of the uppermost electron of this state, E_{N+1}, can be roughly calculated from Eq. (9.3.1) by letting $n = N + 1$. The difference of energy between the S_1 and S_0 states is thus seen to be equal to $E_{N+1} - E_N$, and, according to Eq. (9.3.1) can be shown to decrease with increasing length L of the chain. If the spin is flipped, the total spin is $S = 1$ and the resulting state is a triplet state, labeled T_1 in the figure. Excited singlet, S_2 (not shown in Fig. 9.15), and triplet,

* Molecular systems with unpaired electrons are known as radicals and they tend to react readily, thus forming a more stable system with paired electrons.

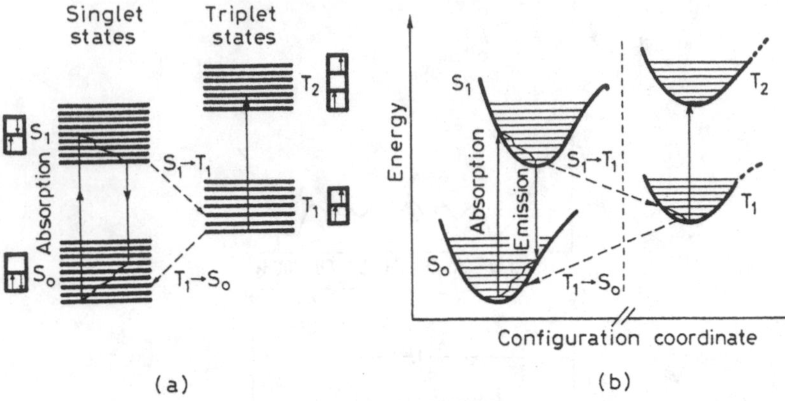

FIG. 9.15. (a) Typical energy levels for a dye in solution. The singlet and triplet states are shown in separate columns. (b) Energy level diagram of a dye, in a configuration-coordinate representation (after,[28] by permission).

T_2, states result when the electron is promoted to the next higher level, and so on. Note that in Fig. 9.15a the corresponding energy levels are indicated by a close set of horizontal lines representing the inclusion of vibrational energy. Note also that, in Fig 9.15b, the energy states and the vibrational structure of a dye molecule is represented as a function of a configuration-coordinate (i.e., a coordinate describing one of the many vibrational modes that a long-chain dye molecule has). Note finally that, due to the large number of vibrational and rotational levels involved and due to the effective line broadening mechanisms in liquids, the rotational-vibrational structure is in fact unresolved at room temperature.

We now look at what happens when the molecule is subjected to electromagnetic radiation. First, we recall that the selection rules require that $\Delta S = 0$. Hence singlet-singlet as well as triplet-triplet transitions are allowed, while singlet-triplet transitions are forbidden. Therefore, the interaction with electromagnetic radiation can raise the molecule from the ground level S_0 to some vibrational levels of the S_1 state, account being taken of the Franck-Condon principle (see Fig. 9.15b) or, more precisely, of the corresponding Franck-Condon factors (see Sect. 3.1.3.). Since the vibrational and rotational structure is unresolved, the absorption spectrum will show a broad and featureless transition as indeed shown in Fig. 9.13 for the case of rhodamine 6G. Note that an important characteristic of dyes is that they have a very large dipole matrix element μ. This is because the π-electrons are free to move over a distance roughly equal to the chain length, L, and, since L is quite large, it follows that μ is also large ($\mu \approx eL$). It then follows that the absorption cross section, σ_a, which is proportional to μ^2, is also large ($\sim 10^{-16}\,\text{cm}^2$, see Fig. 9.13). Once in the excited state, the molecule nonradiatively decays in a very short time ($\tau_{nr} \cong 100\,\text{fs}$, resulting from collisional deactivation) to the lowest vibrational level of the S_1 state (Fig. 9.15b)*. From there it decays radiatively to some vibrational level of S_0 state account being taken again of the Franck-Condon principle (Fig. 9.15b). The fluorescent emission will then take the form of a broad and featureless band, Stokes shifted to the long-wavelength side of the absorption band (see Fig. 9.13). Due to the large value of the dipole moment μ, the stimulated emission cross section is also expected

* More precisely, thermalization among the many rotational-vibrational levels of this state will occur.

to be quite large again ($\sim 10^{-16}\,\text{cm}^2$, see Fig. 9.13). Having dropped to an excited vibrational level of the ground S_0 state, the molecule will then return to the lowest vibrational state by another very fast ($\sim$100 fs) nonradiative decay. Note that, while the molecule is in the lowest level of S_1, it can also decay to the T_1 state. This process is referred to as *intersystem crossing* and, although radiatively forbidden, may occur rather readily via collisions. Similarly, the transition $T_1 \to S_0$ takes place mainly by near-resonant-energy-transfer collisions with species within the solution (e.g., dissolved oxygen) provided that these collisions, according to the Wigner rule (see Sect. 6.4.1.1.), preserve the total spin of the colliding partners. Note finally that, while the molecule is in the lowest level of T_1, it can also absorb radiation to undergo the $T_1 \to T_2$ transition, which is optically allowed. Unfortunately, this absorption tends to occur in the same wavelength region where stimulated emission occurs (see again, for example, Fig. 9.13) and, as we shall see below, it may represent a serious obstacle to laser action.

The three decay processes considered above, occurring from states S_1 and T_1, can be characterized by the following three constants: (1) τ_{sp}, the spontaneous-emission lifetime of the S_1 state. (2) k_{ST}, the intersystem crossing rate (s^{-1}) of the $S_1 \to T_1$ transition. (3) τ_T, the lifetime of the T_1 state. If we let τ be the overall lifetime of the S_1 state, then, according to Eq. (2.6.18), we get

$$\frac{1}{\tau} = \frac{1}{\tau_{sp}} + k_{ST} \qquad (9.3.2)$$

Owing to the large value of the dipole matrix element μ, the radiative lifetime falls in the nanosecond range (e.g. $\tau_{sp} \cong 5$ ns for rhodamine 6G). Since k_{ST}^{-1} is usually much longer (e.g., $\sim$100 ns for rhodamine 6G), it follows that most of the molecules decay from the S_1 state by fluorescence. The fluorescence quantum yield (number of photons emitted by fluorescence divided by number of molecules raised to the S_1 state) is therefore nearly unity. In fact, according to Eq. (2.6.22), one has

$$\phi = \tau/\tau_{sp} \qquad (9.3.3)$$

The triplet lifetime τ_T depends on the dye solution and, particularly, on the amount of dissolved oxygen. The lifetime can range from 10^{-7} s, in an oxygen-saturated solution, to 10^{-3} s or more in a solution that has been deoxygenated.

As a summary, Table 9.6 indicates the typical ranges of various relevant optical and spectroscopic parameters of dye laser media.

9.3.2. *Characteristics of Dye Lasers*

From the discussion above, one can see that these materials have the appropriate characteristics for exhibiting laser action, over the wavelength range of fluorescence, according to a 4-level laser scheme. In fact, the fast nonradiative decay within the excited singlet state S_1 populates the upper laser level very effectively, while the fast nonradiative decay within the ground state is effective in depopulating the lower laser level. It was, however, quite late in the general development of laser devices before the first dye laser was operated (1966),[29,30] and we now look for some reasons for this. One problem which presents itself is the very

TABLE 9.6. Range of optical and spectroscopic parameters of typical dye laser media

Available laser wavelengths [nm]	320÷1,500
Concentration [molar]	$10^{-3}\div10^{-4}$
N_t [10^{19} mol./cm^3]	0.1÷1
σ_e [10^{-16} cm^2]	1÷4
σ_T [10^{-16} cm^2]	0.5÷0.8
$\Delta\lambda$[nm]	25÷30
τ[ns]	2÷5
k_{ST}^{-1}[ns]	≈100
τ_T[s]	$10^{-7}\div10^{-3}$
Refractive index	1.3÷1.4

short lifetime τ of the S_1 state since the required pump power is inversely proportional to τ. Although this is, to some extent, compensated by the comparatively large value of the stimulated emission cross-section, the product $\sigma\tau$ [one should always remember that, for a 4-level laser, the threshold pump power is inversely proportional to $\sigma\tau$, see Eqs. (7.3.12) and (6.3.20) for the space independent and space dependent model, respectively] is still about three orders of magnitude smaller for e.g., rhodamine 6G compared to Nd:YAG. A second problem arises from intersystem crossing. In fact, if τ_T is long compared to k_{ST}^{-1}, then molecules accumulate in the triplet state resulting in absorption at the laser wavelength due to triplet-triplet transition. In fact, it can be readily shown that a necessary condition for laser action is that τ_T is less than some particular value, which depends on other optical parameters of the dye molecule. To obtain this result, let N_2 and N_T be the populations of the upper laser state and of the triplet state respectively. A necessary condition for laser action can then be established by requiring that the gain coefficient, due to stimulated emission, exceeds the intrinsic loss due to triplet-triplet absorption, i.e.,

$$\sigma_e N_2 > \sigma_T N_T \tag{9.3.4}$$

where σ_T is the cross section for triplet-triplet absorption and where the values of both σ_e and σ_T are taken at the wavelength where laser action is considered. In the steady state, the rate of decay of triplet population, N_T/τ_T, must equal the rate of increase due to intersystem crossing $k_{ST}N_2$, i.e.,

$$N_T = k_{ST}\tau_T N_2 \tag{9.3.5}$$

Combining Eqs. (9.3.4) and (9.3.5) we get

$$\tau_T < \sigma_e/\sigma_T k_{ST} \tag{9.3.6}$$

which is a necessary condition for cw laser action [i.e., in a sense equivalent to the condition Eq. (7.3.1) for a simple two-level system]. If this condition is not satisfied, the dye laser can only operate in a pulsed regime. In this case, the duration of the pump pulse must be short enough to ensure that an excessive population does not accumulate in the triplet state. Finally, a third crucial problem comes from the presence of thermal gradients produced in the liquid by the pump. These tend to produce refractive-index gradients and hence optical distortions that can prevent laser action.

Dye lasers can be operated pulsed or, when condition Eq. (9.3.6) is satisfied, also cw. Pulsed laser action has been obtained from very many different dyes by using one of the following pumping schemes: (1) Fast and intense flashlamps, with pulse duration usually less than ~100 μs. (2) Short light pulses from another laser. In both cases, the short pulse duration serves the purpose of producing laser action before an appreciable population has accumulated in the triplet state and before the onset of refractive-index gradients in the liquid.

For flashlamp pumping, linear lamps in an elliptical-cylinder pumping chamber (see Fig. 6.1a) have been used with the liquid containing the active medium flowing through a glass tube placed along the second focal line of the ellipse. To achieve better pumping uniformity and hence more symmetric refractive index gradients, annular flashlamps, consisting of two concentric glass tubes, with the dye solution in a central glass tube, are also used.

For pulsed laser pumping, nitrogen lasers are sometimes used, its UV output beam being suitable for pumping many dyes that oscillate in the visible range*. To obtain more energy and higher average power, the more efficient excimer lasers (in particular KrF and XeF) are being increasingly used as UV pumps, while, for dyes with emission wavelength longer than ~550–600 nm, the second harmonic of a Q-switched Nd:YAG laser ($\lambda = 532$ nm) or the green and yellow emissions of a copper vapor laser are being increasingly used. For these visible pump lasers, the conversion efficiency from pump laser to dye laser output is rather higher (30–40%) than that obtained with UV laser pumping (~10%). Furthermore, dye degradation due to the pump light is considerably reduced. For all the cases considered above where pulsed laser pumping is used, a transverse pump configuration (i.e., direction of the pump beam orthogonal to the resonator axis) is generally adopted (Fig. 9.16). The laser pump beam is focused by the lens L, generally a combination of a spherical and cylindrical lens, to a fine line along the axis of the laser cavity. The length of the line focus is made equal to that of the dye cell (a few millimeters), while the transverse dimensions are generally less than 1 mm. To tune the output wavelength within the wide emission bandwidth of a dye (~30–50 nm),

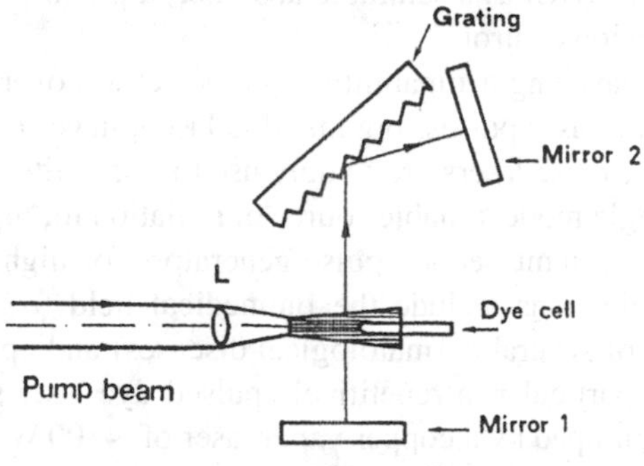

FIG. 9.16. Arrangement for a transversely pumped dye laser. The pumping beam may be that of a nitrogen, excimer, or copper vapor laser, or it may be the second harmonic beam of a Q-switched Nd:YAG laser.

* In this case the pump light is usually absorbed by the $S_0 \rightarrow S_2$ transition of the dye and then rapidly transferred to the bottom of the S_1 state.

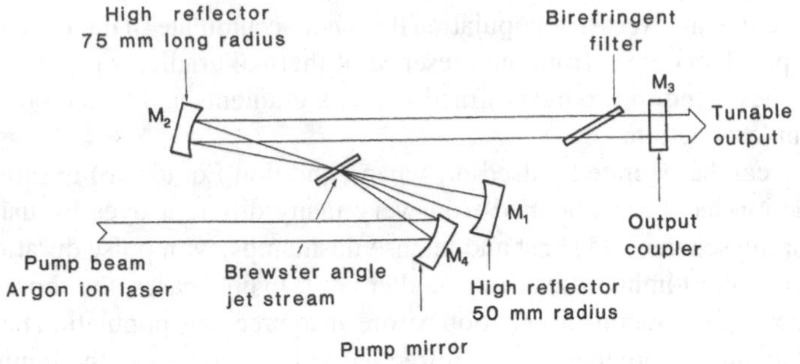

FIG. 9.17. Arrangement of an Ar-ion laser-pumped cw dye laser.

a grazing-incidence diffraction grating is commonly inserted in the laser cavity (see Fig. 9.16) and tuning is achieved by rotating the mirror labeled as mirror 2 in the figure. Grazing incidence is used to increase the resolving power of the grating* and hence to considerably reduce the bandwidth of the emitted radiation (~0.01–0.02 nm). Smaller bandwidths, down to single mode operation, can be obtained by inserting one or more Fabry-Perot etalons, as discussed in Sect. 7.8.2.1.

For continuous laser pumping, Ar^+ lasers (and sometimes also Kr^+ lasers) are often used. To achieve a much lower threshold, as required for cw pumping, the near-longitudinal pumping configuration of Fig. 9.17 is now used. The liquid dye medium is in the form of a thin jet stream (~200 μm thickness) freely flowing in a plane orthogonal to the plane of the figure and inclined at Brewster's angle relative to the dye-laser beam direction. Accordingly, the laser beam is linearly polarized with its electric field in the plane of the figure. Both pump and laser beams have their waist in the jet stream with similar, very small, spot sizes (~10 μm). For laser tuning, a birefringent filter may be inserted within the laser cavity. To achieve single longitudinal mode operation, a birefringent filter and generally two Fabry-Perot etalons in an unidirectional ring cavity are often used (see Fig. 7.25). For femtosecond pulse generation a colliding pulse mode locked (CPM) laser configuration is generally used (see Fig. 8.29). To achieve the shortest pulse duration (~25 fs for a combination of a solution of rhodamine 6G as active medium and of DODCI as a saturable absorber) a prism-pair is also inserted within the laser cavity for dispersion control.

By virtue of their wavelength tunability, wide spectral coverage, and the possibility of generating femtosecond laser pulses, organic dye lasers have found an important role in many fields. In particular, these lasers are widely used in scientific applications, either as a narrow band, down to single mode, tunable source of radiation for high-resolution frequency-domain spectroscopy, or as femtosecond-pulse generators for high resolution time-domain spectroscopy. Other applications include the biomedical field (e.g., treatment of diabetic retinopathy or treatment of several dermatological diseases) and applications in the field of laser photochemistry. In particular, a repetitively pulsed dye laser system, made of 20 dye lasers each transversely pumped by a copper vapor laser of ~100 W average power, has been built for laser isotope separation of ^{235}U.

* The resolving power, $\nu/\Delta\nu$, $\Delta\nu$ being the resolved bandwidth, turns out to be just equal to the number of lines of the diffraction grating illuminated by the laser beam. At grazing incidence, this number increases and thus the resolving power also increases.

9.4. SEMICONDUCTOR LASERS[31,32]

Semiconductor lasers represent one of the most important class of lasers in use today, not only because of the large variety of direct applications in which they are involved but also because they have found a widespread use as pumps for solid state lasers. These lasers will therefore be considered at some length here.

Semiconductor lasers require, for the active medium, a direct gap material and, accordingly, the normal elemental semiconductors (e.g., Si or Ge) cannot be used. The majority of semiconductor-laser materials are based on a combination of elements belonging to the third group of the periodic table (such as Al, Ga, In) with elements of the fifth group (such as N, P, As, Sb) (*III–V compounds*). Examples include the best known, GaAs, as well as some ternary (e.g. AlGaAs, InGaAs) and quaternary (e.g., InGaAsP) alloys. The cw laser emission wavelength of these III–V compounds generally ranges between 630–1,600 nm. Quite recently, however, very interesting InGaN semiconductor lasers, providing cw room-temperature emission in the blue (~410 nm), have been developed and promise to become the best candidates for semiconductor laser emission in the very important blue-green spectral region. Semiconductor laser materials are not limited to III–V compounds, however. For the blue-green end of the spectrum we note that there are wide-gap semiconductors using a combination between elements of the second group (such as Cd and Zn) and of the sixth group (S, Se) (*II–VI compounds*). For the other end of the e.m. spectrum, we mention semiconductors based on some *IV–VI compounds* such as Pb salts of S, Se, and Te, all oscillating in the mid-infrared (4 μm–29 μm). Due to the small band gap, these last lasers reuire cryogenic temperatures, however. In the same wavelength range, we thus mention the recent invention of the *quantum cascade* laser,[33] which promises efficient mid infrared sources without requiring cryogenic temperatures.

9.4.1. Principle of Semiconductor Laser Operation

The principle of operation of a semiconductor laser can be simply explained with the help of Fig. 9.18, where the semiconductor valence band, V, and conduction band, C, separated by the energy gap, E_g, are indicated. For simplicity, let us first assume that the semiconductor is held at $T = 0$ K. Then, for a non-degenerate semiconductor, the valence band will be completely filled with electrons while the conduction band will be completely empty (see Fig. 9.18a, in which the energy states belonging to the hatched area are completely filled by

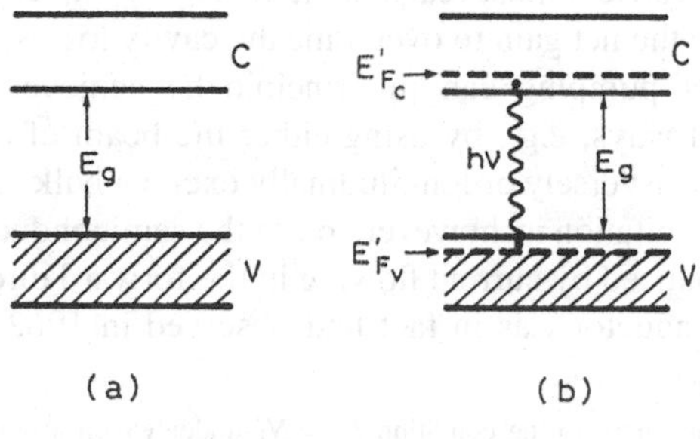

FIG. 9.18. Principle of operation of a semiconductor laser.

electrons). Suppose now that some electrons are raised from the valence band to the conduction band by a suitable pumping mechanism. After a very short time (~1 ps), the electrons in the conduction band drop to the lowest unoccupied levels of this band, and, meanwhile, any electron near the top of the valence band also drops to the lowest unoccupied levels of this band, thus leaving holes at the top of the valence band (Fig. 9.18b). This situation can be described by introducing the quasi-Fermi levels, E'_{F_c}, for the conduction band and, E'_{F_v}, for the valence band (see Sect. 3.2.3.). At $T = 0\,\mathrm{K}$ they define, for each band, the energy below which states are fully occupied by electrons and above which, states are empty. Light emission can now occur when an electron, of the conduction band, falls back to the valence band recombining with a hole. This, so-called recombination-radiation process, is the process by which radiation is emitted in light emitting diodes (LED). Given the appropriate conditions, however, a process of stimulated emission of this recombination radiation, thus leading to laser action, can occur. It was shown in Sect. 3.2.5. that the condition for a photon to be amplified rather than absorbed by the semiconductor is simply given by [see Eq. (3.2.39)]

$$E_g \le h\nu \le E'_{F_c} - E'_{F_v} \tag{9.4.1}$$

In the simple case where $T = 0\,\mathrm{K}$, this condition can be readily understood from Fig. 9.18b, since the non-hatched area in the valence band corresponds to states which are empty, and a conduction band electron can only fall into an empty state of the valence band. However, the detailed treatment of Sect. 3.2.5. shows that condition of Eq. (9.4.1) in fact holds for any temperature and simply means that, for the range of transition energy $h\nu$ defined by Eq. (9.4.1), the gain arising from stimulated emission exceeds the absorption. To achieve condition of Eq. (9.4.1) one must, of course, have $E'_{F_c} - E'_{F_v} \ge E_g$. It is important at this point to realize that the values of both E'_{F_c} and E'_{F_v} depend on the intensity of the pumping process, i.e. on the number density, N, of electrons raised to the conduction band (see Fig. 3.15). Actually $E'_{F_c} = E'_{F_c}(N)$ increases while $E'_{F_v} = E'_{F_v}(N)$ decreases as N is increased. Thus, to obtain $E'_{F_c} - E'_{F_v} \ge E_g$ i.e., to have gain exceeding absorption losses, the electron density N must exceed some critical value established by the condition

$$E'_{F_c}(N) - E'_{F_v}(N) = E_g \tag{9.4.2}$$

The value of the injected carrier density which satisfies Eq. (9.4.2) is referred to as the carrier density at transparency*, N_{tr}. If now the injected carrier density is larger than N_{tr}, the semiconductor will exhibit a net gain and, if this active medium is placed in a suitable cavity, laser action can occur if this net gain is sufficient to overcome the cavity losses. Thus, to obtain laser action, the injected carriers must reach some threshold value, N_{th}, larger than N_{tr} by a sufficient margin to allow the net gain to overcome the cavity losses.

Semiconductor laser pumping can in principle be achieved, and indeed has been achieved, in a number of ways, e.g., by using either the beam of another laser, or an auxiliary electron beam, to transversely or longitudinally excite a bulk semiconductor. By far the most convenient way of excitation is, however, to use the semiconductor laser in the form of a diode with excitation produced by current flowing in the forward direction of the junction.[34] Laser action in a semiconductor was in fact first observed in 1962 by using a *p-n* junction

* Condition (9.4.2) is thus equivalent to the condition $N_2 = N_1$ under which a non-degenerate two level system becomes transparent

diode, the demonstration being made almost simultaneously by four groups,[35–38] three of which were using GaAs. The devices developed during the early stage of semiconductor laser research made use of the same material for both the p and n sides of the junction and are therefore referred to as homojunction lasers. The homojunction laser is now only of historical importance, since it has been essentially superseded by the double heterostructure (DH) laser where the active medium is sandwiched between p and n materials which are different from the active material. Homojunction lasers could in fact operate cw only at cryogenic temperatures ($T = 77$ K), while it was only after the invention of the heterojunction laser that it became possible to operate semiconductor lasers cw at room temperature. This development occurred 7 years after the invention of the homojunction laser (1969)[39–41] and opened up the way to the great variety of applications in which semiconductor lasers are nowadays used. Homojunction semiconductor lasers will nevertheless be discussed briefly in the next section since this discussion helps to understand the great advantages offered by the DH lasers.

9.4.2. The Homojunction Laser

In the homojunction laser, the pumping process is achieved in a p-n junction where both p-type and n-type regions, being of the same material (e.g., GaAs), are in the form of a degenerate semiconductor. This means that the donor and acceptor concentrations are so large ($\approx 10^{18}$ atoms/cm^3) that the Fermi levels fall in the valence band for the p type, E_{F_p}, and in the conduction band for the n type, E_{F_n}. When a junction is formed, and if no voltage is applied, the band structure will be as shown in Fig. 9.19a, where the two Fermi energies are seen to be the same. When a forward bias voltage V is applied, the band structure becomes as shown in Fig. 9.19b and the two Fermi levels become separated by $\Delta E = eV$. We see from this figure that, in the junction region, electrons are injected into the conduction band (from the n-type region) while holes are injected into the valence band (from the p-type region). Thus, under appropriate values of current density, the transparency condition and then the laser threshold condition can be reached. Actually, one of the main limitations of this device comes from the very small potential barrier that an electron, in the conduction band, encounters when it reaches the p-side of the junction. The electron can then penetrate into the p-type material where it becomes a minority carrier thus recombining with a hole. The penetration

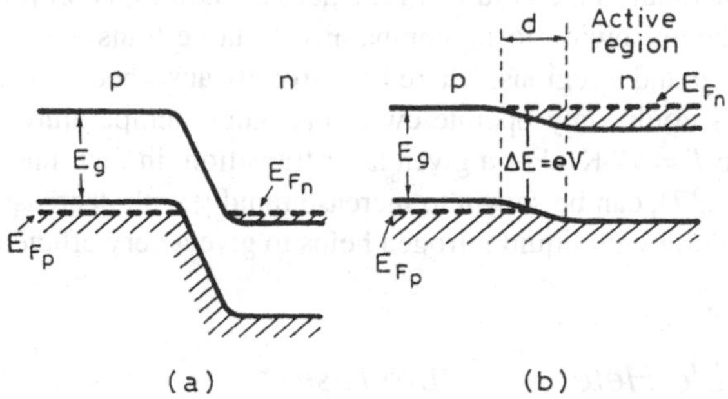

FIG. 9.19. Band structure of a p-n junction semiconductor laser with zero voltage, (a), and forward voltage, (b), applied to the junction.

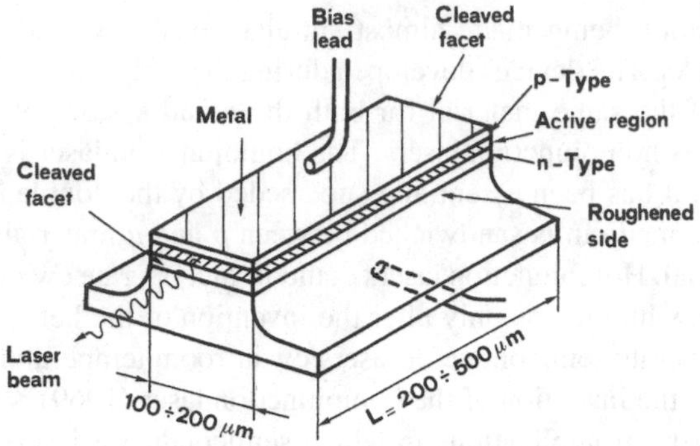

FIG. 9.20. Typical broad-area *p-n* homojunction laser.

depth, d, of the electron will then be given, according to diffusion theory, by $d = \sqrt{D\tau}$, where D is the diffusion coefficient and τ is the electron lifetime, as established by electron-hole recombination. In GaAs, $D = 10\,\mathrm{cm}^2/\mathrm{s}$ and $\tau \cong 3$ ns so that $d \approx 1\ \mu$m which shows that the active region is quite thick being limited by the diffusion length d rather then by the thickness of the depletion layer ($\approx$0.1 μm).

A typical configuration of a p-n junction laser is shown in Fig. 9.20, the shaded region corresponding to the active layer. It is seen that the diode dimensions are very small (some hundreds of microns). To provide feedback for laser action, two parallel end faces are prepared, usually by cleavage along crystal planes. Often the two surfaces are not provided with reflective coatings. In fact, since the refractive index of a semiconductor is very large (e.g., $n = 3.6$ for GaAs), there is already a sufficient high reflectivity ($\sim$32% for GaAs) from the Fresnel reflection at the semiconductor-air interface. Note that, as mentioned earlier, the thickness of the active region in the direction perpendicular to the junction is $d \approx 1\ \mu$m. Because of diffraction, however, the transverse dimension of the laser beam in this direction ($\approx$5 μm) is significantly larger than the active region.

A homojunction laser has a very high threshold current density at room temperature ($J_{th} \cong 10^5\ \mathrm{A/cm}^2$) which prevents the laser from operating cw at room temperature (without suffering destruction in a very short time!). There are two main reasons for this high threshold value: (1) The thickness of the active region ($d \approx 1\ \mu$m) is quite large and the threshold current, being proportional to the volume of the active medium, is proportional to this thickness. (2) The laser beam, owing to its comparatively large transverse dimensions, extends considerably into the p and n regions, where it is strongly absorbed. Given the above reasons, homojunctions lasers could only operate cw at cryogenic temperatures (typically at liquid nitrogen temperature $T = 77$ K). For a given laser transition, in fact, the semiconductor gain, according to Eq. (3.2.37), can be shown to increase rapidly with decreasing temperature and, also, contact of the diode with liquid nitrogen helps to give a very efficient cooling.

9.4.3. The Double-Heterostructure Laser

The limitations discussed in the previous section prevented any widespread use of semiconductor lasers until, first, the single-heterostructure, and, immediately after, the

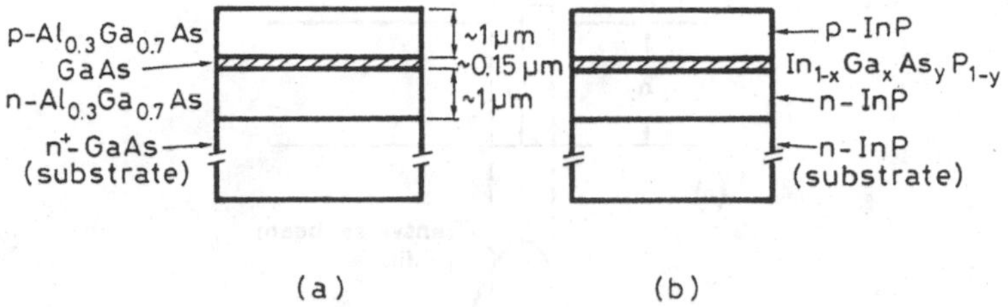

FIG. 9.21. Schematic diagram of a double-heterostructure where the active medium (hatched area) consists of GaAs, (a), and InGaAsP, (b).

double-heterostructure lasers were introduced. We will limit our discussion here to the latter type of laser structure since it is the only one that is now in common use.

Two examples of a double-heterostructure, where the active medium is a thin layer (0.1–0.2 μm) of either GaAs or of the quaternary alloy InGaAsP, are shown in Fig. 9.21a and b respectively. For the two cases considered, the p and n sides are made of $Al_{0.3}Ga_{0.7}As$ and InP, respectively. When properly optimized (see Fig. 9.23), the room-temperature threshold current-density of such a diode structure can be reduced by about two orders of magnitude (i.e., to $\sim 10^3\ A/cm^2$) compared to the corresponding homojunction devices, thus making cw room temperature operation feasible. This strong reduction of threshold current density is due to the combined effect of three circumstances: (1) The refractive index of the active layer n_1 (e.g., $n_1 = 3.6$ for GaAs) is significantly larger than that, n_2, of the p-side and n-side cladding-layers (e.g., $n_2 = 3.4$ for $Al_{0.3}Ga_{0.7}As$), thus providing a guiding structure (see Fig 9.22a). This means that the laser beam will now be mostly confined to the active layer region, i.e., where the gain exists (*photon-confinement*, see Fig. 9.22b). (2) The band gap E_{g_1} of the active layer (e.g., $E_{g_1} \cong 1.5$ eV in GaAs) is significantly smaller than that, E_{g_2}, of the cladding layers* (e.g., $E_{g_2} \cong 1.8$ eV for $Al_{0.3}Ga_{0.7}As$). Energy barriers are thus formed at the two junction planes thus effectively confining the injected holes and electrons within the active layer (*carrier-confinement*, see Fig. 9.22c). Thus, for a given current density, the concentration of holes and electrons in the active layer is increased and therefore the gain is increased. (3) Since E_{g_2} is appreciably larger than E_{g_1}, the laser beam, which has a frequency $\nu \cong E_{g_1}/h$, is much less strongly absorbed in the tails of the beam profile (see Fig. 9.22b) by the cladding layers, the loss arising, in this case, only from free-carriers (*reduced absorption*).

To form a double heterostructure, thus taking advantage of all its favorable properties, a very important requirement must be fulfilled, namely that the lattice period of the active layer must be equal (to within $\sim 0.1\%$) to that of the cladding layers†. In fact, if this condition is not fulfilled, the resulting strain at the two interfaces will result in misfit dislocations being produced there, each dislocation acting as a rather effective center for electron-hole nonradiative recombination. For the GaAs/AlGaAs structure, this requirement of lattice matching does not constitute a limitation because the lattice periods of GaAs (0.564 m) and AlAs (0.566 m) are

* It is a general rule for all III–V compounds that, any change in composition that produces a change, in a given sense e.g. a *decrease*, of band gap also produces a change, in the opposite sense i.e. an *increase*, of the refractive index.

† All III–V compounds crystallize in the cubic structure.

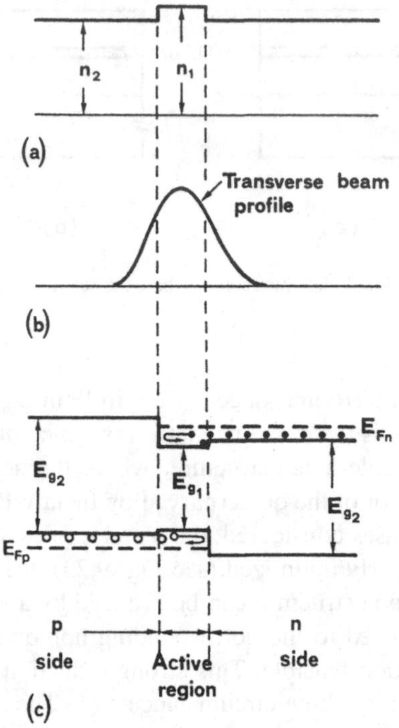

FIG. 9.22. (a) Refractive index profile, (b) transverse beam profile, and (c) band structure (very schematic) of a double-heterostructure diode-laser.

very close in value (the atomic radii of Ga and Al are, in fact, almost the same). In the case of the quaternary compound $In_{1-x}Ga_xAs_yP_{1-y}$, the alloy can be lattice matched to InP for a specific y/x ratio, as one can appreciate from the following argument: suppose that, starting with InP for the active layer, some fraction, x, of Ga is added, substituting for some In in the lattice (which hence becomes In_{1-x}). Since the radius of Ga is *smaller* (by ~19 pm) than that of In, the lattice period of the $In_{1-x}Ga_xP$ will be decreased compared to InP. Suppose now that some fraction, y, of As (As_y) substitutes for P (thus becoming P_{1-y}). Since the radius of As is now *larger* (by ~10 pm) than that of P, this addition will tend to increase the lattice period. So, if the y/x ratio of the two substituents has an appropriate value, the two effects will cancel each other thus resulting in $In_{1-x}Ga_xAs_yP_{1-y}$ being lattice matched to InP. This lattice-matching condition turns out to be given by $y \cong 2.2\,x$. Upon changing x, while keeping the y/x ratio equal to the lattice-matching value, the semiconductor band-gap and hence the emission wavelength can be changed. In this way the emission wavelength of $In_{1-x}Ga_xAs_yP_{1-y}$ can be varied between 1,150 and 1,670 nm, for cw room temperature operation, thus encompassing the so-called second (~1300 nm) and third (~1550 nm) transmission windows of silica optical fibers.

Experimental and theoretical plots of the threshold current density, J_{th}, vs thickness, d, of the active layer, for a DH GaAs laser are shown in Fig. 9.23.[42] Note that, as d decreases, J_{th} first decreases then reaches a minimum value ($J_{th} \cong 1\ \text{kA/cm}^2$ for $d \cong 0.1\ \mu\text{m}$) and thereafter increases. To understand this behavior, we need first to relate the threshold current density, J_{th}, to the threshold carrier density, N_{th}. We begin by defining R_p as the rate at which electrons (and holes) are injected into unit volume of the active layer. We also let η_i, usually referred to as the *internal quantum efficiency*, be the fraction of the carriers which recombine radiatively in the layer, the remaining fraction undergoing nonradiative electron-hole recombination mostly at the junction boundaries. The quantity η_i can also be looked upon as the effective fraction of the injected carriers, while the remaining fraction can be considered as not having been injected into the active region at all. For a given current density, J, flowing through the junction, R_p is then readily seen to be given by $R_p = \eta_i J/ed$ where e is the electron charge and d is the thickness of the active layer. Under steady state conditions, a simple balance condition gives the corresponding expression for the carrier density N as $N = R_p \tau_r$, where τ_r is the radiative recombination time (given the assumptions made before, all carriers recombine radiatively in the active layer). From the previous two expressions one gets $J = edN/\eta_i \tau_r$ so that, at threshold, one has:

$$J_{th} = \left(\frac{ed}{\eta_i \tau_r} \right) N_{th} \tag{9.4.3}$$

With the help of Eq. (9.4.3) we can now qualitatively understand the relevant features of Fig. 9.23. We first note that, for sufficiently large values of d, the threshold carrier density, N_{th}, turns out to be almost the same as the transparency density, N_{tr} (see Example 9.1) and hence it is a constant. Equation (9.4.3) then predicts a linear relation between J_{th} and d as indeed observed in Fig. 9.23 for sufficiently large values of d (larger than ~0.15 μm). However, when the thickness d becomes sufficiently small, the confinement action of the active layer, indicated in Fig. 9.22b, will no longer be so effective and the beam will extend considerably into the p and n sides of the junction. This situation will result in a reduction of the effective gain and, at the same time, an increase of losses experienced in the cladding layers, both of which effects lead to a strongly increased N_{th}. So, at sufficiently small values of d, J_{th} is expected to increase as d is decreased.

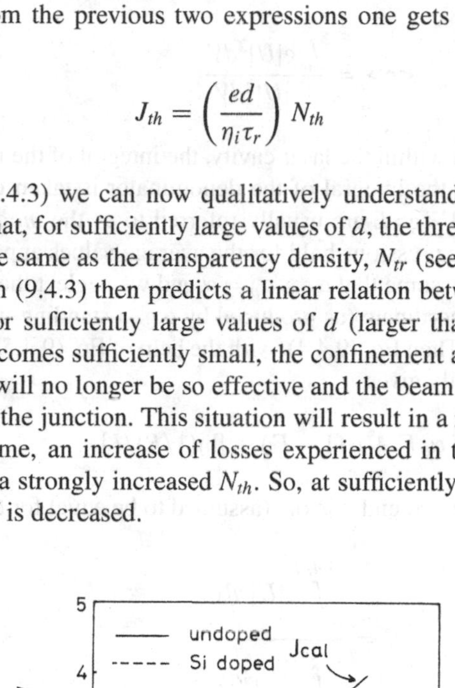

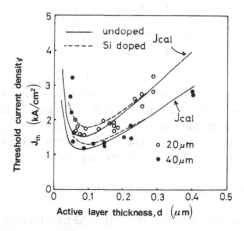

FIG. 9.23. Calculated (*continuous* and *dashed lines*) and experimental (*open* and *closed circles*) values of the threshold current density, J_{th}, vs active layer thickness, d, for a 300 μm long AlGaAs DH laser. *Closed* and *open circles* represent data for a 40 μm and 20 μm stripe width, respectively. The theoretical curves, J_{cal}, refer to the cases of "undoped" and low Si-doped active layers (after[42] by permission).

Example 9.1. *Carrier and current densities at threshold for a DH GaAs laser* Since the laser field is space-dependent, the threshold condition must be derived, as in previous examples (see e.g., Sect. 6.3.4.), by setting the condition that the spatially averaged gain must equal the spatially averaged losses. Thus, in general terms, we write

$$<g>L = <\alpha_a>L + <\alpha_n>L + <\alpha_p>L + \gamma_m \qquad (9.4.4)$$

where L is the length of the active medium, g is the gain coefficient, α_a is the scattering loss of the active layer, α_n and α_p are the losses in the n and p sides of the cladding, respectively, and γ_m is the mirror loss. The average values appearing in Eq. (9.4.4) are to be taken over the field intensity distribution, so that, for example, the average gain is given by

$$<g> = \frac{\int_a g|U|^2 dV}{\int_c |U|^2 dV} \qquad (9.4.5)$$

where $U(x, y, z)$ is the field distribution within the laser cavity, the integral of the numerator is taken over the volume of the active medium and the integral of the denominator is taken over the whole volume of the cavity. The quantities $<g>$ and $<g>L$ are usually referred to as the *modal gain coefficient* and the *modal gain*, respectively. Similar expressions hold for the average values appearing in the right hand side of Eq. (9.4.4). For simplicity, we assume that $\alpha_n \cong \alpha_p = \alpha$ and we neglect the spatial variation of the cavity field along the longitudinal z-coordinate (as produced by e.g. a standing wave pattern) and along the coordinate parallel to the junction. Then Eq. (9.4.4), with the help of Eq. (9.4.5) and the corresponding expressions for $<\alpha_a>$ and $<\alpha_p>$, readily gives

$$g\Gamma = \alpha_a\Gamma + \alpha(1 - \Gamma) + [\ln(1/R)/L] \qquad (9.4.6)$$

where R is the power reflectivity of the two end mirrors (assumed to be equal for the two mirrors) and

$$\Gamma = \frac{\int_{-d/2}^{+d/2} |U|^2 dx}{\int_{-\infty}^{+\infty} |U|^2 dx} \qquad (9.4.7)$$

where x is the coordinate along the direction orthogonal to the junction. The quantity Γ represents the fraction of the beam power which is actually in the active layer and is usually referred to as the *beam confinement factor*. According to the discussion in Sect. 3.2.5, we can now approximate g as $g = \sigma (N - N_{tr})$ where σ is the differential gain coefficient and N_{tr} is the carrier density at transparency. If we now further assume, for simplicity, $\alpha_a = \alpha$, then Eq. (9.4.6) simplifies to

$$\sigma\Gamma(N_{th} - N_{tr}) = \alpha + [\ln(1/R)/L] = \gamma/L \qquad (9.4.8)$$

where N_{th} is the threshold carrier density and $\gamma = \alpha L + \ln(1/R)$ is the total loss per pass. From Eq. (9.4.8) we finally get the desired expression for the threshold carrier density as

$$N_{th} = (\gamma/\sigma L\Gamma) + N_{tr} \qquad (9.4.9)$$

To proceed with the calculation, we now need to evaluate the confinement factor Γ. A fairly accurate and simple expression turns out to be given by given by[43]

$$\Gamma \cong D^2/(2+D^2) \tag{9.4.10}$$

where

$$D = 2\pi(n_1^2 - n_2^2)^{1/2}d/\lambda \tag{9.4.11}$$

is the normalized thickness of the active layer (we recall that n_1 and n_2 are the refractive indices of the active medium and cladding layers, respectively). If we now take $n_1 = 3.6, n_2 = 3.4$, and $\lambda = 850$ nm (as appropriate for a GaAs laser) and we make the calculation for $d = 0.1\ \mu$m, we obtain $D \cong 0.875$ and hence $\Gamma \cong 0.28$. To obtain a numerical estimate of the corresponding value of N_{th}, we take the reflectivity of the two end faces equal to that of the uncoated surfaces ($R \cong 32\%$), and assume a loss coefficient $\alpha \cong 10\ \text{cm}^{-1}$ and a cavity length of $L = 300\ \mu$m. We obtain $\gamma = \ln(1/R) + \alpha L \cong 1.44$. If we now take (see Table 3.1) $\sigma = 1.5 \times 10^{-16}\ \text{cm}^2$ and $N_{tr} = 2 \times 10^{18}$ carriers/cm^3, we obtain from Eq. (9.4.9)

$$N_{th} = (1.14 + 2) \times 10^{18}\ \text{carriers/cm}^3 \tag{9.4.12}$$

where, for convenience, the numerical values of the two terms appearing in the right hand side of Eq. (9.4.9) have been left separated. Equation (9.4.12) thus shows that, in this case, the first term, i.e., the carrier density needed to overcome cavity losses, is comparable to N_{tr}.

The threshold current density is now readily obtained by substituting Eq. (9.4.9) in Eq. (9.4.3). We get

$$J_{th} = \left(\frac{ed}{\eta_i \tau_r}\right)\left[\left(\frac{\gamma}{\sigma L \Gamma}\right) + N_{tr}\right] \tag{9.4.13}$$

We have seen that, for sufficiently large values of d, N_{tr} is the dominant term in the square bracket of Eq. (9.4.13). In this case J_{th} is expected to be proportional to d, as indeed shown in Fig. 9.23 for d larger than ~0.15 μm, and most of the threshold pump current is just used to reach the semiconductor transparency condition. When d becomes very small, however, the confinement factor also becomes very small [according to Eq. (9.4.10), for very small values of d, one has $\Gamma \propto d^2$), the first term in the brackets eventually dominates and J_{th} reaches a point where it increases with decreasing d. To get a numerical evaluation of J_{th} from Eq. (9.4.13), we assume $d = 0.1\ \mu$m, we take $\eta_i \cong 1$, $\tau_r = 4$ ns and use the previously calculated value of N_{th}. We obtain $J_{th} \cong 10^3\ \text{A/cm}^2$ in reasonable agreement with the results of Fig. 9.23.

9.4.4. Quantum Well Lasers[44]

If the thickness of the active layer of a DH laser is greatly reduced to a point where the dimension is comparable to the de-Broglie wavelength ($\lambda \cong h/p$), where p is the electron momentum a quantum well, QW, double heterostructure laser is produced. Such lasers exploit the more favorable optical properties of a QW or of a multiple

Example 9.2. *Carrier and current densities at threshold for a GaAs/AlGaAs quantum well laser* We assume that a functional relation of the form $g = \sigma(N - N_{th})$ still holds approximately for a QW*, so that the relations established in Example 9.1. can still be applied. To make a comparison with the case discussed in Example 9.1, we take the same values of the loss coefficient $\alpha(10\,\text{cm}^{-1})$ and mirror reflectivity R (32%), hence of the total loss $\gamma(\gamma = 1.44)$. We also take the same value of cavity length (300 μm) and carrier density at transparency N_{tr} $(2\times10^{18}\,\text{cm}^{-3})$, while we now take $\sigma \cong 6\times10^{-16}\,\text{cm}^2$ (see Sect. 3.3.5.). To calculate the confinement factor we assume that the beam field profile can be written as $U \propto \exp-(x^2/w_\perp^2)$, where $w_\perp$ is the beam spot size in the direction orthogonal to the junction. From Eq. (9.4.7) we then readily get $\Gamma = (d/0.62d_\perp)$, where d is the well thickness and $d_\perp = 2w_\perp$ so that, taking $d = 10\,\text{nm}$ and $d_\perp = 1\,\mu\text{m}$, we obtain $\Gamma = 1.6\times10^{-2}$. From Eq. (9.4.9) we now get $N_{th} = (5+2)\times10^{18}\,\text{cm}^{-3}$, where the numerical values of the two terms appearing in the right hand side of Eq. (9.4.9) have again been left separated. We see that, due to the much smaller value of the confinement factor, the first term, i.e., the carrier density required to overcome cavity losses, is now appreciably larger than the second term, N_{tr}. The threshold current density is now readily obtained by substituting the previously calculated value of N_{th} into Eq. (9.4.3). Assuming again that $\eta_i = 1$ and $\tau = 4\,\text{ns}$, we obtain $J_{th} \cong 280\,\text{A/cm}^2$ which is about 4 times smaller than the value calculated for a DH laser. Note that, in this case, since it is cavity loss that mainly determines the value of N_{th}, a reduction of this loss is helpful to further reduce J_{th}. If we now take, for example, $\alpha = 3\,\text{cm}^{-1}$ and $R = 80\%$, we get $\gamma = 0.28$, hence $N_{th} = (2.3+2)\times10^{18}\,\text{cm}^{-3}$ and $J_{th} \cong 170\,\text{A/cm}^2$.

quantum well, MQW, structure compared to those of the corresponding bulk material, in particular, the increased differential gain (see Example 3.12), and decreased dependence of this gain on temperature. These favorable properties are essentially related to the completely different density of states of QW materials compared to bulk materials, arising from quantum-confinement along the well direction (see Sect. 3.3). Single QW and also MQW lasers would, however, be seriously affected by the strong reduction of the confinement factor arising from the reduced layer thickness. To limit the beam size along the QW direction, one then needs to use a separate confinement structure.

Several structures have been introduced for this purpose, and a particularly simple example is shown in Fig. 9.24a. Everything in this figure is to scale, except for the bulk GaAs band-gap energy, which has been reduced for clarity. At the center of the structure is the thin ($\sim$10 nm) quantum well (GaAs) and, on both sides of the well, there are two thicker ($\sim$0.1 μm), inner barrier, layers of wider band-gap and hence lower refractive-index material ($Al_{0.2}Ga_{0.8}As$). Outside the inner barrier layers there are two, much thicker ($\sim$1 μm), cladding layers of still wider band-gap material ($Al_{0.6}Ga_{0.4}As$), constituting the p- and n-sides of the diode. Beam confinement is established by the higher refractive index of the inner barrier layers compared to the cladding layers, while the contribution to confinement by the very thin QW is negligible. The resulting beam intensity profile for this waveguide configuration is also shown as a dashed line in Fig. 9.24a. One can see that the full width between the $1/e^2$ points is, in this case, confined to a comparatively small dimension ($\sim$0.8 μm). A somewhat similar and widely used structure is that shown in Fig. 9.24b, where the index composition of the inner barrier layer, $Al_xGa_{1-x}As$, is gradually changed from e.g., $x = 0.2$ at the QW interface to the value $x = 0.6$ at the interfaces between the two cladding layers, where it matches the index of

* This approximation holds with less accuracy for a QW, a plot of g vs N showing a curve which, due to the essentially two-dimensional structure of the density of states, saturates at sufficiently high values of the current injection (see[45]).

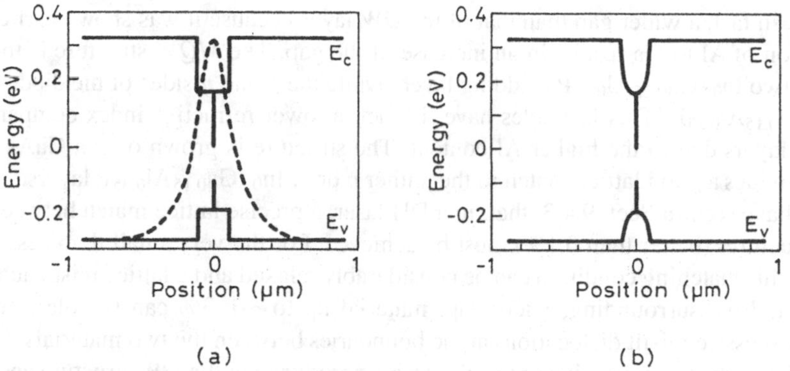

FIG. 9.24. (a) Energy bands of a step-index $Al_xGa_{1-x}As$-GaAs separate confinement *QW* heterostructure. The resulting optical mode intensity profile for this waveguide structure is shown as a *dashed line* (after[46] by permission). (b) Energy bands of a graded-index $Al_xGa_{1-x}As$-GaAs separate confinement *QW* heterostructure (GRINSCH).

the cladding layers. This structure is usually referred to as GRINSCH, from GRaded-INdex Separated-Confinement Heterostructure. Note that, in both structures of Fig. 9.24, the carriers are confined by the QW structure while the beam is confined by the step-index or graded-index profile of Fig. 9.24a and b respectively. Note also that, although the thickness of the QW layer is much smaller than the width of the beam, optical confinement results in a sufficiently high confinement factor to now take advantage of the expected reduction of J_{th} arising from the strong reduction of the active layer thickness d [see Eqs. (9.4.3) and (9.4.13)]. In fact, as shown in Example 9.2, one may now typically obtain values of J_{th} which are ~4–5 times smaller than those of a DH laser (i.e., ~200 A/cm^2). This threshold reduction thus arises from a combination of the following two features: (1) Reduction in threshold as expected from the strong reduction of layer thickness, once the problem of beam confinement has been partially overcome by a separated confinement structure. (2) Increase (by about a factor 2) of the differential gain which occurs in a QW compared to the corresponding bulk material.

The separated confinement structures of Fig. 9.24 may include either a single QW, as shown in the figure, or a multiple quantum well (MQW) structure, the structure in this case consisting of a number of alternating layers of narrow and wide band-gap materials. A noteworthy example is the $In_{0.5}Ga_{0.5}P/In_{0.5}Ga_{0.25}Al_{0.25}P$ MQW structure (see Fig. 9.25) leading to laser emission, in the red, at 670 nm wavelength. The thickness of each $In_{0.5}Ga_{0.5}P$ QW is seen to be 5 nm while the thickness of the $In_{0.5}Ga_{0.25}Al_{0.25}P$ barrier layer is 4 nm. This

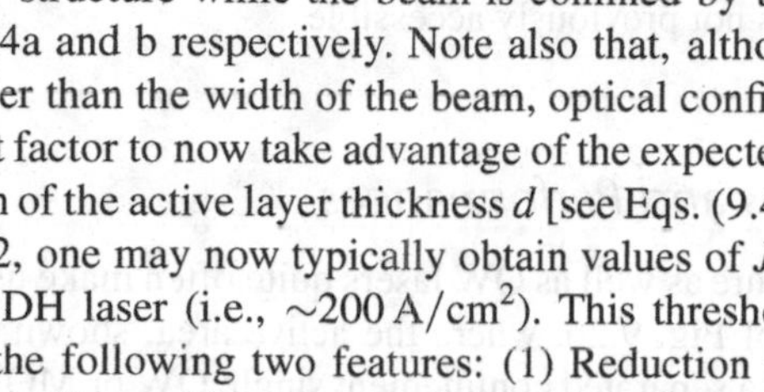

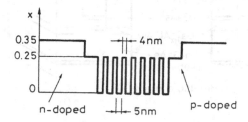

FIG. 9.25. Composition variations in a $In_{0.5}GaP_{0.5}/In_{0.5}(Ga_{0.5-x}Al_x)P$ MQW active layer, where $x = 0.25$ for well barriers and $x = 0.35$ for confinement layers, producing laser emission at 670 nm wavelength. Substrate material is GaAs which provides a good lattice match to the $In_{0.5}Ga_{0.15}Al_{0.35}P$ confinement layer.

layer has, in fact, a wider gap than that of the *QW* layer because it was shown earlier that the substitution of Al to Ga results in an increase of the gap. The *MQW* structure is then terminated by two $In_{0.5}Ga_{0.25}Al_{0.25}P$ cladding layers while the *p* and *n* sides of the diode are made of $In_{0.5}Ga_{0.15}Al_{0.35}P$. This last sides have, in fact, a lower refractive index compared to the cladding layers due to the higher Al content. The structure is grown over a GaAs substrate which provides a good lattice match to the, either *p* or *n*, $In_{0.5}Ga_{0.15}Al_{0.35}P$ layers.

We have seen in Sect. 9.4.3. that, in a DH laser, a precise lattice match between the two heterostructures (better than 0.1%) must be achieved. For the very small thickness of a *QW*, however, this matching condition can be considerably relaxed and a lattice mismatch between the *QW* and the surrounding, wider-gap, material up to ~1–3% can be tolerated without creating excessive misfit dislocations at the boundaries between the two materials. Due to the lattice mismatch, a compressive or tensile strain is produced in the *QW* structure and one thus has a strained *QW*. Strained quantum wells present two main advantages: (1) Structures can be grown which can produce laser action in wavelength ranges not otherwise covered (e.g., 900–1,100 nm for $In_xGa_{1-x}As/GaAs$). (2) As discussed in Sect. 3.3.6., under compressive strain, the effective mass of the hole in a direction parallel to the junction decreases to a value closer to the effective mass of the electron. This situation lowers the transparency density, N_{tr}, and increases the differential gain, compared to an unstrained *QW*. Thus, strained quantum well lasers allow laser action to be obtained, with very low threshold current density and high efficiency, at wavelengths not previously accessible.

9.4.5. Laser Devices and Performances

Double-heterostructure as well as QW lasers quite often make use of the so-called stripe-geometry configuration of Fig. 9.26, where the active area, shown dashed, may be either a double heterostructure or a separated confinement single QW or MQW structure. One can see from both figures that, by introducing a suitable insulating oxide layer, the current from the positive electrode is constrained to flow in a stripe of narrow width s ($s = 3$–$10\ \mu m$). Compared to a broad area device (see Fig. 9.20), this stripe-geometry device has the advantage of considerably reducing the area A ($A = Ls$, where L is the semiconductor length) through

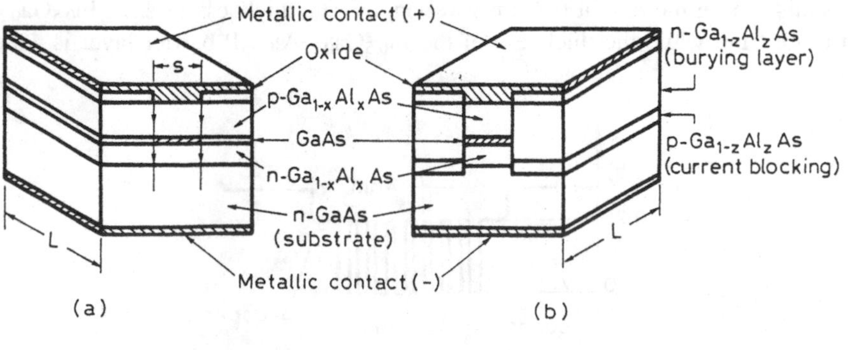

FIG. 9.26. Detail of stripe-geometry double-heterostructure semiconductor lasers: (a) Gain-guided laser; (b) buried-heterostructure index-guided laser.

which the current flows. Thus, for a given current density J, the required total current, $I = JA$, is correspondingly reduced. Furthermore, since the width of the gain region in the junction plane is also roughly equal to s, this mechanism can already be used to confine the beam transverse dimension in the direction parallel to the junction. The corresponding device is referred to as a *gain-guided laser*, see Fig. 9.26a and, if s is made sufficiently small ($s < 10\,\mu$m), gain confinement results in the beam being restricted to the fundamental transverse mode in the direction parallel to the junction. In the direction orthogonal to the junction, on the other hand, the beam is also confined to the fundamental transverse mode by the index guiding effect produced by the double heterostrusture (see Fig. 9.22) or by the separated confinement structure (see Fig. 9.24). The output beam thus consists of a single transverse mode of elliptical cross section ($\sim 1\,\mu$m $\times$ 5 μm). The gain-guided structure of Fig. 9.26a has the disadvantage that the unpumped regions of the active layer are strongly absorbing and the beam confinement action arising from these regions inevitably introduces some loss for the beam. A better solution is to provide lateral confinement via a refractive-index guiding action within the junction plane as well (*index-guided lasers*). A possible solution is to surround the active layer with semiconductor materials of lower refractive index such as in the buried heterostructure laser of Fig. 9.26b. The advantage of an index-guided laser is that the laser beam suffers less absorption by the laterally confining media. In fact, index-guided structures (e.g., buried or ridge waveguide structures) appear to be increasingly favored in commercial devices.

We now consider some properties of the output beam namely output power, beam divergence and spectral content.

Plots of output power versus input current, for two different temperatures, are shown in Fig. 9.27 for a gain-guided DH GaAs semiconductor laser. Note that the threshold current, I_{th}, at room temperature is less than 100 mA as a result of using the stripe geometry. Threshold currents lower than this (~15 mA) are now more typical for both gain-guided and index-guided DH GaAs semiconductor lasers, while much lower values (~1 mA) have been

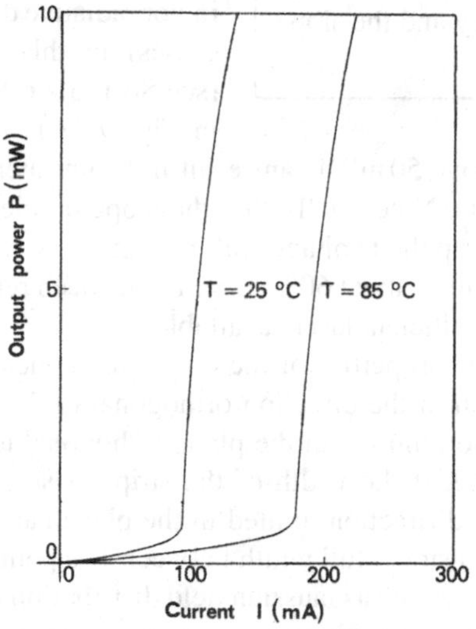

FIG. 9.27. Plot of the output power vs the input current for a DH laser at room temperature and elevated temperature.

Example 9.3. *Output power and external quantum efficiency of a semiconductor laser* To calculate the output power, we first note that, under steady-state conditions, the power emitted by stimulated emission can simply be written as $P_e = (I - I_{th})\eta_i h\nu/e$, where η_i is the internal quantum efficiency, introduced in Sect. 9.4.3., and ν is the frequency of the emitted radiation. Part of this power is dissipated by the internal losses (due to scattering and cladding losses) and part is available as output power from the two cavity ends. This power can then be written as

$$P = \left[\frac{(I - I_{th})\eta_i h\nu}{e}\right]\left(\frac{-\ln R}{\alpha L - \ln R}\right) \tag{9.4.14}$$

where R is the power reflectivity of the two end mirrors, α is the internal loss coefficient and L is the cavity length. One can now define the external quantum efficiency η_{ex} as the ratio between the increase in emitted photons and the corresponding increase in injected carriers, i.e., $\eta_{ex} = d(P/h\nu)/d(I/e)$. From Eq. (9.4.14) we then get

$$\eta_{ex} = \eta_{in}\left(\frac{-\ln R}{\alpha L - \ln R}\right) \tag{9.4.15}$$

which shows that η_{ex} increases by reducing the cavity length. Note also that, following our previous definitions, the relation between the external efficiency and the slope efficiency is simply $\eta_{ex} = \eta_s(eV/h\nu)$.

obtained with particular *QW* devices [indeed, assuming $J_{th} = 200\ \mathrm{A/cm^2}$, see Example 9.2., $s = 4\ \mu$m and $L = 150\ \mu$m, one gets $I_{th} = 1.2$ mA]. Note also from Fig. 9.27 the rapid increase of I_{th} with temperature. In most laser diodes this increase has been found empirically to follow the law $I_{th} \propto \exp(T/T_0)$ where T_0 is a characteristic temperature, dependent on the particular diode, whose value is a measure of the quality of the diode. The ratio between the threshold values at two temperatures differing by ΔT is in fact given by $(I'_{th}/I''_{th}) = \exp(\Delta T/T_0)$. Thus the larger T_0 the less sensitive is I_{th} to temperature. In the case of Fig. 9.27 it can be readily calculated that $T_0 \cong 91$ K. In DH GaAs lasers, T_0 typically ranges between 100 to 200 K, while T_0 is usually larger (> 270 K) for GaAs *QW* lasers. Thus, the increase in characteristic temperature of a *QW* laser is another favorable feature of *QW* devices and results from a weaker dependence of quasi-Fermi energies and hence of the differential gain on temperature (compare Fig. 3.25 with Fig. 3.15). The characteristic temperature for DH InGaAsP/InP lasers is considerably lower than the above values ($50\ \mathrm{K} < T_0 < 70$ K) probably due to the rapid increase in nonradiative decay rate (due to Auger processes) in this narrower bandgap material (see Sect. 3.2.6.). Note that the output power in Fig. 9.27 is limited to $\sim$10 mW. Higher output powers (typically above 50 mW) can result in beam intensities high enough to damage the semiconductor facets. Note finally that the slope efficiency of the laser is given by $\eta_s = dP/VdI$, where V is the applied voltage. Taking $V \cong 1.8$ V we get $\eta_s = 40\%$. Even higher slope efficiencies than this (up to about 60%) have in fact been reported. Thus, semiconductor lasers are currently the most efficient lasers available.

As far as the divergence properties of the output are concerned, we first note that, due to the small beam dimension in the direction orthogonal to the junction ($\sim$1 μm), the output beam is always diffraction limited in the plane orthogonal to the junction. Furthermore, as already pointed out above, if the width of the stripe is smaller that some critical value ($\sim$10 μm), the beam is also diffraction limited in the plane parallel to the junction. Now let $d_\perp$ and $d_\parallel$ be the beam dimensions (full width between $1/e$ points of the electric field) along the two directions and let us assume a Gaussian field distribution in both transverse directions. According to Eq. (4.7.19), the beam divergences in the plane parallel to the junction, $\theta_\parallel$, and in the plane orthogonal to the junction, $\theta_\perp$, will be given by $\theta_\parallel = 2\lambda/\pi d_\parallel$ and $\theta_\perp = 2\lambda/\pi d_\perp$,

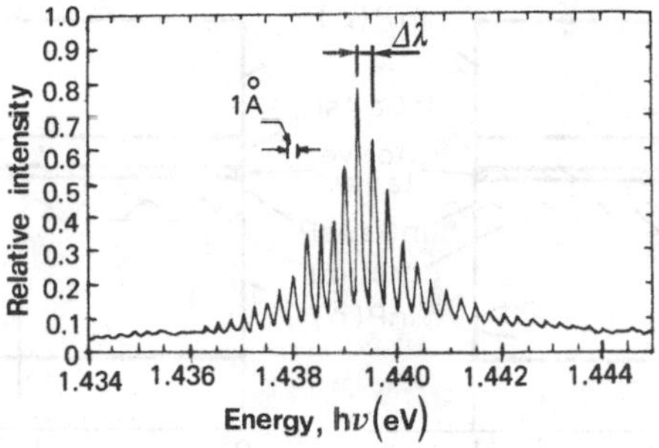

FIG. 9.28. Typical spectral emission of a Fabry-Perot type DH GaAs semiconductor laser with a cavity length of 250 μm.

respectively. For an output beam of elliptical cross-section (e.g., 1 μm × 5 μm) the divergence in the plane orthogonal to the junction will be larger than that in the plane parallel to the junction. Thus the beam ellipticity rotates by 90° at a distance some tens of microns away from the semiconductor exit face (see Fig. 6.9a). As discussed in Sect. 6.3.2.1., optical systems can be developed to compensate for this astigmatic behavior of the beam.

A typical emission spectrum of a diode laser, in which optical feedback is provided by the two end-face reflections, is shown in Fig. 9.28. The equally spaced peaks corresponds to different longitudinal modes of the Fabry-Perot cavity. Two points should be noted from this figure: (1) The relative spectral bandwidth $\Delta\nu_L/\nu$ is sufficiently small ($\sim 1.1\times 10^{-3}$) to justify stating, according to Eq. (9.4.1), that the emission frequency is roughly equal to E_g/h. (2) The absolute value of this bandwidth ($\Delta\nu_L \cong 400$ GHz in Fig. 9.28) is sufficiently large, however, to be a problem for optical fiber communications, due to the chromatic dispersion of an optical fiber, particularly around $\lambda = 1,550$ nm. To obtain much smaller linewidths, the best approach is to use either a distributed feedback laser or a laser with distributed Bragg reflectors. These lasers are briefly considered in the next section.

9.4.6. Distributed Feedback and Distributed Bragg Reflector Lasers

A distributed feedback (DFB) laser consists of an active medium in which a periodic thickness variation is produced in one of the cladding layers forming part of the heterostructure.[47] A schematic example of a DFB laser oscillating at 1,550 nm is shown in Fig. 9.29a, where a InGaAsP active layer ($\lambda = 1,550$ nm) is sandwiched between two InGaAsP cladding layers ($\lambda = 1,300$ nm), one of the two layers showing this periodic thickness variation. Since the refractive index of the InGaAsP cladding layers is larger than that of the InP, *p*- and *n*-type, layers, the electric field of the oscillating mode will see an effective refractive index $n_{eff}(z) = \langle n(x, z)\rangle_x$ which depends on the longitudinal *z*-coordinate. In the previous expression $\langle\rangle_x$ stands for a weighted spatial average taken over the *x*-coordinate, orthogonal to the junction, the weight being determined by the transverse distribution of the beam intensity, $|U(x)|^2$ [see also Eq. (9.4.5)]. We now assume that $n_{eff}(z)$ is a periodic function of *z* i.e.,

$$n_{eff}(z) = n_0 + n_1 \sin[(2\pi z/\Lambda) + \varphi] \tag{9.4.16}$$

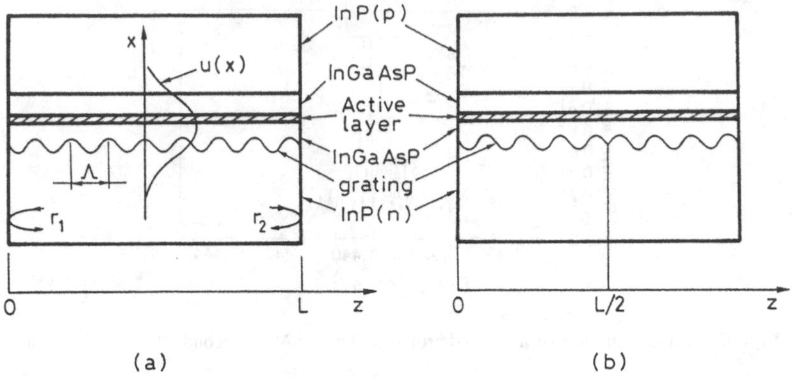

FIG. 9.29. Schematic structure of: (a) A DFB laser with a uniform grating. (b) A $\lambda/4$-shifted DFB laser.

where Λ is the pitch of the periodic thickness change (see Fig. 9.29a). In accordance with the ideas of Bragg, for the scattering from a periodic array of elements, the forward and backward propagating beams of the DFB laser will be effectively coupled to each other if the free space wavelength of the radiation is such that

$$\lambda = \lambda_B = 2\langle n_{eff}\rangle \Lambda \tag{9.4.17}$$

where $\langle n_{eff}\rangle$ is some suitable average value, along the z-coordinate, of n_{eff}, whose value will be discussed later on. To appreciate the significance of the above expression, we assume, for simplicity, that $n_{eff}(z)$ consists of a periodic square-wave function of period Λ. In this case, the structure of Fig. 9.29a is equivalent to a periodic sequence of high and low refractive index layers, the thickness of each layer being equal to $\Lambda/2$. This case is rather similar to a periodic sequence of multilayer dielectric mirrors (see Sect. 4.4) and constructive reflection is expected to occur when $(\langle n_{eff}\rangle\Lambda/2) = \lambda/4$. Equation (9.4.17) then shows that, for a given pitch Λ, there is only one wavelength, i.e., only one mode satisfying the Bragg condition. Only this mode is then expected to oscillate when the appropriate threshold condition is satisfied.

The above simple considerations are very approximate and a better understanding of the DFB laser behavior would require a detailed analytical treatment. In this analysis, the two oppositely traveling waves are assumed to see an effective gain coefficient as established by the active medium and to be coupled by a periodic change of the dielectric constant i.e., of the refractive index. One also generally assumes that there are finite values, r_1 and r_2, of the electric-field reflectivity from the two end faces. We will not go into this analysis here, referring elsewhere for a detailed treatment,[47,48] and only discussing a few important results. First we consider a rather peculiar result, indicated, for the simple case $r_1 = r_2 = 0$, in Fig. 9.30a which shows the intensity transmittance T, e.g., $T = |E_f(0, L)/E_f(0, 0)|^2$ for the forward beam, vs normalized detuning $\delta L = (\beta - \beta_B)L$. In the above expressions $E_f(x, z)$ is the electric field of the forward beam, L is the cavity length, $\beta = 2\pi n_0/\lambda$ and $\beta_B = \pi/\Lambda$. The plots shown in the figure have been obtained for the value $kL = 2$ of the normalized coupling coefficient k $(k \cong 2\pi n_1/\lambda)$ and for several values of the effective gain $\langle g\rangle L$.[49] Figure 9.30a shows that a transmission minimum actually occurs at exact resonance ($\delta = 0$) while several transmission maxima, i.e., several modes are present, symmetrically located from either side of exact

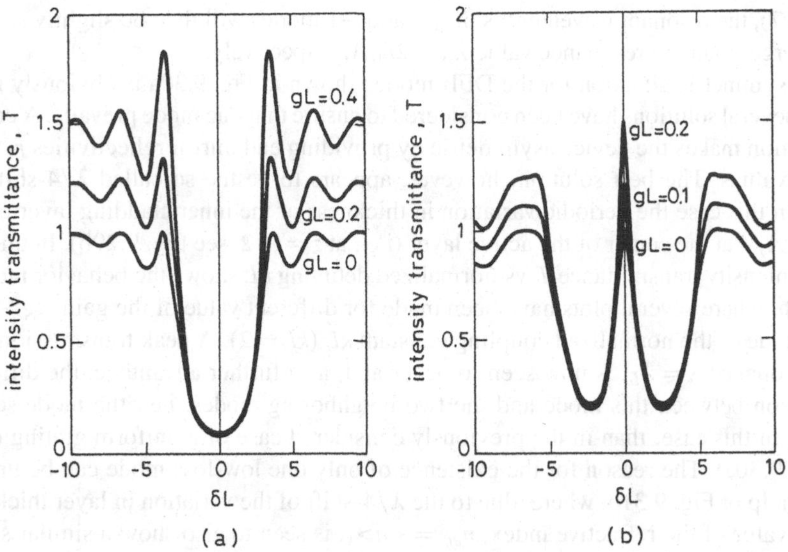

FIG. 9.30. Transmittance T vs normalized detuning δL, at several values of the gain gL, for zero end-face reflectivity and for: (a) A uniform grating. (b) A $\lambda/4$-shifted grating (from,[49] by permission).

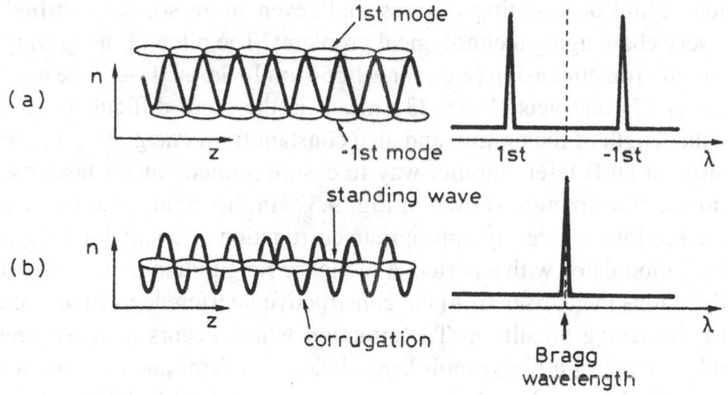

FIG. 9.31. Schematic representation of the refractive index change, mode patterns and corresponding resonance wavelengths for a DFB laser with a uniform grating and for a DFB laser with a $\lambda/4$-shifted grating (after,[49] by permission).

resonance. The reason for the existence of e.g., the first and strongest two resonances can be understood with the help of Fig. 9.31a, where the longitudinal variation of the refractive index, the standing wave patterns of the two modes and the corresponding resonant wavelengths are shown.[49] One can see that the mode labeled +1 is subjected to an effective, i.e., longitudinally averaged, refractive index $<n_{eff}>_1$ which is slightly smaller than that, $<n_{eff}>_{-1}$, of mode −1. In both previous expressions, the spatial average $<>$ is now taken over the longitudinal intensity distribution of the cavity mode. According to the general Bragg condition

Eq. (9.4.17), the resonant wavelengths of +1 and −1 modes will then be slightly smaller and slightly larger than the resonance value $\lambda_B = 2n_0\Lambda$, respectively.

The symmetric situation for the DFB modes shown in Fig. 9.30a is obviously not desirable and several solutions have been considered to ensure that one mode prevails. A commonly used solution makes the device asymmetric by providing end mirror reflectivities r_1 and r_2 of different values. The best solution, however, appears to be the so-called $\lambda/4$-shifted DFB laser.[50] In this case the periodic variation in thickness of the inner cladding layer undergoes a shift of $\Lambda/4$ at the center of the active layer (i.e., at $z = L/2$, see Fig. 9.29b). In this case, in fact, the intensity transmittance T vs normalized detuning δL shows the behavior indicated in Fig. 9.30b, where several plots have been made for different value of the gain $<g>L$ and for a given value of the normalized coupling constant kL ($kL = 2$). A peak transmission, at exact Bragg resonance $\lambda = \lambda_B$, is now seen to occur and, as a further advantage, the difference in transmission between this mode and the two neighboring modes, i.e., the mode selectivity, is higher, in this case, than in the previously considered case of a uniform grating (compare with Fig. 9.30a). The reason for the existence of only one low-loss mode can be understood with the help of Fig. 9.31b, where, due to the $\lambda/4$-shift of the variation in layer thickness, the effective value of the refractive index, $n_{eff} = <n>_x$, is seen to also show a similar shift in its longitudinal variation (see Fig. 9.31b). The standing-wave pattern of the lowest-loss mode is also shown in the figure, and the longitudinal spatial average of the effective refractive index $<n_{eff}>$ is now seen to be equal to n_0. The resonance condition will then be $(\lambda/2n_0) = \Lambda$ and the wavelength λ of the mode will coincide with the Bragg wavelength $\lambda_B = 2n_0\Lambda$.

Fabrication of uniform-grating devices and, even more so, $\lambda/4$-shifted DFB lasers presents some very challenging technological problems. The pitch of the grating Λ has, typically, to be of submicron dimension [e.g., for a 1,550 nm InGaAsP laser, one has $<n_{eff}> \cong 3.4$ and from Eq. (9.4.17) one gets $\Lambda \cong 0.23\ \mu$m]. It is therefore difficult to make this pitch uniform along the length of the grating and also constant from one grating to the next.

Besides using a DFB laser, another way to ensure semiconductor laser oscillation on a single line is to use the structure shown in Fig. 9.32. In this figure, the two cavity ends are made of passive sections where, by appropriate corrugation of a suitable layer, the effective refractive index is modulated with a period Λ along the longitudinal direction. The reflectivity of the two end sections then arises from the constructive interference which occurs in the two sections, under the Bragg condition. The situation which occurs in these sections is then somewhat similar to that of a $(\lambda/4)$ multilayer dielectric mirror and maximum reflectivity is expected to occur at the wavelength $\lambda = 2<n_{eff}>\Lambda$. Compared to DFB lasers, DBR lasers have the advantage that the grating is fabricated in an area separated from the active layer. This introduces some simplification in the fabrication process and makes the DBR structure

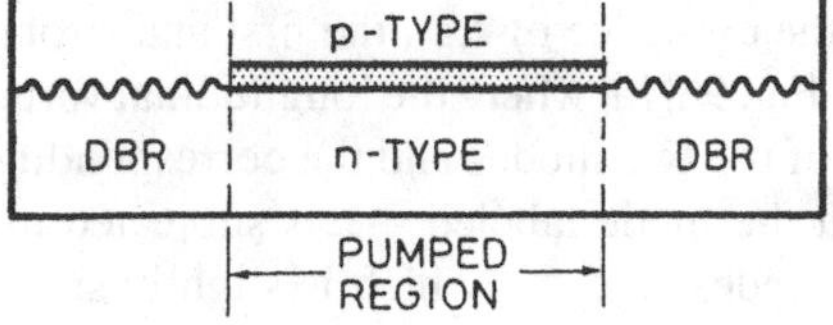

FIG. 9.32. Schematic representation of a distributed Bragg reflector (DBR) semiconductor laser.

more suitable for integration with other devices, such as separated sections for laser tuning or modulation. The wavelength selectivity of a DBR laser is however less than that of a DFB laser owing to the presence of many Fabry-Perot longitudinal modes. Actually, due to the small length of the active section, usually only one mode falls within the high-reflectivity bandwidth of the DBR structure. Temperature variations can however produce jumping between adjacent modes and, for this reason, DBR lasers are much less widely used than DFB lasers.

9.4.7. Vertical Cavity Surface Emitting Lasers

So far, we have considered semiconductor diode lasers that generate light traveling in a direction parallel to the junction plane, and hence emitted from one edge of the device (*edge emitting lasers*). For several applications which we will discuss in the next section, semiconductor lasers emitting normal to the junction plane have been developed. These devices are usually referred to as *surface emitting lasers* and are made using one of the following two approaches: (1) Use of a conventional edge-emitting geometry, but with some optical element, e.g., a 45° mirror, to deflect the output beam vertically (Fig. 9.33a). (2) Use of highly reflective mirrors to clad the active layer thus resulting in a vertical cavity that produces an output beam propagating normal to the junction plane (vertical-cavity surface-emitting laser, VCSEL, see Fig. 9.33b). Surface emitting lasers, of the type shown in Fig. 9.33a, are no different, conceptually, from a conventional edge-emitting laser. A peculiar characteristic of a VCSEL, on the other hand, is the very short length of active medium and thus the very small gain involved. However, once this low-gain limitation is overcome by using sufficiently high reflectivity mirrors, low thresholds can be obtained and these lasers then present some distinct advantages over the corresponding edge-emitting devices, due to the inherently high packaging density and low threshold currents that can be achieved. In the discussion that follows we therefore concentrate on vertical-cavity surface-emitting lasers.[51]

A schematic view of a VCSEL using, as active medium, three $In_{0.2}Ga_{0.8}As/GaAs$ strained QW layers, each 8 nm thick, is shown in Fig. 9.34. The three active layers are sandwiched between two $Ga_{0.5}Al_{0.5}As$ spacers to form an overall thickness of one wavelength. The bottom and top mirrors are made of a 20.5-pairs of *n*-doped, and 16-pairs of *p*-doped,

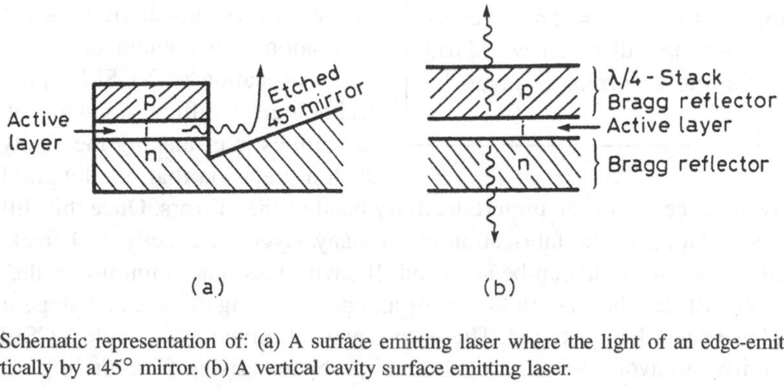

FIG. 9.33. Schematic representation of: (a) A surface emitting laser where the light of an edge-emitting laser is deflected vertically by a 45° mirror. (b) A vertical cavity surface emitting laser.

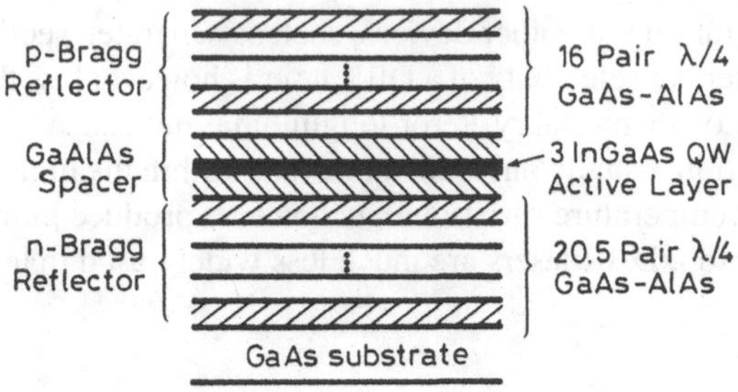

FIG. 9.34. Schematic representation of a bottom emitting VCSEL design. The top-most GaAs layer is a half-wavelength thick to provide phase matching for the metal contact (after,[52] by permission).

quarter-wave GaAs/AlAs stacks, respectively. On account of the relatively high refractive-index change between the two layers (the refractive index is 3.6 for GaAs and 2.9 for AlAs), high reflectivity (~99%) can be achieved for both end mirrors. Due to the short length of the cavity, scattering and absorption losses in the active layer are very small and reasonably low-threshold current densities can be obtained ($J_{th} \cong 4\,\mathrm{kA/cm^2}$). Since the diameter of the circular surface, through which current flows, is usually made very small ($D = 5 \div 10\,\mu\mathrm{m}$), a very small threshold current is obtained (~1 mA). Due to this small cross-sectional area of the active-region, VCSELs tend to oscillate on a TEM_{00} mode even at currents well above (e.g., two times) the threshold value. It should also be noted that, due to the small length of the laser cavity (1–2 μm), consecutive longitudinal modes are widely spaced in wavelength ($\Delta\lambda \cong 100\,\mathrm{nm}$). Thus, if one mode is made to coincide with the peak reflectivity of each of the quarter-wave stacks, the two adjacent modes fall outside the high reflectivity band of the mirrors and single longitudinal mode oscillation is also obtained.

Example 9.4. *Threshold current density and threshold current for a VCSEL* According to Eq. (9.4.9), assuming a confinement factor $\Gamma \cong 1$, we can write in this case $N_{th} = (\gamma/\sigma l) + N_{tr}$, where l is the thickness of the active layer. Following Example 9.2, we will assume the values $N_{tr} = 2 \times 10^{18}\,\mathrm{cm^{-3}}$ and $\sigma \cong 6 \times 10^{-16}\,\mathrm{cm^2}$ for the transparency carrier density and for the differential gain of the strained-layer quantum wells. The single pass loss is then given by $\gamma = -\ln R + \alpha_i L$, where R is the power reflectivity of each of the two mirrors, α_i is the internal loss coefficient and L is the cavity length. If we take $R = 99\%$, $\alpha_i = 20\,\mathrm{cm^{-1}}$ and $L = 2\,\mu\mathrm{m}$, we obtain $\gamma = 1.4 \times 10^{-2}$. If we now assume $l = 24\,\mathrm{nm}$ for the overall thickness of the active layers, we obtain $N_{th} \cong 11 \times 10^{18}\,\mathrm{cm^{-3}}$, which shows that the value of N_{th} is dominated by the loss term $(\gamma/\sigma l)$. From Eq. (9.4.3), taking $\eta_i \cong 1$ and $\tau = 2\,\mathrm{ns}$, we get $J_{th} = 3.84 \times 10^3\,\mathrm{A/cm^2}$. Assuming a diameter $D = 8\,\mu\mathrm{m}$ for the active area, the threshold current is given by $I_{th} = (\pi D^2/4)J_{th} \cong 0.44\,\mathrm{mA}$.

Fabrication of VCSELs presents some technological difficulties. One of the main difficulties is in making the cavity length to such a precision that one longitudinal mode falls exactly at the center of the high reflectivity band of the mirrors. Once this difficulty and the other one, relating to the fabrication of so many layers of exactly $\lambda/4$ thickness, have been solved, a low threshold can be achieved. If cavity losses are minimized, the slope efficiency of a VCSEL can be as high as that of an edge-emitting device, and slope-efficiencies up to 50% have been demonstrated. The output power emitted by a single VCSEL is rather limited (~1 mW) to avoid the onset of thermal problems arising from the injection of high pump powers into such a small volume of active layer. Arrays of e.g., 8×8, independently addressable, lasers as well as matrix-addressable arrays have been made.

9.4.8. Applications of Semiconductor Lasers

Semiconductor lasers lend themselves to a large variety of both low-power and high-power applications, some of which will be briefly reviewed here. We first refer to Table 9.7 where some characteristics of the most common DH or QW lasers are presented. Since all the structures shown are grown on either a GaAs or InP substrate, the laser material is characterized by the active layer-substrate combination. In each case, the laser wavelength is largely determined by the composition index of the active layer and, for a QW laser, also by the thickness of this layer. Most recent lasers use separated-confinement (e.g., GRINSCH) QW and MQW layers in a gain-guiding or, more often, index guiding configuration.

In the $Al_xGa_{1-x}As/Al_yGa_{1-y}As$ laser structures, the composition index y of the cladding layers must be larger than that, x, of the active layer. Depending on the value of this last composition index, the emission wavelength usually ranges between 720 and 850 nm. Low-power ($P = 5$–20 mW), single-stripe devices are widely used in Compact-Disk (CD) players and laser printers. Higher power single-stripe lasers, laser arrays, laser bars and stacks of laser bars (Figs. 6.9 and 6.10) are used to pump solid-state lasers such as Nd (pump wavelength $\lambda_p \cong 800$ nm), Tm:Ho ($\lambda_p \cong 790$ nm) and Cr:LISAF ($\lambda_p \cong 670$ nm). Some of these laser systems and the corresponding fields of applications have been discussed in Sect. 9.2 of this chapter as well as in previous chapters.

In the $In_{1-x}Ga_xAs_yP_{1-y}$/InP lasers, lattice matching is achieved by letting $y \cong 2.2\,x$ and the oscillation wavelength covers the so-called second (centered at $\lambda = 1,310$ nm, which corresponds to $x = 0.27$) and third ($\lambda = 1,550$ nm center wavelength, corresponding to $x = 0.42$) transmission windows of optical fibers. Thus, these lasers find their widest use in optical communications. Most recent optical communication systems use lasers around 1,550 nm wavelength. Due to the relatively large group-delay dispersion of optical fibers around this wavelength, narrow-linewidth ($\Delta\nu_L < 10$ MHz) DFB lasers are now widely used. With these

TABLE 9.7. Some characteristic parameters of the most important semiconductor laser diodes

Material/Substrate	AlGaAs/GaAs	InGaAsP/InP	InGaAs/GaAs	InGaAlP/GaAs
Wavelength [nm]	720–850	1,200–1,650	900–1,100	630–700
Internal loss [cm^{-1}]	4–15	5–10	2–10	~10
Threshold current density J_{th} [A/cm^2]	80–700	200–1,500	50–400	200–3,000
Characteristic Temperature T_0 [K]	120–200	50–70	100–200	60–100

lasers, modulation rates up to a few Gbits/s have been demonstrated by direct modulation of the diode current. Even higher modulation rates (up to a few tens of Gbits/s) have also been demonstrated using external modulators such as $LiNbO_3$ waveguide modulators. For communication systems operating at even larger communication bit-rates (from a few hundreds Gbit/s up to the Tbit/s range) Wavelength-Division-Multiplexing (WDM) systems have increasingly been used. For this application, many DFB lasers tuned at distinct wavelengths in the low transparency region (spanning ~13 THz) around 1,550 nm wavelength are used. Systems based on WDM thus allow exceptionally high bit rate capacity to be achieved.

$In_{1-x}Ga_xAs/GaAs$ strained-layer QW lasers allow oscillation to be obtained over a wide range (900–1,100 nm) at previously inaccessible wavelengths. Lasers with emission around 980 nm wavelength ($x = 0.8$) are of particular interest as pumps for Er-doped fiber amplifiers and lasers as well as for pumping Yb:Er:glass and Yb:YAG lasers. For these applications, output powers up to ~100 mW are available in a diffraction-limited beam (of $1 \times 4\ \mu m$ area) and up to ~1 W from a broader area ($1 \times 30\ \mu m$) device. To obtain higher output powers (~40 W or even larger), diode laser arrays and laser bars are also available. Given the favorable laser properties of strained-layer QW structures, vertical-cavity surface-emitting lasers based on $In_{1-x}Ga_xAS/GaAs$ structures have been particularly actively developed. They promise to offer interesting solutions for optical interconnects, optical communications, and optical signal processing.

InGaAlP/GaAs lasers are particularly interesting since they emit radiation in the visible, red, region of the e.m. spectrum.[53] The $In_{0.5}Ga_{0.5}P/In_{0.5}(Ga_{0.5-x}Al_x)P$ QW or MQW structure (where $x = 0.25$ for the well barriers and $x = 0.35$ for the confinement layers, see Fig. 9.25) oscillating at 670 nm wavelength has been particularly developed. These lasers are commercially available with sufficiently high power (up to ~10 mW) and long lifetime to be used for CD players or as substitutes for red-emitting He-Ne lasers for applications such as bar-code scanners and general alignment uses.

The development of semiconductor lasers is by no means limited to the laser categories shown in Table 9.7. On the short wavelength side (blue-green region) the most interesting category now appears to be the III-V nitride-based diode-lasers, e.g., the $In_{0.2}Ga_{0.8}N/In_{0.05}Ga_{0.95}N$ MQW structure oscillating in the blue (417 nm).[54] In the same wavelength region, wide-gap II-VI lasers such as ZnCdSe/ZnSSe QW lasers have also been demonstrated. After a number of years of intense development, however, these lasers are still suffering from a few technological limitations with particular problems of rather limited operating lifetime (~100 h). Although nitride-lasers also present similar lifetime problems (less than 100 h), their recent development and the rapid progress which is occurring indicate nitride-based diode-lasers as the best candidates for blue-green semiconductor emitters. Potentially, blue-green lasers are of strong interest for e.g. a new generation of CD players where, due to the shorter wavelength, substantially higher bit densities could be achieved on the CD. On the long wavelength side, on the other hand, we just mention the IV–VI compounds such as the Pb and Sn salts (e.g., PbSSe, PbSnTe, and PbSnSe), oscillating in the middle-far infrared (4–29 μm). All these lasers, however, need to operate at cryogenic temperatures ($T < 100$ K) to avoid problems such as the increased free-carrier absorption and the increased rate of non-radiative decay arising from the much narrower band-gaps. Thus, due to this cryogenic requirement, these lasers have found only a limited use (e.g., for spectroscopy). It should be noted, however, that the recent invention of a quantum-cascade laser promises efficient mid-IR sources without the need for cryogenic temperatures.[33]

9.5. CONCLUSIONS

In this chapter, a few of the most notable solid-state, dye and semiconductor lasers have been considered. These lasers use high-density active media and thus share a few common features. A first feature is that they generally show wide and strong absorption band(s) which indicates that they are generally suitable for optical pumping, this type of pumping being in fact always used for solid-state and dye lasers and sometimes also used for semiconductor lasers. The high values of absorption coefficient allow for lasers with dimensions down to a few microns (microlasers). A second relevant feature is that these media generally show wide fluorescence, and hence wide gain bandwidths. On the one hand, this offers tunability over wide (a few to several nm) bandwidths. On the other hand, this also implies that very short pulse durations (femtoseconds) can be obtained in mode-locked operation. A third relevant feature is that the optical-to-optical laser efficiency, for solid-state and dye media, and electrical to optical efficiency, for semiconductors, is generally quite large. It should also be noted that laser pumping involving combination of these three categories of lasers has been used increasingly (e.g., diode-pumped solid-state lasers or solid-state-laser pumped dye lasers) thus allowing compact and efficient lasers to be realized. Thus, as a conclusion, high-density laser media appear to present some of the best solutions to requirements for laser radiation in the visible-to-near-infrared range, even at high power levels.

PROBLEMS

9.1. Make up a diagram in which the tuning ranges of all the tunable solid-state lasers considered in this chapter are plotted vs oscillation wavelength.

9.2. For pollution monitoring, a tunable laser oscillating around 720 nm wavelength is needed. Which kind of solid-state laser would you use?

9.3. For biomedical photo-coagulation purposes, using an endoscopic apparatus, a cw laser with power exceeding 50 W needs to be used. Which laser would you use?

9.4. For material working applications, a laser with an average power of 2 kW, to be transmitted through a ~1 mm diameter optical fiber, is needed. Which laser would you use?

9.5. Consider a 6 mm diameter, 10 cm long Nd:phosphate glass laser rod. Using data from Table 9.3 and the results discussed in Sect. 2.9.2., calculate the maximum inversion and the corresponding maximum value of stored energy which are allowed if the onset of amplified spontaneous emission is to be avoided. Compare the results obtained with those for a Nd:YAG rod of the same dimensions.

9.6. With reference to the energy level diagram of alexandrite shown in Fig. 9.8, assume that the $^4T_2$ and the ^2E states are strongly coupled and take $\tau_T = 1.5$ ms and $\tau_E = 6.6\ \mu$s as the lifetimes of the two states. Knowing that the degeneracy of the two states is the same ($g = 4$), calculate the effective lifetime of the $^4T_2$ state at $T = 300$ K and $T = 400$ K. Knowing that the true cross-section for the $^4T_2 \rightarrow {}^4A_2$ transition is $\sigma \cong 4 \times 10^{-19}$ cm^2, calculate then the effective value of the cross section at the two temperatures. Using these results, consider whether the laser threshold is expected to increase or decrease when the crystal temperature is raised from 300 to 400 K.

9.7. Consider a Cr:LISAF laser longitudinally pumped, at 647.1 nm wavelength, by a Kr ion laser. Assume the linearly polarized beam of the laser to be sent along the *c*-axis of the LISAF and assume an active medium with 1 at. % Cr^{3+} concentration and a length $l = 4$ mm. Also assume both pump and mode spot sizes to be matched to a value of 60 μm, the output coupling transmission to be 1%, and the internal loss per pass to be 1%. Using data from Fig. 9.11 and from Table 9.5 and neglecting both ground-state and excited-state absorption, calculate the expected threshold pump power.

9.8. With reference to the previous problem, how would the expression for the threshold pump power, given by Eq. (6.3.19), need to be modified if ground state absorption, characterized by a loss per pass γ_a, and excited-state absorption, characterized by an excited-state absorption cross-section σ_{ESA}, were taken into account? Compare the result with that given in.[25]

9.9. Derive an expression for the threshold pump power of a longitudinally pumped dye laser, taking into account triplet-triplet absorption (assume Gaussian transverse profiles for both pump and mode beams). Compare this expression with that obtained for Cr:LISAF in the previous problem.

9.10. Using the expression for the threshold pump power obtained in the previous problem and using data from Fig. 9.13, calculate the threshold power for an Ar^+ pumped rhodamine 6G (see Fig. 9.17) laser oscillating at 580 nm wavelength. For this calculation, assume an output coupling of 3%, an internal loss per pass of 1%, assume also that 80% of the pump power is absorbed in the dye jet stream, take the lifetime for the first excited singlet state to be 5 ns, the intersystem crossing rate to be $k_{ST} \cong 10^7\ s^{-1}$ and the triplet lifetime to be $\tau_T \cong 0.1\ \mu s$. Compare this value of P_{th} with that obtained for Cr:LISAF in problem 9.7 and explain the numerical differences.

9.11. At the very small values of the thickness, d, corresponding to the minimum of J_{th} in Fig. 9.23, the expression for the beam confinement factor Γ of a DH semiconductor-laser, given by Eq. (9.4.10), can be approximated by $\Gamma \cong D^2/2$, where D is expressed by Eq. (9.4.11). Under this approximation, calculate the expression for this thickness, d_m, that minimizes J_{th}. From data given in Example 9.1, then calculate the value of d_m and the corresponding value of J_{th}.

9.12. From the expression for the output power of a semiconductor laser given in Example 9.3, derive an expression for the laser slope efficiency. Using data given in Example 9.1, then calculate the predicted slope-efficiency of a DH GaAs/AlGaAs laser by taking an applied voltage of 1.8 V.

9.13. Assume that the beam at the exit face of a semiconductor laser is spatially coherent. Assume that the transverse field distributions have Gaussian profiles along the directions parallel and perpendicular to the junction, with spot sizes $w_{||}$ and $w_{\perp}$ respectively. Assume also that, for both field distributions, the location of the beam-waists occurs at the exit face. Under these assumptions, derive an expression for the propagation distance at which the beam becomes circular. Taking $w_{\perp} = 0.5\ \mu m$ and $w_{||} = 2.5\ \mu m$, calculate the value of this distance for $\lambda = 850$ nm.

9.14. By taking into account the fact that the refractive index of a semiconductor, n, is a relatively strong function of the wavelength λ, derive an expression for the frequency difference between two consecutive longitudinal modes of a Fabry-Perot semiconductor laser [express this frequency difference in terms of the material group index $n_g = n - \lambda(dn/d\lambda)$].

9.15. The calculations leading to Fig. 9.30a were performed by assuming $kL = 2$, where k is the coupling constant between forward and backward propagating beams in a DFB laser and L is its length. From the definition of k given in Sect. 9.4.6., calculate the value n_1, appearing in Eq. (9.4.16), for $\lambda = 1,550$ nm and $L = 600\ \mu m$.

9.16. The two strongest peaks of Fig. 9.30a are seen from the figure to be separated by a normalized frequency difference $\Delta(\delta L) \cong 7.28$. From the definition of the normalized frequency detuning δ

given in Sect. 9.4.6., calculate the frequency difference $\Delta\nu$ between the two modes by taking, for a InGaAsP DFB laser, $L = 600\,\mu$m, $n_0 = 3.4$ and $\lambda = 1,550$ nm. Compare this value with the corresponding one obtained for the frequency separation between two consecutive longitudinal modes of a Fabry-Perot semiconductor laser with the same length and wavelength and with a group index, n_g, equal to n_0.

References

1. A. A. Kaminski, *Crystalline Lasers: Physical Processes and Operating Schemes*, (CRC Press Inc. Boca Raton, FL 1996)
2. T. H. Maiman, Stimulated Optical Radiation in Ruby Masers, *Nature*, **187**, 493 (1960)
3. T. H. Maiman, Optical Maser Action in Ruby, *Brit. Commun. Electron.*, **7**, 674 (1960)
4. W. Koechner, *Solid-State Laser Engineering*, fourth edition (Springer Berlin 1996), Sects. 2.2 and 3.6.1.
5. Ref. [4], Sects. 2.3.1. and 3.6.3.
6. E. Snitzer and G.C. Young, Glass Lasers, in *Lasers*, ed. by A. K. Levine (Marcel Dekker, New York, 1968), Vol. 2, Chap. 2
7. Ref. [4], Sect. 2.3.4
8. T. Y. Fan, Diode-Pumped Solid State Lasers, in *Laser Sources and Applications*, ed. by A. Miller and D. M. Finlayson (Institute of Physics, Bristol, 1996) pp 163–193
9. P. Lacovara *et al.*, Room-Temperature Diode-Pumped Yb:YAG Laser, *Opt. Lett.*, **16**, 1089–1091 (1991)
10. H. Bruesselbach and D. S. Sumida, 69-W-average-power Yb:YAG Laser, *Opt. Lett.*, **21**, 480–482 (1996)
11. G. Huber, Solid-State Laser Materials, in *Laser Sources and Applications*, ed. by A. Miller and D. M. Finlayson (Institute of Physics, Bristol, 1996) pp. 141–162
12. E. V. Zharikov *et al.*, *Sov. J. Quantum Electron.*, **4**, 1039 (1975)
13. S. J. Hamlin, J. D. Myers, and M. J. Myers, High Repetition Rate Q-Switched Erbium Glass Lasers, in *Eyesafe Lasers: Components, Systems, and Applications* ed. by A. M. Johnson, SPIE, **1419**, 100–104 (1991)
14. S. Taccheo, P. Laporta, S. Longhi, O. Svelto, C. Svelto, Diode-Pumped Bulk Erbium-Ytterbium Lasers, *Appl. Phys.*, **B63**, 425–436 (1996)
15. D. Sliney and M. Wolbarsht, *Safety with Lasers and other Optical Sources* (Plenum Press, New York 1980)
16. T. Y. Fan, G. Huber, R. L. Byer, and P. Mitzscherlich, Spectroscopy and Diode Laser-Pumped Operation of Tm, Ho:YAG, *IEEE J. Quantum Electron.*, **QE-24**, 924–933 (1988)
17. D. C. Hanna, Fibre Lasers, in *Laser Sources and Applications*, ed. by A. Miller and D. M. Finlayson (Institute of Physics, Bristol, 1996) pp. 195–208
18. E. Snitzer, Optical Maser Action on Nd^{3+} in a Barium Crown Glass, *Phys. Rev. Lett.*, **7**, 444–446 (1961)
19. J. C. Walling, O.G. Peterson, H. P. Jenssen, R. C. Morris, and E. W. O'Dell, Tunable Alexandrite Lasers, *IEEE J. Quantum Electron.*, **QE-16**, 1302–1315 (1980)
20. L. F. Mollenauer, Color Center Lasers, in *Laser Handbook*, ed. by M.L. Stitch and M. Bass (North Holland, Amsterdam, 1985), Vol. 4, pp. 143–228
21. P. F. Moulton, Spectroscopy and Laser Characteristics of Ti : Al_2O_3, *J. Opt. Soc. Am. B*, **3**, 125–132 (1986)
22. Günther Huber, Solid-State Laser Materials: Basic Properties and New Developments, in *Solid State Lasers: New Developments and Applications*, ed. by M. Inguscio and R. Wallenstein (Plenum Press New York 1993) pp. 67–81
23. P. Albers, E. Stark, and G. Huber, Continuous-wave Laser Operation and Quantum Efficiency of Titanium-Doped Sapphire, *J. Opt. Soc. Am. B*, **3**, 134–139 (1986)
24. S. A. Payne, L. L. Chase, L. K. Smith, W. L. Kway, and H. W. Newkirk, Laser Performance of $LiSrAlF_6 : Cr^{3+}$, *J. Appl. Phys.*, **66**, 1051–1055 (1989)
25. S. A. Payne, L. L. Chase, H. W. Newkirk, L. K. Smith, and W. F. Krupke, $LiCaAlF_6 : Cr^{3+}$: A Promising New Solid-State Laser Material, *IEEE J. Quantum Electron.*, **QE-24**, 2243–2252 (1988)
26. *Dye Lasers*, second edition., ed. by F. P. Schäfer (Springer-Verlag, Berlin, 1977)
27. H. D. Försterling and H. Kuhn, *Physikalische Chemie in Experimenten, Ein Praktikum*, (Verlag Chemie, Weinheim, 1971)

28. J. T. Verdeyen, *Laser Electronics*, third edition (Prentice-Hall International Inc., Englewood Cliffs, N. J., 1995) Fig. 10.19
29. P. P. Sorokin and J. R. Lankard, Stimulated Emission Observed from an Organic Dye, Chloro-Aluminum Phthalocyanine, *IBM J. Res. Dev.*, **10**, 162 (1966)
30. F. P. Schäfer, F. P. W. Schmidth, and J. Volze, Organic Dye Solution Laser, *Appl. Phys. Lett.*, **9**, 306–308 (1966)
31. *Semiconductor Lasers: Past, Present, Future* ed. by G. P. Agrawal (AIP Press, Woodbury, New York 1995).
32. G. P. Agrawal and N.K. Dutta, *Long Wavelength Semiconductor Lasers* (Chapman and Hall, New York, 1986).
33. J. Faist, F. Capasso, D. L. Sivco, C. Sirtori, A. L. Hutchinson, A. Y. Cho, *Science*, **264**, 553 (1994)
34. N. G. Basov, O. N. Krokhin, and Y. M. Popov, Production of Negative Temperature States in *p-n* Junctions of Degenerate Semiconductors, *J. Exp. Theoret. Phys.*, **40**, 1320 (1961)
35. R. N. Hall, G. E. Fenner, J. D. Kinhsley, F. H. Dills, and G. Lasher, Coherent Light Emission from GaAs Junctions, *Phys. Rev. Lett.*, **9**, 366–368 (1962)
36. M. I. Nathan, W. P. Dumke, G. Burns, F. H. Dills, and G. Lasher, Stimulated Emission of Radiation from GaAs p-n Junction, *Appl. Phys. Lett.*, **1**, 62 (1962)
37. N. Holonyak, Jr. and S. F. Bevacqua, Coherent (Visible) Light Emission from $Ga(As_{1-x}P_x)$ Junctions, *Appl. Phys. Lett.*, **1**, 82 (1962)
38. T. M. Quist, R. J. Keyes, W. E. Krag, B. Lax, A. L. McWhorter, R. H. Rediker, and H. J. Zeiger, Semiconductor Maser of GaAs, *Appl. Phys. Lett.*, **1**, 91 (1962)
39. Z. I. Alferov, V. M. Andreev, V. I. Korolkov, E. L. Portnoi, and D. N. Tretyakov, Coherent Radiation of Epitaxial Heterjunction Structures in the AlAs-GaAs System, *Soviet. Phys. Semicond.*, **2**, 1289 (1969)
40. I. Hayashi, M. B. Panish, and P. W. Foy, A Low-Threshold Room-Temperature Injection Laser, *IEEE J. Quantum Electron.*, **QE-5**, 211 (1969)
41. H. Kressel and H. Nelson, Close Confinment Gallium Arsenide p-n Junction Laser with reduced Optical Losses at Room Temperature, *RCA Rev.*, **30**, 106 (1969)
42. N. Chinone, H. Nakashima, I. Ikushima, and R. Ito, Semiconductor Lasers with a Thin Active Layer ($> 0.1\,\mu$m) for Optical Communications, *Appl. Opt.*, **17**, 311–315 (1978)
43. D. Botez, Analytical Approximation of the Radiation Confinement Factor for the TE_0 Mode of a Double Heterojunction Laser, *IEEE J. Quantum Electron.*, **QE-14**, 230–232 (1978)
44. *Quantum Well Lasers*, ed. by Peter S. Zory (Academic Press, Boston 1993)
45. Ref. [32], Fig. 9.8 and 9.10.
46. J. J. Coleman, Quantum-Well Heterostructure Lasers, in *Semiconductor Lasers: Past, Present, Future* ed. by G. P. Agrawal (AIP Press, Woodbury, New York 1995) Fig. 1.6.
47. H. Kogelnik and C. V. Shank, Stimulated Emission in a Periodic Structure, *Appl. Phys. Lett.*, **18**, 152–154 (1971)
48. Ref. [32] Chap. 7
49. N. Chinone and M. Okai, Distributed Feed-Back Semiconductor Lasers, in *Semiconductor Lasers: Past, Present, Future* ed. by G. P. Agrawal (AIP Press, Woodbury, New York 1995), Chap. 2, pp. 28–70.
50. H. A. Haus and C. V. Shank, Antisymmetric Taper of Distributed Feedback Lasers, *IEEE J. Quantum Electron.*, **QE-12**, 532 (1976)
51. C. J. Chang-Hasnain, Vertical-Cavity Surface-Emitting Lasers, in *Semiconductor Lasers: Past, Present, Future* ed. by G. P. Agrawal (AIP Press, Woodbury, New York 1995), Chap. 4, pp. 110–144.
52. C. J. Chang-Hasnain, J. P. Harbison, C.-H. Zah, M. W. Maeda, L. T. Florenz, N. G. Stoffel, and T.-P. Lee, Multiple Wavelength Tunable Surface-Emitting Laser Array, *IEEE J. Quantum Electron.*, **QE-27**, 1368 (1991)
53. G.-I. Hatakoshi, Visible Semiconductor Lasers, in *Semiconductor Lasers: Past, Present, Future* ed. by G. P. Agrawal (AIP Press, Woodbury, New York 1995), Chap. 6, pp. 181–207
54. S. Nakamura *et al.*, *Jpn. J. Appl. Phys.*, **35**, L74 (1994)

10

Gas, Chemical, Free Electron, and X-Ray Lasers

10.1. INTRODUCTION

In this chapter, the most important types of lasers involving low density active media are considered, namely gas, chemical and free electron lasers. Some considerations on X-ray lasers involving highly ionized plasmas will also be presented. The main emphasis, again, is to stress the physical behavior of the laser and to relate this behavior to the general concepts developed in the previous chapters. Some engineering details are also presented with the main intention again of providing for a better physical insight into the behavior of the particular laser. To complete the picture, some data relating to laser performances (e.g., oscillation wavelength(s), output power or energy, wavelength tunability, etc.) are also included to help provide some indication of the laser's field of application. For each laser, after some introductory comments, the following items are generally covered: (1) Relevant energy levels; (2) excitation mechanisms; (3) characteristics of the laser transition(s); (4) engineering details relating to the laser structure(s); (5) characteristics of the output beam; (6) applications.

10.2. GAS LASERS

In general, for gases, the broadening of the energy levels is rather small (of the order of a few GHz or less), since the line-broadening mechanisms are weaker than in solids. For gases at the low pressures often used in lasers (a few tens of torr), in fact, the collision-induced broadening is very small, and the linewidth is essentially determined by Doppler broadening. Thus, no broad absorption bands are present in the active medium and optical pumping by c.w. or pulsed lamps is not used for gases. This would, in fact, be very inefficient since the emission spectrum of these lamps is more or less continuous. Therefore, gas lasers are usually excited

O. Svelto, *Principles of Lasers*,
DOI: 10.1007/978-1-4419-1302-9_10,

by electrical means, i.e., by passing a sufficiently large current (which may be continuous, at radiofrequency or pulsed) through the gas. The principal pumping mechanisms occurring in gas lasers have been discussed in Sect. 6.4. It should also be pointed out that some gas lasers can in addition be pumped by mechanisms other than electrical pumping. In particular, we mention pumping by gas-dynamic expansion, chemical pumping, and optical pumping by means of another laser (particularly used for far-infrared lasers).

Once a given species is in its excited state, it can decay to lower states, including the ground state, by four different processes: (1) Collisions between an electron and the excited species, in which the electron takes up the excitation energy as kinetic energy (super-elastic collision); (2) near-resonant collisions between the excited species and the same or a different species in the ground state; (3) collisions with the walls of the container; (4) spontaneous emission. Regarding this last case, one should always take into account the possibility of radiation trapping, particularly for the usually very strong UV or VUV transitions. This process slows down the effective rate of spontaneous emission (see Sect. 2.9.1).

For a given discharge current, these various processes of excitation and de-excitation lead eventually to some equilibrium distribution of population among the energy levels. Thus it can be seen that, due to the many processes involved, the production of a population inversion in a gas is a more complicated matter that, e.g., in a solid-state laser. In general we can say that a population inversion between two given levels will occur when either (or both) of the following circumstances occur: (1) The excitation rate is greater for the upper laser level (level 2) than for the lower laser level (level 1); (2) the decay of level 2 is slower than that of level 1. In this regard we recall that a necessary condition for cw operation is that the rate of the $2 \rightarrow 1$ transition be smaller than the decay rate of level 1 [see Eq. (7.3.1)]. If this condition is not satisfied, however, laser action can still occur under pulsed operation provided the condition (1), above, is fulfilled (self-terminating lasers).

10.2.1. Neutral Atom Lasers

These lasers make use of neutral atoms in either gaseous or vapor form. Neutral atom gas lasers constitute a large class of lasers and include in particular most of the noble gases. All these lasers oscillate in the infrared (1–10 μm), apart from the notable exceptions of green and red emission from the He-Ne laser. Metal vapor lasers also constitute a large class of lasers, including, for example, Pb, Cu, Au, Ca, Sr, and Mn. These lasers generally oscillate in the visible, the most important example being the copper vapor laser oscillating on its green (510 nm) and yellow (578.2 nm) transitions. All metal vapor lasers are self-terminating and therefore operate in a pulsed regime.

10.2.1.1. Helium-Neon Lasers[1,2]

The He-Ne laser is certainly the most important of the noble gas lasers. Laser action is obtained from transitions of the neon atom, while helium is added to the gas mixture to greatly facilitate the pumping process. The laser oscillates on many wavelengths, by far the most popular being at $\lambda = 633$ nm (red). Other wavelengths include the green (543 nm) and the infrared ones at $\lambda = 1.15\ \mu$m and $\lambda = 3.39\ \mu$m. The helium-neon laser oscillating on its $\lambda = 1.15\ \mu$m transition was the first gas laser and the first cw laser ever to be operated.[3]

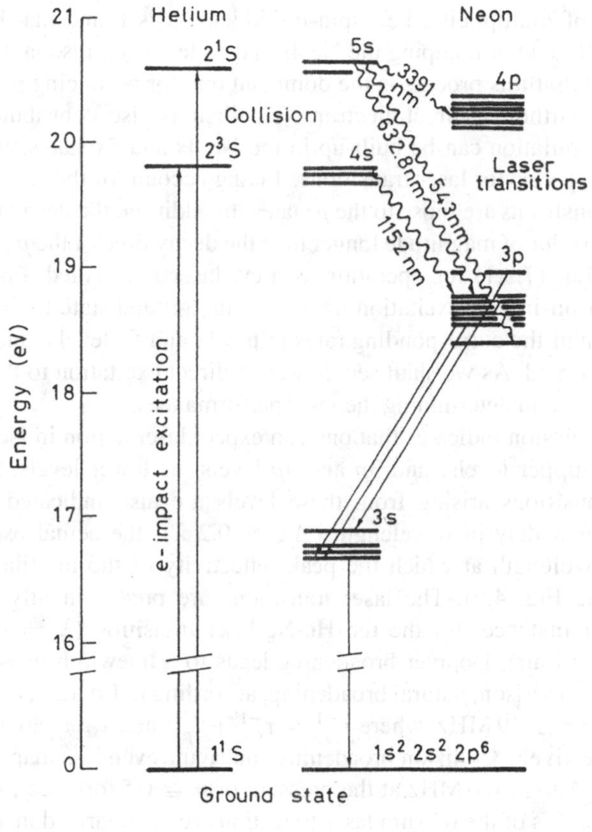

FIG. 10.1. Relevant energy levels of the He-Ne laser.

The energy levels of the He-Ne system that are relevant for laser action are shown in Fig. 10.1. The level notation for He is according to Russell-Saunders coupling with the principal quantum number of the given level also indicated as the first number. Thus the 1^1S state corresponds to the case where the two electrons of He are both in the $1s$ state with opposite spins. The 2^3S and 2^1S states correspond to a situation where one of the two electrons is raised to the $2s$ state with its spin either in the same or opposite direction to that of the other electron, respectively. Neon, on the other hand, has an atomic number of ten and a number of ways have been used, such as Paschen or Racah notations, to indicate its energy levels. For simplicity, however, we will limit ourselves here to simply indicating the electron configuration corresponding to each level. So, the ground state is indicated by the $1s^22s^22p^6$ configuration, while the excited states shown in the figure correspond to the situation where one $2p$ electron is raised to excited s states ($3s$, $4s$, and $5s$) or excited p states ($3p$ and $4p$). One should also notice that, due to the interaction with the remaining five electrons in the $2p$ orbitals, these s and p states are split into four and ten sub-levels, respectively.

It is apparent form Fig. 10.1 that the levels 2^3S and 2^1S of He are nearly resonant with the $4s$ and $5s$ states of Ne. Since the 2^3S and 2^1S levels are metastable ($S \rightarrow S$ transitions are electric dipole forbidden and, furthermore, the $2^3S \rightarrow 1^1S$ transition is also forbidden due to the change of multiplicity, i.e., spin-forbidden), it is found that He atoms in these states prove very efficient at pumping the Ne $4s$ and $5s$ levels by resonant energy transfer. It has been confirmed that this process is the dominant one for producing population inversion in the He-Ne laser, although direct electron-Ne collisions also contribute to the pumping. Since significant population can be built-up in the Ne $4s$ and $5s$ states, they prove suitable candidates as upper levels for laser transitions. Taking account of the selection rules, we see that the possible transitions are those to the p states. In addition, the decay time of the s states ($\tau_s \cong 100\,\mathrm{ns}$) is an order of magnitude longer than the decay time of the p states ($\tau_p \cong 10\,\mathrm{ns}$). So, the condition Eq. (7.3.1) for operation as a cw laser is satisfied. Finally, it should be noted that the electron-impact excitation rates from the ground state to the $3p$ and $4p$ levels are much smaller than the corresponding rates to the $4s$ and $5s$ levels, due to smaller values of cross section involved. As we shall see, however, direct excitation to the $3p$ and $4p$ levels plays an important role in determining the laser performances.

The above discussion indicates that one can expect laser action in Ne to occur between $5s$ or $4s$ levels, as upper levels, and $3p$ and $4p$ levels, as lower levels. Some of the most important laser transitions arising from these levels are also indicated in Fig. 10.1. For transitions differing widely in wavelength ($\Delta\lambda > 0.2\,\lambda$), the actual oscillating transition depends on the wavelength at which the peak reflectivity of the multilayer dielectric mirror is centered (see Fig. 4.9). The laser transitions are predominantly broadened by the Doppler effect. For instance, for the red He-Ne laser transition ($\lambda = 633\,\mathrm{nm}$ in vacuum and $\lambda = 632.8\,\mathrm{nm}$ in air), Doppler broadening leads to a linewidth of $\sim$1.5 GHz (see also example 2.6). By comparison, natural broadening, according to Eq. (2.5.13), can be estimated to be $\Delta\nu_{nat} = 1/2\pi\tau \cong 19\,\mathrm{MHz}$, where $\tau^{-1} = \tau_s^{-1} + \tau_p^{-1}$ and τ_s, τ_p are the lifetimes of the s and p states, respectively. Collision broadening contribute even less than natural broadening [e.g., for pure Ne, $\Delta\nu_c \cong 0.6\,\mathrm{MHz}$ at the pressure of $p \cong 0.5$ torr, see example 2.2]. Some spectroscopic properties of the 633 nm laser transition are summarized in Table 10.1.

The basic design of a He-Ne laser is shown in Fig. 10.2. The discharge is produced between a ring anode and a large tubular cathode, which can thus withstand the collisions from positive ions. The discharge is confined to a capillary for most of the tube length and high inversion is only achieved in the region where the capillary is present. The large volume of gas available in the tube surrounding the capillary acts as a reservoir to replenish the He-Ne mixture in the capillary. When a polarized output is needed, a Brewster angle plate is also inserted inside the laser tube. The laser mirrors are directly sealed to the two tube ends. The most commonly used resonator configuration is nearly hemispherical since this is easy

TABLE 10.1. Spectroscopic properties of laser transitions and gas-mixture composition in some relevant atomic and ionic gas lasers

Laser type	He-Ne	Copper Vapor	Argon Ion	He-Cd
Laser wavelength [nm]	633	510.5	514.5	441.6
Cross-section [$10^{-14}\,\mathrm{cm}^2$]	30	9	25	9
Upper-state lifetime [ns]	150	500	6	700
Lower-state lifetime [ns]	10	$\approx 10^4$	~ 1	1
Transition Linewidth [GHz]	1.5	2.5	3.5	1
Partial pressures of gas mixture [torr]	4 (He) 0.8 (Ne)	40 (He) 0.1–1 (Cu)	0.1 (Ar)	10 (He) 0.1 (Cd)

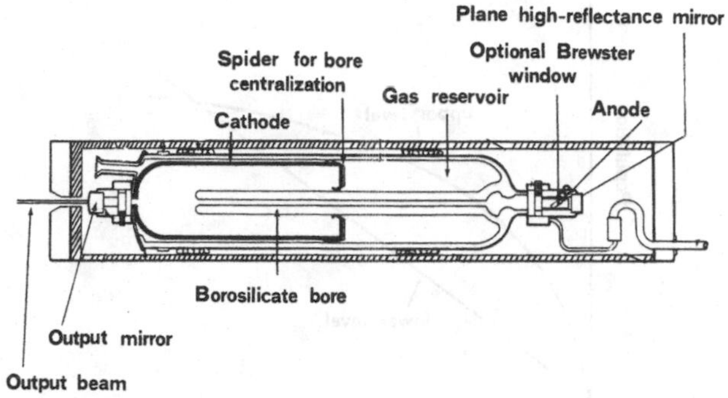

FIG. 10.2. Internal design of a hard-sealed helium-neon laser (courtesy of Melles-Griot).

to adjust, is very stable against misalignment, and readily gives TEM_{00} mode operation. The only disadvantage of this configuration is that it does not fully utilize the volume of the plasma discharge since the mode spot size is much smaller at the plane mirror than at the concave mirror. If, however, the left-hand mirror in Fig. 10.2 is chosen to be the plane mirror, the region of smaller spot size for the near-hemispherical TEM_{00} mode will be outside the capillary, i.e., in a region of low inversion.

One of the most characteristic features of the He-Ne laser is that the output power does not increase monotonically with the discharge current but reaches a maximum and thereafter decreases. For this reason, commercially available He-Ne lasers are provided with a power supply designed to give only the optimum current. The fact that there is an optimum value of current, i.e., of current density J within the capillary is because (at least for the 633 nm and 3.39 μm transitions), at high current densities, de-excitation of the He (2^3S and 2^1S) metastable states takes place not only by collision at the walls but also by super-elastic collision processes such as

$$\mathrm{He}(2^1S) + e \rightarrow \mathrm{He}(1^1S) + e \tag{10.2.1}$$

Since the rate of this process is proportional to the electron density N_e, and hence to J, the overall rate of excitation can be written as $k_2 + k_3J$. In this expression k_2 is a constant that represents de-excitation due to collisions with the walls and k_3J, where k_3 is also a constant, represents the superelastic collision rate of process Eq. (10.2.1). The excitation rate, on the other hand, can be expressed as k_1J, where k_1 is again a constant. Under steady state conditions we can then write that $N_tk_1J = (k_2 + k_3J)N^*$, where N_t is the ground-state He atom population and N^* is the excited (2^1S) state population. The equilibrium 2^1S population is then given by

$$N^* = N_t \frac{k_1J}{k_2 + k_3J} \tag{10.2.2}$$

which can be seen to saturate at high current densities. Since the steady-state population of the 5*s* state of Ne is established by near-resonant energy transfer from He 2^1S state, the population of the upper, 5*s*, laser level will also show a similar saturation behavior as J increases (see Fig. 10.3). On the other hand, in the absence of laser action, the population of the lower

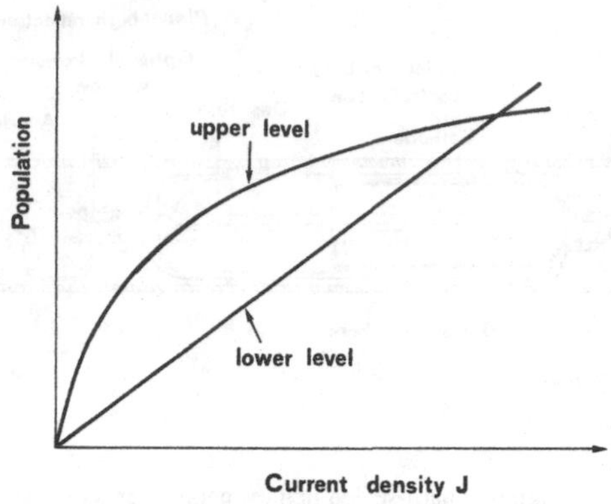

FIG. 10.3. Schematic dependence on current density of upper-level and lower-level populations in a He-Ne laser.

laser level ($3p$ or $4p$), produced by direct pumping from ground-state Ne atoms and radiative cascading from higher levels, is found experimentally to increase linearly with J (see Fig. 10.3). Therefore, as the discharge current is increased, the population difference and hence the output power rises to some optimum value and thereafter falls.

Besides this optimum value of current density, the He-Ne laser has optimum values for other operational parameters, namely: (1) An optimum value for the product of total gas pressure p and capillary diameter $D(pD = 3.6\text{–}4$ torr × mm). This optimum value of the pD product is a consequence of the fact that the electron temperature needs to be optimized (see Sect. 6.4.5). (2) An optimum value of the ratio between the partial pressures of He and Ne ($\sim 5:1$ at $\lambda = 632.8$ nm and $\sim 9:1$ at $\lambda = 1.15\ \mu$m). (3) An optimum value of the capillary diameter ($D \cong 2$ mm). This can be understood when it is realized that, for a constant value of pD i.e., for a given electron temperature, all electron-collision excitation processes scale simply as the number of atoms available for excitation. Since both upper and lower laser levels are ultimately populated by electron-collision processes, their populations and hence the laser gain will be proportional to p i.e., to D^{-1} at constant pD. On the other hand, diffraction losses of the laser cavity will increase as D is decreased and a value of the capillary diameter which optimizes the net gain (gain minus diffraction losses) is therefore expected.

According to the discussion of the behavior presented in Fig. 10.3, He-Ne lasers are typically low power devices (under optimized conditions the available output power at the 633 nm transition may range between 1 and 10 mW for tube lengths ranging between 20 and 50 cm, while the output power on the green transition is typically an order of magnitude less). The efficiency of a He-Ne laser, on any of its laser transitions, is always very low ($<10^{-3}$) a major cause of this low efficiency being the low quantum efficiency. In fact, from Fig. 10.1, one can readily see that each elementary pumping cycle requires an energy of ~20 eV while the energy of the laser photon is less than 2 eV. The narrow gain linewidth, on the other hand, is an advantage when single longitudinal mode is required. In fact, if the cavity length is short enough ($L<15$–20 cm), single longitudinal mode is readily achieved when the cavity length is tuned (by a piezo-electric translator) to bring a cavity mode into coincidence with the peak of the gain line (see Sect. 7.8.2). Single mode He-Ne lasers can then be frequency stabilized to a high degree $[(\Delta\nu/\nu) = 10^{-11} \div 10^{-12}]$ against a frequency reference (such as a high finesse

Fabry-Perot interferometer or, for absolute stabilization, against an $^{129}I_2$ absorption line, for the 633 nm transition).

He-Ne lasers oscillating on the red transition are still widely used for many applications where a low power visible beam is needed (such as alignment or bar-code scanners). Most supermarkets and other stores use red He-Ne lasers to read the coded information contained in the bar code on each product. For some of these applications, however, He-Ne lasers are facing a very strong competition from red-emitting semiconductor lasers, these lasers being smaller and more efficient. Given the better visibility of a green beam to the eye, green-emitting He-Ne lasers are increasingly being used for alignment and cell cytometry. In this last application, individual cells (e.g., red-blood cells) stained by suitable fluorochromes are flowed, rapidly, through a capillary, onto which the He-Ne laser is focused, and are characterized by their subsequent scattering or fluorescent emission. Single mode He-Ne lasers are also often used for metrological applications (e.g., very precise, interferometric, distance measurements) and for holography.

10.2.1.2. Copper Vapor Lasers[4]

The relevant energy levels of the copper vapor laser are shown in Fig. 10.4, where Russell-Saunders notation is again used. The $^2S_{1/2}$ ground state of Cu corresponds to the electron configuration $3d^{10}4s^1$, while, the excited $^2P_{1/2}$ and $^2P_{3/2}$ levels correspond to the outer $4s$ electron being promoted to the next higher $4p$ orbital. The $^2D_{3/2}$ and $^2D_{5/2}$ levels arise from the electron configuration $3d^94s^2$ in which an electron has been promoted from the $3d$ to the $4s$ orbital.

The relative values of the corresponding cross-sections are such that the rate of electron-impact excitation to the P states is greater than that to the D states. Thus, the P states are preferentially excited by electron-impact. On the other hand, the $^2P \rightarrow {}^2S_{1/2}$ transition is

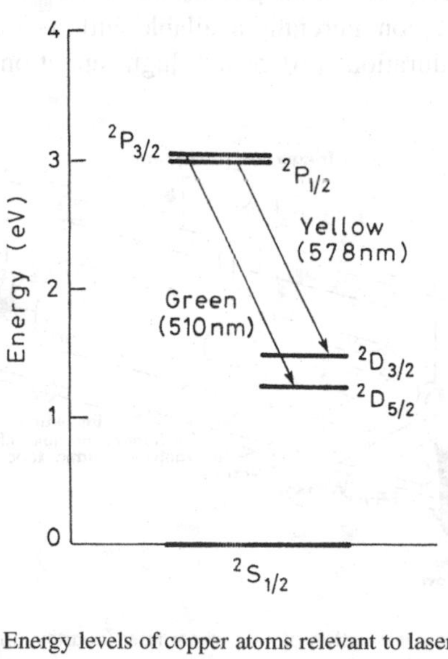

FIG. 10.4. Energy levels of copper atoms relevant to laser operation.

strongly electric-dipole allowed (we recall that the selection rules for optically allowed transitions require that $\Delta J = 0$ or ± 1) and the corresponding absorption cross section turns out to be quite large. At the temperature used for Cu lasers ($T = 1500\,°C$), the vapor pressure is then sufficiently high (~ 0.1 torr) that the $^2P \to {}^2S_{1/2}$ transition is completely trapped. Thus, the only effective decay route of the ^2P state is via the ^2D states and the corresponding decay time is rather long ($\sim 0.5\,\mu s$) since the transition is only weakly allowed. It then follows that the ^2P states can accumulate a large population and are therefore good candidates to act as upper laser levels. Thus laser action in Cu can occur on both the $^2P_{3/2} \to {}^2D_{5/2}$ (green) and $^2P_{1/2} \to {}^2D_{3/2}$ (yellow) transitions. One should note that the $^2D \to {}^2S$ transition is electric-dipole forbidden and the lifetime of the ^2D state is very long (a few tens of microseconds). It then follows that the laser transition is self-terminating and the laser can only operate on a pulsed basis with pulse duration of the order of or shorter than the lifetime of the ^2P state. One should also note that the $^2D \to {}^2S$ decay occurs mainly via super-elastic collisions with cold electrons remaining after the pump pulse, and that the corresponding decay rate sets an upper limit to the repetition rate of the laser. Some relevant spectroscopic properties of the copper-vapor green transition are indicated in Table 10.1 as a representative example.

The construction of a metal vapor laser is based on the arrangement shown schematically in Fig. 10.5, where the vapor is contained in an alumina tube which is thermally isolated in a vacuum chamber. The necessary high temperature is usually maintained by the power dissipated in the tube due to the repetitively pulsed pumping-current. Anode and cathode are in the form of ring electrodes and are placed at the ends of the alumina tube. A 25–50 torr neon buffer gas is used to provide enough electron density, after the passage of the discharge pulse, to allow for de-excitation of the lower ^2D state by superelastic collisions. The neon gas is also helpful in reducing the diffusion length of the Cu vapor thus preventing metal-vapor deposition on the (cold) end-windows. More recently, the so-called Copper-HyBrID lasers have also been introduced, which use HBr in the discharge. Since, in this case, CuBr molecules are formed in the discharge region and these molecules are much more volatile than Cu atoms, the temperatures required in the gas discharge are lower.

Copper vapor lasers are commercially available with average output powers in excess of 100 W, with short pulse durations (30–50 ns), high repetition rates (up to $\sim$10 kHz) and

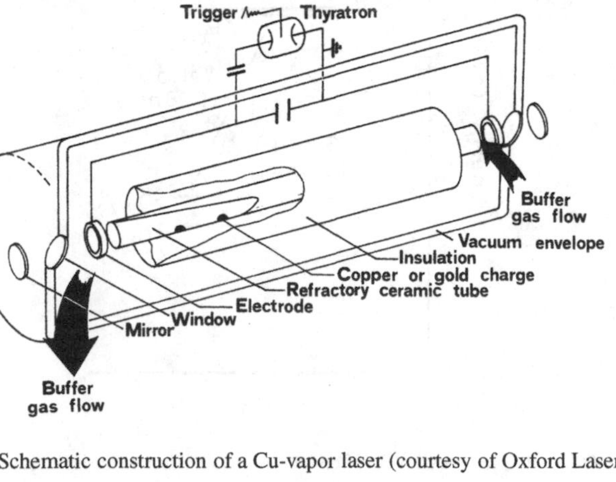

FIG. 10.5. Schematic construction of a Cu-vapor laser (courtesy of Oxford Lasers, Ltd).

relatively high efficiency (~1%). The latter is the result of both the high quantum efficiency of the copper laser (~55%, see Fig. 10.4) and the high electron-impact cross section for the $^2S \rightarrow {}^2P$ transition. Even higher output powers (~200 W) and higher efficiencies (~3%) have recently been obtained with Copper-HyBrID lasers.

Copper vapor lasers are used for some industrial applications (such as high-speed photography, resistor trimming, and more recently, micromachining) and as a pump for dye lasers. In particular, in high-speed flash photography, the short pulse (tens of ns) and high-repetition-rate (10–20 kHz) are exploited in stroboscopic illumination of various, rapidly moving, objects (e.g., a bullet in flight). A large facility based on copper-laser pumped dye lasers (using many copper lasers, each with average power up to 100 W) is currently in use in a pilot plant, in the United States, for ^{235}U isotope separation.

10.2.2. Ion Lasers

In the case of an ionized atom, the scale of energy levels is expanded in comparison with neutral atoms. In this case, in fact, an electron in the outermost orbital(s) experiences the field due to the positive charge Ze of the nucleus (Z being the atomic number and e the electronic charge) screened by the negative charge $(Z-2)e$ of the remaining electrons. Assuming, for simplicity, the screening to be complete, the net effective charge is then $2e$ rather than simply e for the corresponding neutral atom. This expansion in energy scale means that ion lasers typically operate in the visible or ultraviolet regions. As in the case of neutral atom lasers, ion lasers can be divided into two categories: (1) Ion gas lasers, involving most of the noble gases, the most notable example being the Ar^+ laser, which we consider below, and the Kr^+ laser. Both lasers oscillate on many transitions, the most common being in the green and blue (514.5 nm and 488 nm) for the Ar^+ laser and in the red (647.1 nm) for the Kr^+ laser. (2) Metal-ion vapor lasers, involving many metals (Sn, Pb, Zn, Cd, and Se), the most notable example being the He-Cd laser, discussed below, and the He-Se laser.

10.2.2.1. Argon Laser[5,6]

A simplified scheme for the relevant energy levels in an argon laser is shown in Fig. 10.6. The Ar^+ ground state is obtained by removing one electron out of the six electrons of the, $3p$, outer shell of Ar. The excited $4s$ and $4p$ states are then obtained by promoting one of the remaining $3p^5$ electrons to the $4s$ or $4p$ state, respectively. As a consequence of the interaction with the other $3p^4$ electrons, both the $4s$ and $4p$ levels, indicated as single levels in Fig. 10.6, actually consist of many sublevels.

Excitation of the Ar ion to its excited states occurs by a two-step process involving collisions with two distinct electrons. The first collision, in fact, ionizes Ar i.e., raises it to the Ar^+ ground state, while the second collision excites the Ar ion. Since the lifetime of the $4p$ level ($\sim 10^{-8}$ s, set by the $4p \rightarrow 4s$ radiative transition) is about ten times longer than the radiative lifetime of the $4s \rightarrow 3p^5$ transition, excited Ar ions accumulate predominantly in the $4p$ level. This means that the $4p$ level can be used as the upper laser level, for the $4p \rightarrow 4s$ laser transition, and that, according to Eq. (7.3.1), cw laser action can be achieved. It should be noted that excitation of the Ar ion can lead to ions in the $4p$ state by three distinct processes

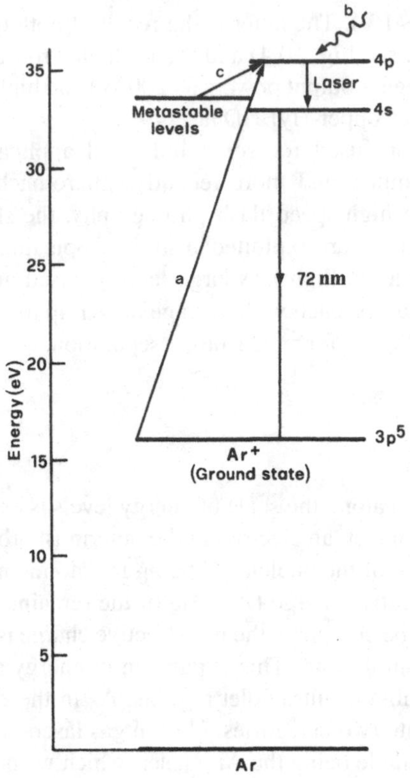

FIG. 10.6. Energy levels of Ar relevant for laser action.

(see Fig. 10.6): (a) direct excitation to the $4p$ level starting from the Ar^+ ground level; (b) excitation to higher-lying states followed by radiative decay to the $4p$ level; (c) excitation to metastable levels followed by a third collision leading to excitation to the $4p$ state. Considering, for simplicity, only processes (a) and (b) one can readily see that the pumping process to the upper state is expected to be proportional to the square of the discharge current density. In fact, since processes (a) and (b) involve a two-steps electron collision, the rate of upper state excitation, $(dN_2/dt)_p$, is expected to be of the form

$$(dN_2/dt)_p \propto N_e N_i \cong N_e^2 \tag{10.2.3}$$

where N_e and N_i are the electron and ion density in the plasma ($N_e \cong N_i$ in the positive column of the plasma). Since the electric field of the discharge is independent of the discharge current, the drift velocity, v_{drift}, will also be independent of the discharge current. From the standard equation $J = ev_{drift}N_e$, one then see that the electron density N_e is proportional to the discharge current density, and, from Eq. (10.2.3) it follows that $(dN_2/dt)_p \propto J^2$. Laser pumping thus increases rapidly with current density and high current densities ($\sim$1 kA/cm^2) are required if the inherently inefficient two-step processes, considered above, are to pump enough ions to the upper state. This may explain why the first operation of an Ar^+ laser occurred some three-years after the first He-Ne laser.(7)

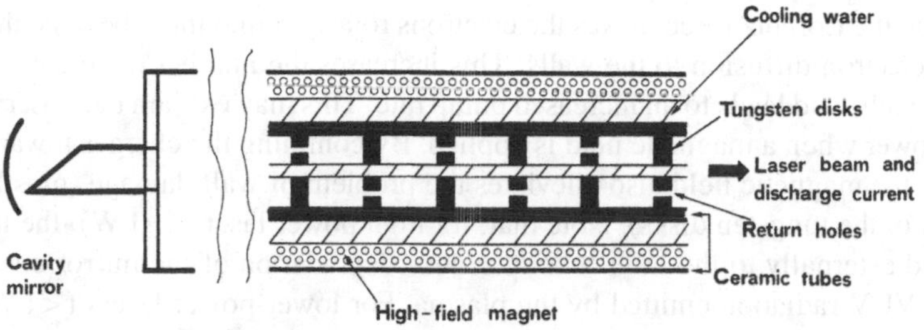

FIG. 10.7. Schematic diagram of a high-power water-cooled Ar^+ laser tube.

From the discussion above, one expects laser action in an Ar laser to occur on the $4p \rightarrow 4s$ transition. Since both the $4s$ and $4p$ levels actually consist of many sub-levels, the argon laser is found to oscillate on many lines, the most intense being in the green ($\lambda = 514.5$ nm) and in the blue ($\lambda = 488$ nm). From spectral measurements of the spontaneously emitted light it is found that the Doppler linewidth $\Delta\nu_0^*$, on e.g. the green transition, is about 3500 MHz. From Eq. (2.5.18) it is seen that this implies an ion temperature of $T \cong 3000$ K. The ions are therefore very hot, a result of ion acceleration by the electric field of the discharge. Some relevant spectroscopic properties of the Ar ion green laser transition are summarized in Table 10.1.

A schematic diagram of a high power (>1 W) argon laser is shown in Fig. 10.7. One sees that both the plasma current and the laser beam are confined by metal disks (of tungsten) inserted in a larger bore tube of ceramic material (BeO). The use of this thermally conductive and resistant metal-ceramic combination is necessary to ensure a good thermal conductivity of the tube and, at the same time, to reduce the erosion problems that arise from the high ion temperature. The diameter of the central holes in the disks is kept small (~2 mm) to confine oscillation to a TEM_{00} mode (long-radius of curvature mirrors are commonly used for the resonator) and to reduce the total current required. A problem that must be overcome in an argon laser is that which arises from the cataphoresis of the argon ions. Due to the high current density, in fact, a substantial migration of Ar ions occurs toward the cathode, where they are neutralized upon combining with electrons emitted by the cathode surface*. Thus an accumulation of neutral atoms tends to build up at this electrode with a corresponding reduction of the Ar pressure in the discharge capillary, below its optimum value. To overcome this problem, off-center holes are also made in the disk to provide return paths for the atoms, from cathode to anode, by diffusion. The holes are arranged in such a way that no current flows along this return path on account of the longer path lengths involved compared to that of the central path. The inner ceramic tube is water cooled to remove the large amount of heat that it is inevitably dissipated in the tube (some kW/m). Note also that a static magnetic field is also applied in the discharge region, parallel to the tube axis, by a solenoid. With this

* With reference to the discussion of Sect. 6.4.4., we recall that direct electron ion recombination cannot occur in the discharge volume since the process cannot satisfy, simultaneously, both energy and momentum conservation. Electron-ion recombination can therefore only occur in the presence of a third partner e.g., at the tube walls or at the cathode surface.

arrangement, the Lorentz force makes the electrons rotate around the tube axis, thus reducing the rate of electron diffusion to the walls. This increases the number of free electrons at the center of the tube and leads to an increased pump rate. This may explain the observed increase in output power when a magnetic field is applied. By confining the charges toward the center of the tube, the magnetic field also alleviates the problem of wall damage (mostly occurring at the holes of the tungsten disks). Note that, for high power lasers (>1 W), the laser mirrors are mounted externally to the laser tube to reduce degradation of the mirror coatings due to the intense VUV radiation emitted by the plasma. For lower-power lasers (<1 W), the laser tube is often simply made of a block of ceramic (BeO) provided with a central hole for the discharge current. In this case, no magnetic field is applied, the tube is air cooled, and the mirrors are directly sealed to the ends of the tube as in a He-Ne laser.

Water-cooled argon lasers are commercially available with power ranging between 1 and 20 W, operating on both blue and green transitions simultaneously or, by using the configuration of Fig. 7.16b, operating on a single line. Air-cooled argon lasers of lower power (~100 mW) and much simpler design are also commercially available. In both cases, once above threshold, the output power increases rapidly with current density ($\propto J^2$) since, by contrast with the behavior of He-Ne lasers, in this case there is no process leading to saturation of inversion. The laser efficiency is, however, very low ($<10^{-3}$), because the laser quantum efficiency is rather low (~7.5%, see Fig. 10.6) and because the electron impact excitation involves many levels that are not coupled effectively to the upper laser level. Argon lasers are often operated in mode-locked regime using an acousto-optic modulator. In this case, fairly short mode-locked laser pulses (~150 ps) are achievable by virtue of the relatively large transition line-width (~3.5 GHz) which, furthermore, is inhomogeneously broadened.

Argon lasers are widely used in ophthalmology (particularly to cure diabetic retinopathy), and in the field of laser entertainment (laser light shows). In a more scientifically-oriented contest, argon lasers are also widely used for a variety of studies of light-matter interaction (particularly in mode-locked operation) and also as a pump for solid-state lasers (particularly Ti:sapphire) and dye lasers. For many of these applications, argon lasers are tending to be superseded now by cw diode-pumped Nd : YVO_4 lasers in which a green beam, $\lambda = 532$ nm, is produced by intracavity second harmonic generation. Lower-power Ar lasers are extensively used for high-speed laser printers and cell cytometry.

10.2.2.2. He-Cd Laser

The energy levels of the He-Cd system that are relevant for laser action are shown in Fig. 10.8, where, again, level labeling is according to Russell-Saunders notation. Pumping of Cd^+ upper laser levels ($^2D_{3/2}$ and $^2D_{5/2}$) is achieved with the help of He through the Penning ionization process. This process can be written in the general form

$$A^* + B \rightarrow A + B^+ + e \tag{10.2.4}$$

where the ion B^+ may or may not be left in an excited state. Of course, the process can only occur if the excitation energy of the excited species A^* is greater than or equal to the ionization energy of species B (plus the excitation energy of B^+ if the ion is left in an excited state). Note that, unlike near-resonant energy transfer, Penning ionization is a non-resonant process, any surplus energy being in fact released as kinetic energy of the ejected electron. In the case of the

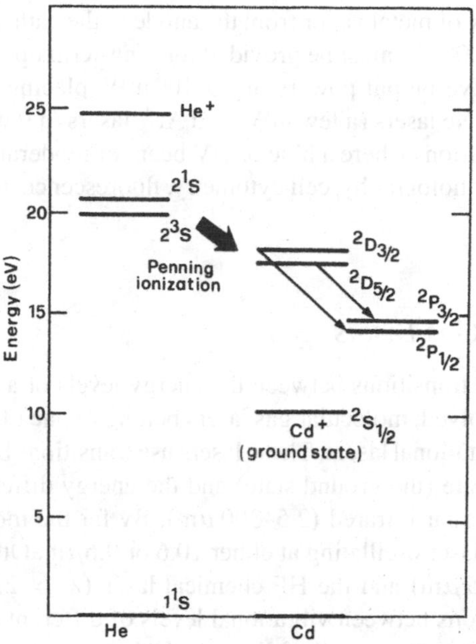

FIG. 10.8. Relevant energy levels of the He-Cd laser.

He-Cd system, the 2^1S and 2^3S metastable states of He act as the species A^* in Eq. (10.2.4), and, upon collision, this excitation energy is given up to ionize the Cd atom and to excite the Cd^+ ion. Although the process is not resonant, it has been found that the rate of excitation to the D states is about three times larger than that to the P states*. What is more important, however, is that the lifetime of the D states ($\sim 0.1\ \mu s$) is much longer than the lifetime of the P states (~ 1 ns). Population inversion between the D and P states can, therefore, be produced readily and cw laser action can be produced. Indeed, in accordance with the selection rule $\Delta J = 0, \pm 1$, laser action is obtained on the ${}^2D_{3/2} \rightarrow {}^2P_{1/2}$ ($\lambda = 325$ nm, UV) and the ${}^2D_{5/2} \rightarrow {}^2P_{3/2}$ ($\lambda = 416$ nm, blue) lines. The Cd^+ ion then drops to its ${}^2S_{1/2}$ ground state by radiative decay.

A typical construction for the He-Cd laser is in the form of a tube, terminated by two Brewster's angle windows, with the two laser mirrors mounted separate from the tube. In one possible arrangement, the tube, which is filled with helium, also has a small reservoir containing the Cd metal, located near the anode. The reservoir is raised to a high enough temperature ($\sim$250 °C) to produce the desired vapor pressure of Cd atoms in the tube. The discharge itself then produces enough heat to prevent condensation of the vapor on the tube walls along the discharge region. Due to ion cataphoresis, however, the ions migrate to the cathode where they recombine with the electrons emitted by the cathode. Neutral Cd atoms then condense around the cathode region, where there is no discharge and the temperature is low. The net

* According to e.g., (2.6.2), the rate of excitation, k_{A^*B}, of the general process Eq. (10.2.4) can be defined via the relation $(dN/dt)_{AB^+} = k_{A^*B} N_{A^*} N_B$, where $(dN/dt)_{B^+}$ is the number of species B^+, which are produced per unit volume per unit time, and N_{A^*} and N_B are the concentrations of the colliding partners.

result is a continuous flow of metal vapor from the anode to the cathode. Therefore a sufficient supply of Cd (~1 g per 1,000 h) must be provided for long-term operation of the laser.

He-Cd lasers can give output powers of 50–100 mW, placing them in an intermediate position between red He-Ne lasers (a few mW) and Ar^+ lasers (a few W). Thus, He-Cd lasers are used for many applications where a blue or UV beam of moderate power is required (e.g., high-speed laser printers, holography, cell cytometry, fluorescence analysis of e.g., biological specimens).

10.2.3. Molecular Gas Lasers

These lasers exploit transitions between the energy levels of a molecule. Depending on the type of transition involved, molecular gas lasers belong to one of the three following categories: (1) Vibrational-rotational lasers. These lasers use transitions between vibrational levels of the same electronic state (the ground state) and the energy difference between the levels falls in the middle- to the far-infrared (2.5–300 μm). By far the most important example of this category is the CO_2 laser oscillating at either 10.6 or 9.6μm. Other noteworthy examples are the CO laser ($\lambda \cong 5\,\mu$m) and the HF chemical laser ($\lambda \cong 2.7$–$3.3\,\mu$m). (2) Vibronic lasers, which use transitions between vibrational levels of different electronics states: In this case the oscillation wavelength generally falls in the UV region. The most notable example of this category of laser is the nitrogen laser ($\lambda = 337$ nm). A special class of lasers, which can perhaps be included in the vibronic lasers, is the excimer laser. These lasers involve transitions between different electronic states of special molecules (excimers) with corresponding emission wavelengths generally in the UV. Excimer lasers, however, involve not only transitions between bound states (bound-bound transitions) but also, and actually more often, transitions between a bound upper state and a repulsive ground state (bound-free transitions). It is more appropriate therefore to treat these lasers as being in a category of their own. (3) Pure rotational lasers, which use transitions between different rotational levels of the same vibrational state (usually an excited vibrational level of the ground electronic state). The corresponding wavelength falls in the far infrared (25 μm to 1 mm). Since these pure rotational lasers are relatively less important than the other categories, we shall not discuss them further in the sections that follow. We therefore limit ourselves to pointing out here that laser action is more difficult to achieve in this type of laser since the relaxation between rotational levels is generally very fast. Therefore these lasers are usually pumped optically, using the output of another laser as the pump (commonly a CO_2 laser). Optical pumping excites the given molecule (e.g., CH_3F, $\lambda = 496\,\mu$m) to a rotational level belonging to some vibrational state above the ground level. Laser action then takes place between rotational levels of this upper vibrational state.

10.2.3.1. The CO_2 Laser[8,9]

The laser utilizes, as active medium, a suitable mixture of CO_2, N_2, and He. Oscillation takes place between two vibrational levels of the CO_2 molecule, while, as we shall see, the N_2 and He greatly improve the efficiency of laser action. The CO_2 laser is actually one of the most powerful lasers (output powers of more than 100 kW have been demonstrated from a CO_2 gas-dynamic laser) and one of the most efficient (15–20% slope efficiency).

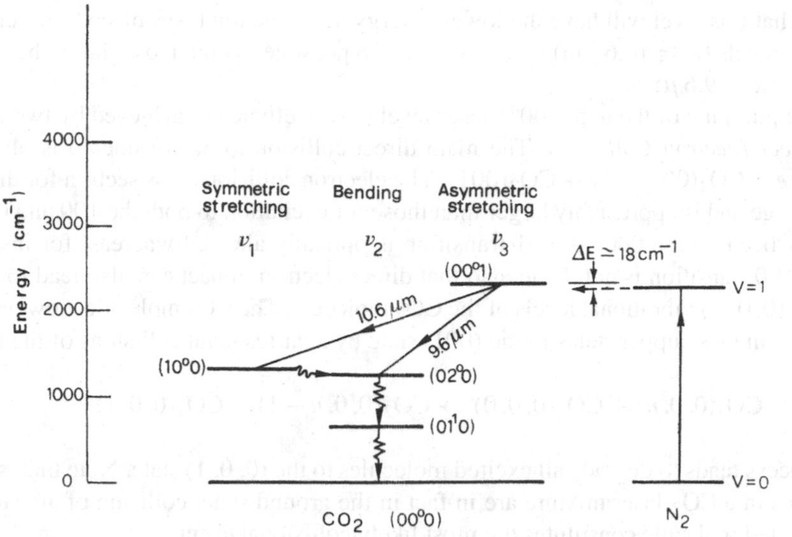

FIG. 10.9. The lowest vibrational levels of the ground electronic state of a CO_2 molecule and a N_2 molecule (for simplicity, the rotational levels are not shown).

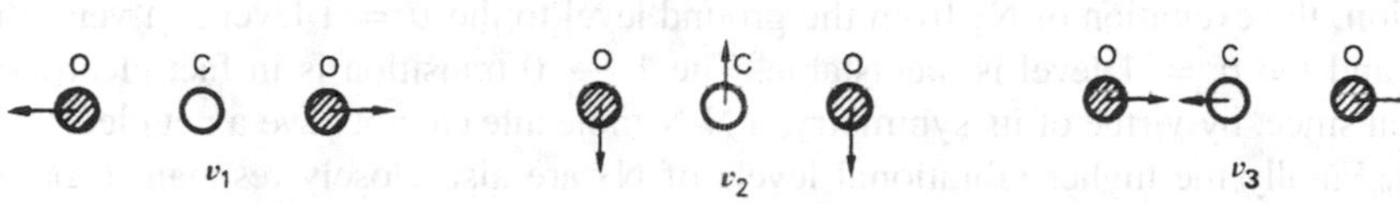

FIG. 10.10. The three fundamental modes of vibration for a CO_2 molecule: (ν_1) symmetric stretching mode, (ν_2) bending mode, (ν_3) asymmetric stretching mode.

Figure 10.9 shows the relevant vibrational-energy levels for the electronic ground states of the CO_2 and N_2 molecules. N_2, being a diatomic molecule, has only one vibrational mode whose lowest two energy levels ($\upsilon = 0$, $\upsilon = 1$) are indicated in the figure. The energy levels for CO_2 are more complicated since CO_2 is a linear triatomic molecule. In this case, there are three nondegenerate modes of vibration (Fig. 10.10): (1) Symmetric stretching mode, (2) bending mode, and (3) asymmetric stretching mode. The oscillation behavior and the corresponding energy levels are therefore described by means of three quantum numbers n_1, n_2 and n_3, which give the number of quanta in each vibrational mode. This means that, apart from zero-point energy, the energy of the level is given by $E = n_1 h\nu_1 + n_2 h\nu_2 + n_3 h\nu_3$, where ν_1, ν_2, and ν_3 are the resonance frequencies of the three modes. For example, the 01^10 level* corresponds to an oscillation in which there is one vibrational quantum in mode 2. Since mode 2 has the smallest force constant of the three modes (the vibrational motion is transverse), it

* The superscript (which we will denote by l) on the bending quantum number arises from the fact that the bending vibration is, in this case, doubly degenerate. In fact, it can occur both in the plane of Fig. 10.10 and in the orthogonal plane. A bending vibration therefore consists of a suitable combination of these two vibrations and the superscript l characterizes this combination; more precisely, $l\hbar$ gives the angular momentum of this vibration about the axis of the CO_2 molecule. For example, in the 02^00 state ($l = 0$), the two degenerate vibrations combine in such a way to give an angular momentum $l\hbar = 0$.

follows that this level will have the lowest energy. Laser action takes place between the 00^01 and 10^00 levels ($\lambda \cong 10.6\,\mu$m) although it is also possible to obtain oscillation between 00^01 and 02^00 ($\lambda \cong 9.6\,\mu$m).

The pumping of the upper 00^01 laser level is very efficiently achieved by two processes:

Direct Electron Collisions. The main direct collision to be considered is obviously as follows: $e + CO_2(000) \rightarrow e + CO_2(001)$. The electron collision cross section for this process is very large and is appreciably larger than those for excitation to both the 100 and 020 levels, probably because the 000 $\rightarrow$ 001 transition is optically allowed whereas, for instance, the 000 $\rightarrow$ 100 transition is not. Note also that direct electron impact can also lead to excitation of upper $(0, 0, n)$ vibrational levels of the CO_2 molecule. The CO_2 molecule, however, rapidly relaxes from these upper states to the (001) state by near resonant collisions of the type*

$$CO_2(0,0,n) + CO_2(0,0,0) \rightarrow CO_2(0,0,n-1) + CO_2(0,0,1) \tag{10.2.5}$$

This process tends to degrade all excited molecules to the (0, 0, 1) state. Note that, since most molecules in a CO_2 laser mixture are in fact in the ground state, collision of an excited with an unexcited molecule constitutes the most likely collisional event.

Resonant Energy Transfer from N_2 Molecule. This process is also very efficient due to the small energy difference between the excited levels of the two molecules ($\Delta E = 18\,\text{cm}^{-1}$). In addition, the excitation of N_2 from the ground level to the $\upsilon = 1$ level is a very efficient process and the $\upsilon = 1$ level is metastable. The $1 \rightarrow 0$ transition is in fact electric-dipole forbidden since, by virtue of its symmetry, a N-N molecule cannot have a net electric dipole moment. Finally the higher vibrational levels of N_2 are also closely resonant ($\Delta E < kT$) with the corresponding CO_2 levels (up to 00^05), and transitions between the excited levels, $00n$, and the 001 level of the CO_2 molecule occur rapidly through the process indicated in Eq. (10.2.5).

The next point to consider is the decay of both upper and lower laser levels. We note that, although the transitions $00^01 \rightarrow 10^00$, $00^01 \rightarrow 02^00$, $10^00 \rightarrow 01^00$, and $02^00 \rightarrow 01^00$ are optically allowed, the corresponding decay times τ_{sp} for spontaneous emission are very long (we recall that $\tau_{sp} \propto 1/\nu^3$). The decay of these various levels is therefore determined essentially by collisions. Accordingly, the decay time τ_s of the upper laser level can be obtained from a formula of the type

$$(1/\tau_s) = \Sigma a_i p_i \tag{10.2.6}$$

where p_i are the partial pressures and a_i the rate constants that are corresponding to the gases in the discharge. Taking, for example, the case of a total pressure of 15 torr (in a 1:1:8 $CO_2 : N_2 :$ He partial pressure ratio) one finds that the upper level has a lifetime $\tau_s \cong 0.4$ ms. As far as the relaxation rate of the lower level is concerned, we begin by noting that the 100 $\rightarrow$ 020 transition is very fast and it occurs even in a isolated molecule. In fact the energy difference between the two levels is much smaller than kT. Furthermore, a coupling between the two states is present (Fermi resonance) because a bending vibration tends to induce a change of distance between the two oxygen atoms (i.e., induce a symmetric stretching). Levels

* Relaxation processes in which vibrational energy is given up as vibrational energy of another like or unlike molecule are usually referred to as *VV relaxations*.

10^00 and 02^00 are then effectively coupled to the 01^10 level by the two near-resonant collision processes, involving CO_2 molecules in the ground state (VV relaxation), as shown below:

$$CO_2(10^00) + CO_2(00^00) \rightarrow CO_2(01^10) + CO_2(01^10) + \Delta E \quad (10.2.7a)$$

$$CO_2(02^00) + CO_2(00^00) \rightarrow CO_2(01^10) + CO_2(01^10) + \Delta E' \quad (10.2.7b)$$

The two processes have a very high probability since ΔE and $\Delta E'$ are much smaller than kT. It follows, therefore, that the three levels 10^00, 02^00 and 01^10 reach thermal equilibrium in a very short time. We are therefore left with the decay from the 01^10 to the ground level 00^00. If this decay were slow, it would lead to an accumulation of molecules in the 01^10 level during laser action. This in turn would produce an accumulation in the (10^00) and (02^00) levels since these are in thermal equilibrium with the (01^10) level. Thus a slowing down of the decay process of all three levels would occur, i.e., the $01^10 \rightarrow 00^00$ transition would constitute a "bottleneck" in the overall decay process. It is, therefore, important to look into the question of the lifetime of the 01^10 level. Note that, since the $01^10 \rightarrow 00^00$ transition is the least energetic transition in any of the molecules in the discharge, relaxation from the 01^10 level can only occur by transferring this vibrational energy to translational energy of the colliding partners (VT relaxation). From the theory of elastic collisions we then know that energy is most likely to be transferred to the lighter atoms, i.e., to helium in this case. This means that the lifetime is again given by an expression of the type of Eq. (10.2.6) where the coefficient a_i for He is much larger than that of the other species. For the same partial pressures as considered in the example above, one obtains a lifetime of about 20 μs. It follows from the above discussion that this will also be the value of the lifetime of the lower laser level. So, by virtue of the much larger value of the upper state lifetime, population accumulates in the upper laser level and the condition for cw laser action is also fulfilled. Note that He has another valuable effect: The He, because of its high thermal conductivity, helps to keep the CO_2 cool by conducting heat away to the walls of the container. A low translational temperature for CO_2 is necessary to avoid population of the lower laser level by thermal excitation. The energy separation between the levels is, in fact, comparable to kT. In conclusion, the beneficial effects of nitrogen and helium can be summarized as follows: Nitrogen helps to produce a large population in the upper laser level while helium helps to empty population from the lower laser level.

From the above considerations it is seen that laser action in a CO_2 laser may occur either on the $(00^01) \rightarrow (10^00)(\lambda{=}10.6\,\mu\text{m})$ transition or on the $(00^01) \rightarrow (02^00)$ transition ($\lambda{=}9.6\mu$ m). Since the first of these transitions has the greater cross section and since both transitions share the same upper level, it follows that it is usually the $00^01 \rightarrow 10^00$ transition that oscillates. To obtain oscillation on the 9.6μ m line, some appropriate frequency-selective device is placed in the cavity to suppress laser action on the line with highest gain (the system of Fig. 7.16a is often used). Our discussion has so far ignored the fact that both upper and lower laser levels actually consist of many closely spaced rotational levels. Accordingly, the laser emission may occur on several equally spaced rotational-vibrational transitions belonging to either P or R branches (see Fig. 3.7) with the P branch exhibiting the largest gain. To complete our discussion one must now also take into account the fact that, as a consequence of the Boltzmann distribution between the rotational levels, the $J' = 21$ rotational level of the upper 00^01 state happens to be the most heavily populated (see Fig. 10.11)*. Laser oscillation

* For symmetry reasons, only levels with odd values of J are occupied in a CO_2 molecule.

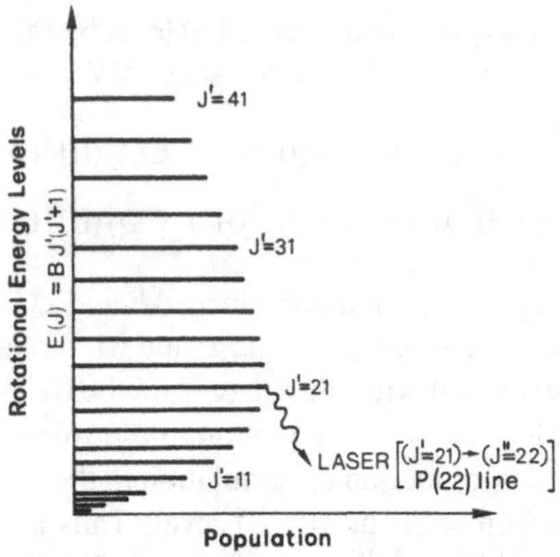

FIG. 10.11. Relative population of the rotational levels of the upper laser level of CO_2.

will then occur on the rotational-vibrational transition with the largest gain, i.e., originating from the most heavily populated level. This is because, in a CO_2 laser, the rate of thermalization of the rotational levels ($10^7\ s^{-1}\ torr^{-1}$) is faster than the rate of decrease of population (due to spontaneous and stimulated emission) of the rotational level from which laser emission is occurring. Therefore, the entire population of rotational levels will contribute to laser action on the rotational level with highest gain. Consequently, and as a conclusion to this discussion, we can say that laser action in a CO_2 laser normally occurs on the $P(22) \rightarrow$ i.e., $(J' = 21)$ $(J'' = 22)$ line of the $(00^01) \rightarrow (10^00)$ transitions. Other lines of the same transition as well as lines belonging to the $(00^01) \rightarrow (02^00)$ transition (the separation between rotational lines in a CO_2 laser is about $2\,cm^{-1}$) can be selected with, e.g., the scheme of Fig. 7.16a.

A major contribution to the laser linewidth in a CO_2 laser comes from the Doppler effect. Compared with, e.g., a visible gas laser, the corresponding value of the Doppler linewidth is, however, rather small (about 50 MHz) on account of the low frequency ν_0 of the laser transition [see example 3.2]. Collision broadening is, however, not negligible [see example 3.3]. and actually becomes the dominating line-broadening mechanism for CO_2 lasers operating at high pressures ($p > 100$ torr).

From the point of view of their constructional design, CO_2 lasers can be separated into eight categories: (1) Lasers with slow axial flow, (2) sealed-off lasers, (3) waveguide lasers, (4) lasers with fast axial flow, (5) diffusion-cooled area-scaling lasers, (6) transverse-flow lasers, (7) transversely excited atmospheric pressure (TEA) lasers, and (8) gas-dynamic lasers. We will not consider the gas-dynamic laser, here, since its operating principle has already been described in Sect. 6.1. Before considering the other categories, it is worth pointing out here that, although the corresponding lasers differ from one another as far as many of their operating parameters are concerned (e.g., the output power), they all share a very important characteristic feature namely a high slope efficiency (15–25%). This high efficiency is a consequence of the high laser quantum efficiency (~40%; see Fig. 10.10) and of the highly efficient pumping processes that occur in a CO_2 laser at the optimum electron temperature of the discharge (see Fig. 6.27).

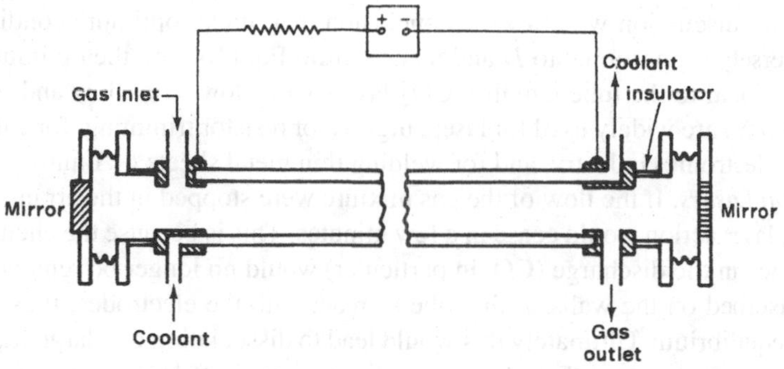

FIG. 10.12. Schematic diagram of a CO_2 laser with longitudinal gas flow.

Lasers with Slow Axial Flow. Operation of the first CO_2 laser was achieved in a laser of this type (C. K. N. Patel, 1964[10]). The gas mixture is flowed slowly along the laser tube (see Fig. 10.12) simply to remove the dissociation products, in particular CO, that would otherwise contaminate the laser. Heat removal is provided by radial conduction of heat to the tube walls (usually made of glass), which are cooled externally by a suitable coolant (usually water). An internal mirror arrangement is often used, and, at least in the design of Fig. 10.12, one of the metal mounts holding the cavity mirrors needs to be held at high voltage. One of the main limitations of this laser arises from the fact that there is a maximum laser output power per unit length of the discharge (50–60 W/m) that can be obtained, independently of the tube diameter. In fact, from Eq. (6.4.24), we find that the number of molecules pumped into the upper laser level per unit volume and unit time can be written as

$$\left(\frac{dN_2}{dt}\right)_p = N_t \frac{J}{e}\left[\frac{<\upsilon\sigma>}{\upsilon_{drift}}\right] \tag{10.2.8}$$

where J is the current density, σ is a suitable electron-impact cross section, which takes into account both direct and energy-transfer excitations, and N_t is the total CO_2 ground-state population. For pump rates well above threshold, the output power, P, is proportional to $(dN_2/dt)_p$ and to the volume of the active medium, V_a. From Eq. (10.2.8) we can therefore write

$$P \propto JN_tV_a \propto JpD^2l \tag{10.2.9}$$

where D is the diameter of the active medium, l its length, and p the gas pressure. Under optimum operating conditions we now have the following: (1) The product pD must be constant (~22.5 torr × cm, e.g., 15 torr for $D = 1.5$ cm) to keep the discharge at the optimum electron temperature. (2) Due to thermal limitations imposed by the requirement of heat conduction to the tube walls, an optimum value of the current density exists, and this value is inversely proportional to the tube diameter D. The existence of an optimum value of J can be understood when one notices that excessive current density leads to excessive heating of the mixture (even with an efficiency of 20%, some 80% of the electrical power is dissipated in the discharge as heat) with consequent thermal population of the lower laser levels. The fact that the optimum value of J is inversely proportional to D can now be understood when one realizes that, for larger diameters, the generated heat has more difficulty in escaping to the walls. From this discussion we draw the conclusion that, under optimum conditions, both J and p are inversely proportional to D and, hence, from Eq. (10.2.9), the optimum value of P is only proportional to the tube length l. CO_2 lasers with slow axial flow and relatively low power (50–100 W) are widely used for laser surgery, for resistor trimming, for cutting ceramic plates for the electronics industry, and for welding thin metal sheets (<1 mm).

Sealed-off Lasers. If the flow of the gas mixture were stopped in the arrangement shown in Fig. 10.12, laser action would cease in a few minutes. This is because the chemical reaction products formed in the discharge (CO, in particular) would no longer be removed and would instead be adsorbed on the walls of the tube or react with the electrodes, thus upsetting the CO_2-CO-O_2 equilibrium. Ultimately this would lead to dissociation of a large fraction of CO_2 molecules in the gas mixture. For a non-circulating sealed-off laser, some kind of catalyst must be present in the gas tube to promote the regeneration of CO_2 from the CO. A simple way to achieve this is to add a small amount of H_2O (1%) to the gas mixture. This leads to regeneration of CO_2, probably through the reaction

$$CO^* + OH \rightarrow CO_2^* + H \tag{10.2.10}$$

involving vibrationally excited CO and CO_2 molecules. The relatively small amount of H_2O vapor required may be added in the form of hydrogen and oxygen gas. Actually, since oxygen is produced during the dissociation of CO_2, it is found that only hydrogen needs be added. Another way of inducing the recombination reaction relies on the use of a hot (300 °C) Ni cathode, which acts as a catalyst. With these techniques, lifetimes for sealed-off tubes in excess of 10,000 h have been demonstrated.

Sealed-off lasers have produced output powers per unit length of ~60 W/m, i.e., comparable to those of longitudinal-flow lasers. Low-power (~1 W) sealed-off lasers of short length and hence operating in a single mode are often used as local oscillators in optical heterodyne experiments. Sealed-off CO_2 lasers of somewhat higher power (~10 W) are attractive for laser microsurgery and for micromachining.

Capillary Waveguide Lasers. If the diameter of the laser tube in Fig. 10.12 is reduced to around a few millimeters (2–4 mm), a situation is reached where the laser radiation is guided by the inner walls of the tube. Such waveguide CO_2 lasers have a low diffraction loss. Tubes of BeO or SiO_2 have been found to give the best performance. The main advantage of a waveguide CO_2 laser stems from the fact that, owing to the small bore diameter, the pressure of the gas mixture needs to be considerably increased (100–200 torr). With this increase in pressure, the laser gain per unit length is correspondingly increased. This means that one can make short CO_2 lasers ($L < 50$ cm) without having to face a difficult requirement on reduction of cavity losses. On the other hand, however, the power available per unit length of the discharge suffers the same limitation discussed earlier for the slow axial flow laser (~50 W/m). Therefore waveguide CO_2 lasers are particularly useful when short and compact CO_2 lasers of low power ($P < 30$ W) are needed (e.g., for laser microsurgery). To fully exploit their compactness these lasers usually operate as sealed-off devices. The laser configuration may either be similar to that of Fig. 10.12, in which the discharge current flows longitudinally along the laser tube, or as shown in Fig. 10.13, where the current (usually provided by a rf source) flows transversely across the tube. For a given value of discharge electric field $\mathcal{E}$, owing to the fact that the quantity $\mathcal{E}/p$ must be constant, the transverse pumping configuration offers a significant advantage over longitudinal pumping by way of a much reduced (one to two orders of

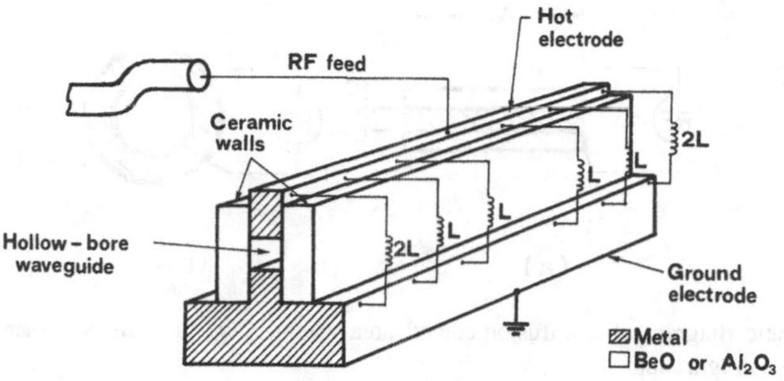

FIG. 10.13. Schematic diagram of a rf excited waveguide CO_2 laser.

magnitude) electrode voltage. Radiofrequency ($\nu \cong 30$ MHz) excitation presents many advantages, the most important of which are perhaps:[11] (1) The avoidance of permanent anodes and cathodes, which eliminates the associated gas-chemistry problem at the cathode; (2) a stable discharge with the help of simple non-dissipative elements (e.g., a dielectric slab) in series in the discharge circuit (see e.g., Fig. 6.20). As a result of these various advantages, rf discharge lasers are being used increasingly not only for waveguide lasers but also for fast axial flow lasers and for the transverse flow lasers which we consider next. We note finally that the laser tube of a waveguide CO_2 laser is either not cooled or, for the largest power units, cooled by forced air.

Lasers with Fast Axial Flow. To overcome the output power limitation of a CO_2 laser with slow-axial-flow, as discussed with the help of Eqs. (10.2.8) and (10.2.9), a possible solution, and a very interesting one, practically, is to flow the gas mixture through the tube at very high supersonic speed (about 50 m/s). In this case the heat is removed simply by removing the hot mixture, which is then cooled outside the tube by a suitable heat exchanger before being returned to the tube as shown in the schematic diagram of Fig. 10.14. Excitation of each of the two laser tubes indicated in the figure may be provided either by a dc longitudinal discharge or, more frequently, by a rf transverse discharge across the glass tube (see Fig. 6.20). When operated in this way there is no optimum for the current density, the power actually increases linearly with J, and much higher output powers per unit discharge length can be obtained (~1 kW/m or even greater). While outside the laser tube, the mixture, besides being cooled, is also passed over a suitable catalyst to let CO recombine with O_2 and thus achieve the

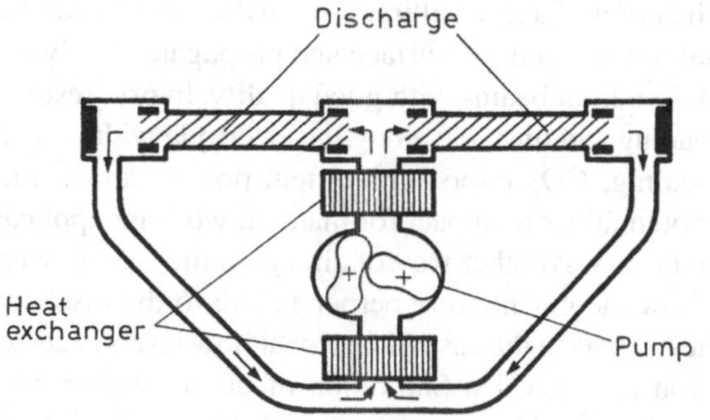

FIG. 10.14. Schematic diagram of a fast axial flow CO_2 laser.

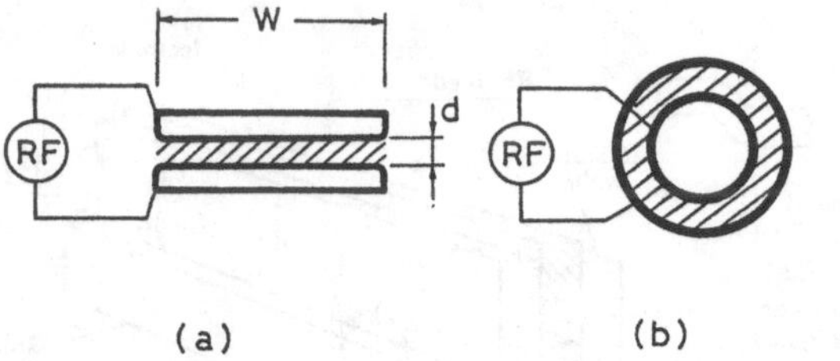

FIG. 10.15. Schematic diagram of a diffusion-cooled area-scaling CO_2 laser using either a planar, (a), or an annular,(b), electrode configuration.

required regeneration of CO_2 molecules (some O_2 is already present in the mixture owing to dissociation of the CO_2 in the discharge region). In this way either completely sealed-off operation can be achieved or at least replenishment requirements for the mixture are kept to a minimal level. Fast axial flow CO_2 lasers with high power (1–6 kW) are now commonly used for many material working applications and in particular for laser cutting of metals (with a thickness of up to a few millimeters).

Diffusion-Cooled Area-Scaling Lasers. An alternative way of circumventing the power limitation of a slow-axial-flow laser is to use a transverse discharge with electrodes spacing, d, much smaller than the electrode width, W (see Fig. 10.15a). In this case the gas mixture is cooled very effectively by one-dimensional heat flow toward the electrodes which are water cooled. It can be shown that the laser output power scales, in this case, as $P_{out} = C(Wl)/d$, where C is a constant ($C \cong 50\,\mathrm{W/m}$) and l is the electrode length.[12] Thus, for a given electrode spacing, the output power scales as the electrode area, Wl, rather than electrode length like in e.g., CO_2 lasers with slow axial flow [see Eq. (10.2.9)]. For sufficiently small electrode spacing, large powers per unit electrode area can then be obtained [e.g., $(P_{out}/Wl) \cong 20\,\mathrm{kW/m^2}$ for $d = 3\,\mathrm{mm}$]. Instead of the planar configuration of Fig. 10.15a, the annular configuration of Fig. 10.15b can also be used, this latter configuration being technically more complicated but allowing more compact devices to be achieved.

It should be stressed again that the above results hold if the electrode width is appreciably larger (by $\sim$ an order of magnitude) than the electrode spacing. To produce a good-quality discharge and an output beam with good divergence properties, from a gain medium with such a pronounced elongation, poses some difficult problems. Stable and spatially uniform discharges can however be obtained exploiting the advantages of radiofrequency excitation. For electrode spacing of the order of a few millimeters, on the other hand, the laser beam is guided in the direction normal to the electrode surface and propagates freely in the direction parallel to this surface. To obtain output beams with good quality, hybrid resonators, which are stable along the electrode spacing and unstable along the electrode width, have been developed.[13]

Compact, area-scaling, CO_2 lasers with output powers above 1 kW are now commercially available with potential large impact for material working applications.

Transverse-Flow Lasers. Another way of circumventing the power limitations of a slow axial flow laser is to flow the gas mixture perpendicular to the discharge (Fig. 10.16). If the flow is fast enough, the heat, as in the case of fast axial flow lasers, gets removed by convection rather than by diffusion to the walls. Saturation of output power versus discharge current does not then occur, and, as in the case of fast axial flow, very high output powers per unit discharge length can be obtained (a few kW/m; see also Fig. 7.7). It should be noted that the

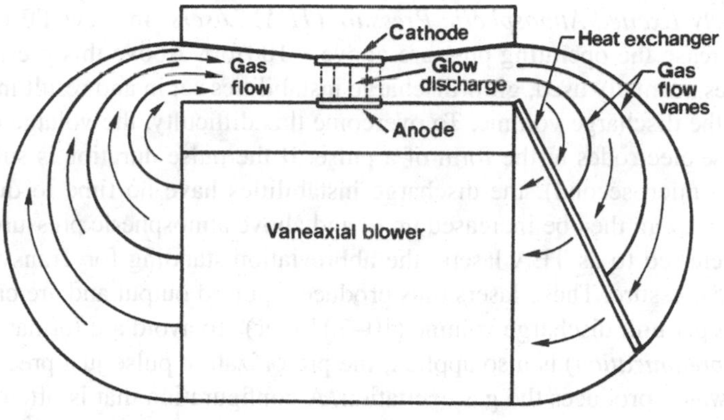

FIG. 10.16. Schematic diagram for a transverse-flow CO_2 laser.

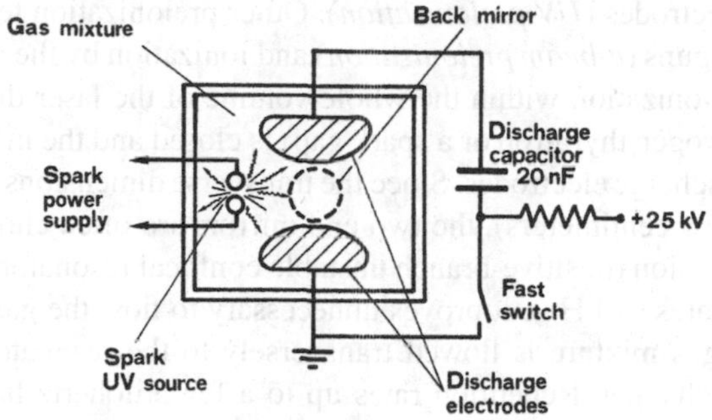

FIG. 10.17. Schematic diagram (viewed along the laser axis) of a CO_2 TEA laser. The laser uses UV radiation from several spark sources placed along the tube direction, to provide for gas preionization.

optimum total pressure (~100 torr) is now typically an order of magnitude greater than that of large-bore longitudinal flow systems. The increase in total pressure p requires a corresponding increase of the electric field $\mathcal{E}$ in the discharge. In fact, for optimum operating conditions, the ratio $\mathcal{E}/p$ must remain approximately the same for all these cases since this ratio determines the temperature of the discharge electrons [see Eq. (6.4.20)]. With this higher value of electric field, a longitudinal-discharge arrangement such as in Fig. 10.12 would be impractical since it would require very high voltages (100–500 kV for a 1-m discharge). For this reason, the discharge is usually applied perpendicular to the resonator axis (TE lasers, i.e., lasers with transverse electric field).

TE CO_2 lasers with fast transverse flow and high output power (1–20 kW) are used in a great variety of metal-working applications (cutting, welding, surface hardening, surface metal alloying). Compared to the fast axial flow lasers, these lasers turn out to be simpler devices in view of the reduced flow speed requirement for transverse flow. However, due to the cylindrical symmetry of their discharge current distribution, the beam quality of fast axial flow lasers is considerable better and this makes them particularly interesting for cutting applications.

Transversely Excited Atmospheric Pressure (TEA) Lasers. In a cw TE CO_2 laser, it is not easy to increase the operating pressure above ~100 torr. Above this pressure and at the current densities normally used, glow discharge instabilities set in and result in the formation of arcs within the discharge volume. To overcome this difficulty, the voltage can be applied to the transverse electrodes in the form of a pulse. If the pulse duration is sufficiently short (a fraction of a microsecond), the discharge instabilities have no time to develop and the operating pressure can then be increased up to and above atmospheric pressure. These lasers are therefore referred to as TEA lasers, the abbreviation standing for Transversely Excited at Atmospheric pressure. These lasers thus produce a pulsed output and are capable of large output energies per unit discharge volume (10–50 J/liter). To avoid arc formation, some form of ionization (*preionization*) is also applied, the preionization pulse just preceding the main voltage pulse which produces the gas excitation. A configuration that is often used is shown in Fig. 10.17, where the ionization is produced by the strong UV emission of a row of sparks which runs parallel to the tube length. The deep UV emission of these sparks produces the required ionization by both photoionization of the gas constituents and UV-induced electron emission from the electrodes (*UV-preionization*). Other preionization techniques include the use of pulsed *e*-beam guns (*e-beam preionization*) and ionization by the corona effect (*corona preionization*). Once ionization within the whole volume of the laser discharge is produced, the fast switch (a hydrogen thyratron or a spark gap) is closed and the main discharge pulse is passed through the discharge electrodes. Since the transverse dimensions of the laser discharge are usually large (a few centimeters), the two end mirrors are often chosen to give an unstable resonator configuration (positive-branch unstable confocal resonator; see Fig. 5.18b). For low pulse repetition rates (~1 Hz), it proves unnecessary to flow the gas mixture. For higher repetition rates, the gas mixture is flowed transversely to the resonator axis and is cooled by a suitable heat exchanger. Repetition rates up to a few kilohertz have been achieved in this way. Another interesting characteristic of these lasers is their relatively broad linewidths (~4 GHz at $p = 1$ atm, due to collision broadening). Thus, optical pulses with less than 1-ns duration have been produced with mode-locked TEA lasers. Transverse-flow CO_2 TEA lasers of relatively high repetition rate (~50 Hz) and relatively high average output power ($\langle P_{out} \rangle \cong 300$ W) are commercially available. Besides being used in scientific applications, these lasers find a number of industrial uses for those material working applications in which the pulsed nature of the beam presents some advantage (e.g., pulsed laser marking or pulsed ablation of plastic materials).

10.2.3.2. The CO Laser

The second example of a gas laser using vibrational-rotational transitions that we will briefly consider is that of the CO laser. This laser has attracted considerable interest on account of its shorter wavelength ($\lambda \cong 5\mu$ m) than the CO_2 laser, combined with high efficiencies and high power. Output powers in excess of 100 kW and efficiencies in excess of 60% have been demonstrated.[14] However, to achieve this sort of performance the gas mixture must be kept at cryogenic temperature (77–100 K). Laser action, in the 5μ-m region, arises from several rotational-vibrational transitions [e.g., from $\upsilon'(11) \rightarrow \upsilon(10)$, to $\upsilon'(7) \rightarrow \upsilon(6)$ at $T = 77$ K] of the highly excited CO molecule.

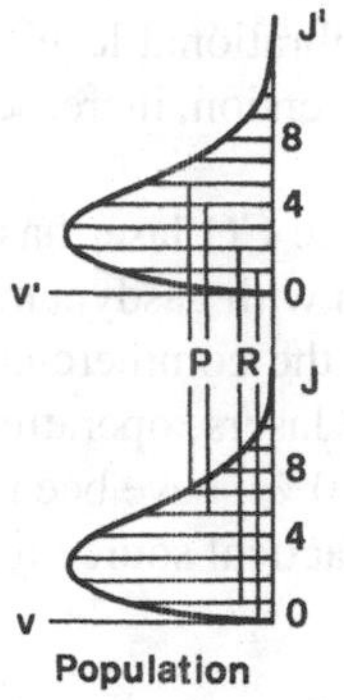

FIG. 10.18. Partial inversion between two vibrational transitions (υ and υ') having the same total population.

Pumping of the CO vibrational levels is achieved by electron-impact excitation. Like the isoelectronic N_2 molecule, the CO molecule has an unusually large cross section for electron-impact excitation of its vibrational levels. Thus, nearly 90% of the electron energy in a discharge can be converted into vibrational energy of CO molecules. Another important feature of the CO molecule is that VV relaxation proceeds at a much faster rate than VT relaxation (which is unusually low). As a consequence of this, a non-Boltzmann population buildup in higher vibrational levels, by a process known as *anharmonic pumping*, plays a very important role*. Although this phenomenon does not allow a total inversion in the vibrational population of a CO molecule, a situation known as *partial inversion* may occur. This is illustrated in Fig. 10.18, in which the rotational populations of two neighboring vibrational states are indicated. One sees from the figure that, although the total population for the two vibrational states is equal, an inversion exists for the two P transitions $[(J' = 5) \rightarrow (J = 6),\ (J' = 4) \rightarrow (J = 5)]$ and also for the two R-branch transitions indicated in the figure. Under these conditions of partial inversion, laser action can thus take place, and a new phenomenon, called cascading, can then play an important role. The effect of laser action is in fact to depopulate a rotational level of the upper state and populate a rotational level of the lower vibrational state. The latter level can then accumulate enough population to result in population inversion with respect to a rotational level of a still lower vibrational state. At the same time, the rotational level of the upper state may become sufficiently depopulated to result in population inversion with respect to a rotational level of a still higher vibrational state. This process of cascading coupled with the very low VT rate results in most of the vibrational energy being extracted as laser output energy. This, together with the very high excitation efficiency, accounts for the high efficiency of the CO laser. The low-temperature requirement arises from the need for very efficient anharmonic pumping. In

* Anharmonic pumping arises from a collision of the type $CO(\upsilon = n) + CO(\upsilon = m) \rightarrow CO(\upsilon = n+1) + CO(\upsilon = m-1)$ with $n > m$. Because of anharmonicity (a phenomenon shown by all molecular oscillators), the separation between vibrational levels becomes smaller for levels higher up in the vibrational ladder (see also Fig. 3.1). This means that, in a collision process of the type indicated above, with $n > m$, the total vibrational energy of the two CO molecules after collision is somewhat smaller than before collision. The collision process therefore has a greater probability of proceeding in this direction rather than the reverse direction. This means that the hottest CO molecule $[CO(\upsilon = n)]$ can climb up the vibrational ladder and this leads to a non-Boltzmann distribution of the population among the vibrational levels.

fact, the overpopulation of the high vibrational levels compared to the Boltzmann distribution, and hence the degree of partial inversion, increases rapidly with decreasing translational temperature.

As in the case of the CO_2 laser, the CO laser has been operated with longitudinal flow, with *e*-beam preionized pulsed TE, and with gasdynamic excitation. The requirement of cryogenic temperatures has so far limited the commercial development of CO lasers. Recently, however, high power ($P > 1$ kW) CO lasers, operating at room temperature while retaining a reasonably high slope efficiency (~10%), have been introduced commercially. This laser is now under active consideration as a practical source for material-working applications.

10.2.3.3. The N_2 Laser[15]

As a particularly relevant example of vibronic lasers, we will consider the N_2 laser. This laser has its most important oscillation at $\lambda = 337$ nm (UV), and belongs to the category of self-terminating lasers.

The relevant energy level scheme for the N_2 molecule is shown in Fig. 10.19. Laser action takes place in the so-called second positive system, i.e., in the transition from the $C^3\Pi_u$ state (henceforth called the C state) to the $B^3\Pi_g$ state (B state)*. The excitation of the C state is believed to arise from electron-impact collisions with ground-state N_2 molecules. Since both C and B states are triplet states, transitions from the ground state are spin-forbidden. On the basis of the Franck-Condon principle, we can however expect the excitation cross section, due to e.m. wave interaction and hence due to the electron-impact cross section, to the $\upsilon = 0$ level of the C state to be larger than that to the $\upsilon = 0$ level of the B state. The potential minimum

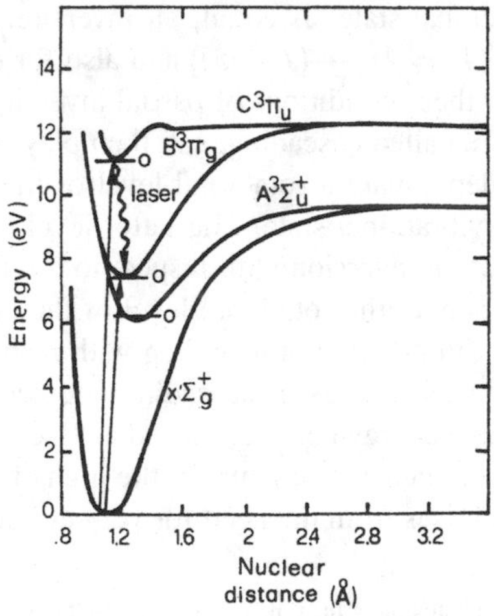

FIG. 10.19. Energy states of the N_2 molecule. For simplicity only the lowest vibrational level ($\upsilon = 0$) is shown for each electronic state.

* Under different operating conditions laser action can also take place, in the near infrared (0.74–1.2 μm), in the first positive system involving the $B^3\Pi_g \rightarrow A^3\Sigma_u^+$ transition.

TABLE 10.2. Spectroscopic properties of UV laser transitions and gas-mixture composition in nitrogen and KrF lasers

Laser type	N_2	$(KrF)^*$
Laser wavelength [nm]	337.1	248
Cross-section [10^{-14} cm^2]	40	0.05
Upper-state lifetime [ns]	40	10
Lower-state lifetime [ns]	10 s	
Transition Linewidth [THz]	0.25	3
Partial press. of gas mixture [mbar]	40 (N2)	120 (Kr)
	960 (He)	6 (F_2)
		2,400 (He)

of the B state is in fact shifted (relative to the ground state) to larger internuclear separation than that of the C state. The lifetime (radiative) of the C state is 40 ns, while the lifetime of the B state is 10 μs. Clearly the laser cannot operate cw since condition Eq. (7.3.1) is not satisfied. The laser can, however, be excited on a pulsed basis provided the electrical pulse is appreciably shorter than 40 ns. Laser action takes place predominantly on several rotational lines of the $\upsilon''(0) \rightarrow \upsilon'(0)$ transition ($\lambda = 337.1$ nm) because this transition exhibits the largest stimulated-emission cross-section. Oscillation on the $\upsilon''(1) \rightarrow \upsilon'(0)(\lambda = 357.7$ nm) transition and the $\upsilon''(0) \rightarrow \upsilon'(1)(\lambda = 315.9$ nm) transition also occurs, although with lower intensity. Some spectroscopic data for the N_2 laser are summarized in Table 10.2.

Given the high pressure of the gas mixture (~40 mbar of N_2 and 960 mbar of He) and the correspondingly high electric field (~10 kV/cm), a TEA configuration (see Fig. 10.17) is normally used for a nitrogen laser. To obtain the required fast current pulse (5–10 ns), the discharge circuit must have as low an inductance as possible. Owing to the high gain of this self-terminating transition, oscillation takes place in the form of amplified spontaneous emission (ASE) and the laser can be operated even without mirrors. Usually, however, a single mirror is placed at one end of the laser since this reduces the threshold gain and hence the threshold electrical energy for ASE emission (see Sect. 2.9.2). The mirror also ensures a unidirectional output and reduces the beam divergence. With this type of laser, it is possible to obtain laser pulses, of high peak power (up to ~1 MW) and short duration (~10 ns), at high repetition rate (up to ~100 Hz). Nitrogen lasers with nitrogen pressure up to atmospheric pressure and without helium have also been developed. In this case, the problem of arcing is alleviated by further reducing (to ~1 ns) the duration of the voltage pulse. The increased gain per unit length, due to the higher N_2 pressure, and the fast discharge, leads to this type of laser usually being operated without any mirrors. The length can be kept very short (10–50 cm) and, as a consequence, output pulses of shorter time duration can be obtained (~100 ps with 100 kW peak power). Nitrogen lasers of both long (~10 ns) and short (~100 ps) time duration are widely used as pumps for dye lasers, since most dyes absorb strongly in the UV.

10.2.3.4. Excimer Lasers[17]

Excimer lasers represent an interesting and important class of molecular lasers involving transitions between different electronic states of special molecules referred to as excimers. Consider a diatomic molecule A_2 with potential energy curves as in Fig. 10.20, for the ground

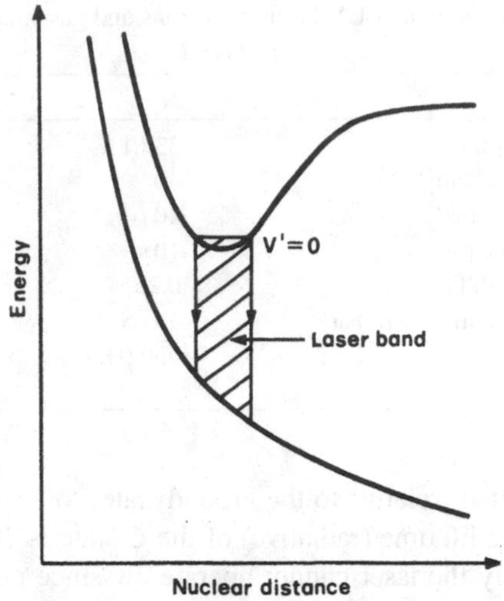

FIG. 10.20. Energy states of an excimer laser.

and excited states, respectively. Since the ground state is repulsive, the molecule cannot exist in this state, i.e., the species only exists in the monomer form A in the ground state. Since, however, the potential energy curve shows a minimum for the excited state, the species is bound in this state i.e., it exists in dimer form. Such a molecule A_2^* is called an *excimer*, a contraction of the words *exc*ited d*imer*. Now suppose that a large fraction of excimers are somehow produced in the given volume of the medium. Under appropriate conditions, laser action can then be produced on the transition between the upper (bound) state and the lower (free) state (bound-free transition). This is referred to as an excimer laser, a classical example being the Ne_2^* laser, the first excimer laser to be operated ($\lambda = 170\,\text{nm}$).[16]

Excimer lasers have three notable and important properties: (1) Since the transition occurs between different electronic states of a molecule, the corresponding transition wavelength generally falls in the UV spectral region. (2) Once the molecule, after undergoing stimulated emission, reaches the ground state, it rapidly dissociates due to the repulsive potential of this state. This means that the lower laser level can be considered to be empty and the laser operates as a four-level laser. (3) Due to the lack of energy levels in the ground state, no rotational-vibrational transitions exist, and the transition is observed to be featureless and relatively broad ($\Delta\nu = 20\text{–}100\,\text{cm}^{-1}$). It should be noted, however, that, in some excimer lasers, the energy curve of the ground state does not correspond to a pure repulsive state but features a (shallow) minimum. In this case the transition occurs between an upper bound state and a lower (weakly) bound state (bound-bound transition). However, since the ground state is only weakly bound, a molecule in this state undergoes rapid dissociation either by itself (a process referred to as predissociation) or as a result of the first collision with another species of the gas mixture. Thus, in this case also, the light emission produces a continuous spectrum.

We now consider a particularly important class of excimer laser in which a rare gas atom (notably Kr, Ar, Xe) is combined, in the excited state, with a halogen atom (notably

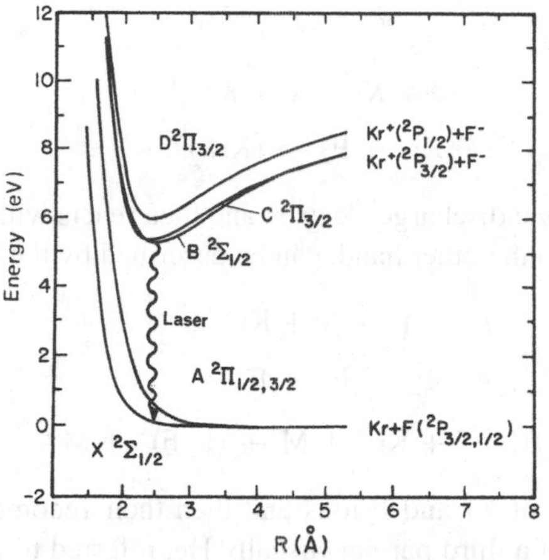

FIG. 10.21. Potential energy diagram showing the energy states of KrF (by permission from ref.[17]).

F, Cl) to form a rare-gas-halide excimer*. Specific examples are ArF ($\lambda = 193$ nm), KrF ($\lambda = 248$ nm), XeF ($\lambda = 351$ nm) and XeCl ($\lambda = 309$ nm), all oscillating in the UV. The reason why rare-gas-halides are readily formed in the excited state can be understood if we consider an excited rare-gas atom reacting with a ground-state halogen. In fact, an excited rare-gas atom becomes chemically similar to an alkali atom, and such an atom is known to react readily with halogens. This analogy also indicates that the bonding in the excited state has ionic character: In the bonding process, the excited electron is transferred from the rare gas atom to the halogen atom. This bound state is therefore also referred to as a charge-transfer state.

We will now consider the KrF laser in some detail, as it represents one of the most important lasers of this category (Fig. 10.21). The upper laser level is an ionically bound charge-transfer state which, at large internuclear distances ($R \to \infty$), corresponds to the ^2P state of the Kr positive ion and to the ^1S state of the negative F ion. Thus, for large values of the internuclear distance, the energy curves of the upper state obey the Coulomb law. The interaction potential between the two ions therefore extends to much greater distances (0.5–1 nm) than those occurring when covalent interactions predominate (compare with e.g., Fig. 10.19). The lower state, on the other hand, is covalently bonded and, at large internuclear distances ($R \to \infty$), corresponds to the ^1S state of Kr atom and ^2P state of the F atom. As a result of interaction of the corresponding orbitals, both upper and lower states are then split, for short internuclear distance, into the well-known $^2\Sigma$ and $^2\Pi$ states of molecular spectroscopy. Laser action occurs on the $B^2\Sigma \to X^2\Sigma$ transition, since it exhibits the largest cross section. Note that, during the transition, the radiating electron transfers from the F^- to the Kr^+ ion. Some relevant spectroscopic data for this transition are shown in Table 10.2.

The two main excitation mechanisms responsible for producing KrF excimers arise either from excited Kr atoms or from Kr ions. The route involving excited atoms can be described

* Strictly speaking these should not be referred to as excimers since they involve binding between unlike atoms. In fact, the word exciplex, a contraction from excited complex, has been suggested as perhaps being more appropriate for this case. However, the word excimer is now widely used in this context, and we will follow this usage.

by the following reactions

$$e + Kr \rightarrow e + \mathrm{Kr}^* \tag{10.2.11a}$$

$$\mathrm{Kr}^* + \mathrm{F}_2 \rightarrow (\mathrm{KrF})^* + \mathrm{F} \tag{10.2.11b}$$

where Kr is first excited by a discharge electron and then reacts with a fluorine molecule. The route involving Kr ions, on the other hand, can be described by the following three reactions

$$e + \mathrm{Kr} \rightarrow 2e + \mathrm{Kr}^+ \tag{10.2.12a}$$

$$e + \mathrm{F}_2 \rightarrow \mathrm{F}^- + \mathrm{F} \tag{10.2.12b}$$

$$\mathrm{F}^- + \mathrm{Kr}^+ + \mathrm{M} \rightarrow (\mathrm{KrF})^* + \mathrm{M} \tag{10.2.12c}$$

involving first production of Kr and F ions and then their recombination, in the discharge volume, in the presence of a third partner (usually He, referred to as a buffer gas), to satisfy both energy and momentum conservation. Note that reaction Eq. (10.2.12b) is a peculiar one, usually referred to as dissociative attachment, resulting from the high electron affinity of F atoms. Note also that, due to the long-range interaction between the two reacting ions, the reaction Eq. (10.2.12c) can proceed at very fast rate provided that the buffer gas pressure is sufficiently high. Indeed, He partial pressures well above atmospheric pressure are normally used (a typical gas mixture may contain ~120 mbar of Kr, 6 mbar of F_2 and 2,400 mbar of He). Under this condition, the reaction pathway described by Eqs. (10.2.12) becomes the dominant mechanism of $(KrF)^*$ production.

Since the pressure of the gas mixture is above atmospheric pressure, excimer lasers can only be operated in a pulsed regime and the general TEA configuration of Fig. 10.17 is used. In the case of excimer lasers, however, the components of the laser tube and laser flow system must be compatible with the highly reactive F_2 gas. Furthermore, owing to the shorter lifetime of the upper state and to avoid the onset of arc formation, faster pumping is usually provided for excimer lasers compared to TEA CO_2 lasers (pump durations of 10–20 ns are typical). For standard systems, preionization is achieved, as in Fig. 10.17, by a row of sparks. However, for the largest systems, more complex preionization arrangements are adopted, using either an auxiliary electron-beam or an X-ray source.

Excimer lasers with high repetition rate (up to ~500 Hz) and high average power (up to ~100 W) are commercially available, while larger, laboratory, systems exist with higher average power (in excess of a few kW). The efficiency of these lasers is usually quite high (2–4%) as a result of the high quantum efficiency (see Fig. 10.21) and the high efficiency of the pumping processes.

Excimer lasers are used to ablate plastic as well as biological materials with great precision since these material exhibit strong absorption at UV wavelengths. In fact, in some of these materials, the penetration depth, for each laser pulse, may be only a few μm. Due to the strong absorption and short pulse duration, a violent ablation process is produced wherein these materials are directly transformed into volatile components. Applications include drilling of very precise holes in thin plastic films (as used e.g., for the ink-jet printer head) and corneal sculpturing to change the refractive power of the eye and hence correct myopia. In the field of lithography, the 193 nm UV light provides a good illumination source for achieving submicron-size features in semiconductor microchips. Excimer lasers can also be used as dye-laser pumps since most dye absorb strongly in the UV.

10.3. CHEMICAL LASERS[18, 19]

A *chemical laser* is usually defined as one in which the population inversion is "directly" produced by an exothermic chemical reaction*. Chemical lasers usually involve either an associative or a dissociative chemical reaction between gaseous elements.

An associative reaction can be described by an equation of the form $A + B \to AB$. For an exothermic reaction, some of the heat of reaction will appear as either rotational-vibrational or electronic energy of the molecule AB. Thus, if a population inversion is achieved, the associative reaction can in principle lead to either a vibrational-rotational or a vibronic transition. In spite of much effort, however, only chemical lasers operating on vibrational-rotational transitions have been demonstrated so far. For this kind of transition, the range of oscillation wavelengths, achieved so far, lies between 3 and 10 μm, with the HF and DF lasers, considered in the next section, being the most notable examples.

A dissociative reaction, on the other hand, can be described by an equation of the general form $ABC \to A + BC$. If the reaction is exothermic, some of the heat of reaction can be left either as electronic energy of the atomic species A or as internal energy of the molecular species BC. The most notable example of a laser exploiting this type of reaction is the atomic iodine laser, in which iodine is chemically excited to its $^2P_{1/2}$ state and laser action occurs between the $^2P_{1/2}$ state and the $^2P_{3/2}$ ground state ($\lambda = 1.315\,\mu$m). Excited atomic iodine may be produced via the exothermic dissociation of CH_3I (or CF_3I, C_3F_7I), the dissociation being produced by means of UV light (~300 nm) from powerful flashlamps. More recently, excited iodine has been produced via generation of excited molecular oxygen by reacting molecular chlorine with hydrogen peroxide. The molecular oxygen, excited to its long-lived singlet-state (the ground electronic state of oxygen molecule happens in fact to be a triplet state), in turn transfers its energy to atomic iodine (oxygen-iodine chemical laser).

Chemical lasers are important mainly for two main reasons: (1) They provide an interesting example of direct conversion of chemical energy into electromagnetic energy. (2) They are potentially able to provide either high output power (in cw operation) or high output energy (in pulsed operation). This is because the amount of energy available in an exothermic chemical reaction is usually quite large†.

10.3.1. The HF Laser

HF chemical lasers can be operated using either SF_6 or F_2 as compounds donating the atomic fluorine, and, from a practical view-point, the two lasers are very different. In commercial devices, the inert SF_6 molecule is used as fluorine donor and the gas mixture also contains H_2 and a large amount of He. An electrical discharge is then used to dissociate the SF_6 and excite the reaction. The overall pressure of the mixture is around atmospheric pressure, the

* According to this definition, the gas-dynamic CO_2 laser, briefly considered in Sect. 6.1., should not be regarded as a chemical laser even though the upper state population arises ultimately from a combustion reaction.

† For example, a mixture of H_2, F_2, and other substances (16% of H_2 and F_2 in a gas mixture at atmospheric pressure) has a heat of reaction equal to 2,000 J/liter, of which 1,000 J is left as vibrational energy of HF (a large value in terms of available laser energy).

laser is pulsed, and the laser configuration is very similar to that of a CO_2 TEA laser. The output energy of this type of device is, however, appreciably smaller than the input electrical energy. Thus, the laser derives only a small part of its output energy from the energy of the chemical reaction and it can only be considered marginally as a chemical laser. On the other hand, when a F_2 and H_2 gas mixture is used, the laser operates cw and it derives most its power from the chemically reacting species. So, only the latter type of HF chemical laser will be considered below.

In a $F_2 + H_2$ chemical laser a certain amount of atomic fluorine gets produced from the fluorine molecules. This atomic fluorine can then react with molecular hydrogen according to the reaction

$$F + H_2 \rightarrow HF^* + H \tag{10.3.1}$$

which produces atomic hydrogen. This atomic hydrogen can then react with molecular fluorine according to the second reaction

$$H + F_2 \rightarrow HF^* + F \tag{10.3.2}$$

Atomic fluorine is then restored after this second reaction and this fluorine atom can then repeat the same cycle of reaction, and so on. We thus have a classical chain-reaction which can result in a large production of excited HF molecules. It should be noted that the heats of reaction of Eqs. (10.3.1) and (10.3.2) are 31.6 kcal/mole and 98 kcal/mole respectively, the two reactions being therefore referred to as the cold and hot reaction, respectively. It should also be noted that, in the case of the cold reaction, the energy released, $\Delta H = 31.6\,\text{kcal/mole}$, can easily be shown to correspond to an energy of $\Delta H \cong 1.372\,\text{eV}$ for each molecular HF produced. Since the energy difference between two vibrational levels of HF, corresponding to a transition wavelength of $\lambda \cong 3\,\mu\text{m}$, is about $\Delta E \cong 0.414\,\text{eV}$, one then understands that, if all this energy were released as vibrational excitation, vibrationally excited HF molecules up to the $\upsilon = \Delta H/\Delta E \cong 3$ vibrational quantum number could be produced (see Fig. 10.22a). It is found, however, that the fraction of the reaction energy which goes into vibrational energy depends on the relative velocity of the colliding partners and on the orientation of this velocity compared to the H-H axis. In a randomly oriented situation such as in a gas, one can then calculate the fraction of molecules found in the $\upsilon = 0, 1, 2$ or 3 vibrational states respectively. The relative numbers $N(\upsilon)$ of excited HF molecules are also indicated in the same figure. For instance, one can see that 5 out of 18 molecules are found in the $\upsilon = 3$ state and thus take up almost all the available energy as vibrational energy. On the other hand, 1 out of 18 molecules are found in the ground ($\upsilon = 0$) vibrational state and, in this case, all the reaction enthalpy is found as kinetic energy of the reaction products (mostly H, this being the lightest product). From the figure one thus sees that, if this were the only reaction, a population inversion would be established, particularly for the ($\upsilon = 2$) $\rightarrow$ ($\upsilon = 1$) transition. In the case of the hot reaction, on the other hand, excited HF up to the ($\upsilon = 10$) vibrational level can be produced (Fig. 10.22b). The relative populations of these vibrational levels, $N(\upsilon)$, can be calculated and is also shown in the same figure. One now sees that a pretty strong population inversion exists particularly for the ($\upsilon = 5$) $\rightarrow$ ($\upsilon = 4$) transition. From the above considerations one can then easily calculate that, e.g., for the cold reaction of Eq. (10.3.1), more than 60%

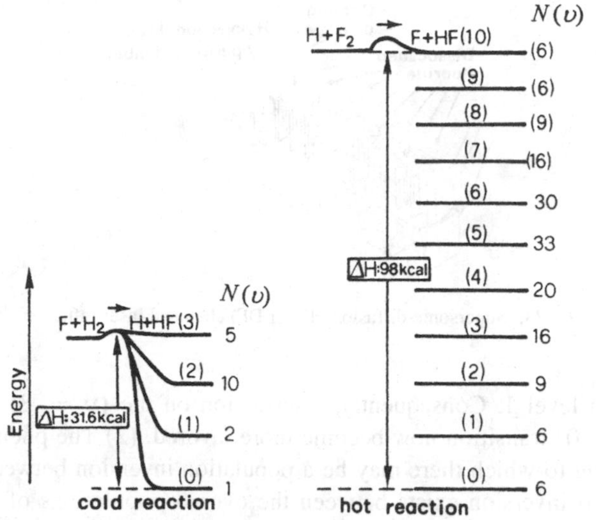

FIG. 10.22. Pumping of the vibrational levels of the HF molecule by the two reactions $F + H_2 \rightarrow HF^* + H$ (a) and $H + F_2 \rightarrow HF^* + F$ (b). The relative populations, $N(\upsilon)$, of each vibrational state, of quantum number υ, are also indicated in the two figures.

of the reaction energy is, on average, released as vibrational energy. The reason why the HF molecule is left in an excited vibrational state, after chemical reaction, can be understood in a simple way. Consider an F atom colliding with a H_2 molecule. As a result of the high electron affinity of a fluorine atom, the interaction is strongly attractive and leads to a considerable polarization of the H_2 charge distribution even at large F-H_2 distances. As a consequence of the electron's low inertia, an electron can be transferred to the fluorine atom from the nearest hydrogen atom, hence forming the HF ionic bond, before the spacing between the hydrogen and fluorine has adjusted itself to the internuclear separation corresponding to the HF equilibrium distance. This classical picture indicates that, after reaction has occurred, the HF molecule is left in an excited vibrational state.

As one can see from the above discussion, a consequence of the combined effect of both the cold and hot reactions, described by Eqs. (10.3.1) and (10.3.2), is that a population inversion between several vibrational levels of HF is produced in an HF laser. If the active medium is placed in a suitable resonator, laser action on a number of transitions from vibrationally excited HF molecules is thus expected to occur. Laser action has indeed been observed on several rotational lines of the $(\upsilon = 1) \rightarrow (\upsilon = 0)$ transition up to the $(\upsilon = 6) \rightarrow (\upsilon = 5)$ transition. In fact, due to the anharmonicity of the interaction potential, the vibrational energy levels of Fig. 10.22 are not equally spaced and the laser spectrum actually consists of many roto-vibrational lines encompassing a rather wide spectral range ($\lambda = 2.7$–$3.3\ \mu$m). It should also be noted that the number of observed laser transitions is larger than one would have expected according to the population inversion situation described in Fig. 10.22. As already discussed in the case of a CO laser, there are two reasons why oscillation can occur on so many lines: (1) The phenomenon of cascading: if, in fact, the $(\upsilon = 2) \rightarrow (\upsilon = 1)$ transition (usually the strongest one) lases, the population will be depleted from level 2 and

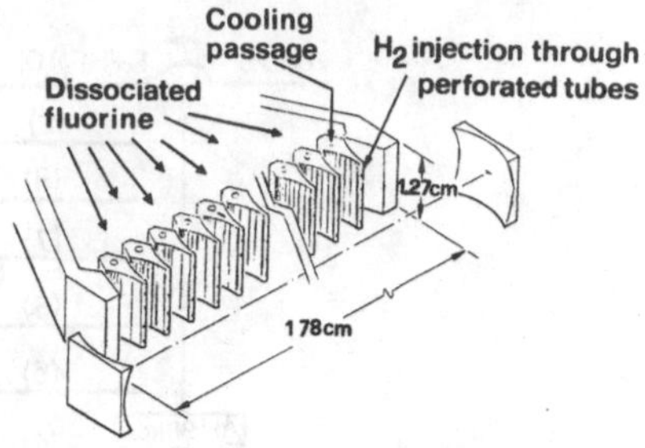

FIG. 10.23. Supersonic-diffusion HF (or DF) chemical laser (after ref.[18]).

will accumulate in level 1. Consequently, laser action on the $(\upsilon = 3) \rightarrow (\upsilon = 2)$ and $(\upsilon = 1) \rightarrow (\upsilon = 0)$ transition now become more favored. (2) The phenomenon of partial inversion, according to which there may be a population inversion between some rotational lines even when no inversion exists between the overall populations of the corresponding vibrational levels. Finally, it should be noted that, besides laser action in HF, laser action can also be achieved in the analogous compounds DF, HCl, and HBr thus providing oscillation on a large number of transitions in the 3.5–5 μm range.

A possible configuration for a high-power cw HF or DF laser is shown in Fig. 10.23. Fluorine is thermally dissociated by an arc jet heater and then expanded to supersonic velocity (~ Mach 4) through some appropriate expansion nozzles. Molecular hydrogen is then mixed downstream through some appropriate perforated tubes inserted in the nozzles. Downstream in the expansion regions, excited HF molecules are produced by the chain reactions Eqs. (10.3.1) and (10.3.2) and a suitable resonator, with its axis orthogonal to the flow direction, is placed around this region. To handle the large power available in the expanding beam, usually of large diameter, unstable resonators exploiting metallic water-cooled mirror are often utilized. Chemical lasers of this type can produce very large c.w. output powers (in the MW range!) with good chemical efficiency.

Pulsed TEA type HF lasers are commercially available and have found a limited use when an intense source of middle infrared radiation is needed (e.g., in spectroscopy). HF and DF chemical lasers of the type described in Fig. 10.23 are used exclusively for military applications. Safety considerations have indeed prevented the use of these lasers for commercial applications. In fact, the F_2 molecule is one of the most corrosive and reactive element known, the waste products are difficult to dispose of, and, under certain conditions, the chain reaction Eqs. (10.3.1) and (10.3.2) may even become explosive. In the military field, due to the large available output powers, these lasers can be used as directed energy weapons to e.g., destroy enemy missiles. A military c.w. device named MIRACL (Mid-Infrared Advanced Chemical Laser), using DF, has given the largest c.w. power of any laser (2.2 MW). DF rather than HF molecule was used because the system was intended for use from a ground station and the DF emission wavelengths fall in a region of relatively good atmospheric transmission. More recently, HF lasers of somewhat higher power (~5 MW) have been constructed. The laser is intended to be used either from a high altitude plane, for destruction of missile during their ballistic flight, or from a space station, for missile destruction during their lift-off phase (where the rocket has much lower speed and hence is much more vulnerable).

10.4. THE FREE-ELECTRON LASER[20]

In a free-electron laser (FEL) an electron beam moving at a speed close to the speed of light is made to pass through the magnetic field generated by a periodic structure (called the wiggler or the undulator) (Fig. 10.24). The stimulated emission process comes about through the interaction of the e.m. field of the laser beam with these relativistic electrons moving in the periodic magnetic structure. As in any other laser, two end mirrors are used to provide feedback for laser oscillation. The electron beam is injected into the laser cavity and then deflected out the cavity using suitable bending magnets.

To understand how this interaction comes about, we first consider the case of spontaneously emitted radiation, i.e., when no mirrors are used. Once injected into the periodic structure, the electrons acquire a wiggly, or undulatory, motion in the plane orthogonal to the magnetic field direction (Fig. 10.24). The resulting electron acceleration produces a longitudinal emission of the synchrotron radiation type. The frequency of the emitted radiation can be derived, heuristically, by noting that the electron oscillates in the transverse direction at an angular frequency $\omega_q = (2\pi/\lambda_q)\upsilon_z \cong (2\pi/\lambda_q)c$, where λ_q is the magnet period and υ_z is the (average) longitudinal velocity of the electron (which is almost equal to the vacuum light velocity c). Let us now consider a reference frame that is moving longitudinally at velocity υ_z. In this frame, the electron will be seen to oscillate essentially in the transverse direction and thus looks like an oscillating electric dipole. In this reference frame, due to the Lorentz time-contraction, the oscillation frequency will then be given by

$$\omega' = \frac{\omega_q}{[1-(\upsilon_z/c)^2]^{1/2}} \tag{10.4.1}$$

and this will therefore be the frequency of the emitted radiation. If we now go back to the laboratory frame, the radiation frequency undergoes a (relativistic) Doppler shift. The observed frequency ω_0 and the corresponding wavelength λ_0 will then be given by

$$\omega_0 = \left[\frac{1+\upsilon_z/c}{1-(\upsilon_z/c)}\right]^{1/2} \omega' \cong \frac{2\omega_q}{1-(\upsilon_z/c)^2} \tag{10.4.2}$$

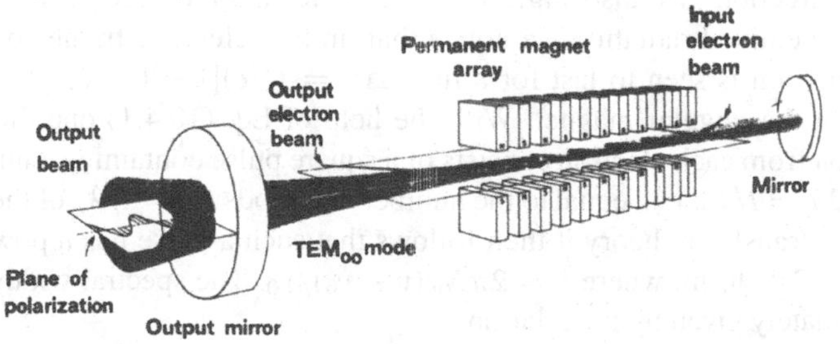

FIG. 10.24. Basic structure of a free-electron laser (courtesy of Luis Elias, University of California, at Santa Barbara Quantum Institute).

and

$$\lambda_0 = \frac{\lambda_q}{2}\left[1 - \left(\frac{\upsilon_z}{c}\right)^2\right] \tag{10.4.3}$$

respectively. Note that, since $\upsilon_z \cong c$, λ_0 is generally much smaller than the magnet period. To calculate the quantity $[1 - (\upsilon_z/c)^2]$ appearing in both Eqs. (10.4.2) and (10.4.3), we begin by noting that, for a completely free electron moving with velocity υ_z along the z axis, one would have $[1 - (\upsilon_z/c)^2] = (m_0c^2/E)^2$, where m_0 is the rest mass of the electron and E its energy. However, for a given total energy, the wiggling motion reduces the value of υ_z, i.e., it increases the value of $[1 - (\upsilon_z/c)^2]$. A detailed calculation then shows that this quantity is given by

$$1 - \left(\frac{\upsilon_z}{c}\right)^2 = (1 + K^2)\left(\frac{m_0c^2}{E}\right)^2 \tag{10.4.4}$$

where the numerical constant K, which is usually smaller than 1, is referred to as the undulator parameter. Its value is obtained from the expression $K = e\langle B^2\rangle^{1/2}\lambda_q/2\pi m_0c^2$, where B is the magnetic field of the undulator and where the average is taken along the longitudinal direction. From Eqs. (10.4.2) and (10.4.3), with the help of Eq. (10.4.4), we get our final result

$$\omega_0 = \frac{4\pi c}{\lambda_q}\left(\frac{1}{1 + K^2}\right)\left(\frac{E}{m_0c^2}\right)^2 \tag{10.4.5}$$

and

$$\lambda_0 = \frac{\lambda_q}{2}\left(\frac{m_0c^2}{E}\right)^2(1 + K^2) \tag{10.4.6}$$

which shows that e.g., the emission wavelength can be changed by changing the magnet period λ_q, and/or the energy E of the electron beam. Assuming, as an example, $\lambda_q = 10\,\text{cm}$ and $K = 1$, we find that the emitted light can range from the infrared to the ultraviolet by changing the electron energy from 10^2 to 10^3 MeV. Note that, according to our earlier discussion, the emitted radiation is expected to be polarized in the plane orthogonal to the magnetic field direction (see also Fig. 10.24). To calculate the spectral line shape and the bandwidth of the emitted radiation we notice that, in the reference frame considered above, the electron emission is seen to last for a time $\Delta t' = (l/c)[1 - (\upsilon_z/c)^2]^{1/2}$, where l is the overall length of the wiggler magnet. With the help of Eq. (10.4.1) one then sees that the emitted radiation from each electron consists of a square pulse containing a number of cycles, $N_{cyc} = \omega'\Delta t'/2\pi = l/\lambda_q$, i.e., equal to the number of periods $N_w = l/\lambda_q$ of the wiggler. From standard Fourier-transform theory it then follows that such a pulse has a power spectrum of the $[\sin(x/2)/(x/2)]^2$ form, where $x = 2\pi N_w(\nu - \nu_0)/\nu_0$. The spectral width $\Delta\nu_0$ (FWHM) is then approximately given by the relation

$$\frac{\Delta\nu_0}{\nu_0} = \frac{1}{2N_w} \tag{10.4.7}$$

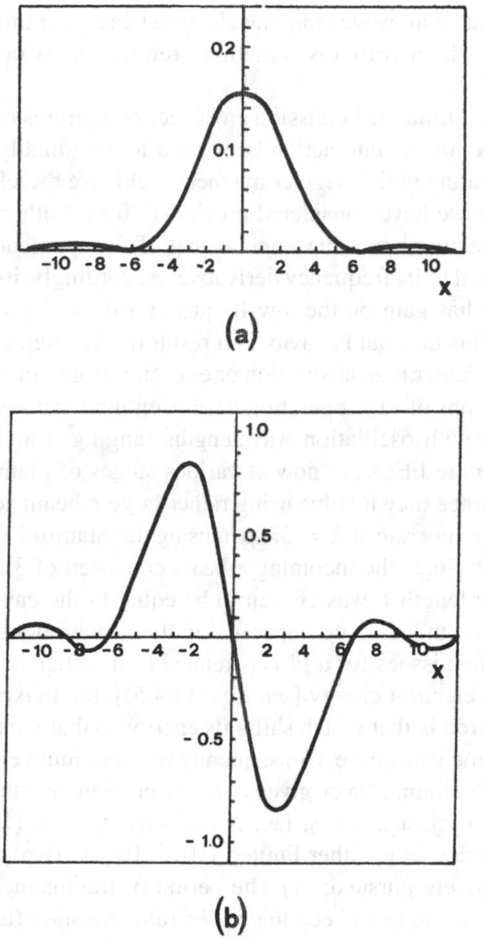

FIG. 10.25. Spectrum (a) of the spontaneously emitted radiation, and (b) of the stimulated emission cross section in a free- electron laser as a function of the normalized quantity $x = 2\pi N_w(\nu - \nu_0)/\nu_0$.

This spectrum is shown in Fig. 10.25a as a function of the dimensionless quantity x. Since all electrons, if injected with the same velocity and in the same direction, will show the same line shape, this corresponds to a homogeneous line shape for the FEL laser. Inhomogeneous effects arise from such factors as spread in electron energy, angular divergence of the electron beam, and variation in magnetic field over the beam cross section. Note that, since the number of undulator periods may typically be $N_w \sim 10^2$, we have from Eq. (10.4.7) $\Delta\nu_0/\nu_0 \cong 5 \times 10^{-3}$. Note also that there is an alternative way of considering the behavior of the emitted radiation. In the rest frame of the electron that we considered earlier, the magnetic field of the undulator appears to move at nearly the velocity of light. It can be shown that the static magnetic field then appears to the electrons essentially like a counter-propagating e.m. wave. The synchrotron emission can therefore be considered as arising from Compton back scattering of this "virtual" e.m. wave from the electron beam. For this reason the corresponding type of free electron laser (FEL) is sometimes referred to as operating in the Compton regime (Compton FEL).

A calculation of the stimulated emission cross section requires a detailed analysis, which we do not consider here, of the interaction between a longitudinally propagating e.m. wave and the electron propagating in the wiggler magnetic field. We merely wish to point out that, unlike the situation that we have considered previously for all other lasers, the spectral distribution of this cross section is not the same as that of the spontaneously emitted radiation but is instead proportional to its frequency derivative. Accordingly, its shape will be as shown in Fig. 10.25b and one has gain on the low-frequency side and loss on the high-frequency side of the transition. This unusual behavior is a result of the interaction being based upon a light-scattering process rather than absorption or emission from bound states.

So far, demonstrations of FEL operation have been made on several devices (more than ten) around the world, with oscillation wavelengths ranging from millimeter waves up to the UV region. Many more FELs are now at various stages of planning. All of these lasers require large facilities since they involve using rather large e-beam accelerators. Historically, the first FEL was made to operate at $\lambda = 3.4\,\mu$m using the Stanford University superconducting linear accelerator.[21] Since the incoming e-beam consisted of 3.2-ps pulses separated by $\tau = 84.7$ ns, the cavity length L was chosen to be equal to the cavity round trip time (i.e., $L = c\tau/2 = 12.7$ m) and the laser was operated in the synchronously mode-locked regime. One of the most important issues for a FEL is related to its efficiency. Since the emitted frequency depends on the electron energy [see Eq. (10.4.5)], the maximum energy that can be extracted from the electron is that which shifts its energy so that the corresponding operating frequency falls outside the gain curve. Consequently the maximum efficiency η_{max}, defined as the ratio between the maximum energy given to the laser beam and the initial electron energy, is approximately given by just $\Delta\nu_0/\nu_0$, i.e., is given by $\eta_{max} = (1/2N_w)$. This means that the efficiency in such a device is rather limited (10^{-2}–10^{-3}). Two ways of obtaining higher efficiency are being actively pursued: (1) The period of the magnet is gradually decreased along the e-beam direction so as to keep the λ_q/E^2 ratio constant (tapered wiggler). (2) The energy remaining in the electron beam, after leaving the wiggler, is recovered by decelerating the electrons. Much higher efficiencies are predicted to be achievable using these approaches and indeed have been achieved to some degree. As a final comment we point out that the FELs described so far all use e-beam machines of high energy ($E > 10$ MeV) and low current ($I \sim 1$–100 A). Under these conditions, as previously discussed, the light emission can be described as arising from Compton scattering of the virtual quanta of the magnetic field from individual electrons (*Compton regime FEL*). Free electron lasers using e-beams of lower energy ($E = 1$–2 MeV) and much higher currents ($I \sim 10$–20 kA) have also been made to operate. In this case, the electron-electron interaction becomes so strong that collective oscillatory motions (plasma waves) are induced in the e-beam when interacting with the e.m. wave in the wiggler. The emission can then be looked upon as arising from scattering of the virtual quanta of the magnetic field from these collective motions rather than from single electrons. The emitted frequency, $\nu_0 = 2\pi/\omega_0$, is then no longer given by Eq. (10.4.5) but in fact downshifted by the frequency of this collective motion. The phenomenon is analogous to Raman scattering of light from molecular vibrations, and the corresponding laser is said to operate in the *FEL-Raman regime*. Because of the lower value of electron energy involved, these lasers all oscillate in the millimeter wave region.

As a conclusion to this section we can say that the most attractive properties of FELs are: (1) Wide tunability; (2) excellent beam quality, close to the diffraction limit; and potentially, (3) very high efficiency, and thus very high laser power (the average power of the e-beam of the Stanford Linear Accelerator is about 200 kW). Free electron lasers are, however, inherently large and expensive machines, and interest in their applications is likely to be strongest in the frequency ranges where more conventional lasers are not so readily available e.g., in the far IR (100–400 μm) or in the vacuum ultraviolet ($\lambda < 100$ nm).

10.5. X-RAY LASERS[22]

The achievement of coherent oscillation in the x-ray region has been a long sought dream that is slowly but steadily coming true. The potential applications of x-ray lasers are indeed very important. They include in fact such possibilities as: (i) X-ray holography or x-ray microscopy of e.g., living cells or cell constituents, allowing, respectively, three-dimensional or two-dimensional pictures with sub-nanometer resolution to be obtained. (ii) X-ray lithography, where patterns with extremely high resolution could be produced.

Before discussing what has been achieved so far in this wavelength region, let us indicate the difficulties that have to be overcome to obtain x-ray laser operation. Starting with fundamental considerations, we recall that the threshold pump power of a four-level laser is given by Eq. (7.3.12) which is here reported for convenience

$$P_{th} = \frac{h\nu_{mp}}{\eta_p}\frac{\gamma A}{\sigma \tau} \tag{10.5.1}$$

The minimum threshold, P_{mth}, is of course attained for $\sigma = \sigma(\nu = \nu_0) = \sigma_p$, where σ_p is the cross section at the peak of the transition. Furthermore, one must take into account that, in the x-ray region, the upper state lifetime τ is established by the spontaneous lifetime τ_{sp}. From Eqs. (2.4.29) and (2.3.15) one then gets $1/\sigma_p\tau_{sp} \propto \nu_0^2/g_t(0)$, independently of the transition matrix element $|\mu|$. For either Eq. (2.4.9b) (homogeneous line) or Eq. (2.4.28) (inhomogeneous line) one then finds that $g_t(0) \propto 1/\Delta\nu_0$, where $\Delta\nu_0$ indicates here the transition linewidth for either a homogeneous or inhomogeneous line. Thus, in either case, from Eq. (10.5.1), with $h\nu_{mp} \cong h\nu_0$, we obtain $P_{mth} \propto \nu_0^3\Delta\nu_0$. At frequencies in the VUV to soft x-rays and at moderate pressures, we may assume that the linewidth is dominated by Doppler broadening. Hence, [see Eq. (2.5.18)], $\Delta\nu_0 \propto \nu_0$ and P_{mth} is expected to increase as ν_0^4. At the higher frequencies corresponding to the x-ray region, the linewidth is dominated by natural broadening since the radiative lifetime becomes very short (down to the femtosecond region). In this case one has $\Delta\nu_0 \propto 1/\tau_{sp} \propto \nu_0^3$ and P_{mth} is expected to increase as ν_0^6. Thus, if we go from e.g., the green ($\lambda = 500$ nm) to the soft x-ray region ($\lambda \cong 10$ nm) the wavelength decreases by a factor of 50 and P_{mth} increases by very many orders of magnitude. From a more practical viewpoint we note that multilayer dielectric mirrors for the x-ray region are lossy and difficult to make. A basic problem is that the difference in refractive index between various materials become very small in this region. Dielectric multi layers with a large number of layers (hundreds) are therefore needed to achieve a reasonable reflectivity. Scattering of light at the many interfaces then makes the mirrors very lossy and, furthermore, the mirrors have difficulty to withstand the high intensity of a x-ray-laser beam. For these reasons, the x-ray lasers that have been operated so far, have operated without mirrors as ASE (amplified spontaneous emission) devices.

As a representative example, we will consider the soft x-ray laser based on 24-times ionized selenium (Se^{24+}) as the active medium,(23) this being the first laser of a kind (the *x-ray recombination laser*) which now includes a large number of materials. Pumping is achieved by the powerful second harmonic beam ($\lambda = 532$ nm) of the Novette laser (pulse energy ~1 KJ, pulse duration ~1 ns), consisting of one arm of the Nova laser, at Lawrence Livermore Laboratory in the U.S.. The beam is focused to a fine line ($d \cong 200\,\mu$m, $l = 1.2$ cm) on a thin stripe (75 nm thick) of selenium evaporated on a 150-nm-thick foil of Formvar (Fig. 10.26). The foil could be irradiated from one or both sides. Exposed to the high intensity of this pump beam (~5×10^{13} W/cm^2), the foil explodes and a highly ionized selenium plasma is formed, whose shape is approximately cylindrical with a diameter $d \cong 200\,\mu$m. During the electron-ion recombination process, a particularly long-lived constituent of this plasma is formed, consisting of Se^{24+}. This ion has the same ground-state electronic configuration as neutral Ne ($1s^2 2s^2 2p^6$, see Fig. 10.1) and, accordingly, it is usually referred to as neon-like selenium. Impact collisions with the hot plasma electrons ($T_e \cong 1$ keV) then raises Se^{24+} from its ground state to excited states and population inversion between the states $2p^5 3p$ and $2p^5 3s$ is achieved because the lifetime of the $3s \rightarrow$ ground state transition is much shorter than the lifetime of the $3p \rightarrow 3s$ transition (both transitions are electric-dipole allowed). With a pump configuration as in Fig. 10.26, a strong longitudinal emission due to ASE is observed on two lines ($\lambda_1 = 20.63$ nm and $\lambda_2 = 20.96$ nm) of the $2p^5 3p \rightarrow 2p^5 3s$ transition (see Fig. 10.1). Owing to the much higher nuclear charge of Se compared to Ne, these lines fall in the soft x-ray region. From the length dependence of the emitted energy it is deduced that a maximum single-pass gain, $G = \exp(\sigma_p N l)$, of about 700 has been obtained. Note that this gain is still well below the "threshold" for ASE as defined in Sect. 2.9.2. In fact, for the experimental situation described here, the emission solid angle is $\Omega \cong 10^{-4}$ sterad and the linewidth is expected to be still dominated by Doppler broadening. From Eq. (2.9.4b) one then gets $G_{th} \cong 4.5 \times 10^5$. This means that the emitted intensity due to ASE is still much smaller than the saturation intensity of the amplifier. Indeed, the x-ray output energy produced was an extremely small fraction (~10^{-10}) of the pump energy.

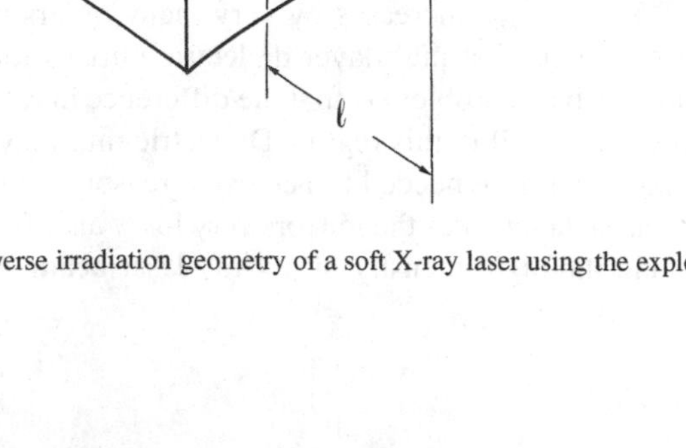

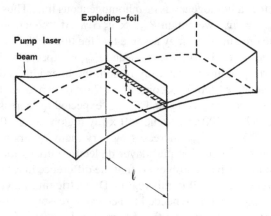

FIG. 10.26. Transverse irradiation geometry of a soft X-ray laser using the exploding-foil technique.

Since this first laser was demonstrated, research activity in this field has been very strong.[24] Thus many new active media have been made to operate namely many more neon-like ions (from Ag^{37+} to Ar^{8+}), many hydrogen-like ions (from Al^{12+} to C^{5+}), lithium-like ions (from Si^{11+} to Al^{10+}) and nickel-like ions (from Au^{51} to Eu^{35+}) ions. The range of the oscillation wavelength now extends from ~3.6 to 47 nm while the single pass gain, G, generally ranges from 10 to 10^3. To achieve the high pump powers required [see Eq. (10.5.1)] and, at the same time, reduce the corresponding pump energy, picosecond or even femtosecond laser pulses are now also used for pumping. Amplified spontaneous emission (at $\lambda = 46.9$ nm) has also been observed in neon-like Ar^{8+} by passing a strong current pulse of short duration through a $1 \div 10$ cm long capillary filled with Ar.

10.6. CONCLUDING REMARKS

In this chapter, the most notable examples of lasers involving low-density media have been considered. In general these lasers tend to be more bulky, and often less efficient than the lasers considered in the previous chapter (notably semiconductor and diode-pumped solid-state lasers). For these reasons, whenever possible, these low-density lasers are tending to be superseded by the solid-state-laser counterparts. This is for instance the case of the Argon laser, which is facing a strong competition from the green beam emitted by a diode-pumped Nd laser (e.g., Nd:YVO_4) with intracavity second-harmonic generation. This is also the case, at least for quite a few applications, of the red-emitting He-Ne laser, which is facing strong competition from red-emitting InGaAlP semiconductor lasers. Low-density lasers will however survive in those frequency ranges which are not covered effectively by semiconductor or diode-pumped solid-state lasers. This is for instance the case for middle-infrared lasers, the CO_2 lasers being the most notable example here, and for the lasers oscillating in the UV (e.g. excimer lasers) down to the X-ray spectral region. Another field of applications where low-density lasers will continue to perform well is where very high powers are required, with CO_2 lasers, excimer lasers, and chemical lasers as important examples. Thus, as a conclusion, one can foresee that lasers based on low-density media will still maintain an important role in the laser field.

PROBLEMS

10.1. List at least four lasers, using a low-density active medium, whose wavelengths fall in the infrared.

10.2. List at least four lasers, using a low-density active medium, whose wavelengths fall in the UV to soft-x-ray region. What are the problems to be faced in achieving laser action in the UV or x-ray region?

10.3. For metalworking applications, a laser with cw output power >1 kW is required. Which lasers meet this requirement?

10.4. The 514.5 nm transition of an argon-ion laser is Doppler broadened to a line-width of ~3.5 GHz. The cavity length of the laser is 100 cm and, when pumped three times above threshold, the laser emits a power of 4 W in a TEM_{00} mode-profile. Assuming that the frequency of one of the oscillating TEM_{00} modes coincides with the center of the gain line, calculate the number of TEM_{00} modes which are expected to oscillate.

10.5. Consider the argon-ion laser described in the previous problem and assume that the laser is mode-locked by an acousto-optic modulator. Calculate: (i) The duration and the peak power of the mode-locked pulses. (ii) The drive frequency of the rf oscillator.

10.6. Assume that the bond between the two nitrogen atoms of the N_2 molecule can be simulated by a spring of suitable elastic constant. Knowing the vibrational frequency (Fig. 10.9) and the atomic mass calculate the elastic constant. Compare this constant with that obtainable from the ground state curve of Fig. 10.19.

10.7. Show that, if the elastic constant of the N-N bond is taken to be the same as that of the iso-electronic CO molecule, the $(\upsilon' = 1) \rightarrow (\upsilon = 0)$ transition wavelength of the N_2 molecule is approximately the same as that of the CO molecule.

10.8. Assume that each of the two oxygen-carbon bonds of the CO_2 molecule can be simulated by a spring with elastic constant k. With the assumption that there is no interaction between the two oxygen atoms and knowing the ν_1 frequency ($\nu_1 = 1337\,\text{cm}^{-1}$), calculate this constant.

10.9. Knowing the elastic constant k between the two oxygen-carbon bonds, obtained in the previous problem, calculate the expected frequency ν_3 of the asymmetric stretching mode and compare the result with the value shown in Fig. 10.9.

10.10. Show that each C-O bond of the CO_2 molecule cannot be simulated by elastic springs if the harmonic oscillation corresponding to the bending mode of frequency ν_2 has to be calculated.

10.11. From the knowledge that, for a Boltzmann distribution, the maximum population of the upper laser level of a CO_2 molecule occurs for the rotational quantum number $J' = 21$ (see Fig. 10.11), calculate the rotational constant B [assume $T = 400\,\text{K}$, which corresponds to an energy kT such that $(kT/h) \cong 280\,\text{cm}^{-1}$]. From this value then calculate the equilibrium distance between the C atom and each O atom.

10.12. Using the result of the previous problem, calculate the frequency spacing (in cm^{-1}) between the rotational lines of the CO_2 laser transition [assume that the rotational line constant of the lower laser level is the same as that of the upper laser level, and remember that only levels with odd values of J are occupied in a CO_2 molecule].

10.13. The linewidth, due to collision broadening, of the CO_2 laser transition is given by $\Delta\nu_c = 7.58\,(\psi_{CO_2} + 0.73\,\psi_{N_2} + 0.6\,\psi_{He})p(300/T)^{1/2}$ MHz, where the ψ are the fractional partial pressures of the gas mixture, T is the gas temperature and p is the total pressure (in torr) (see example 3.3). Taking a ratio of the partial pressures of CO_2, N_2, and He molecules of 1:1:8 and assuming a separation between rotational lines of the CO_2 laser transition of $\Delta\nu_r \cong 2\text{cm}^{-1}$, calculate the total gas pressure needed to make all the rotational lines merge together. What would the width of the gain curve be?

10.14. Consider a CO_2 laser with high enough pressure to have all its rotational lines merged together. If this laser were mode-locked, what would be the order of magnitude of the corresponding laser pulse-width?

10.15. Show that a reaction energy of 31.6 kcal/mole, as in the HF cold reaction [see Eq. (10.3.1)], is equivalent to an energy of 1.372 eV released for each molecular reaction.

10.16. Consider the cold reaction of Fig. 10.22a and take the values shown in the figure for the relative populations of HF molecules which, after reaction, are left in the first three vibrational levels. Calculate the fraction, η, of heat released in the reaction which goes into vibrational energy.

10.17. Repeat the previous calculation for the hot reaction of Fig. 10.22b.

10.18. To reach the final end products of the cold reaction, see Eq. (10.3.1), one can choose a reaction path-way where one first dissociates the H_2 molecule, to obtain the single atoms F, H, and H, and then let the fluorine and one hydrogen atom recombine together. Similarly, for the hot reaction, see Eq. (10.3.2), one can first dissociate molecular fluorine and then recombine one fluorine with the hydrogen atom. Given the two possibilities, relate the difference in heat of reaction of these two reactions to the difference in dissociation energy between the fluorine and hydrogen molecules.

References

1. R. Arrathoon, Helium-Neon Lasers and the Positive Column, in *Lasers*, ed. by A. K. Levine and A. J. De Maria (Marcel Dekker, New York, 1976), Vol. 4, Chap. 3.
2. W. B. Bridges, Atomic and Ionic Gas Lasers, in *Methods of Experimental Physics*, ed. by C. L. Tang (Academic, New York, 1979), Vol. 15, pp. 33–151.
3. A. Javan, W. R. Bennett, and D. H. Herriott, Population Inversion and Continuous Optical Maser Oscillation in a Gas Discharge Containing a He-Ne Mixture, *Phys. Rev. Lett.*, **6**, 106 (1961).
4. C. E. Webb, Metal Vapor Lasers: Recent Advances and Applications, in *Gas Flow and Chemical Lasers*, Springer Proceedings in Physics N. 15, ed. by S. Rosenwork (Springer-Verlag, Berlin, 1987), pp. 481–494.
5. C. C. Davis and T. A. King, Gaseous Ion Lasers, in *Advances in Quantum Electronics*, ed. by D. W. Goodwin (Academic, New York, 1975), Vol. 3, pp. 170–437.
6. D. H. Dunn and J. N. Ross, The Argon Ion Laser, in *Progress in Quantum Electronics*, ed. by J. H. Sanders and S. Stenholm (Pergamon, London, 1977), Vol. 4, pp. 233–270.
7. W. B. Bridges, Laser Oscillation in Singly Ionized Argon in the Visible Spectrum, *Appl. Phys. Letters* **4**, 128 (1964).
8. P. K. Cheo, CO_2 Lasers, in *Lasers*, ed. by A. K. Levine and A. J. De Maria (Marcel Dekker, New York, 1971), Vol. 3, Chap. 2.
9. A. J. De Maria, Review of High-Power CO_2 Lasers, in *Principles of Laser Plasmas*, ed. by G. Bekefi (Wiley-Interscience, New York, 1976), Chap. 8.
10. C. K. N. Patel, W. L. Faust, and R. A. Mc Farlane, CW Laser Action on Rotational Transitions of the $\Sigma_u^+ \rightarrow \Sigma_g^+$ Vibrational Band of CO_2, *Bull. Am. Phys. Soc.* **9**, 500 (1964).
11. D. R. Hall and C. A. Hill, Radiofrequency-Discharge-Excited CO_2 Lasers, in *Handbook of Molecular Lasers* ed. by P. Cheo (Marcel Dekker, New York, 1987) Chapt. 3.
12. K. M. Abramski, A. D. Colley, H. J. Baker, and D. R. Hall, Power Scaling of Large-Area Transverse Radiofrequency Discharge CO_2 Lasers, *Appl. Phys. Letters*, **54**, 1833–1835 (1989).
13. P. E. Jackson, H. J. Baker, and D. R. Hall, CO_2 Large-Area Discharge Laser Using an Unstable-Waveguide Hybrid Resonator, *Appl. Phys. Letters*, **54**, 1950–1952 (1989).
14. R. E. Center, High-Power, Efficient Electrically-Excited CO Laser, in *Laser Handbook*, ed. by M. L. Stitch (North-Holland, Amsterdam, 1979), Vol. 3, pp. 89–133.
15. C. S. Willet, *An Introduction to Gas Lasers: Population Inversion Mechanisms* (Pergamon-Press, Oxford, 1974), Secs. 6.2.1 and 6.2.3.
16. N. G. Basov, V. A. Danilychev, and Yu. M. Popov, Stimulated Emission in the Vacuum Ultraviolet Region, *Soviet J. Quantum Electron.*, **1**, 18 (1971).
17. J. J. Ewing, Excimer Lasers, in *Laser Handbook*, ed. by M. L. Stitch (North-Holland, Amsterdam, 1979), Vol. 3, pp. 135–197.

18. A. N. Chester, Chemical Lasers, in *High-Power Gas Lasers*, ed. by E. R. Pike (The Institute of Physics, Bristol and London, 1975), pp. 162–221.
19. C. J. Ultee, Chemical and Gas-Dynamic Lasers, in *Laser Handbook*, ed. by M. L. Stitch and M. Bass (North-Holland, Amsterdam, 1985), Vol. 3, pp. 199–287.
20. G. Dattoli and R. Renieri, Experimental and Theoretical Aspects of the Free-Electron Lasers, in *Laser Handbook*, ed. by M. L. Stitch (North-Holland, Amsterdam, 1979), Vol. 4, pp. 1–142.
21. D. A. G. Deacon, L. R. Elias, J. M. J. Madey, G. J. Ramian, H. A. Schwettman, and T. I. Smith, First Operation of a Free-Electron Laser, *Phys. Rev. Lett.* **38**, 892 (1977).
22. R. C. Elton, *X-Ray Lasers* (Acedemic, Boston 1990).
23. D. L. Matthews et. al., Demonstration of a Soft X-ray Amplifier, *Phys. Rev Lett.* **54**, 110 (1985).
24. *X-Ray Lasers 1996*, ed. by S. Svanberg and C. G. Wahlstrom, Institute of Physics Conference Series N. 151 (Institute of Physics, Bristol 1996).

11

Properties of Laser Beams

11.1. INTRODUCTION

In Chap. 1 it was stated that the most characteristic properties of laser beams are (1) monochromaticity, (2) coherence (spatial and temporal), (3) directionality, (4) brightness. The material presented in earlier chapters allows us to now examine these properties in more detail and compare them with the properties of conventional light sources (thermal sources).

In most cases of interest to us, the spectral bandwidth of the light source $\Delta\omega$ is much smaller than the mean frequency $<\omega>$ of the spectrum (*quasi-monochromatic wave*). In this case, the electric field of the wave, at position $\mathbf{r}$ and time t, can be written as

$$E(\mathbf{r}, t) = A(\mathbf{r}, t) \exp j\,[<\omega>t - \phi(\mathbf{r}, t)] \tag{11.1.1}$$

where $A(\mathbf{r},\ t)$ and $\phi(\mathbf{r},\ t)$ are both slowly varying over an optical period, i.e.,

$$\left[\left|\frac{\partial A}{A\partial t}\right|, \left|\frac{\partial \phi}{\partial t}\right|\right] \ll \langle\omega\rangle \tag{11.1.2}$$

We then define the intensity of the beam as

$$I(\mathbf{r}, t) = E(\mathbf{r}, t)E^*(\mathbf{r}, t) = |A(\mathbf{r}, t)|^2 \tag{11.1.3}$$

11.2. MONOCHROMATICITY

We have seen in Sects. 7.9–7.11 that the frequency fluctuations of a c.w. single-mode laser mostly arise from phase fluctuations rather than amplitude fluctuations. Amplitude fluctuations come in fact from pump or cavity loss fluctuations. They are usually very

O. Svelto, *Principles of Lasers*,
DOI: 10.1007/978-1-4419-1302-9_11, © Springer Science+Business Media LLC 2010

small (~1%) and can be further reduced by suitable electronically-controlled feedback loops. To first order we can thus take $A(t)$ to be constant for a single-mode laser and so consider its degree of monochromaticity to be determined by frequency fluctuations. The theoretical limit to this monochromaticity arises from zero-point fluctuations and is expressed by Eq. (7.9.2). However, this limit generally corresponds to a very low value for the oscillating bandwidth, $\Delta\nu_L$, which is seldom reached in practice. In the example 7.9, for instance, the value of $\Delta\nu_L$ is calculated to be ~0.4 mHz for a He-Ne laser with 1 mW output power. A notable exception to this situation occurs for a semiconductor laser, where, due to the short length and high loss of the laser cavity, this limit is very much higher ($\Delta\nu_L \cong 1$ MHz) and the actual laser linewidth is indeed often determined by these quantum fluctuations. Otherwise, in most other cases, technical effects such as vibrations and thermal expansion of the cavity are dominant in determining the laser linewidth $\Delta\nu_L$. If a monolithic structure is used for the laser cavity configuration, as in the non-planar ring oscillator of Fig. 7.26, typical values of $\Delta\nu_L$ may fall in the 10–50 kHz range. Much smaller linewidths (down to ~0.1 Hz) can be obtained by stabilizing the laser frequency against an external reference, as discussed in Sect. 7.10. In pulsed operation the minimum linewidth is obviously limited by the inverse of the pulse duration τ_p. For example, for a single-mode Q-switched laser, assuming $\tau_p \cong 10$ ns, one has $\Delta\nu_L \cong 100$ MHz.

In the case of a laser oscillating on many modes, the monochromaticity is obviously related to the number of oscillating modes. We recall, for example, that in mode-locked operation, pulses down to a few tens of femtoseconds have been obtained. In this case, actually, the corresponding laser bandwidth is in the range of a few tens of THz and the condition for quasi-monochromatic radiation no longer holds well.

The degree of monochromaticity required depends, of course, on the given application. Actually, the very narrowest laser linewidths are only needed for the most sophisticated applications dealing with metrology and with fundamental measurements in physics (e.g. gravitational wave detection). For other more common applications, such as interferometric measurements of distances, coherent laser radar, and coherent optical communications, the required monochromaticity falls in the 10–100 kHz range while a monochromaticity of ~1 MHz is typical of what is needed for much of high resolution spectroscopy and it is certainly enough for optical communications using wavelength-division-multiplexing (WDM). For some applications, of course, laser monochromaticity is not relevant. This is certainly the case for important applications such as laser material working and for most applications in the biomedical field.

11.3. FIRST-ORDER COHERENCE[1]

In Chap. 1 the concept of coherence of an e.m. wave was introduced in an intuitive fashion, with two types of coherence being distinguished: (1) spatial and (2) temporal coherence. In this section we give a more detailed discussion of these two concepts. In fact, as will be better appreciated by the end of this chapter, it turns out that spatial and temporal coherence describe the coherence properties of an e.m. wave only to first order.

11.3.1. Degree of Spatial and Temporal Coherence

In order to describe the coherence properties of a light source, we can introduce a whole class of correlation functions for the corresponding field. For the moment, however, we will limit ourselves to looking at the first-order functions.

Suppose that a measurement of the field is performed at some point $\mathbf{r}_1$ in a time interval between 0 and T. We can then obtain the product $E(\mathbf{r}_1,\ t_1)E^*(\mathbf{r}_1,\ t_2)$ where t_1 and t_2 are given time instants within the time interval $0-T$. If the measurement is now repeated a large number of times, we can calculate the average of the above product over all the measurements. This average is called the ensemble average and written as

$$\Gamma^{(1)}(\mathbf{r}_1, \mathbf{r}_1, t_1, t_2) = \langle E(\mathbf{r}_1, t_1)E^*(\mathbf{r}_1, t_2)\rangle \tag{11.3.1}$$

For the remainder of this section as well as in the next two sections, we will consider the case of a stationary beam,* as would for instance apply to a single-mode c.w. laser or to a c.w. laser oscillating on many modes that are not locked in phase, or to a c.w. thermal light source. In these cases, by definition, the ensemble average will only depend upon the time difference $\tau = t_1 - t_2$ and not upon the particular times t_1 and t_2. We can then write

$$\Gamma^{(1)}(\mathbf{r}_1, \mathbf{r}_1, t_1, t_2) = \Gamma^{(1)}(\mathbf{r}_1, \mathbf{r}_1, \tau) = \langle E(\mathbf{r}_1, t+\tau)E^*(\mathbf{r}_1, t)\rangle \tag{11.3.2}$$

where we have set $t = t_2$ and $\Gamma^{(1)}$ only depends upon τ. If now the field, besides being stationary is also ergodic (a condition that also usually applies to the cases considered above), then, by definition, the ensemble average is the same as the time average. We can then write

$$\Gamma^{(1)}(\mathbf{r}_1, \mathbf{r}_1, \tau) = \lim_{T\to\infty} \frac{1}{T}\int_0^T E(\mathbf{r}_1, t+\tau)\, E^*(\mathbf{r}_1, t)dt \tag{11.3.3}$$

Note that, the definition of $\Gamma^{(1)}$ in terms of a time average is perhaps easier to deal with than that based on ensemble averages. However, the definition of $\Gamma^{(1)}$ in terms of an ensemble average is more general and, in the form given by Eq. (11.3.1), can be applied also to non stationary beams, as we shall see in Sect. 11.3.4.

Having defined the first-order correlation function $\Gamma^{(1)}$ at a given point $\mathbf{r}_1$, we can define a normalized function $\gamma^{(1)}(\mathbf{r}_1, \mathbf{r}_1, \tau)$ as follows

$$\gamma^{(1)} = \frac{\langle E(\mathbf{r}_1,\ t+\tau)\, E^*(\mathbf{r}_1,\ t)\rangle}{\langle E(\mathbf{r}_1,\ t)\, E^*(\mathbf{r}_1,\ t)\rangle^{1/2}\langle E(\mathbf{r}_1,\ t+\tau)\, E^*(\mathbf{r}_1,\ t+\tau)\rangle^{1/2}} \tag{11.3.4}$$

Note that, for a stationary beam, the two ensemble averages in the denominator of Eq. (11.3.4) are equal to each other and, according to Eq. (11.1.3), are both equal to the average beam intensity $\langle I(\mathbf{r}_1,\ t)\rangle$. The function $\gamma^{(1)}$, as defined by Eq. (11.3.4), is referred to as the *complex degree of temporal coherence* while its magnitude, $|\gamma^{(1)}|$, as the *degree of temporal coherence*. Indeed $\gamma^{(1)}$ gives the degree of correlation between the fields at the same point $\mathbf{r}_1$ at two

* A process is said to be stationary when the ensemble average of any variable that describes it (e.g., the analytic signal or the beam intensity, in our case), is independent of time.

instants separated by a time τ. The function $\gamma^{(1)}$ has the following main properties: (1) $\gamma^{(1)} = 1$ for $\tau = 0$, as apparent from Eq. (11.3.4); (2) $\gamma^{(1)}(\mathbf{r}_1, \mathbf{r}_1, -\tau) = \gamma^{(1)*}(\mathbf{r}_1, \mathbf{r}_1, \tau)$ as can readily be seen from Eq. (11.3.4) with the help of Eq. (11.1.1); (3) $|\gamma^{(1)}(\mathbf{r}_1, \mathbf{r}_1, \tau)| \leq 1$, which follows from applying the Schwarz inequality to Eq. (11.3.4).

We can now say that a beam has perfect temporal coherence when $\gamma^{(1)} = 1$ for any τ. For a c.w. beam this essentially implies that both amplitude and phase fluctuations of the beam are zero so that the signal reduces to that of a sinusoidal wave, i.e., $E = A(\mathbf{r}_1)\,\exp j[\omega t - \phi(\mathbf{r}_1)]$. Indeed, the substitution of this expression into Eq. (11.3.4) shows that $|\gamma^{(1)}| = 1$ in this case. The opposite case of complete absence of temporal coherence occurs when $\langle E(\mathbf{r}_1,\ t+\tau)E^*(\mathbf{r}_1,\ t)\rangle$ and hence $\gamma^{(1)}$ vanishes for $\tau > 0$. Such would be the case for a thermal light source of very large bandwidth (e.g., a blackbody source, see Fig. 2.3). In more realistic situations $|\gamma^{(1)}|$ is generally expected to be a decreasing function of τ as indicated in Fig. 11.1. Note that, according to the property stated as point (2) above, $|\gamma^{(1)}|$ is a symmetric function of τ. We can therefore define a characteristic time τ_{co} (referred to as the *coherence time*) as, for instance, the time for which $|\gamma^{(1)}| = 1/2$. For a perfectly coherent wave, one obviously has $\tau_{co} = \infty$, while for a completely incoherent wave one has $\tau_{co} = 0$. Note that we can also define a *coherence length* L_c as $L_c = c\tau_{co}$.

In a similar way, we can define a first-order correlation function between two points $\mathbf{r}_1$ and $\mathbf{r}_2$ at the same time as

$$\Gamma^{(1)}(\mathbf{r}_1, \mathbf{r}_2, 0) = \langle E(\mathbf{r}_1,\ t)\ E^*(\mathbf{r}_2,\ t)\rangle = \lim_{T\to\infty} \frac{1}{T}\int_0^T E(\mathbf{r}_1,\ t)\ E^*(\mathbf{r}_2,\ t)dt \tag{11.3.5}$$

We can also define the corresponding normalized function $\gamma^{(1)}(\mathbf{r}_1,\ \mathbf{r}_2, 0)$ as

$$\gamma^{(1)} = \frac{\langle E(\mathbf{r}_1,\ t)\ E^*(\mathbf{r}_2,\ t)\rangle}{\langle E(\mathbf{r}_1,\ t)\ E^*(\mathbf{r}_1,\ t)\rangle^{1/2}\ \langle E(\mathbf{r}_2,\ t)\ E^*(\mathbf{r}_2,\ t)\rangle^{1/2}} \tag{11.3.6}$$

The quantity $\gamma^{(1)}(\mathbf{r}_1,\ \mathbf{r}_2,\ 0)$ is referred to as the *complex degree of spatial coherence* and its magnitude, the *degree of spatial coherence*. Indeed $\gamma^{(1)}$ gives, in this case, the degree of correlation between the fields at the two space points $\mathbf{r}_1$ and $\mathbf{r}_2$ at the same time. Note that, from the Schwarz inequality, we again find that $|\gamma^{(1)}| \leq 1$. A wave will be said to have a

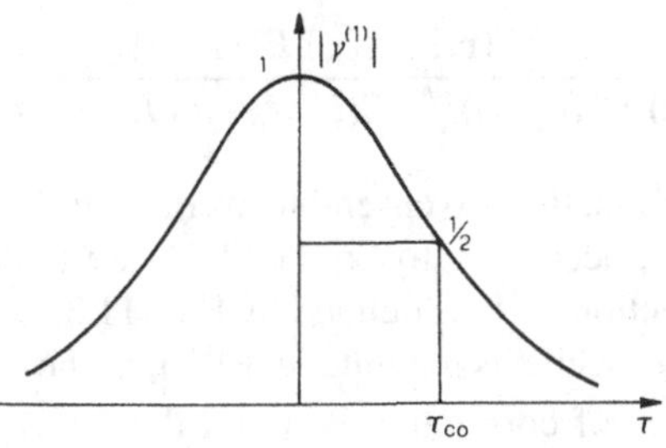

FIG. 11.1. Example of possible behavior of the degree of temporal coherence $|\gamma^{(1)}(\tau)|$. The coherence time, τ_{co}, can be defined as the half-width of the curve at half-height.

perfect spatial coherence if $|\gamma^{(1)}| = 1$ for any two points $\mathbf{r}_1$ and $\mathbf{r}_2$ (provided that they lie on the same wave front or on wave fronts whose separation is much smaller than the coherence length L_c). Often, however, one has a situation of partial spatial coherence. This means that, for a fixed value of $\mathbf{r}_1$, the degree of spatial coherence, $|\gamma^{(1)}|$, as a function of $|\mathbf{r}_2 - \mathbf{r}_1|$, decreases from the value 1 (which occurs for $\mathbf{r}_2 = \mathbf{r}_1$) to zero as $|\mathbf{r}_2 - \mathbf{r}_1|$ increases. This situation is illustrated in Fig. 11.2 where the function $|\gamma^{(1)}(\mathbf{r}_2 - \mathbf{r}_1)|$ is plotted vs $\mathbf{r}_2$ for a given position of point P_1 (of coordinate $\mathbf{r}_1$) on the wavefront. One sees that $|\gamma^{(1)}|$ will be larger than some prescribed value (e.g., 1/2) over a certain characteristic area, referred to as the *coherence area* of the beam at point P_1 of the wave-front.

The concepts of spatial and temporal coherence can be combined by means of the so-called mutual coherence function, defined as

$$\Gamma^{(1)}(\mathbf{r}_1, \mathbf{r}_2, \tau) = \langle E(\mathbf{r}_1, t+\tau) E^*(\mathbf{r}_2, t) \rangle \tag{11.3.7}$$

which can also be normalized as follows

$$\gamma^{(1)}(\mathbf{r}_1, \mathbf{r}_2, \tau) = \frac{\langle E(\mathbf{r}_1,\ t+\tau)\ E^*(\mathbf{r}_2,\ t)\rangle}{\langle E(\mathbf{r}_1,\ t)\ E^*(\mathbf{r}_1,\ t)\rangle^{1/2}\ \langle E(\mathbf{r}_2,\ t)\ E^*(\mathbf{r}_2,\ t)\rangle^{1/2}} \tag{11.3.8}$$

This function, referred to as the *complex degree of coherence*, provides a measure of the coherence between two different points of the wave at two different times. For a quasi-monochromatic wave, it follows from Eqs. (11.1.1) and (11.3.8) that we can write

$$\gamma^{(1)}(\tau) = |\gamma^{(1)}| \exp\{j[\langle\omega\rangle\tau - \phi(\tau)]\} \tag{11.3.9}$$

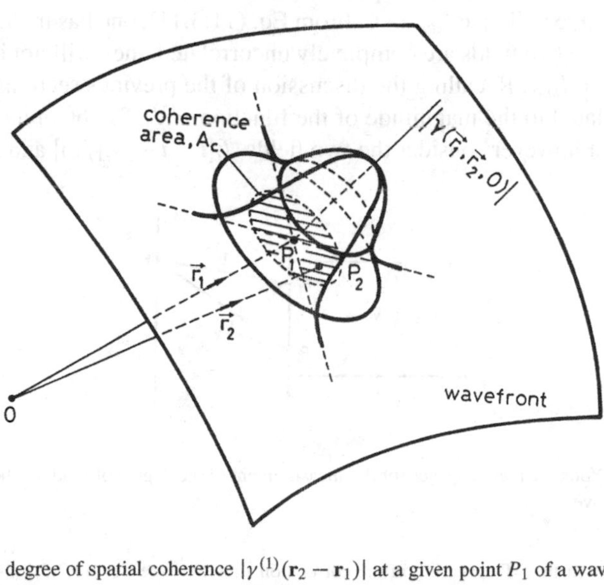

FIG. 11.2. Plot of the degree of spatial coherence $|\gamma^{(1)}(\mathbf{r}_2 - \mathbf{r}_1)|$ at a given point P_1 of a wavefront, to illustrate the concept of coherence area.

where $|\gamma^{(1)}|$ and $\phi(\tau)$ are both slowly varying functions of τ, i.e.,

$$\left[\frac{d\,|\gamma^{(1)}|}{|\gamma^{(1)}|\,d\tau}, \left|\frac{d\phi}{d\tau}\right|\right] \ll \langle\omega\rangle \tag{11.3.10}$$

11.3.2. Measurement of Spatial and Temporal Coherence

One very simple way of measuring the degree of spatial coherence between two points, P_1 and P_2, on the wave front of a light wave is by using Young's interferometer (Fig. 11.3). This simply consists of a screen 1, in which two small holes have been made at positions P_1 and P_2, and a screen 2 on which an interference pattern is produced by the light diffracted from the two holes. More precisely, the interference at point P and time t will arise from the superposition of the waves emitted from points P_1 and P_2 at times $[t-(L_1/c)]$ and $[t-(L_2/c)]$, respectively. One will therefore see interference fringes on screen 2, around point P, that are more distinct the better the correlation between the two fields of the light wave, $E[\mathbf{r}_1,\ t-(L_1/c)]$ and $E[\mathbf{r}_2,\ t-(L_2/c)]$, where $\mathbf{r}_1$ and $\mathbf{r}_2$ are the coordinates of points P_1 and P_2*. If we now let I_{max} and I_{min} represent, respectively, the maximum intensity of a bright fringe and the minimum intensity of a dark fringe in the region of the screen around P, we can define a visibility, V_P, of the fringes as

$$V_P = \frac{I_{max} - I_{min}}{I_{max} + I_{min}} \tag{11.3.11}$$

One can now see that, if the diffracted fields, at point P, from the two holes 1 and 2 have the same amplitude and if the two fields are perfectly coherent, their destructive interference at the point of the dark fringe will give $I_{min} = 0$. From Eq. (11.3.11), one has in this case $V_P = 1$. If, on the other hand, the two fields are completely uncorrelated, they will not interfere at all and one will have $I_{min} = I_{max}$. Recalling the discussion of the previous section, one can now see that V_P must be related to the magnitude of the function $\gamma^{(1)}$. To obtain the degree of spatial coherence one must however consider the two fields, $E[\mathbf{r}_1,\ t-(L_1/c)]$ and $E[\mathbf{r}_2,\ t-(L_2/c)]$,

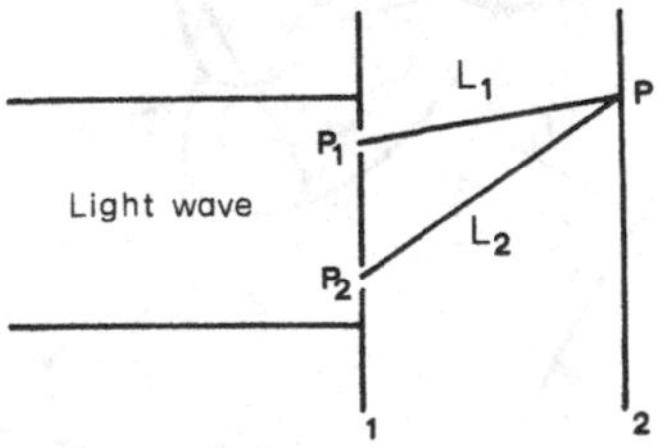

FIG. 11.3. The use of Young's interferometer for the measurement of the degree of spatial coherence between points P_1 and P_2 of an e.m. wave.

* Note that the integration time T appearing in the correlation function [see Eq. (11.3.5)] is now equal to the time taken for the measurement of the fringes (e.g., the exposure time of a photographic plate).

at the same time. This then requires that the point P be chosen so that $L_1 = L_2$. In this case, it will be shown in example 11.1 that

$$V_P = |\gamma^{(1)}(\mathbf{r}_1, \mathbf{r}_2, 0)| \tag{11.3.12}$$

On the other hand, if the two holes 1 and 2 do not produce the same amplitudes, i.e., the same illumination, at point P, instead of Eq. (11.3.12) one has

$$V_P = \frac{2[\langle I_1\rangle \langle I_2\rangle]^{1/2}}{\langle I_1\rangle + \langle I_2\rangle}\left|\gamma^{(1)}(\mathbf{r}_1, \mathbf{r}_2, 0)\right| \tag{11.3.13}$$

where $\langle I_1\rangle$ and $\langle I_2\rangle$ are the average intensities of the ligth diffracted to point P from the two holes. Note also that, for the general point P shown in Fig. 11.3, the visibility V_P can be shown to be equal to $\left|\gamma^{(1)}(\mathbf{r}_1, \mathbf{r}_2, \tau)\right|$, where $\tau = (L_2 - L_1)/c$.

Example 11.1. *Calculation of the fringe visibility in Young's interferometer.* The field $E(\mathbf{r}_P, t')$, at point P of Fig. 11.3 and time t', can be expressed as a superposition of the fields diffracted from holes 1 and 2 at times $(t' - L_1/c)$ and $(t' - L_2/c)$, respectively. We can then write

$$E(\mathbf{r}_P, t') = K_1\, E(\mathbf{r}_1, t' - L_1/c) + K_2\, E(\mathbf{r}_2, t' - L_2/c) \tag{11.3.14}$$

where $E(\mathbf{r}_1, t' - L_1/c)$ and $E(\mathbf{r}_2, t' - L_2/c)$ are the fields at points P_1 and P_2 while K_1 and K_2 represent the fractions of these two fields which are diffracted to position P, respectively. The factors K_1 and K_2 are expected to be inversely proportional to L_1 and L_2 and to also depend on the hole dimensions and the angle between the incident wave and the wave diffracted from P_1 and P_2. Since the diffracted secondary wavelets are always a quarter of a period out of phase with the incident wave [see also the discussion of "Huygens wavelets" appearing in Sect. 4.6] it follows that

$$K_1 = |K_1| \exp[-(j\pi/2)] \tag{11.3.15a}$$

$$K_2 = |K_2| \exp[-(j\pi/2)] \tag{11.3.15b}$$

If we now define $t_1 = t' - L_2/c$ and $\tau = (L_2/c) - (L_1/c)$, Eq. (11.3.14) can be written as

$$E = K_1\, E(\mathbf{r}_1, t+\tau) + K_2\, E(\mathbf{r}_2, t) \tag{11.3.16}$$

According to Eq. (11.1.3), the intensity at the point P can be written as $I = EE^* = |K_1\, E(\mathbf{r}_1, t+\tau) + K_2\, E(\mathbf{r}_2, t)|^2$, where Eq. (11.3.16) has been used. From this expression, with the help of Eq. (11.3.15), one obtains

$$I = I_1(t+\tau) + I_2(t) + 2\text{Re}\,[K_1\, K_2^*\, E(\mathbf{r}_1, t+\tau) E^*(\mathbf{r}_2, t)] \tag{11.3.17}$$

where Re stands for real part. In the previous expressions, I_1 and I_2 are the intensities, at point P, due to the emission from point P_1 alone and point P_2 alone, respectively, and are given by

$$I_1 = |K_1|^2 |E(\mathbf{r}_1, t+\tau)|^2 = |K_1|^2\, I(\mathbf{r}_1, t+\tau) \tag{11.3.18a}$$

$$I_2 = |K_2|^2 |E(\mathbf{r}_2, t)|^2 = |K_2|^2\, I(\mathbf{r}_2, t) \tag{11.3.18b}$$

where $I(\mathbf{r}_1, t+\tau)$ and $I(\mathbf{r}_2, t)$ are the intensities at points P_1 and P_2. Taking the time average of both sides of Eq. (11.3.17) and using Eq. (11.3.7), we find

$$\langle I\rangle = \langle I_1\rangle + \langle I_2\rangle + 2|K_1||K_2|\mathrm{Re}[\Gamma^{(1)}(\mathbf{r}_1, \mathbf{r}_2, \tau)] \tag{11.3.19}$$

where equations (11.3.15) have also been used. Equation (11.3.19) can be expressed in terms of $\gamma^{(1)}$ by noting that from Eq. (11.3.8) we have

$$\Gamma^{(1)} = \gamma^{(1)}[\langle I(\mathbf{r}_1, t+\tau)\rangle\langle I(\mathbf{r}_2, t)\rangle]^{1/2} \tag{11.3.20}$$

The substitution of Eq. (11.3.20) in Eq. (11.3.19) with the help of Eq. (11.3.18) gives $\langle I\rangle = \langle I_1\rangle + \langle I_2\rangle + 2(\langle I_1\rangle\langle I_2\rangle)^{1/2}\,\mathrm{Re}[\gamma^{(1)}(\mathbf{r}_1, \mathbf{r}_2, \tau)]$. From Eq. (11.3.9) we then get

$$\langle I\rangle = \langle I_1\rangle + \langle I_2\rangle + 2(\langle I_1\rangle\langle I_2\rangle)^{1/2}|\gamma^{(1)}|\cos[\langle\omega\rangle\tau - \phi(\tau)] \tag{11.3.21}$$

Now, since both $|\gamma^{(1)}|$ and $\phi(\tau)$ are slowly varying as a function of τ, it follows that the variation of intensity $\langle I\rangle$ as P is changed, i.e., the fringe pattern, is due to the rapid variation of the cosine term with its argument $\langle\omega\rangle\tau$. So, in the region around P, we have

$$I_{max} = \langle I_1\rangle + \langle I_2\rangle + 2(\langle I_1\rangle\langle I_2\rangle)^{1/2}|\gamma^{(1)}| \tag{11.3.22a}$$

$$I_{min} = \langle I_1\rangle + \langle I_2\rangle - 2(\langle I_1\rangle\langle I_2\rangle)^{1/2}|\gamma^{(1)}| \tag{11.3.22b}$$

and therefore, from equation (11.3.11)

$$V_P = \frac{2\,(\langle I_1\rangle\langle I_2\rangle)^{1/2}}{\langle I_1\rangle + \langle I_2\rangle}\left|\gamma^{(1)}(r_1, r_2, \tau)\right| \tag{11.3.23}$$

For the case $\tau = (L_2/c) - (L_1/c) = 0$, Eq. (11.3.23) reduces to Eq. (11.3.13).

The measurement of the degree of temporal coherence is usually performed by means of the Michelson interferometer (Fig. 11.4a). Let P be the point where the temporal coherence of the wave is to be measured. A combination of a suitably small diaphragm placed at P and a lens with its focus at P transforms the incident wave into a plane wave (see Fig. 11.12). This wave then falls on a partially reflecting mirror S_1 (of reflectivity $R = 50\%$) which splits the beam into the two beams A and B. These beams are reflected back by mirrors S_2 and S_3 (both of reflectivity $R = 100\%$) and recombine to form the beam C. Since the waves A and B interfere, the illumination in the direction of C will be either light or dark according to whether $2(L_3 - L_2)$ is an even or odd number of half wavelengths. Obviously this interference will only be observed as long as the difference $L_3 - L_2$ does not become so large that the two beams A and B become uncorrelated in phase. Thus, for a partially coherent wave, the intensity I_c of beam C as a function of $2(L_3 - L_2)$ will behave as shown in Fig. 11.4b. At a given value of the $L_3 - L_2$ difference between the lengths of the interferometer arms, i.e., at a given value of the delay $\tau = 2(L_3 - L_2)/c$ between the two reflected waves, one can again

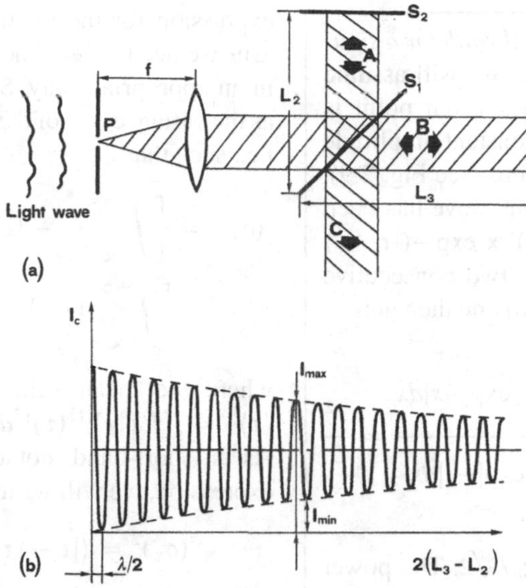

FIG. 11.4. (a) Michelson interferometer for the measurement of the degree of temporal coherence of an e.m. wave at point P. (b) Behavior of the output intensity, along the direction C of the interferometer, as a function of the difference $L_3 - L_2$ between the lengths of the interferometer arms.

define a fringe visibility V_P as in Eq. (11.3.11), where the quantities $I_{\max}$ and $I_{\min}$ are as shown in Fig. 11.4b. One then expects V_P to be a function of the time delay τ and, as in the case of Young's interferometer, related to the degree of temporal coherence. One can show, in fact, that one has, in this case,

$$V_P(\tau) = |\gamma^{(1)}(\mathbf{r}, \mathbf{r}, \tau)| \tag{11.3.24}$$

where $\mathbf{r}$ is the coordinate of point P. Once $V_P(\tau) = |\gamma^{(1)}(\tau)|$ has been measured, the value of the coherence time τ_{co} defined as, e.g., the time at which $V_P(\tau_{co}) = 1/2$ (see Fig. 11.1) can be determined. The corresponding coherence length will then be given by $L_c = c\tau_{co}$ and, given the definition τ_{co} adopted in Fig. 11.1, we then see that L_c is equal to twice the difference $L_3 - L_2$ between the interferometer arms for which the fringe visibility falls to $V_P = 1/2$.

11.3.3. Relation Between Temporal Coherence and Monochromaticity

From the paragraphs above it is clear that, for a stationary beam, the concept of temporal coherence is intimately connected with the monochromaticity. For example, the more monochromatic the wave is, the greater its temporal coherence, i.e. the coherence time, τ_{co}, has an inverse dependence on the laser oscillation bandwidth, $\Delta\nu_L$. In this section we wish to discuss this relationship in more depth.

We start by noting that the spectrum of an e.m. wave, as measured by, e.g., a spectrograph, is proportional to the power spectrum $W(\mathbf{r}, \omega)$ of the field $E(\mathbf{r}, t)$. Since the power spectrum W is equal to the Fourier transform of the auto correlation function $\Gamma^{(1)}$, either one of these quantities can be obtained once the other is known. To give a precise

Example 11.2. *Coherence time and bandwidth for a sinusoidal wave with random phase jumps.* We will assume that the time behavior of the field at a given point is described by a sinusoidal wave with constant amplitude and with phase undergoing random jumps (see Fig. 2.9). The correlation function $\Gamma^{(1)}(\tau)$ for this wave has been calculated in Appendix B to be $\Gamma^{(1)}(\tau) \propto \exp -(|\tau|/\tau_c)$ where τ_c is the average time between two consecutive phase jumps. According to Eq. (11.3.26) one then gets

$$\sigma_\tau^2 = \frac{\int_0^\infty \tau^2 \exp[-(2\tau/\tau_c)]d\tau}{\int_0^\infty \exp[-(2\tau/\tau_c)]d\tau} = \frac{\tau_c^2}{4}\frac{\int_0^\infty x^2 \exp[-x]dx}{\int_0^\infty \exp[-x]dx} = \frac{\tau_c^2}{2}$$

where $x = 2\tau/\tau_c$. We thus get $\sigma_\tau = \tau_c/\sqrt{2}$. The power spectrum, $W(\nu - \nu_0)$, of this signal is then described by a Lorentzian function (see again Appendix B) so that, according to Eq. (11.3.27) we can write

$$\sigma_\nu^2 = \frac{\int_0^\infty (\nu-\nu_0)^2\left[\frac{1}{1+4\pi^2(\nu-\nu_0)^2\tau_c^2}\right]^2 d(\nu-\nu_0)}{\int_0^\infty \left[\frac{1}{1+4\pi^2(\nu-\nu_0)^2\tau_c^2}\right]^2 d(\nu-\nu_0)} =$$

$$= \frac{1}{4\pi^2\tau_c^2}\frac{\int_0^\infty x^2\left[\frac{1}{1+x^2}\right]^2 dx}{\int_0^\infty \left[\frac{1}{1+x^2}\right]^2 dx}$$

where $x = 2\pi(\nu - \nu_0)\tau_c$. It turns out that the two integrals on the right-hand side of the above equation are both equal to $(\pi/4)$[11] so that $\sigma_\nu = (1/2\pi\tau_c)$. From the previous calculations one then finds that the product $\sigma_\tau\sigma_\nu$ is equal to $1/(2\sqrt{2})\pi$, i.e., it is $\sqrt{2}$ times larger than the minimum value, $1/4\pi$, that holds for a Gaussian spectrum.

expression for the relation between τ_{co} and $\Delta\nu_L$ we need to redefine these two quantities in an appropriate way. So we will define τ_{co} as the variance, σ_τ, of the function $|\Gamma^{(1)}(\tau)|^2$, i.e., such that

$$(\sigma_\tau)^2 = \left[\int_{-\infty}^{+\infty}(\tau - \langle\tau\rangle)^2|\Gamma^{(1)}(\tau)|^2 d\tau\right] \Big/ \left[\int_{-\infty}^{+\infty}|\Gamma^{(1)}(\tau)|^2 d\tau\right],$$

where the mean value $\langle\tau\rangle$ is defined by $\langle\tau\rangle = [\int \tau|\Gamma^{(1)}(\tau)|^2 d\tau]/[\int |\Gamma^{(1)}(\tau)|^2 d\tau]$. As a short-hand notation for the above expression, we will write

$$(\sigma_\tau)^2 = \langle[\tau - \langle\tau\rangle]^2\rangle \qquad (11.3.25)$$

Since $|\Gamma^{(1)}(-\tau)|^2 = |\Gamma^{(1)}(\tau)|^2$, one has $\langle\tau\rangle = 0$ and Eq. (11.3.25) reduces to

$$(\sigma_\tau)^2 = \langle\tau^2\rangle \qquad (11.3.26)$$

The coherence time defined in this way is conceptually simpler, although sometimes involving lengthier calculation, than that defined earlier, i.e., the half-width at half-height of the curve $|\Gamma^{(1)}(\tau)|^2$, see Fig. 11.1. In fact, if $|\Gamma^{(1)}(\tau)|^2$ were an oscillatory function of the coherence time, τ_{co}, as defined in Fig. 11.1, would not be uniquely determined. Similarly we define the laser oscillation bandwidth $\Delta\nu_L$ as the variance of $W^2(\nu)$, i.e., such that

$$(\Delta\nu_L)^2 = (\sigma_\nu)^2 = \langle[\nu - \langle\nu\rangle]^2\rangle \qquad (11.3.27)$$

where $\langle\nu\rangle$, the mean frequency of the spectrum, is given by $\langle\nu\rangle = [\int \nu W^2 d\nu]/[\int W^2 d\nu]$. Now, since W and $\Gamma^{(1)}$ are related by a Fourier transform, it can be shown that σ_ν and σ_τ, as we have just defined them, satisfy the condition

$$\sigma_\tau\,\sigma_\nu \geq (1/4\pi) \qquad (11.3.28)$$

The relation (11.3.28) is closely analogous to the Heisenberg uncertainty relation and can be proved using the same procedure as used to derive the uncertainty relation.[2] The equality sign in Eq. (11.3.28) applies when $|\Gamma^{(1)}(\tau)|$ [and hence $W(\nu)$] are Gaussian functions. This case is obviously the analogue of the minimum uncertainty wave packets of quantum mechanics.[2]

11.3.4. Nonstationary Beams*

We will now briefly consider the case of a nonstationary beam. In this case, by definition, the function $\Gamma^{(1)}$ in Eq. (11.3.1) depends on both t_1 and t_2 and not only on their difference $\tau = t_1 - t_2$. Examples would include an amplitude-modulated laser, an amplitude-modulated thermal light source, a Q-switched or a mode-locked laser. For a nonstationary beam, the correlation function can be obtained as the ensemble average of many measurements of the field in a time interval between 0 and T, where the origin of the time interval is synchronized to the driving signal (e.g., synchronized to the amplitude modulator for a mode-locked laser or the Pockels cell driver for a Q-switched laser). The degree of temporal coherence at a given point $\mathbf{r}$ can then be defined as

$$\gamma^{(1)}(t_1, t_2) = \frac{\langle E(t_1)\, E^*(t_2)\rangle}{\langle E(t_1)\, E^*(t_1)\rangle^{1/2}\,\langle E(t_2)\, E^*(t_2)\rangle^{1/2}} \tag{11.3.29}$$

where t_1 and t_2 are two given times, in the interval 0-T, and where all signals are measured at point $\mathbf{r}$. We can now say that the beam has a perfect temporal coherence if $|\gamma^{(1)}(t_1,\ t_2)| = 1$ for all times t_1 and t_2. According to this definition we can see that a nonstationary beam *without amplitude and phase fluctuations* has a *perfect temporal coherence*. In the absence of fluctuations, in fact, the products $E(t_1)E^*(t_2)$, $E(t_1)E^*(t_1)$, and $E(t_2)E^*(t_2)$ appearing in Eq. (11.3.29) remain the same for all measurements of the ensemble. These products are thus equal to the corresponding ensemble averages and $\gamma^{(1)}(t_1,\ t_2)$ reduces to

$$\gamma^{(1)}(t_1, t_2) = \frac{E(t_1)E^*(t_2)}{|E(t_1)||E(t_2)|} \tag{11.3.30}$$

From Eq. (11.3.30) we then immediately see that $|\gamma^{(1)}| = 1$. According to this definition of temporal coherence, the coherence time of a nonstationary beam, e.g., of a mode-locked laser, is infinite if the beam does not fluctuate in amplitude or phase. This shows that the coherence time of a nonstationary beam is not related to the inverse of the oscillating bandwidth. In a practical situation, however, if we correlate, e.g., one pulse of a mode-locked train with some other pulse of the train, i.e., if we choose $t_1 - t_2$ to be larger than the pulse repetition time, some lack of correlation will be found due to fluctuations. This means that $|\gamma^{(1)}|$ will decrease as $t_1 - t_2$ increases beyond the pulse repetition time. Thus the coherence time is expected to be finite although not related to the inverse of the oscillating bandwidth but to the *inverse of the fluctuation bandwidth*.

11.3.5. Spatial and Temporal Coherence of Single-Mode and Multimode Lasers

Consider first a cw laser oscillating on a single transverse and longitudinal mode. Above threshold, as explained in Sects. 7.10 and 7.11, the amplitude fluctuations may, to first order, be neglected and the fields of the wave, at the two points $\mathbf{r}_1$ and $\mathbf{r}_2$, can be written as

$$E(\mathbf{r}_1, t) = a_0 u(\mathbf{r}_1)\exp\{j[\omega t - \phi(t)]\} \tag{11.3.31a}$$

$$E(\mathbf{r}_2, t) = a_0 u(\mathbf{r}_2)\exp\{j[\omega t - \phi(t)]\} \tag{11.3.31b}$$

* The author wishes to acknowledge some enlightening discussion on this topic with Professor V. Degiorgio.

where a_0 is a constant, $u(\mathbf{r})$ describes the mode amplitude and ω is the angular frequency at band center. The substitution of Eq. (11.3.31) into Eq. (11.3.6) then gives $\gamma^{(1)} = u(\mathbf{r}_1)u^*(\mathbf{r}_2)/|u(\mathbf{r}_1)||u(\mathbf{r}_2)|$, which shows that $|\gamma^{(1)}| = 1$. Thus, a laser oscillating on a single mode has perfect spatial coherence. Its temporal coherence, on the other hand, is established by the laser bandwidth $\Delta\nu_L$. As an example, a single mode monolithic Nd:YAG laser (see Fig. 7.26) may have $\Delta\nu_L \cong 20$ kHz. The coherence time will then be $\tau_{co} \cong 1/\Delta\nu_L \cong 0.05$ ms and the coherence length $L_c = c\tau_{co} \cong 15$ km (note the very large value of this coherence length).

Let us now consider a laser oscillating on a single transverse mode and on l longitudinal modes. In terms of the cavity mode amplitudes $u(\mathbf{r})$, the fields at two points $\mathbf{r}_1$ and $\mathbf{r}_2$ belonging to the same wave-front can generally be written as

$$E(\mathbf{r}_1, t) = \sum_1^l {}_k\, a_k\, u(\mathbf{r}_1) \exp j[\omega_k t - \phi_k(t)] \tag{11.3.32a}$$

$$E(\mathbf{r}_2, t) = \sum_1^l {}_k\, a_k\, u(\mathbf{r}_2) \exp j[\omega_k t - \phi_k(t)] \tag{11.3.32b}$$

where a_k are constant factors, ω_k and ϕ_k are, respectively, the frequency and the phase of the k-th mode. Note that, since the transverse field configuration is the same for all modes (e.g., that of a TEM_{00} mode), the amplitude u has been taken to be independent of the mode index k. The function $u(\mathbf{r})$ can thus be taken out of the summation in both equations (11.3.32) so that one gets the following result

$$E(\mathbf{r}_2, t) = [u(\mathbf{r}_2)/u(\mathbf{r}_1)]E(\mathbf{r}_1, t) \tag{11.3.33}$$

This means that, whatever time variation $E(\mathbf{r}_1, t)$ is observed in $\mathbf{r}_1$, the same time variation will be observed at $\mathbf{r}_2$ except for a proportionality constant. Substitution of Eq. (11.3.33) into Eq. (11.3.6) then readily gives $|\gamma^{(1)}| = 1$. Thus a laser beam made of many longitudinal modes with the same transverse profile (e.g., corresponding to a TEM_{00} mode) still has a perfect spatial coherence. The temporal coherence, if all the mode phases are random, is again equal to the inverse of the oscillating bandwidth. If no frequency-selecting elements are used in the cavity, the oscillating bandwidth may now be comparable to the laser gain bandwidth and hence the coherence time may be much shorter than in the example considered previously generally in the range of nanoseconds to picoseconds. When these modes are locked in phase, however, the temporal coherence may become very long, as discussed in the previous section. Thus a mode-locked laser can in principle have perfect spatial and temporal coherence.

The last case we should consider is that of a laser oscillating on many transverse modes. It will be shown in the following example that, in this case, the laser has only partial spatial coherence.

Example 11.3. *Spatial coherence for a laser oscillating on many transverse-modes.* For a laser oscillating on l transverse modes, the fields of the output beam, at two points $\mathbf{r}_1$ and $\mathbf{r}_2$ on the same wave-front, can be written as [compare with Eq. (11.3.32)]

$$E(\mathbf{r}_1, t) = \sum_1^l {}_k\, a_k\, u_k(\mathbf{r}_1) \exp j[\omega_k t - \phi_k(t)] \tag{11.3.34a}$$

$$E(\mathbf{r}_2, t) = \sum_1^l {}_k\, a_k\, u_k(\mathbf{r}_2) \exp j[\omega_k t - \phi_k(t)] \tag{11.3.34b}$$

If the product $E(\mathbf{r}_1, t)E^*(\mathbf{r}_2, t)$ is carried out, one obtains a set of terms proportional to $\exp j[(\omega_k - \omega_{k'})\, t - (\phi_k - \phi_k')]$ with $k' \neq k$. These terms can be neglected because, performing the time average, they will average out to zero. We are thus left only with terms for which $k' = k$, so that we obtain

$$\langle E(\mathbf{r}_1, t)E^*(\mathbf{r}_2, t)\rangle = \sum_1^l {}_k\, a_k\, a_k^* u_k(\mathbf{r}_1) u_k^*(\mathbf{r}_2) \tag{11.3.35}$$

For $\mathbf{r}_2 = \mathbf{r}_1$, one gets from Eq. (11.3.35)

$$\langle E(\mathbf{r}_1, t)E^*(\mathbf{r}_1, t)\rangle = \sum_1^l {}_k\, |a_k|^2\, |u_k(\mathbf{r}_1)|^2 \tag{11.3.36}$$

Likewise, for $\mathbf{r}_1 = \mathbf{r}_2$, one gets from Eq. (11.3.35)

$$\langle E(\mathbf{r}_2, t)E^*(\mathbf{r}_2, t)\rangle = \sum_1^l {}_k\, |a_k|^2\, |u_k(\mathbf{r}_2)|^2 \tag{11.3.37}$$

The substitution of Eqs. (11.3.35)–(11.3.37) in Eq. (11.3.6) then gives

$$\gamma^{(1)} = \frac{\sum_1^l {}_k\, a_k a_k^* u_k(\mathbf{r}_1) u_k^*(\mathbf{r}_2)}{\left[\sum_1^l {}_k\, |a_k|^2\, |u_k(\mathbf{r}_1)|^2\right]^{1/2} \left[\sum_1^l {}_k\, |a_k|^2\, |u_k(\mathbf{r}_2)|^2\right]^{1/2}} \tag{11.3.38}$$

If we now let $\mathbf{R}_1$ represent a complex vector characterized by the components $a_k u_k(\mathbf{r}_1)$ in an l-dimensional space, and similarly for $\mathbf{R}_2$, one writes their scalar product as

$$\mathbf{R}_1 \cdot \mathbf{R}_2 = \sum_1^l {}_k\, a_k a_k^* u_k(\mathbf{r}_1) u_k^*(\mathbf{r}_2) \tag{11.3.39}$$

The magnitude of the two vectors is, on the other hand, given by

$$R_1 = |\mathbf{R}_1| = \left[\sum_1^l {}_k\, |a_k|^2\, |u_k(\mathbf{r}_1)|^2\right]^{1/2} \tag{11.3.40}$$

and

$$R_2 = |\mathbf{R}_2| = \left[\sum_1^l {}_k\, |a_k|^2\, |u_k(\mathbf{r}_2)|^2\right]^{1/2} \tag{11.3.41}$$

respectively. The substitution of Eqs. (11.3.39)–(11.3.41) in Eq. (11.3.38) shows that

$$\gamma^{(1)} = \frac{\mathbf{R}_1 \cdot \mathbf{R}_2}{R_1 R_2} \tag{11.3.42}$$

From the Schwartz inequality it then follows that, since now $\mathbf{R}_1 \neq a\mathbf{R}_2$, where a is a constant, one always has $|\gamma^{(1)}|<1$.

11.3.6. Spatial and Temporal Coherence of a Thermal Light Source

We shall now discuss, briefly, the coherence properties of the light emitted by an ordinary lamp, which may be either a filament lamp or a lamp filled with a gas at a suitable pressure. Since the emitted light now arises from spontaneous emission of atoms essentially under thermal equilibrium conditions, these sources will generally be referred to as *thermal sources*.

As far as the temporal coherence is concerned, one notes that the light emitted by a c.w. gas lamp generally consists of several emission lines (see, e.g., Fig. 6.6b), the width of these lines being rather broad ($\Delta\nu = 1\text{–}10$ THz) due to the high pressures generally used. On the other hand, in the case of a flashlamp (see, e.g., Fig. 6.6a) or a filament source, the emission is much broader, resembling that of a blackbody radiator (see Fig. 2.3). Thus the coherence time, $\tau_{co} \cong 1/\Delta\nu$, of a thermal light source is, generally, very short (τ_{co}<1 ps).

As far as the spatial coherence is concerned, one may note that, since the e.m. wave originates from spontaneous emission by independent emitters, the wave is completely incoherent at a location very near to the source while it acquires an increasing degree of spatial coherence as the considered location moves away from the source. This situation can be understood with the help of Fig. 11.5, where uncorreletated emitters (dots) are assumed to be present within the aperture of a hole, of diameter d, in a screen S. Emission will occur over a 4π solid angle and the degree of spatial coherence between points P_1 and P_2 is measured at some distance z from the emitters. For simplicity, let P_1 be the point on the symmetry axis of the system and let r be the distance between the two points. It is clear from the figure that, for very small values of z, points P_1 and P_2 ($r < d/2$) will mostly see the emission of the emitter just facing to it and the fields at the two points will then be completely uncorrelated. As z is increased, however, each of the two points will receive an increasing contribution from all other emitters

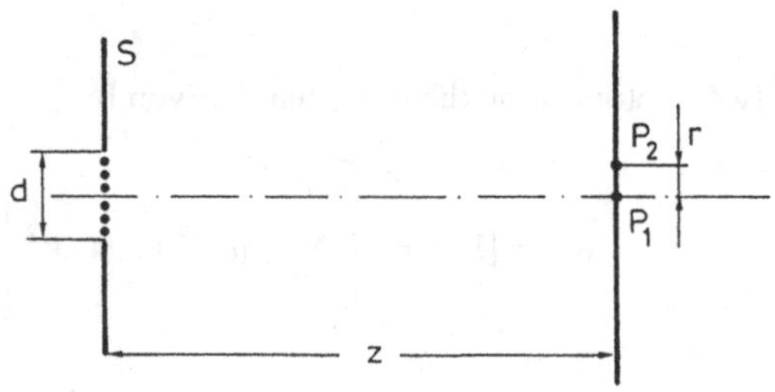

FIG. 11.5. Degree of spatial coherence in a plane at a distance z from independent emitters, covering an area of diameter d.

and the fields at the two points will become more and more correlated. A calculation of the degree of spatial coherence, $|\gamma^{(1)}|$, as a function of the coordinate r is beyond the scope of this book and we refer elsewhere for this treatment.[4] We limit ourselves to pointing out that $|\gamma^{(1)}|$ turns out to be a function of the dimensionless quantity $(rd/\lambda z)$ and, as an example, one has $|\gamma^{(1)}| = 0.88$ for

$$r \cong 0.16(\lambda z/d) \tag{11.3.43}$$

a result which will be used in a following section.

11.4. DIRECTIONALITY

There are usually two ways by which the directionality, i.e., the divergence, of a laser beam, or more generally of any light source, can be measured, namely: (i) By measuring the degree of beam spreading at very large distances from the source. In fact, if we let W represent a suitably defined radius of the beam at a very large distance z, the half-angle beam divergence can be obtained from the relation

$$\theta_d = W/z \tag{11.4.1}$$

(ii) By measuring the radial intensity distribution, $I(r)$, of the focused beam in the focal plane of a lens. To understand the basis of the latter measurement, let r be the coordinate of a general point in this focal plane, relative to an origin at the beam center. According to the discussion made in relation to Fig. 1.8, the beam can be thought of as composed of a set of plane waves propagating along slightly different directions.[5] The wave inclined at an angle θ to the propagation axis will then be focused, in the focal plane, to the point of radial coordinate r given by (for small θ)

$$r = f\theta \tag{11.4.2}$$

Hence, a knowledge of the intensity distribution $I(r)$ in the focal plane of a lens gives the angular distribution of the original beam. In what follows, we will use either one of the above methods, as convenient.

The directional properties of a laser beam are strictly related to its spatial coherence. We will therefore discuss first the case of an e.m. wave with perfect spatial coherence and then the case of partial spatial coherence.

11.4.1. Beams with Perfect Spatial Coherence

Let us first consider a beam with perfect spatial coherence consisting of a plane wave front of circular cross section, with diameter D, and constant amplitude over this cross section. Following the discussion above, the beam divergence can be calculated from a knowledge of the intensity distribution, $I(r)$, in the focal plane of a lens. This calculation was performed in

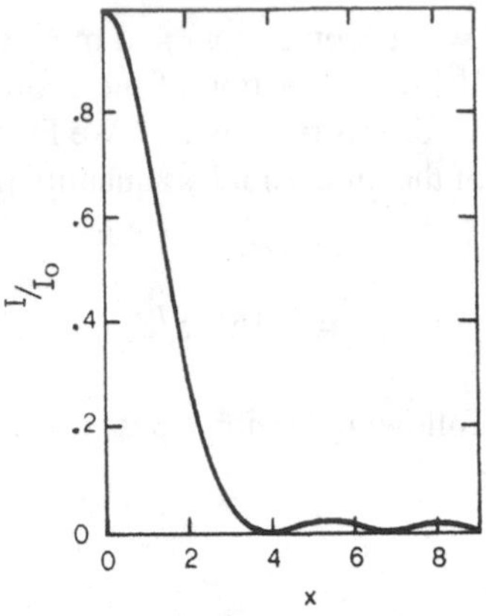

FIG. 11.6. Distribution of light intensity in the focal plane of a lens, of diameter D, as a function of radial distance r in terms of the normalized variable $x = krD/2f$.

the 19th century, using diffraction theory, by Airy.[6] The expression obtained for $I(r)$, known as *Airy formula*, can be written as

$$I = \left[\frac{2J_1\,(krD/2f)}{krD/2f}\right]^2 I_0 \tag{11.4.3}$$

where $k = 2\pi/\lambda$, J_1 is the first-order Bessel function, and I_0 is given by

$$I_0 = P_i\left(\frac{\pi D^2}{4\lambda^2 f^2}\right)_r \tag{11.4.4}$$

P_i being the power of the beam incident at the lens. Note that, since the value of the expression in the square brackets of Eq. (11.4.3) becomes unity when $r = 0$, I_0 represents the beam intensity at the center of the focal spot.

The behavior of the intensity, I, as a function of the dimensionless quantity

$$x = krD/2f \tag{11.4.5}$$

is shown in Fig. 11.6. From this figure one can see that the diffraction pattern consists of a circular central zone (the *Airy disk*) surrounded by a series of rings of rapidly decreasing intensity. The divergence angle θ_d of the original beam is defined as corresponding to the radius of the first minimum shown in Fig. 11.6. While this is an arbitrary definition, it is convenient, and is the conventionally accepted definition. So, from the value of x which corresponds to this minimum, with the help of Eqs. (11.4.5) and (11.4.2) one obtains

$$\theta_d = 1.22\,\lambda/D \tag{11.4.6}$$

As a second example of the propagation of a spatially coherent beam, we consider the case of a Gaussian beam (TEM_{00}) such as can be obtained from a stable laser cavity consisting of two spherical mirrors. If we let w_0 be the spot size at the beam waist, the spot size w and the radius of curvature R of the equiphase surface at a distance z from the waist can be obtained from Eqs. (4.7.17a) and (4.7.17b), respectively. At a large distance from the waist [i.e., for $(\lambda z/\pi w_0^2) \gg 1$] one then see that $w \cong \lambda z/\pi w_0$ and $R \cong z$. Since both w and R increase linearly with distance, the wave can be considered to be a spherical wave having its origin at the waist. Following the convention of identifying the radius of the beam with the spot size w, the beam divergence is obtained as

$$\theta_d = w/z = \lambda/\pi w_0 \tag{11.4.7}$$

A comparison of Eqs. (11.4.7) and (11.4.6) indicates that one can set $D = 2w_0$, implying that for the same diameter, a Gaussian beam has a divergence about half that of a plane beam.

As a conclusion to this section we can say that the divergence θ_d of a spatially coherent beam can generally be written as

$$\theta_d = \beta\lambda/D \tag{11.4.8}$$

where D is a suitably defined beam diameter and β is a numerical factor of the order of unity whose exact value depends on the field amplitude distribution as well as on the way by which both θ_d and D are defined. Such a beam is commonly referred to as being *diffraction-limited.*

11.4.2. Beams with Partial Spatial Coherence

For an e.m. wave with partial spatial coherence the divergence is greater than for a spatially coherent wave having the same intensity distribution. This can, for example, be understood following the argument used to explain the divergence of a beam of uniform amplitude in Fig. 1.6. In fact, if the wave considered in Fig. 1.6 is not spatially coherent, the secondary wavelets emitted over its cross section would no longer be in phase and the wave front produced by diffraction would have a larger divergence than that given by Eq. (11.4.6). A rigorous treatment of this problem (i.e., the propagation of partially coherent waves) is beyond the scope of this book, and the reader is referred to more specialized texts.[3] We will limit ourselves to considering first a particularly simple case of a beam of diameter D (Fig. 11.7a), which is made up of many smaller beams (shaded in the figure) of diameter d. We will assume that each of these smaller beams is diffraction-limited (i.e., spatially coherent). Now, if the various beams are mutually uncorrelated, the divergence of the beam as a whole will be equal to $\theta_d = \beta\lambda/d$. If, on the other hand, the various beams were correlated, the divergence would be $\theta_d = \beta\lambda/D$. The latter case is actually equivalent to a number of antennas (the small beams) all emitting in phase with each other. After this simple case, we can now consider the general case where the partially coherent beam has a given intensity distribution over its diameter D and a coherence area A_c at a given point P (Fig. 11.7b). By analogy with the previous case one can readily understand that one has, in this case,

$$\theta_d = \beta\lambda/D_c \tag{11.4.9}$$

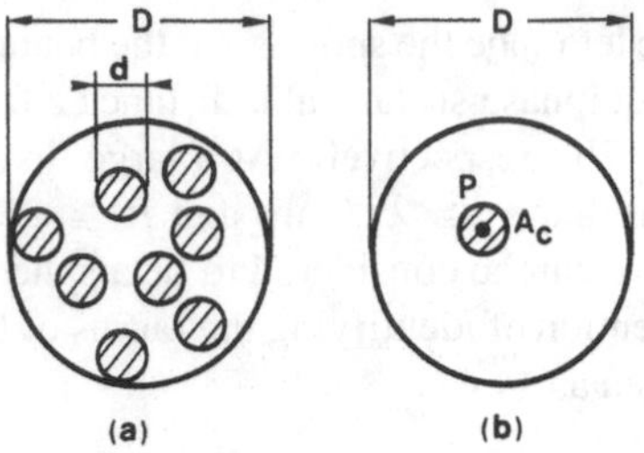

FIG. 11.7. Examples to illustrate the different divergence properties of coherent and partially coherent waves: (a) Beam of diameter D made of the superposition of several smaller and coherent beams of diameter d. (b) Beam of diameter D and coherence area A_c at point P.

where D_c is the diameter of the coherence area and β is a numerical factor of the order of unity whose value depends on how θ_d and D_c are defined. The concept of directionality is thus intimately related to that of spatial coherence.

11.4.3. The M^2 Factor and the Spot-Size Parameter of a Multimode Laser Beam

The expressions for the beam divergence given above [see Eqs. (11.4.9) and (11.4.8)] involve a degree of imprecision which has its origin in the arbitrary definition of beam diameter. We now present some precise definitions of beam radius and beam divergence which allow to describe, in a more general way, beam propagation both for a diffraction-limited laser beam with arbitrary transverse profile and for a non-diffraction-limited, multi-transverse-mode, partially-coherent laser beam.[6]

Let $I(x, y, z)$ be the, time averaged, intensity profile of the laser beam at the longitudinal coordinate z. Note that, for great generality, we do not restrict ourserlves to radially symmetric beams and, accordingly, the intensity I is written as a function of both transverse coordinates x and y, separately. One can now define a beam variance, $\sigma_x^2(z)$, along, e.g., the x-coordinate so that

$$\sigma_x^2(z) = \frac{\iint (x - \langle x \rangle)^2\, I(x, y, z)dxdy}{\iint I(x, y, z)dxdy} \tag{11.4.10}$$

where $\langle x \rangle = [\iint x\, I(x, y, z)dxdy]/[\iint I(x, y, z)dxdy]$, and equivalent definition applies for the y coordinate. To introduce a similar definition for the beam divergence, let $\hat{I}(s_x, s_y)$ be the wave intensity at the normalized angular coordinates $s_x = \theta_x/\lambda$, and $s_y = \theta_y/\lambda$. The s_x and s_y coordinates are usually referred to as the spatial-frequency coordinates of the wave and their use is quite common in diffraction optics.[7] If, for instance, the divergence is obtained by measuring the beam intensity $I(x', y')$ at a plane, x', y', at a large distance, z, from the source, the angular intensity $\hat{I}(s_x, s_y)$ is obtained from $I(x', y')$ using the relations

$$x' = \theta_x z = s_x\, \lambda z \tag{11.4.11a}$$

and

$$y' = \theta_y z = s_y \lambda z \tag{11.4.11b}$$

Having defined the intensity $\hat{I}(s_x,\ s_y)$, the variance of the spatial frequency s_x can now be readily defined as

$$\sigma_{s_x}^2 = \frac{\iint (s_x - \langle s_x \rangle)^2\, \hat{I}(s_x,\ s_y) ds_x ds_y}{\iint \hat{I}(s_x, s_y) ds_x ds_y} \tag{11.4.12}$$

where $\langle s_x \rangle = [\iint s_x \hat{I}(s_x, s_y) ds_x ds_y]/[\iint \hat{I}(s_x, s_y) ds_x ds_y]$, and similarly for the y-coordinate. If we now let $u(x, y, z)$ be the transverse amplitude profile of the beam (so that $I \propto |u|^2$) and $\hat{u}(s_x, s_y)$ be the spatial-frequency profile (so that $\hat{I} \propto |\hat{u}|^2$), we know that, for any arbitrary optical beam, the two functions are related through a Fourier transform.[8] For any arbitrary real laser beam, one can then prove that $\sigma_x^2(z)$ obeys the free space propagation equation

$$\sigma_x^2(z) = \sigma_{x0}^2 + \lambda^2 \sigma_{s_x}^2 (z - z_{0x})^2 \tag{11.4.13}$$

where σ_{x0} is the minimum value of σ_x and z_{0x} is the coordinate at which this minimum is attained. One can also show from the same treatment that [compare with Eq. (11.3.28)]

$$\sigma_{x0}\sigma_{s_x} \geq 1/4\pi \tag{11.4.14}$$

the equality holding only for a coherent Gaussian beam. In this case, in fact, one has $I(x, y, z) \propto \exp[-2(x^2 + y^2)/w^2(z)]$ while it can be readily shown, using the coordinate transformations Eq. (11.4.11), that $\hat{I}(s_x, s_y) \propto \exp\left[-2\pi^2 w_0^2 \left(s_x^2 + s_y^2\right)\right]$. From Eqs. (11.4.10) and (11.4.12) one then gets

$$\sigma_x(z) = w(z)/2 \tag{11.4.15a}$$

$$\sigma_{s_x} = 1/2\pi w_0 \tag{11.4.15b}$$

and obviously, setting $z = 0$ in (11.4.15a),

$$\sigma_{x0} = w_0/2 \tag{11.4.15c}$$

From Eqs. (11.4.15b) and (11.4.15c) one then obtains

$$(\sigma_{x0}\sigma_{s_x})_G = 1/4\pi \tag{11.4.16}$$

Following the previous argument one can now define an M_x^2 factor as the ratio between the $(\sigma_{x0}\sigma_{s_x})$ product of the beam and the corresponding product, $(\sigma_{x0}\sigma_{s_x})_G$, for a Gaussian beam i.e.,

$$M_x^2 = (\sigma_{x0}\sigma_{s_x})/(\sigma_{x0}\sigma_{s_x})_G = 4\pi(\sigma_{x0}\sigma_{s_x}) \tag{11.4.17}$$

and similarly for the y-coordinate. Note that, according to Eq. (11.4.14) one has $M_x^2 \geq 1$. The term M_x^2 is usually referred to as the beam quality. Since larger values of M_x^2 correspond to lower beam quality, it has also been referred to (less commonly) as the inverse beam quality factor. Note also that, if a comparison is made with a Gaussian beam having the same variance, i.e., if $(\sigma_{x0})_G = (\sigma_{x0})$, M_x^2 then specifies how much the beam divergence exceeds that of a Gaussian beam.

An alternative way to Eq. (11.4.13) for describing the propagation of a multimode laser beam can be obtained by noting that, for a Gaussian beam, according to Eqs. (11.4.15a) and (11.4.15c) one has $w_x(z) = 2\sigma_x(z)$ and $w_{x0} = 2\sigma_{x0}$. Consequently, for a laser beam of general transverse profile, one can define the spot-size parameters $W_x(z)$ and W_{x0} as

$$W_x(z) = 2\sigma_x(z) \tag{11.4.18a}$$

$$W_{x0} = 2\sigma_{x0} \tag{11.4.18b}$$

Note that we are using the upper-case symbols $W_x(z)$ and W_{x0} to indicate the spot-size parameters of an arbitrary laser beam. The substitution, in Eq. (11.4.13), of $\sigma_x(z)$ and σ_{x0} from Eqs. (11.4.18) and of σ_{sx} from Eq. (11.4.17) then gives

$$W_x^2(z) = W_{x0}^2 + M_x^4 \frac{\lambda^2}{\pi^2 W_{x0}^2}(z - z_{0x})^2 \tag{11.4.19}$$

For a Gaussian beam, one obviously has $W_x(z) = w_x(z)$, $W_{x0} = w_{x0}$, and $M_x^2 = 1$ and Eq. (11.4.19) reduces to Eq. (4.7.13a). For a multimode laser beam, on the other hand, Eq. (11.4.19) represents an equation formally similar to that of a Gaussian beam except that the second term on the right hand side, expressing the spreading due to beam diffraction, is multiplied by a factor M_x^4.

Equation (11.4.19) expresses the propagation of a multimode laser beam as a function of a precisely defined spot-size parameter $W_x(z)$. Note that the beam propagation is determined by the three parameters W_{x0}, M_x^2 and z_{0x}. Their values can in principle be determined by performing a measurement of the beam size $W_x(z)$ at three different z-coordinates. Note also that, at large distances from the waist position, z_{0x}, one obtains from Eq. (11.4.19) $W_x(z) \cong \left(M_x^2\lambda/\pi W_{x0}\right)(z - z_{0x})$. For a multimode laser beam, one can then define a beam divergence as

$$\theta_{dx} = W_x(z)/(z - z_{0x}) = M_x^2(\lambda/\pi W_{x0}) \tag{11.4.20}$$

The beam divergence of a multimode laser beam is thus M_x^2 times that of a Gaussian beam of the same spot size (i.e., such that $w_{x0} = W_{x0}$). A comparison of Eq. (11.4.20) with Eq. (11.4.9) also allows one to establish a relation between M_x^2, W_{x0} and the diameter D_c of the coherence area.

Example 11.4. *M^2-factor and spot-size parameter of a broad area semiconductor laser.* We will consider a broad area AlGaAs/GaAs semiconductor laser with output beam dimensions, at the exit face of the laser i.e., in the near-field, of $d_\perp = 0.8\ \mu$m and $d_\parallel = 100\ \mu$m and with beam divergences $\theta_\perp = 20°$ and $\theta_\parallel = 10°$. The labels $\perp$ and $\parallel$ represent directions perpendicular and parallel to the junction plane, respectively. The diameters, d, are measured between the half-intensity points and, likewise, the divergences, θ, are measured as half-angles between half-maximum-intensity points (HWHM). Since the exit face of the semiconductor is plane, the waist positions for both axes can be taken to coincide with this face. The near-field intensity distribution perpendicular to the junction (the 'fast-axis') can be taken to be approximately Gaussian. The corresponding spot-size $w_{0\perp} = W_{0\perp}$ must then be such that $\exp[-2(d_\perp/2w_{0\perp})^2] = (1/2)$. We get $w_{0\perp} = W_{0\perp} = d_\perp/[2\ln 2]^{1/2} \cong 0.68\,\mu$m. The far-field intensity profile along the same direction may also be taken to be Gaussian. According to Eq. (11.4.11) its intensity profile, in terms of the transverse coordinate θz, can be written, for large z, as $\propto \exp[-2(\theta z/W_\perp)^2]$ where $W_\perp(z) = w_\perp(z)$ is the spot-size. Its value can be obtained by setting $\propto \exp[-2(\theta_\perp z/W_\perp)^2] = (1/2)$. Since the beam divergence is now defined as $\theta_{d\perp} = W_\perp/z$, one obtains $\theta_{d\perp} = [2/\ln 2]^{1/2}\theta_\perp \cong 0.59$ rad and from Eq. (11.4.20) $M_\perp^2 = \pi W_{0\perp}\theta_{d\perp}/\lambda = \pi\theta_\perp d_\perp/(\ln 2)\lambda = 1.5$, where $\lambda \cong 850$ nm. As expected, the $M_\perp^2$-factor is close to that of a true Gaussian beam. The near-field intensity distribution in the direction parallel to the junction (the 'slow-axis') can be taken to be approximately constant. From Eq. (11.4.10) one then gets $W_{0\|} \cong d_\|/2 = 50\,\mu$m. The far-field intensity distribution, on the other hand, is a bell-shaped function that can be approximated by a Gaussian function. As before, we then get $\theta_{d\|} = [2/\ln 2]^{1/2}\theta_\| \cong 0.148$ rad and, using Eq. (11.4.20) $M_\|^2 = \pi W_{0\|}\theta_{d\|}/\lambda = \pi\theta_\| d_\|/[2\ln 2]^{1/2}\lambda = 55$. Thus, in the slow-axis direction the beam divergence is much larger than the diffraction-limit of a Gaussian beam, i.e. the beam is many time diffraction limited.

11.5. LASER SPECKLE[9, 10]

Following the discussion on first-order coherence given in Sect. 11.3, we now briefly consider a very striking phenomenon, characteristic of laser light, known as laser speckle. Laser speckle is apparent when one looks at the scattered ligth from a laser beam, of sufficiently large diameter, incident on, for example, the surface of a wall or a transparent diffuser. The scattered light is then seen to consist of a random collection of alternately bright and dark spots (or *speckles*) (Fig. 11.8a). Despite the randomness, one can distinguish an average speckle (or grain) size. This phenomenon was soon recognized by early workers in the field as being due to constructive and destructive interference of radiation coming from the small scattering centers within the area where the laser beam is incident. Since the phenomenon depends on there being a high degree of first-order coherence, it is an inherent feature of laser light.

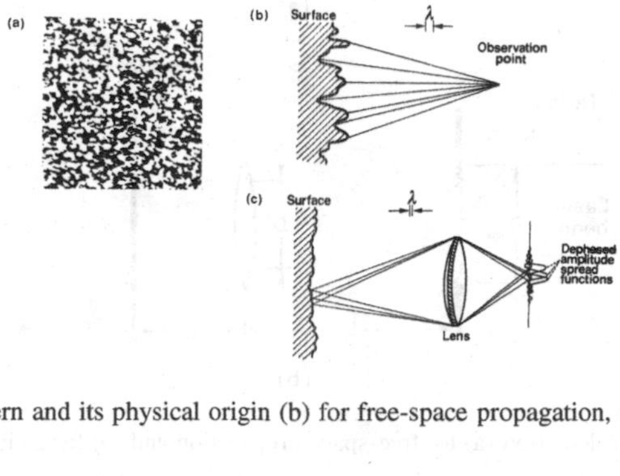

FIG. 11.8. (a) Speckle pattern and its physical origin (b) for free-space propagation, and (c) for an image-forming system.

The physical origin of the observed granularity can be readily understood, both for free-space propagation (Fig. 11.8b) and for an imaging system (Fig. 11.8c), when it is realized that the surfaces of most materials are extremely rough on the scale of an optical wavelength. For free-space propagation, the resulting optical wave at any moderately distant point from the scattering surface consists of many coherent components or wavelets, each arising from a different microscopic element of the surface. Referring to Fig. 11.8b, one notes that the distances traveled by these various wavelets may differ by many wavelengths. Interference of the phase-shifted but coherent wavelets results in the granular intensity (or *speckle pattern*, as it is usually referred to). When the optical arrangement is that of an imaging system (Fig. 11.8c), an explanation of the observed pattern must take account of diffraction as well as interference. In fact, due to the finite resolving power of even a perfectly corrected imaging system, the intensity at a given image point can result from the coherent addition of contributions from many independent parts of the surface. This situation occurs in practice when the point-spread function of the imaging system is broad in comparison to the microscopic surface variations.

One can readily obtain an order-of-magnitude estimate for the grain size d_g (i.e., the average size of the spots in the speckle pattern) for the two cases just considered. In the first case (Fig. 11.9a) the scattered light is assumed to be recorded on a photographic film at a distance L from the diffuser with no lens between film and diffuser. Suppose now that a bright speckle is present at some point P in the recording plane. This means that the light diffracted by all points of the diffuser will interfere at point P in a predominantly constructive way so as to give an overall peak for the field amplitude. In a heuristic way we can then say that the diffractive contributions, at point P, from the wavelets scattered from points P_1, P'_1, P''_1, etc. add (on the average) in phase with those from points P_2, P'_2, P''_2, etc. We now ask how far the point P must be moved along the x-axis in the recording plane for this constructive interference to become a destructing interference. This situation will occur when the contributions of, e.g.,

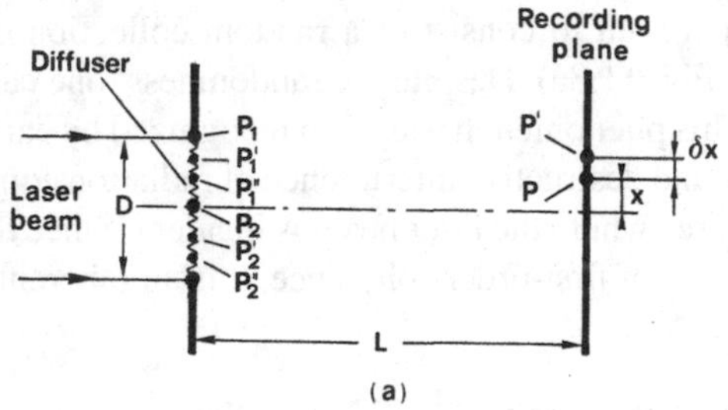

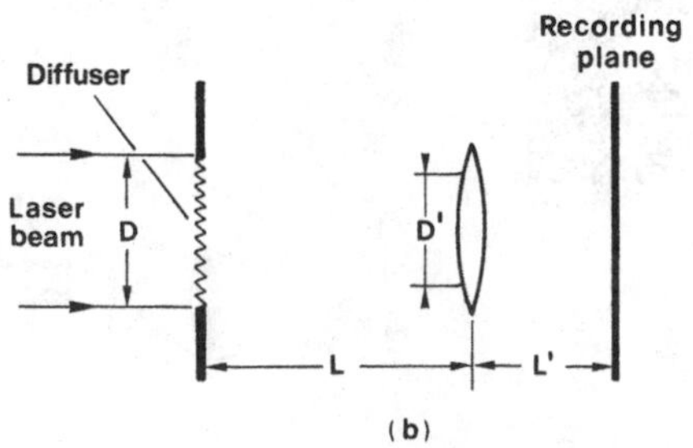

FIG. 11.9. Grain-size calculation (a) for free-space propagation and (b) for an image-forming system.

the diffracted waves from points P_1 and P_2 interfere, at the new point P', destructively rather than constructively. In this case, in fact, one can show that the contributions from points P'_1, and P'_2 will also interfere destructively, the same circumstance occurring also for points P''_1 and P''_2 etc., and the overall light intensity will have a minimum value. Taking, e.g., points P_1 and P_2, one then requires that the change δx, in the x-coordinate of point P, be such that the corresponding change $\delta(P_2P - P_1P)$ in the path difference $P_2P - P_1P$ be equal to $\lambda/2$. Since $P_2P = (x^2 + L^2)^{1/2}$ and $P_1P = \{[(D/2) - x]^2 + L^2\}^{1/2}$, one obtains (for $D \ll L$) the result that $\delta(P_2P - P_1P) \cong (D/2L)\delta x$. The requirement $\delta(P_2P - P_1P) = \lambda/2$ then gives

$$\delta x = \lambda L/D \tag{11.5.1}$$

Following a similar calculation, one can readily show that the same result is obtained by considering points P'_1 and P'_2 (or points P''_1 and P''_2 etc.) rather than points P_1 and P_2. All the corresponding contributions will now (on average) combine destructively rather than constructively and one can thus write the following approximate expression for the grain size, d_g:

$$d_g \cong 2\delta x = 2\lambda L/D \tag{11.5.2}$$

To obtain an approximate expression of the grain size for the imaging system of Fig. 11.8b, we first note that a similar argument to that presented above can be used to calculate the beam diameter of the Airy spot in the focal plane of a lens. Consider in fact the case where the diffuser in Fig. 11.9a is replaced by a lens of focal length $f = L$. Following the argument above one now realizes that an intensity maximum must be present at $x = 0$, i.e., at the center of the recording plane. In fact, as a result of the spherical wave front produced by the lens, the contributions from points P_1, P'_1, P''_1, etc. add in phase, there, with the contributions from points P_2, P'_2, P''_2 etc. The approximate size of the spot in the focal plane is then expected to be given again by Eq. (11.5.2), i.e., equal to $d_g \cong 2\lambda L/D$. This result should be compared with the value $d_g = 2.44\lambda L/D$ which can be obtained from the Airy function shown in Fig. 11.6. From this example, one can now understand the following general property of a diffracted wave: whenever the whole aperture, of diameter D, of an optical system contributes coherently to the diffraction to one or more spots in a plane located at a distance L, the minimum spot size in this recording plane is always approximately given by $2\lambda L/D^*$. Note that, in the case of a diffuser, this coherent contribution from the whole aperture D occurs provided that: (1) the size d_s of the individual scatters is much smaller than the aperture D; (2) there is an appreciable overlap, at the recording plane, between wavelets diffracted from various scattering centers. This implies that the dimension of each of these wavelets at the recording plane ($\sim\lambda L/d_s$) is larger than their mean separation ($\sim D$). The length L must therefore be such that $L > d_s D/\lambda$. Thus, for instance, with $d_s = 10\,\mu$m and $\lambda = 0.5\,\mu$m one has $L > 20D$.

We now go on to consider the case where the scattered light is recorded on a photographic plate after passing through a lens which images the diffuser onto the plate (Fig. 11.9b). We assume that the diameter, D', of the lens aperture is fully illuminated by the light diffracted by

* Since, for $D \gg L$, the field distribution in the recording plane is the Fourier transform of that in the input plane,[7] this property emerges as a general property of the Fourier transform.

each individual scatterer, i.e., that $(2\lambda L/d_s) \geq D'$. Given this condition, the whole aperture contributes via diffraction to each spot on the photographic plate and the grain size, d_g, at the plate is given by

$$d_g = 2\lambda L'/D' \tag{11.5.3}$$

Example 11.5. *Grain size of the speckle pattern as seen by a human observer.* We will consider a red, $\lambda = 633$ nm, He-Ne laser beam illuminating, e.g., an area of diameter $D = 2$ cm on a scattering surface for which the individual scatterers are taken to have a dimension $d_s = 50$ μm. The scattered ligth is observed by a human eye at a distance $L = 50$ cm from the diffuser. We will take $L' = 2$ cm as the distance between the retina and the lens of the eye and assume a pupil diameter $D' = 2$ mm. Since $(2\lambda L/d_s) \cong 12.7$ mm is much greater than D', the whole aperture of the eye is illuminated by the light diffracted by each individual scatterer. The apparent size of the speckle at the illuminated region of diffuser is obtained from Eq. (11.5.4) as $d_{ag} \cong 316$ μm. Note that, if the observer moves to a distance $L = 100$ cm from the diffuser, the apparent grain-size on the $D = 2$ cm illuminated spot of the diffuser will double to ~ 632 μm.

It should be noted that the arrangement of Fig. 11.9b also corresponds to the case where one looks directly at a scattering surface. In this case the lens and the recording plane correspond to the lens of the eye and the retina, respectively. Accordingly, d_g, as given by Eq. (11.5.3), can be taken as the expression for the grain size on the retina. Note that the apparent grain size on the scattering surface, d_{ag}, is $d_g(L/L')$ so that

$$d_{ag} = 2\lambda L/D' \tag{11.5.4}$$

This expression, which actually gives the eye's resolution for objects at a distance L, shows that d_{ag} is expected to increase with increasing L, i.e., with increasing distance between the observer and the diffuser and to decrease with increasing aperture of the iris (i.e., when the eye is dark-adapted). Both these predictions are readily confirmed by experimental observations.

Speckle noise often constitutes an undesirable feature of coherent light. The spatial resolution of the image of an object made via illumination with laser light is in fact often limited by speckle noise. Speckle noise is also apparent in the reconstructed image of a hologram, again limiting the spatial resolution of this image. Some techniques have therefore been developed to reduce speckle from coherently illuminated objects.[10] Speckle noise is not always a nuisance, however. In fact techniques have been developed that exploit the presence of speckle to show up, in a rather simple way, the deformation of large objects arising, e.g., from stresses or vibrations (*speckle interferometry*).

11.6. BRIGHTNESS

The brightness B of a light source or of a laser source has already been introduced in Sect. 1.4.4. [see Eqs. (1.4.3) and (1.4.4)]. We note again that the most significant parameter of a laser beam (and in general of any light source) is not simply its power or its intensity, but its brightness. This was already pointed out in Sect. 1.4.4. where it was shown that the maximum peak intensity, which can be obtained by focusing a given beam, is proportional to the beam brightness [see Eq. (1.4.6)]. This is further emphasized by the fact that, although the intensity of a beam can be increased, its brightness cannot. In fact, the simple arrangement of confocal lenses shown in Fig. 11.10 can be used to decrease the beam diameter, if $f_2 < f_1$,

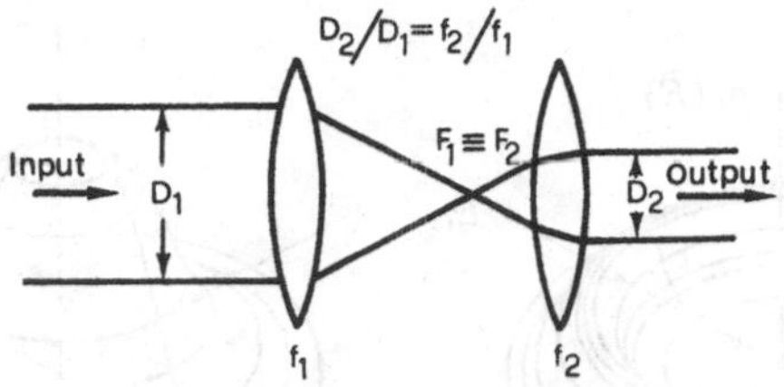

FIG. 11.10. Method for increasing the intensity of a laser beam.

and hence increase its intensity. However, the divergence of the output beam ($\sim \lambda/D_2$) is also correspondingly increased relative to that ($\sim \lambda/D_1$) of the input beam, and so one can see that the brightness remains invariant. This property, illustrated here for a particular case, is, in fact, of general validity (for incoherent sources also): given some light source and an optical imaging system, the image cannot be brighter that the original source (this is true provided the source and image are in media of the same refractive index).

The brightness of laser sources is typically several orders of magnitude greater than that of the most powerful incoherent sources. This is due to the extreme directionality of a laser beam. Let us compare, for example, an Ar laser oscillating on its green, $\lambda = 514$ nm, transition with a power of $P = 1$ W, with what is probably the brightest conventional source. This would be a high-pressure mercury vapor lamp (PEK Labs type 107/109), driven by an electrical power of $\sim$100 W and with an optical output power of $P_{out} \cong 10$ W and a brightness B of $\sim$95 W/cm$^2\times$sr in its most intense green line at $\lambda = 546$ nm wavelength. We will assume that the laser is oscillating on a TEM$_{00}$ mode so that we can take $A = \left(\pi w_0^2/2\right)$ as the beam area, where w_0 is the spot-size at the beam waist. Note the factor 2 in the denominator of the above expression which arises from the fact that w_0 actually represents the $(1/e)$ spot-size of the laser field rather than of the laser intensity. Likewise, since the $(1/e)$ field-divergence is given by $\theta = (\lambda/\pi w_0)$, the emission solid angle can be taken to be $\Omega = (\pi\theta^2/2)$. The brightness of this laser source can now be written as $B = (P/A\Omega)$, where P is the power. From the previous two expressions for the beam area and for the emission solid angle we then obtain

$$B = (4P/\lambda^2) \tag{11.6.1}$$

Inserting the appropriate values, for P and λ, for this laser gives $B \cong 1.6\times10^9$ W/cm$^2\times$sr. The brightness of the Ar laser is thus more than 7 orders of magnitude larger than that of the lamp. Since this will also be ratio of the two peak intensities obtained by focusing the corresponding sources, we now have a more quantitative appreciation of why a focused laser beam can be used, while a focused lamp cannot be used, in applications like, e.g., welding and cutting.

11.7. STATISTICAL PROPERTIES OF LASER LIGHT AND THERMAL LIGHT[11]

The temporal fluctuations of the field generated by a laser source or by a thermal light source can be described effectively in terms of the corresponding statistical behavior. Let $E(t) = A(t)\exp j[\omega t - \phi(t)]$ be the field generated by the source at some given space point. Writing $E(t) = \tilde{E}(t)\exp j(\omega t)$, where $\tilde{E} = A\exp -j(\phi)$ we then concentrate on just the slowly

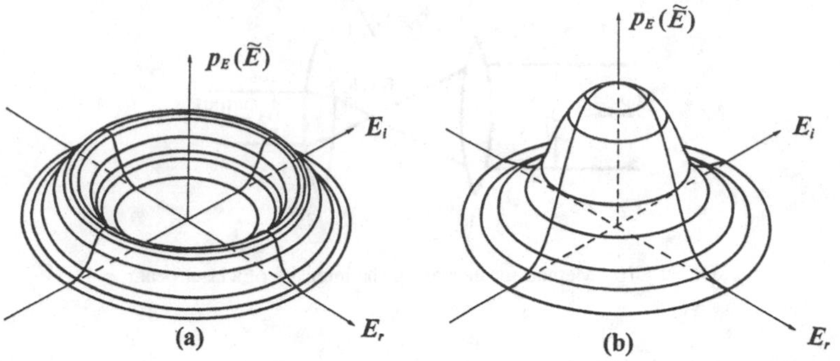

FIG. 11.11. Plot of the probability density $p_E(\tilde{E})$ of the field $\tilde{E}$ vs the real, E_r, and imaginary, E_i, parts of $\tilde{E}$. (a) Case of a coherent signal such as that of a single-mode laser. (b) Case of a thermal light such as that emitted by a conventional light source.

varying (over an optical cycle) complex-amplitude $\tilde{E}(t)$. Suppose now that a number of measurements are made, at different times t, of $\tilde{E}$, e.g., of its real, E_r, and imaginary, E_i, parts. In the limit of a very large number of measurements, we can obtain the bidimensional probability density $p_E(\tilde{E}) = p_E(E_r, E_i)$ defined such that $dp = p_E(\tilde{E})dE_r dE_i$ represents the elemental probability that a field measurement gives a value for the real part between E_r and $E_r + dE_r$ and imaginary part between E_i and $E_i + dE_i$. Alternatively, we can represent $p_E(\tilde{E})$ as a function of the amplitude, A, and phase, ϕ, and thus write $dp = p_E(\tilde{E})AdAd\phi$ as the elemental probability that a measurement gives a value of the amplitude between A and $A+dA$ and phase between ϕ and $\phi + d\phi$. Once $p_E(\tilde{E})$ is known, the average intensity of the wave, according to Eq. (11.1.3), can be written as

$$<I> = \frac{\iint |E|^2 p_E(\tilde{E})dE_r dE_i}{\iint p_E(\tilde{E})dE_r dE_i} = \frac{\iint |A|^2 p_E(\tilde{E})AdAd\phi}{\iint p_E(\tilde{E})AdAd\phi} \tag{11.7.1}$$

The probability density can be represented, in a very effective way, in three-dimensional space as a function of E_r and E_i.

The plot of $p_E(\tilde{E})$ vs (E_r, E_i) for a single mode laser source is shown in Fig. 11.11a. As already pointed out in Sect. 7.11, the output intensity and hence the field amplitude of this laser is fixed, for a given pump rate, by the condition that upward transitions, due to pumping, must be balanced by downward transitions due to both stimulated emission and spontaneous decay. Small amplitude fluctuations may then arise from fluctuations of both pump rate and cavity loss. On the other hand, the phase $\phi(t)$ is not controlled by such a balancing process and is thus free to take any value between 0 and ∞. Since, now, one has $A = \left[E_r^2 + E_i^2\right]^{1/2}$ and $\phi = -\tan^{-1}(E_i/E_r)$, the expected plot will be as shown in Fig. 11.11a. Note that the amplitude fluctuations of $A = A(t)$ have been greatly exaggerated in the figure, relative amplitude fluctuations of a few percent or less being typical in free-running operation for, e.g., a good diode-pumped solid-state laser source (see Fig. 7.30). To first order we can then assume

$$p_E(\tilde{E}) \propto \delta(A - A_0) \tag{11.7.2}$$

where A_0 is a constant determined by the average intensity value. The substitution of Eqs. (11.7.2) into (11.7.1) gives in fact $<I> = A_0^2$. In the time domain, the point that represents $\tilde{E}(t)$ in the (E_r, E_i) plane will then travel essentially along the circumference of radius $|\tilde{E}| = A_0$. As a consequence of the statistical nature of the phase fluctuations, this movement will be random and the corresponding angular speed, $d\phi/dt$, will establish the laser bandwidth.

The plot of $p_E(\tilde{E})$ vs (E_r, E_i) for a thermal light source is shown in Fig. 11.11b. In this case, the field is due to the superposition of the uncorreleted light emitted, by spontaneous emission, from the individual atoms of the light source. Since the number of these emitters is very large, it follows from the central limit theorem of statistics that the probability distribution of both real and imaginary parts of $\tilde{E}$ must follow a Gaussian law. We can thus write

$$p_E(\tilde{E}) \propto \exp - \left(\frac{E_r^2 + E_i^2}{A_0^2}\right) = \exp - \left(\frac{A^2}{A_0^2}\right) \tag{11.7.3}$$

where A_0 is a constant again determined by the average light intensity. The substitution of Eqs. (11.7.3) into (11.7.1) gives, in fact, $<I> = A_0^2$. Note that the average values of both E_r and E_i are now zero. Thus, the time-domain movement of the point representing $\tilde{E}$ in the (E_r, E_i) plane consists now of a random movement around the origin. The speed of this movement in terms of both amplitude and phase (i.e., in term of both dA/Adt and $d\phi/dt$) establishes, in this case, the bandwidth of this thermal source.

11.8. COMPARISON BETWEEN LASER LIGHT AND THERMAL LIGHT

A comparison will now be made between a red-emitting, $\lambda = 633$ nm, He-Ne laser oscillating, on a single mode, with a "modest" output power of 1 mW with what is probably the brightest conventional source, already considered in Sect. 11.6. (PEK Labs type 107/109), giving an optical output power of $P_{out} \cong 10$ W and a brightness B of $\sim 95\,\text{W/cm}^2 \times \text{sr}$ in its most intense green line at $\lambda = 546$ nm wavelength. To obtain a beam with good spatial coherence from this lamp, one can use the arrangement of Fig. 11.12, where a lens of focal length, f, and suitable aperture, D, collects a fraction of the emitted light. The lamp is simulated by individual emitters, located within an aperture, of diameter d, made in a screen S. Following the discussion in Sect. 11.3.6, to obtain a beam with a spatial coherence which approaches the ideal, the aperture D must be chosen so that [see Eq. (11.3.43)]

$$D \cong 0.32\lambda f/d \tag{11.8.1}$$

This beam, although having a degree of spatial coherence somewhat less than unity $[\gamma^{(1)}(P_1, P_2) \cong 0.88$, in the case considered], may be regarded, generously, as having the same degree of spatial coherence as that of the He-Ne laser [whose value for $\gamma^{(1)}$ may be taken to be essentially unitary]. The output power of the beam obtained after the lens is then given by $P_{out} = B A \Omega$, where B is the brightness of the lamp, A is its emitting area ($A = \pi d^2/4$) and Ω is the acceptance solid angle of the lens ($\Omega = \pi^2 D^2/4f^2$). With the help of Eq. (11.8.1)

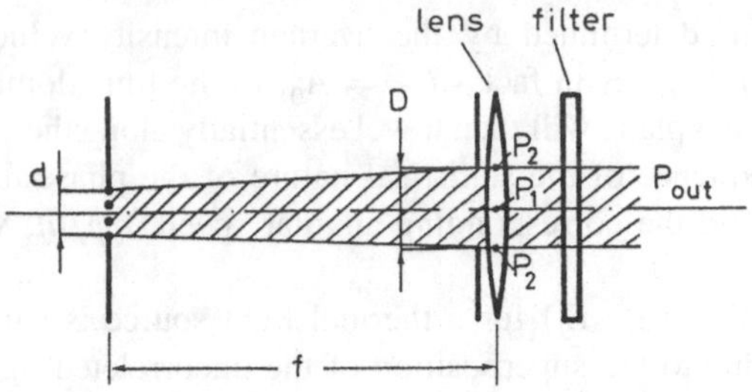

FIG. 11.12. Spatial and frequency filter to obtain a, first-order, coherent output beam from an incoherent lamp.

one then obtains [compare with Eq. (11.6.1)]

$$P_{out} \cong (\lambda/4)^2 B \tag{11.8.2}$$

Using the value $B = 95\ W/\text{cm}^2 \times \text{sr}$ for the lamp brightness one obtains $P_{out} \cong 1.8 \times 10^{-8}$ W. Note that this power is now ~5 orders of magnitude smaller than that emitted by the He-Ne laser and ~9 orders of magnitude smaller than that emitted by the lamp. Note also that, according to Eq. (11.8.2), the power that can be obtained in a spatially coherent beam depends only on the brightness of the lamp and this further illustrates the importance of this quantity.

Having paid such a high penalty in terms of output power, one has now a beam which has about the same degree of spatial coherence as that of our He-Ne laser. The degree of temporal coherence, however, is still much worse because the emission bandwidth of the lamp is certainly much larger than that of the He-Ne laser. In fact, the linewidth of the lamp is broadened, by the high pressure of its internal gas, to a width of $\Delta\nu \cong 10^{13}$ Hz. The linewidth of the He-Ne laser, on the other hand, even using a modest frequency stabilization scheme, may be taken to be $\Delta\nu_L \cong 1$ kHz. To achieve the same degree of time coherence we must then arrange for the bandwidths of the two sources to be the same. In principle this can be done by inserting a frequency filter with the (exceptionally) narrow linewidth of 1 kHz in the output beam obtained from the lamp (see Fig. 11.12). However, this filter further reduces the output power of this source by about ten orders of magnitude [corresponding to $(\Delta\nu_L/\Delta\nu) \cong 10^{-10}$] so that the final output power of the, spatially- and frequency-filtered lamp would be $P_{out} \cong 10^{-18}$ W.

Thus, for a penalty of about 19 orders of magnitude on the original green output power emitted by the lamp, we can now say that the He-Ne laser beam and the filtered output beam of Fig. 11.12 show, approximately, the same degree of spatial and temporal coherence. To compare the two beams at the same output power, one may now place an attenuator with a factor 10^{15} attenuation in front of the He-Ne laser. At this stage the power and the degree of coherence for the two beams is the same and it is therefore natural to ask the question as to whether these two sources would now be effectively the same, i.e., indistinguishable. The answer is however negative. In fact, a detailed comparison between the two sources show that the two beams remain basically different and that, notably, the He-Ne laser beam still remains more coherent.

A first comparison can readily be made in terms of the statistical properties of the two light sources. One may notice, in fact, that the filtering operation, applied to the lamp output, and an attenuator, placed in front of the He-Ne laser beam, do not alter the statistical properties of the corresponding light. Thus Fig. 11.11a and Fig. 11.11b can still be used to describe the statistical properties of the two sources and, from these figures, a first basic difference between the two beams is now apparent. Note that, since the output powers of the two beams have been equalised, the quantity A_0 appearing in Eqs. (11.7.2) and (11.7.3) is the same for the two sources. Note also that, if the degree of temporal coherence of the two beams is made equal, this simply means that the speed of movement of the point representing $\tilde{E}$ in the (E_r, E_i) plane is the same for the two cases. Note lastly that, if the degree of spatial coherence of the two beams is equalised, this simply means that, for each beam, the movement in the (E_r, E_i) plane will be the same at any point of the wavefront. Nonetheless, the statistical properties of the two beams, as represented in Fig. 11.11, still remain fundamentally different.

A second comparison between the two sources can be made in terms of the coherence properties to higher-order (see Appendix H). To this purpose, we recall that the coherence function $\Gamma^{(1)}$ was introduced in Sect. 11.3 in terms of the product $E(x_1)E^*(x_2)$ between fields taken at two different space-time points $x_i = (\mathbf{r}_i, t_i)$. For this reason, a superscript (1) was used for Γ as a reminder of the fact that one is actually performing a, first-order, correlation between the two fields. To higher order, one can in fact introduce a whole class of correlation functions, e.g., $<E(x_1)E(x_2)E^*(x_3)E^*(x_4)>$ involving the four distinct space-time points x_1, x_2, x_3, and x_4, and so on to yet higher order*. One can then introduce some suitable definition of a higher-order coherence, $\Gamma^{(n)}$, in terms of these higher-order correlation functions. When this is done, it turns out that the higher-order coherence of our monomode laser source is still larger than the filtered beam taken from a lamp (see Appendix H). It also turns out that, fundamentally, one can at best arrange for the two sources to exhibit the same, first-order, coherence i.e., the same degree of spatial and temporal coherence, as indeed achieved by the filtering system of Fig. 11.12.

Thus, as a conclusion, despite having paid such a heavy penalty in terms of output power, our filtered thermal source remains basically different from a laser.

PROBLEMS

11.1. Calculate $\Gamma^{(1)}(\mathbf{r}_1, \mathbf{r}_1, \tau)$ for a sinusoidal wave.

11.2. Prove Eq. (11.3.9).

11.3. For the Michelson interferometer of Fig. 11.4a, find the analytical relation between the intensity along the C direction, I_c, and $\Gamma^{(1)}(\mathbf{r}, \mathbf{r}, \tau)$, where $\tau = 2(L_3 - L_2)/c$.

11.4. Assume that the field at point P of the Michelson interferometer of Fig. 11.4a is made of a sinusoidal wave with constant amplitude and random phase jumps (see Fig. 2.9). Using the expression for $\Gamma^{(1)}(\tau)$ calculated in Appendix B for this field and the relation between I_c and $\Gamma^{(1)}(\tau)$ calculated in the previous problem, find the analytical expression of $V_p = V_p(\tau)$.

11.5. The shape of the spectral output of a CO_2 laser beam operating at $\lambda = 10.6\ \mu m$ is approximately Gaussian with a bandwidth of 10 kHz [$\Delta\nu_L$ is defined according to Eq. (11.3.27)]. Calculate the coherence length L_c and the distance ΔL between two successive maxima of the intensity curve of Fig. 11.4b.

* Actually, it can be shown that a given field $E(x)$ can, in principle. be completely characterized by either the infinite set of values for E obtained by changing x, or by the infinite set of correlation functions as indicated above.

11.6. A plane e.m. wave of circular cross section, uniform intensity, and perfect spatial coherence is focused by a lens. What is the increase in intensity at the focus compared to that of the incident wave?

11.7. A Gaussian beam is focused by a lens of focal length f. Assuming that the waist of the incident beam is located at the lens position and that the corresponding spot-size w_0 is appreciably smaller than the lens diameter, relate the peak intensity at the focal spot to the power, P_i, of the incident beam. Compare then the resulting expression with Eq. (11.4.4).

11.8. The near-field transverse-intensity profile of a Nd:YAG laser beam, at $\lambda = 1.064\ \mu m$ wavelength, is, to a good approximation, Gaussian with a diameter (FWHM) $D \cong 4$ mm. The half-cone beam divergence, measured at the half-maximum point of the far-field intensity distribution, is $\theta_d \cong 3$ mrad. Calculate the corresponding M^2 factor.

11.9. The near-field transverse intensity profile of a pulsed TEA CO_2 laser beam, at $\lambda = 10.6\ \mu m$ wavelength, is, to a good approximation, constant over its 1 cm × 4 cm dimension. The laser is advertised to have a M^2 factor of 16 along both axes. Assuming the waist position to be located at the position of the output mirror, calculate the spot-size parameters at a distance from this mirror of $z = 3$ m.

References

1. M. Born and E. Wolf, *Principles of Optics*, 6th edn. (Pergamon Press, Oxford, 1980), pp. 491–544.
2. W. H. Louisell, *Radiation and Noise in Quantum Electronics* (McGraw-Hill, New York, 1964), pp. 47–53.
3. A. Jeffrey, *Handbook of Mathematical Formulas and Integrals* (Academic, San Diego 1995) p. 244.
4. Reference [1], pp. 508–518.
5. J. W. Goodman, *Introduction to Fourier Optics* (McGraw-Hill, New York, 1968).
6. Reference [1], pp. 395–398.
7. A. E. Siegman, Defining and Measuring Laser Beam Quality, in *Solid-State Lasers-New Developments and Applications*, ed. by M. Inguscio and R. Wallenstein (Plenum, New York, 1993) pp. 13–28.
8. Reference [4], Chapter 5.
9. *Laser Speckle and Related Phenomena*, ed. by J. C. Dainty (Springer-Verlag, Berlin 1975).
10. M. Françon, *Laser Speckle and Applications in Optics* (Academic, New York, 1979).
11. R. J. Glauber, Optical Coherence and Photon Statistics, in *Quantum Optics and Electronics*, ed. by C. De Witt, A. Blandin, and C. Cohen-Tannoudji (Gordon and Breach, New York, 1965), pp. 71, 94–98, 103, 151–155.

12

Laser Beam Transformation: Propagation, Amplification, Frequency Conversion, Pulse Compression and Pulse Expansion

12.1. INTRODUCTION

Before it is put to use, a laser beam is generally transformed in some way. The most common type of transformation is that which occurs when the beam is simply made to propagate in free space or through a suitable optical system. Since this produces a change in the spatial distribution of the beam (e.g., the beam may be focused or expanded), we shall refer to this as a *spatial transformation* of the laser beam. A second type of transformation, also rather frequently encountered, is that which occurs when the beam is passed through an amplifier or chain of amplifiers. Since the main effect here is to alter the beam amplitude, we shall refer to this as *amplitude transformation*. A third, rather different, case occurs when the wavelength of the beam is changed as a result of propagating through a suitable nonlinear optical material (*wavelength transformation* or *frequency conversion*). Finally the temporal behavior of the laser beam can be modified by a suitable optical element. For example, the amplitude of a cw laser beam may be temporally modulated by an electro-optic or acousto-optic modulator or the time duration of a laser pulse may be increased (pulse expansion) or decreased (pulse compression) using suitably dispersive optical systems or nonlinear optical elements. This fourth and last case will be referred to as *time transformation*. It should be noted that these four types of beam transformation are often interrelated. For instance, amplitude transformation and frequency conversion often result in spatial and time transformations occurring as well.

In this chapter the four cases of laser beam transformation introduced above will be briefly discussed. In the case of frequency conversion, of the various nonlinear optical effects

O. Svelto, *Principles of Lasers*,
DOI: 10.1007/978-1-4419-1302-9_12, © Springer Science+Business Media LLC 2010

that can be used[1] to achieve this, only the so-called parametric effects will be considered here. These in fact provide some of the most useful techniques so far developed for producing new sources of coherent light. Time transformation will be considered only for the cases of pulse expansion or pulse compression, while we refer elsewhere for the case of amplitude modulation.[2] We also omit a number of aspects of amplitude and time transformation that arise from the nonlinear phenomenon of self-focusing and the associated phenomenon of self-phase-modulation,[3] although it should be noted that they can play a very important role in limiting, for instance, the performance of laser amplifiers.

12.2. SPATIAL TRANSFORMATION: PROPAGATION OF A MULTIMODE LASER BEAM[4,5]

The free-space propagation of a Gaussian beam and of a multi-transverse-mode beam has already been considered in Sects. 4.7.2 and 11.4.3, respectively. In Sect. 4.7.2 it was shown that a Gaussian beam, of, e.g., circular cross-section, is characterized by two parameters, namely the coordinate, z_0, of the beam-waist and the corresponding spot-size, w_0. By contrast to this, it was shown in Sect. 11.4.3 that a multi-mode beam, of, e.g., circular cross section again, is characterized by three parameters, namely the coordinate, z_0, of the beam-waist, the spot-size parameter W_0, and M^2, the beam-quality factor. The propagation of a Gaussian beam through a general optical system characterized by a given *ABCD* matrix, has, on the other hand, been considered in Sect. 4.7.3. It was shown there that the complex q-parameter of the beam, after passing through the optical system, can be readily obtained from the q-parameter of the input beam once the matrix elements, *A, B, C, D*, are known. To complete this picture, we will consider, in this Section, the propagation of a multi-mode laser beam through a general optical system characterized by a given *ABCD* matrix.

Consider first the free-space propagation of a multimode laser beam. The spot-size parameter, along, e.g., the transverse direction x and at a longitudinal coordinate z, is described by Eq. (11.4.19) which, for convenience, is reproduced here

$$W_x^2(z) = W_{0x}^2 + M_x^4 \frac{\lambda^2}{\pi^2 W_{0x}^2}(z - z_{0x})^2 \tag{12.2.1}$$

We can now see that, as far as free-space propagation is concerned, the multimode laser beam behaves as if it contained an "embedded Gaussian beam" having the same waist location, z_{0x}, as the multimode beam and a spot-size, at any coordinate z, given by

$$w_x(z) = W_x(z)/M_x \tag{12.2.2}$$

where $M_x = \sqrt{M_x^2}$ is a constant. In fact, the substitution of Eq. (12.2.2) in Eq. (12.2.1) readily gives Eq. (4.7.13a). It can also be shown that the radius of curvature, $R(z)$, of this embedded Gaussian beam equals, at any z, that of the multimode beam.

This notion of an embedded Gaussian beam can be shown to hold also for propagation through a general optical system described by, e.g., its *ABCD* matrix. Accordingly, the propagation of the multi-mode beam can be obtained by the following simple procedure involving steps (a), (b), (c), and (d): (a) Starting with the multimode laser beam characterized by given values of W_{0x}, M_x^2, and z_{0x}, one defines the embedded Gaussian beam with $w_{0x} = W_{0x}/M_x$ and beam-waist at the location of the multimode beam-waist. (b) One then calculates the propagation of the embedded Gaussian beam through the optical system by, e.g., using the *ABCD* law of Gaussian-beam propagation. (c) At any location within the optical system, the wavefront radius of curvature of the multimode beam will then coincide with that of the embedded Gaussian beam. This means, in particular, that any waist will have the same location for the two beams. (d) The spot-size parameter, W_x, of the multimode beam, at any location, will then be given by $W_x(z) = M_x w_x(z)$.

Example 12.1. *Focusing of a multimode Nd:YAG beam by a thin lens* Consider a multimode beam from a repetitively pulsed Nd:YAG laser, at $\lambda \cong 1.06\,\mu$m wavelength, such as used for welding or cutting metallic materials. The near-field transverse-intensity profile may be taken to be approximately Gaussian with a diameter (FWHM) of $D = 4$ mm, while the M^2 factor may be taken to be ~40. We want now to see what happens when the beam is focused by a spherical lens of $f = 10$ cm focal length. We assume that the waist of this multimode beam coincides with the output mirror, this being a plane mirror. We will also assume that the lens is located very near to this mirror so that the waist of the multimode beam and hence of the embedded Gaussian beam can be taken to coincide with the lens location. For a Gaussian intensity profile, the spot size parameter of the input beam, $W = W_0$, is then related to the beam diameter, D, by the condition $\exp -2(D/2W_0)^2 = (1/2)$. We get $W_0 = D/[2\ln 2]^{1/2} \cong 3.4$ mm, so that $w_0 = W_0/\sqrt{M^2} \cong 0.54$ mm. According to Eq. (4.7.26), since the Rayleigh range corresponding to this spot size, $z_R = \pi w_0^2/\lambda \cong 85$ cm, is much larger than the focal length of the lens, the waist formed beyond the lens will approximately be located at the lens focus. From Eq. (4.7.28), the spot-size of the embedded Gaussian beam at this focus, w_{0f}, is then given by $w_{0f} \cong \lambda f/\pi w_0 \cong 63\,\mu$m and the spot-size parameter of the multimode beam by $W_{0f} = \sqrt{M^2} w_{0f} \cong 400\,\mu$m.

12.3. AMPLITUDE TRANSFORMATION: LASER AMPLIFICATION[(6–8)]

In this section we consider the rate-equation treatment of a laser amplifier. We assume that a plane wave of uniform intensity I enters (at $z = 0$) a laser amplifier extending for a length l along the z direction. We limit our considerations to the case where the incoming laser beam is in the form of a pulse (pulse amplification) while we refer elsewhere[(8)] for the amplification of a c.w. beam (steady-state amplification).

We consider first the case of an amplifier medium working on a four-level scheme and further assume that pulse duration, τ_p, is such that $\tau_1 \ll \tau_p \ll \tau$, where τ_1 and τ are the lifetime of the lower and upper levels of the amplifier medium, respectively. In this case the population of the lower level of the amplifier can be set equal to zero. This is perhaps the most relevant case to consider as it would apply, for instance, to the case of a *Q*-switched laser pulse from a Nd: YAG laser being amplified. We will also assume that pumping to the amplifier upper-level and subsequent spontaneous decay can be neglected during the passage of the pulse and that the transition is homogeneously-broadened. Under these conditions and with the help of Eq. (2.4.17) [in which we set $F = I/h\nu$], the rate of change of population inversion $N(t, z)$ at a point z within the amplifier can be written as

$$\frac{\partial N}{\partial t} = -WN = -\frac{NI}{\Gamma_s} \tag{12.3.1}$$

where

$$\Gamma_s = (h\nu/\sigma) \tag{12.3.2}$$

is the saturation energy fluence of the amplifier [see Eq. (2.8.29)]. Note that a partial derivative is required in Eq. (12.3.1) since we expect N to be a function of both z and t, i.e., $N = N(t, z)$, on account of the fact that $I = I(t, z)$. Note also that Eq. (12.3.1) can be solved for $N(t)$ to yield

$$N(\infty) = N_0 \exp -(\Gamma/\Gamma_s) \tag{12.3.3}$$

where $N_0 = N(-\infty)$ is the amplifier's upper-level population before the arrival of the pulse, as established by the combination of pumping and spontaneous decay, and where

$$\Gamma(z) = \int_{-\infty}^{+\infty} I(z, t)\, dt \tag{12.3.4}$$

is the total fluence of the laser pulse.

Next we derive a differential equation describing the temporal and spatial variation of intensity I. To do this we first write an expression for the rate of change of e.m. energy within unit volume of the amplifier. For this we refer to Fig. 12.1 where an elemental volume of the amplifier medium of length dz and cross-section S is indicated by the shaded area. We can then write

$$\frac{\partial \rho}{\partial t} = \left(\frac{\partial \rho}{\partial t}\right)_1 + \left(\frac{\partial \rho}{\partial t}\right)_2 + \left(\frac{\partial \rho}{\partial t}\right)_3 \tag{12.3.5}$$

where $(\partial\rho/\partial t)_1$ accounts for stimulated emission and absorption in the amplifier, $(\partial\rho/\partial t)_2$ for the amplifier loss (e.g., scattering losses), and $(\partial\rho/\partial t)_3$ for the net photon flux which flows into the volume. With the help again of Eq. (2.4.17) [$F = I/h\nu$] we obtain

$$\left(\frac{\partial \rho}{\partial t}\right)_1 = WNh\nu = \sigma NI \tag{12.3.6}$$

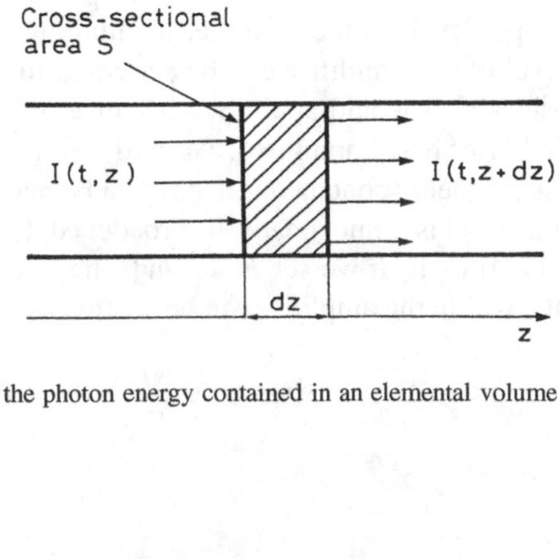

FIG. 12.1. Rate of change of the photon energy contained in an elemental volume of length dz and cross sectional area S of a laser amplifier.

and from Eqs. (2.4.17) and (2.4.32) we obtain

$$\left(\frac{\partial \rho}{\partial t}\right)_2 = - W_a N_a h\nu = -\alpha I \tag{12.3.7}$$

where N_a is the density of the loss centers, while W_a is the absorption rate, and α the absorption coefficient associated with the loss centers. To calculate $(\partial\rho/\partial t)_3$, we refer again to Fig. 12.1, and note that $(\partial\rho/\partial t)_3 S dz$ is the rate of change of e.m. energy in this volume due to the difference between the incoming and outgoing laser power. We can then write $(\partial\rho/\partial t)_3 S dz = S[I(t, z) - I(t, z + dz)]$, which readily gives

$$\left(\frac{\partial \rho}{\partial t}\right)_3 dz = -\frac{\partial I}{\partial z} dz \tag{12.3.8}$$

Equation (12.3.5), with the help of Eqs. (12.3.6)–(12.3.8) and with the observation that $(\partial\rho/\partial t) = (\partial I/c\partial t)$, gives

$$\frac{1}{c}\frac{\partial I}{\partial t} + \frac{\partial I}{\partial z} = \sigma N I - \alpha I \tag{12.3.9}$$

This equation, together with Eq. (12.3.1), completely describes the amplification process. Note that Eq. (12.3.9) has the usual form of a time-dependent transport equation.

Equations (12.3.1) and (12.3.9) must be solved with the appropriate boundary and initial conditions. As the initial condition we take $N(0, z) = N_0$, where N_0 is the amplifier upper-level population before the arrival of the laser pulse. The boundary condition is obviously established by the intensity $I_0(t)$ of the light pulse injected into the amplifier, i.e., $I(t, 0) = I_0(t)$. For negligible amplifier losses (i.e., neglecting the term $-\alpha I$), the solution to Eqs. (12.3.1) and (12.3.9) can be written as

$$I(z, \tau) = I_0(\tau)\left\{1 - [1 - \exp(-gz)]\exp\left[-\int_{-\infty}^{\tau} I_0(\tau')\, d\tau'/\Gamma_s\right]\right\}^{-1} \tag{12.3.10}$$

where $\tau = t - (z/c)$ and where $g = \sigma N_0$ is the unsaturated gain coefficient of the amplifier.

From Eqs. (12.3.1) and (12.3.9), we can also obtain a differential equation for the total fluence of the pulse, $\Gamma(z)$, given by Eq. (12.3.4). Thus, we first integrate both sides of Eq. (12.3.1) with respect to time, from $t = -\infty$ to $t = +\infty$, to obtain $\left(\int_{-\infty}^{+\infty} N I dt/\Gamma_s\right) = N_0 - N(+\infty) = N_0[1 - \exp(-\Gamma/\Gamma_s)]$, where Eq. (12.3.3) has been used. We then integrate both sides of Eq. (12.3.9) with respect to time, on the same time interval, and use the above expression for $\left(\int_{-\infty}^{+\infty} N I dt/\Gamma_s\right)$ and the fact that $I(+\infty, z) = I(-\infty, z) = 0$. We obtain

$$\frac{d\Gamma}{dz} = g\Gamma_s[1 - \exp(-\Gamma/\Gamma_s)] - \alpha\Gamma \tag{12.3.11}$$

Again neglecting amplifier losses, Eq. (12.3.11) gives

$$\Gamma(l) = \Gamma_s \ln\left\{1 + \left[\exp\left(\frac{\Gamma_{in}}{\Gamma_s}\right) - 1\right] G_0\right\} \tag{12.3.12}$$

where $G_0 = \exp(gl)$ is the unsaturated gain of the amplifier and Γ_{in} is the energy fluence of the input beam. As a representative example the ratio Γ/Γ_s is plotted in Fig. 12.2 versus Γ_{in}/Γ_s for $G_0 = 3$. Note that, for $\Gamma_{in} \ll \Gamma_s$, Eq. (12.3.12) can be approximated as

$$\Gamma(l) = G_0 \Gamma_{in} \tag{12.3.13}$$

and the output fluence increase linearly with the input fluence (linear amplification regime). Equation (12.3.13) is also plotted in Fig. 12.2 as a dashed straight line starting from the origin. At higher input fluences, however, Γ increases with Γ_{in} at a lower rate than that predicted by Eq. (12.3.13) i.e., amplifier saturation begins to occur. For $\Gamma_{in} \gg \Gamma_s$ (deep saturation regime) Eq. (12.3.12) can be approximated to

$$\Gamma(l) = \Gamma_{in} + gl\Gamma_s \tag{12.3.14}$$

Equation (12.3.14) has also been plotted in Fig. 12.2 as a dashed straight line. Note that Eq. (12.3.14) shows that, for high input fluences, the output fluence is linearly dependent on the length l of the amplifier. Since $\Gamma_s gl = N_0 lh\nu$, one then realizes that every excited atom undergoes stimulated emission and thus contributes its energy to the beam. Such a condition obviously represents the most efficient conversion of stored energy to beam energy, and for this reason amplifier designs operating in the saturation regime are used wherever practical.

It should be pointed out again that the previous equations have been derived for an amplifier having an ideal four-level scheme. For a quasi-three-level scheme, on the other hand, one can see from the considerations developed in Sect. 7.2.2 that Eq. (12.3.1) still applies provided that Γ_s is now given by

$$\Gamma_s = h\nu/(\sigma_e + \sigma_a) \tag{12.3.15}$$

where σ_e and σ_a are the effective cross-sections for stimulated emission and absorption, respectively. One can also show that Eq. (12.3.9) still applies provided that σ is replaced by σ_e.

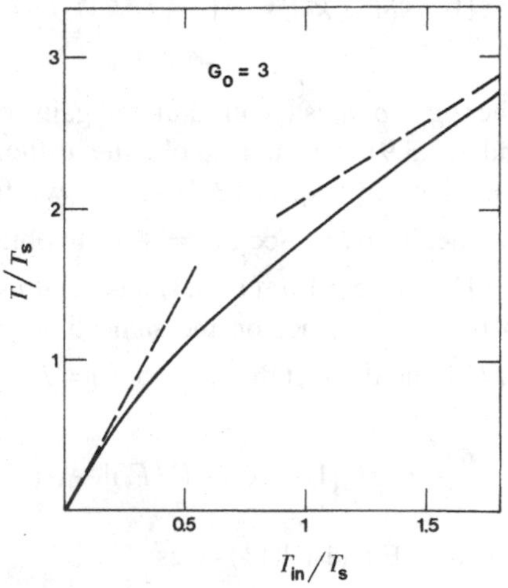

FIG. 12.2. Output laser energy fluence Γ versus input fluence Γ_{in} for a laser amplifier with a small signal gain $G_0 = 3$. The energy fluence is normalized to the laser saturation fluence $\Gamma_s = h\nu/\sigma$.

It then follows that Eq. (12.3.12) still remain valid provided that Γ_s is given by Eq. (12.3.15) and G_0 given by $G_0 = \exp \sigma_e N_0 l$. Similar considerations can be made for an amplifier operating on a four-level scheme when the pulse duration becomes much shorter than the lifetime of the lower level of the transition. In this case the population driven to the lower level by stimulated emission remains in this level during the pulse and one can show that Eq. (12.3.12) still remains valid provided that σ is replaced by σ_e, and Γ_s is given by Eq. (12.3.15), where σ_a is the effective absorption cross-section of the lower level of the transition.

If amplifier losses cannot be neglected, the above picture has to be modified somewhat. In particular the output fluence $\Gamma(l)$ does not continue increasing with input fluence, as in Fig. 12.2, but reaches a maximum and then decreases. This can be understood by noting that, in this case, the output as a function of amplifier length tends to grow linearly due to amplification [at least for high input fluences, see Eq. (12.3.14)] and to decrease exponentially due to loss [on account of the term $-\alpha\Gamma$ in Eq. (12.3.11)]. The competition of these two terms then gives a maximum for the output fluence Γ. For $\alpha \ll g$ this maximum value of the output fluence, Γ_m, turns out to be

$$\Gamma_m \cong g\Gamma_s/\alpha \tag{12.3.16}$$

It should be noted, however, that, since amplifier losses are typically quite small, other phenomena usually limit the maximum energy fluence that can be extracted from an amplifier. In fact, the limit is usually set by the amplifier damage fluence Γ_d (10 J/cm^2 is a typical value for a number of solid-state media). From Eq. (12.3.14) we then get the condition

$$\Gamma \cong gl\Gamma_s < \Gamma_d \tag{12.3.17}$$

Another limitation to amplifier performance arises from the fact that the unsaturated gain $G_0 = \exp(gl)$ must not be made too high, otherwise two undesirable effects can occur in the amplifier: (1) parasitic oscillations, (2) amplified spontaneous emission (ASE). Parasitic oscillation occurs when the amplifier starts lasing by virtue of some internal feedback which will always be present to some degree, (e.g., due to the amplifier end faces). The phenomenon of ASE has already been discussed in Sect. 2.9.2. Both these phenomena tend to depopulate the available inversion and hence decrease the laser gain. To minimize parasitic oscillations one should avoid elongated amplifiers and in fact ideally use amplifiers with roughly equal dimensions in all directions. Even in this case, however, parasitic oscillations set an upper limit $(gl)_{max}$ to the product of gain coefficient, g, with amplifier length, l, i.e.,

$$gl < (gl)_{max} \tag{12.3.18}$$

where, for typical cases, $(gl)_{max}$ may range between three and five. The threshold for ASE has already been given in Sect. 2.9.2 [see Eq. (2.9.4a), for a Lorentzian line]. For an amplifier in the form of a cube (i.e., for $\Omega \cong 1$) and for a unitary fluorescence quantum yield we get $G \cong 8$ [i.e., $gl \cong 2.1$] which is comparable to that established by parasitic oscillations. For smaller values of solid angle Ω, which are more typical, the value of G for the onset of ASE is expected to increase [Eq. (2.9.4a)]. Hence parasitic oscillations, rather than ASE, usually determine the maximum gain that can be achieved.

Example 12.2. *Maximum energy which can be extracted from an amplifier.* It is assumed that the maximum value of gl is limited by parasitic oscillations such that $(gl)^2_{max} \cong 10$ and the rather low gain coefficient of $g = 10^{-2}\,\text{cm}^{-1}$ is also assumed. For a damage energy-fluence of the amplifier medium of $\Gamma_d = 10\,\text{J/cm}^2$, we get from Eq. (12.3.19) $E_m \cong 1$ MJ. It is however worth noting that this represents an upper limit to the energy since it would require a somewhat impractical amplifier dimension of the order of $l_m \cong (gl)_m/g \cong 3$ m.

When both limits, due to damage, Eq. (12.3.17), and parasitic oscillations, Eq. (12.3.18), are taken into account, one can readily obtain an expression for the maximum energy E_m, that can be extracted from an amplifier, as

$$E_m = \Gamma_d l_m^2 = \Gamma_d (gl)_m^2/g^2 \qquad (12.3.19)$$

where l_m is the maximum amplifier dimension (for a cubic amplifier) implied by Eq. (12.3.18). Equation (12.3.19) shows that E_m is increased by decreasing the amplifier gain coefficient g. Ultimately, a limit to this reduction in gain coefficient would be established by the amplifier losses α.

So far we have concerned ourselves mostly with the change of laser pulse energy as the pulse passes through an amplifier. In the saturation regime, however, important changes in both the temporal and spatial shape of the input beam also occur. The spatial distortions can be readily understood with the help of Fig. 12.2. For an input beam with a bell-shaped transverse intensity profile (e.g., a Gaussian beam), the beam center, as a result of saturation, will experience less gain than the periphery of the beam. Thus, the width of the beam's spatial profile is enlarged as the beam passes through the amplifier. The reason for temporal distortions can also be seen quite readily. Stimulated emission caused by the leading edge of the pulse implies that some of the stored energy has already been extracted from the amplifier by the time the trailing edge of the pulse arrives. This edge will therefore see a smaller population inversion and thus experience a reduced gain. As a result, less energy is added to the trailing edge than to the leading edge of the pulse, and this leads to considerable pulse reshaping. The output pulse shape can be calculated from Eq. (12.3.10), and it is found that the amplified pulse may either broaden or narrow (or even remain unchanged), the outcome depending upon the shape of the input pulse.[7]

12.3.1. Examples of Laser Amplifiers: Chirped-Pulse-Amplification

One of the most important and certainly the most spectacular example of laser pulse amplification is that of Nd:glass amplifiers used to produce pulses of high energy (10–100 kJ) for laser fusion research.[8] Very large Nd:glass laser systems have, in fact, been built and operated at a number of laboratories throughout the world, the one having the largest output energy being operated at the Lawrence Livermore National Laboratory in the USA (the NOVA laser). Most of these Nd:glass laser systems exploit the master-oscillator power-amplifier (MOPA) scheme. This scheme consists of a master oscillator, which generates a well controlled pulse of low energy, followed by a series of power amplifiers, which amplify the pulse to high energy. The clear aperture of the power amplifiers is increased along the chain to avoid optical damage as the beam energy increases. A schematic diagram of one of the ten arms of the NOVA system is shown in Fig. 12.3. The initial amplifiers in the chain consist of phosphate-glass rods (of 380 mm length and with a diameter of 25 mm for the first amplifiers, 50 mm for the last). The final stage of amplification is achieved via face-pumped disk amplifiers (see

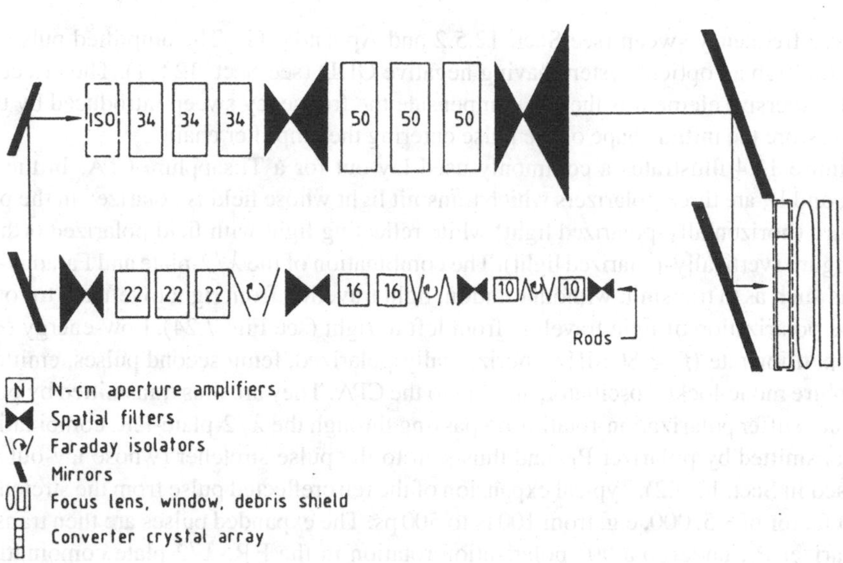

FIG. 12.3. Schematic layout of the amplification system, utilizing Nd:glass amplifiers, for one arm of the Nova system [after ref.,[8] by permission].

Fig. 6.3b) with large clear-aperture diameter (10 cm for the first amplifiers, 20 cm for the last). Note the presence in Fig. 12.3 of Faraday isolators (see Fig. 7.23) whose purpose is to avoid reflected light counter-propagating through the amplifier chain and thus damaging the initial stages of the system. Note also the presence of spatial filters consisting of two lenses, in a confocal arrangement (Fig. 11.10), with a pinhole at the common focus. These filters serve the double purpose of removing the small-scale spatial irregularities of the beam, as well as matching the beam profile between two consecutive amplifiers of different aperture. The laser system of Fig. 12.3 delivers an output energy of ~10 kJ in a pulse of duration down to 1 ns, which gives a total energy of the 10-arms NOVA system of ~100 kJ. Laser systems based on this layout concept and delivering an overall output energy of ~1 MJ are now being built in USA (National Ignition Facility, NIF, Livermore) and in France (Megajoule project, Limeil) [see also Sect. 9.2.2.2].

A second class of laser amplifiers which has revolutionized the laser field in terms of focusable beam intensity, relies on the Chirped Pulse Amplification (CPA) concept[9] and is used to amplify picosecond or femtosecond laser pulses. At such short pulse durations, in fact, the maximum energy which can be obtained from an amplifier depends on the onset, either of self-focusing, which is related to the beam peak power, or multi-photon-ionization, which is related to the beam peak intensity. To overcome these limitations, one can adopt a technique, already used in radar technology, of pulse expansion (or pulse-stretching), before amplification, followed by pulse compression, to its original shape, after the amplification process. In this way the peak power and hence the peak intensity of the pulse, in the amplifier chain, may be reduced by a few orders of magnitude (10^3–10^4). This allows a corresponding increase in the maximum energy which can be safely extracted from a given amplifier. Pulse expansion is achieved via an optical system which provides, e.g., a positive group-delay dispersion (GDD). In this way the pulse may be considerably expanded in time while acquiring a positive frequency sweep (see Sect. 12.5.2 and Appendix G). The amplified pulse is then passed through an optical system having negative GDD (see Sect. 12.5.1). The effect of this second dispersive element is then to compensate the frequency sweep introduced by the first and so restore the initial shape of the pulse entering the amplifier chain.

Figure 12.4 illustrates a commonly used lay-out for a Ti:sapphire CPA. In the figure, P_1, P_2, and P_3 are three polarizers which transmit light whose field is polarized in the plane of the figure (horizontally-polarized light) while reflecting light with field polarized orthogonal to the figure (vertically-polarized light). The combination of the $\lambda/2$-plate and Faraday-rotator (F.R.) is such as to transmit, without rotation, light traveling from right to left and to rotate, by 90°, the polarization of light traveling from left to right (see Fig. 7.24). Low-energy (~1 nJ), high-repetition rate ($f \cong 80$ MHz), horizontally-polarized, femtosecond pulses, emitted by a Ti:sapphire mode-locked oscillator, are sent to the CPA. They are thus transmitted by polarizer P_2, do not suffer polarization-rotation on passing through the $\lambda/2$-plate-F.R. combination, are then transmitted by polarizer P_1, and thus sent to the pulse stretcher (whose lay-out will be discussed in Sect. 12.5.2). Typical expansion of the retroreflected pulse from the stretcher may be by a factor of ~5,000, e.g. from 100 fs to 500 ps. The expanded pulses are then transmitted by polarizer P_1, undergo a 90° polarization rotation in the F.R.-$\lambda/2$-plate combination and are reflected by polarizer P_2. With the help of polarizer P_3, the expanded pulses are then injected into a so-called regenerative amplifier which consists of a Ti:sapphire amplifier and a Pockels cell (P.C.) located in a three-mirror (M_1, M_2, and M_3) folded resonator. The Pockels cell is oriented so as to produce a static $\lambda/4$ retardation. The cavity Q is thus low before the pulse arrival and the regenerative amplifier is below the oscillation threshold. In this situation, any injected pulse become horizontally polarized after a double passage through the P.C., and is thus transmitted by polarizer P_3 toward the Ti:sapphire amplifier. After returning from

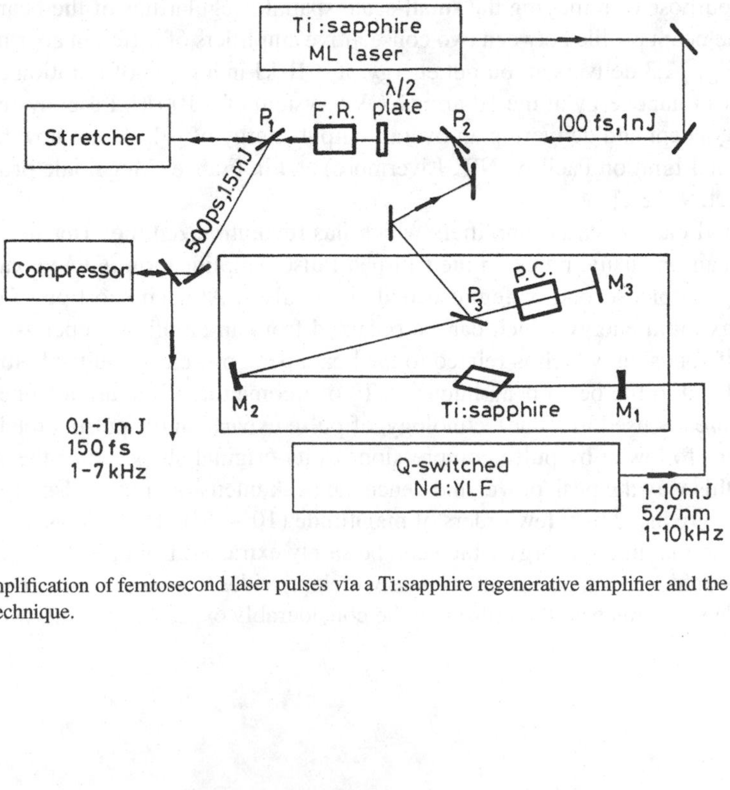

FIG. 12.4. Amplification of femtosecond laser pulses via a Ti:sapphire regenerative amplifier and the chirped-pulse-amplification technique.

the amplifier, the pulse is transmitted by polarizer P_3, and, after again double-passing the P.C., becomes vertically polarized and thus reflected out of the cavity by polarizer P_3. So, in this double transit through the regenerative amplifier, very little amplification is obtained for the output pulse. If however, while the pulse is between the polarizer and mirror M_1, a $\lambda/4$-voltage is applied to the P.C., the cell becomes equivalent to a $\lambda/2$-plate, and the pulse does not change its polarization state after each double passage through the cell. Therefore, the pulse gets trapped in the regenerative amplifier and, on each pass through the amplifying medium, it is amplified. After a suitable number of round-trips in the cavity (typically 15–20), the pulse energy reaches its maximum value and is then extracted from the cavity by applying an additional $\lambda/4$ voltage to the P.C.. In this case, in fact, after a double pass through the cell, the pulse becomes vertically polarized and is reflected by polarizer P_3 back in the direction of the incoming pulses. This, high-energy, vertically-polarized pulse is reflected by polarizer P_2, does not suffer polarization-rotation on passing through the $\lambda/2$-plate-F.R. combination, and is reflected by polarizer P_1 toward the pulse compressor (whose lay-out will be discussed in Sect. 12.5.1). The retroreflected beam from the compressor then consists of a train of high energy pulses, each with a duration approximately equal to that of the original pulses emitted by the oscillator, and with a repetition rate equal to that at which the Ti:sapphire amplifier is pumped (1–10 kHz, usually by the second-harmonic green-beam of a repetitively Q-switched Nd:YLF laser).

Systems of this type, exploiting the CPA technique, have allowed the development of lasers with ultra-high peak-power.(10) For instance, using Ti:sapphire active media, table-top CPA systems with peak power of ~20 TW have already been demonstrated while systems with peak powers approaching 100 TW (e.g., 2 J in a 20 fs pulse) are under construction. The largest peak power, so far achieved by exploiting the CPA technique, is actually ~1.25 PW (1 PW = 10^{15} W),(11) obtained using a chain of amplifiers taken from one arm of the NOVA laser (so as to obtain an amplified pulse with ~580 J energy and ~460 fs duration). The peak intensity obtained by focusing these ultra-high-power pulses is extremely high (10^{19}–10^{20} W/cm^2), representing an increase of four to five orders of magnitude compared to that available before introducing the CPA technique. When these ultra-high intensity beams interact with a solid target or with a gas, a highly ionized plasma is obtained and a completely new class of nonlinear optical phenomena is produced. Applications of these high intensities cover a broad area of science and technology including ultrafast x-ray and high-energy electron sources, as well as novel fusion concepts and plasma astrophysics.(12)

A third class of amplifier, widely used in optical fiber communications, is represented by the Er-doped optical-fiber amplifier (EDFA).(13) This amplifier is diode-pumped either in the 980 nm or 1480 nm pump bands of the Er^+ ion [see Fig. 9.4] and is used to amplify pulses at wavelengths corresponding to the so-called third transmission window of silica optical fibers ($\lambda \cong 1550$ nm). Since, usually, the pulse repetition rate of a communication system is very high (~ GHz) and the upper-state lifetime of Er^+ is very long (~10 ms, see Table 9.4) the saturation behavior of the Er^+ population is a cumulative result of many laser pulses, i.e., determined by the average beam intensity. The rate-equation treatment of this type of amplification can then be made in terms of average beam intensity and, in principle, is very simple. Complications however arise from several factors, namely: (1) The Er^+ system works on an almost pure three-level scheme (see Sect. 9.2.4) and therefore, the effective cross-sections of stimulated-emission and absorption, both covering a large spectral bandwidth, must be taken into account. (2) Transverse variation, within the fiber, of both the Er^+ population profile and the intensity profile of the propagating mode, must be taken into account. (3) Account must also be taken of the simultaneous presence of bidirectional noise arising from amplified spontaneous emission (ASE). We therefore make no attempt to cover this subject in any detail, limiting ourselves to pointing out that a vast literature exists,[14] that very high small-signal gain (up to ~50 dB), relatively large saturated average output powers (~100 mW), and low noise are achieved via these amplifiers. Thus, Erbium-doped fiber amplifiers must be considered a major break-through in the field of optical fiber communications, with applications regarding both long-haul systems as well as distribution networks.

12.4. FREQUENCY CONVERSION: SECOND-HARMONIC GENERATION AND PARAMETRIC OSCILLATION[1,15]

In classical linear optics one assumes that the induced dielectric polarization of a medium is linearly related to the applied electric field, i.e.,

$$\mathbf{P} = \varepsilon_0 \chi \mathbf{E} \tag{12.4.1}$$

where χ is the dielectric susceptibility. With the high electric fields involved in laser beams the above linear relation is no longer a good approximation, and further terms in which **P** is related to higher-order powers of **E** must also be considered. This nonlinear response can lead to an exchange of energy between e.m. waves at different frequencies.

In this section we will consider some of the effects produced by a nonlinear polarization term that is proportional to the square of the electric field. The two effects that we will consider are: (1) Second-harmonic generation (SHG), in which a laser beam at frequency ω is partially converted, in the nonlinear material, to a coherent beam at frequency 2ω [as first shown by Franken *et al.*[16]]. (2) Optical parameter oscillation (OPO), in which a laser beam at frequency ω_3 causes the simultaneous generation, in the nonlinear material, of two coherent beams at frequency ω_1 and ω_2 such that $\omega_1 + \omega_2 = \omega_3$ [as first shown by Giordmaine and Miller[17]]. With the high electric fields available in laser beams the conversion efficiency of both these processes can be very high (approaching 100% in SHG). Therefore, these techniques are increasingly used to generate new coherent waves at different frequencies from that of the incoming wave.

12.4.1. Physical Picture

We will first introduce some physical concepts using the simplifying assumption that the induced nonlinear polarization P^{NL} is related to the electric field E of the e.m. wave by a scalar equation, i.e.,

$$P^{NL} = 2\varepsilon_0 d E^2 \tag{12.4.2}$$

where d is a coefficient whose dimension is the inverse of an electric field.* The physical origin of Eq. (12.4.2) resides in the nonlinear deformation of the outer, loosely bound, electrons

* We use $2\varepsilon_0\, dE^2$ rather than dE^2 (as often used in other textbooks) to make d conform to increasingly accepted practice.

of an atom or atomic system when subjected to high electric fields. This is analogous to a breakdown of Hooke's law for an extended spring, resulting in the restoring force no longer being linearly dependent on the displacement from equilibrium. A comparison of Eqs. (12.4.2) and (12.4.1) shows that the nonlinear polarization term becomes comparable to the linear one for an electric field $E \cong \chi/d$. Since $\chi \cong 1$, we see that $(1/d)$ represents the field strength for which the linear and nonlinear terms become comparable. At this field strength, a sizable nonlinear deformation of the outer electrons must occur and $(1/d)$ is then expected to be of the order of the electric field that an electronic charge produces at a distance corresponding to a typical atomic dimension a, i.e., $(1/d) \cong e/4\pi\varepsilon_0 a^2$ [thus $(1/d) \cong 10^{11}$ V/m for $a \cong 0.1$ nm]. We note that d must be zero for a centrosymmetric material, such as a centrosymmetric crystal or the usual liquids and gases. For symmetry reasons, in fact, if we reverse the sign of E, the sign of the total polarization $P_t = P + P^{NL}$ must also reverse. Since, however, $P^{NL} \propto dE^2$, this can only occur if $d = 0$. From now on we will therefore confine ourselves to a consideration of non-centrosymmetric materials. We will see that the simple Eq. (12.4.2) is then able to account for both SHG and OPO.

12.4.1.1. Second-Harmonic Generation

We consider a monochromatic plane wave of frequency ω propagating along some direction, denoted as the z-direction, within a nonlinear crystal, the origin of the z-axis being taken at the entrance face of the crystal. For a plane wave of uniform intensity we can write the following expression for the electric field $E_\omega(z, t)$ of the wave

$$E_\omega(z, t) = (1/2)\{E(z, \omega)\exp[j(\omega t - k_\omega z)] + c.c.\} \tag{12.4.3}$$

In the above expression $c.c.$ means the complex conjugate of the other term appearing in the brackets and

$$k_\omega = \frac{\omega}{c_\omega} = \frac{n_\omega \omega}{c} \tag{12.4.4}$$

where c_ω is the phase velocity, in the crystal, of a wave of frequency ω, n_ω is the refractive index at this frequency, and c is the velocity of light *in vacuum*. Substitution of Eq. (12.4.3) into Eq. (12.4.2) shows that P^{NL} contains a term* oscillating at frequency 2ω, namely,

$$P_{2\omega}^{NL} = (\varepsilon_0 d/2)\{E^2(z, \omega)\exp[j(2\omega t - 2k_\omega z)] + c.c.\} \tag{12.4.5}$$

Equation (12.4.5) describes a polarization wave oscillating at frequency 2ω and with a propagation constant $2k_\omega$. This wave is then expected to radiate at frequency 2ω, i.e., to generate an e.m. wave at the second harmonic (SH) frequency 2ω. The analytical treatment, given later, involves in fact substitution of this polarization in the wave equation for the e.m. field. The radiated SH field can be written in the form

$$E_{2\omega}(z, t) = (1/2)\{E(z, 2\omega)\exp[j(2\omega t - k_{2\omega} z)] + c.c.\} \tag{12.4.6}$$

* The quantity P^{NL} also contains a term at frequency $\omega = 0$ which leads to development of a dc voltage across the crystal (optical rectification).

where

$$k_{2\omega} = \frac{2\omega}{c_{2\omega}} = \frac{2n_{2\omega}\omega}{c} \tag{12.4.7}$$

is the propagation constant of a wave at frequency 2ω. The physical origin of SHG can thus be traced back to the fact that, as a result of the nonlinear relation Eq. (12.4.2), the e.m. wave at the fundamental frequency ω will beat with itself to produce a polarization at 2ω. A comparison of Eq. (12.4.5) with Eq. (12.4.6) reveals a very important condition that must be satisfied if this process is to occur efficiently, viz., that the phase velocity of the polarization wave ($v_P = 2\omega/2k_\omega$) be equal to that of the generated e.m. wave ($v_E = 2\omega/k_{2\omega}$). This condition can thus be written as

$$k_{2\omega} = 2k_\omega \tag{12.4.8}$$

In fact, if this condition is not satisfied, the phase of the polarization wave at coordinate $z = l$ into the crystal, $2k_\omega l$, will be different from that, $k_{2\omega} l$, of the wave generated at $z = 0$ which has subsequently propagated to $z = l$. The difference in phase, $(2k_\omega - k_{2\omega})l$, would then increase with distance l and the generated wave, being driven by a polarization which does not have the appropriate phase, will then not grow cumulatively with distance l. Equation (12.4.8) is therefore referred to as the *phase-matching* condition. Note that, according to Eqs. (12.4.4) and (12.4.7), equation (12.4.8) implies that

$$n_{2\omega} = n_\omega \tag{12.4.9}$$

Now, if the polarization directions of E_ω and P^{NL} (and hence of $E_{2\omega}$) were indeed the same [as implied by Eq. (12.4.2)] condition Eq. (12.4.9) could not be satisfied owing to the dispersion ($\Delta n = n_{2\omega} - n_\omega$) of the crystal. This would then set a severe limit to the crystal length l_c over which P^{NL} can give contributions which keep adding cumulatively to form the second harmonic wave. This length l_c (*the coherence length*) must in fact correspond to the distance over which the polarization wave and the *SH* wave get out of phase with each other by an amount π. This means that $k_{2\omega} l_c - 2k_\omega l_c = \pi$, from which, with the help of Eqs. (12.4.4) and (12.4.7), one gets

$$l_c = \frac{\lambda}{4\Delta n} \tag{12.4.10}$$

where $\lambda = 2\pi c/\omega$ is the wavelength in *vacuum* of the fundamental wave. Taking, as an example, $\lambda \cong 1\ \mu\text{m}$ and $\Delta n = 10^{-2}$, we get $l_c = 25\ \mu\text{m}$. Note that, at this distance into the crystal, the polarization wave becomes 180° out of phase compared to the *SH* wave and the latter begins to decrease with increased distance rather than continuing to grow. Since, as seen in the previous example, l_c is usually very small, only a very small fraction of the incident power can then be transformed into the second harmonic wave.

At this point it is worth pointing out another useful way of visualizing the SHG process, in terms of photons rather than fields. First we write the relation between the frequency of the fundamental (ω) and second-harmonic (ω_{SH}) wave, viz.,

$$\omega_{SH} = 2\omega \tag{12.4.11}$$

If we now multiply both sides of Eqs. (12.4.11) and (12.4.8) by $\hbar$, we get

$$\hbar\omega_{SH} = 2\hbar\omega \tag{12.4.12a}$$

$$\hbar k_{2\omega} = 2\hbar k_{\omega} \tag{12.4.12b}$$

respectively. For energy to be conserved in the SHG process, we must have $dI_{2\omega}/dz = -dI_{\omega}/dz$, where $I_{2\omega}$ and I_{ω} are the intensities of the waves at the two frequencies. With the help of Eq. (12.4.12a) we get $dF_{2\omega}/dz = -(1/2)dF_{\omega}/dz$, where $F_{2\omega}$ and F_{ω} are the photon fluxes of the two waves. From this equation we can then say that, whenever, in the SHG process, one photon at frequency 2ω is produced, correspondingly two photons at frequency ω disappear. Thus the relation Eq. (12.4.12a) can be regarded as a statement of conservation of photon energy. Remembering that $\hbar k$ is the photon momentum, Eq. (12.4.12b) is then seen to correspond to the requirement that photon momentum is also conserved in the process.

We now reconsider the phase-matching condition Eq. (12.4.9) to see how it can be satisfied in a suitable, optically anisotropic, crystal.(18,19) To understand this we will first need to make a small digression to explain the propagation behavior of waves in an anisotropic crystal, and also to show how the simple nonlinear relation Eq. (12.4.2) should be generalized for anisotropic media.

In an anisotropic crystal it can be shown that, for a given direction of propagation, there are two linearly polarized plane waves that can propagate with different phase velocities. Corresponding to these two different polarizations one can then associate two different refractive indices, the difference of refractive index being referred to as the crystal's birefringence. This behavior is usually described in terms of the so-called index ellipsoid which, for a uniaxial crystal, is an ellipsoid of revolution around the optic axis (the z axis of Fig. 12.5). Given this ellipsoid, the two allowed directions of linear polarization and their corresponding refractive indices are found as follows: Through the center of the ellipsoid one draws a line in the direction of beam propagation (line OP of Fig. 12.5) and a plane perpendicular to this line. The intersection of this plane with the ellipsoid is an ellipse. The direction of the two axes of the ellipse then give the two polarization directions and the length of each semiaxis gives the refractive index corresponding to that polarization. One of these directions is necessarily perpendicular to the optic axis and the wave having this polarization is referred to as the ordinary wave. Its refractive index, n_o, can be seen from the figure to be independent of the direction of propagation. The wave with the other direction of polarization is referred to as the extraordinary wave and the corresponding index, $n_e(\theta)$, depends of the angle θ and ranges in value from that of the ordinary wave n_0 (when OP is parallel to z) to the value n_e, referred to as the extraordinary index, which occurs when OP is perpendicular to z. Note now that one defines a positive uniaxial crystal as corresponding to the case $n_e > n_o$ while a negative uniaxial crystal corresponds to the case $n_e < n_o$. An equivalent way to describe wave propagation is through the so-called normal (index) surfaces for the ordinary and extraordinary waves (Fig. 12.6). In this case, for a given direction of propagation OP and, for either ordinary or extraordinary waves, the length of the segment between the origin O and the point of interception of the ray OP with the surface gives the refractive index of that wave. The normal surface for the ordinary wave is thus a sphere, while the normal surface for the extraordinary wave is an ellipsoid of revolution around the z axis. In Fig. 12.6 the intersections of these two normal surfaces with the y-z plane are indicated for the case of a positive uniaxial crystal.

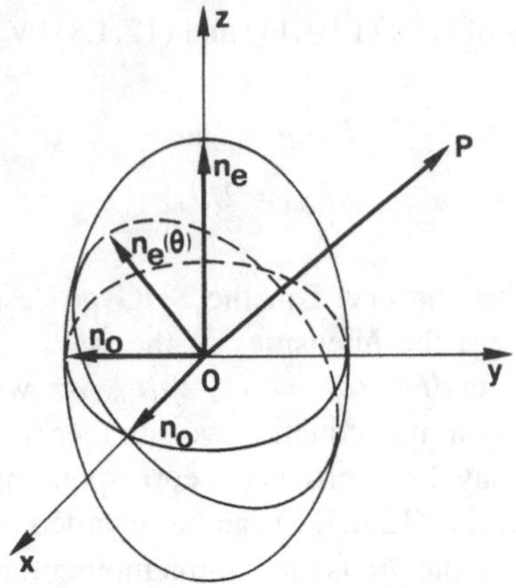

FIG. 12.5. Index ellipsoid for a positive uniaxial crystal.

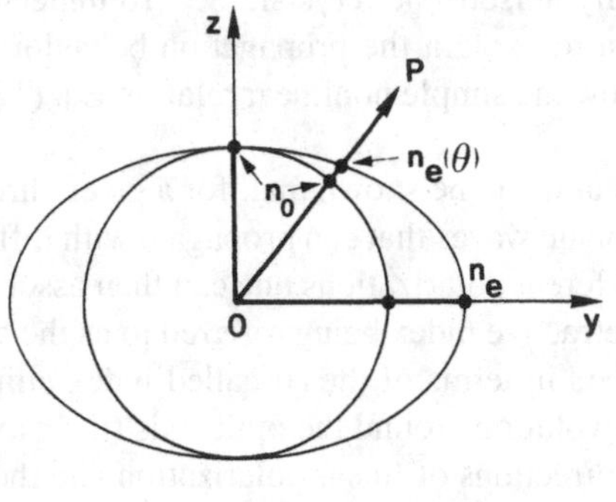

FIG. 12.6. Normal (index) surface for both the ordinary and extraordinary waves (for a positive uniaxial crystal).

After this brief discussion of wave propagation in anisotropic crystals, we now return to the problem of the induced nonlinear polarization. In general, in an anisotropic medium, the scalar relation Eq. (12.4.2) does not hold and a tensor relation needs to be introduced. First, we write the electric field $\mathbf{E}^{\omega}(\mathbf{r},\ t)$ of the e.m. wave at frequency ω and at a given point $\mathbf{r}$ and the nonlinear polarization vector at frequency 2ω, $\mathbf{P}^{2\omega}_{NL}(\mathbf{r},\ t)$ in the form

$$\mathbf{E}^{\omega}(\mathbf{r}, t) = (1/2)[\mathbf{E}^{\omega}(\mathbf{r}, \omega)\exp(j\omega t) + c.c.] \tag{12.4.13a}$$

$$\mathbf{P}^{2\omega}_{NL}(\mathbf{r}, t) = (1/2)[\mathbf{P}^{2\omega}(\mathbf{r}, 2\omega)\exp(2j\omega t) + c.c.] \tag{12.4.13b}$$

A tensor relation can then be established between $\mathbf{P}^{2\omega}$ $(\mathbf{r},\ 2\omega)$ and $\mathbf{E}^{\omega}(\mathbf{r},\ \omega)$. In fact, the second harmonic polarization component, along, e.g., the i-direction of the crystal, can be written as

$$P_i^{2\omega} = \sum_{j,k=1,2,3} \varepsilon_0 d_{ijk}^{2\omega} E_j^{\omega} E_k^{\omega} \tag{12.4.14}$$

Note that Eq. (12.4.14) is often written in condensed notation as

$$P_i^{2\omega} = \sum_1^6 {}_m\varepsilon_0 d_{im}^{2\omega} (EE)_m \tag{12.4.15}$$

where m runs from one to six. The abbreviated field notation is that $(EE)_1 \equiv E_1^2 \equiv E_x^2, \equiv (EE)_2 \equiv E_2^2 \equiv E_y^2$, $(EE)_3 \equiv E_3^2 \equiv E_z^2$, $(EE)_4 \equiv 2E_2E_3 \equiv 2E_yE_z$, $(EE)_5 \equiv 2E_1E_3 \equiv 2E_xE_z$, and $(EE)_6 \equiv 2E_1E_2 \equiv 2E_xE_y$, where both the 1, 2, 3 and the x, y, z notation for axes have been indicated. Note that, expressed in matrix form, d_{im} is a 3×6 matrix that operates on the column vector $(EE)_m$. Depending on the crystal symmetry, some of the values of the d_{im} matrix may be equal and some may be zero. For the $\bar{4}2m$ point group symmetry, which includes the important nonlinear crystals of the KDP type and the chalcopyrite semiconductors, only d_{14}, d_{25}, and d_{36} are non-zero and these three d coefficients, are themselves equal. Therefore only one coefficient, e.g., d_{36}, needs to be specified, and one can write

$$P_x = 2\varepsilon_0 \, d_{36} \, E_yE_z \tag{12.4.16a}$$

$$P_y = 2\varepsilon_0 \, d_{36} \, E_zE_x \tag{12.4.16b}$$

$$P_z = 2\varepsilon_0 \, d_{36} \, E_xE_y \tag{12.4.16c}$$

where the z-axis is again taken along the optic axis of the uniaxial crystal. The nonlinear optical coefficients, the symmetry class, the transparency range, and the damage threshold of some selected nonlinear materials are indicated in Table 12.1. Except for cadmium germanium arsenate and $AgGaSe_2$, which are commonly used around the 10μm range, all the other crystals listed are used in the near UV to near IR range. The table includes the more recent crystals of KTP (potassium titanyl phosphate), and BBO (beta-barium borate), the former being commonly used for second harmonic generation at, e.g., the Nd:YAG wavelength. The nonlinear d-coefficients are normalized to that of KDP, whose actual value is $d_{36} \cong 0.5 \times 10^{-12}$ m/V.

Following this digression on the properties of anisotropic media we can now go on to show how phase matching can be achieved for the particular case of a crystal of $\bar{4}2m$ point group symmetry. From Eq. (12.4.16) we note that, if $E_z = 0$, only P_z will be non-vanishing and will thus tend to generate a second-harmonic wave with a non-zero z-component. We recall (see Fig. 12.5) that a wave with $E_z = 0$ is an ordinary wave while a wave with $E_z \neq 0$ is an extraordinary wave. Thus an ordinary wave at the fundamental frequency ω tends, in this case, to generate an extraordinary wave at 2ω. To satisfy the phase-matching condition one can then propagate the fundamental wave at an angle θ_m to the optic axis, in such a way that

$$n_e(2\omega, \theta_m) = n_o(\omega) \tag{12.4.17}$$

This can be better understood with the help of Fig. 12.7 which shows the intercepts of the normal surfaces $n_o(\omega)$ and $n_e(2\omega, \theta)$ with the plane containing the z axis and the propagation direction. Note that, since crystals usually show a normal dispersion, one has $n_o(\omega) < n_o(2\omega)$, while for a negative uniaxial crystal one has $n_e(2\omega) < n_o(2\omega)$, where, as a short-hand notation (see Fig. 12.7), we have set $n_e(2\omega) \equiv n_e(2\omega, 90°)$ and $n_o(2\omega) \equiv n_e(2\omega, 0)$. Thus the ordinary circle, corresponding to the wave at frequency, intersects the extraordinary ellipse,

TABLE 12.1. Nonlinear optical coefficients for selected materials

Material	Formula	Nonlinear d coefficient (relative to KDP)	Symmetry class	Transparency range (μm)	Damage threshold (GW/cm^2)
KDP	KH_2PO_4	$d_{36} = d_{14} = 1$	$\bar{4}2m$	0.22–1.5	0.2
KD*P	KD_2PO_4	$d_{36} = d_{14} = 0.92$	$\bar{4}2m$	0.22–1.5	0.2
ADP	$NH_4H_2PO_4$	$d_{36} = d_{14} = 1.2$	$\bar{4}2m$	0.2–1.2	0.5
CDA	CsH_2AsO_4	$d_{36} = d_{14} = 0.92$	$\bar{4}2m$	0.26–1.4	0.5
Lithium iodate	$LiIO_3$	$d_{31} = d_{32} = d_{24}$ $d_{15} = 12.7$	6	0.3–5.5	0.5
Lithium niobate	$LiNbO_3$	$d_{31} = 12.5$ $d_{22} = 6.35$	$3m$	0.4–5	0.05
KTP	$KTiOPO_4$	$d_{31} = 13$ $d_{32} = 10$ $d_{33} = 27.4$ $d_{24} = 15.2$ $d_{15} = 12.2$	$mm2$	0.35–4.5	1
BBO	$\beta - BaB_2O_4$	$d_{22} = 4.1$	$3m$	0.19–3	5
Cadmium germanium arsenide	$CdGeAs_2$	$d_{36} = d_{14} = 538$	$\bar{4}2m$	2.4–20	0.04
Silver-gallium selenide	$AgGaSe_2$	$d_{36} = d_{14} = 66$	$\bar{4}2m$	0.73–17	0.05

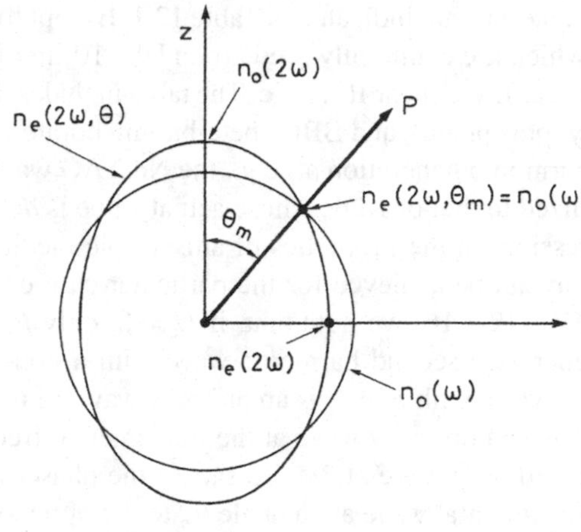

FIG. 12.7. Phase-matching angle θ_m for type I second-harmonic generation in a negative uniaxial crystal.

corresponding to the wave at frequency 2ω, at some angle θ_m.* For light propagating at this angle θ_m to the optic axis (i.e., for all ray directions lying in a cone around the z axis, with cone angle θ_m), Eq. (12.4.17) is satisfied and hence the phase-matching condition is satisfied.

* It should be noted that for this intersection to occur at all it is necessary for $n_e(2\omega, 90^\circ)$ to be less than $n_o(\omega)$, otherwise the ellipse for $n_e(2\omega)$ (see Fig 12.7) will lie wholly outside the circle for $n_o(\omega)$. Thus $n_e(2\omega, 90^\circ) = n_e(2\omega) < n_o(\omega) < n_o(2\omega)$, which shows that crystal birefringence $n_o(2\omega) - n_e(2\omega)$ must be larger than crystal dispersion $n_o(2\omega) - n_o(\omega)$.

It should be noted that, if $\theta_m \neq 90°$, the phenomenon of double refraction will occur in the crystal, i.e., the direction of the energy flow for the extraordinary (SH) beam will be at an angle slightly different from θ_m. Thus the fundamental and SH beams will travel in slightly different directions (although satisfying the phase-matching condition). For a fundamental beam of finite transverse dimensions this will put an upper limit on the interaction length in the crystal. This limitation can be overcome if it is possible to operate with $\theta_m = 90°$, i.e., $n_e\,(2\omega,\ 90°) = n_o(\omega)$. This is referred to as the 90° phase matching condition. Since n_e and n_o generally undergo different changes with temperature, it turns out that 90° phase matching condition can, in some cases, be reached by changing the crystal temperature. To summarize the above discussion, we can say that phase matching is possible in a (sufficiently birefringent) negative uniaxial crystal when an ordinary ray at ω [E_x beam of Eq. (12.4.16c)] combines with an ordinary ray at ω [E_y beam of Eq. (12.4.16c)] to give an extraordinary ray at 2ω, or, in symbols, $o_\omega + o_\omega \rightarrow e_{2\omega}$. This is referred to as *type I second-harmonic generation*. In a negative uniaxial crystal another scheme for phase-matched SHG, called type II, is also possible. In this case an ordinary wave at ω combines with an extraordinary wave ω at to give an extraordinary wave at 2ω, or, in symbols, $o_\omega + e_\omega \rightarrow e_{2\omega}$.*

Example 12.3. *Calculation of the phase-matching angle for a negative uniaxial crystal.* With reference to Fig. 12.7, we label the horizontal-axis as the y-axis. If we then let z and y represent the cartesian coordinates of general point of the ellipse describing the extraordinary index, $n_e(2\omega,\ \theta)$, one can write

$$\frac{z^2}{\left(n_2^o\right)^2} + \frac{y^2}{\left(n_2^e\right)^2} = 1$$

where, as a short-hand notation, we have set $n_2^o = n_o(2\omega)$ and and $n_2^e = n_e(2\omega)$. If the coordinates z and y are now expressed as a function of $n_e(2\omega,\ \theta)$ and of the angle θ, the previous equation transforms to

$$\frac{[n_e(2\omega,\ \theta)]^2}{\left(n_2^o\right)^2}\cos^2\theta + \frac{[n_e(2\omega,\ \theta)]^2}{\left(n_2^e\right)^2}\sin^2\theta = 1$$

For $\theta = \theta_m$, Eq. (12.4.17) must hold. Substitution of this equation into the equation above then gives

$$\left(\frac{n_1^o}{n_2^o}\right)^2 (1 - \sin^2\theta_m) + \left(\frac{n_1^o}{n_2^e}\right)^2 \sin^2\theta_m = 1$$

where, again as a short-hand notation we have set $n_1^o = n_o(\omega)$. This last equation can be solved for $\sin^2\theta_m$ to obtain

$$\sin^2\theta_m = \frac{1 - \left(\frac{n_1^o}{n_2^o}\right)^2}{\left(\frac{n_1^o}{n_2^e}\right)^2 - \left(\frac{n_1^o}{n_2^o}\right)^2} = \frac{\left(\frac{n_2^o}{n_1^o}\right)^2 - 1}{\left(\frac{n_2^o}{n_2^e}\right)^2 - 1}$$

Second-harmonic generation is currently used to provide coherent sources at new wavelengths. The nonlinear crystal may be placed either outside or inside the cavity of the laser producing the fundamental beam. In the latter case one takes advantage of the greater e.m. field strength inside the resonator to increase the conversion efficiency. Very high conversion efficiencies (approaching 100%) have been obtained with both arrangements. Among the most frequent applications of SHG are frequency doubling the output of a Nd:YAG laser (thus producing a green beam, $\lambda = 532$ nm, from an infrared one, $\lambda = 1.064\ \mu$m) and generation of tunable UV radiation (down to $\lambda \cong 205$ nm) by frequency doubling a tunable dye laser. In both of these cases either cw

* More generally, interactions in which the polarizations of the two fundamental waves are the same are termed type I (e.g., also $e_\omega + e_\omega \rightarrow o_{2\omega}$), and interactions in which the polarization of the fundamental waves are orthogonal are termed type II.

or pulsed laser sources are used. The nonlinear crystals most commonly used as frequency doublers for Nd:YAG lasers are KTP and $\beta - BaB_2O_4$ (BBO), while BBO, due to its more extended transparency toward the UV, is particularly used when a SH beam at UV wavelengths down to ~200 nm have to be generated. Efficient frequency conversion of infrared radiation from CO_2 or CO lasers is often produced in chalcopyrite semiconductors (e.g., $CdGeAs_2$).

12.4.1.2. Parametric Oscillation

We now go on to discuss the process of parametric oscillation. We begin by noticing that the previous ideas introduced in the context of SHG can be readily extended to the case of two incoming waves at frequencies ω_1 and ω_2 combining to give a wave at frequency $\omega_3 = \omega_1 + \omega_2$ (sum-frequency generation). Harmonic generation can in fact be thought of as a limiting case of sum-frequency generation with $\omega_1 = \omega_2$ and $\omega_3 = 2\omega$. The physical picture is again very similar to the SHG case: By virtue of the nonlinear relation Eq. (12.4.2) between P^{NL} and the total field $E[E = E_{\omega_1}(z, t) + E_{\omega_2}(z, t)]$, the wave at ω_1 will beat with that at ω_2, to give a polarization component at $\omega_3 = \omega_1 + \omega_2$. This will then radiate an e.m. wave at ω_3. Thus for sum-frequency generation we can write

$$\hbar\omega_1 + \hbar\omega_2 = \hbar\omega_3 \tag{12.4.18a}$$

which, according to a description in terms of photons rather than fields, implies that one photon at ω_1 and one photon at ω_2 disappear while a photon at ω_3 is created. We therefore expect the photon momentum to be also conserved in the process, i.e.,

$$\hbar\,\mathbf{k}_1 + \hbar\,\mathbf{k}_2 = \hbar\,\mathbf{k}_3 \tag{12.4.18b}$$

where the relationship is put in its general form, with the $\mathbf{k}$ denoted by vectors. Equation (12.4.18b), which expresses the phase-matching condition for sum-frequency generation, can be seen to be a straightforward generalization of that for SHG [compare with Eq. (12.4.12b)].

Optical parametric generation is in fact just the reverse of sum-frequency generation. Here, instead, a wave at frequency ω_3 (the pump frequency) generates two waves (called the idler and signal waves) at frequencies ω_1 and ω_2, in such a way that the total photon energy and momentum is conserved, i.e.,

$$\hbar\omega_3 = \hbar\omega_1 + \hbar\omega_2 \tag{12.4.19a}$$

$$\hbar\,\boldsymbol{k}_3 = \hbar\,\boldsymbol{k}_1 + \hbar\,\boldsymbol{k}_2 \tag{12.4.19b}$$

The physical process occurring in this case can be visualized as follows. Imagine first that a strong wave at ω_3 and a weak wave at ω_1 are both present in the nonlinear crystal. As a result of the nonlinear relation Eq. (12.4.2), the wave at ω_3 will beat with the wave at ω_1 to give a polarization component at $\omega_3 - \omega_1 = \omega_2$. If the phase-matching condition Eq. (12.4.19b) is satisfied, a wave at ω_2 will thus build up as it travels through the crystal. Then the total E field will in fact be the sum of three fields $[E = E_{\omega_1}(z,t) + E_{\omega_2}(z,t) + E_{\omega_3}(z,t)]$ and the wave at ω_2 will in turn beat with the wave at ω_3 to give a polarization component at

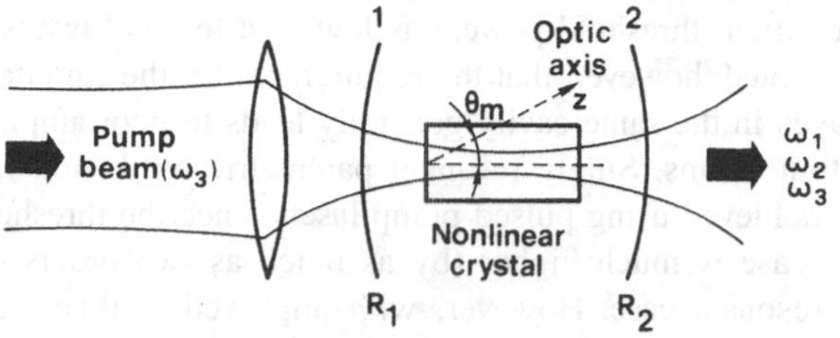

FIG. 12.8. Schematic diagram of an optical parametric oscillator.

$\omega_3 - \omega_2 = \omega_1$. This polarization will cause the ω_1 wave to grow also. Thus power will be transferred from the beam at ω_3 to those at ω_1 and ω_2, and the weak wave at ω_1 which was assumed to be initially present will be amplified. From this picture one sees a fundamental difference between parametric generation and SHG. In the latter case only a strong beam at the fundamental frequency is needed for the SHG process to occur. In the former case, however, a weak beam at ω_1 is also needed and the system behaves like an amplifier at frequency ω_1 (and ω_2). In practice, however, the weak beam need not be supplied by an external source (such as another laser) since it is generated, internally to the crystal, as a form of noise (so-called parametric noise). One can then generate coherent beams from this noise in a way analogous to that used in a laser oscillator. Thus, the nonlinear crystal, which is pumped by an appropriately focused pump beam, is placed in an optical resonator (Fig. 12.8). The two mirrors (1 and 2) of this parametric oscillator have high reflectivity (e.g., $R_1 = 1$ and $R_2 \cong 1$) either at ω_1 only (singly resonant oscillator, SRO) or at both ω_1 and ω_2 (doubly resonant oscillator, DRO). The mirrors are ideally transparent to the pump beam. Oscillation will start when the gain arising from the parametric effect just exceeds the losses of the optical resonator. Some threshold power of the input pump beam is therefore required before oscillation will begin. When this threshold is reached, oscillation occurs at both ω_1 and ω_2, and the particular pair of values of ω_1 and ω_2 is determined by the two Eq. (12.4.19). For instance, with type I phase matching involving an extraordinary wave at ω_3 and ordinary waves at ω_1 and ω_2 (i.e., $e_{\omega_3} \to o_{\omega_1} + o_{\omega_2}$), Eq. (12.4.19b) would give

$$\omega_3 n_e(\omega_3, \theta) = \omega_1 n_o(\omega_1) + \omega_2 n_o(\omega_2) \tag{12.4.20}$$

For a given θ, i.e., for a given inclination of the nonlinear crystal with respect to the cavity axis, Eq. (12.4.20) provides a relation between ω_1 and ω_2 which, together with the relation Eq. (12.4.19a), determines the values of both ω_1 and ω_2. Phase-matching schemes of both type I and type II (e.g., $e_{\omega 3} \to o_{\omega_1} + e_{\omega 2}$ for a negative uniaxial crystal) are possible and tuning can be achieved by changing either the crystal inclination (angle tuning) or its temperature (temperature tuning). As a final comment, we note that, if the gain from the parametric effect is large enough, one can dispense with the mirrors altogether, and an intense emission at ω_1 and ω_2, grows from parametric noise in a single pass through the crystal. This behavior is often referred to as superfluorescent parametric emission and such a device is referred to as an optical parametric generator (OPG).

Singly resonant and doubly resonant optical parametric oscillators have both been used. Doubly resonant parametric oscillation has been achieved with both c.w. and pulsed pump lasers. For c.w. excitation, threshold powers as low as a few milliwatts have been demonstrated. It should be noted, however, that the requirement for the simultaneous resonance of both parametric waves in the same cavity generally leads to poor amplitude and frequency stability of the output beams. Singly resonant parametric oscillation had, until relatively recently, only been achieved using pulsed pump lasers since the threshold pump power for the singly resonant case is much higher (by as much as two orders of magnitude) than that of the doubly resonant case. However, with improved nonlinear crystals, c.w. oscillation is now readily achieved. Singly resonant oscillators produce a much more stable output and, for this reason, is the most frequently used. Optical parametric oscillators producing coherent radiation from the visible to the near infrared (0.5–5 μm) are now well developed, with the most successful devices being based on BBO, LBO and lithium niobate ($LiNbO_3$). Optical parametric oscillators can also generate coherent radiation at longer infrared wavelengths (to $\sim$14 μm) using crystals such as silver-gallium selenide ($AgGaSe_2$) and cadmium selenide (CdSe). Synchronous pumping of OPOs, using a mode-locked pump, is also proving very attractive as a means of generating short pulses with very wide tunability. A notable feature of these devices is that their gain is determined by the peak power of the pump pulse, so that thresholds corresponding to very low average powers (a few milliwatts) can be achieved even for a singly resonant oscillator. It should last be observed that the efficiency of an OPO can be very high, approaching the theoretical 100% photon efficiency.

12.4.2. Analytical Treatment

To arrive at an analytical description of both SHG and parametric processes, we need to see how the nonlinear polarization [e.g., Eq. (12.4.2)], which acts as the source term to drive the generated waves, is introduced into the wave equation. The fields within the material obey Maxwell's equations

$$\nabla \times \mathbf{E} = -\frac{\partial \mathbf{B}}{\partial t} \tag{12.4.21a}$$

$$\nabla \times \mathbf{H} = \mathbf{J} + \frac{\partial \mathbf{D}}{\partial t} \tag{12.4.21b}$$

$$\nabla \cdot \mathbf{D} = \rho \tag{12.4.21c}$$

$$\nabla \cdot \mathbf{B} = 0 \tag{12.4.21d}$$

where ρ is the free-charge density. For the media of interest here we can assume the magnetization $\mathbf{M}$ to be zero; thus

$$\mathbf{B} = \mu_0 \mathbf{H} + \mu_0 \mathbf{M} = \mu_0 \mathbf{H} \tag{12.4.22}$$

Losses within the material (e.g., scattering losses) can be simulated by the introduction of a fictitious conductivity σ_s such that

$$\mathbf{J} = \sigma_s \mathbf{E} \tag{12.4.23}$$

Finally we can write

$$\mathbf{D} = \varepsilon_0 \mathbf{E} + \mathbf{P}^L + \mathbf{P}^{\mathrm{NL}} = \varepsilon \mathbf{E} + \mathbf{P}^{\mathrm{NL}} \tag{12.4.24}$$

where $\mathbf{P}^L$ is the linear polarization of the medium and is taken account, in the usual way, by introducing the dielectric constant ε. Upon applying the $\nabla\times$ operator to both sides of Eq. (12.4.21a), interchanging the order of $\nabla\times$ and $\partial/\partial t$ operators on the right-hand side of the resulting equation, and making use of Eqs. (12.4.22), (12.4.21b), (12.4.23), and (12.4.24), we obtain

$$\nabla\times\nabla\times\mathbf{E} = -\mu_0\left(\sigma_s\frac{\partial\mathbf{E}}{\partial t} + \varepsilon\frac{\partial^2\mathbf{E}}{\partial t^2} + \frac{\partial^2\mathbf{P}^{\mathrm{NL}}}{\partial t^2}\right) \tag{12.4.25}$$

Using the identity $\nabla\times\nabla\times\mathbf{E} = \nabla(\nabla\cdot\mathbf{E}) - \nabla^2\mathbf{E}$ and under the assumption that $\nabla\cdot\mathbf{E} \cong 0$, we get from Eq. (12.4.25)

$$\nabla^2\mathbf{E} - \frac{\sigma_s}{\varepsilon c^2}\frac{\partial\mathbf{E}}{\partial t} - \frac{1}{c^2}\frac{\partial^2\mathbf{E}}{\partial t^2} = \frac{1}{\varepsilon c^2}\frac{\partial^2\mathbf{P}^{\mathrm{NL}}}{\partial t^2} \tag{12.4.26}$$

where $c = (\varepsilon\mu_0)^{-1/2}$ is the phase velocity in the material. Equation (12.4.26) is the wave equation with the nonlinear polarization term included. Note that the linear part of the medium polarization has been transferred to the left-hand side of Eq. (12.4.26) and its effect is included in the dielectric constant. The nonlinear part $\mathbf{P}^{\mathrm{NL}}$ has been kept on the right-hand side since it will be shown to act as a source term for the waves being generated at new frequencies as well as a loss term for the incoming wave. Confining ourselves to the simple scalar case of plane waves propagating along the z-direction, one sees that Eq. (12.4.26) reduces to

$$\frac{\partial^2 E}{\partial z^2} - \frac{\sigma_s}{\varepsilon c^2}\frac{\partial E}{\partial t} - \frac{1}{c^2}\frac{\partial^2 E}{\partial t^2} = \frac{1}{\varepsilon c^2}\frac{\partial^2 P^{\mathrm{NL}}}{\partial t^2} \tag{12.4.26a}$$

We now write the field of the wave at frequency ω_i as

$$E_{\omega_i}(z,t) = (1/2)\,\{E_i(z)\exp[j(\omega_i t - k_i z)] + c.c.\} \tag{12.4.27a}$$

where E_i is taken to be complex in general. Likewise, the amplitude of the nonlinear polarization at frequency ω_i will be written as

$$P^{\mathrm{NL}}_{\omega_i} = (1/2)\,\{P^{\mathrm{NL}}_i(z)\exp[j(\omega_i t - k_i z)] + c.c.\} \tag{12.4.27b}$$

Since Eq. (12.4.26a) must hold separately for each frequency component of the waves present in the medium, Eqs. (12.4.27a) and (12.4.27b) can be substituted into the left- and right-hand sides of Eq. (12.4.26a), respectively. Within the slowly varying amplitude approximation, we can neglect the second derivative of $E_i(z)$, i.e., assume that $d^2E_i/dz^2 \ll k_i\,(\mathrm{d}E_i/dz)$. Equation (12.4.26a) then yields

$$2\frac{dE_i}{dz} + \frac{\sigma_i}{n_i\varepsilon_0 c}E_i = -j\left(\frac{\omega_i}{n_i\varepsilon_0 c}\right)P^{\mathrm{NL}}_i \tag{12.4.28}$$

where σ_i, n_i, and ε_i are, respectively, the loss, the refractive index, and the dielectric constant of the medium at frequency ω_i, and where use has been made of the relations $k_i = n_i\omega_i/c$ and $\varepsilon_i = n_i^2\varepsilon_0$.

Equation (12.4.28) is the basic equation that will be used in the next sections. Note that it has been obtained subject to the assumption of a scalar relation between $\mathbf{P}^{NL}$ and $\mathbf{E}$ [see Eq. (12.4.2)]. This assumption is not correct, and actually a tensor relation should be used [see Eq. (12.4.15)]. However, it can be shown that one can still use this scalar equation provided that E_i now refers to the field component along an appropriate axis and an effective coefficient, d_{eff}, is substituted for d in Eq. (12.4.2). In general, d_{eff} is a combination of one or more of the d_{im} coefficients appearing in Eq. (12.4.15) multiplied by appropriate trigonometric functions of the angles θ and ϕ that define the direction of wave propagation in the crystal[20] (θ is the angle between the propagation vector and the z axis while ϕ is the angle that the projection of the propagation vector in the x-y plane makes with the x axis of the crystal). For example, for a crystal of $\bar{4}2m$ point group symmetry and for type I phase matching one obtains $d_{eff} = d_{36}$ sin 2ϕ sin θ. As a short-hand notation, however, we will still retain the symbol d in Eq. (12.4.2) while bearing in mind that it means the effective value of the d coefficient, d_{eff}.

12.4.2.1. Parametric Oscillation

We now consider three waves at frequencies ω_1, ω_2 and ω_3 [where $\omega_3 = \omega_1 + \omega_2$] interacting in the crystal. Thus we write the total field $E(z, t)$ as

$$E(z, t) = E_{\omega_1}(z, t) + E_{\omega_2}(z, t) + E_{\omega_3}(z, t) \tag{12.4.29}$$

where each of the fields can be written in the form of Eq. (12.4.27a). Upon substituting Eq. (12.4.29) into Eq. (12.4.2) and using Eq. (12.4.27a) we obtain an expression for the components $P_i^{NL}(z)$ [as defined by Eq. (12.4.27b)] of the nonlinear polarization at the frequency ω_i. After some lengthy but straightforward algebra we find that, for instance, the component P_1^{NL} at frequency ω_1 is given by

$$P_1^{NL} = 2\varepsilon_0 d E_3(z) E_2^*(z) \exp[j(k_1 + k_2 - k_3)z] \tag{12.4.30}$$

The components of P^{NL} at ω_2 and ω_3 are obtained in a similar way. For each of the three frequencies, the field equation is then obtained by substituting into Eq. (12.4.28) the expression for P^{NL} corresponding to the appropriate frequency. We thus arrive at the following three equations:

$$\frac{dE_1}{dz} = -\left(\frac{\sigma_1}{2n_1\varepsilon_0 c}\right)E_1 - j\left(\frac{\omega_1}{n_1 c}\right) dE_3E_2^* \exp[-j(k_3 - k_2 - k_1)z] \tag{12.4.31a}$$

$$\frac{dE_2}{dz} = -\left(\frac{\sigma_2}{2n_2\varepsilon_0 c}\right)E_2 - j\left(\frac{\omega_2}{n_2 c}\right) dE_3E_1^* \exp[-j(k_3 - k_1 - k_2)z] \tag{12.4.31b}$$

$$\frac{dE_3}{dz} = -\left(\frac{\sigma_3}{2n_3\varepsilon_0 c}\right)E_3 - j\left(\frac{\omega_3}{n_3 c}\right) dE_1E_2 \exp[-j(k_1 - k_2 - k_3)z] \tag{12.4.31c}$$

These are the basic equations describing the nonlinear parametric interaction. One sees that they are coupled to each other via the nonlinear coefficient d.

It is convenient at this point to define new field variables A_i as

$$A_i = (n_i/\omega_i)^{1/2}/E_i \tag{12.4.32}$$

Since the intensity of the wave is $I_i = n_i\varepsilon_0 c|E_i|^2/2$, the corresponding photon flux F_i will be given by $F_i = I_i/\hbar\omega_i = (\varepsilon_0 c/2\hbar)|A_i|^2$. Thus $|A_i|^2$ is seen to be proportional to the photon flux F_i with a proportionality constant independent of both n_i and ω_i. When reexpressed in terms of these new field variables, equations (12.4.31) transform to

$$\frac{dA_1}{dz} = -\frac{\alpha_1 A_1}{2} - j\delta\, A_3 A_2^* \exp[-j(\Delta kz)] \tag{12.4.33a}$$

$$\frac{dA_2}{dz} = -\frac{\alpha_2 A_2}{2} - j\delta\, A_3 A_1^* \exp[-j(\Delta kz)] \tag{12.4.33b}$$

$$\frac{dA_3}{dz} = -\frac{\alpha_3 A_3}{2} - j\delta\, A_1 A_2 \exp[j(\Delta kz)] \tag{12.4.33c}$$

where we have put $\alpha_i = \sigma_i/n_i\varepsilon_0 c$, $\Delta k = k_3 - k_2 - k_1$, and

$$\delta = \frac{d}{c}\left(\frac{\omega_1\omega_2\omega_3}{n_1 n_2 n_3}\right)^{1/2} \tag{12.4.34}$$

The advantage of using A_i instead of E_i is now apparent since, unlike Eq. (12.4.31), relations Eq. (12.4.33) now involve a single coupling parameter δ.

If losses are neglected (i.e., when $\alpha_i = 0$), we can obtain from Eq. (12.4.33) some very useful conservation laws. For instance, if we multiply both sides of Eq. (12.4.33a) by A_1^* and both sides of Eq. (12.4.33b) by A_2^*, and compare the resulting expressions, we arrive at the following relation: $d|A_1|^2/dz = -d|A_3|^2/dz$. Similarly from Eqs. (12.4.33b) and (12.4.33c) we get $d|A_2|^2/dz = -d|A_3|^2/dz$. We can therefore write

$$\frac{d\,|A_1|^2}{dz} = \frac{d\,|A_2|^2}{dz} = -\frac{d\,|A_3|^2}{dz} \tag{12.4.35}$$

which are known as the Manley-Rowe relations. Note that, since $|A_i|^2$ is proportional to the corresponding photon flux, Eq. (12.4.35) implies that whenever a photon at ω_3 is destroyed, a photon at ω_1 and a photon at ω_2 are created. This is consistent with the photon model for the parametric process which was discussed in Sect. 12.4.1.2. Note also that Eq. (12.4.35) means, for instance, that $(dP_1/dz) = -(\omega_1/\omega_3)(dP_3/dz)$, where P_1 and P_3 are the powers of the two waves. Thus only a fraction (ω_1/ω_3) of the power at frequency ω_3 can be converted into power at frequency ω_1.

Strictly speaking, equations (12.4.33) apply to a traveling wave situation in which an arbitrarily long crystal is being traversed by the three waves at ω_1, ω_2, ω_3. We now want to see how these equations might be applied to the case of an optical parametric oscillator as in Fig. 12.8. Here we will first consider the DRO scheme. The waves at ω_1 and ω_2 will therefore travel back and forth within the cavity, and the parametric process will only occur

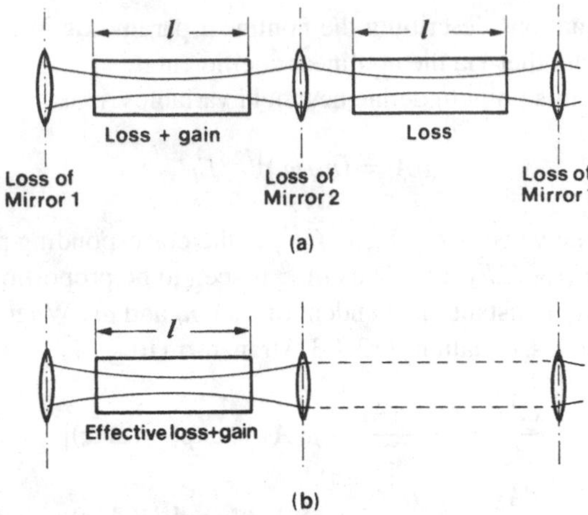

FIG. 12.9. (a) Unfolded path for an optical parametric oscillator; (b) Reduction to a single-pass scheme with mirror losses incorporated into the distributed losses of the crystal.

when their propagation direction is the same as that of the pump wave (since it is only under these circumstances that phase matching can be satisfied). If we unfold the optical path, it will look like that of Fig. 12.9a, and it can be seen that loss occurs on every pass while parametric gain occurs only once in every second pass. This situation can be reduced to that of Fig. 12.9b if one chooses an appropriate definition of the effective loss coefficient $\alpha_i (i = 1, 2)$. The loss due to a crystal of length l in Fig. 12.9b must in fact equal the losses incurred in a double pass in Fig. 12.9a. The latter losses must account for the actual losses in the crystal, as well as the mirror and diffraction losses. Thus the coefficients α_1 and α_2 in Eq. (12.4.33) must be appropriately defined so as to incorporate these various losses. From Eq. (12.4.33), neglecting the parametric interaction [i.e., setting $\delta = 0$], one sees that, after the beam at frequency ω_i $(i = 1, 2)$ traverses the length l of the crystal, its power will be reduced by a fraction $\exp(-\alpha_i l)$. This reduction must then account for the round-trip losses of the cavity, which requires that

$$\exp(-\alpha_i l) = R_{1i} R_{2i} (1 - L_i)^2 \tag{12.4.35a}$$

where R_{1i} and R_{2i} are the power-reflectivities of the two mirrors at frequency ω_i, and L_i is the crystal loss (plus diffraction loss) per pass. If we now define [compare with Eq. (1.2.4)] $\gamma_{1i} = -\ln R_{1i}$, $\gamma_{2i} = -\ln R_{2i}$, $\gamma_i' = -\ln(1 - L_i)$ and $\gamma_i = [(\gamma_{1i} + \gamma_{2i})/2] + \gamma_i'$, we can rewrite Eq. (12.4.35a) as

$$\alpha_i l = 2\gamma_i \tag{12.4.36}$$

where γ_i is the overall cavity loss per pass at frequency ω_i. Note that this amounts to simulating the mirror losses by losses distributed through the crystal and then including them into an effective absorption coefficient of the crystal $\alpha_i (i = 1, 2)$. The loss α_3, on the other hand, only involves crystal losses and can in general be neglected. Thus we can say that Eq. (12.4.33) can be applied to a DRO provided that α_1 and α_2 are given by Eq. (12.4.36) and provided one sets

$\alpha_3 \cong 0$. To obtain the threshold condition of a DRO, Eq. (12.4.33) can be further simplified if we neglect depletion of the pump wave by the parametric process. This assumption together with the assumption $\alpha_3 = 0$ means that we can take $A_3(z) \cong A_3(0)$, where $A_3(0)$, the field amplitude of the incoming pump wave, is taken to be real. With the further assumption of $\Delta k = 0$ (perfect phase matching), Eq. (12.4.33) is considerably simplified and becomes

$$\frac{dA_1}{dz} = -\frac{\alpha_1 A_1}{2} - j\frac{g}{2}A_2^* \tag{12.4.37a}$$

$$\frac{dA_2}{dz} = -\frac{\alpha_2 A_2}{2} - j\frac{g}{2}A_1^* \tag{12.4.37b}$$

where we have set

$$g = 2\delta\, A_3(0) = 2d\frac{E_3(0)}{c}\left(\frac{\omega_1\omega_2}{n_1 n_2}\right)^{1/2} \tag{12.4.38}$$

The threshold condition for a DRO is then readily obtained from Eq. (12.4.37) by imposing the condition $dA_1/dz = dA_2/dz = 0$. This leads to

$$\alpha_1 A_1 + jgA_2^* = 0 \tag{12.4.39a}$$

$$jgA_1 - \alpha_2 A_2^* = 0 \tag{12.4.39b}$$

where the complex conjugate of Eq. (12.4.37b) has been taken. The solution of this homogeneous system of equations will yield non-zero values for A_1 and A_2 only if

$$g^2 = \alpha_1\alpha_2 = 4(\gamma_1\gamma_2/l^2) \tag{12.4.40}$$

where Eq. (12.4.36) has been used. It should be noted that, according to Eq. (12.4.38), g^2 is proportional to $E_3^2(0)$, i.e., to the intensity of the pump wave. Thus condition Eq. (12.4.40) means that a certain threshold intensity of the pump wave is needed in order for parametric oscillation to start. As will be shown in the following example, the threshold intensity is proportional to the product of the single-pass (power) losses, γ_1 and γ_2 of the two waves at ω_1 and ω_2, and inversely proportional to d^2 and l^2.

Example 12.4. *Calculation of the threshold intensity for the pump beam in a doubly resonant optical parametric oscillator.* From Eqs. (12.4.38) and (12.4.40) one readily gets the following value for expression $E_3^2(0)$, the square of the pump-beam amplitude at threshold

$$E_{3th}^2(0) = \frac{c^2}{d^2}\frac{n_1 n_2}{\omega_1\omega_2}\frac{\gamma_1\gamma_2}{l^2}$$

Since the pump-beam intensity is given by $I_3 = n_3\varepsilon_0 c|E_3|^2/2$, one then obtains the following expression for the threshold pump intensity

$$I_{3th} = \left[\frac{n_3}{2Zd^2}\right]\left[\frac{n_1 n_2 \lambda_1 \lambda_2}{(2\pi\, l)^2}\right][\gamma_1\gamma_2]$$

where $Z = 1/\varepsilon_0 c = 377$ Ohms, is the free-space impedance while λ_1 and λ_2 are the wavelengths of the signal and idler waves, respectively. Note that the term in the first bracket on the right hand-side of the above equation has the dimension of an intensity while the factors in the remaining two brackets are dimensionless.

The SRO case is somewhat more involved. If the laser cavity is resonant only at ω_1, then α_1 can again be written as in Eq. (12.4.36). Since the wave at ω_2 is no longer fed back, α_2 will involve only the crystal losses and it can therefore be neglected. Again, neglecting depletion of the pump wave and assuming perfect phase matching, Eq. (12.4.37) will still be applicable provided we now set $\alpha_2 = 0$. For small parametric conversion we can set $A_1^*(z) \cong A_1^*(0)$ on the right-hand side of Eq. (12.4.37b). We thus get

$$A_2(z) = -jgA_1^*(0)z/2 \tag{12.4.41}$$

where the condition $A_2(0) = 0$ has been assumed (i.e., no field, at frequency ω_2, is fed back into the crystal by the resonator). If one then substitutes Eq. (12.4.41) in Eq. (12.4.37a) and put $A_1(z) \cong A_1(0)$ in the right-hand side of the latter equation, one gets

$$\frac{dA_1}{dz} = \left(-\frac{\alpha_1}{2} + \frac{g^2 z}{4}\right) A_1(0) \tag{12.4.42}$$

Integration of Eq. (12.4.42) over the length, l, of the crystal gives the following expression for the field at frequency ω_1

$$A_1(l) = A_1(0)\left(1 - \frac{\alpha_1 l}{2} + \frac{g^2 l^2}{8}\right) \tag{12.4.43}$$

The threshold condition is then reached when $A_1(l) = A_1(0)$, i.e., when

$$g^2 = \frac{4\alpha_1}{l} = \frac{8\gamma_1}{l^2} \tag{12.4.44}$$

Since g^2 is proportional to intensity I of the pump wave, a comparison of Eq. (12.4.44) with Eq. (12.4.40) gives the ratio of threshold pump intensities as

$$\frac{I_{SRO}}{I_{DRO}} = \frac{2}{\gamma_2} \tag{12.4.45}$$

For example, if one takes a loss per pass of $\gamma_2 = 0.02$ (i.e., 2% loss), one finds from Eq. (12.4.45) that the threshold power for the SRO is 100 times larger than for the DRO case.

12.4.2.2. Second-Harmonic Generation

In the case of SHG we take

$$E(z,t) = (1/2)\{E_\omega \exp[j(\omega t - k_\omega z)] + E_{2\omega}\exp[j(2\omega t - k_{2\omega} z)] + c.c.\} \tag{12.4.46}$$

$$P^{\mathrm{NL}}(z,t) = (1/2)\{P_\omega^{\mathrm{NL}} \exp[j(\omega t - k_\omega z)] + P_{2\omega}^{\mathrm{NL}}\exp[j(2\omega t - k_{2\omega} z)] + c.c.\} \tag{12.4.47}$$

Substitution of Eqs. (12.4.46) and (12.4.47) into Eq. (12.4.2) gives

$$P_{2\omega}^{\mathrm{NL}} = \varepsilon_0 d E_\omega^2 \exp[-j(2k_\omega - k_{2\omega})z] \tag{12.4.48a}$$

$$P_\omega^{\mathrm{NL}} = 2\varepsilon_0 d E_{2\omega} E_\omega^* \exp[-j(k_{2\omega} - 2k_\omega)z] \tag{12.4.48b}$$

If one then substitutes Eq. (12.4.48) into Eq. (12.4.28) and neglects crystal losses (i.e., one takes $\sigma_i = 0$), one gets

$$\frac{dE_{2\omega}}{dz} = -j\frac{\omega}{n_{2\omega}c} dE_\omega^2 \exp(j\Delta kz) \tag{12.4.49a}$$

$$\frac{dE_\omega}{dz} = -j\frac{\omega}{n_\omega c} dE_{2\omega}E_\omega^* \exp(-j\Delta kz) \tag{12.4.49b}$$

where $\Delta k = k_{2\omega} - 2k_\omega$. These are the basic equations describing SHG.

To solve these equations, it is first convenient to define new field variables E'_ω and $E'_{2\omega}$ as

$$E'_\omega = (n_\omega)^{1/2} E_\omega \tag{12.4.50a}$$

$$E'_{2\omega} = (n_{2\omega})^{1/2} E_{2\omega} \tag{12.4.50b}$$

Since the intensity I_ω of the wave at ω is proportional to $n_\omega|E_\omega|^2$, the quantity $|E'_\omega|^2$ is seen to be proportional to I_ω with a proportionality constant independent of refractive index. The substitution of Eq. (12.4.50) into Eq. (12.4.49) then gives

$$\frac{dE'_{2\omega}}{dz} = -\frac{j}{l_{SH}}\frac{E'^2_\omega}{E'_\omega(0)} \exp[j(\Delta kz)] \tag{12.4.51a}$$

$$\frac{dE'_\omega}{dz} = -\frac{j}{l_{SH}}\frac{E'_{2\omega}E'^*_\omega}{E'_\omega(0)} \exp[-j(\Delta kz)] \tag{12.4.51b}$$

In the above equations $E'_\omega(0)$ is the value of E'_ω at $z = 0$, assumed to be a real quantity, and l_{SH} is a characteristic length, for second-harmonic interaction, given by

$$l_{SH} = \frac{\lambda\,(n_\omega n_{2\omega})^{1/2}}{2\pi dE_\omega(0)} \tag{12.4.52}$$

where λ is the wavelength and $E_\omega(0)$, the field amplitude of the incident wave at frequency ω, is also a real quantity. Note that the advantage of using the new field variables E'_ω and $E'_{2\omega}$ is now apparent from Eq. (12.4.51) since they involve a single coupling parameter, l_{SH}. From Eq. (12.4.51) one readily finds

$$\frac{d\left|E'_{2\omega}\right|^2}{dz} = -\frac{d\left|E'_\omega\right|^2}{dz} \tag{12.4.53}$$

which represents the Manley-Rowe relation for SHG. Note that the equation shows that, e.g., a decrease of beam power, or intensity, at frequency ω, must correspond to an increase, by the same amount, of the power or intensity at frequency 2ω. Thus 100% conversion of fundamental-beam power into second-harmonic power is possible in this case.

As a first example of the solution of Eq. (12.4.51), we consider the case where there is an appreciable phase mismatch (i.e., $l_{SH}\Delta k \gg 1$) so that little conversion of fundamental into SH is expected to occur. We therefore set $E'_\omega(z) \cong E'_\omega(0)$ on the right-hand side of

Eq. (12.4.51a). The resulting equation can then be readily integrated and, using the boundary condition $E'_{2\omega}(0) = 0$, one obtains

$$E'_{2\omega}(1) = -\frac{E'_{\omega}(0)}{l_{SH}}\left[\frac{\exp(-j\Delta kl) - 1}{\Delta k}\right] \tag{12.4.54}$$

One can then readily show that

$$\left|\frac{E'_{2\omega}(l)}{E'_{\omega}(0)}\right|^2 = \frac{\sin^2(\Delta kl/2)}{(\Delta k l_{SH}/2)^2} \tag{12.4.55}$$

Since $|E'_{2\omega}|^2$ is proportional to the SH intensity $I_{2\omega}$, the variation of this intensity with crystal length is immediately obtained from Eq. (12.4.55). The corresponding behavior of I_ω vs l can then be obtained from the condition $I_\omega + I_{2\omega} = I_\omega(0)$, this condition being an immediate consequence of Eq. (12.4.53). The plots of $[I_\omega/I_\omega(0)]$ and $[I_{2\omega}/I_\omega(0)]$ vs (l/l_{SH}), obtained in this way for $l_{SH}\Delta k = 10$, are shown as dashed curves in Fig. 12.10. Note that, due to the large phase mismatch, only a small conversion to second harmonic occurs. Note also that, as can be readily shown from Eq. (12.4.55), the first maximum of $[I_{2\omega}/I_\omega(0)]$ occurs at $l = l_c$, where l_c, the coherence length, is given by Eq. (12.4.10).

As a second example of a solution to Eq. (12.4.51), we consider the case of perfect phase matching ($\Delta k = 0$). In this case appreciable conversion to second harmonic may occur and the depletion of the fundamental beam must therefore be taken into account. This means that Eq. (12.4.51) must now be solved without the approximation $E'_\omega(z) \cong E'_\omega(0)$. If $\Delta k = 0$, however, it can be shown from Eq. (12.4.51) that, if $E'_\omega(0)$ is taken to be real, then $E'_\omega(z)$ and $E'_{2\omega}(z)$ turn out to be real and imaginary, respectively. We can therefore write

$$E'_\omega = |E'_\omega| \tag{12.4.56a}$$

$$E'_{2\omega} = -j\,|E'_{2\omega}| \tag{12.4.56b}$$

and Eq. (12.4.51) then gives

$$\frac{d\,|E'_\omega|}{dz} = -\frac{1}{l_{SH}}\frac{|E'_{2\omega}|\,|E'_\omega|}{E'_\omega(0)} \tag{12.4.57a}$$

$$\frac{d\,|E'_{2\omega}|}{dz} = \frac{1}{l_{SH}}\frac{|E'_\omega|^2}{E'_\omega(0)} \tag{12.4.57b}$$

The solution of Eq. (12.4.57) with the boundary conditions $E'_\omega(z=0) = E'_\omega(0)$ and $E'_{2\omega}(0) = 0$ is

$$|E'_{2\omega}| = E'_\omega(0)\,\tanh(z/l_{SH}) \tag{12.4.58a}$$

$$|E'_\omega| = E'_\omega(0)\,\operatorname{sech}(z/l_{SH}) \tag{12.4.58b}$$

Since the intensity of the wave, at a given frequency, is proportional to $|E'|^2$, one has $I_{2\omega}/I_\omega(0) = |E'_{2\omega}|^2/E'^2_{2\omega}(0)$ and $I_\omega/I_\omega(0) = |E'_\omega|^2/E'^2_\omega(0)$. The dependence of $I_{2\omega}/I_\omega(0)$ and

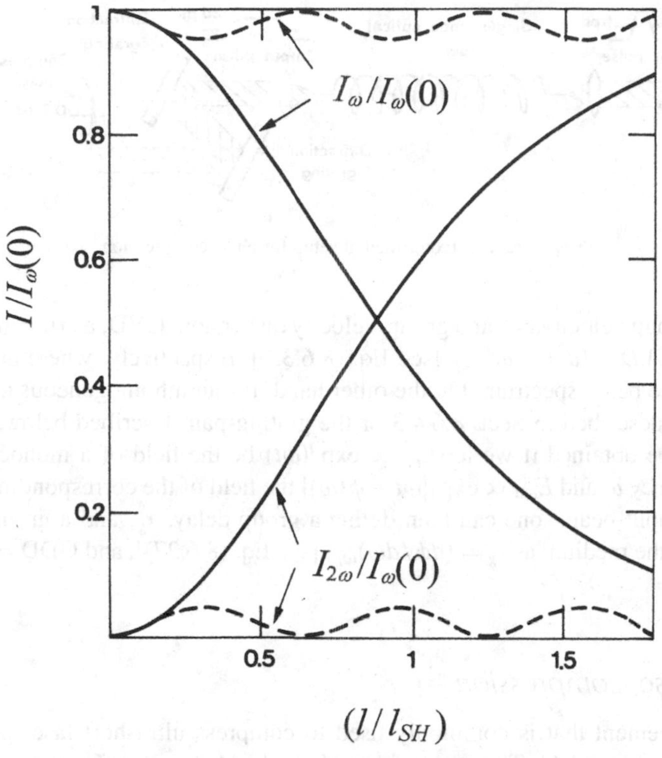

FIG. 12.10. Normalized plots of second-harmonic intensity $I_{2\omega}$ and fundamental intensity I_ω versus crystal length l for perfect phase-matching (continuous curves) and for a finite phase mismatch (dashed curves).

$I_\omega/I_\omega(0)$ on crystal length, as predicted by Eq. (12.4.58), is shown by solid curves in Fig. 12.9. Note that, for $l = l_{SH}$, an appreciable fraction (~59%) of the incident wave has been converted into a SH wave. This illustrates the role of l_{SH} as a characteristic length for the second-harmonic interaction. Note that, according to Eq. (12.4.52) the value of l_{SH} is inversely proportional to the square root of beam-intensity at the fundamental frequency ω. Note also that, for $l \gg l_{SH}$, the radiation at the fundamental frequency can be completely converted into second-harmonic radiation, in agreement with the Manley-Rowe relation Eq. (12.4.53).

12.5. TRANSFORMATION IN TIME: PULSE COMPRESSION AND PULSE EXPANSION

The phenomena of pulse compression and pulse expansion, now widely used in the field of ultrashort laser pulses, are considered in this section. Before entering into a detailed discussion of these phenomena, it is worth recalling that, given a homogeneous medium, such as a piece of glass or an optical fiber, characterized by the dispersion relation $\beta = \beta(\omega)$, one

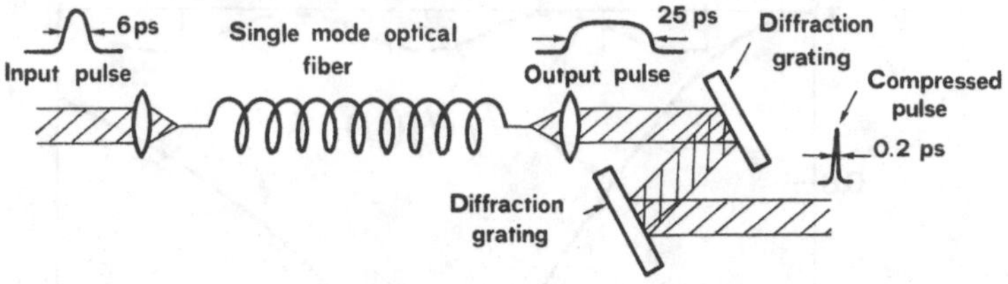

FIG. 12.11. Experimental setup for pulse compression.

can define a group velocity, υ_g, and group-velocity dispersion, GVD, as $\upsilon_g = (d\omega/d\beta)_{\omega_L}$ [see (8.6.26)] and $GVD = (d^2\beta/d\omega^2)_{\omega_L}$ [see Eq. (8.6.33)], respectively, where ω_L is the central frequency of the beam spectrum. On the other hand, for an inhomogeneous medium such as the prism-pair described in Sect. 8.6.4.3 or the grating-pair described below, a more useful approach can be obtained if we let $E_{in} \propto \exp j(\omega t)$ be the field of a monochromatic input beam at frequency ω and $E_{out} \propto \exp j[\omega t - \phi(\omega)]$ the field of the corresponding output beam. For a pulsed input beam, one can then define a group delay, τ_g, and a group-delay dispersion, GDD, of the medium as $\tau_g = (d\phi/d\omega)_{\omega_L}$, [see Eq. (8.6.27)], and $\text{GDD} = (d^2\phi/d\omega^2)_{\omega_L}$, respectively.

12.5.1. Pulse Compression

An arrangement that is commonly used to compress ultrashort laser pulses is shown schematically in Fig. 12.11. The pulse of a mode-locked laser, of sufficient power (in practice a relatively modest peak-power of, e.g., $P_p = 2\,\text{kW}$) and long time duration (e.g., $\tau_p = 6\,\text{ps}$), is sent through a single-mode silica optical fiber of suitable length (e.g., $L = 3\,\text{m}$). The wavelength of the pulse (e.g., $\lambda \cong 590\,\text{nm}$) falls in the region of positive GVD of the fiber ($\lambda < 1.32\,\mu\text{m}$, for non-dispersion-shifted fibers). After leaving the fiber, the output is collimated and passed through an optical system consisting of two identical gratings aligned parallel to each other and whose tilt and spacing are appropriately chosen, as described below. Under appropriate conditions, the output beam then consists of a light pulse with a much shorter duration (e.g., $\tau_p = 200\,\text{fs}$) than that of the input pulse and, hence, of much higher peak power (e.g., $P_p = 20\,\text{kW}$). Thus, the arrangement of Fig. 12.11 can readily provide a large compression factor (e.g., ~30 in the illustrated case) of the input pulse. The rather subtle phenomena involved in this pulse compression scheme are discussed below.[21,22]

We start by considering what happens when the pulse propagates in the optical fiber. First we recall that, due to the phenomenon of self-phase modulation, a light pulse of uniform intensity profile, which travels a distance z in a material exhibiting the optical Kerr effect, acquires a nonlinear phase term given by Eq. (8.6.38). In an optical fiber, however, the situation is somewhat more complicated due to the non-uniform transverse intensity profile of its fundamental mode (EH_{11}). In this case, it can be shown that the whole mode profile acquires a phase term given by[22]

$$\phi(t, z) = \omega_L t - \frac{\omega_L n_0}{c} z - \frac{\omega_L n_2 P}{c A_{eff}} z \tag{12.5.1}$$

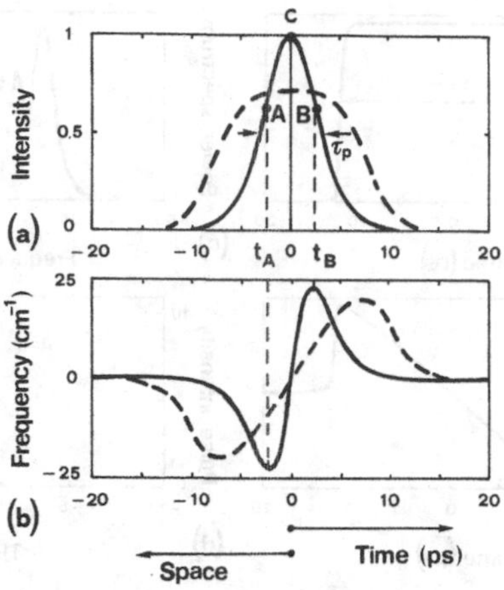

FIG. 12.12. Time behavior (a) of the pulse intensity, and (b) of the pulse frequency, when propagating through a single mode fiber of suitable length. The solid and dashed curves refer, respectively, to the cases of no group velocity dispersion and positive group velocity dispersion in the fiber.

where n_0 is the low-intensity refractive index, n_2 is the coefficient of the nonlinear index of the medium [see Eq. (8.6.23)], $P = P(t, z)$ is the power of the beam traveling in the fiber and A_{eff} is a suitably defined effective area of the beam in the fiber. The instantaneous carrier frequency of the light pulse is then obtained from Eq. (12.5.1) as

$$\omega(t, z) = \frac{\partial \phi}{\partial t} = \omega_L - \frac{\omega_L}{cA_{eff}} z n_2 \frac{\partial P}{\partial t} \tag{12.5.2}$$

and it is seen to be linearly dependent on the negative time derivative of the corresponding power, P. Thus, for a bell-shaped pulse as in Fig. 12.12a, the carrier frequency will vary with time as indicated by the solid curve in Fig. 12.12b. Notice that, around the peak of the pulse, the time behavior of the power can be described by a parabolic law and the instantaneous carrier frequency then increases linearly with time (i.e., the pulse is said to show a positive frequency chirp). Note however that the frequency chirp becomes negative after the pulse inflection points, i.e., for $t < t_A$ or $t > t_B$ in Fig. 12.12b.

It should be noted that the physical situation described so far has neglected the presence of GVD in the fiber. In the absence of GVD the pulse shape does not change with propagation i.e. the field amplitude remains a function of the variable $(z - \upsilon_g t)$, where υ_g is the group velocity (see Appendix G). The z dependence of the pulse, at any given time, is then obtained from the corresponding time dependence by reversing the positive direction of the axis and multiplying the time scale by υ_g (see Fig. 12.12). This means that a point such as A of Fig. 12.12a is actually in the leading edge while a point such as B is in the trailing edge of the pulse. Note now that, according to Fig. 12.12b, the carrier frequency of the pulse around point A will be smaller than at C, where it is roughly equal to ω_L. On the other hand, the carrier frequency of the pulse around point B will be higher than at C.

Assume now that the fiber has a positive GVD. That part of the pulse, in the vicinity of point A, will move faster than that corresponding to point C and this will, in turn, move faster than the region around point B. This means that the central part of the pulse, while traveling in

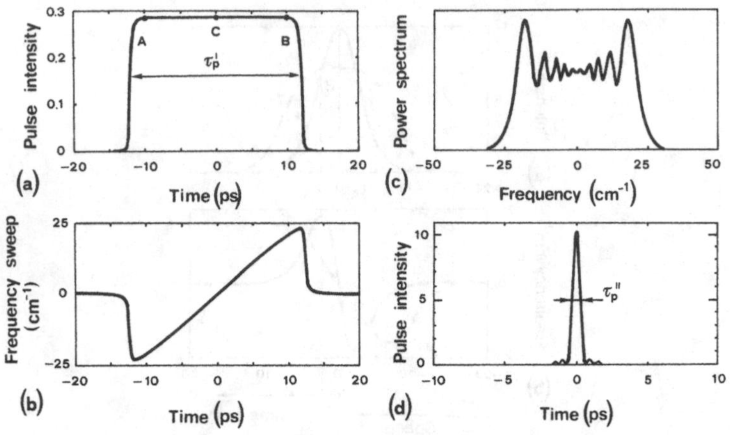

FIG. 12.13. Plots of calculated values of (a) self-broadening and (b) self-phase-modulation of an initial 6-ps pulse after propagating through 30 m of single-mode fiber with positive group velocity dispersion; (c) output pulse spectrum; (d) compressed pulse produced by an optical system having a negative group velocity dispersion with linear dispersion (after ref.,[21] by permission).

the fiber, will be expanded. Following a similar argument one sees that the outer parts of the pulse will be compressed rather than expanded, because there the frequency chirp is negative. Thus, when a positive GVD is considered, the actual shape of the pulse intensity as a function of time, at a given z position, will look like the dashed curve of Fig. 12.12a. The corresponding behavior for the frequency change will then be as shown by the dashed curve of Fig. 12.12b. Note from Fig. 12.12a that, owing to the pulse broadening produced by the GVD, the peak intensity of the dashed curve is lower than that of the solid curve. Note also that, since the parabolic part of the pulse now extends over a wider region around the peak, the linear part of the positive frequency chirp will now occur over a larger fraction of the pulse.

Having established these general features of the interplay between SPM and GVD, we are now able to understand how, for a long enough fiber, the time behavior of the pulse amplitude and of the pulse frequency, at the exit of the fiber, can actually develop into those shown in Figs. 12.13a and 12.13b. Note that the pulse has been squared and expanded to a duration of $\tau_p' \cong 23$ ps while the positive frequency chirp is linear with time over most of this light pulse duration. The corresponding pulse spectrum is shown in Fig. 12.13c and one can see that, due to the strong SPM occurring for such a small beam in the fiber (the core diameter for the condition depicted in Fig. 12.13 was $d \cong 4\,\mu$m), the spectral extension of the output pulse, $\Delta\nu_L' = 50\,\text{cm}^{-1}$, is considerably larger than that of the input pulse to the fiber. The latter is, in fact, established by the inverse of the pulse duration and, for the case considered of $\tau_p \cong 6$ ps, corresponds to $\Delta\nu_L \cong 0.45/\tau_p \cong 2.5\,\text{cm}^{-1}$. This means that the bandwidth of the output pulse is predominantly established by the phase modulation of the pulse rather than by the duration of its envelope.

Suppose now that the pulse of Fig. 12.13a and 12.13b is passed through a medium of negative GDD. With the help of a similar argument to that used in relation to Fig. 12.12, we can now see that the region of the pulse around point A will move more slowly than that around C and this in turn will move more slowly than that around B. This implies that the pulse will now be compressed. Let us next suppose that the GDD of the medium, besides being negative, is also independent of frequency. According to Eq. (8.6.27), this means that the dispersion in group-delay, $d\tau_g/d\omega$, will also be negative and independent of frequency. Thus τ_g decreases linearly with frequency and, since the frequency chirp of the pulse increases linearly with time (see Fig. 12.13b), all points of the pulse of Fig. 12.13a will tend to be compressed together at the same time if the GDD has the appropriate value. According to Eq. (8.6.31), this optimum value of GDD must be such that

$$\left(\frac{d^2\phi}{d\omega^2}\right)_{\omega_L} \Delta\omega_L' = \tau_p' \tag{12.5.3}$$

where $\Delta\omega_L' = 2\pi\Delta\nu_L'$ is the total frequency sweep of the pulse of Fig. 12.13b and τ_p' is the duration of the expanded pulse of Fig. 12.13a. It should be noted, however, that this compression mechanisms cannot produce an indefinitely sharp pulse as, at first sight, one may be led to believe. In fact, the system providing the negative GDD is a linear medium and this implies that the pulse spectrum must remain unchanged on passing through such a system. This means that the spectrum of the compressed pulse still remains as that shown in Fig. 12.13c. Even under optimal conditions, the duration of the compressed pulse, τ_p'', cannot then be shorter than approximately the inverse of the spectral bandwidth, i.e., $\tau_p'' \cong 1/\Delta\nu_L' \cong 0.75$ ps. Since the time duration τ_p of the pulse originally entering the optical fiber was ~6 ps (see Fig. 12.11a), the above result indicates that a sizable compression of the incoming light pulse has been achieved.*

The above heuristic discussion is based on the assumption that a chirped pulse can be subdivided into different temporal regions with different carrier frequencies. Although this idea is basically correct and allows a description of the phenomena in simple physical terms, a more critical detailed examination of this approach would reveal some conceptual difficulties. To validate this analysis, however, the analytical treatment of the problem can be performed in a rather straightforward way although the intuitive physical picture of the phenomenon gets somewhat obscured. For this analytical treatment, in fact, one merely takes the Fourier transform, $E_\omega(\omega)$, of the pulse of Figs. 12.13a and 12.13b and then multiplies, $E_\omega(\omega)$, by the transmission $t(\omega)$ of the medium exhibiting negative GDD. The resulting pulse, in the time domain, is then obtained by taking the inverse Fourier transform of $E(\omega)t(\omega)$. Note that, for a lossless medium, $t(\omega)$ must be represented by a pure phase term, i.e., it can be written as

$$t(\omega) = \exp(-j\phi) \tag{12.5.4}$$

where $\phi = \phi(\omega)$. If the medium has a constant GDD, the Taylor-series expansion of $\phi(\omega)$ around the central carrier frequency ω_L gives

$$\phi(\omega) = \phi(\omega_L) + \left(\frac{d\phi}{d\omega}\right)_{\omega_L} (\omega - \omega_L) + \frac{1}{2}\left(\frac{d^2\phi}{d\omega^2}\right)_{\omega_L} (\omega - \omega_L)^2 \tag{12.5.5}$$

where $(d\phi/d\omega)_{\omega_L}$ is the group delay and $(d^2\phi/d\omega^2)_{\omega_L}$ is the group-delay dispersion. By substituting Eq. (12.5.5) into Eq. (12.5.4) and taking the inverse Fourier transform of $E(\omega)t(\omega)$

* Techniques of this type to produce shorter pulses by first imposing a linear frequency chirp followed by pulse compression have been extensively used in the field of radar (chirped radars) since World War II.

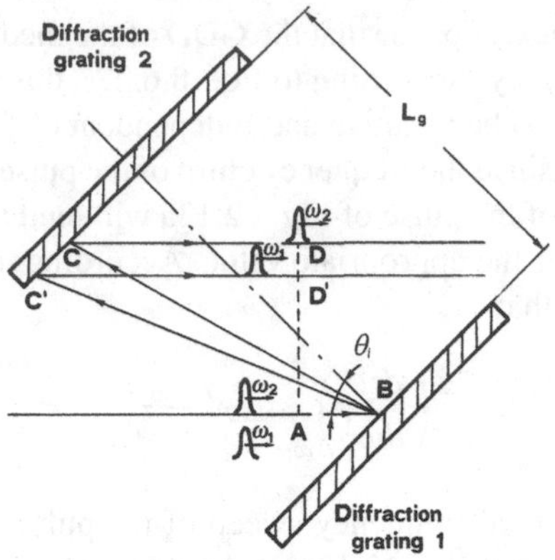

FIG. 12.14. Grating-pair for pulse compression.

one then finds that, if $(d^2\phi/d\omega^2)_{\omega_L}$ is negative and satisfies condition Eq. (12.5.3), the optimum pulse compression occurs. The optimally compressed pulse, calculated in this way, is shown in Fig. 12.13d. The resulting pulse duration turns out to be $\tau_p'' \cong 0.6$ ps, rather than the approximate value (0.75 ps) estimated before.

We are now left with the problem of finding a suitable optical system that can provide the required negative GDD. Notice that, since one can write $\mathrm{GDD} = d\tau_g/d\omega$, a negative GDD implies that the group delay must decrease with increasing ω. As discussed in Sect. 8.6.4.3, one such system consists of the two-prism couple shown in Fig. 8.26. Another such system is the pair of parallel and identical gratings shown in Fig. 12.11.[23] To understand the main properties of this last system we refer to Fig. 12.14, which shows a plane wave, represented by the ray AB, incident on the grating 1 with a propagation direction at an angle θ_i to the grating normal. We now assume that the incident wave consists of two synchronous pulses, at frequency ω_2 and ω_1, with $\omega_2 > \omega_1$. As a result of the grating dispersion, the two pulses will then follow paths ABCD and ABC$'$D$'$, respectively, and one sees that the delay suffered by the pulse at frequency ω_2, $\tau_{d_2} = ABCD/\upsilon_g$, is smaller than that, $\tau_{d_1} = ABC'D'/\upsilon_g$, for the pulse at frequency ω_1. Since $\omega_2 > \omega_1$, this means that the pulse delay dispersion is negative. A detailed calculation then shows that the GDD can be expressed as[23]

$$\mathrm{GDD} = \frac{d^2\phi}{d\omega^2} = -\frac{4\pi^2 c}{\omega^3 d^2}\frac{1}{\{[1-[\sin\theta_i - (\lambda/d)]^2\}^{3/2}} L_g \tag{12.5.6}$$

where ω is the frequency of the wave, λ its wavelength, d the grating period, and L_g the distance between the two gratings. Note the minus sign on the right-hand side of Eq. (12.5.6) indicating indeed a negative GDD. Note also that the value of the dispersion can be changed by changing L_g and/or the incidence angle θ_i. It should lastly be observed that the two-grating system shown in Fig. 12.14 has the drawback that a lateral walk-off is present in the output beam, the amount of walk-off depending upon the difference in frequency between the beam components (as it occurs e.g. between rays CD and C$'$D$'$). For beams of finite size, this walk-off can represent a problem. This problem can however be circumvented by retroreflecting the output beam back to itself by a plane mirror. In this case the overall dispersion resulting from the double pass through the diffraction-grating-pair is, of course, twice that given by Eq. (12.5.6).

The system of Fig. 12.11 has been used to produce compression of both picosecond and femtosecond laser pulses over a wide range of conditions.[24] For example, pulses of ~6 ps duration (and ~2 kW peak power), from a synchronously-pumped mode-locked dye laser, have been compressed, using a 3-m-long fiber, to about 200 fs ($P_p = 20$ kW). These pulses have again been compressed, by a second system as in Fig. 12.11, using a 55-cm-long fiber, to optical pulses of 90 fs duration. One of the most interesting results achieved involves the compression of 50 fs pulses, from a colliding-pulse mode-locked dye laser down to ~6 fs, using a 10-mm-long fiber.[25] To achieve this record value of pulse duration, for such a configuration, second-order group-delay dispersion [GDD = $(d^2\phi/d\omega^2)_{\omega_L}$], and third-order group-delay dispersion [TOD = $(d^3\phi/d\omega^3)_{\omega_L}$] were compensated using both two consecutive grating pairs (each pair as in Fig. 12.14) and a four-prism sequence as in Fig. 8.26. In fact, the TOD of the two compression system could be arranged to be of opposite sign, so as to cancel each other.

A limitation of the optical-fiber compression scheme of Fig. 12.11 arises from the small diameter ($d \cong 5\,\mu$m) of the core of the fiber. Accordingly, the pulse energy that can be launched into the fiber is necessarily limited to a low value (~10 nJ). A recently introduced guiding configuration to produce wide-bandwidth SPM spectra, uses a hollow-silica fiber filled with noble gases (Kr, Ar) at high pressures (1–3 atm).[26] With an inner diameter for the hollow-fiber of 150–300 μm, a much higher-energy input pulse (~2 mJ) could be launched into the fiber. Using a fiber length of ~1 m, wide SPM spectra (~200 nm) have been obtained starting with input pulses of femtosecond duration (20–150 fs). With the help of a specially designed two-prism sequence, in a double-pass configuration, and also using two reflections from a specially designed chirped mirror,[27] 20 fs pulses from the amplified beam of a mode-locked Ti:sapphire laser were compressed to ~4.5 fs.[28] These pulses containing ~1.5 cycle of the carrier-frequency, are the shortest pulses generated to date and have a relative large amount of energy (~100 μJ).

12.5.2. Pulse Expansion

It was already pointed out in Sect. 12.3 that, for chirped-pulse amplification, one needs first to subject the pulse to a large expansion in time. In principle this expansion can be achieved by a single-mode fiber of suitable length (see Fig. 12.11 and Fig. 12.12a). However, the linear chirp, produced in this way (see Fig. 12.13b), cannot be exactly compensated by a grating-pair compressor Fig. 12.14, due to the higher order dispersion exhibited by this compressor. For pulses of short duration (subpicosecond), this system would thus only provide a partial compression of the expanded pulse to its original shape. A much better solution[29] involves using an expander which also consists of a grating-pair, but in an anti-parallel configuration and with a 1:1 inverting telescope between the two gratings, as shown in Fig. 12.15.[30] To achieve the desired positive GDD, the two gratings must be located outside the telescope but within a focal length of the lens, i.e., one must have $(s_1, s_2) < f$, where f is the focal length of each of the two lenses. In this case, under the ideal paraxial wave-propagation conditions

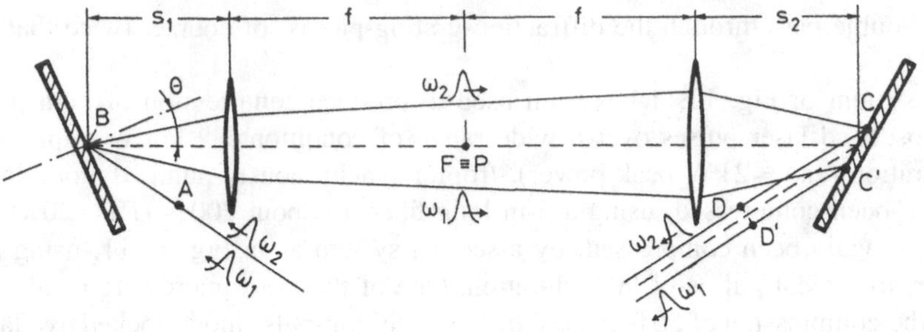

FIG. 12.15. Pulse expander consisting of two diffraction gratings, in an antiparallel configuration, with a 1:1 inverting telescope between them.

and negligible dispersion of the lens material, the GDD can be shown to be given by[30]

$$GDD = \frac{d^2\phi}{d\omega^2} = \frac{4\pi^2 c}{\omega^3 d^2 \cos^2\theta}(2f - s_1 - s_2) \tag{12.5.7}$$

where ω is the frequency of the wave, d is the grating period, and θ is the angle shown in Fig. 12.15. Equation (12.5.7) indeed shows that one has, in this case, a positive value of GDD. To understand this result we refer again to Fig. 12.15, where the plane wave incident on the first grating, represented by the ray AB, is assumed to consist of two synchronous pulses, at frequency ω_2 and ω_1, with $\omega_2 > \omega_1$. As a result of the grating dispersion, the two pulses will then follow paths ABCD and ABC′D′, respectively, and one sees that the delay suffered by the pulse at frequency ω_2, $\tau_{d_2} = ABCD/\upsilon_g$, is now larger than that, $\tau_{d_1} = ABC'D'/\upsilon_g$, for the pulse at frequency ω_1. Since $\omega_2 > \omega_1$, this means that the pulse delay dispersion is now positive. It should be observed that the two-grating telescopic-system shown in Fig. 12.15 has the drawback that a lateral walk-off is present in output the beam, the amount of walk-off depending upon the difference in frequency between the beam components (e.g., between rays CD and C′D′). For beams of finite size, this walk-off can represent a problem. This problem can however be circumvented by retroreflecting the output beam back to itself by a plane mirror. In this case the overall dispersion resulting from the double pass through the system of Fig. 12.15 is, of course, twice that given by Eq. (12.5.7)

To compare the positive GDD of this pulse expander with the negative GDD of the grating-pair of Fig. 12.14, we first remember that, according to the grating equation, one has $\sin\theta_i - (\lambda/d) = \sin\theta'$, where θ_i is the angle of incidence at the grating and θ' is the corresponding diffraction angle. One can now substitute this grating equation into Eq. (12.5.6) and compare the resulting expression with Eq. (12.5.7). One then readily sees that, if $\theta' = \theta$, the two expressions become identical, apart from having opposite sign, provided that

$$(L_g/cos\,\theta) = 2f - s_1 - s_2 \tag{12.5.8}$$

It should be stressed that this equivalence only holds under the ideal conditions considered above. In this case, and when Eq. (12.5.8) applies, the expander of Fig. 12.15 is said to be conjugate to the compressor of Fig. 12.14. Physically, the conjugate nature of this expander comes about from the fact that the telescope produces an image of the first grating which is located beyond the second grating and parallel to it. The expander of Fig. 12.15 is thus equivalent to a two-parallel-grating system with negative separation and, under condition Eq. (12.5.8), this system has exactly opposite dispersion, for all orders, to that of the compressor of Fig. 12.14. In practice, due to lens aberrations and dispersion, the expander of Fig. 12.15 works well for pulse durations larger than ∼100 fs and for expansion ratios less than a few thousands. For shorter pulses and larger expansion ratios, the 1:1 telescope of the expander is usually realized via a suitably designed cylindrical-[31] or spherical-mirror configuration.[32] In particular, the use of the cylindrical-mirror configuration has resulted in an expander with expansion ratio greater than 10^4 and with second-, third-, and fourth-order dispersion being matched, for a suitable choice of material dispersion in the amplification chain, to that of the compressor.[31]

PROBLEMS

12.1. The Nd:YAG laser beam of example 12.1. is first propagated in free-space for a distance of 1 m, starting from its waist, and then focused by a positive lens with $f = 10$ cm focal length. Calculate the waist position after the lens and the spot-size parameter at this waist. [Hint: To calculate this waist position, the lens can be considered to consist of two positive lenses, f_1, and f_2 ($f_1^{-1} + f_2^{-1} = f^{-1}$), the first lens compensating the curvature of the incoming wavefront, thus producing a plane wave front, while the second focuses the beam....]

12.2. The output of a Q-switched Nd:YAG laser ($E = 100$ mJ, $\tau_p = 20$ ns) is to be amplified by a 6.3-mm-diameter Nd:YAG amplifier having a small signal gain of $G_0 = 100$. Assume that: (1) The lifetime of the lower level of the transition is much shorter than τ_p. (2) The beam transverse intensity profile is uniform. (3) The effective peak cross-section for stimulated emission is $\sigma \cong 2.8 \times 10^{-19}$ cm^2. Calculate the energy of the amplified pulse, the corresponding amplification, and the fraction of the stored energy in the amplifier that is extracted by the incident pulse.

12.3. A large Nd: glass amplifier, to be used for amplifying 1-ns laser pulses for fusion experiments, consists of a disk-amplifier with disk clear-aperture of $D = 9$ cm and overall length of the disks of 15 cm. Assume: (1) A measured small signal gain, G_0, for this amplifier of ∼4. (2) An effective, stimulated-emission peak cross-section for Nd:glass of $\sigma = 4 \times 10^{-20}$ cm^2 (see Table 9.3). (3) That the lifetime of the lower level of the transition is much shorter than the laser pulse. Calculate the total energy available in the amplifier and the required energy of the input-pulse to generate an output energy of $E_{out} = 450$ J.

12.4. Following the analysis made in deriving Eqs. (12.3.1) and (12.3.9) (assume $\alpha = 0$) as well as the rate-equation calculation for a quasi-three level laser [see Eqs. (7.2.21)–(7.2.24)], prove Eq. (12.3.15).

12.5. With reference to problem 12.2., assume now that the input pulse duration is much shorter than the lifetime, τ_1, of the lower laser level ($\tau_1 \cong 100$ ps). Using data obtained in example 2.10. and knowing that the fractional population of the lower laser sub-level of the $^4I_{11/2}$ state is $f_{13} \cong 0.187$, calculate the energy of the amplified pulse and the corresponding amplification. Compare the results with those obtained in problem 12.2.

12.6. A large CO_2 TEA amplifier (with a gas mixture CO_2:N_2:He in the proportion 3:1.4:1) has dimensions of 10 × 10 × 100 cm. The small signal gain coefficient for the $P(22)$ transition has been measured to be $g = 4 \times 10^{-2}$ cm^{-1}. The duration of the input light pulse is 200 ns, which can therefore be assumed much longer than the thermalization time of the rotational levels of both the upper and lower vibrational states. The laser pulse is however much shorter than the decay time of the lower laser level. The true peak cross-section for the $P(22)$ transition is $\sigma \cong 1.54 \times 10^{-18}\,\text{cm}^2$, while, for $T = 300\,\text{K}$, the fractional population of both initial and final rotational states can be calculated to be $f = 0.07$. Calculate the output energy and the gain available from this amplifier for an input energy of 17 J. Also calculate the energy per unit volume available in the amplifier.

12.7. Prove Eq. (12.3.12).

12.8. The frequency of a Nd:YAG laser beam ($\lambda = 1.06\,\text{m}$) is to be doubled in a KDP crystal. Knowing that, for KDP, $n_o(\lambda = 1.06\,\mu\text{m}) = 1.507$, $n_o(\lambda = 532\,\text{nm}) = 1.5283$, and $n_e(\lambda = 532\,\text{nm}) = 1.48222$, calculate the phase-matching angle, θ_m.

12.9. Prove Eq. (12.4.30).

12.10. Calculate the threshold pump intensity for parametric oscillation at $\lambda_1 \cong \lambda_2 = 1\,\mu\text{m}$ in a 5-cm-long $LiNbO_3$ crystal pumped at $\lambda_3 = 0.5\,\mu\text{m}$ [$n_1 = n_2 = 2.16$, $n_3 = 2.24$, $d \cong 6 \times 10^{-12}\,\text{m/V}$, $\gamma_1 = \gamma_2 = 2 \times 10^{-2}$]. If the pumping beam is focused in the crystal to a spot of $\sim$100 μm diameter, calculate the resulting threshold pump power.

12.11. Calculate the second-harmonic conversion efficiency via type I second-harmonic generation in a perfectly phase-matched 2.5-cm-long KDP crystal for an incident beam at $\lambda = 1.06\,\text{m}$ having an intensity of $100\,\text{MW/cm}^2$ [for KDP $n \cong 1.5$, $d_{eff} = d_{36}\,\sin\theta_m = 0.28 \times 10^{-12}\,\text{m/V}$, where $\theta_m \cong 50°$ is the phase-matching angle].

References

1. R. W. Boyd, *Nonlinear Optics* (Academic Press, New York, 1992).
2. A. Yariv, *Optical Electronics* fourth edn. (Holt Rinehart and Winston, New York, 1991), Chaps. 9 and 12.
3. O. Svelto, Self-Focusing, Self-Steepening and Self-Phase-Modulation of Laser Beams, in *Progress in Optics*, ed. by E. Wolf (North-Holland, Amsterdam 1974), Vol. XII, pp. 3–50.
4. A. E. Siegman, New Developments in Laser Resonators, in *Laser Resonators* ed. by D. A. Holmes, *Proc. SPIE*, **1224**, 2–14 (1990).
5. A. E. Siegman, Defining and Measuring Laser Beam Quality, in *Solid State Lasers-New Develpments and Applications*, ed. by M. Inguscio and R. Wallenstein (Plenum, New York 1993) pp 13–28.
6. L. M. Franz and J. S. Nodvick, Theory of Pulse Propagation in a Laser Amplifier, *J. Appl. Phys.*, **34**, 2346–2349 (1963).
7. P. G. Kriukov and V. S. Letokhov, Techniques of High-Power Light-Pulse Amplification, in *Laser Handbook*, ed. by F. T. Arecchi and E. O. Schultz-Dubois (North-Holland, Amsterdam, 1972), Vol. 1, pp. 561–595.
8. W. Koechner, *Solid-State Laser Engineering*, fourth edn. (Springer, Berlin 1996), Chap. 4.
9. D. Strickland and G. Mourou, Compression of Amplified Chirped Optical Pulses, *Opt. Commun.*, **56**, 219–221 (1985).
10. G. Mourou, The Ultra-High-Peak-Power Laser: Present and Future, *Appl. Phys. B*, **65**, 205–211 (1997).
11. M. D. Perry *et al.*, The Petawatt Laser and its Application to Inertial Confinement Fusion, *CLEO '96 Conference Digest* (Optical Society of America, Whashington) paper CWI4.
12. Proceedings of the *International Conference on Superstrong Fields in Plasmas*, ed. by M. Lontano, G. Mourou, F. Pegoraro, and E. Sindoni (American Institute of Physics Series, New York 1998).
13. R. J. Mears, L. Reekie, I. M. Jauncey, and D. N. Payne, Low Noise Erbium-Doped Fiber Amplifier at 1.54 μm, *Electron. Lett.*, **23**, 1026–1028 (1987).
14. Emmanuel Desurvire, *Erbium-Doped Fiber Amplifiers* (John Wiley and Sons, New York, 1994).

15. R. L. Byer, Optical Parametric Oscillators, in *Quantum Electronics*, ed. by H. Rabin and C. L. Tang (Academic, New York, 1975), Vol. 1, Part B, pp. 588–694.
16. P. A. Franken, A. E. Hill, C. W. Peters, and G. Weinreich, Generation of Optical Harmonics, *Phys. Rev. Lett.* 7, 118 (1961).
17. J. A. Giordmaine and R. C. Miller, Tunable Optical Parametric Oscillation in $LiNbO_3$ at Optical Frequencies, *Phys. Rev. Lett.* **14**, 973 (1965).
18. J. A. Giordmaine, Mixing of Light Beams in Crystals, *Phys. Rev. Lett.* **8**, 19 (1962).
19. P. D. Maker, R. W. Terhune, M. Nisenhoff, and C. M. Savage, Effects of Dispersion and Focusing on the Production of Optical Harmonics, *Phys. Rev. Lett.* **8**, 21 (1962).
20. F. Zernike and J. E. Midwinter, *Applied Nonlinear Optics* (Wiley, New York, 1973), Sec. 3.7.
21. D. Grischkowsky and A. C. Balant, Optical Pulse Compression Based on Enhanced Frequency Chirping, *Appl. Phys. Lett.* **41**, 1 (1982).
22. G. P. Agrawal, *Nonlinear Fiber Optics*, second edn. (Academic, San Diego, 1995) Chapter 2.
23. E. B. Treacy, Optical Pulse Compression with Diffraction Gratings, *IEEE J. Quantum Electron.* **QE-5**, 454 (1969).
24. Reference [22] Chapter 6.
25. R. L. Fork et al., Compression of Optical Pulses to Six Femtosecond by Using Cubic Phase Compensation, *Opt. Lett.*, **12**, 483–485 (1987).
26. M. Nisoli, S. De Silvestri and O. Svelto, Generation of High Energy 10 fs Pulses by a New Compression Technique, *Appl. Phys. Letters*, **68**, 2793–2795 (1996).
27. R. Szipöcs, K. Ferencz, C. Spielmann, F. Krausz, Chirped Multilayer Coatings for Broadband Dispersion Control in Femtosecond Lasers, *Opt. Letters*, **19**, 201–203 (1994).
28. M. Nisoli et al., Compression of High-Energy Laser Pulses below 5 fs, *Opt. Letters*, **22**, 522–524 (1997).
29. M. Pessot, P. Maine and G. Mourou, 1000 Times Expansion-Compression Optical Pulses for Chirped Pulse Amplification, *Opt. Comm.*, **62**, 419–421 (1987).
30. O. E. Martinez, 3000 Times Grating Compressor with Positive Group Velocity Dispersion: Application to Fiber Compensation in 1.3–1.6 μm Region, *IEEE J. Quantum Electron.*, **QE-23**, 59–64 (1987).
31. B. E. Lemoff and C. P. J. Barty, Quintic-Phase-Limited, Spatially Uniform Expansion and Recompression of Ultrashort Optical Pulse, *Opt. Letters*, **18**, 1651–1653 (1993).
32. Detao Du *et al.*, Terawatt Ti:Sapphire Laser with a Spherical Reflective-Optic Pulse Expander, *Opt. Letters*, **20**, 2114–2116 (1995).

A

Semiclassical Treatment of the Interaction of Radiation with Matter

The calculation that follows will make use of the so-called semiclassical treatment of the interaction of radiation with matter. In this treatment the atomic system is assumed to be quantized and it is therefore described quantum mechanically while the e.m. radiation is treated classically, i.e., by using Maxwell's equations.

We will first examine the phenomenon of absorption. We therefore consider the usual two-level system where we assume that, at time $t = 0$, the atom is in its ground state 1 and that a monochromatic e.m. wave at frequency ω made to interact with it. Classically, the atom has an additional energy H' when interacting with the e.m. wave. For instance this may be due to the interaction of the electric dipole moment of the atom $\boldsymbol{\mu}_e$ with the electric field $\mathbf{E}$ of the e.m. wave ($H' = \boldsymbol{\mu}_e \cdot \mathbf{E}$), referred to as as an electric dipole interaction. This is not the only type of interaction through which the transition can occur, however. For instance, the transition may result from the interaction of the magnetic dipole moment of the atom $\boldsymbol{\mu}_m$ with the magnetic field $\mathbf{B}$ of the e.m. wave ($\boldsymbol{\mu}_m \cdot \mathbf{B}$, magnetic dipole interaction). To describe the time evolution of this two-level system, we must now resort to quantum mechanics. Thus, just as the classical treatment involves an interaction energy H', so the quantum mechanical approach introduces an interaction Hamiltonian $\mathcal{H}'$. This Hamiltonian can be obtained from the classical expression for H' according to the well-known rules of quantum mechanics. The precise expression for $\mathcal{H}'$ need not concern us at this point, however. We only need to note that $\mathcal{H}'$ is a sinusoidal function of time with frequency equal to that of the incident wave. Accordingly we put

$$\mathcal{H}' = \mathcal{H}_{0'} \sin \omega t \tag{A.1}$$

The total Hamiltonian $\mathcal{H}$ for the atom can then be written as

$$\mathcal{H} = \mathcal{H}_0 + \mathcal{H}' \tag{A.2}$$

where $\mathcal{H}_0$ is the atomic Hamiltonian in the absence of the e.m. wave. Once the total Hamiltonian $\mathcal{H}$ for $t > 0$ is known, the time evolution of the wave function ψ of the atom is obtained from the time-dependent Schrödinger equation

$$\mathcal{H}\psi = j\hbar\frac{\partial\psi}{\partial t} \tag{A.3}$$

To solve Eq. (A.3) for the wave-function $\psi(t)$ we begin by introducing, according to Eq. (2.3.1), $\psi_1 = u_1 \exp[-(jE_1 t/\hbar)]$ and $\psi_2 = u_2 \exp[-(jE_2 t/\hbar)]$, as the unperturbed eigenfunctions of levels 1 and 2, respectively. Thus u_1 and u_2 satisfy the time-independent Schrödinger wave-equation

$$\mathcal{H}_0 u_i = E_i u_i \ (i = 1, 2) \tag{A.4}$$

Under the influence of the e.m. wave, the wave-function of the atom can be written as

$$\psi = a_1(t)\psi_1 + a_2(t)\psi_2 \tag{A.5}$$

where a_1 and a_2 will be time-dependent complex numbers that, according to quantum mechanics, obey the relation

$$|a_1|^2 + |a_2|^2 = 1 \tag{A.6}$$

Since, according to Eq. (1.1.6), one has $W_{12} = -d|a_1(t)|^2/dt = d|a_2(t)|^2/dt$, to calculate W_{12}, we must calculate the function $|a_2(t)|^2$. To do this we will first generalize Eq. (A.5) as

$$\psi = \sum_1^m a_k\psi_k = \sum_1^m a_k\, u_k \exp[-j(E_k/\hbar)\, t] \tag{A.7}$$

where k denotes a general state of the atom and m gives the number of these states. By substituting Eq. (A.7) into Eq. (A.3) we obtain

$$\begin{aligned}\sum_k(\mathcal{H}_0 + \mathcal{H}')a_k\, u_k \exp[-j(E_k/\hbar)\, t] = \sum_k \{&j\hbar(da_k/dt)u_k \exp[-j(E_k/\hbar)\, t] \\ &+ a_k u_k E_k \exp[-j\,(E_k/\hbar)\, t]\}\end{aligned} \tag{A.8}$$

This equation, with the help of Eq. (A.4), reduces to

$$\sum j\hbar(da_k/dt)u_k \exp[-j(E_k/\hbar)\, t] = \sum a_k\mathcal{H}' u_k \exp[-j(E_k/\hbar)\, t] \tag{A.9}$$

By multiplying each side of this equation by the arbitrary eigenfunction u_n^* and then integrating over the whole space, we obtain

$$\begin{aligned}&\sum j\hbar(da_k/dt) \exp[-j(E_k/\hbar)\, t] \int u_k\, u_n^* dV \\ &= \sum a_k \exp[-j(E_k/\hbar)t] \int u_n^* \mathcal{H}' u_k\, dV\end{aligned} \tag{A.10}$$

Since the wave functions u_k are orthogonal (i.e., $\int u_n^* u_k \, dV = \delta_{kn}$), Eq. (A.10) gives

$$(da_n/dt) = \frac{1}{(j\hbar)} \sum_1^m {}_k H'_{nk} a_k \exp\left[-j\frac{(E_k - E_n)t}{\hbar}\right] \tag{A.11}$$

where $H'_{nk} = H'_{nk}(t)$ is given by

$$H'_{nk}(t) = \int u_n^* \mathcal{H}' \, u_k \, dV \tag{A.12}$$

Equation (A.11) comprises a set of m differential equations for the m variables $a_k(t)$, and these equations can be solved once the initial conditions are known. For the simpler case of a two-level system the wave-function ψ is given by Eqs. (A.5) and (A.11) reduces to the two equations

$$\left(\frac{da_1}{dt}\right) = \left(\frac{1}{j\hbar}\right) \left\{H'_{11} \, a_1 + H'_{12} \, a_2 \, \exp\left[-j(E_2 - E_1)\frac{t}{\hbar}\right]\right\} \tag{A.13a}$$

$$\left(\frac{da_2}{dt}\right) = \left(\frac{1}{j\hbar}\right) \left\{H'_{21} \, a_1 \exp\left[-j(E_1 - E_2)\,\frac{t}{\hbar}\right] + H'_{22} \, a_2\right\} \tag{A.13b}$$

which are to be solved with the initial condition $a_1(0) = 1$, $a_2(0) = 0$.

So far, no approximations have been made. Now, to simplify the solution of the Eq. (A.13), we will make use of a perturbation method. We will assume that on the right-hand side of Eq. (A.13) we can make the approximations that $a_1(t) \cong 1$ and $a_2(t) \cong 0$. By solving Eq. (A.13) subject to this approximation, we obtain the first-order solutions for $a_1(t)$ and $a_2(t)$. For this reason, the theory that follows is known as *first-order perturbation theory*. The solutions $a_1(t)$ and $a_2(t)$ obtained in this way can then be substituted in the right-hand side of Eq. (A.13) to get a solution which could then be a second-order approximation, and so on to higher orders. To first order, therefore, Eq. (A.13) gives

$$\left(\frac{da_1}{dt}\right) = \left(\frac{1}{j\hbar}\right) H'_{11} \tag{A.14a}$$

$$\left(\frac{da_2}{dt}\right) = \left(\frac{1}{j\hbar}\right) H'_{21} \exp\,(j\omega_0 t) \tag{A.14b}$$

where we have written $\omega_0 = (E_2 - E_1)/\hbar$ for the transition frequency of the atom. To calculate the transition probability, we need only to solve Eq. (A.14b). Thus, making use of Eqs. (A.1) and (A.12), we write

$$H'_{21} = H'^0_{21} \sin \omega \, t = H'^0_{21}[\exp(j\,\omega\, t) - \exp(-j\,\omega\, t)]/2j \tag{A.15}$$

where H'^0_{21} is given by

$$H'^0_{21} = \int u_2^* \, \mathcal{H}_{0'} \, u_1 \, dV \tag{A.16}$$

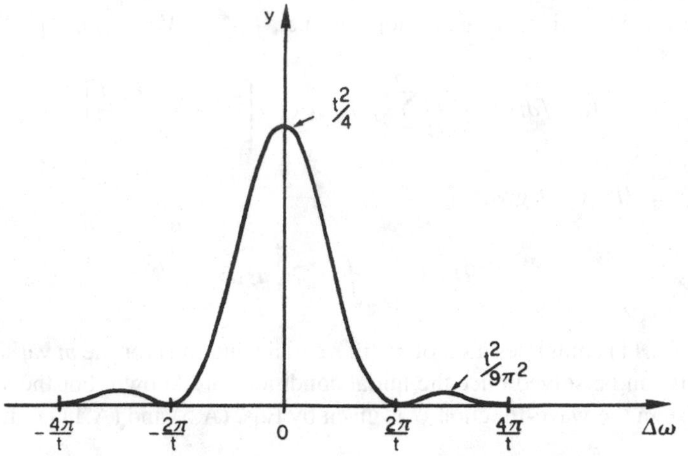

FIG. A.1. Plot of the function $y = [\sin(\Delta\omega\, t\,/\,2)\,/\,\Delta\omega]^2$ versus $\Delta\omega$.

and is, in general, a complex constant. By substituting Eq. (A.15) into Eq. (A.14b) and integrating with the initial condition $a_2(0) = 0$, we obtain

$$a_2(t) = \frac{H_{21}'^0}{2j\hbar}\left[\frac{\exp[j(\omega_0-\omega)t]-1}{\omega_0-\omega} - \frac{\exp[j(\omega_0+\omega)t]-1}{\omega_0+\omega}\right] \quad \text{(A.17)}$$

If we now assume that $\omega \cong \omega_0$, we see that the first term in the square brackets is much larger than the second. We can then write

$$a_2(t) \cong -\frac{H_{21}'^0}{2j\hbar}\left[\frac{\exp(-j\Delta\omega\, t)-1}{\Delta\omega}\right] \quad \text{(A.18)}$$

where $\Delta\omega = \omega - \omega_0$. One can then readily calculate that

$$|a_2(t)|^2 = \frac{|H_{21}'^0|^2}{\hbar^2}\left[\frac{\sin(\Delta\omega\, t\,/\,2)}{\Delta\omega}\right]^2 \quad \text{(A.19)}$$

The function $y = [\sin(\Delta\omega t/2)/\Delta\omega]^2$ is plotted in Fig. A.1 vs $\Delta\omega$ and one can then see that the peak value of the function becomes greater and its width narrower as t increases. Furthermore, one can show that

$$\int_{-\infty}^{+\infty}\left[\frac{\sin(\Delta\omega\, t\,/\,2)}{\Delta\omega}\right]^2 d\Delta\omega = \frac{\pi\, t}{2} \quad \text{(A.20)}$$

For large enough values of t we can then put

$$\left[\frac{\sin(\Delta\omega\, t\,/\,2)}{\Delta\omega}\right]^2 \cong \frac{\pi\, t}{2}\delta(\Delta\omega) \quad \text{(A.21)}$$

where δ is the Dirac δ function. Within this approximation Eq. (A.19) gives

$$|a_2(t)|^2 = \frac{|H'^0_{21}|^2}{\hbar^2}\frac{\pi}{2}t\delta(\Delta\omega) \tag{A.22}$$

which shows that, for long enough times, the probability $|a_2(t)|^2$ of finding the atom in level two is proportional to the time t. Consequently, the transition probability W_{12} is obtained as

$$W_{12} = \frac{d|a_2(t)|^2}{dt} = \frac{\pi}{2}\frac{|H'^0_{21}|^2}{\hbar^2}\delta(\Delta\omega) \tag{A.23}$$

To calculate W_{12} explicitly, we must now calculate the quantity $\left|H'^0_{21}\right|^2$. Let us assume that the interaction responsible for the transition occurs between the electric field of the e.m. wave and the electric dipole moment of the atom (electric-dipole interaction). Classically, if we let $\mathbf{r}$ be the vector that specifies the position of the electron with respect to the nucleus and e the magnitude of the electron charge, the corresponding dipole moment of the atom will be $\boldsymbol{\mu}_e = -e\mathbf{r}$. The classical interaction energy H' is then given by $H' = \boldsymbol{\mu}_e \cdot \mathbf{E} = -e\mathbf{E}(\mathbf{r}, t) \cdot \mathbf{r}$, where $\mathbf{E}$ is the electric field of the incident e.m. wave at the electron position. Following the well-known rules of quantum mechanics, the interaction Hamiltonian is then simply given by

$$\mathcal{H}' = -e\mathbf{E}(\mathbf{r}, t) \cdot \mathbf{r} \tag{A.24}$$

Substitution of Eq. (A.24) into Eq. (A.12) with $n = 2$ and $k = 1$, gives

$$H'_{21} = -e\int u_2^* \mathbf{E} \cdot \mathbf{r}\, u_1\, dV \tag{A.25}$$

Let us now suppose that the wavelength of the e.m. wave is much larger than the atomic dimension. This is satisfied very well for e.m. waves in the visible. One has in fact $\lambda = 500$ nm for green light while typical atomic dimensions are $\sim$0.1 nm). In this case, we can assume that $\mathbf{E}(\mathbf{r}, t)$ does not change appreciably over an atomic dimension and thus remains equal to its value, $\mathbf{E}(0, t)$, at $\mathbf{r} = 0$, i.e., at the center of the nucleus (*electric-dipole approximation*). We can thus write

$$\mathbf{E}(\mathbf{r}, t) \cong \mathbf{E}(0, t) = \mathbf{E}_0 \sin\omega t \tag{A.26}$$

where $\mathbf{E}_0$ is a constant. If Eq. (A.26) is now substituted into Eq. (A.25) and the resulting expression for H'_{21} compared with that given in Eq. (A.15), one finds that H'^0_{21} can be expressed as

$$H'^0_{21} = \mathbf{E}_0 \cdot \boldsymbol{\mu}_{21} \tag{A.27}$$

where $\boldsymbol{\mu}_{21}$ is given by

$$\boldsymbol{\mu}_{21} = -\int u_2^* e\mathbf{r} u_1\, dV \tag{A.28}$$

and is called the matrix element of the electric dipole moment. If we now let as $\cos\theta$ be the angle between $\boldsymbol{\mu}_{21}$ and $\mathbf{E}_0$ we obtain from Eq. (A.27) that

$$\left|H'^{0}_{21}\right|^2 = E_0^2\left|\mu_{21}\right|^2\cos^2\theta \tag{A.29}$$

where $|\mu_{21}|$ is the magnitude of the complex quantity $\boldsymbol{\mu}_{21}$. If we now assume that the e.m. wave interacts with several atoms whose vectors $\boldsymbol{\mu}_{21}$ are randomly oriented with respect to $\mathbf{E}_0$, the average value of $\left|H'^{0}_{21}\right|^2$ will be obtained by averaging Eq. A.29 over all possible angles θ and ϕ (in two dimensions). To carry this out, let $p(\theta)$ represent the probability density for the orientation of the atomic dipoles, so that $p(\theta)d\Omega$ gives the elemental probability that the vector $\boldsymbol{\mu}_{21}$ is within the solid angle $d\Omega$ making an angle θ with the direction of $\mathbf{E}_0$. For randomly oriented dipoles, one has $p(\theta) = \text{const.}$ and accordingly one has $< \cos^2\theta >= 1/3$, where the brackets $<>$ indicate the average value over all dipole orientations. From Eq. (A.29) we then get

$$\left\langle |H'^{0}_{21}|^2 \right\rangle = E_0^2|\mu_{21}|^2/3 \tag{A.30}$$

The substitution of this last expression into Eq. (A.23) then gives

$$W_{12} = \frac{\pi}{6}\frac{(2\pi)^2E_0^2|\mu_{21}|^2}{h^2}\delta(\omega-\omega_0) \tag{A.31}$$

If the function $\delta(\nu-\nu_0)$ is used instead of $\delta(\omega-\omega_0)$ in Eq. (A.31) then, since $\delta(\nu-\nu_0) = 2\pi\delta(\omega-\omega_0)$, Eq. (A.31) transforms to Eq. (2.4.5).

Having calculated the rate of absorption, we can now go on to calculate the rate of stimulated emission. To do this, we should start again from Eq. (A.13), this time using the initial conditions $a_1(0) = 0$ and $a_2(0) = 1$. We immediately see, however, that the required equations in this case are obtained from the corresponding Eqs. [(A.13)–(A.31)] for the case of absorption simply by interchanging the indexes 1 and 2. Since it can be seen from Eq. (A.28) that $|\mu_{12}| = |\mu_{21}|$, it follows from Eq. (A.31) that $W_{12} = W_{21}$, which shows that the probabilities of absorption and stimulated emission are equal.

B

Lineshape Calculation for Collision Broadening

As explained in Sect. 2.5.1, the calculation of the lineshape for collision broadening can be obtained from the normalized spectral density, $g(\nu' - \nu)$, of the sinusoidal waveform with random phase-jumps shown in Fig. 2.9. The signal wave of Fig. 2.9 will be written as

$$E(t) = E_0 \exp j[\omega t - \phi(t)] \tag{B.1}$$

where E_0 is taken to be a real constant, $\omega = 2\pi\nu$ is the angular frequency of the radiation and where the phase $\phi(t)$ is assumed to undergo random jumps at each atom's collision. We will assume that the probability density, $p_\tau(\tau)$, of the time τ between two consecutive collisions can be described by Eq. (2.5.7). The calculation of $g(\nu' - \nu)$ is best done in terms of the angular frequency ω, i.e., in terms of the distribution, $g(\omega' - \omega)$. Since obviously one has $g(\nu' - \nu)d\nu' = g(\omega' - \omega)d\omega'$, it follows that $g(\nu' - \nu) = 2\pi g(\omega' - \omega)$. Apart from a proportionality constant, the function $g(\omega' - \omega)$ is then given by the power spectrum $W(\omega' - \omega)$ of the waveform shown in Fig. 2.9. For this proportionality constant to be unity, we require, according to Eq. (2.5.4), that $W(\omega' - \omega)$ be such that $\int W(\omega' - \omega)d\omega' = 1$. From Parseval's theorem one then has

$$\int_{-\infty}^{+\infty} W(\omega' - \omega)d\omega' = \lim_{T\to\infty} \frac{\pi}{T} \int_{-T}^{+T} |E(t)|^2 dt = 2\pi E_0^2 \tag{B.2}$$

The condition $\int W(\omega' - \omega)d\omega' = 1$ then leads to the result $2\pi E_0^2 = 1$. This means that $g(\omega' - \omega)$ can be obtained as the power spectrum of the signal $E(t)$ given by Eq. (B.1) and with amplitude

$$E_0 = (2\pi)^{-1/2} \tag{B.3}$$

To calculate the power spectrum $W(\omega' - \omega)$ of the signal given in Eq. (B.1) we make use of the Wiener-Kintchine theorem so that $W(\omega' - \omega)$ is obtained as the Fourier transform of the signal autocorrelation function, $\Gamma(\tau)$. We can thus write

$$W(\omega' - \omega) = \int_{-\infty}^{+\infty} \Gamma(\tau) \exp -(j\omega'\tau)\, d\tau \tag{B.4}$$

where the autocorrelation function $\Gamma(\tau)$ is given by

$$\Gamma(\tau) = \lim_{T\to\infty} \frac{1}{2T} \int_{-T}^{+T} E^*(t)E(t+\tau)dt \tag{B.5}$$

For the waveform shown in Fig. 2.9, we can then write

$$\Gamma(\tau) = \lim_{T\to\infty} \frac{1}{2T} E_0^2 \exp(j\omega\tau) \left\{ \int_{corr.} dt + \int_{uncorr.} \exp[-j(\Delta\phi)]dt \right\} \tag{B.6}$$

The first integral in the right hand side of Eq. (B.6) is calculated over the time intervals, between $-T$ and $+T$, in which no phase-jumping collision has occurred and thus the signals $E(t)$ and $E(t+\tau)$ have the same phase (*correlated intervals*). The second integral in the right hand side of Eq. (B.6) is calculated over the time intervals in which a collision has occurred and thus the two signals $E(t)$ and $E(t+\tau)$ have a random phase difference $\Delta\phi$ (*uncorrelated intervals*). This situation can be illustrated schematically as in Fig. B.1, where the vertical bars (continuous-line) indicate the instants of phase jumps for both $E(t)$ and $E(t+\tau)$. On projecting the vertical bars of e.g., $E(t+\tau)$ onto the plot of $E(t)$ (dashed vertical lines), the correlated time intervals, shown as hatched areas, are obtained.

To calculate $\Gamma(\tau)$ from Eq. (B.6) we first notice that the contribution of the second integral in the right hand side of Eq. (B.6) vanishes as T tends to infinite because the integrand $\exp[-j(\Delta\phi)]$ is a random number with zero average. Equation (B.6) then reduces to

$$\Gamma(\tau) = \lim_{T\to\infty} \frac{E_0^2}{2T} \exp(j\omega\tau) \left[\int_{corr.} dt \right] \tag{B.7}$$

The integral in Eq. (B.7) is equal to the total time of phase correlation, i.e. to the sums of the time intervals corresponding to the hatched areas in Fig. B.1. If we let τ'_n represent the time interval between the n-th phase-jump and the next one (see Fig. B.1), this sum can be expressed as $\sum_n (\tau'_n - \tau)$, where summation is extended over the values τ'_n for which $\tau'_n > \tau$. Equation (B.7) can then be written as

$$\Gamma(\tau) = E_0^2 \exp(j\omega\tau) \lim_{T\to\infty} \frac{\sum_n (\tau'_n - \tau)}{2T} \tag{B.8}$$

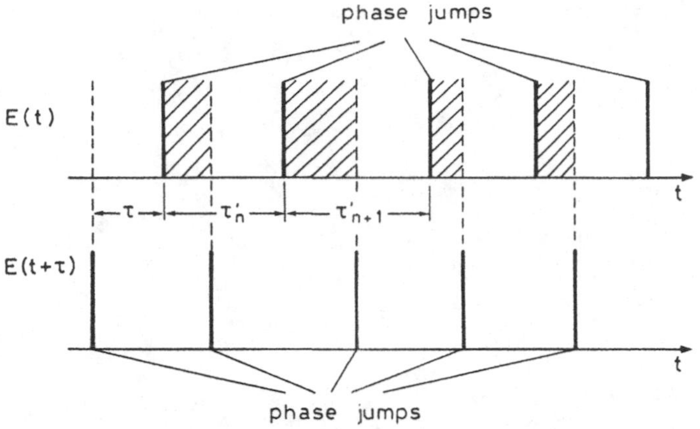

FIG. B.1. Plots of the phase jumps of the two functions $E(t)$ and $E(t+\tau)$ with the time intervals where the two functions are correlated indicated as hatched areas.

If we now let N be the number of phase-jumps between $-T$ and $+T$, we can write

$$\lim_{T\to\infty} \frac{\sum_n (\tau_n' - \tau)}{2T} = \lim_{T\to\infty} \frac{[\sum_n (\tau_n' - \tau)]/N}{2T/N} = \frac{<(\tau' - \tau)>}{\tau_c} \tag{B.9}$$

where $\tau_c = 2T/N$ is the average time between two consecutive phase-jumps and $<(\tau' - \tau)>$ is the average value of $(\tau' - \tau)$ (with the constraint $\tau' > \tau$). Using the probability density of the time intervals between consecutive jumps given by Eq. (2.5.7), the value $<(\tau' - \tau)>$ can be obtained as

$$<\tau' - \tau> = (1/\tau_c) \int_{\tau}^{+\infty} (\tau' - \tau) \exp -(\tau'/\tau_c)\, d\tau' = \tau_c \exp -(\tau/\tau_c) \tag{B.10}$$

From Eqs. (B.8), (B.9) and (B.10) we get the expression for the correlation function as $\Gamma(\tau) = E_0^2 \exp[j\omega\tau - (\tau/\tau_c)]$. If $\Gamma(\tau)$ is now extended also to the case $\tau < 0$, then, since one must have $\Gamma(-\tau) = \Gamma^*(\tau)$, we can write our final result as

$$\Gamma(\tau) = E_0^2 \exp[j\omega\tau - (|\tau|/\tau_c)] \tag{B.11}$$

With the help of the Wiener-Kintchine theorem, see Eq. (B.4), we can now readily calculate $W(\omega' - \omega)$ and hence, using Eq. (B.3) for E_0, also $g(\omega' - \omega)$. We obtain

$$g(\omega' - \omega) = \frac{\tau_c}{\pi} \frac{1}{[1 + (\omega' - \omega)^2 \tau_c^2]} \tag{B.12}$$

Since $g(\nu' - \nu) = 2\pi g(\omega' - \omega)$, we can also write [see Eq. (2.5.9)]

$$g(\nu' - \nu) = 2\tau_c \frac{1}{[1 + (\nu' - \nu)^2 4\pi^2 \tau_c^2]} \tag{B.13}$$

from which the lineshape function given by Eq. (2.5.10) is readily obtained.

C

Simplified Treatment of Amplified Spontaneous Emission

We will assume that ASE emission occurs in both directions within the active medium and therefore refer to the geometry shown in Fig. C.1. We will further assume that the amplifier behaves as an ideal four-level system, so that the lower level population can be neglected. We will consider both homogeneously and inhomogeneously broadened transitions. A detailed theory that holds for these conditions has been developed by Casperson.[1] The theory is rather complicated, however, and as a result one cannot readily obtain the main aspects of the physical behavior of ASE from this treatment. Following some very recent work,[2] we present here a simplified treatment of ASE aimed at obtaining some asymptotic expressions for ASE behavior in the low saturation regime.

In the low saturation regime, we assume that the upper state population, N_2, and hence the population inversion $N \cong N_2$, are not appreciably saturated by the ASE intensity. With reference to Fig. C.1, we let $I_\nu(z, \nu)$ be the spectral ASE intensity at coordinate z for the beam propagating in the positive z-direction. The elemental variation, dI_ν, along the z-coordinate, must account not only for the stimulated emission but also for the spontaneous-emission contribution arising from the element dz. We can thus write

$$\frac{\partial I_\nu}{\partial z} = \sigma N I_\nu + N A_\nu \frac{\Omega(z)}{4\pi} \tag{C.1}$$

where σ is the transition cross section at frequency ν, A_ν is the rate of spontaneous emission at frequency ν and $\Omega(z)$ is the solid angle subtended by the exit face as seen from the element dz. Note that the factor $\Omega(z)/4\pi$, in the right hand side of Eq. (C.1), accounts for the fact that the spontaneous radiation from the element dz is emitted uniformly over the whole solid angle, 4π, while we are only interested in that fraction emitted within the solid angle $\Omega(z)$. To calculate the ASE spectral intensity $I_\nu(l, \nu)$ at the $z = l$ exit face of the medium, we must integrate Eq. (C.1) over the z-coordinate. Noting that most of ASE emission arises from emitting elements near $z = 0$, which experience the largest gain, we can assume $\Omega(z) \cong \Omega$,

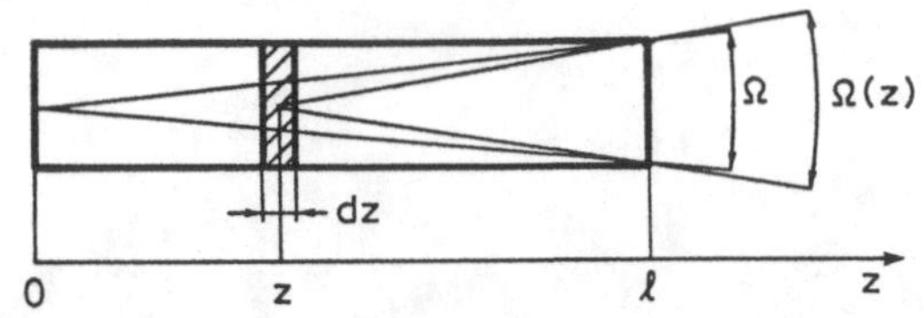

FIG. C.1. Calculation of the ASE spectral emission along the z-direction.

where Ω is the solid angle subtended by one face of the active medium as seen from the other face (see Fig. C.1). Equation (C.1) can then be readily integrated and, using the boundary condition $I_\nu(0, \nu) = 0$, one obtains

$$I_\nu(l, \nu) = \frac{\Omega}{4\pi} A_\nu \frac{h\nu[\exp(\sigma Nl) - 1]}{\sigma} \tag{C.2}$$

Equation (C.2) can be rearranged by noting that $A_\nu = Ag(\nu - \nu_0)$ and $\sigma = \sigma_p g(\nu - \nu_0)/g_p$, where A is the rate of spontaneous emission, $g(\nu - \nu_0)$ is the transition lineshape function, g_p is its peak value, σ_p is the peak cross-section and ν_0 is the frequency of the transition peak. Equation (C.2) can then be written as

$$I_\nu(l, \nu) = \phi I_s \frac{\Omega}{4\pi} g_p[\exp(\sigma Nl) - 1] \tag{C.3}$$

where $\phi = \tau/\tau_r = \tau A$ is the fluorescence quantum yield [see Eq. (2.6.22)] and $I_s = h\nu/\sigma_p\tau$ is the saturation intensity of the amplifier [see Eq. (2.8.24)].

The ASE emission at a general frequency ν, normalized to its peak value, is readily obtained from Eq. (C.3) as $[I_\nu(l, \nu)/I_\nu(l, \nu_0)] = [\exp(\sigma Nl) - 1]/[\exp(\sigma_p Nl) - 1]$. For both Lorentzian and Gaussian lines, this spectral emission can be readily computed for a given value of the peak gain, $G = \exp(\sigma_p Nl)$. As an example, Fig. 2.24 shows, as solid lines, the computed spectral profiles vs the normalized frequency offset, $2(\nu-\nu_0)/\Delta\nu_0$, for a Lorentzian line with peak gains of 10^3 and 10^6, respectively. On the other hand, an approximate expression for the ASE linewidth, $\Delta\nu_{ASE}$, can be obtained from Eq. (C.3) if we assume that the ASE spectrum can be approximated by a Gaussian function. We accordingly write

$$[\exp(\sigma Nl) - 1] \cong [\exp(\sigma_p Nl) - 1] \times \exp(-kx^2) \tag{C.4}$$

where k is a constant and x represents the normalized frequency offset i.e.,

$$x = 2(\nu - \nu_0)/\Delta\nu_0 \tag{C.5}$$

where $\Delta\nu_0$ is the transition linewidth (for either a homogeneously or inhomogeneously broadened transition). From Eq. (C.4) we readily obtain the expression for kx^2 as

$$kx^2 \cong \ln[\exp(\sigma_p Nl) - 1] - \ln[\exp(\sigma Nl) - 1] \tag{C.6}$$

If we now let $f(x)$ represent the function on the right hand side of Eq. (C.6), the constant k can be readily obtained from the relation

$$k = \frac{1}{2}\left[\frac{d^2 f}{dx^2}\right]_{x=0} \tag{C.7}$$

For a Lorentzian line, we now write $\sigma = \sigma_p/(1 + x^2)$ in Eq. (C.6). After some lengthy but straightforward calculations, we then get from Eq. (C.7)

$$k = \frac{G \ln G}{(G-1)} \tag{C.8}$$

Similarly, for the case of a Gaussian line, we write $\sigma = \sigma_p \exp-(x^2 \ln 2)$ in Eq. (C.6) and thus obtain from Eq. (C.7)

$$k = (\ln 2)\frac{G \ln G}{(G-1)} \tag{C.9}$$

With the above Gaussian approximation, the normalized ASE linewidth, in terms of the normalized frequency offset x, is now readily obtained as $\Delta x_{ASE} = 2(\ln 2/k)^{1/2}$. Since, from Eq. (C.5), one has $\Delta\nu_{ASE} = \Delta x_{ASE} \times (\Delta\nu_0/2)$, the ASE linewidth can be found from the two previous expressions once the value for k given by Eq. (C.8) or (C.9) is used. We thus obtain

$$\Delta\nu_{ASE} = [\ln 2]^{1/2}\left[\frac{G-1}{G \ln G}\right]^{1/2} \Delta\nu_0 \tag{C.10}$$

for a Lorentzian line and

$$\Delta\nu_{ASE} = \left[\frac{G-1}{G \ln G}\right]^{1/2} \Delta\nu_0 \tag{C.11}$$

for a Gaussian line. As an example, expression Eq. (C.10) has been used to obtain the normalized ASE linewidth, $\Delta\nu_{ASE}/\Delta\nu_0$, vs peak gain G which is plotted in Fig. 2.25 as a dashed line.

With the help of these considerations on the spectral behaviour of ASE, we can now also obtain an approximate expression for the total ASE intensity, $I = \int I_\nu d\nu$, at the exit face of the active medium. To obtain this we first integrate both sides of Eq. (C.3) over frequency ν and obtain

$$I = \phi I_s \left(\frac{\Omega}{4\pi}\right) g_p \int_{-\infty}^{+\infty} [\exp(\sigma N l) - 1] d\nu \tag{C.12}$$

With the help of the Gaussian approximation for the ASE spectrum given by Eq. (C.4), the integral in Eq. (C.12) is readily calculated, using Eqs. (C.8) and (C.9), to give [see Eq. (2.9.3)]

$$I = \phi I_s \left(\frac{\Omega}{4\pi^{3/2}}\right) \frac{(G-1)^{3/2}}{[G \ln G]^{1/2}} \tag{C.13}$$

for a Lorentzian and

$$I = \phi I_s \left(\frac{\Omega}{4\pi}\right) \frac{(G-1)^{3/2}}{[G \ln G]^{1/2}} \tag{C.14}$$

for a Gaussian line.

References

1. L.W. Casperson, Threshold Characteristics of Mirrorless Lasers, *J. Appl. Phys.* **48**, 256 (1977)
2. O. Svelto, S. Taccheo, and C. Svelto, Analysis of Amplified Spontaneous Emission: Some Corrections to the Lindford Formula, *Opt. Comm.* **149**, 277–282 (1998)

D

Calculation of the Radiative Transition Rates of Molecular Transitions

A simplified treatment is considered here merely to show how the selection rules arise for a molecular transition.

The transition probability can be expressed in the form of Eq. (2.4.9) provided that the appropriate value for the amplitude of the oscillating dipole moment $|\mu|$ is used. Thus we begin by recalling that, for an ensemble of negative charges (the electrons of the molecule), each of value e (with the sign included) and of positive charges of value e_h (the nuclei of the molecule) the classical electrical dipole moment is given by $\boldsymbol{\mu} = \Sigma_i\, e\mathbf{r}_i + \Sigma_j\, e_h\, \mathbf{R}_j$. Here, $\mathbf{r}_i$ and $\mathbf{R}_j$ specify the positions of the electrons and of the nuclei, respectively, relative to some given reference point, and the sum is taken over all electrons and nuclei of the molecule. If the reference point is taken to be center of the positive charges, then $\Sigma_j\, e_h\, \mathbf{R}_j = 0$ and $\boldsymbol{\mu}$ reduces to

$$\boldsymbol{\mu} = \Sigma_i\, e\mathbf{r}_i \tag{D.1}$$

To simplify matters, we will now consider a diatomic molecule. In this case the nuclear coordinates may be reduced to the magnitude R of the internuclear spacing $\mathbf{R}$ and to the angular coordinates, θ, and, ϕ, of $\mathbf{R}$ relative to a given reference system. According to quantum mechanics the oscillating dipole moment of the molecule is then given by [see also Eq. (2.3.6)]

$$\boldsymbol{\mu}_{osc} = 2\,\mathrm{Re} \int \psi_2^*(\mathbf{r}_i, R, \mathbf{r}_r)\, \boldsymbol{\mu}\, \psi_1(\mathbf{r}_i, R, \mathbf{r}_r)\, d\mathbf{r}_i\, dR\, d\mathbf{r}_r \tag{D.2}$$

where ψ_2 and ψ_1 are, respectively, the wave functions of the final and initial states of the transition. Note that both ψ_1 and ψ_2 are taken to be functions of the positions of all electrons, of the internuclear distance R, and of the rotational coordinates $\mathbf{r}_r$ (a shorthand notation for θ and ϕ), and the integral is taken over all these coordinates. Following the Born-Oppenheimer approximation, the molecular wave functions ψ can now be written as

$$\psi(\mathbf{r}_i, R, \mathbf{r}_r) = u_e(\mathbf{r}_i, R)\, u_v(R)\, u_r(\mathbf{r}_r)\, \exp[-j\,(E/\hbar)\,t] \tag{D.3}$$

where u_e, u_v, and u_r are the electronic, vibrational, and rotational wave functions, respectively, and $E = E_e + E_v + E_r$ is the total energy of the given state. The substitution of Eq. (D.3) into Eq. (D.2) then shows that μ_{osc} oscillates at the frequency $\nu_{21} = (E_2 - E_1)/h$ with a complex amplitude μ_{21} given by [compare with Eq. (2.3.7)]

$$\mu_{21} = \left(\int u_{v2}^*\, \mu_e\, u_{v1}\, dR\right)\left(\int u_{r2}^*\, u_{r1}\, d\mathbf{r}_r\right) \tag{D.4}$$

In the above expression one has

$$\mu_e = \mu_e(R) = \int u_{e2}^*(\mathbf{r}_i, R)\mu\, u_{e1}(\mathbf{r}_i, R)\, d\mathbf{r}_i \tag{D.5}$$

where μ is the dipole moment given by Eq. (D.1). Since the electronic wave functions are slowly varying functions of R, $\mu_e(R)$ can then be expanded in a power series around the equilibrium internuclear distance R_0 in the form

$$\mu_e(R) = \mu_e(R_0) + \frac{d\mu_e}{dR}(R - R_0) + \ldots \tag{D.6}$$

Let us first consider pure rotational transitions. In this case one has $u_{e2} = u_{e1}$ and $u_{v_2} = u_{v_1}$ and, from Eq. (D.5), the dipole moment $\mu_e(R_0)$ is seen to be given by

$$\mu_e(R_0) = \int \mu\, |u_{e1}(\mathbf{r}_i, R_0)|^2\, d\mathbf{r}_i \tag{D.7}$$

which is the permanent electric dipole moment μ_{ep} of the molecule. If we assume $\mu = \mu_e(R_0)$ in the first integral on the right hand side of Eq. (D.4) and remember that $\int u_{v2}^*\, u_{v1}\, dR = \int |u_{v1}|^2\, dR = 1$, we get the following expression for $|\mu_{21}|^2 = |\mu_{21}|^2$ to be used in Eq. (2.4.9):

$$|\mu_{21}|^2 = |\mu_{ep}|^2 \left|\int u_{r2}^* u_{r1} d\mathbf{r}_r\right|^2 \tag{D.8}$$

The first factor in the right-hand side of Eq. (D.8) indicates that pure rotational transitions are possible only in molecules possessing a permanent dipole moment μ_{ep}. This can be easily understood because the stimulated emission process can be considered to arise from the interaction of the incident e.m. wave with this rotating dipole moment. For molecules with a permanent dipole moment, $|\mu_{21}|^2$ is then proportional to the second factor on the right-hand side of Eq. (D.8). From the symmetry properties of the rotational wave functions, it then follows that this factor is nonzero only if the quantum jump ΔJ between the rotational numbers of the two states obeys the selection rule $\Delta J = \pm 1$.

Let us next consider rotational-vibrational transitions. One has again $u_{e2} = u_{e1}$ and, to first order, we again put $\mu_e(R) \cong \mu_e(R_0) = \mu_{ep}$ into Eq. (D.4). We readily see that μ_{21}

reduces to $\left(\mu_{ep}\int u^*_{v2}u_{v1}dR\right)\left(\int u^*_{r2}u_{r1}d\mathbf{r}_r\right)$ which is zero on account of the orthogonality of the vibrational wave functions belonging to the same electronic state. To calculate the transition rate we need therefore to consider the second term in the expansion Eq. (D.6). The substitution of this term into Eq. (D.4) then gives the following expression for $|\mu_{21}|^2$:

$$|\mu_{21}|^2 = \left|\frac{d\mu_e}{dR}\right|^2 \left|\int u^*_{v2}(R-R_0)u_{v1}dR\right|^2 \left|\int u^*_{r2}u_{r1}d\mathbf{r}_r\right|^2 \tag{D.9}$$

The third factor in the right hand side of Eq. (D.9) again gives the selection rule $\Delta J = \pm 1$ for the rotational quantum jump. As far as the second factor is concerned, we remember that, if the potential energy curve $U(R-R_0)$ of the molecule is approximated by a parabola, i.e. for harmonic restoring force, the wave functions u_v are given by the well-known harmonic-oscillator functions i.e., by the product of Hermite polynomials with a Gaussian function. As a result of the symmetry properties of these functions, $|\mu_{21}|^2$ then turns out to be non-zero only if $\Delta v = \pm 1$. Overtones arise if the above parabolic assumption is relaxed (i.e., in the case of anharmonicity of the potential-energy) or if the higher-order terms in Eq. (D.6) are taken into account (electrical anharmonicity). Note finally that, under certain symmetry conditions for the ground state electronic wave-function, the first factor in Eq. (D.9) turns out to be zero and the transition is said to be infrared inactive. An obvious case for this circumstance occurs when the two atoms of the molecule are identical as in the case of e.g. a N_2 molecule involving the same isotopic species for the two atoms. In this case in fact, for symmetry reasons, the molecule cannot have a net dipole moment, $\mu_e(R)$, for any value of the internuclear distance R, and $|\mu_{21}|^2$ in Eq. (D.9) is always zero.

Lastly we consider the case of vibronic transitions. If we again take only the first term in the expansion Eq. (D.6), then $|\mu_{21}|^2$, from Eq. (D.4), is seen to be given by

$$|\mu_{21}|^2 = |\mu_e(R_0)|^2 \left|\int u^*_{v2}u_{v1}dR\right|^2 \left|\int u^*_{r2}u_{r1}d\mathbf{r}_r\right|^2 \tag{D.10}$$

As a result of the symmetry properties of the electronic wave functions of the two states, the first factor in the right hand side of Eq. (D.10) may turn out to be zero. In this case the vibronic transition is said to be electric dipole forbidden. For a dipole-allowed transition, the third factor on the right-hand side of Eq. (D.10) leads again to the selection rule $\Delta J = \pm 1$. Within this selection rule and again for a dipole-allowed transition, Eq. (D.10) then shows that $|\mu_{21}|^2$ is proportional to the second factor appearing in the right hand side of the equation, known as the Franck-Condon factor. Note that, in this case, this factor is non-zero because u_{v2} and u_{v1} belong to different electronic states. The transition probability W is thus determined by the degree of overlap between the nuclear wave functions, as discussed in chap. 3 in connection with Fig. 3.6.

E

Space Dependent Rate Equations

The purpose of this appendix is to develop a rate-equation treatment, and to solve these equations for the c.w. case, when the spatial variation of both the pump-rate and the cavity field are taken into account. As a result of these spatial variations, the population inversion will also be space dependent. In all cases, we will assume the laser oscillating on a single mode.

E.1. FOUR-LEVEL LASER

For an ideal four-level laser, we can neglect the population, N_1, of the lower laser level and thus let $N \cong N_2$ represent the population inversion. We can then write

$$\frac{\partial N}{\partial t} = R_p - WN - \frac{N}{\tau} \tag{E.1.1a}$$

$$\frac{d\phi}{dt} = \int_a WNdV - \frac{\phi}{\tau_c} \tag{E.1.1b}$$

where the integral in Eq. (E.1.1b) is taken over the volume of the active medium and where the meaning of all other symbols has already been given in Chap. 7. Equation (E.1.1a) expresses a local balance between pumping, stimulated emission, and spontaneous decay processes. Note that a partial derivative has been used on the left-hand side of the equation on account of the expected spatial variation of N. The integral term on the right-hand side of Eq. (E.1.1b) accounts for the contribution of stimulated processes to the total number of cavity photons ϕ. This term has been written on the basis of a simple balance, using the fact that each individual stimulated process produces a photon. For a plane wave, we can now write $W = \sigma F = \sigma I/h\nu$ and $I = c\rho/n$, where σ is the stimulated-emission cross-section, F is photon flux, I is the intensity of the wave, ρ is the energy density in the active medium and n is its refractive index.

From the previous expressions, we can then relate W to the energy density of the wave as

$$W = \frac{c\sigma}{nh\nu}\rho \tag{E.1.2}$$

Although this equation has been derived, for simplicity, for a plane wave, one immediately recognizes that it merely establishes a local relation between the transition rate and the energy density of the e.m. field. It then follows that this equation has a general validity and can thus represent the relation between W and ρ for e.g. the e.m. field of the cavity. In this case ρ is expected to be dependent on both space, $\mathbf{r}$, and, for a transient case, also on time, t, the spatial dependence accounting for the spatial variation of the cavity mode. Equations (E.1.1), with the help of Eq. (E.1.2) give

$$\frac{\partial N}{\partial t} = R_p - \frac{c\sigma}{nh\nu}\rho N - \frac{N}{\tau} \tag{E.1.3a}$$

$$\frac{d\phi}{dt} = \frac{c\sigma}{nh\nu}\int_a \rho N dV - \frac{\phi}{\tau_c} \tag{E.1.3b}$$

Note that, since both R_p and ρ are assumed to depend on position (and time for a transient case), this will also apply to N, which cannot therefore be taken out of the integral in Eq. (E.1.3b). Note also that the total number of cavity photons ϕ may be related to the energy density of the e.m. wave by

$$\phi = \frac{1}{h\nu}\int_c \rho\, dV \tag{E.1.4}$$

where the integral is taken over the whole volume of the cavity. We will consider a laser cavity of length L in which is inserted an active medium of length l and refractive index n and we will assume that the beam waist is located somewhere in the active medium. Under these conditions, the energy density of the mode outside, ρ_{out}, and inside, ρ_{in}, the active medium can be written, respectively, as

$$\rho_{out} = \rho_0|u(\mathbf{r})|^2 \tag{E.1.5a}$$

$$\rho_{in} = n\rho_0|u(\mathbf{r})|^2 \tag{E.1.5b}$$

where $u(\mathbf{r})$ is the field amplitude, at the general coordinate, $\mathbf{r}$, normalized to its peak value (occurring at the waist), and $n\rho_0$ is the energy density at the waist location. From Eqs. (E.1.4) and (E.1.5) we then obtain

$$\phi = \frac{\rho_0}{h\nu}\left(n\int_a |u|^2 dV + \int_r |u|^2 dV\right) \tag{E.1.6}$$

where the two integrals are taken over the active medium and the remaining volume of the cavity, respectively. The form of Eq. (E.1.6) suggests that we define an effective volume of the mode in the cavity, V, as

$$V = \left(n \int_a |u|^2 dV + \int_r |u|^2 dV \right) \tag{E.1.7}$$

and a volume V_a of the mode in the active medium as

$$V_a = \int_a |u|^2 dV \tag{E.1.8}$$

With the help of Eqs. (E.1.5b), (E.1.6), and (E.1.7), Eq. (E.1.3) transform to

$$\frac{\partial N}{\partial t} = R_p - \frac{c\sigma}{V}\phi N|u|^2 - \frac{N}{\tau} \tag{E.1.9a}$$

$$\frac{d\phi}{dt} = \left[\frac{c\sigma}{V} \int_a N|u|^2 dV - \frac{1}{\tau_c} \right] \phi \tag{E.1.9b}$$

which represents our final result describing the space dependent rate equations of a four-level laser.

We will now solve Eq. (E.1.9) for the case of a c.w. laser oscillating on a single TEM_{00} mode. For simplicity, we will further assume the e.m. cavity field, $u(\mathbf{r})$, to be independent of the longitudinal coordinate z. This implies that we are neglecting both the spot size variation and the mode standing-wave-pattern along the laser cavity. We will also take the cladded-rod model discussed in Sect. 6.3.3. so that one needs not to worry about the aperturing effect caused by a finite rod-diameter. Under these assumptions we can write the following very simple expression for $|u(r)|$, holding for any value of the longitudinal coordinate z in the active medium, and for any value of the radial coordinate, r, between 0 and ∞:

$$|u| = \exp -(r / w_0)^2 \tag{E.1.10}$$

where w_0 is the spot size at the beam waist. Equation (E.1.7) then gives

$$V = \frac{\pi w_0^2}{2} L_e \tag{E.1.11}$$

where the equivalent length of the cavity, L_e, can be expressed as [see Eq. (7.2.11)]

$$L_e = L + (n-1)l \tag{E.1.12}$$

Likewise, from Eq. (E.1.8), we obtain

$$V_a = \frac{\pi w_0^2}{2} l \tag{E.1.13}$$

The threshold condition for the population inversion is now obtained from Eq. (E.1.9b) by letting $(d\phi/dt) = 0$. Upon defining a spatially averaged population inversion as [see Eq. (7.3.20)]

$$<N> = \frac{\int_a N|u|^2 dV}{\int_a |u|^2 dV} = \frac{\int_a N|u|^2 dV}{V_a} \tag{E.1.14}$$

Equation (E.1.9b) gives [see Eq. (7.3.19)]

$$<N>_c = \frac{1}{c\sigma\tau_c}\frac{V}{V_a} = \frac{\gamma}{\sigma\, l} \tag{E.1.15}$$

where Eqs. (E.1.11), (E.1.13) and (7.2.14) have been used. The threshold expression for the pump rate, on the other hand, can be obtained from Eq. (E.1.9a) by letting $(\partial N/\partial t) = 0$ and $\phi = 0$. We get

$$R_p(r,z) = N(r,z)/\tau \tag{E.1.16}$$

A spatially averaged pump rate $<R_p>$ can now be defined as

$$<R_p> = \frac{\int_a R_p|u|^2 dV}{\int_a |u|^2 dV} = \frac{\int_a R_p|u|^2 dV}{V_a} \tag{E.1.17}$$

The substitution of Eq. (E.1.16) on the right hand side of Eq. (E.1.17) with the further help of Eq. (E.1.15) gives

$$<R_p>_c = \frac{<N>_c}{\tau} = \frac{\gamma}{\sigma\, l\tau} \tag{E.1.18}$$

Above threshold, the steady state average population, $<N>_0$, is obtained from Eq. (E.1.9b) by letting $(d\phi/dt) = 0$. This gives

$$<N>_0 = <N>_c = \gamma/\sigma l \tag{E.1.19}$$

The steady state photon number, ϕ_0, on the other hand, is obtained from Eq. (E.1.9a) by letting $(\partial N/\partial t) = 0$. One gets

$$N\left[1 + \frac{c\sigma\tau}{V}\phi_0|u|^2\right] = R_p\tau \tag{E.1.20}$$

To proceed further we relate ϕ_0, appearing in Eq. (E.1.20), to the output power, P_{out}, with the help of Eq. (7.2.18). Equation (E.1.20) can then be transformed to

$$N = \frac{R_p\tau}{\left[1 + \frac{P_{out}}{P_s}|u|^2\right]} \tag{E.1.21}$$

where use has been made of Eq. (E.1.11) for the cavity volume and of Eq. (7.2.14) for the photon decay time and where a saturation power, P_s, has been defined as [see Eq. (7.3.28)]

$$P_s = \frac{\gamma_2}{2} \frac{\pi w_0^2}{2} \frac{h\nu}{\sigma\tau} \tag{E.1.22}$$

Upon multiplying both sides of Eq. (E.1.21) by $|U|^2$ and integrating over the volume of the active medium, we obtain

$$<N>_0 = \frac{1}{V_a} \int \frac{R_p |u|^2 \tau}{\left[1 + \frac{P_{out}}{P_s}|u|^2\right]} dV \tag{E.1.23}$$

where Eq. (E.1.14) has been used and where the spatially-averaged inversion has been denoted by $<N>_0$ since the laser is operating c.w.. With the help of Eqs. (E.1.19), and (E.1.13), the previous expression gives

$$\frac{\gamma}{\sigma} = \frac{2}{\pi w_0^2} \int \frac{R_p |u|^2 \tau}{\left[1 + \frac{P_{out}}{P_s}|u|^2\right]} dV \tag{E.1.24}$$

To proceed further we need to specify the spatial variation of R_p and to relate its value to the pump power P_p. This will be done below for a pumping profile which is either uniform or has a Gaussian transverse distribution.

In the case of uniform pumping, R_p is a constant and, for both lamp pumping and electrical pumping, is given by [see Eqs. (6.2.6) and (6.4.26)]

$$R_p = \eta_p \frac{P_p}{\pi a^2 l h\nu_{mp}} \tag{E.1.25}$$

Note that Eq. (E.1.25) holds only for $0 \leq r \leq a$, where a is the radius of the medium, while one has $R_p(r) = 0$ for $r > a$. Note also that, according to the discussion in Sect. 6.3.3., the expression valid for diode pumping, with a uniform illumination, is simply obtained from Eq. (E.1.25) by replacing ν_{mp}, the minimum pump frequency described in Fig. 6.18, with ν_p, the frequency of the pumping diode. The substitution of Eq. (E.1.25) in the integral of Eq. (E.1.24) then gives

$$\frac{\gamma}{\sigma} = \eta_p \left[\frac{P_p \tau}{\pi a^2 h\nu_{mp}}\right] \left[\frac{2}{\pi w_0^2}\right] \int_0^a \frac{|u|^2}{\left[1 + \frac{P_{out}}{P_s}|u|^2\right]} 2\pi r dr \tag{E.1.26}$$

where the integration along the longitudinal coordinate, z, of the active medium has already been performed. We now define a minimum pump threshold P_{mth}, and the dimentionless variables x and y as in Eqs. (7.3.26), (7.3.25), and (7.3.27), respectively. Equation (E.1.26) then transforms to

$$\frac{1}{x} = \left[\frac{2}{\pi w_0^2}\right] \int_0^a \frac{\exp -2(r/w_0)^2}{[1 + y \exp -2(r/w_0)^2]} 2\pi r dr \tag{E.1.27}$$

where Eq. (E.1.10) has been used for $|U|^2$. Equation (E.1.27) can be readily integrated by the substitution

$$t = \exp -2(r / w_0)^2 \tag{E.1.28}$$

One obtains [compare with Eq. (7.3.30)]

$$\frac{1}{x} = \int_{\beta}^{1} \frac{dt}{1+yt} = \frac{1}{y} \ln \left[\frac{1+y}{1+\beta y}\right] \tag{E.1.29}$$

where

$$\beta = \exp -(a / w_0)^2 \tag{E.1.30}$$

In the case of a Gaussian distribution for the transverse pumping profile, as can be produced by e.g., longitudinal diode-pumping, R_p may be related to the pump power, P_p, by Eq. (6.3.7). From Eq. (E.1.24), using the expression for $|U|$ given by Eq. (E.1.10), we then get

$$\frac{\gamma}{\sigma} = \eta_r \eta_t \left(\frac{2}{\pi w_p^2}\right)\left(\frac{P_p \tau}{h\nu_p}\right)\left(\frac{2}{\pi w_0^2}\right)\left[\int_0^{\infty} \frac{\exp -2r^2\left[\left(w_0^2+w_p^2\right) / w_0^2 w_p^2\right]}{1+(P_{out} / P_s)\exp -\left[2r^2 / w_0^2\right]} 2\pi\, r dr\right] \times \int_0^{l} \alpha \exp -[\alpha z]\, dz \tag{E.1.31}$$

According to Eq. (6.3.11), the second integral in the right hand side of Eq. (E.1.31) gives the pump-absorption efficiency η_a. We now define a minimum pump threshold as [see Eq. (7.3.32)]

$$P_{mth} = \left(\frac{\gamma}{\eta_p}\right)\left(\frac{h\nu_p}{\tau}\right)\left(\frac{\pi w_p^2}{2\sigma_e}\right) \tag{E.1.32}$$

where $\eta_p = \eta_r \eta_t \eta_a$ is the pump efficiency. We also define the dimensionless variables x and y as in Eqs. (7.3.25) and (7.3.27), respectively. Equation (E.1.31) then gives

$$\frac{1}{x} = \left(\frac{2}{\pi w_0^2}\right)\left[\int_0^{\infty} \frac{\exp -2r^2\left[\left(w_0^2+w_p^2\right) / w_0^2 w_p^2\right]}{1+y\exp -\left[2r^2 / w_0^2\right]} 2\pi\, r dr\right] \tag{E.1.33}$$

If we now define a variable t as in Eq. (E.1.28) and a quantity δ as $\delta = (w_0 / w_p)^2$, the previous equation simplifies to

$$\frac{1}{x} = \int_0^1 \frac{t^{\delta}}{1+y\delta} dt \tag{E.1.34}$$

The integral in Eq. (E.1.34) can be calculated analytically for integer values of δ. In particular, for $\delta = 1$, one gets

$$\frac{1}{x} = \left[\frac{t}{y} - \frac{1}{y^2}\ln(1 + yt)\right]_0^1 \tag{E.1.35}$$

from which Eq. (7.3.34) is immediately obtained.

E.2. QUASI-THREE-LEVEL LASER

The procedure to establish the space-dependent rate equations for a quasi-three-level laser and to solve them for the c.w. case follows the same path as that of a four-level laser. According to Eq. (7.2.19) we now write

$$N_1 + N_2 = N_t \tag{E.2.1a}$$

$$\frac{\partial N_2}{\partial t} = R_p - (W_e N_2 - W_a N_1) - \frac{N_2}{\tau} \tag{E.2.1b}$$

$$\frac{d\phi}{dt} = \int_a (W_e N_2 - W_a N_1) dV - \frac{\phi}{\tau_c} \tag{E.2.1c}$$

In the previous equations, following Eq. (E.1.2), the rates for stimulated emission, W_e, and for absorption, W_a, can be written as

$$W_e = \frac{c\sigma_e}{nh\nu}\rho \tag{E.2.2a}$$

and

$$W_a = \frac{c\sigma_a}{nh\nu}\rho \tag{E.2.2b}$$

where σ_e and σ_a are the effective cross sections for emission and absorption, respectively. One can now proceed following the same steps, from Eq. (E.1.4) to (E.1.8), as for a four-level laser, to obtain [compare with Eq. (7.2.24)]

$$\frac{\partial N}{\partial t} = R_p(1 + f) - \frac{c(\sigma_e + \sigma_a)}{V}\phi N|u|^2 - \frac{fN_t + N}{\tau} \tag{E.2.3a}$$

$$\frac{d\phi}{dt} = \left[\frac{c\sigma_e}{V}\int_a N|u|^2 dV - \frac{1}{\tau_c}\right]\phi \tag{E.2.3b}$$

where $N = N_2 - fN_1$ [see Eq. (7.2.23)] and $f = \sigma_a/\sigma_e$ [see Eq. (7.2.22)]. These equations represent our final result describing the space-dependent rate equations of a quasi-three-level laser.

We will now solve Eq. (E.2.3) for the case of a c.w. laser oscillating on a single TEM_{00} mode. Under the assumption, again, that $|u(r)|$ is described by Eq. (E.1.10) for $0<r<\infty$, we obtain again Eqs. (E.1.11), (E.1.12) and (E.1.13) for V, L_e, and V_a, respectively. We also define the spatially averaged values, $<N>$, and, $<R_p>$, as in Eqs. (E.1.14) and (E.1.17), respectively.

The threshold value of $<N>$ is obtained from Eq. (E.2.3b) by letting $(d\phi/dt) = 0$. This gives

$$<N>_c = \gamma/\sigma_e l \tag{E.2.4}$$

The threshold value of $<R_p>$ is then obtained from Eq. (E.2.3a) by letting $(\partial N/\partial t) = 0$ and $\phi = 0$. One gets

$$<R_p>_c = \frac{f<N_t> + <N>_c}{(1+f)\tau} = \frac{\sigma_a <N_t> l + \gamma}{(\sigma_e + \sigma_a) l\tau} \tag{E.2.5}$$

Above threshold, the steady state value, $<N>_0$, is again obtained from Eq. (E.2.3b) by letting $(d\phi/dt) = 0$. One gets

$$<N>_0 = <N>_c = \gamma/\sigma_e l \tag{E.2.6}$$

From Eq. (E.2.3a), under the condition $(\partial N/\partial t) = 0$, one obtains

$$N = \frac{R_p(1+f)\tau - f N_t}{1 + [c(\sigma_e + \sigma_a)\tau/V]\phi_0 |u|^2} \tag{E.2.7}$$

and the steady-state number of photons, ϕ_0, can again be related to the output power, P_{out}, by Eq. (7.2.18). Equation (E.2.7) can then be transformed to

$$N = \frac{R_p(1+f)\tau - f N_t}{1 + y|u|^2} \tag{E.2.8}$$

In the previous expression we have again defined $y = P_{out}/P_s$, where the saturation power, P_s, is now given by

$$P_s = \frac{\gamma_2}{2}\frac{\pi w_0^2}{2}\frac{h\nu}{(\sigma_e + \sigma_a)\tau} \tag{E.2.9}$$

We now multiply both sides of Eq. (E.2.8) by $|U|^2$ and integrate over the volume of the active medium. With the help of Eqs. (E.1.13), (E.1.14) and (E.2.6) we obtain [compare with Eq. (E.1.24)]

$$\frac{\gamma}{\sigma} = \frac{2}{\pi w_0^2} \int \frac{[R_p(1+f)\tau - f N_t]|u|^2}{[1 + y|u|^2]} dV \tag{E.2.10}$$

To proceed further, the spatial dependence of R_p needs to be specified. In the case of longitudinal pumping with a pump beam of Gaussian radial profile we use Eq. (6.3.7). With the help of Eq. (E.1.10) for $|u|$, Eq. (E.2.10) can be written as

$$\frac{\gamma}{\sigma_e l} = \eta_r \eta_t (1+f) \left(\frac{P_p \tau}{h\nu_p}\right) \left(\frac{2}{\pi w_p^2 l}\right) \int_0^1 \frac{t^\delta}{1+yt} dt \int_0^l \alpha \exp-(\alpha z) dz -$$

$$-f N_t \int_\beta^1 \frac{dt}{1+yt} \tag{E.2.11}$$

where t and β are expressed by Eqs. (E.1.28) and (E.1.30), respectively, and

$$\delta = (w_0 / w_p)^2 \tag{E.2.12}$$

Again recognizing that the integral over the longitudinal coordinate z is the absorption efficiency η_a, and under the simplifying assumption $w_0 << a$ (i.e. $\beta \cong 0$), Eq. (E.2.11) gives

$$\gamma = \eta_p (\sigma_e + \sigma_a) \left(\frac{P_p \tau}{h\nu_p}\right) \left(\frac{2}{\pi w_p^2}\right) \int_0^1 \frac{t^\delta}{1+yt} dt - \sigma_a N_t l \frac{\ln(1+y)}{y} \tag{E.2.13}$$

and this equation can be solved for P_p to get

$$P_p = \gamma \left(\frac{h\nu_p}{\eta_p \tau}\right) \left[\frac{\pi w_p^2}{2(\sigma_e + \sigma_a)}\right] \frac{1}{\int_0^1 [t^\delta / 1 + yt]\, dt} \left[1 + B\frac{\ln(1+y)}{y}\right] \tag{E.2.14}$$

where $B = \sigma_a N_t l / \gamma$. The minimum pump threshold is obtained from Eq. (6.3.25) when $w_0 << w_p$ and $\sigma_a N_t l << \gamma$. One obtains [see Eq. (7.4.16)]

$$P_{mth} = \gamma \left(\frac{h\nu_p}{\eta_p \tau}\right) \left[\frac{\pi w_p^2}{2(\sigma_e + \sigma_a)}\right] \tag{E.2.15}$$

By taking the ratio between Eqs. (E.2.14) and (E.2.15) we obtain our final result [see Eq. (7.4.18)]

$$x = \frac{1}{\int_0^1 \frac{t^\delta}{1+yt}\, dt} \left[1 + B\frac{\ln(1+y)}{y}\right] \tag{E.2.16}$$

F

Theory of Mode-Locking: Homogeneous Line

According to the discussion presented in Sect. 8.6.3, a theory of mode-locking can be developed, in the time domain, by requiring that the pulse reproduce itself after each cavity round trip. We will limit our discussion here to the case of a homogeneous line and we will further assume that the lifetime of the upper laser level is much longer than the cavity round-trip time. Under these conditions, the saturated single-pass power gain of the amplifier, at the transition peak, is given by $g_0 = \sigma_p N_0 l$, where σ_p is the peak cross-section, l is the length of the active medium and N_0 is the steady-state inversion, as established by the cumulative effect of the passage of many pulses. This means that this saturated gain will be determined by the average intracavity beam intensity $<I>$ and can thus be related to the unsaturated gain, g, by

$$g_0 = \frac{g}{1 + (<I>/I_s)} \tag{F.1}$$

where $I_s = h\nu_0/\sigma_p\tau$ is the saturation intensity of the amplifier, at the transition peak.

F.1. ACTIVE MODE-LOCKING

This theory was developed by Kuizenga and Siegman[1] and, later on, put in a more general framework by Haus [2,3]. We will follow the latter treatment and for brevity we will limit our considerations to mode-locking by an amplitude modulator. Thus, we consider the laser configuration of Fig. F1 and assume that the modulator is very thin and placed as near as possible to mirror 2. Under these conditions, a single light pulse is expected to be traveling back and forth within the cavity (see Fig. 8.19). At any given position within the cavity, the

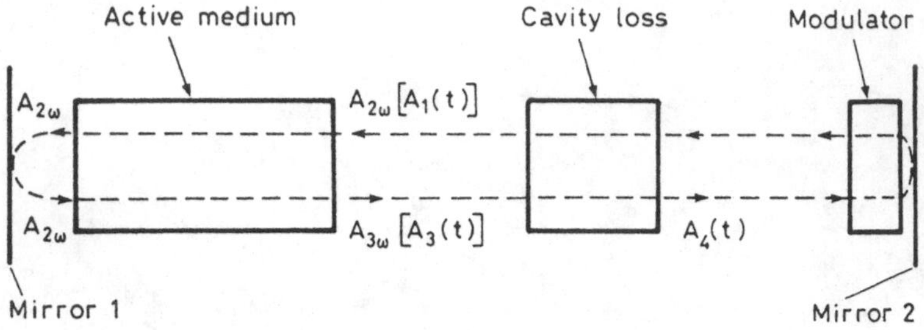

FIG. F.1. Schematic diagram of the laser cavity considered for the theoretical analysis of mode-locking.

electric field of this pulse can be written as

$$E(t) = A(t) \exp j(\omega_0 t - \phi) \tag{F.1.1}$$

where t represents a suitable local time to account for pulse propagation. The spectral amplitude $A_\omega(\omega-\omega_0)$ of the pulse is then obtained by taking the Fourier transform of $E(t)$ i.e.,

$$A_\omega(\omega - \omega_0) = \int_{-\infty}^{+\infty} E(t) \exp(-j\omega t) dt = \int_{-\infty}^{+\infty} A(t) \exp[-j(\omega - \omega_0)t] dt \tag{F.1.2}$$

The field amplitude $A(t)$ is then related to $A_\omega(\omega - \omega_0)$ by the inverse Fourier transform

$$A(t) = \int_{-\infty}^{+\infty} A_\omega(\omega - \omega_0) \exp[j(\omega - \omega_0)t] \, d(\omega - \omega_0) \tag{F.1.3}$$

We first consider the passage of the light pulse through the amplifier. If we let $A_{1\omega}$ and $A_{2\omega}$ represent the spectral amplitude of the light pulse before and after a single passage (see Fig. F1), we can write $A_{2\omega} = t_g A_{1\omega}$, where the single-pass electric-field transmission, t_g, through the amplifier can be shown to be given by[(4)]

$$t_g = \frac{A_{2\omega}}{A_{1\omega}} = [\exp -j(\omega \, nl/c)] \times \exp \left\{ \frac{(g_0/2)}{[1 + 2j(\omega - \omega_0)/\Delta\omega_0]} \right\} \tag{F.1.4}$$

where n is the refractive index of the active medium and $\Delta\omega_0$ is the width (FWHM) of the laser line. Note that, according to Eq. (F.1.4), the power gain is given by

$$G(\omega) = |t_g|^2 = \exp g_0(\omega) \tag{F.1.5}$$

where the frequency dependent gain $g_0(\omega)$ is given by

$$g_0(\omega) = \frac{g_0}{\left\{1 + [2(\omega - \omega_0)/\Delta\omega_0]^2\right\}} \tag{F.1.6}$$

i.e., it shows the expected Lorentzian shape of a homogeneous line. If we assume that the spectral width of the light pulse is appreciably narrower than $\Delta\omega_0$, the expression appearing as the argument of the second exponential function in Eq. (F.1.4) can be expanded as a power series in $(\omega - \omega_0)$. To first order, this gives

$$t_g = \exp -j\{(\omega\, nl/c) + [g_0(\omega - \omega_0)/\Delta\omega_0]\} \times \\ \times \exp(g_0/2)\left\{1 - [2(\omega - \omega_0)/\Delta\omega_0]^2\right\} \tag{F.1.7}$$

The imaginary terms in the first exponential function correspond to a phase delay $\phi = (\omega nl/c) + [g_0(\omega - \omega_0)/\Delta\omega_0]$, from which, according to Eq. (8.6.27), the time delay, τ_d, experienced by the pulse after traveling in the active medium is obtained as

$$\tau_d = \frac{d\phi}{d\omega} = \frac{nl}{c} + \frac{g_0}{\Delta\omega_0} \tag{F.1.8}$$

Note that this delay is not simply nl/c because the gain line gives an additional finite contribution. This delay should be taken into account when considering the requirement that the round-trip pulse propagation time be set equal to the period of the amplitude modulator. For simplicity we will not consider any further the effect of this delay as well as that of all other cavity elements, the pulse amplitude being referred, in any case, to a local time where these delays are taken into account. We therefore ignore the phase term in Eq. (F.1.7) and write

$$t_g = \exp\left\{(g_0/2)\left\{1 - [2(\omega - \omega_0)/\Delta\omega_0]^2\right\}\right\} \tag{F.1.9}$$

We also ignore the loss introduced by mirror 1, since this will be taken into account in the overall cavity loss. After passing once more through the active medium, the spectral amplitude of the light pulse will experience another transmission factor t_a as given by Eq. (F.1.9). The round trip transmission through the amplifier is then given by $t_g^2 = (A_{3\omega}/A_{1\omega}) = \exp\left\{g_0\left\{1 - [2(\omega - \omega_0)/\Delta\omega_0]^2\right\}\right\}$, where $A_{3\omega}$ is the spectral amplitude of the light pulse after one round trip through the amplifier. Under the assumption $g_0 \ll 1$, this last equation gives

$$A_{3\omega} = t_g^2 A_{1\omega} = A_{1\omega}\left(1 + g_0\left\{1 - [2(\omega - \omega_0)/\Delta\omega_0]^2\right\}\right) \tag{F.1.10}$$

To proceed further we need to calculate the effect of this transmission in the time-domain rather than in the frequency-domain. For this we note the following property of a Fourier transform ($F.T.$):

$$F.T.\left[\frac{d^n A(t)}{dt^n}\right] = [j(\omega - \omega_0)]^n A_\omega(\omega - \omega_0) \tag{F.1.11}$$

This relation can be readily proved by taking first the n-th derivative and then the Fourier transform of both sides of Eq. (F.1.3). Equation (F.1.11) shows that the multiplication of the spectral amplitude A_ω by $k(\omega - \omega_0)^n$, where k is a constant, is equivalent, in the time domain, of taking (k/j^n) times the n-th derivative of the amplitude $A(t)$. We can apply this rule to each individual term on the right hand side of Eq. (F.1.10) to obtain

$$A_3(t) = \left\{1 + g_0\left[1 + \left(\frac{2}{\Delta\omega_0}\right)^2 \frac{d^2}{dt^2}\right]\right\} A_1(t) \tag{F.1.12}$$

where $A_1(t)$ and $A_3(t)$ are, respectively, the amplitudes of the light pulse entering the amplifier and after one round trip (see Fig. F1). Equation (F.1.12) shows that the effect on the light pulse amplitude of a round trip passage through the gain medium, can be described by a round-trip operator

$$\hat{T}_g = \left\{1 + g_0\left[1 + \left(\frac{2}{\Delta\omega_0}\right)^2 \frac{d^2}{dt^2}\right]\right\} \tag{F.1.13}$$

We next consider the effect of cavity losses arising from finite mirror reflectivities and internal losses. These losses are represented by the central box in Fig. F1. If we then let γ be the logarithmic power loss per pass, we can write, for a single passage in the cavity,

$$A_4(t) = [\exp(-\gamma/2)]A_3(t) \tag{F.1.14}$$

In fact, according to Eq. (F.1.14), the ratio of corresponding intensities $(I_4/I_3) = (A_4/A_3)^2$ indeed shows the expected value $\exp(-\gamma)$. From Eq. (F.1.14) one finds that the transmission accounting for the round trip losses is given by $\exp(-\gamma)$ and this expression, for $\gamma \ll 1$, can be approximated by $1 - \gamma$. This means that the operator corresponding to the round trip loss in the cavity is simply

$$\hat{T}_l = 1 - \gamma \tag{F.1.15}$$

Lastly, we consider the effect of the amplitude modulator. We let $\gamma_m[1 - \cos\omega_m t]$, represent the single-pass logarithmic power-loss introduced by the modulator. In this expression ω_m is the modulator frequency which is assumed to be such that the modulator period is equal to the round trip time of the light pulse in the laser cavity. The single pass transmission of the field amplitude through the modulator will then be given by

$$t_m = \exp\{-(\gamma_m/2)[1 - \cos\omega_m t]\} \tag{F.1.16}$$

The transmission for a double pass through the modulator is then $t_m^2 = \exp\{-(\gamma_m)[1 - \cos\omega_m t]\}$ and this expression, for $\gamma_m \ll 1$, can be approximated to $t_m^2 \cong 1 - (\gamma_m)[1 - \cos\omega_m t]$. We now assume that the pulse passes through the modulator when the modulator-loss is zero (see Fig. 8.20), i.e. at time $t = 0$, and that the pulsewidth is much smaller that the modulator period $2\pi/\omega_m$. Under these conditions, the round trip transmission can be further approximated to $t_m^2 = 1 - (\gamma_m/2)(\omega_m t)^2$. The operator corresponding to a double-passage of the pulse through the modulator is then simply given by

$$\hat{T}_m = 1 - \frac{\gamma_m}{2}(\omega_m t)^2 \tag{F.1.17}$$

Having established the operators which describe the time-domain evolution of the pulse upon a double passage through the three components considered, we now require that, in the steady state regime, the pulse amplitude $A(t)$ reproduces itself after a round trip. We thus write

$$\hat{T}_m\hat{T}_l\hat{T}_g\, A(t) = A(t) \tag{F.1.18}$$

Using the previous expressions for $\hat{T}_m$, $\hat{T}_l$, and $\hat{T}_g$ and again using the condition that $[g_0, \gamma, \gamma_m] \ll 1$ we obtain the following differential equation

$$\left\{g_0\left[1+\left(\frac{2}{\Delta\omega_0}\right)^2\frac{d^2}{dt^2}\right]-\gamma-\frac{\gamma_m}{2}\omega_m^2 t^2\right\}A(t)=0 \tag{F.1.19}$$

which is the final result of our calculation. This equation is readily seen to be equivalent to the Schrödinger equation for a particle in a parabolic potential (harmonic oscillator), whose solution is well known. The solution, in our case, can then be written as

$$A(t)=H_n(\omega_p t)\exp-\left(\frac{\omega_p^2 t^2}{2}\right) \tag{F.1.20}$$

where H_n is the Hermite polynomial of order n, and where

$$\omega_p=\left[\frac{\gamma_m}{2g_0}\right]^{1/4}\left[\frac{\omega_m\Delta\omega_0}{2}\right]^{1/2} \tag{F.1.21}$$

and

$$1-\frac{\gamma}{g_0}=\frac{4\omega_p^2}{\Delta\omega_0^2}(2n+1) \tag{F.1.22}$$

It can also be shown, however, that, out of all these solutions, only the first order Gaussian solution ($n=0$) is stable.

Equations (F.1.21) and (F.1.22) represent a set of two relations for the unknown parameters ω_p and g_0. From a knowledge of ω_p one then gets the width of the mode-locked pulse. The full width at half maximum intensity of the pulse, $\Delta\tau_p$, is in fact given by $\Delta\tau_p = 2\,[\ln 2]^{1/2}/\omega_p$ and from Eq. (F.1.21) one gets

$$\Delta\tau_p=\left[\frac{2\sqrt{2}\ln 2}{\pi^2}\right]^{1/2}\left(\frac{g_0}{\gamma_m}\right)^{1/4}\left(\frac{1}{\nu_m\Delta\nu_0}\right)^{1/2} \tag{F.1.23}$$

where $\nu_m=\omega_m/2\pi$ and $\Delta\nu_0=\Delta\omega_0/2\pi$. We note that the first factor on the right hand side of Eq. (F.1.23) is approximately equal to 0.45 while the second factor, as a result of the 1/4th power, is approximately equal to unity. The values of $\Delta\tau_p$ and, hence, ω_p are then only weakly dependent on g_0 and from Eq. (F.1.23) one obtains the following approximate expression for $\Delta\tau_p$ [see Eq. (8.6.19)]: $\Delta\tau_p \cong 0.45/(\nu_m\Delta\nu_0)^{1/2}$. From Eq. (F.1.22), with $n=0$, the value of g_0 can now be obtained. Note that, according to Eq. (F.1.22), g_0 turns out to be larger than γ on account of the presence of the modulator loss. Once the value of g_0 is calculated, the average intracavity laser intensity, $< I >$, is obtained from (F.1) since one has $g=\sigma_p Nl = x\gamma$ where $x=N/N_c=R_p/R_{pc}$ is the amount by which threshold is exceeded. Knowing the average intracavity laser intensity, the laser pulse duration and the pulse repetition rate, one then obtains the laser peak intensity.

F.2. PASSIVE MODE-LOCKING

We will consider the theory of passive mode-locking by a saturable absorber with a lifetime much shorter than the pulse duration (fast saturable absorber).[5] We will again refer to Fig. F1, where the modulator is now replaced by this fast absorber.

According to Eq. (F.1.13), the effect on the pulse amplitude due to a round-trip passage through the gain medium can be described by the operator

$$\hat{T}_g = \left\{1 + g_0' \left[1 + \left(\frac{2}{\Delta\omega_0}\right)^2 \frac{d^2}{dt^2}\right]\right\} \tag{F.2.1}$$

where, to conform with the discussion presented in Sect. 8.6.3.2., the saturated gain has been denoted by g_0'. According to Eq. (F.1.15), the effect of the unsaturable cavity losses, can be described by

$$\hat{T}_l = 1 - \gamma_c \tag{F.2.2}$$

where γ_c is the cavity loss without the saturable absorber. According to Eq. (F.1.14), the single-pass amplitude-transmission of the saturable absorber is then written as

$$t_{sa} = \exp -(\gamma_{sa}/2) \tag{F.2.3}$$

In the previous expression γ_{sa} represents the, saturated, single-pass power-loss of the absorber and is given by

$$\gamma_{sa} = \frac{\gamma'}{1 + (I/I_s)} \tag{F.2.4}$$

where γ' is the unsaturated loss, $I = I(t)$ is the pulse intensity, and I_s is the saturation intensity of the absorber. The amplitude-transmission for a double passage through the saturable absorber is then given by $t_{sa}^2 = \exp -(\gamma_{sa})$ and, under the assumptions $\gamma' \ll 1$ and $(I/I_s) \ll 1$, this last equation, with the help of Eq. (F.2.4) gives $t_{sa}^2 \cong 1 - \gamma_{sa} \cong 1 - \gamma'[1 - (I/I_s)] = 1 - \gamma'[1 - (|A|^2/I_s)]$, where the amplitude $A(t)$ is now normalized so that $|A|^2$ is the beam intensity. From the previous expression for t_{sa}^2 we obtain the operator corresponding to a double pass through the saturable absorber as

$$\hat{T}_{sa} = 1 - \gamma' + \gamma' \frac{|A|^2}{I_s} \tag{F.2.5}$$

Self-consistency now requires that

$$\hat{T}_m \hat{T}_l \hat{T}_{sa} A(t) = A(t) \tag{F.2.6}$$

From Eqs. (F.2.1), (F.2.2) and (F.2.5), again assuming $[g_0', \gamma, \gamma'] \ll 1$, we get

$$\left\{g_0' \left[1 + \left(\frac{2}{\Delta\omega_0}\right)^2 \frac{d^2}{dt^2}\right] - \gamma_c - \gamma' + \gamma' \frac{|A|^2}{I_s}\right\} A(t) = 0 \tag{F.2.7}$$

The solution of Eq. (F.2.7) can be written as

$$A(t) = \frac{A_0}{\cosh(t/\tau_p)} \qquad \text{(F.2.8)}$$

where

$$\tau_p = \left[\frac{2g'_0}{\gamma'}\right]^{1/2} \left[\frac{2}{\Delta\omega_0}\right] \left[\frac{I_s}{|A_0|^2}\right]^{1/2} \qquad \text{(F.2.9)}$$

and

$$\gamma_c + \gamma' - g'_0 = \frac{4g'_0}{\Delta\omega_0^2 \tau_p^2} \qquad \text{(F.2.10)}$$

Since the width of the pulse intensity, $\Delta\tau_p$, (FWHM) is given by $\Delta\tau_p = 1.76\ \tau_p$ and $\Delta\omega_0 = 2\pi\Delta\nu_0$, from Eq. (F.2.9) one obtains Eq. (8.6.22) by recognizing that $|A_0|^2$ is the peak laser intensity. Equation (F.2.10), on the other hand, shows that $g'_0 < \gamma_c + \gamma' = \gamma$, where γ is the overall unsaturated loss of the cavity. This means that, in the absence of the pulse, the laser has a net loss while a net gain only exists during the passage of the mode-locked pulse (see Fig. 8.22).

References

1. D. J. Kuizenga and A. E. Siegman, FM and AM Mode Locking of the Homogeneous Laser- Part I: Theory, *IEEE J. Quantum Electron.*, **QE-6**, 694–708 (1970).
2. H. A. Haus, A Theory of Forced Mode-Locking, *IEEE J. Quantum Electron.*, **QE-11**, 323–330 (1975).
3. H. A. Haus, *Waves and Fields in Optoelectronics*, Prentice Hall, Inc., Englewoods Clifts, 1984, Section 9.3.
4. A. E. Siegman, *Lasers* (Oxford University Press, Oxford, 1986) Sec. 7.4.
5. H. A. Haus, Theory of Mode-Locking with a Fast Saturable Absorber, *J. Appl. Phys.*, **46**, 3049–3058 (1975).

G
Propagation of a Laser Pulse Through a Dispersive Medium or a Gain Medium

Consider first a light pulse traveling in a dispersive medium, and let ω_L and $\Delta\omega_L$ be, respectively, the center frequency and the width of the corresponding spectrum (Fig. 8.25a). The electric field, $E(t, z)$, of the corresponding waveform, at coordinate z along the propagation direction, can generally be expressed in terms of a Fourier expansion as

$$E(t,z) = \int_{-\infty}^{+\infty} A_\omega(\omega - \omega_L) \exp[j(\omega t - \beta z)] d\omega \tag{G.1}$$

where $A_\omega = A_\omega(\omega - \omega_L)$ is the complex amplitude of each field component and $\beta = \beta(\omega - \omega_L)$ describes the dispersion relation of the medium.

Let us now assume that the dispersion relation of the medium, over the bandwidth $\Delta\omega_L$, can be approximated by a linear relation, i.e.,

$$\beta = \beta_L + \left(\frac{d\beta}{d\omega}\right)_{\omega_L} (\omega - \omega_L) \tag{G.2}$$

where β_L is the propagation constant corresponding to the frequency ω_L. On substituting Eq. (G.2) in Eq. (G.1) one can see that this last equation can be written as

$$E(t,z) = \exp[j(\omega_L t - \beta_L z)] \times \int_{-\infty}^{+\infty} A_\omega(\Delta\omega) \exp\left\{ j\Delta\omega \left[t - \left(\frac{d\beta}{d\omega}\right)_{\omega_L} z \right] \right\} d\Delta\omega \tag{G.3}$$

where $\Delta\omega = \omega - \omega_L$. From Eq. (G.3) we can now see that the integration over $\Delta\omega$ leads to a function of the single variable $[t - (d\beta/d\omega)_{\omega_L} z]$. Equation (G.3) can therefore be put in the form

$$E(t,z) = A[t - (z/\upsilon_g)] \exp[j(\omega_L t - \beta_L z)] \tag{G.4}$$

where A is the pulse amplitude, $\exp[j(\omega_L t - \beta_L z)]$ is the carrier wave, and υ_g is given by

$$\upsilon_g = \left(\frac{d\omega}{d\beta}\right)_{\beta=\beta_L} \tag{G.5}$$

The fact that the pulse amplitude is a function of the variable $t - (z/\upsilon_g)$ means that the pulse propagates without changing its shape and at a speed υ_g. This velocity is called the *group velocity* of the pulse and, according to Eq. (G.5), is given by the slope of the ω vs β relation at $\omega = \omega_L$ (i.e. $\upsilon_g = \tan\,\theta'$, see Fig. 8.25a).

Consider next the case of a light pulse with a bandwidth $\Delta\omega_L$ so large that it is no longer a good approximation to describe the dispersion relation by a linear law (Fig. 8.25c). In this case, different spectral regions of the pulse will travel with different group velocities and, consequently, the pulse will broaden as it propagates. After traversing the length l of the medium, the broadening of the pulse, $\Delta\tau_d$, will then be given approximately by the difference in group delay between the slowest and the fastest spectral components, respectively. We can then write

$$\Delta\tau_d = \left(\frac{l}{\upsilon'_g} - \frac{l}{\upsilon''_g}\right) = l\left[\left(\frac{d\beta}{d\omega}\right)_{\omega'} - \left(\frac{d\beta}{d\omega}\right)_{\omega''}\right] \tag{G.6}$$

where υ'_g and υ''_g are the two group velocities of these components and ω' and ω'' are the corresponding frequencies. Let us now assume that the dispersion relation, within the bandwidth $\Delta\omega_L$, can be approximated by a parabolic (or quadratic) law, i.e.,

$$\beta = \beta_L + \left(\frac{d\beta}{d\omega}\right)_{\omega_L}(\omega - \omega_L) + \frac{1}{2}\left(\frac{d^2\beta}{d\omega^2}\right)_{\omega_L}(\omega - \omega_L)^2 \tag{G.7}$$

From Eqs. (G.6) and (G.7) one gets

$$\Delta\tau_d \cong l\left|\left(\frac{d^2\beta}{d\omega^2}\right)_{\omega_L}\right|\Delta\omega_L = |\phi''(\omega_L)|\,\Delta\omega_L \tag{G.8}$$

where we have defined $\phi = \beta l$, $\phi'' = d^2\phi/d\omega^2$ and ϕ'' is taken at the central laser frequency ω_L. Given the form of Eq. (G.8), the quantity $\phi''(\omega_L)$, is referred to as the *group delay dispersion* (GDD) while the quantity

$$(d^2\beta/d\omega^2)_{\omega_L} = \text{GVD} = [d(1/\upsilon_g)/d\omega]_{\omega_L} \tag{G.9}$$

is referred to as the *group velocity dispersion* (GVD) at frequency ω_L.

The calculation leading to Eq. (G.8) is open to criticism because, to obtain Eq. (G.8), we have in fact been considering the propagation of limited spectral components of the pulse, each of which actually correspond to a different, and in fact, longer pulse than the original one. A more precise and instructive calculation can be performed by assuming that the pulse, upon entering the medium at coordinate $z = 0$, has a Gaussian amplitude profile i.e.,

$$E(t) = A_0 \exp - \left(t^2/2\tau_p^2\right) \exp(j\omega_L t) \tag{G.10}$$

where τ_p is the 1/e half-width of the pulse intensity. Since the spectral amplitude, $A_\omega(\omega - \omega_L)$, is a Gaussian function of $(\omega - \omega_L)$, the electric field after a distance z in the medium can be easily calculated from Eq. (G.1) if we assume that the dispersion relation can be developed in a Taylor expansion up to second order of $(\omega - \omega_L)$, as shown in Eq. (G.7). In this case, in fact, the integrand in Eq. (G.1) can be readily expressed in terms of the inverse Fourier transform of a generalized Gaussian function of complex argument, whose integral is well known. To show the final result for the pulse amplitude, A(t, z), we refer to a new coordinate system given by

$$t' = t - (z/\upsilon_g) \tag{G.11a}$$

$$z' = z \tag{G.11b}$$

υ_g being the group velocity. This means that the pulse amplitude is referred to a local time which takes into account the group delay of the pulse. Using this new coordinate system, the pulse amplitude turns out to be given by[(1)]

$$A(t', z) = \frac{A_0 \tau_p}{\left(\tau_p^2 + jb_2 z\right)^{1/2}} \exp - \left[\frac{(t')^2}{2\left(\tau_p^2 + jb_2 z\right)}\right] \tag{G.12}$$

where, for simplicity, we have written $b_2 = (d^2\beta/d\omega^2)_{\omega_L} = \text{GVD}$, and where, on account of Eq. (G.11b), we have written, again for simplicity, $z' = z$. According to Eq. (G.12), $A(t', z)$ then turns out to be given by a Gaussian function of t' with complex argument so that we can write

$$A(t', z) = |A(t', z)| \exp -j\varphi(t', z) \tag{G.13}$$

From Eq. (G.12), the pulse magnitude $|A(t', z)|$ can be readily calculated as

$$|A(t', z)| = \frac{A_0 \tau_p}{[\tau_p^4 + b_2^2 z^2]^{1/4}} \exp - \left[\frac{t'^2 \tau_p^2}{2(\tau_p^4 + b_2^2 z^2)}\right] \tag{G.14}$$

Equation (G.14) shows that the Gaussian pulse maintains its shape on propagation and comparison with Eq. (G.10) shows that the width of the pulse at coordinate z, $\tau_p(z)$, is such that $\tau_p^2(z) = \left(\tau_p^4 + b_2^2 z^2\right)/\tau_p^2$. This expression can then be transformed to

$$\tau_p(z) = \tau_p \left[1 + \left(\frac{z}{L_D}\right)^2\right]^{1/2} \tag{G.15}$$

where $L_D = \tau_p^2/|b_2|$ is called the *dispersion length* of the pulse in the medium. Note from Eq. (G.15) the analogy of time-broadening of a Gaussian pulse in a dispersive medium and spot-size increase of a Gaussian-beam due to diffraction [compare with Eq. (4.7.17a)], the dispersive length, in the former case, being the equivalent of the Rayleigh range in the latter case. This analogy can be traced back to a formal analogy between the diffraction equation, in the paraxial approximation, and the differential equation describing pulse propagation in a quadratic dispersive medium.[(1)]

From Eq. (G.12), the phase, $\varphi(t', z)$, can also be readily calculated as

$$\varphi(t', z) = -\frac{\mathrm{sgn}(b_2)(z/L_D)}{1 + (z/L_D)^2}\frac{t'^2}{\tau_p^2} + \frac{1}{2}\tan^{-1}\left(\frac{z}{L_D}\right) \tag{G.16}$$

where $\mathrm{sgn}(b_2)$ stands for the sign of b_2 i.e., of GVD. Equation (G.16) shows that, $\varphi(t', z)$ contains, besides the constant term $(1/2)\tan^{-1}(z/L_D)$, another term which is quadratic in t'. This means that the instantaneous frequency of the pulse, $\omega(t') = \partial[(\omega_L t') - \varphi(t', z)]/\partial t'$, given by

$$\omega = \omega_L + \mathrm{sgn}(b_2)\frac{(z/L_D)}{1 + (z/L_D)^2}\frac{2t'}{\tau_p^2} \tag{G.17}$$

now has a term which changes linearly with time. Thus the pulse acquires a linear frequency *chirp* whose sign depends on the sign of b_2. In particular, the instantaneous frequency will decrease in time for a negative GVD.

If the length l of the medium is much smaller than the dispersion length, L_D, the relative pulse broadening, $(\delta\tau_p/\tau_p)_D = [\tau_p(l) - \tau_p]/\tau_p$, can be readily obtained from Eq. (G.15), as

$$\left(\frac{\delta\tau_p}{\tau_p}\right)_D \cong \frac{1}{2}\left(\frac{l}{L_D}\right)^2 = \frac{1}{2}\left(\frac{\phi''}{\tau_p^2}\right)^2 \tag{G.18}$$

where $\phi'' = \phi''(\omega_L) = [d^2\phi/d\omega^2]_{\omega_L}$. The width of a Gaussian pulse intensity (FWHM) is then related to the quantity τ_p appearing in Eq. (G.10) by $\Delta\tau_p = 2(\ln 2)^{1/2}\tau_p$. From Eq. (G.18) we then obtain

$$\left(\frac{\delta\tau_p}{\tau_p}\right)_D = (8\ln^2 2)\frac{\phi''^2}{\Delta\tau_p^4} \tag{G.19}$$

Let us now consider a Gaussian pulse, as in Eq. (G.10), entering a homogeneously broadened gain medium. The spectral amplitude of the pulse $A_\omega(\omega - \omega_0)$, while entering the medium, is obtained by taking the inverse Fourier transform of Eq. (G.10) and it is readily shown to be given by

$$A_\omega(\omega - \omega_L) \propto \exp -\left[\frac{(\omega - \omega_L)^2\tau_p^2}{2}\right] \tag{G.20}$$

where ω_L is the central laser frequency. If the spectral width of the pulse is much smaller than the gain linewidth, the gain for the electric field amplitude can be approximated as (see Appendix F)

$$G_e(\omega - \omega_0) = \exp\left\{\left(\frac{g_0}{2}\right)\{1 - [2(\omega - \omega_0)/\Delta\omega_0]^2\}\right\} \tag{G.21}$$

where $g_0 = N_0\sigma l$ is the saturated single pass power gain through the amplifier and $\Delta\omega_0$ is the linewidth of the transition. The spectral amplitude of the pulse, after passing through the gain medium, is then given by

$$A_{g\omega}(\omega - \omega_0) = G_e(\omega - \omega_0) \times A_\omega(\omega - \omega_L) \tag{G.22}$$

From Eq. (G.22) with the help of Eqs. (G.20) and (G.21) and assuming $\omega_L = \omega_0$, we obtain

$$A_{g\omega}(\omega - \omega_0) \propto \exp - \left\{ (\omega - \omega_0)^2 \left[\frac{\tau_p^2}{2} + \frac{g_0}{2} \left(\frac{2}{\Delta\omega_0} \right)^2 \right] \right\} \tag{G.23}$$

Equation (G.23) shows that, within the approximation made, the spectrum remains Gaussian after the pulse has traversed the gain medium. A comparison of Eq. (G.23) with Eq. (G.20) then indicates that the pulse duration has been broadened to a value τ_p' where $\tau_p'^2/2$ corresponds to the term appearing in the square brackets of Eq. (G.23). We thus get

$$\tau_p' = \tau_p \left[1 + g_0 \left(\frac{2}{\tau_p \Delta\omega_0} \right)^2 \right]^{1/2} \tag{G.24}$$

For small changes of pulse duration, the relative pulse broadening, $(\delta\tau_p/\tau_p)_g = \left(\tau'_p - \tau_p\right)/\tau_p$, after the gain medium, is obtained from Eq. (G.24) as

$$\left(\frac{\delta\tau_p}{\tau_p} \right)_g = \frac{1}{2} \left(\frac{2}{\tau_p \Delta\omega_0} \right)^2 g_0 \tag{G.25}$$

Equation (G.25) can be readily recast in terms of the gain linewidth $\Delta\nu_0 = \Delta\omega_0/2\pi$ and laser pulsewidth $\Delta\tau_p = 2(\ln 2)^{1/2}\tau_p$. We obtain

$$\left(\frac{\delta\tau_p}{\tau_p} \right)_g = \left(\frac{2\ln 2}{\pi^2} \right) \left(\frac{1}{\Delta\tau_p^2 \Delta\nu_0^2} \right) g_0 \tag{G.26}$$

References

1. Hermann A. Haus, *Waves and Fields in Optoelectronics*, Prentice Hall, Inc., Englewoods Clifts, 1984, Sect. 6.6

H

Higher-Order Coherence

The degree of coherence $\Gamma^{(1)}$ introduced in Sect. 11.3 involves the first order correlation function $< E(x_1)E^*(x_2) >$, see Eq. (11.3.7), where, as a short-hand notation, the symbol $x_i = (\mathbf{r}_i, t_i)$ has been used to denote both space and time coordinates of the field. Likewise we can define

$$\Gamma^{(n)}(x_1, x_2, \ldots, x_{2n}) = \langle E(x_1)\ldots E(x_n)E^*(x_{n+1})\ldots E^*(x_{2n})\rangle \tag{H.1}$$

which involves the product of $2n$ terms, these being the functions E evaluated at the $2n$ space-time points $x_1, x_2, \ldots, x_{2n}$. The corresponding normalized quantity, $\gamma^{(n)}$, can then be defined as

$$\gamma^{(n)}(x_1, x_2, \ldots, x_{2n}) = \frac{\langle E(x_1)\ldots E(x_n)E^*(x_{n+1})\ldots E^*(x_{2n})\rangle}{\prod_1^{2n}{}_r\langle E(x_r)E^*(x_r)\rangle^{1/2}} \tag{H.2}$$

where $\prod$ is the symbol for product. Obviously these expression reduce to Eqs. (11.3.7) and (11.3.8) for the case $n = 1$.

In terms of these higher-order correlation functions, we need now to define what we mean by a completely coherent beam. Thus, we begin by noting that, if a wave is perfectly coherent to first order (i.e., if $\gamma^{(1)}(x_1, x_2) = 1$), then one must have

$$\Gamma^{(1)}(x_1, x_2) = E(x_1)\, E^*(x_2) \tag{H.3}$$

i.e., $\Gamma^{(1)}$ must factorize into a product of the fields at x_1 and x_2. Indeed, if field fluctuations are completely absent, the ensemble averages of e.g., Eq. (11.3.7) or (11.3.8) simply reduce to the product of the corresponding signals. By analogy one can define a perfectly coherent e.m. wave as one for which $\Gamma^{(n)}$ factorizes to all orders n. This means that

$$\Gamma^{(n)}(x_1, x_2, \ldots, x_{2n}) = \prod_1^{n}{}_r E(x_r) \prod_{n+1}^{2n}{}_k E^*(x_k) \tag{H.4}$$

Indeed, when field fluctuations are completely absent, the ensemble average of Eq. (H.1) will again be simply the product of the corresponding fields. If now Eq. (H.4) is substituted in the numerator of the right hand-side of Eq. (H.2), one readily finds that

$$\gamma^{(n)}(x_1, x_2, \ldots, x_{2n}) = 1 \tag{H.5}$$

for all orders n. It should be noted that the field of a cw laser oscillating in a single mode with narrow linewidth can be considered, for all practical purposes, to satisfy condition Eq. (H.4) to all orders. In fact, as discussed in Sect. 11.7, this field can be considered as showing only phase fluctuations. For a narrow linewidth laser the rate of change of this phase is rather slow, however. For example, in the case of the He–Ne laser considered in Sect. 11.8, having a bandwidth of $\Delta\nu_L \cong 1\,\text{kHz}$, the phase change will occur in $\tau_{co} \cong 1/\Delta\nu_L = 1\,\text{ms}$. This means that, for time intervals much smaller that τ_{co} i.e., for separations between the equiphase surfaces of the $2n$ space-time points much smaller than $c\tau_{co} = 300\,\text{km}$, phase fluctuations will be the same for all $2n$ space-time points and one readily gets Eqs. (H.4) and (H.5).

The difference, to the n-th order, between a completely coherent beam, e.g., the single-mode He–Ne laser just considered, and a thermal light source, is easily illustrated in the case $x_1 = x_2 = \ldots = x_{2n} = x$, i.e., by considering field correlations at the same point and at the same time. The correlation function $\Gamma^{(n)}(x, x, \ldots, x)$ can then be obtained from Eq. (H.1) as

$$\Gamma^{(n)} = \langle |E|^{2n} \rangle = \frac{\iint A^{2n} p_E(\tilde{E}) A\, dA\, d\phi}{\iint p_E(\tilde{E}) A\, dA\, d\phi} \tag{H.6}$$

where the field amplitude $A = A(x)$ is given by Eq. (11.1.1) and $p_E(\tilde{E})$ is the probability density introduced in Sect. 11.7. In particular, for $n = 1$, one has

$$\Gamma^{(1)}(x, x) = \langle |E|^2 \rangle = \langle I \rangle = \frac{\iint A^2 p_E(\tilde{E}) A\, dA\, d\phi}{\iint p_E(\tilde{E}) A\, dA\, d\phi} \tag{H.7}$$

In the case of a coherent field one can use the expression Eq. (11.7.2) for $p_E(\tilde{E})$. From Eq. (H.6) we then get $\Gamma^{(n)} = A_0^{2n}$ while from Eq. (H.7) we get $\Gamma^{(1)} = A_0^2$ so that one can write

$$\Gamma^{(n)}(x, x, \ldots, x) = [\Gamma^{(1)}(x, x)]^n \tag{H.8}$$

In the case of a thermal light source, on the other hand, Eqs. (H.6) and (H.7), with the help of Eq. (11.7.3) for $p_E(\tilde{E})$, give

$$\Gamma^{(n)}(x, x, \ldots, x) = n![\Gamma^{(1)}(x, x)]^n \tag{H.9}$$

To obtain the normalized n-th order coherence function $\gamma^{(n)}$ using Eq. (H.2), we just note that the denominator of the fractional expression on the right-hand side of this equation is, in any case, equal to $[\Gamma^{(1)}(x, x)]^n$. From Eqs. (H.8) and (H.9) one then obtains

$$\gamma^{(n)}(x, x, \ldots, x) = 1 \tag{H.10}$$

and

$$\gamma^{(n)} = n! \tag{H.11}$$

for the single-mode laser source and for the thermal source, respectively. Equation (H.10) obviously shows that the laser beam satisfies the general coherent condition Eq. (H.5). Equation (H.11) then shows that a thermal source can satisfy the coherence condition only for $n = 1$, i.e., only to first order. It follows that one can, at best, arrange for a thermal light source to have perfect, first-order, coherence i.e., perfect spatial and temporal coherence, as indeed discussed in Sect. 11.8.

I

Physical Constants and Useful Conversion Factors

Physical constant	Value	Units
Planck constant (h)	$6.6260755(40) \times 10^{-34}$	$J \cdot s$
($\hbar = h/2\pi$)	$1.05457266(63) \times 10^{-34}$	$J \cdot s$
Electronic charge (e)	$1.60217733(49) \times 10^{-19}$	C
Electron rest mass (m_e)	$9.1093897 \times 10^{-31}$	kg
Proton rest mass (m_p)	$1.6726231(10) \times 10^{-27}$	kg
Neutron rest mass (m_n)	$1.6749286(10) \times 10^{-27}$	kg
Velocity of light in vacuum (c)	2.99792458×10^{8}	m/s
Boltzmann constant (k)	$1.380658(12) \times 10^{-23}$	$J \cdot K^{-1}$
Bohr magneton (β)	$9.2740154(30) \times 10^{-24}$	$A \cdot m^2$
Permittivity of vacuum (ε_0)	$8.854187817\ldots \times 10^{-12}$	F/m
Permeability of vacuum (μ_0)	$4\pi \times 10^{-7}$	H/m
Avogadro's constant (N_A)	$6.0221367(36) \times 10^{23}$	mol^{-1}
Ideal-gas constant ($R = N_A \cdot k$)	8.31451	$J \cdot K^{-1} \cdot mol^{-1}$
Radius of first Bohr orbit [$a_o = (4\pi^2\varepsilon_0/me^2)$]	$0.529177249(24) \times 10^{-10}$	m
Stefan-Boltzmann constant (σ_{SB})	$5.67051(19) \times 10^{-8}$	$W \cdot m^{-2} \cdot K^{-4}$
Free-space impedance ($Z = 1/\varepsilon_0 c$)	376.73...	Ω
Ratio of the mass of the proton to the mass of the electron (m_p/m_e)	1836.152...	

Physical constant	Value	Units
Energy corresponding to 1 eV	$1.602\ldots \times 10^{-19}$	J
Energy of a photon with wavelenght $\lambda = 1\ \mu m$	$1.986\ldots \times 10^{-19}$	J
Wavenumbers corresponding to an energy spacing of $kT (T = 300\,K)$ ($\tilde{\nu} = kT/hc$)	$208.512\ldots$	cm^{-1}
Atmospheric pressure (1 atm = 760 torr = 1.013 . . . bar)	$1.013\ldots \times 10^{5}$	Pa

Answers to Selected Problems

Chapter 1

1.1. Far infrared: 1 mm – 50 μm. Medium infrared: 50 – 2.5 μm. Near infrared: 2.5 μm – 750 nm. Visible: 750 – 380 nm. Ultraviolet: 380 – 180 nm. Vacuum ultraviolet: 180 – 40 nm. Soft x-ray: 40 – 1 nm. X-ray: 1 – 0.01 nm.

1.4. For $g_1 = g_2$ one gets from Eq. (1.2.2) $E_2 - E_1 = kT = 208.5\,\text{cm}^{-1}$ so that $\lambda = (1/208.5)$ cm $\cong 48\,\mu$m, falling in the medium infrared.

1.5. $\gamma_1 = 1$, $\gamma_2 = -\ln R_2 \cong 0.693$, $\gamma_i \cong 0.01$, $\gamma = \gamma_i + (\gamma_1 + \gamma_2)/2 \cong 0.357$, $N_c = \gamma/\sigma l \cong 1.7 \times 10^{17}\,\text{cm}^{-3}$.

1.6. $D_m \cong (2\lambda/D)L \cong 533$ m, where D_m is the beam diameter on the moon, D is the telescope aperture, and L is the distance between earth and moon. The first earth-moon ranging experiment was achieved under these conditions, using a Q-switched ruby-laser. Owing to the large beam diameter on the moon and to the surface variations over this diameter, the precision of this ranging experiment was rather limited (~1 m). Using special mirrors, as beam reflectors, placed on the moon surface by visiting astronauts, the earth moon distance can now be measured with an accuracy of roughly a few mm.

Chapter 2

2.1. $N(\Delta\lambda) = 8\pi V\,\Delta\lambda/\lambda^4 \cong 1.9 \times 10^{12}$ modes !

2.2. $\rho_\lambda = \rho_\nu|d\lambda/d\nu| = (c_n/\lambda^2)\,\rho_\nu$ where the relation $\lambda\nu = c_n$ (c_n being the light velocity in the medium filling the black-body cavity) has been used. From Eq. (2.2.22) with the substitution $\nu = c_n/\lambda$ one then obtains

$$\rho_\lambda = \frac{8\pi c_n h}{\lambda^5}\frac{1}{\exp(hc_n/\lambda kT) - 1}$$

2.3. By imposing the condition $(d\rho_\lambda/d\lambda) = 0$ and using the expression for ρ_λ given in the answer of Problem 2.2, one gets the equation $5 \cdot [\exp(hc_n/\lambda kT) - 1] - (hc_n/\lambda kT)\exp(hc_n/\lambda kT) = 0$. If one writes $y = (hc_n/\lambda kT)$ in the previous expression one sees that the value of y corresponding to the peak of ρ_λ must satisfy the equation $5 \cdot [1 - \exp(y_M)] = y_M$. The solution of this equation can be obtained, by a fast converging iterative procedure, as $y_M \cong 4.965$. For $c_n = c$, c being the light velocity in vacuum, the wavelength, λ_M, at which the maximum value of ρ_λ occurs, must then satisfy the relation (Wien's law): $\lambda_M T = hc_n/y_M k \cong 2.3 \times 10^{-3}$ m × K.

2.6. The density of the Nd^{3+} ions, N, expressed in ions/cm^3, and hence the Nd^{3+} concentration in the $^4I_{9/2}$ manifold is given by $N = 1 \times 10^{-2}\ 3(\rho/\text{M.W.})N_A$ where ρ is the density, expressed in g/cm^3, M.W. is the molecular weight of YAG, and N_A is Avogadro's number. The factor 3 in the previous expression accounts for the presence of three yttrium atoms per molecule. Since the YAG molecular weight is 594 g/mol we obtain $N \cong 1.38 \times 10^{20}$ ions/cm^{-3}. According to Eq. (1.2.2), the fraction, f, of this population belonging to the lowest sub-level of the $^4I_{9/2}$ state is then given by

$$f = \frac{1}{1 + \sum_{i=1}^{4} \exp -(E_i / kT)}$$

where $E_i (i = 1\text{–}4)$ is the energy separation between the higher sublevels and the ground sublevel. Given the values of E_i for these sublevels one gets $f = 46\%$.

2.7. From Eqs. (2.4.25) and (2.3.15) one gets $\sigma_{in} = (\lambda_n^2/8\pi)\,[g(\nu - \nu_0)/\tau_{sp}]$ where $\lambda_n = c/n \cdot \nu_0$ is the wavelength, in the medium, of an e.m. wave of frequency ν_0. For $\nu = \nu_0$ and for a pure inhomogeneous broadening, using Eq. (2.4.28), we obtain the following expression for the peak cross-section $\sigma_p = 0.939\,(\lambda_n^2/8\pi)\,(1/\Delta\nu_0^*\tau_{sp})$. For $\lambda_n = 1.15\,\mu$m ($n \cong 1$), $\Delta\nu_0^* = 9 \times 10^8$ Hz, and $\tau_{sp} \cong 10^{-7}$ s we then get $\sigma_p \cong 5.5 \times 10^{-12}$ cm^2.

2.8. Consider a plane wave, of uniform intensity I, crossing a surface, of area S, in a medium of refractive index n. The e.m. energy flux through the surface S in a time Δt is $E = IS\Delta t$ and this energy is found, uniformly distributed, in a volume $V = S(c/n) \cong \Delta t$. The energy density in the medium is then $\rho_n(E/V) = (n/c)I$.

2.11. The answer is readily obtained following Example 2.13. One gets $(\Omega/4\pi) = (D/4l)^2 \cong 4.4 \times 10^{-4}$ and from Eq. (2.9.4a), by a fast iterative procedure, one finds $G = 1.24 \times 10^4$. The threshold inversion is then $N_{th} = \ln G/\sigma_p l = 4.49 \times 10^{18}$ cm^{-3} and the maximum stored energy $E_M = N_{th}(\pi D^2 l/4) \cdot h\nu = 1.96$ J.

2.13. Under thermal equilibrium the two processes Eqs. (2.6.9) and (2.6.10) must balance each other. Thus the relation $\kappa_{B^*A} N_{B^*} N_A = \kappa_{BA^*} N_B N_{A^*}$ must hold. At exact resonance and again at thermal equilibrium one has $(N_{A^*}/N_A) = (N_{B^*}/N_B) = \exp -(E/kT)$ where E is the energy level separation between either one of the two-level system. We then get $\kappa_{B^*A} = \kappa_{BA^*}$.

2.14. For a Lorentzian line one has

$$I_s = I_{so}\{1 + [2(\nu - \nu_0)/\Delta\nu_0]^2\}$$

We then get

$$\alpha(\nu - \nu_0) = \frac{\alpha_0(0)}{1 + [2(\nu - \nu_0)/\Delta\nu_0]^2} \cdot \frac{1}{1 + (I/I_s)} =$$

$$= \frac{\alpha_0(0)}{1 + [2(\nu - \nu_0)/\Delta\nu_0]^2} \cdot \frac{1}{1 + \frac{I}{I_{so}} \frac{1}{1 + [2(\nu - \nu_0)/\Delta\nu_0]^2}} =$$

$$= \frac{\alpha_0(0)}{1 + [2(\nu - \nu_0) / \Delta\nu_0]^2 + (I / I_{so})}$$

2.15. On setting $\nu = \nu_o$ in the expression obtained in Problem 2.14, we get

$$\alpha_p(0) = \frac{\alpha_0(0)}{1 + (I / I_{s0})}$$

According to this equation, the saturation intensity, I_{s0}, is the intensity of a resonant ($\nu \cong \nu_o$) e.m. wave at which the peak saturated absorption coefficient, $\alpha_p(0)$, is half of the corresponding unsaturated value, $\alpha_0(0)$. The (1/2) power points of $\alpha = \alpha(\nu - \nu_o)$, in the expression obtained in the previous problem, occur at frequency ν' so that $[2(\nu' - \nu_0)/\Delta\nu_0]^2 = 1 + (I/I_{s0})$. The saturated linewidth (FWHM) is then readily obtained as $\Delta\nu_{sat} = \Delta\nu_0[1 + (I/I_{s0})]^{1/2}$.

Chapter 3

3.1. The center of mass is midway between the two atoms. Considering the x-axis to be along the vibration direction with the origin at the center of mass, the restoring force on each atom is given by $F = -2k_0(x - x_0)$, where x_0 is the equilibrium coordinate of each atom. The equation of motion can then be written as $[Md^2(x - x_0)dt^2] = -2k_0(x - x_0)$ so that the resonance frequency is $\omega = [2k_0/M]^{1/2}$.

3.2. Using the result of the previous problem we obtain $k_0 = (2\pi\tilde{\nu}\, c)^2 \cdot (M/2) = 2{,}314\,\mathrm{J \cdot m^{-2}}$ where the atomic weight of the N atom has been taken to be $\cong 14$. The potential energy of the system is then given by $U = k_0(R - R_0)^2/2$, where R is the internuclear separation and R_0 is the equilibrium value. For $R - R_0 = 0.03$ nm one obtains $U \cong 6.5$ eV.

3.6. $B = \hbar^2/2I$. In this expression one has $I = 2M_oR_0^2$, where M_o is the oxygen mass and R_0 is the equilibrium distance between oxygen and carbon. For $B = 0.37\,\mathrm{cm^{-1}}$ and $M_o = 16$ g/mol (atomic weight) we obtain $R_0 = \hbar/2 \cdot [M_o \cdot B]^{1/2} = 0.0515$ nm.

3.8. From Fig. 3.15b, for $N = 1.6 \times 10^{-18}\,\mathrm{cm^{-3}}$, one gets $E_{Fc} = 2.35\,kT$ and $E_{Fv} = -1.45\,kT$. The overall gain bandwidth is then $\Delta\tilde{\nu} = (E_{Fc} + E_{Fv})/hc = 0.9(kT/hc) = 187.65\,\mathrm{cm^{-1}}$. [we recall that $(kT/hc) = 208.5\,\mathrm{cm^{-1}}$ at $T = 300$ K].

3.9. $E_2 + E_1 = 0.45\,kT$. From Eq. (3.2.2) we get $(E_2/E_1) = m_v/m_c = 6.865$ where m_v is the hole mass and m_c is the electron mass in the conduction band. From the previous two equations we get $E_2 = 0.392\,kT$ and $E_1 = 0.0572\,kT$.

3.10. From Fig. 3.16 and for $E - E_g = 0.45\,kT \cong 12$ meV we obtain $\alpha = \alpha_0 = 1.8 \times 10^3\,\mathrm{cm^{-1}}$. Since the probabilities of occupation of the upper and lower laser level are given by [see Eq. (3.2.10)]

$$f_c(E_2) = \frac{1}{1 + \exp\left[(E_2 - E_{Fc}) / kT\right]}$$

$$f_v(E_1) = \frac{1}{1 + \exp\left[(E_{Fv} - E_1) / kT\right]}$$

from the results of Problems 3.8 and 3.9, we obtain $f_c(E_2) = 0.877$ and $f_v(E_1) = 0.8186$. From Eq. (3.2.37) we then get $g = \alpha_0[f_c(E_2) - f_v(E_1)] \cong 104\,\mathrm{cm^{-1}}$.

3.11. $\sigma = (dg/dN) = [g/(N - N_{tr})] = 2.6 \times 10^{-16}\,\text{cm}^2$, where $g = 104\,\text{cm}^{-1}$, $N = 1.6 \times 10^{18}\,\text{cm}^{-3}$ and, see Example 3.7, $N_{tr} = 1.2 \times 10^{18}\,\text{cm}^{-3}$.

3.13. From Fig. 3.26, for $N = 2 \times 10^{18}\,\text{cm}^{-3}$, one gets $E_{Fc} - E_{1c} \cong 2.8\,kT \cong 72\,\text{meV}$ and $E_{Fv} - E_{1v} \cong -1.1\,kT \cong -28.6\,\text{meV}$. Using the results of Example 3.9 for E_{1c} and E_{1v} one then obtains $E_{Fc} = 128.2\,\text{meV}$ and $E_{Fv} = -20.6\,\text{meV}$. According to Eq. (3.3.26) the gain bandwidth occurs for E-E_g ranging between $\Delta E_1 = E_{1c} + E_{1v} = 64.2\,\text{meV}$ and $E_{Fc} + E_{Fv} = 107.6\,\text{meV}$.

Chapter 4

4.3. One has $T = 1 - R - A = 5 \times 10^{-3}$. From Eq. (4.5.6a) the peak transmission is obtained as $(T_{FP})_p = [T/(1-R)]^2 = 25\%$. Note the strong reduction in peak transmission even for such a small mirror loss. A comparison between Eqs. (4.5.6a) and (4.5.6) [with $R_1 = R_2 = R$] shows that the expression for the finesse still remains that given by Eq. (4.5.14); thus $F = \pi R^{1/2}/(1-R) \cong 312.4$.

4.4. From Eqs. (4.5.8) and (4.5.3), assuming normal incidence ($\theta = 0$) and unit refractive index ($n_r = 1$), one gets $L = c/2\Delta\nu_{fsr} = 5\,\text{cm}$. From Eq. (4.5.13) the finesse is then obtained as $F = \Delta\nu_{fsr}/\Delta\nu_c = 50$. From Eq. (4.5.14), with $R_1 = R_2 = R$, we readily obtain the value of the mirror reflectivity as $R \cong 94\%$. From Eq. (4.5.6a), for a peak transmission of 50%, we then obtain $T = 4.24 \times 10^{-2}$ and hence $A = 1 - R - T = 1.76 \times 10^{-2}$.

4.7. The wavefront radius of curvature, at the lens position, is given by [see Eq. (4.7.17b)] $R = d[1 + (z_R/d)^2]$, where $z_R = \pi w_0^2/\lambda$ is the Rayleigh length. To compensate this curvature, the focal length of the lens must just equal R.

4.9. From Eq. (4.7.19) one gets $w_0 = \lambda/\pi\theta_d \cong 201\,\mu\text{m}$. Assuming that the output beam forms a waist, the peak intensity will occur at this waist position and will be given by $I_p = P/\left(\pi w_0^2/2\right) \cong 7.85\,\text{W/cm}^2$. From Eqs. (2.4.10) and (2.4.6), the intensity of an e.m. wave is seen to be related to the field amplitude E_0 by $I = nE_0^2/2Z$ where $Z = 1/\varepsilon_0 c \cong 377$ ohms is the impedance of free space. For $n = 1$ we then get $E_0 = [2IZ]^{1/2} \cong 77\,\text{V/cm}$.

4.11. From Eq. (4.7.27) one readily gets $w_{02} = \lambda f/D$. The numerical aperture of a lens is defined as (see Sect. 1.4.4) $N.A. = \sin(\theta)$, where θ is the half-angle of the cone formed by the aperture D seen from the central point in the focal plane. Since $\tan(\theta) = D/2f$, one gets $w_{02} = \lambda/2\tan(\sin^{-1} N.A.)$. For a small numerical aperture one has $w_{02} \cong \lambda/2N.A.$

4.13. According to Eq. (4.7.4), the beam parameter q, after the plate, is related to the input beam parameter q_0 by

$$\frac{1}{q} = \frac{C + (D/q_0)}{A + (B/q_0)}$$

In our case $q_0 = -jz_R$, where $z_R = \pi w_0^2/\lambda$ is the Rayleigh length. Using the *ABCD* matrix elements of a plate of length L and refractive index n [see Table 4.1] we obtain

$$\frac{1}{q} = \frac{-jz_R}{1 - j(L/nz_R)}$$

From the real and imaginary parts of the above equation, with the help of Eq. (4.7.8), one obtains

$$w^2 = w_0^2[1 + (L'/z_R)^2]$$
$$R = L'[1 + (z_R/L')^2]$$

where $L' = L/n$. This proves the statement of Problem 4.13. After the plate, the spot size at distance z from the waist would then be equal to that which would occur, without the plate, at a distance $z' = (z-L)+(L/n) = z-[(n-1)/n]L$. At large distances from the waist, i.e. for $z \gg z_R$, one then has $w(z) = (\lambda/\pi w_0)\cdot\{z-[(n-1)/n]L\}$. The beam divergence is then $\theta_d = w(z)/z$ and, if $L \leq z_R$, one has $z \gg L$ and the beam divergence remains equal to $(\lambda/\pi w_0)$ i.e. it is unaffected by the presence of the plate.

4.14. Equation (4.7.26) can be written as

$$z_m = \frac{1}{\frac{1}{f} + \frac{f}{z_{R_1}}}$$

For a given value of z_{R_1}, the denominator of the right hand side of the above equation has a minimum for $f = z_{R_1}$. At this value of focal length, z_m then reaches its maximum value given by $z_m = z_{R_1}/2$.

Chapter 5

5.2. $w_o = \sqrt{L\lambda/2\pi} = 0.29\,\text{nm}$; $w_s = \sqrt{2}\,w_o \cong 0.4\,\text{mm}$; $\Delta\nu_c = c/2L = 150\,\text{MHz}$. Number of modes $N = \Delta\nu_o^*/(c/4L) = 47$, where $(c/4L)$ is the frequency spacing between two consecutive non-degenerate modes of a confocal resonator (see Fig. 5.10a).

5.4. The curvature of the wavefront must coincide with that of the mirror, at the mirror's location. From Eq. (4.7.13b), setting $z = L/2$ where L is the cavity length, one then gets

$$\frac{\pi w_0^2}{\lambda} = \frac{L}{2}\left[\frac{2R}{L} - 1\right]^{1/2}$$

The above expression gives $z_R = \pi w_0^2/\lambda \cong 1.32\,\text{m}$ and $w_0 \cong 0.466\,\text{mm}$. The spot-size at the mirror is then obtained from Eq. (4.7.13a) as $w = w_0[1 + (L/2z_R)^2)]^{1/2} = 0.498\,\text{mm}$.

5.6. Since the equiphase surfaces, at the two mirror positions, must coincide with the mirror surfaces, from Eq. (4.7.17b) one writes

$$-R_1 = z_1 + \left(\frac{z_R^2}{z_1}\right)$$

$$R_2 = z_2 + \left(\frac{z_R^2}{z_2}\right)$$

$$L = z_2 - z_1$$

where z_1 and z_2 are the coordinates of the mirrors as measured from the waist. Note the minus sign in front of the term R_1 in the first equation. It arises from the fact that, if e.g. mirror 1 is a concave mirror, R_1 is positive while the sign of the wavefront is negative because the center of curvature is to the right of the wavefront. From the above three equations one can then eliminate z_2 to obtain

$$-R_1 z_1 = z_1^2 + z_R^2$$

$$R_2(L + z_1) = (L + z_1) + z_R^2$$

From these two equations one can eliminate z_R^2, to obtain $z_1[2L - R_1 - R_2] = L(R_2 - L)$. Note that, for $R_1 = R_2 = R$, this equation gives $z_1 = -L/2$, which shows that, in this case, the waist is located at the cavity center. From Eq. (5.4.10) one now finds $R_1 = L/(1-g_1)$ and $R_2 = L/(1-g_2)$. The substitution of these two expressions for R_1 and R_2 into the previous equation then readily gives $-z_1(g_1 + g_2 - 2g_1g_2) = g_2(1 - g_1)L$.

5.7. We have $g_1 = 0.333$, $g_2 = 0.75$. From the expression for z_1 obtained in the previous problem we then find $z_1 = -0.857$ m. From Eqs. (5.5.8) and (5.5.9) we also find $w_1 = 0.533$ mm, $w_2 = 0.355$ mm and $w_0 = 0.349$ mm.

5.9. For symmetry reasons the waist must be located at a distance $L_p/2$ from the lens. From Eq. (4.7.17b) one then finds that the absolute value of the wavefront's radius of curvature, on both sides of the lens, is given by $R = (L_p/2)[1 + (2z_R/L_p)^2]$. The lens must then transform one wavefront into the other. One must then have $f = R/2$. Since $z_R = \pi w_0^2/\lambda$, one obtains from the previous expressions $w_0^2 = (\lambda/2\pi)[L(4f - L_p)]^{1/2}$. Note that the previous expression gives a real value of w_0^2 only when $L_p \le 4f$, which represents the stability condition for our case.

5.11. One has $g_1 = 1$ and $g_1 = 1 - L/(L + \Delta) \cong \Delta/L$. From Eq. (5.5.8b) one then has $w_2 = w_m = (L\lambda/\pi)^{1/2}\{1/(\Delta/L)[1 - (\Delta/L)]\}^{1/4}$. For $w_m = 0.5$ mm, $L = 30$ cm and $\lambda = 633$ nm, the above expression gives $\Delta = 1.85$ cm. We then have $R_2 = L + \Delta = 31.85$ cm, and $g_2 = 0.058$, and, from Eq. (5.5.8a), the spot size at the plane mirror is obtained as $w_1 \cong 0.122$ mm.

5.13. In this case, from Eqs. (5.5.5) and (5.4.6) the stability condition is seen to be given by $-2 < 2(2A_1D_1 - 1) < 2$ i.e. by $0 < A_1D_1 < 1$. Since $B_1D_1 - A_1C_1 = 1$, one then finds $-1 < B_1C_1 < 0$.

5.17. (1) $g_1 = 1$, $g_2 = 1.25$; (2) from Eq. (5.6.1): $r_1 = 2.24$ and $r_2 = 1.24$; (3) $a_1 > M_{21}a_2 = 1.8\ a_2$; (4) $M = M_{12}M_{21} = 2.62$, so that $\gamma = (M^{2-1})/M^2 = 0.85$.

5.18. Positive branch confocal unstable resonator. From Fig. 5.22, for $\gamma = 0.2$ and $N_{eq} = 7.5$, one finds $M = 1.35$. It then follows that $2a_2 = 2[2L\lambda N_{eq}/(M - 1)]^{1/2} = 4.26$. To achieve a single-ended resonator one must have $a_1 > 2Ma_2 = 5.75$ cm. The radii of the two mirrors, on the other hand, must be such that $L = (R_1 + R_2)/2$ and $M = -R_1/R_2$ (note that $R_2 < 0$). We obtain $R_1 = 7.7$ m and $R_2 = -5.7$ m.

5.20. (1) From Eq. (5.6.20) one gets $\exp -2(a/w)^6 = 2 \times 10^{-2}$ i.e. $w = 2.94$ mm. (2) From Eq. (5.6.21) $w_m = w/(M^6 - 1)^{1/6} = 2.32$ mm. (3) $\gamma = 1 - (R_0/M^2) \cong 0.744$. (4) Since $g_2 = 1$, from Eqs. (5.6.3) and (5.6.1) one gets $M = g_1\{1 + [1 + (1/g_1)]^{1/2}\}^2$ from which one obtains $g_1 = 1.0285$. The radius of curvature of the convex mirror is then $R_1 = L/(1 - g_1) \cong -17.5$ m.

Chapter 6

6.1. For radial propagation, the power absorbed in the laser rod can be written as $P_a = \int S[1 - \exp -(2\alpha R)]I_{e\lambda}\mathrm{d}\lambda$, where S is the rod's lateral surface area. Since the power entering the rod is $P_e = \int SI_{e\lambda}\mathrm{d}\lambda$, we obtain $\eta_a = P_a/P_e = \int[1 - \exp -(2\alpha R)]I_{e\lambda}\mathrm{d}\lambda / \int I_{e\lambda}\mathrm{d}\lambda$.

6.4. From Eq. (6.2.6) with $h\nu_{mp} \cong 2.11 \times 10^{-19}$ J [$\lambda_{mp} = 940$ nm] one readily gets $R_{cp} \cong 2.01 \times 10^{20}$ cm^{-3}s^{-1}.

6.6. The pump efficiency is, in this case, equal to $\eta_p = \eta_t\eta_a\eta_{pq} = 5.3\%$. The laser of Problem 6.4, with a pump efficiency of $\eta'_p = 4.5\%$, has a threshold of $P'_{th} = 2.36$ kW. With the present pump configuration, the threshold pump power will then be $P_{th} = \left(\eta_p/\eta'_p\right)P'_{th} = 2.36$ kW. To pump the laser 2 times above threshold we then need a pump power of $P_p \cong 4.72\,\text{kW}$. The area of the collecting optics is then given by $A = P_p/I \cong 4.72\,\text{m}^2$, where I is the solar intensity. If we let D and l be the rod diameter and rod length respectively, the focal lengths of the two lenses must be such that $f_1\alpha = D$ and $f_2\alpha = l$, where α is the angle which the sun's disc subtends at the earth. One gets $f_1 = 0.64\,\text{m}$ and $f_2 = 8.05\,\text{m}$. A cheaper focusing scheme could be made using a spherical mirror, with focal length $f_2 = 8.05\,\text{m}$ and diameter $D = (4A/\pi)^{1/2} = 2.45\,\text{m}$, followed by a cylindrical lens.

6.7. $R_{cp} = \eta_p E_{th}/Vh\nu_{mp}\Delta t$, where Δt is the pump duration. Since $h\nu_{mp} = 2.11 \times 10^{-19}\,\text{J}$ one obtains $R_{cp} = 5.75 \times 10^{21}\,\text{cm}^{-3}\text{s}^{-1}$. The rate equation involving pumping and spontaneous decay is $(dN_2/dt) = R_p - (N_2/\tau)$ whose solution, for $R_p = \text{const}$ and $t \geq 0$, is $N_2(t) = R_p\tau\,[1-\exp-(t/\tau)]$. Assuming $\tau = 230\,\mu\text{s}$, $t = \Delta t = 100\,\mu\text{s}$, and with the help of the previously calculated value of R_{cp}, one gets the critical inversion as $N_{2c} = R_{cp}\tau[1-\exp-(\Delta t/\tau)] = 4.66 \times 10^{17}\,\text{cm}^{-3}$. If the pulse duration is increased to $\Delta t' = 300\,\mu\text{s}$, the pump rate, R'_{cp}, to achieve the same inversion must be such that $R'_{cp}[1-\exp-(\Delta t'/\tau)] = R_{cp}[1-\exp-(\Delta t/\tau)]$. One gets $R'_{cp} \cong 0.48\,R_{cp} \cong 2.78 \times 10^{21}\,\text{cm}^{-3}\text{s}^{-1}$. The new threshold pump energy is then $E'_p = (R'_{cp}\Delta t'/R_{cp}\Delta t)\,E_p \cong 1.44\,E_p \cong 4.9\,\text{J}$.

6.10. From Table 6.2 one gets $N_t = 9 \times 10^{20}\,\text{cm}^{-3}$ so that $\sigma_a N_t l + \gamma = 0.169$. Assuming a $\sim$80% efficiency for the transfer optics, the pump efficiency can be taken to be $\eta_p = \eta_t\eta_a = \eta_t[1-\exp(\alpha l)] = 0.424$ where $\alpha = 5\,\text{cm}^{-1}$ is the absorption coefficient at the pump wavelength (see Table 6.2). From Eq. (6.3.25), with $w_p = w_0$, one then readily gets $P_{th} \cong 177\,\text{mW}$.

6.12. For a Maxwell–Boltzmann distribution one has $kT_e = (2/3)\,(mv_{th}^2/2)$. Since $(mv_{th}^2/2) = 10\,\text{eV}$, we then get $kT_e = 6.67\,\text{eV}$.

6.15. Using Gauss's theorem, the radially oriented electric field, at the radial coordinate r within the medium, can be expressed as $E(r) = N_i er/2\varepsilon_0$. By integration, the potential drop between wall and center is readily obtained as $V = N_i eR^2/4\varepsilon_0 \cong 4.56 \times 10^6\,\text{V}$, where R is the tube radius. The very high value of the voltage obtained shows that there is a negligible probability for electrons to disappear at a different rate from that of the ions.

6.18. The thermal velocity is given by $\upsilon_{th} = (2E/m)^{1/2} = 1.87 \times 10^8\,\text{cm/s}$, where m is the mass of the electron. For an ideal gas, the molar volume is given by $V = RT/p$, where $R = 8.314\,\text{J} \cdot \text{mol}^{-1} \cdot \text{K}^{-1}$ is the gas constant. For $p = 1.3\,\text{torr} \cong 1.73 \times 10^2\,\text{Pa}$ and $T = 400\,\text{K}$ one gets $V = 19.2 \times 10^6\,\text{cm}^3$. The atomic density in the gas is then given by $N = N_A/V = 3.14 \times 10^{16}\,\text{cm}^{-3}$ where N_A is Avogadro's constant (see Appendix *I*). We then obtain the electron mean free path as $l = 1/N\sigma = 638\,\mu\text{m}$. From Eqs. (6.4.14) and (6.4.15), the drift velocity can then be obtained as $\upsilon_{drift} = e\mathcal{E}l/m\upsilon_{th} \cong 1.8 \times 107\,\text{cm/s}$ ($\upsilon_{drift}/\upsilon_{th} \cong 9.6 \times 10^{-2}$).

Chapter 7

7.3. $L_e = L + (n-1)l = 56.15\,\text{cm}$, $\gamma = 0.12$, $\tau_c = 15.6\,\text{ns}$.

7.4. The overall lifetime of the upper laser level is such that $(1/\tau) = (1/\tau_{21}) + (1/\tau')$, where $(1/\tau_{21})$ is the rate of the $2 \to 1$ transition and $(1/\tau')$ is the rate of all other spontaneous transitions originating from level 2. Since the branching ratio, β, is given by $\beta = (1/\tau_{21})/(1/\tau) = \tau/\tau_{21}$, we get $\tau_{21} = \tau/\beta = 451\,\mu\text{s}$. Below threshold, at steady state, one has $(N_1/\tau_1) = (N_2/\tau_{21})$, where τ_1 is the lifetime of the lower laser level. Thus, for $(N_1/N_2) < 1\%$, one must have $\tau_1 < 10^{-2}\,\tau_{21} \cong 4.5\,\mu\text{s}$. For an output power of $P_{out} = 200\,\text{W}$, the rate of emission of photons from the active medium is $(d\phi/dt) = P_{out}(2\gamma/\gamma_2)/h\nu = 1.58 \times 10^{21}$ photons/s. The population ending up in the lower laser level, per unit time, will then be $dN'/dt = (d\phi/dt)/A_b l = 9.16 \times 10^{20}\,\mathrm{cm^{-3}s^{-1}}$, where A_b is the effective beam area in the rod and l is the rod length. The steady-state population N_1 is then given by $N_1 = \tau_1(dN'/dt)$ while the upper-state population, N_2, is now equal to the threshold population, i.e., $N_2 = N_c \cong 5.7 \times 10^{16}\,\mathrm{cm^{-3}}$. For $(N_1/N_2) < 1\%$ one must now have $\tau_1 < 10^{-2} \cdot N_c/(dN'/dt) \cong 0.6\,\mu\mathrm{s}$. (The actual lifetime τ_1 for Nd:YAG is ~ 100 ps).

7.7. The minimum threshold power is given by $P_{mth} = P_{th} \cdot \gamma_i/\gamma = 2.75\,\mathrm{kW}$ where γ_i is the internal loss ($\gamma_i = 0.02$) and γ is the total loss ($\gamma = 0.32$). At a pump power of $P_p = 140\,\mathrm{kW}$ we then have $x_m = P_p/P_{mth} = 50.9$. From Eq. (7.5.5) one finds $S_{op} = 6.135$, so that $\gamma_{2op} = 2S_{op}\gamma_i = 0.25$. The corresponding optimum output power is obtained from Eq. (7.5.6) as $P_{op} = 16.78\,\mathrm{kW}$. We have $T_{2op} = 1 - \exp(-\gamma_{2op}) \cong 0.25$. Since the peak intensity in the focal plane of a lens is proportional to $(M^{2^{-1}})/M^2$ (see Sect. 5.6.3) the ratio between the two intensities is $(I_{op}/I) = P_{op}T_{2op}/PT_2 = 16.78 \times 0.22/12 \times 4.45 = 0.777$.

7.8. The lens of focal length f can be divided into two, closely spaced, lenses each of focal length $2f$. The radius of curvature of the wavefront at the lens position is given by $R = (L/2)\cdot\{1 + [z_R/(L/2)]^2\}$, with $z_R = \pi\, w_0^2/\lambda$, where w_0 is the spot size at each of the two mirrors. For symmetry reasons, the wavefront between the two lenses must be plane. Thus, one must have $R = 2f = 50\,\mathrm{cm}$. Using this value of R in the previous expression one finds $z_R = 25\,\mathrm{cm}$ i.e. $w_0 = 290\,\mu\mathrm{m}$. The spot size at the lens position is then $w = \sqrt{2}w_0 = 410\,\mu\mathrm{m}$.

7.10. The minimum threshold pump power for zero output coupling is $P'_{mth} = P_{th}\gamma_i/\gamma = 12.5\,\mathrm{mW}$ [note that P'_{mth} should not be confused with P_{mth} given by Eq. (7.3.32)]. At $P_p = 1.14\,\mathrm{W}$ one has $x_m = 91.2$. From Eq. (7.5.5) one then gets $S_{op} = 8.54$ i.e. $\gamma'_{2op} = 2S_{op}\gamma_i = 8.5\%$. To calculate the output power we note that the total loss is now given by $\gamma' = \left(\gamma'_{2op}/2\right) + \gamma_i = 4.75\%$ and the amount by which threshold is exceeded is $x' = x\gamma/\gamma'$, where x and γ are the corresponding values obtained in Example 7.4 ($x = 30$ and $\gamma = 3\%$). We get $x' = 19$ so that, from Eq. (7.3.34), we find $y' = 15.6$. The expected power is then $P'_{out} = P_{out}(y'/y)\left(\gamma'_{2op}/\gamma_2\right) = 510\,\mathrm{mW}$ where P_{out}, y, and γ_2 are the corresponding values obtained in Example 7.4 ($P_{out}, = 500\,\mathrm{mW}$, $y = 26$, $\gamma_2 = 5\%$). We should note that, indeed, our optimization procedure has resulted in a value of P'_{out} (slightly) larger than P_{out}.

7.12. The lens of focal length f can be divided into two, closely spaced, lenses each of focal length $f' = 2f$. The spot size between the two lenses is w_a and, for symmetry reasons, the wavefront is plane. The position of the two plane mirrors must also correspond to a beam waist. From Eq. (4.7.26) we then find that the distance between each plane mirror and the corresponding lens f' is given by $z_m = f'/[1 + (f'/z_R)^2]$ where, in our case, $z_R = \pi\, w_a^2/\lambda \cong 581\,\mathrm{cm}$. For $f' = 2f = 42\,\mathrm{cm}$ one finds $z_m \cong 41.8\,\mathrm{cm}$. The spot size on each mirror is then obtained from Eq. (4.7.28) as $w_0 = (\lambda/\pi w_a)f' \cong 100\,\mu\mathrm{m}$.

7.15. One has $\gamma_2 = -\ln(1 - T_2) \cong 5.1\%$. To avoid oscillation on the TEM_{01} mode, the required diffraction loss per pass for this mode must be $\gamma \geq 7.45\%$. From Fig. 5.13b and for $g = 1 - (L/R) = 0.8$ one then finds $N = a^2/\lambda L \leq 2$, i.e. $a \leq \sqrt{2\lambda L} \cong 1\,\mathrm{mm}$.

7.17. $N_c = \gamma/\sigma_e l \cong 4 \times 10^9\,\mathrm{ions/cm^3}$; $R_{cp} = N_c/\tau \cong 8 \times 10^{17}\,\mathrm{cm^{-3}s^{-1}}$; $\Delta\nu = c/2L \cong 1.5\times 10^8\,\mathrm{Hz}$; $(R_p/R_{cp}) = \exp\left[(2\Delta\nu/\Delta\nu_0^*)^2 \ln 2\right] = 1.005$.

7.19. One has $L = n(\lambda/2)$, where n is an integer number. According to this equation, if L is increased by $\lambda/2$, the oscillation wavelength will increase by $\Delta\lambda = \lambda/n$. Since $\lambda\nu = c$ we then get $\Delta\nu \cong -(\Delta\lambda/\lambda)\nu = -\nu/n$. From the relation $\nu = n(c/2L)$ we then obtain $\Delta\nu \cong -(c/2L)$.

7.21. Assume that a transmission peak of the FP etalon is made coincident with the central mode. The two adjacent longitudinal modes, which are frequency spaced by $\Delta\nu = c/2L$, will not oscillate if the FP transmission at these two frequencies, $T(\Delta\nu)$, is such that $T(\Delta\nu)\exp(\sigma_p Nl - \gamma) \leq 1$. We obtain $T(\Delta\nu) \leq 0.8$. From Eqs. (4.5.6) and (4.5.14), with $R_1 = R_2 = R$, one gets $T(\Delta\nu) = 1/[1 + (2F/\pi)^2 \sin^2\phi]$ where $\phi = 2\pi L'\Delta\nu/c$ with $L' \cong n_r L_{et} = 2.9$ cm (L_{et} is the thickness and n_r is the refraction index of the etalon). We get $\phi = \pi L'/L = 9.1 \times 10^{-1}$ rad and since $T(\Delta\nu) \leq 0.8$, we obtain $(2F/\pi)^2 \sin^2\phi \cong (2F/\pi)^2\phi^2 \geq 0.25$, i.e. $F \geq (0.5\pi/2\phi) = 8.63$. Equation (4.5.14), for $R_1 = R_2 = R$, can then be written as $(1 - R) = \pi R^{1/2}/F$ from which, by an iterative procedure, one gets $R \cong 0.7$. We must also ensure that the mode near the next peak of the FP etalon is below threshold. This occurs if $\exp[-(2\Delta\nu_{fsr}/\Delta\nu_0^*)^2 \ln 2] \times \exp(\sigma_p Nl - \gamma) \leq 1$ i.e. for $(2\Delta\nu_{fsr}/\Delta\nu_0^*)^2 \ln 2 \geq 0.223$, where $\Delta\nu_{fsr} \cong (c/2n_r L_{et})$ is the FP free-spectral-range. We obtain $L_{et} \geq [\ln 2/0.223]^{1/2}\, c/n_r\Delta\nu_0^* \cong 10.4$ cm which shows that the above condition is satisfied in our case.

Chapter 8

8.2. Since $k\nu L' = (\pi/2)$, one finds $V = \lambda/4n_0^3 r_{63}$.

8.4. From Fig. 8.14, for $f^* = f\tau = 2.3$ and $x = 10\,\text{kW}/2.2\,\text{kW} = 4.55$, we obtain $N_i/N_p \cong 1.89$, so that, from Fig. 8.11, we find $\eta_E \cong 0.76$. Since $\gamma_2 = 0.162$ and $A_b = 0.23\,\text{cm}^2$ (see Example 7.2), from Eq. (8.4.20) one finds $E \cong 18$ mJ which gives an average output power of $< P >= Ef = 180$ W, i.e., very close to the c.w. value (202 W, see Fig. 7.5). Since $\gamma = 0.12$ (see Example 7.2) and $L_e = L(n-1)l \cong 56$ cm (where $n = 1.8$ is the refractive index of the YAG crystal), we get $\tau_c = L_e/c\gamma = 15.6$ ns and, from Eq. (8.4.21), $\Delta\tau_p \cong 90$ ns.

8.7. (1) Since t_p is much shorter than the upper state lifetime, one has $(dN/dt) = R_p$ i.e. $N = R_p t$. Since $R_p t_p = 4N_{th}$, where N_{th} is the threshold inversion, the time at which threshold is reached is $t_{th} = t_p/4$. (2) The time behavior of the net gain is then given by $g_{net} = \sigma(N - N_{th})l = (4\sigma N_{th} l)(t - t_{th})/t_p$ where $t_{th} = t_p/4$. (3) Neglecting saturation, one has $(d\phi/dt) = g_{net}\phi/t_T$, where t_T is the transit time. Using the expression for net gain just derived, we find, by integration, $\phi(t') = \phi_i \exp[(4\sigma N_{th} l)\cdot(t'^2/2t_p t_T)]$ where $t' = t - t_{th}$ and $\phi_i \cong 1$. (4) At the end of the pump pulse one has $t' = 3t_p/4$ and, from the above expression for $\phi(t')$ one gets $(9\sigma N_{th} l/8)\cdot(t_p/t_T) = \ln(\phi_p/20)$ where ϕ_p is given by Eq. (8.4.14). Using this last expression one finds $t_p = t_T(8/9\gamma)\ln(\phi_p/20)$, where $\gamma = (-\ln T_2)/2 = 0.35$. To calculate ϕ_p from Eq. (8.4.14) one notes that $N_i/N_p = 4$ and $V_a N_p = \gamma A/\sigma$ where A is the beam area. Thus one finds that $\phi_p = 5.54 \times 10^{10}$ and, since $t_T = 22.7$ ps, it follows $t_p = 1.25$ ns.

8.12. Equation (8.6.14) can be expressed more conveniently as $E(t) \propto \exp(-\Gamma t^2)\exp(j\omega_0 t)$ where $\Gamma = \alpha - j\beta$, i.e. it can be transformed into a Gaussian pulse with a complex Gaussian parameter Γ. Its Fourier transform can then be written as $E(\omega - \omega_0) \propto \exp -(\omega - \omega_0)^2/4\Gamma = \exp -\{[(\omega - \omega_0)^2/4(\alpha^2 + \beta^2)]\cdot(\alpha + j\beta)\}$. The power spectrum will then be $|E(\omega - \omega_0)|^2 \propto \exp -[(\omega - \omega_0)^2\alpha/2(\alpha^2 + \beta^2)]$. If we now write $|E(\omega - \omega_0)|^2 \propto \exp -[4(\omega - \omega_0)^2 \ln 2/\Delta\omega_L^2]$, where $\Delta\omega_L$ is the bandwidth, a comparison of the two above expressions gives $\Delta\omega_L^2 = (8\ln 2)\alpha[1 + (\beta^2/\alpha^2)]$. This equation, with the help of expression Eq. (8.16.15) for α and using the relation $\Delta\nu_L = \Delta\omega_L/2\pi$, then leads to Eq. (8.6.16).

8.14. The average intensity is $< I >= \int_0^\infty I p_I dI / \int_0^\infty p_I dI = I_0$, and the required probability is given by

$$p = \int_{2I_0}^\infty p_I dI / \int_0^\infty p_I dI = \exp(-2) = 0.135.$$

8.16. In our case we have $2\gamma_t = 2\gamma - kP$ while, in the fast saturable absorber case, one can write [see Eq. (8.6.20)] $2\gamma_t = 2\gamma - 2\gamma' \, (P/P_s)$. A comparison of these two expressions shows that k is equivalent to $2\gamma'/P_s$. According to Eq. (8.6.22) the pulse duration can then be written as $\Delta\tau_p \cong (0.79/\Delta\nu_0)\left(2g_0'/kP_p\right)^{1/2}$ where P_p is the peak power. For a hyperbolic secant function, the peak power is related to the pulse energy by $E = 1.13\, P_p\Delta\tau_p$. From the above two expressions we get $\Delta\tau_p \cong (0.79/\Delta\nu_0)^2\left(2g_0'/k\right)(1.13/E) \cong 3.5$ fs.

8.17. From Eq. (8.6.35) with $\phi'' = \beta'' l$ one gets $l = (\delta\tau_p/\tau_p)^{1/2} \times \left(\Delta\tau_p^2/\beta''\right)/2\sqrt{2}\ln 2 \cong 0.46$ mm, where β'' is the GVD.

Chapter 9

9.5. The emission solid angle is $\Omega = \pi D^2/4l^2 \cong 2.83 \times 10^{-3}$ sr. Assuming a Gaussian line, from Eq. (2.9.4b), with $\phi \cong 1$, one gets $G \cong 1.37 \times 10^4$, i.e. $N_{th} = \ln G/\sigma_p l \cong 2.38 \times 10^{19}$ cm^{-3}, where $\sigma_p \cong 4 \times 10^{-20}$ cm^2 for Nd:glass (see Table 9.3). We then obtain $E = N_{th} V h\nu = 12.7$ J where $V = 2.83$ cm^3. For Nd:YAG, at the same value of the solid angle and assuming a Lorentzian line, we get from Eq. (2.9.4a), with $\phi = 1$, $G \cong 2.5 \times 10^4$ i.e. $N_{th} = 3.61 \times 10^{18}$ cm^{-3} so that $E \cong 1.91$ J.

9.7. Neglecting ground state and excited state absorption and under mode-matching conditions ($w_0 \cong w_p$) the threshold pump power, from Eq. (6.3.20), is seen to be given by $P_{th} = (\gamma/\eta_p)(h\nu_p/\tau)\left(\pi w_0^2/\sigma_e\right)$. We have $\gamma = (\gamma_2/2) + \gamma_i = 1.5 \times 10^{-2}$, $\eta_p \cong 1 - \exp -(\alpha_p l) \cong 0.86$ where $\alpha_p = 5$ cm^{-1} is the pump absorption coefficient, $h\nu_p \cong 3.1 \times 10^{-19}$ J, $\tau = 67\,\mu$s, $w_0 = 60\,\mu$m and $\sigma_e = 4.8 \times 10^{-20}$ cm^2. We thus obtain $P_{th} = 190$ mW.

9.8. $P_{th} = [(\gamma + \gamma_a)/\eta_p](h\nu_p/\tau)[\pi\left(w_0^2 + w_p^2\right)/2(\sigma_e - \sigma_{ESA})]$.

9.9. $P_{th} = (\gamma/\eta_p)(h\nu_p/\tau)[\pi\left(w_0^2 + w_p^2\right)/2(\sigma_e - k_{ST}\tau_T\sigma_T)]$.

9.12. $\eta_s = dP/VdI = (h\nu/eV)[-\ln R/(\alpha L - \ln R)]$. For $\lambda = 850$ nm, $h\nu/e = 1.46$ eV, and $V = 1.8$ V one gets $\eta_s \cong 64\%$.

9.13. One must have $w_{0\|}^2\left[1 + \left(z\lambda/\pi w_{0\|}^2\right)2\right]\Big) = w_{0\perp}^2\left[1 + \left(z\lambda/\pi w_{0\perp}^2\right)^2\right]\Big)$, which gives $z = \pi w_{0\|} w_{0\perp}/\lambda$. For the given values of $w_{0\|}$, $w_{0\perp}$ and λ, one obtains $z = 4.6\,\mu$m (note the very short distance).

9.15. $(2\pi n_1 L/\lambda) = 2$ i.e. $n_1 = \lambda/\pi L \cong 8.22 \times 10^{-4}$.

Chapter 10

10.4. If we let $\sigma(\nu - \nu_0)$ be the unsaturated cross section of Ar^+, oscillation will occur up to the n-th mode, away from the central mode, such that $\sigma(n\Delta\nu)\cdot Nl \geq \gamma$, where $\Delta\nu$ is the frequency spacing between consecutive longitudinal modes, N is the unsaturated inversion, l is the length of the active medium, and γ is the cavity loss. The unsaturated cross section is then given by $\sigma(n\Delta\nu) = \sigma_p \exp -[(2n\Delta\nu/\Delta\nu_0^*)^2 \ln 2]$. In this expression σ_p is the peak cross section and, with the laser pumped three-times above threshold, one has $\sigma_p N l = 3\gamma$. From the above three expressions one finds $3\exp -[(2n\Delta\nu/\Delta\nu_0^*)^2 \ln 2] \geq 1$, from which one gets $n \leq [\ln 3/\ln 2]^{1/2}\left(\Delta\nu_0^*/2\Delta\nu\right)$.

Since $\Delta\nu_0^* = 3.5$ GHz and $\Delta\nu = c/2L = 150$ MHz (L is the cavity length), one finds $n \le 14.7$. The number of oscillating modes is then $N_{osc} = 2n + 1 = 29$.

10.6. For a homonuclear molecule consisting of two atoms of mass M, the vibrational frequency, according to Eq. (3.1.3), is given by $\nu_0 = (1/2\pi)[2k_0/M]^{1/2}$ where k_0 is the elastic constant. For $M \cong 14$ a.u. $\cong 2.32 \times 10^{-26}$ kg and $\tilde{\nu}_0 = 2300\,\text{cm}^{-1}$ one finds $k_0 = 2180\,\text{Nm}^{-1}$.

10.8. For the symmetric stretching mode, the carbon position is fixed and the force acting on each oxigen atom is $F = -k(x-x_0)$, where k is the elastic constant and x_0 is the equilibrium separation between carbon and oxygen. It then follows that the resonance frequency of this mode is given by $\omega_1 = [k/M_O]$ where M_O is the mass of the oxygen atom. For $\tilde{\nu}_1 = 1337\,\text{cm}^{-1}$ and $M_O = 16$ a.u. $\cong 2.65 \times 10^{-26}$ kg one finds $k = 1683\,\text{Nm}^{-1}$.

10.10. Let x_0 be the equilibrium distance between one of the oxygen atoms and the carbon atom. A transverse displacement of the carbon atom by Δy would correspond to an elongation Δd of the spring given by $\Delta d = \left(x_0^2 + \Delta y^2\right)^{1/2} - x_0$. For $\Delta y \ll x_0$ one then gets $\Delta d \cong \Delta y^2/2x_0$ from which one sees that the force produced by the spring would be proportional to Δy^2. This implies that the harmonic oscillator model, for oscillation along the y-direction, cannot be derived via the simplified spring-model considered in this problem.

10.13. All ro-vibrational lines will merge together when the collision-broadened linewidth, $\Delta\nu_c$, becomes comparable with the frequency separation between rotational lines. Assuming $\Delta\nu_c = \Delta\nu_r = 60$ GHz, from the given value of $\Delta\nu_c$ one gets a total pressure of $p \cong 13\,997$ torr $= 18.4$ atm. From Fig. 10.11 one then sees that the width of the gain curve, $\Delta\nu_0$, corresponds to J' values ranging between $J' \cong 11$ and $J' \cong 41$, i.e., $\Delta J' \cong 30$. From the solution of Problem 10.11 one finds that the rotational constant, B, of a CO_2 molecule is $B \cong 0.3\,\text{cm}^{-1}$. The width $\Delta\nu_0$ of the gain curve is then given by $\Delta\nu_0 = 2B\Delta J' \cong 60B \cong 18\,\text{cm}^{-1}$.

10.16. The energy which is left, after reaction, as vibrational energy is $E_\upsilon = \sum_0^3{}_\upsilon N(\upsilon)\upsilon\Delta E$ where $N(\upsilon)$ is the population of the vibrational level, with vibrational quantum number υ, and ΔE is the energy spacing between vibrational levels (assumed the same for all levels). On the other hand, the total energy of reaction, E_t, is, on the other hand, given by $E_t = \Delta H \sum_0^3{}_\upsilon N(\upsilon)$ where $\Delta H \cong 3\Delta E$ is the reaction energy. From the above equations one finds that $\eta = (E_\upsilon/E_t) = \sum_0^3{}_\upsilon N(\upsilon)\upsilon \,/\, \sum_0^3{}_\upsilon N(\upsilon) = 68.5\%$.

Chapter 11

11.3. The field of the beam, along the C direction, due to the superposition of the two beams of the interferometer, can be written as $E_c = K_A E(t) + K_B E(t+\tau)$. If the power reflectivity of mirror S_1 is 50% and neglecting, for simplicity, any phase shift arising from reflection at mirrors S_1, S_2, and S_3, we can assume $K_A = K_B = K$. We then get $< I_c(t) > = < E_c(t)E_c^*(t) > = 2\,|K|^2 \{ < I > + \text{Re}[\Gamma^{(1)}(\tau)]\}$ where $< I > = < E(t)E^*(t) > = < E(t+\tau)E^*(t+\tau) >$ and Re stands for real part. With the help of (11.3.4) and (11.3.9) one then gets $< I_c(t) > = 2\,|K|^2 < I > \{1 + \left|\gamma^{(1)}\right| \cos [< \omega > \tau - \phi(\tau)]\}$. Around a given time delay τ, since both $\left|\gamma^{(1)}\right|$ and ϕ are slowly-varying functions of τ, one then has $I_{max} = < I_c(t) >_{max} = 2\,|K|^2 < I > [1 + \left|\gamma^{(1)}(\tau)\right|]$, $I_{min} = < I_c(t) >_{min} = 2\,|K|^2 < I > [1 - \left|\gamma^{(1)}(\tau)\right|]$, and $V_p = \left|\gamma^{(1)}(\tau)\right|$.

11.5. For a Gaussian spectral output, $\gamma^{(1)}(\tau)$ will also be a Gaussian function, i.e. it can be written as $\gamma^{(1)} = \exp-[(\tau/\tau_{co})^2 \ln 2]$, where τ_{co}, the coherence time, is defined as in Fig. 11.1. According to (11.3.28) one then has $\sigma\tau = 1/4\pi\sigma_\nu$. In our case we have $\sigma_\nu = \Delta\nu_L$ while the variance σ_τ of the function $[\gamma^{(1)}]^2 = \exp-[2(\tau/\tau_{co})^2 \ln 2]$ is $\sigma_\tau = \tau_{co}/2\sqrt{\ln 2}$. From the above expressions we obtain $\tau_{co} = \sqrt{\ln 2}/2\pi\sigma_\nu \cong 13.25\,\mu s$ and $L_{co} = c \cdot \tau_{co} \cong 3.98$ km.

11.7. $I_0 = 2P_i/\pi(\lambda f/\pi w_0)^2$. To avoid excessive diffraction losses and creation of diffraction rings from beam truncation by the finite lens aperture, D_L, one should choose a large enough D_L, typically $D_L = \pi w_0$ [see Eq. (5.5.31)]. From the above expressions we then find $I_0 = (2/\pi)P_i D_L^2/(\lambda f)^2$ while, from (11.4.4), with $D = D_L$, we find $I_0 = (\pi/4)P_i D_L^2/(\lambda f)^2$.

11.9. If we let x and y be the coordinates along the smaller and larger dimensions respectively of the near-field pattern, one has $W_{x0} = 0.5$ cm and $W_{y0} = 2$ cm. From (11.4.19) one then has $W_x(z = 3\,\text{m}) \cong 3.28$ cm, while from the equivalent equation along the y-direction, one gets $W_y(z = 3\,\text{m}) \cong 2.16$ cm.

Chapter 12

12.1. Since $w_0 = 0.54$ mm, one has $w(z = 1\,\text{m}) = w_0\,[1 + (z/z_R)]^{1/2} = 0.83$ mm and $R(z = 1\,\text{m}) = z[1 + (z/z_R)^2]^{1/2} \cong 1.74$ m, where $z_R = \pi w_0^2/\lambda \cong 86.1$ cm. The lens of focal length f can be divided into a first lens, of focal length $f_1 = R = 1.74$ m, to compensate for the wavefront curvature, and a second lens, of focal length $f_2 = f_1 \cdot f/(f_1 - f) \cong 10.61$ cm, to focus the beam. To a good approximation, the waist position then occurs at a distance of $z_m \cong f_2 \cong 10.61$ cm from the original lens. The spot-size of the embedded Gaussian beam is $w_0' \cong (\lambda/\pi w) \cdot f_2 \cong 0.043$ mm and the corresponding spot-size parameter is $W_0' = \sqrt{M^2} w_0' \cong 0.274$ mm.

12.3. One has $\Gamma_s = h\nu/\sigma \cong 4.71\,\text{J/cm}^2$ and $S = \pi D^2/4 \cong 63.6\,\text{cm}^2$, so that $\Gamma_{out} = E_{out}/S \cong 7.07\,\text{J/cm}^2$. The total energy available in the amplifier is $E_{av} = h\nu NV = S\Gamma_s \ln G_0 = 415$ J, where N is the initial inversion and V is the volume of the amplifier. To calculate the required input energy, (12.3.12) can be solved for Γ_{in} to give $\Gamma_{in} = [\{[\exp(\Gamma_{out}/\Gamma_s) - 1]/G_0\} + 1] \cong 2.95\,\text{J/cm}^2$ which results in $E_{in} = \Gamma_{in}\,S = 187.8$ J. Thus, out of an available energy of 415 J, the energy extracted from the amplifier is $E_{ex} = E_{out} - E_{in} \cong 262.2$ J. Note that the length of the amplifier does not enter into this calculation.

12.9. With the help of (12.4.27a), substitution of (12.4.29) into (12.4.2) gives $P^{NL} = (\varepsilon_0 d/2)\left\{\sum_1^3{}_i\, E_i(z)\exp[j(\omega_i t - k_i z)] + c.c.\right\}^2$. After manipulation of the right hand side of the above equation, it can easily be seen that, since $\omega_1 = \omega_3 - \omega_2$, the only term at frequency ω_1 is $P^{NL}_{\omega_1} = (\varepsilon_0 d/2)\left\{E_2^*(z)E_3(z)\exp[j(\omega_3 - \omega_2)t - j(k_3 - k_2)z] + c.c.\right\}$. Using the relation $\omega_1 = \omega_3 - \omega_2$, and with the help of (12.4.27b) one then readily obtains (12.4.30).

12.11. From (12.4.58a) the second harmonic conversion efficiency is obtained as $\eta = I_{2\omega}/I_\omega(0) = |E'_{2\omega}|^2 / |E'_\omega(0)|^2 = [\tanh(z/l_{SH})]^2$. From (12.4.52), taking into account the fact that that $E_\omega(0)$ is related to the incident intensity $I = I_\omega(0)$ by $E_\omega(0) = (2ZI)^{1/2}$, where $Z = 1/\varepsilon_0 c \cong 377$ ohms is the free-space impedance, one gets $l_{SH} \cong \lambda n_o/[2\pi d_{eff}(2ZI)^{1/2}] = 2.75$ cm where n_o is the ordinary refractive index of KDP at frequency ω. Substituing this value of l_{SH} into the above expression for η, assuming $z = 2.5$ cm, one gets $\eta = 51.9\%$.

Index

Printed in the United States
By Bookmasters